# The Palgrave Handbook of Environmental Labour Studies

Nora Räthzel • Dimitris Stevis
David Uzzell
Editors

# The Palgrave Handbook of Environmental Labour Studies

palgrave
macmillan

*Editors*
Nora Räthzel
Department of Sociology
Umeå University
Umeå, Sweden

Dimitris Stevis
Department of Political Science
Colorado State University
Fort Collins, CO, USA

David Uzzell
School of Psychology
University of Surrey
Guildford, UK

ISBN 978-3-030-71908-1     ISBN 978-3-030-71909-8 (eBook)
https://doi.org/10.1007/978-3-030-71909-8

© The Editor(s) (if applicable) and The Author(s), under exclusive licence to Springer Nature Switzerland AG 2021
This work is subject to copyright. All rights are solely and exclusively licensed by the Publisher, whether the whole or part of the material is concerned, specifically the rights of translation, reprinting, reuse of illustrations, recitation, broadcasting, reproduction on microfilms or in any other physical way, and transmission or information storage and retrieval, electronic adaptation, computer software, or by similar or dissimilar methodology now known or hereafter developed.
The use of general descriptive names, registered names, trademarks, service marks, etc. in this publication does not imply, even in the absence of a specific statement, that such names are exempt from the relevant protective laws and regulations and therefore free for general use.
The publisher, the authors and the editors are safe to assume that the advice and information in this book are believed to be true and accurate at the date of publication. Neither the publisher nor the authors or the editors give a warranty, expressed or implied, with respect to the material contained herein or for any errors or omissions that may have been made. The publisher remains neutral with regard to jurisdictional claims in published maps and institutional affiliations.

Cover illustration: @Abstract Aerial Art/GettyImages

This Palgrave Macmillan imprint is published by the registered company Springer Nature Switzerland AG.
The registered company address is: Gewerbestrasse 11, 6330 Cham, Switzerland

*For Mike Cooley*
*An inspirational trade unionist whose concern for nature, the ingenuity and skills of workers and the needs of people were inseparable.*

# Praise for *The Palgrave Handbook of Environmental Labour Studies*

"This pathbreaking, impressive collection presents and defines the crucially important emerging discipline of environmental labour studies. A sustainable future for our planet cannot be achieved without engaging workers. Its wide-ranging and thought-provoking essays showcase the interdisciplinary, international, engaged research that focuses on the ecological agency of working people: rural and urban; waged and unwaged; subsistence, service, industrial and extractive. It is an excellent guide to assist effective efforts to mitigate climate change."

—Verity Burgmann, *Monash University, Australia*, author of *Green Bans, Red Union*

"Inspiring reading for trade unionists and all activists engaging in struggles against climate catastrophe, and building transformative models for social, economic and climate justice. This book challenges us to think about Just Transition in radical ways, contesting existing unequal power relations and laying the foundations for new forms of democratic control."

—Alana Dave, *International Transport Workers Federation, Urban Transport Director*

"What an impressive and invaluable resource, providing major theoretical insights into the relation between labour, the environment and 'nature', on the basis of detailed studies of historical developments and current struggles across the globe. The book is remarkable for its range of international coverage and offers a striking contribution to this emerging field."

—Miriam Glucksmann, *Department of Sociology, University of Essex*, author of *Women on the Line*

"I highly recommend this nuanced and engaging Handbook that analyses in-depth how capitalism produces nature and nature produces capitalism. It brings together a remarkable range of theoretical approaches and experiences that embraces Marxism, feminism, post-materialism, environmentalism as well trade unionism making this a truly cutting-edge collection. It is a landmark in the field of environmental labour studies."

—Wendy Harcourt, *International Institute of Social Studies, Erasmus University, Rotterdam, The Netherlands*, editor of *Feminist Political Ecology and the Economics of Care—In Search of Economic Alternatives*

"The global pandemic has revealed a gaping wound at the centre of capitalism—a systemic insecurity in the lives of those whose work matters more than we ever

imagined. This timely collection of impassioned essays does more than simply archive the ongoing struggle. It paints a rich narrative of the many possibilities for change: providing a colourful canvas of 'real utopias' from communities across the world."
—Tim Jackson, *University of Surrey, UK*, author of *Post Growth, Life After Capitalism*

"Climate change and the loss of biodiversity pose existential threats for our species, but the transformation needed to address these environmental challenges poses major tests for our movement of organised labour and social justice around the globe. Just as no country can address environmental degradation alone, neither can one trade union solve the conundrums posed. We need to learn from each other's experiences within the global labour movement and find common collective responses. Only in this way, can we ensure that social justice and worker participation is at the centre of our response to cleaning our planet. I hope this collection of essays will contribute to that common endeavour."
—Judith Kirton-Darling, Deputy General Secretary, *IndustriAll European Trade Union*

"In the course of colonial and neo-colonial economic globalisation, founded on structures of statism, capitalism, patriarchy, racism, and anthropocentrism, both workers and the environment have been marginalised and exploited. Yet, there has been extremely insufficient attention given to the interface between the two, and indeed often they have been seen as opposed to each other due to narrow notions of environmentalism or of labour rights. This book does an enormous service by collating an impressive range of essays analysing these aspects, and the possibilities of integrating worker and ecological interests and rights, towards fundamental transformation towards a just society."
—Ashish Kothari, Co-founder of the *Environmental Action Group Kalpavriksh*, author of *Alternative Futures: India Unshackled*

"When the late Tony Mazzocchi, a leader of the Oil, Chemical and Atomic Workers, introduced the idea of just transition to labor and environmental activists in the late 1970s, very few took him seriously. This stunning compendium of research and analysis shows just how far this idea has traveled and how richly it has developed around the globe. Clearly, now more than ever, we need to study and pursue the paths outlined in this book to sustain both the planet and its working people."
—Les Leopold, Executive Director of the *Labor Institute*, author of *The Man Who Hated Work and Loved Labor: The Life and Times of Tony Mazzocchi*

"This handbook is an outstanding intellectual and political tool. It contains a remarkable collection of essays on the multiple dimensions of the labour environmental struggles against the destructive logic of the (capitalist) system, including not only trade-unions, but also, among others, farmers, indigenous communities and

intellectuals, both in the Global North and the Global South—as well as a discussion of various anti-systemic alternatives, from the Green New Deal to Ecosocialism."
—Michael Löwy, *CNRS (National Centre of Scientific Research), Paris, France*, author of *Ecosocialism: A Radical Alternative to Capitalist Catastrophe*

"Compared to indigenous and other rural peoples at the vanguard of socio-environmental resistance against mining, fossil fuel extraction, hydropower and oil palm plantations, the industrial trade unions are not often counted among the environmentalists. They are sometimes reluctant to accept the 'decarbonization' of the economy. We know however that the industrial and rural working class has long fought against asbestosis, pneumoconiosis and other occupational illnesses. This impressive and magnificent book analyses many kinds of labour, waged and unwaged, and its variety of experiences in India, South America, the United States, Europe, Africa, in a new research field on 'environmental labour studies' uncovering many practical opportunities for 'red-green' alliances."
—Joan Martinez-Alier, *ICTA, Universitat Autònoma de Barcelona, Spain*, author of *The Environmentalism of the Poor*

"A comprehensive exploration of the struggles, conflicts, achievements and potential of working class, indigenous, gender and other grassroots movements in relation to ecological sustainability that re-thinks conventional notions of labour to embrace unpaid, social and nonmarket work."
—Mary Mellor, *Northumbria University, UK*, author of *Money: Myths, Truths and Alternatives*

"The editors have delivered to us an essential resource in the midst of unprecedented planetary crisis. After decades of being told we must choose between 'jobs' and 'environment,' this collection points towards an extraordinary alternative synthesis: of planetary justice and the work of humans and the rest of nature. Revealing the historical, geographical, and economic tissues that bind class, labor, and webs of life, the handbook reminds us that, when it comes to the 'proletariat' and 'biotariat,' an injury to one is an injury to all."
—Jason W. Moore, *Binghamton University, USA*, author of *Capitalism in the Web of Life*

"As somebody who pioneered the incorporation of the environment into the trade union movement and of the social dimension into the international environmental and climate agenda, which culminated in the incorporation of the 'just transition' demand into the Paris Agreement, I see this work as essential in order to understand one of the most creative transformations of our time."
—Joaquín Nieto, Director of the *Spanish Office of the International Labour Organisation*

"A pioneering and valuable international study. Shows how trade unions are advancing just transition, environmental justice and a future of work that is fair for all. A thought-provoking read for everyone who wants to build a greener, more equal global economy."
—Frances O'Grady, General Secretary, *Trades Union Congress, UK*

"In outlining the emerging trans-discipline of Environmental Labour Studies, a complex global field of lived tensions and structural contradictions, the editors have assembled a pioneering anthology. Not least, their innovative inclusion of essays builds on the materialist ecofeminist argument that relations between gender, labour, and nature, are sociologically entangled."
—Ariel Salleh, activist, *Australia*, author of *From Eco-Sufficiency to Global Justice*

"Environmental Labour Studies help us to understand a way forward towards our common future. This globe-spanning volume marks the coming-of-age of this crucial theoretical and empirical field. The original and insightful contributions gathered here dramatically expand our understanding of how work, workers, and trade unions interact with the environment, nature, and environmentalists."
—Victor Silverman, *Pomona College, Department of History, USA*, Emmy-winning filmmaker and author of *Imagining Internationalism*

"Capitalism and its international corporations know no limits in their quest to put profit above life and nature. The Movement of Landless Workers (MST) in Brazil, along with Via Campesina all over the world, have defended what is urgent and necessary now: land, water, forest and minerals must be used in harmony with nature to protect biodiversity and for the common good. This volume brings together timely reflections on these serious problems that our planet is facing."
—João Pedro Stédile, leader of the *Brazilian Movement of Landless Workers*

"In order to win the fight to address our manifold sustainability challenges we need the labor and environmental movements to come up out of their silos and work together. The editors have offered a pathway for a greater confluence of these movements in this first Handbook of Environmental Labour Studies. They have assembled a great group of authors who explore the pitfalls and possibilities and show us how we can expand the boundaries of our movements and begin to take down the forces that keep us apart. So long as humanity views the planet as something for us to dominate, we will continue to destroy it. When we realize that living in harmony with nature is first and foremost, then we can begin to repair the damage we've done."
—Joe Uehlein, Founder and President of the *Labour Network for Sustainability (USA)*

"In this timely handbook, leading scholars and practitioners from around the world engage in key issues of environmental labour studies. The handbook is an invaluable contribution with its intellectual depth, breadth of issues, and geographical spread of cases. It's the first of its kind and should be read widely not only by labour and environmental activists and scholars, but also anyone interested in understanding the challenges and possibilities of overcoming the tensions between labour and the environment."

—Michelle Williams, Chairperson of the *Global Labour University, and Wits University, South Africa*, author of *The Roots of Participatory Democracy: Democratic Communists in South Africa and Kerala, India*

"The environment must be protected and sustainable industries must create decent, safe and healthy work. The perspectives in this Handbook can dispel the notion that environmental protection and decent work are in conflict with each other: we must have both, or we will have neither. To build a bridge to the future we need a 'Just Transition' for workers, their families, and their communities. Integrating labour and environmental studies is a crucial step forward on that bridge."

—Brian Kohler, Director - *Health Safety and Sustainability* (retired), *Industrial Global Union*

# Contents

1 Introduction: Expanding the Boundaries of Environmental
  Labour Studies  1
  *Nora Räthzel, Dimitris Stevis, and David Uzzell*

Part I   Histories  33

2 Labour and the Environment in India  35
  *S. Ravi Rajan*

3 Energy Transitions in the Global South: The Precarious
  Location of Unions  59
  *Dinga Sikwebu and Woodrajh Aroun*

4 The New Struggles to Be Born: The Difficult Birth of a
  Democratic Ecosocialist Working-Class Politics  83
  *Devan Pillay*

5 The Green New Deal and Just Transition Frames within the
  American Labour Movement  105
  *Todd E. Vachon*

6 Working-Class Environmentalism: The Case of Northwest
  Timber Workers  127
  *Erik Loomis*

xiii

xiv Contents

7 Trade Unions and Environmental Justice 149
*Darryn Snell*

Part II Seeking Common Ground 175

8 'Beware of the Crocodile's Smile': Labour-Environmentalism in the Struggle to Achieve a Just Transition in South Africa 177
*Jacklyn Cock*

9 Fighting in the Name of Workers: Exploring the Dynamics of Labour-Environmental Conflicts in Kerala 199
*Silpa Satheesh*

10 Trade Union Politics for a Just Transition: Towards Consensus or Dissensus? 225
*Diego A. Azzi*

11 Climate Jobs Plans: A Mobilizing Strategy in Search of Agency 249
*Andreas Ytterstad*

12 The Role of Ecuadorian Working-Class Environmentalism in Promoting Environmental Justice: An Overview of the Hydrocarbon and Agricultural Sectors 271
*Sara Latorre*

13 A Just Transition for All? A Debate on the Limits and Potentials of a Just Transition in Canada 295
*Bruno Dobrusin*

Part III Farmers, Commoners, Communities 317

14 Labouring the Commons: Amazonia's 'Extractive Reserves' and the Legacy of Chico Mendes 319
*Stefania Barca and Felipe Milanez*

15 Connecting Individual Trajectories and Resistance Movements in Brazil 339
*Beatriz Leandro, Patrícia Vieira Trópia, and Nora Räthzel*

16  Whose Labour, Whose Land? Indigenous and Labour
    Conflicts and Alliances over Resource Extraction       365
    *Erik Kojola*

17  Commoning Labour, Labouring the Commons: Centring the
    Commons in Environmental Labour Studies                389
    *Gustavo A. García-López*

18  Agroecological Farmer Movements and Advocacy Coalitions
    in Sub-Saharan Africa: Between De-Politicization and
    Re-Politicization                                      415
    *Patrick Bottazzi and Sébastien Boillat*

19  Working-Class Environmentalism in the UK: Organising for
    Sustainability Beyond the Workplace                    441
    *Karen Bell*

## Part IV  Trade Unions and the State                      465

20  A Just Transition Towards Environmental Sustainability for All  467
    *Catherine Saget, Trang Luu, and Tahmina Karimova*

21  Labour Resistance Against Fossil Fuel Subsidies Reform:
    Neoliberal Discourses and African Realities            493
    *Camilla Houeland*

22  Challenges and Prospects for Trade Union Environmentalism  517
    *Adrien Thomas and Valeria Pulignano*

23  From 'Just Transition' to the 'Eco-Social State'       539
    *Béla Galgóczi*

24  Environment, Labour and Health: The Ecological-Social
    Debts of China's Economic Development                  563
    *Juan Liu*

## Part V  Organic Intellectuals 581

**25  Introduction: Trade Union Environmentalists as Organic Intellectuals in the USA, the UK, and Spain** 583
*Nora Räthzel, Dimitris Stevis, and David Uzzell*

**26  Embedding Just Transition in the USA: The Long Ambivalence** 591
*Dimitris Stevis*

**27  Caring for Nature, Justice for Workers: Worldviews on the Relationship Between Labour, Nature and Justice** 621
*David Uzzell*

**28  Individuals Transforming Organisations: Spanish Environmental Policies in Comisiones Obreras** 647
*Nora Räthzel*

## Part VI  Rethinking and Broadening Concepts 675

**29  The Commodification of Human Life: Labour, Energy and Money in a Deteriorating Biosphere** 677
*Alf Hornborg*

**30  Workers, Trade Unions, and the Imperial Mode of Living: Labour Environmentalism from the Perspective of Hegemony Theory** 699
*Markus Wissen and Ulrich Brand*

**31  André Gorz's Labour-Based Political Ecology and Its Legacy for the Twenty-First Century** 721
*Emanuele Leonardi and Maura Benegiamo*

**32  Rethinking Labour/Work in a Degrowth Society** 743
*Anna Saave and Barbara Muraca*

| | | |
|---|---|---|
| 33 | **Labour and Societal Relationships with Nature. Conceptual Implications for Trade Unions**<br>*Thomas Barth and Beate Littig* | 769 |
| 34 | **Society–Labour–Nature: How to Think the Relationships?**<br>*Nora Räthzel* | 793 |
| 35 | **Labour-Centred Design for Sustainable and Just Transitions**<br>*Damian White* | 815 |
| 36 | **Technology and the Future of Work: The Why, How and What of Production**<br>*David Elliott* | 839 |

**Index**     863

# Notes on Contributors

**Woodrajh Aroun** is the former Parliamentary Officer of the National Union of Metalworkers of South Africa (NUMSA). He played an active role in the union's energy research group and was part of the International Trade Union Confederation (ITUC) labour delegation that participated in the ILO Tripartite Meeting of Experts on Sustainable Development, Decent Work and Green Jobs (Geneva, 2015).

**Diego A. Azzi** is Professor of International Relations, Federal University of ABC (UFABC), Brazil. Azzi's research interests explore international civil society mobilizations around issues of global political economy, world of labour and climate change, as well as Brazilian Foreign Policy. Background of work as international relations advisor for a number of trade unions, NGOs and social movements.

**Stefania Barca** is a senior researcher at the Centre for Social Studies of the University of Coimbra (Portugal). She writes on the environmental history and political ecology of labour from a feminist perspective.

**Thomas Barth** is a senior researcher at the Institute for Sociology at the Ludwig Maximilian University in Munich, Germany. He is co-editor of the volume (2016) *Sustainable Work: Sociological Contributions to the Redefinition of Societal Relationships with Nature* with Beate Littig and Georg Jochum. His teaching and research focus lies in the fields of environmental sociology, labour and sustainability.

**Karen Bell** is a senior lecturer at the University of the West of England, UK. Her research focuses on fair and inclusive transitions to sustainability. She is the author of *Working-Class Environmentalism: An Agenda for a Just and*

*Fair Transition to Sustainability* (2019, Palgrave). Before becoming an academic, she was a community development worker in disadvantaged areas in and around Bristol.

**Maura Benegiamo** holds a postdoctoral position at the University of Trieste (Italy) and is an affiliated researcher at the Collège d'Études Mondiales of Paris (France). Her research interests include: biocapitalism and non-human value, capitalist transformations in the context of the ecological crisis, decolonial ecologies and technoscientific innovation.

**Sébastien Boillat** is an environmental scientist and human geographer at the Institute of Geography of the University of Bern, Switzerland. His fields of research include land systems, natural resource governance, environmental justice, traditional ecological knowledge, political ecology, agroecology and climate change adaptation. His regional focus covers mainly Latin America and Africa.

**Patrick Bottazzi** is an assistant professor at the Institute of Geography and a member of the Centre for Development and Environment, University of Bern. As an environmental social scientist, he has been working in Latin America and Africa mainly on the local impacts of global environmental changes and related policies. He recently set up a research group Labour and Social-Ecological Transition (LASET).

**Ulrich Brand** is Professor of International Politics at the University of Vienna. In 2018–2019 he was a fellow of the research group on 'Post-growth Societies' at the University of Jena. He co-authored *The Imperial Mode of Living. Everyday Life and the Ecological Crisis of Capitalism* (2021; with Markus Wissen).

**Jacklyn Cock** is Professor Emerita, University of the Witwatersrand, South Africa, and honorary research professor, SWOP Institute. As an academic/activist, she has published widely on social and environmental justice issues. Her latest book is *Writing the Ancestral River. A Biography of the Kowie* (2018).

**Bruno Dobrusin** is a labour and climate justice organizer in Toronto, Canada. He is working on transnational labour solidarity with a focus on Latin America. He is the former coordinator of the Green Economy Network, an alliance of unions, environmental, community and faith-based groups fighting for ambitious climate action in Canada. Previously, he was an advisor at the Workers' Confederation of Argentina.

## Notes on Contributors

**David Elliott** is Emeritus Professor of Technology Policy at The Open University, where he carried out research and developed courses on technological innovation, focusing on renewable energy technology. He was involved with the Lucas worker's campaign for alternative products and production systems and with subsequent local government initiatives in the UK.

**Béla Galgóczi** is an economist working as a senior researcher at the European Trade Union Institute, Brussels, since 2003, and has been working on capital and labour mobility in the EU. His research focus is a just transition towards a carbon-neutral economy with a focus on fair labour market transitions in carbon-intensive sectors and regions.

**Gustavo A. García-López** is a scholar-activist from Puerto Rico working on/with environmental justice movements and commoning initiatives for transformational paths towards more just and ecological worlds. He is an assistant researcher at the Centre for Social Studies, University of Coimbra (Portugal), and the 2019–2021 Prince Claus Chair at the International Institute of Social Studies, The Hague (Netherlands).

**Alf Hornborg** is an anthropologist and Professor of Human Ecology at Lund University (Sweden). He has co-edited several volumes at the intersections of anthropology, environmental history, political ecology and ecological economics. His most recent book is *Nature, Society, and Justice in the Anthropocene* (2019).

**Camilla Houeland** is a postdoctoral fellow, Department of Sociology and Human Geography, University of Oslo (Norway). Her PhD was on Nigerian trade unions. Her project explores how Nigerian and Norwegian oil workers and their unions engage with the green transition. She previously worked as Africa-advisor to the Norwegian Confederation of Trade Unions.

**Tahmina Karimova** is a lawyer specialized in public international law, sustainable development, human rights law, international labour law and arms control. She works at the ILO (Switzerland) on a research programme on the legal framework to the just transition. Previously, she worked as a Human Rights Officer for the Office of the High Commissioner for Human Rights.

**Erik Kojola** is an assistant professor of sociology at Texas Christian University (USA). His research on labour environmentalism, the cultural and class politics of resource extraction, and environmental justice appeared in numerous journal articles.

**Sara Latorre** is a professor at the Latin American Faculty of Social Sciences, Ecuador. Her research work has focused on the study of environmental popular social movements and conflicts in relation to Ecuadorian environmental inequalities from a political ecology and ecological economic perspective.

**Beatriz Leandro** holds a Master's degree in International Humanitarian Law and in Social Sciences (Brazil). She was the researcher responsible for Brazil in the project 'Moments of Danger, Moments of Opportunity' funded by the Swedish Research Council. She edited the book *Water for Life: Ecumenical Action for Rights and Common Goods in Brazil and Latin America* (2020).

**Emanuele Leonardi** is lecturer in Sociology at the University of Parma (Italy) and an affiliated researcher at the Centre for Social Studies of the University of Coimbra (Portugal). His research interests include climate justice movements and their critique of carbon trading, logics of exploitation in contemporary capitalism, and working-class environmentalism.

**Beate Littig** is a sociologist and senior researcher at the Institute for Advanced Studies in Vienna, Austria, and a lecturer at the University of Vienna. The focus of her research and teaching lies in socio-ecological transformation, the future of (sustainable) work, gender studies and practice theories.

**Juan Liu** is associate professor of political ecology and agrarian studies at the College of Humanities and Development Studies, China Agricultural University, Beijing, and researcher at the Institute of Environmental Science and Technology, Universitat Autònoma de Barcelona. Her research includes political economy/ecology of agriculture, food and environment, migration and populations left behind.

**Erik Loomis** is associate professor of history at the University of Rhode Island (USA). Among his books are *A History of America in Ten Strikes* (2020) and *Empire of Timber: Labor Unions and the Pacific Northwest Forests* (2017).

**Trang Luu** is a PhD candidate in Economics at the University of Geneva (Switzerland). Her research interests focus on the impact of climate change on labour markets, including the need to adapt skill development policies and the relevance of industrial policies to just transition.

**Felipe Milanez** is an associate professor at the Institute for Humanities, Arts and Sciences, at the Federal University of Bahia (Brazil). He holds a PhD in Social Studies from the University of Coimbra, as part of the European Network of Political Ecology, and coordinates the network *Political Ecology from Abya Yala* at Clacso.

**Barbara Muraca** is assistant professor of environmental philosophy at the University of Oregon (USA). Her research focuses on human-nature relationships, degrowth, process philosophy and sustainability theory. In 2014 she published *Gut Leben: Eine Gesellschaft jenseits des Wachstums* (Berlin: Wagenbach).

**Devan Pillay** is Head of the Department of Sociology, University of the Witwatersrand, South Africa; previously, Head of Research, National Union of Mineworkers; Director of Policy, Government Communication and Information System; Director of Social Policy Programme, University of Durban-Westville; Managing Editor, *Work In Progress*; and Staff Writer, *SA Labour Bulletin*.

**Valeria Pulignano** is professor of sociology at CESO—KU Leuven (Belgium). Her research interests are labour sociology, comparative industrial (employment) relations, precarious work, labour markets inequality and voice at work. Her recent book is *Reconstructing Solidarity, Labour Unions, Precarious Work, and the Politics of Institutional Change in Europe*, with Virginia Doellgast and Nathan Lillie (2018).

**S. Ravi Rajan** is professor of environmental studies at the University of California, Santa Cruz. His research addresses the political economy of environment—development conflicts; environmental human rights and environmental justice; and environmental risks and disasters. He also contributes to civil society organizations and is a member of the Board of Directors of Greenpeace International.

**Nora Räthzel** is professor emerita at the Umeå University (Sweden). Her research includes environmental labour studies, gender and ethnicity and latest publications are *Marxist-Feminists Theories and Struggles Today* (with Khayaat Fakier, Diana Mulinari); *Transnational Corporations from the Standpoint of Workers* (with Diana Mulinari, Aina Tollefsen); and *Trade Unions in the Green Economy* (with David Uzzell).

**Anna Saave** is a PhD student in social theory specializing in political economy at Friedrich Schiller University Jena (Germany). Her research focuses on appropriation, externalization and the interconnections of social reproduction and ecological processes with the capitalist mode of production. She teaches courses on feminist economics.

**Catherine Saget** is a Chief of Unit with the Research department of the International Labour Office (ILO). She was the lead author of the ILO annual flagship report *World and Employment Social Outlook, 2018: Greening with Jobs* and the ILO 2019 report on the impact of heat stress on productivity and decent work.

**Silpa Satheesh** holds a PhD in Sociology from the University of South Florida (USA). She is an assistant professor at Azim Premji University (School of Development), Bangalore. Her research focuses on social movements, labour and the environment, sociology of development, political economy and ethnographic methods. Her PhD explored the conflicts between trade unions and green movements in Kerala.

**Dinga Sikwebu** is a researcher with the National Union of Metalworkers of South Africa (NUMSA). In 2011–2013, Sikwebu coordinated the union's energy research group and was responsible for the formulation of the organization's climate change policy. He has written and published on labour and politics in South Africa.

**Darryn Snell** is an associate professor in the School of Management at RMIT University, Melbourne (Australia). As the co-coordinator of the Skills, Training and Industry Research Network, he has developed a stream of research focused on labour and economic transitions in carbon-exposed regions. He works closely with unions and governments on finding practical 'just transition' solutions for workers disadvantaged by environmental policies.

**Dimitris Stevis** is professor of politics at Colorado State University (USA). His research focuses on global labour and environmental politics, with particular attention to labour environmentalism and social and ecological justice. He is a founder of the Centre for Environmental Justice and recently published (with Dunja Krause and Edouard Morena) *Just Transitions: Social Justice in the Shift Towards a Low-Carbon World* (Pluto Press 2020).

**Adrien Thomas** is a research scientist in political science at the Luxembourg Institute of Socio-Economic Research. His research interests focus on the sociology of trade unions, labour environmentalism and labour migrations. He has published *Les frontières de la solidarité. Les syndicats et les immigrés au coeur de l'Europe* (Presses universitaires de Rennes, 2015).

**Patrícia Vieira Trópia** is a professor at the Institute of Social Sciences of the Federal University of Uberlândia (Brazil), researching trade unionism and the middle classes. She completed postdoctoral studies at the University Lumière Lyon 2 (France). She is the author of *Força Sindical: Politics and Ideology in Brazilian Syndicalism* (2009).

**David Uzzell** is professor emeritus of environmental psychology at the University of Surrey (UK). His research interests include critical psychological approaches to changing consumption and production practices. He edited (with Nora Räthzel) *Trade Unions in the Green Economy: Working for the Environment* (2013).

**Todd E. Vachon** is a Fellow with the Center for Innovation in Worker Organization and is the Faculty Coordinator of the Labor Education Action Research Network in the School of Management and Labor Relations at Rutgers, The State University of New Jersey (USA).

**Damian White** is dean of liberal arts and professor of sociology and environmental studies at the Rhode Island School of Design (USA). His publications include *Bookchin: A Critical Appraisal* (2008), *Autonomy, Solidarity, Possibility: The Colin Ward Reader* (2011; with Chris Wilbert) and *Environments, Natures and Social Theory* (Palgrave, 2016; with Alan Rudy and Brian Gareau).

**Markus Wissen** is professor of social sciences at Berlin School of Economics and Law (Germany). In 2018 he was a fellow of the research group on 'Postgrowth Societies' at the University of Jena. He co-authored *The Limits to Capitalist Nature. Theorizing and Overcoming the Imperial Mode of Living* (2018, with Ulrich Brand).

**Andreas Ytterstad** is a professor in the Department of Journalism and Media Studies at Oslo Metropolitan University (Norway). His book *100,000 Climate Jobs and Green Workplaces Now! For a Climate Solution from Below* (2013) spawned the Bridge to the Future alliance. Ytterstad hosted a four-part TV series on this alliance in 2018.

# Abbreviations

| | |
|---|---|
| ACFTU | All-China Federation of Trade Unions |
| AFL | American Federation of Labor (USA) |
| AFL-CIO | American Federation of Labor and Congress of Industrial Organizations (USA) |
| ANC | African National Congress (South Africa) |
| CCMPPTF | China Coal Miner Pneumoconiosis Prevention and Treatment Foundation |
| CCOO | Comisiones Obreras (Spain) |
| CCP | Chinese Communist Party |
| CEB | Comunidades Eclesiais de Base (Basic Ecclesial Communities, Brazil) |
| CFDT | Confédération Française Démocratique du Travail (French Democratic Confederation of Labour) |
| CGTP | Confederação Geral dos Trabalhadores Portugueses (General Confederation of Portuguese Workers) |
| CIO | Congress of Industrial Organizations |
| CLC | Canadian Labour Congress |
| CONTAG | Confederação Nacional dos Trabalhadores na Agricultura (National Confederation of Agricultural Workers, Brazil) |
| COP | UN Climate Change Conference of the Parties |
| COSATU | Congress of South African Trade Unions |
| CSD | Commission on Sustainable Development (United Nations) |
| CUPW | Canadian Union of Postal Workers |
| Defra | Department of Environment, Food and Rural Affairs (UK) |
| ECWU | Energy and Chemical Workers Union (Canada) |
| EIUG | Energy Intensive Users Group |
| EPO | European Patent Organisation |
| ETUC | European Trade Union Confederation |

## Abbreviations

| | |
|---|---|
| ETUI | European Trade Union Institute |
| EU | European Union |
| EU ETS | European Union Emissions Trading System |
| FO | General Confederation of Labor-Workers' Force (Confédération Générale du Travail–Force Ouvrière) |
| GDP | Gross Domestic Product |
| GHG | Greenhouse Gas |
| GMO | Genetically modified organisms (GMOs) |
| GND | Green New Deal (USA) |
| GNH | Gross National Happiness |
| GUF | Global Union Federation |
| GWEC | Global Wind Energy Council |
| HDI | Human Development Index |
| IAM | International Association of Machinists and Aerospace Workers (USA) |
| IBEW | International Brotherhood of Electrical Workers (USA) |
| IBT | International Brotherhood of Teamsters (USA) |
| ICEM | International Federation of Chemical, Energy, Mine and General Workers' Unions (now part of IndustriALL) |
| ICFTU | International Confederation of Free Trade Unions |
| IG BCE | IG Bergbau, Chemie, Energie (Germany) |
| ILO | International Labour Organization |
| ILWU | International Longshore and Warehouse Union (USA, Canada) |
| IMF | International Metalworkers' Federation (now part of IndustriALL) |
| IndustriALL | IndustriALL Global Union |
| IOE | International Organisation of Employers |
| IPCC | Intergovernmental Panel on Climate Change |
| IRENA | International Renewable Energy Agency |
| ISTAS | Instituto Sindical de Trabajo, Ambiente y Salud (Union Institute of Work, Environment and Health, Spain) |
| ITF | International Transport Workers' Federation |
| ITUC | International Trade Union Confederation |
| IUD | Industrial Union Department (USA) |
| IWA | International Woodworkers of America |
| LNS | Labor Network for Sustainability |
| MMC | Movimento de Mulheres Camponesas (Women Farmers' Movement, Brazil) |
| MMTR | Movimiento de Trabajadores de las Mujeres Rurales (Rural Women's Workers' Movement, Brazil) |
| MST | Movimiento des Trabalhadores Rurais Sem Terra (Landless Workers' Movement, Brazil) |
| NAACP | National Association for the Advancement of Colored People (USA) |
| NAFTA | North American Free Trade Agreement |

| | |
|---|---|
| NEDLAC | National Economic Development and Labour Council (South Africa) |
| NEHAWU | National Education Health and Allied Workers Union (South Africa) |
| NGO | Non-governmental Organization |
| NSZZ | NSZZ Solidarność (Poland) |
| NUM | National Union of Mineworkers (South Africa; UK) |
| NUMSA | National Union of Metalworkers of South Africa |
| OCAW | Oil, Chemical and Atomic Workers' International Union (USA) |
| OECD | Organisation for Economic Co-operation and Development |
| PACE | Paper, Allied-Industrial, Chemical and Energy Workers International Union (USA) |
| PCS | Public and Commercial Services Union (UK) |
| PSOE | Partido Socialista Obrero Español (Spanish Socialist Workers' Party) |
| PT | Partido dos Trabalhadores (Workers' Party, Brazil) |
| REIPPPP | Renewable Energy Independent Power Producers Procurement Programme |
| SAFSC | South African Food Sovereignty Campaign |
| SAFTU | South African Federation of Trade Unions |
| SAPP | Southern African Power Pool (SAPP) |
| SEIU | Service Employees International Union (USA) |
| TUAC | Trade Union Advisory Committee to the OECD |
| TUC | Trades Union Congress (UK) |
| TUCA | Trade Union Confederation of the Americas |
| TUED | Trade Unions for Energy Democracy |
| TUSDAC | Trade Union Sustainable Development Advisory Committee (UK) |
| UAW | United Auto Workers (USA) |
| UBC | United Brotherhood of Carpenters and Joiners of America (USA and Canada) |
| UCU | University and College Union (UK) |
| UGT | Unión General de Trabajadores (General Union of Workers, Spain) |
| UMWA | United Mine Workers of America (USA, Canada) |
| UN | United Nations |
| UNEP | United Nations Environment Programme |
| UNFCCC | United Nations Framework Convention on Climate Change |
| UNIFOR | Canada's largest private sector union |
| USFS | United States Forest Service |
| USW | United Steelworkers Union (USA, Canada, Caribbean) |

# List of Figures

| | | |
|---|---|---|
| Fig. 15.1 | Agrarian protests per type of claim, Brazil (2000–2014). (Source: DATALUTA 2015. Our elaboration) | 349 |
| Fig. 15.2 | Type of claims from environmental protests (2000–2014) Brazil. (Source: DATALUTA 2015. Our elaboration) | 350 |
| Fig. 18.1 | Timeline of the main milestones in the Senegalese agricultural policies and farmer movements since 1970 | 421 |
| Fig. 20.1 | Percentage of working hours lost due to heat stress, ten most affected countries, 1995 and 2030 (projections). (Source: Kjellstrom et al. 2019) | 469 |
| Fig. 20.2 | Correlation between labour productivity loss due to heat stress and social security coverage, selected countries. (Source: Kjellstrom et al. 2019) | 469 |
| Fig. 29.1 | Industrial technology as capital accumulation: how energy sources (vertical arrows) obscure social exchange relations (horizontal arrows) | 682 |

# List of Tables

| | | |
|---|---|---|
| Table 3.1 | Employment in key sectors of the South African mining industry | 65 |
| Table 3.2 | Policy adjusted IRP 2010–2030 and IRP2019 electricity energy share | 66 |
| Table 12.1 | Timeline of the emergence of Ecuadorian working-class environmental organisations | 291 |
| Table 18.1 | Main organizations cited and their characteristics | 424 |
| Table 20.1 | Sectors most affected by the transition to sustainability in the energy sector | 471 |

# 1

# Introduction: Expanding the Boundaries of Environmental Labour Studies

Nora Räthzel, Dimitris Stevis, and David Uzzell

## Introduction

When we edited our first book about the relationship between labour and nature we focussed on the environmental policies of trade unions across the world (Räthzel and Uzzell 2013). Not much had been published at the time and we sought to bridge the gap that separated research on the environment from research on labour, arguing that labour and environmental movements, though often fighting against each other, had one thing in common: defining nature as labour's 'Other' (ibid., 2). For environmentalists this meant prioritising nature in conflicts between production and the protection of nature, and for unions this meant prioritising workers' jobs. The same gap, we argued, that separated environmental movements from labour movements was

---

N. Räthzel
Department of Sociology, Umeå University, Umeå, Sweden
e-mail: nora.rathzel@umu.se

D. Stevis
Department of Political Science, Colorado State University, Fort Collins, CO, USA
e-mail: dimitris.stevis@colostate.edu

D. Uzzell (✉)
School of Psychology, University of Surrey, Guildford, UK
e-mail: d.uzzell@surrey.ac.uk

replicated in the social sciences, where environmental studies and labour studies rarely took notice of each other. Few researchers explored how the labour movement went beyond caring for workers' health and safety at the workplace and addressed forms of environmental degradation such as biodiversity, various forms of pollution or the climate (Silverman 2006; Adkin 1998; Obach 2004). Given that labour and nature are inextricably linked, and since all labour includes a transformation of nature while without non-human nature humans could not exist, work and survive, we proposed that there should be a discipline analysing this relationship, suggesting the term *environmental labour studies*.

Five years later, when we edited a special issue of *Globalizations* on the theme of 'Labour in the Web of Life', things had developed much further. In many countries of the world researchers were studying the environmental politics of trade unions and in the introduction to this special issue we defined environmental labour studies in a broader way, aiming to reach beyond the way in which unions addressed global environmental degradation:

> [E]nvironmental labour studies includes all research that analyses how workers in any kind of workplace and community are involved in environmental policies/practices and/or how they are affected by environmental degradation in the broadest sense. (Stevis et al. 2020, 4)

However, the papers in that issue were still largely concerned with the environmental policies and practices of trade unions across the globe. This remains a central topic that needs to be studied, since the deep transformations that are needed to develop societies with a just and caring relation towards nature and working people cannot be achieved without the active engagement of workers and their representatives as especially the authors in Parts I, II, IV and V of this volume show. Nevertheless, there is a divide that workers need to overcome in order to muster sufficient strength to effectively counter the capitalist destruction of workers and nature, namely the divide between industrial workers, subsistence workers, unwaged workers and workers extracting the materials on which every kind of production depends (see for this subject the authors in Parts I–III).

It is obvious that these areas of production could not exist without each other. Satheesh (Chap. 9), analysing the relations between industrial workers, peasants and fisherfolk in India, shows that they can become locked in fierce fights against each other when it comes to environmental issues. The ways in which industrial workers may be oblivious of the living and working conditions of those who supply them with the necessary resources for their work and life are reflected in the divide between two areas of research: the study of the 'environmentalism of the poor' as Martínez-Alier and others have coined it, concerned with environmental struggles of workers in rural areas of predominantly the Global South, and the study of 'labour environmentalism' concerned mostly with environmental struggles of industrial workers, predominantly in the Global North. In both areas we find research on mining: in labour environmentalism specifically on miners and in the environmentalism of the poor on the devastating effects of mining on rural communities.

Our aim in this book has been to include contributions from both areas of research in order to encourage and facilitate a conversation between them. Language can obscure these connections letting us forget that 'the poor' are workers and that 'workers' are often poor. Both need to fight for global environmental justice to survive. This does not mean that their struggles, strategies, conflicts and aims are the same. Leandro, Vieira Trópia and Räthzel (Chap. 15), Satheesh (Chap. 9), Latorre (Chap. 12), Bottazzi/Boillat (Chap. 18) and García-López (Chap. 17) explore the living and working conditions of agricultural workers and fisherfolk and the specific environmental struggles resulting from them. Barca/Milanez (Chap. 14) present the legacy of Chico Mendes, a trade unionist and environmentalist, whose vision of a common and respectful usage of the forests is today fought for by a coalition of waged workers, Indigenous peoples and peasants. Comparative research investigating the differences and commonalities of environmental struggles might help to support alliances on the ground. Cock (Chap. 8), Kojola (Chap. 16) and Dobrusin (Chap. 13) provide insights into the possibilities and difficulties of such alliances.

While we broadened the spectrum of environmental labour studies by including research on workers' environmentalism in the agricultural sector worldwide, there is one area of work that we have not been able to cover sufficiently, the huge area of service work. Some aspects of it are covered by Leonardi and Benegiamo (Chap. 31), Houeland (Chap. 21) and White (Chap. 35). If we go by the numbers, this area employs the largest amount of workers worldwide, that is, 50% of all employees (26% of waged labour is in agriculture, a percentage that is steadily shrinking, 23% is in industry, a percentage that is slowly rising (World Bank 2021). However, numbers are not everything. An important reason why service work has not featured prominently in research on labour environmentalism and the environmentalism of the poor is because the main sources of environmental destruction are industrial mining and manufacturing, transport as well as industrialised agriculture, and the main victims of global environmental destruction are workers in rural areas and in marginalised urban areas, populated by working-class people, including those who are migrants, and people of colour.

If we think about environmental destruction more comprehensively, not only in terms of $CO_2$ emissions but include the loss of biodiversity, the acidification of the seas, the deterioration of the soil, the amount of plastic waste, air pollution and other environmental toxins, we will find that the service sector, like any other sector of production needs to transform. In the future, environmental labour studies need to explore this sector, especially, since some of the most environmentally active unions organise workers in the public service sector.

There are two overarching themes that are central to environmental labour studies which all authors in this handbook deal with, explicitly or implicitly: (a) the relationship between society and nature. The way in which it is thought of is crucial for the way in which pathways out of the environmental crisis are imagined; (b) the labour-nature relationship—whether conceptualised as an alliance or a tension defines the priorities of action. In what follows we elaborate on these themes.

## Ways of Understanding the Environment and the Society-Nature Relationship

When people talk about 'the environment' it covers a spectrum of settings. Places and spaces that we refer to as 'environment' include not only mountains, fields and forests but also landscapes of factories and polluting chimneys, exemplified by what William Blake referred to in his poem as the 'dark Satanic mills',[1] which became symbols of the destruction of nature and human relationships. These were replaced in the following century by ex-urban industrial estates and vast out of town shopping malls, and even the oceans around the coasts are now home to the flaring of 'waste' gases from oil and gas platforms which draw carbon-based fuels for our transport, industry and domestic heating systems. At the micro-scale, the environment can be a significant health and safety issue, for example in the workplace such as an uneven floor leading to tripping, stairs without handrails or other ergonomic failings. In

---

[1] These words from a poem by William Blake (1810, 1993) have an iconic place in English culture, extolling a romantic representation of the traditional English landscape: 'And did those feet in ancient time/ Walk upon England's mountains green: /And was the holy Lamb of God,/On England's pleasant pastures seen!/And did the Countenance Divine,/Shine forth upon our clouded hills?/And was Jerusalem builded here,/Among these dark Satanic Mills?' These words later entered popular British culture in a hymn known as 'Jerusalem' (1916) by Sir Hubert Parry.

some cases, what we might think of as 'the environment' is really a referent to 'objects', that is, we want to protect the environment by recycling; thus, the objects (i.e., objects to be recycled) come to 'stand for' the environment. In this sense, the word 'environment' has become an 'empty signifier' (Laclau and Mouffe 2014), so much so that we typically now have to preface certain 'kinds' of environments with the word 'green' in order to contrast them with other spaces and places to which we refer and which are apparently untouched—nature and the natural environment. Such places have been seen as the antithesis of the world of work and thus places of refuge and restoration from labour (Hartig et al. 2011; Korpela and Kinnunen 2010). Interestingly, some of the first examples of workers 'reclaiming' nature as a response to the industrialised lives derive from protests stretching back to the nineteenth century. In 1895, the *International Friends of Nature* were founded in Vienna by a group of socialists, coming together through an advertisement in the *Arbeiter Zeitung* (Räthzel and Uzzell 2013, 1). In 1932, workers and 'environmentalists' joined together for an act of mass trespass on the grouse upland moors of Kinder Scout in Derbyshire which were owned by the aristocracy and wealthy industrialists to protest at the lack of access to green spaces around the industrial cities of the north of England (Walton 2013).

This kaleidoscope of meanings of the word 'environment' leads to fuzzy thinking about the object of our concern. Indeed, this is part of the problem because simply by talking about the *object* of concern as the environment, implies that there is a *subject*—for example, an actor, people and workers and the acted upon, the environment and nature. This sets up the binary relationship between people and especially the natural environment in which human agency is placed in active power over the 'natural' world. But even here, we have a problem with the word 'natural'. Drawing on Lévi-Strauss' (Lévi-Strauss 1992) idea of 'the raw and the cooked' and using the example of food, Lévi-Strauss argues that cooking is a cultural transformative process that turns nature into culture, and so what was nature now becomes 'natural', that is, nature as the processed, altered, commodified natural world. For example, think of how milk is turned into yoghurt. This can be achieved through entirely slow and natural processes or, once in the factory, through the addition of chemicals and sweeteners. Yet it is still advertised as natural with images of cows grazing on green alpine slopes as if the product has come straight from this rural environment to the plastic pot. By this process, categorical opposites drawn from everyday experience provide people with conceptual tools for the formation of abstract notions and for combining these into understandings of the world and how to deal with it. Thus, we can talk about commercial forests as if they are natural even though they are industrial sites, like a feedlot. When we look at the countryside we see a cultural environment. Following Lévi-Strauss we see nature processed into the natural. When we talk about 'green jobs' it makes us feel that we are getting back to nature and the natural order of things. But it is nature in this Lévi-Straussian way.

However, does this mean that all nature can be conceptually absorbed into societal practices? Not least the climate crisis is teaching us that as humans we are not only part of nature but our and the existence of other species is dependent on a specific constellation of the society-nature relationship. Is nature 'inescapably social' as social constructionist geographers would have it (Castree and Braun 2001), or do we need to think of the society-nature relationship as one in which societies work *on* nature and are able to destroy it as a support system for humans and other species? This is for instance the view of theorists of the Anthropocene (Chakrabarty 2009). Marxist ecologists like John Bellamy Foster maintain that it is the capitalist mode of production that alienates humans from nature, creates a 'metabolic rift' between societal life and nature and thereby destroys nature as a basis for human life. Barth and Littig (Chap. 33) discuss the issue using the term: Societal Relations with Nature (gesellschaftliche Naturverhältnisse, a concept widely used in German debates, predominantly by Marxists). They argue that while nature is also a product of society it has nevertheless its own materiality which can be experienced not least by the current environmental crises as a result of human intervention.

The authors in this handbook who discuss trade union environmental policies implicitly adhere to the view that society, more specifically, capitalist forms of production are acting *on* nature, where nature is understood as an object being destroyed and in need of protection as much as workers. This is true mostly for authors exploring trade unions' environmental policies, which can be found in the first, second and fourth parts of this volume. A different view is taken by those authors, who analyse environmental struggles of peasants and fisherfolk, mainly in the third part of the book. Here, following the workers they are investigating, the views are more akin to notions of a co-production.

In the 1940s the Marxist philosopher Ernst Bloch argued for an alliance of humans with non-human nature, more specifically for their co-productivity (Bloch 1973). Räthzel (Chap. 28) argues that the policies of trade union environmentalists in Spain can be understood as aiming at an alliance between society and nature in the way Bloch discussed it. Uzzell (Chap. 27) shows how the life histories of environmental unionists in the UK can explain their views on the society-nature relationship and how this in turn influences the ways in which they pursue environmental policies. Similarly, Stevis (Chap. 26) presents a history of environmental unionism in the USA using life histories of influential unionists. He shows how their experiences of a destructive work-nature relationship shaped their environmental engagement and how some sought a socioecological synthesis while others sought to protect nature through a green economy.

To overcome the difficulty of capturing the society-nature relationship without resorting to a language confined to a subject-object relationship, the feminist biologist Donna Haraway suggests the term 'sympoiesis', co-production as opposed to autopoiesis, self-production (Haraway 2016). Drawing on feminist theories Jason W. Moore seeks to overcome the society-nature dualism, analysing that capitalism produces nature and nature produces capitalism in an unequal process. Saave/Muraca (Chap. 32) discuss feminist analyses of the society-nature relationship who investigate how patriarchal capitalism constructs women as nature, while nature, in turn, is feminised, and thereby both are objectified. Criticising the idea of human dominance over nature, posthumanists aim to level the differences between human and non-human species. Hornborg (Chap. 29) criticises that posthumanism's relativism makes it impossible to critically analyse imperialism and the unequal ecological exchange between the Global North and the Global South as the condition of capitalist industrialisation.

## Ways of Understanding the Work/Labour: Nature Relationship

One might wonder how the different ways of understanding the society-nature relationship could be relevant for environmental labour studies. There appears to be an unbridgeable gap between the abstractions of an 'inescapably social nature', a nature alliance, a sympoiesis, an oikeios and the political language of just transition, climate justice and the everyday efforts of working people and their representatives to fight environmental degradation. However, theorisations can be useful to understand and suggest political practices. For instance, as argued elsewhere (Räthzel and Uzzell 2013, 2019) if workers conceptualise nature as 'labour's other' a space of recreation, rather than an integrated part of the production process in which they are being involved, this is conducive of understanding the protection of workers and the protection of nature as mutually exclusive. This is not to say that there is a necessary link between conceptions and practices, because people have different, contradictory views and understandings and therefore connections between worldviews and practices are loose. In Part V this is discussed more in detail, using life histories of environmental unionists.

Thinking the society-nature relationship is not identical with thinking the labour-nature relationship. In contrast with many other languages that know only one, English knows two words for the transformation of nature by humans: labour and work. Labour can signify workers themselves, as well as the act of producing, often connoting manual 'hard labour' or giving birth. Work can signify all varieties of the act of producing, including office work, service work and the work of professionals. When we speak about the labour-nature relationship we use labour to mean all practices in which human beings produce the means, food, clothing, shelter, transport and the tools and machines necessary for that production, as well as the practices that produce human beings themselves, including care work in the broadest sense. Labour is in this context the mediator between human and non-human nature, acting always within specific 'societal relations with nature'. This term is further developed by Barth/Littig (Chap. 33) who argue that in capitalist societies what is counted as work is paid work, while an environmentally sustainable society would have to include the largely unpaid care work of predominantly women (72.6% globally according to the International Labour Office 2018). Feminist scholars have investigated this issue for decades, at times asking for 'wages for housework' (Federici 1975). While the feminist sociologist Frigga Haug conceptualises both, the production of the means of life and the production of life, as processes of production and reproduction, most feminists, including ecofeminists, define the latter as the area of social reproduction. In Chap. 32 on the role of labour in the degrowth community, Saave and Muraca maintain that it is only recently that degrowth scholarship has acknowledged the role of feminist scholars regarding social reproduction and the perspective it offers for a future environmentally and socially caring society. In the context of the commons and commoning García-López (Chap. 17) uses the concept of social reproduction to mean all freely engaged activities performed outside waged labour without the aim of creating a profit. This includes, for example, the protection of forests, and the production of food and housing.

Authors in Parts II, III and VI refer to the concept of social reproduction (or to reproduction) arguing that we need a new concept of work that includes unpaid care and subsistence work as essential dimensions of the labour-nature relationship. This is argued either in the context of gender relations or referring to the work of farmers and peasants in the Global South or by Indigenous populations. When it comes to researching industrial work, the unpaid care work in households becomes invisible.

Assembling studies of the environmental policies of industrial workers and peasants and fisherfolk we aim to create a space of synergy in which the concepts of work and workers can be reconsidered also by those investigating environmental policies of industrial workers. For instance, for research on environmental union policies to include unpaid domestic work would make it necessary to include gender relations as an integral part of environmental policies. Broadening the concept of work can lead to the conclusion that environmentally and socially just societies would need to reduce the hours of waged work to free men and women for other activities like caring for others, developing their capabilities and taking part in constructing a better society as is argued by Wissen/Brand (Chap. 30), Barth/Littig (Chap. 33), White (Chap. 35) and Elliott (Chap. 36).

Studying the work of farmers and peasants does not necessarily lead to rethinking the concept of work in relation to unpaid work. As Bottazzi/Boillat (Chap. 18) show, another essential question is whether a new form of work, agroecology, has a chance to simultaneously feed agricultural workers and work the earth in a way that improves rather than destroys it. We are used to hear positive accounts of agroecology predominantly from South America but these authors and also Leandro, Vieira Trópia and Räthzel (Chap. 15) demonstrate the difficulties and contradictions that exist for introducing and sustaining it.

## Labour and the Environment: Elements of History

Historically, the relationship between labour and the environment has been subject to several key turning points. Here we focus on US and global labour environmentalism to highlight some of the important questions faced by labour environmentalism over the past several decades. Thus, this is not intended to be a systematic in-depth historical account, but a brief introduction to draw out some salient issues.

As Dewey notes with respect to US environmentalism, until the 1980s 'Many workers viewed environmentalists as elitist and aloof, as "extremists" who were callously indifferent to the economic growth and job opportunities essential to the well-being of ordinary working people' (Dewey 1998, 45), a perception of antagonism, mistrust and hostility that was shared with others including the public, politicians and the media. But Dewey and other environmental historians (e.g., Gordon 2004) argue that organised labour supported environmental initiatives since the 1940s, and their collaboration was one reason for the rise of a more social environmental movement in the 1960s (see also Rome 2003; Gottlieb 1993). The environmental movement in the post-war decades was focused on two issues. On the one hand, there were those who were concerned about the conservation of the natural environment, independent of social impacts, and were typically represented in the USA by the Sierra Club and the Audubon Society and by similar organisations across the industrialised world such as in the UK by the National Trust, the Royal Society for the Protection of Birds and the county naturalist trusts. On the other hand, there was a concern about industrial pollution impacting on the air, watercourses and rivers, and the sea through chemical and nuclear waste emissions as well as the unregulated extraction of natural resources, such as wood (Loomis, Chap. 6). These environmental concerns were more closely allied to health and safety issues, and perhaps therefore it was natural for workers to take a keen and personal interest in these because they were emanating from their workplaces and affecting them and their communities (Snell, Chap. 7, Liu, Chap. 24). But as Dewey states, labour environmentalists were some of the first to bring these two domains together within a single organisational structure, such that.

> Environmentally conscious labor leaders and union members, situated as they were at the crossroads of progressive politics, again preceded most of the rest of the environmental movement in conceptually linking environmental problems with wider social and economic issues. (Ibid., 46)

Dewey notes that labour's role in the environmental movement in this period is rarely acknowledged in narratives concerning labour's history which instead tends to focus on the civil rights, anti-war and feminist movements of the 1960s. US labour environmentalism, however, was an integral part of the fermentation that was taking place during the 1960s and 1970s (Rome 2003; Gordon 2004; Dewey 2019; Vachon, Chap. 5; Stevis Chap. 26). There is insufficient space here to document many examples of individual unions taking on environmental interests, but one can find examples in Barca (2014) who discusses early examples of working-class environmentalism in Brazil and Italy. In Chap. 31 Leonardi and Benegiamo place André Gorz's labour environmentalism within its historical context from the late 1960s to the 1990s.

In the UK, Farnhill (2014) carried out a review of the environmental policies of trade unions since the 1960s and concluded that the involvement of trade unions in the UK at a policy-making level has taken four forms: commented on, engaged and given traction to key environmental arguments and campaigns (e.g., a low-carbon economy and campaigning for lead-free petrol); used environmental arguments to reinforce traditional union demands (e.g., greater investment in public transport); enhanced the environmental component of various policy domains, for example, arguing for ethical public procurement policies; and injected their collectivist values and democratic principles into environmental decision-making debates.

It is worth remarking that Gottlieb (1993) also challenged the conventional perspective on the rise of environmentalism and argued that during the past four decades of the twentieth century the environmental movement was marked by a growing awareness of the issues of social justice, the necessity to take a global perspective and the need to unite 'natural and human environments' which would be more democratic and inclusive. Interestingly, Gottlieb talked of an 'environmentalism of transformation' (ibid., 320), a phrase which would not be out of place today in current discussions of just transition.

As Silverman wrote, just after the turn of the millennium,

> Labour thinking about the environment today grows from a profound, if halting, reorientation in trade union ideology about human beings' place in the natural world. At the same time, environmentalists within labour movements have been engaged in an essentially political process, one that involves compromise and modification of their ideas and programs, especially as they became increasingly involved in international institutions like the UN. The end result of these two intertwined evolutions has been the development of a labour version of sustainable development around which trade unionists in many countries are organizing their environmental activities. (Silverman 2004, 119)

When we wrote the introductory chapter to the book on *Trade Unions and the Green Economy* (Räthzel and Uzzell 2013), it felt as if the relationship between trade unions and the environment was at another turning point. The first conference on trade unions and the environment had been held in Nairobi in 2006, and the United Nations Climate Change Conference in Copenhagen in 2009 (COP15) had seen a significant presence of international trade union bodies arguing for a 'just transition'. Since then, trade unions in the Global North and South have sought to both bring their members on board in terms of supporting union 'green' policies and put pressure on national and international governments to recognise the importance of the environment and the critical role for workers in achieving a sustainable future for the planet (Stevis et al. 2020).

It would be wrong to infer that until this moment there had been little union interest or activism. For example, there was an early engagement by the ICFTU European Regional Organization (the predecessor of the European Trade Union Confederation [ETUC]) (Silverman 2004), and in 1971 the

ILO made the link between the working environment and the wider environment at the International Labour Conference:

> [T]he working environment is an important and integral part of the human environment as a whole since those factors that harm the working environment are also among the major pollutants of nature of people's living environment. (Olsen and Kemter 2013)

Notwithstanding this, Silverman points out that governments (and environmentalists) did not really consider labour interests in formulating international instruments. It is still difficult to disagree with the view Silverman put forward in 2004 that many subsequent intergovernmental agreements have been 'long on rhetoric but short on substantive commitments' (Silverman 2004, 120). Perhaps a notable exception to this has been the work and advocacy of the ILO (Olsen and Kemter 2013).

There are at least four reasons why it may be appropriate to refer to the years 2006–2020 as a turning point in the policies and actions of trade unions with regards to the environment. Prior to 2005, there were many trade unions engaged in environmental actions, most typically active at the local and regional level. These actions were often directed at single issues such as radiation from nuclear power plants and chemical spills but these were often framed, at least in part, in terms of health and safety. However, in Spain, as discussed by Räthzel (Chap. 28), environmental policies of unions were more advanced. Gereluk and Royer (2001) argue that 'trade union achievements in the field of health and safety demonstrate how the fight for sustainable forms of production has been central to workers' historical struggles against unjust conditions of work and community life' (ibid., 3). Consequently, the processes involved in protecting workers' health and safety rights and conditions are relevant to the expansion of their interests in environmental concerns and sustainable development, that is, occupational health and safety.

> [P]rocesses are typically participatory and transformative because they are based on the shop floor, incorporate dialogue between workplace parties, critically analyse organizational and system-wide causes of problems, and promote worker action and empowerment. (Ibid., 4)

The campaigns and actions by some unions in respect of occupational health and safety issues in the context of environmental issues were timely in as much they were sometimes couched in terms of sustainable development, and again were framed in the context and discourse of the social and economic element of Brundtland's sustainable development model (World Commission on Environment and Development 1987). Although this model was seen as strategic and general, it was often invoked only at a community level. This is illustrated in Chapter 29 of the Agenda 21 Report, the principal output of the Rio Earth Summit in 1992, entitled 'Strengthening the Role Workers and their Trade Unions' (UN Department of Economic and Social Affairs 1992). It advocated that not only should workers be participants in the implementation and evaluation of activities related to Agenda 21, but also that unions should be involved in improvements to both the work environment and the production process, as well as working within the local community. As Gereluk and Royer (2001) point out, the 'scientific management' processes promoted by F.W. Taylor which argued for the separation of conception and execution in the labour process with workers only involved in the latter leads to 'an unhealthy relationship of workers to their work, giving them little or no say in the terms and conditions of work and no responsibility for the product or process' (ibid., 3). Giving workers a say in the production process, deciding about its process, its output and usefulness for society was precisely what the Lucas Plan achieved and why it was both innovatory and provides a valuable model. However, as Elliott argues, (Chap. 36) this is also why it was not accepted by management.

But with the climate crisis came two additional challenges. The first of these has been not so much the environmental consequences themselves (although these are fully recognised for the impact they will have on their members), but rather governments' and industries' responses to the climate crisis and how these may affect jobs. The cure, in the form of reducing carbon emissions by means of legislation, tougher regulations, tax incentives and penalties, will have a significant impact on jobs in all sectors of the economy. This has required trade unions to think about the environment in a more comprehensive way and engage with concepts like 'green jobs' and 'just transition'. Authors in Parts I, II, IV and V discuss its importance for ILO policies and regulations. While the concept has not entered the research on the 'environmentalism of the poor' significantly, García-López (Chap. 17) discusses how this predominantly 'Northern' concept could be integrated into policies and practices of the commons, and Bell (Chap. 19) suggests that it might be used to overcome the borders between environmental struggles at work and environmental struggles of working-class communities in their living areas.

While some unions are embracing just transition policies as responses to the crisis, they are also aware that the crises facing labour are global. As it does not make sense to distinguish between the local and the global in respect of the climate and other forms of environmental destruction, it makes even less sense to distinguish between the local and the global in a globalised world when it comes to capital. While policy responses such as the greening of jobs and securing a just transition sound comforting, other responses to the environmental crisis may present serious challenges to trade unions and workers, such as the offshoring of jobs to countries where emission regulations and economies are weaker, and where labour is cheaper. Globalisation has set workers in the Global South and North in competition with each other. There is a necessity for a global response to the environmental crisis and the crisis of capital. This means that the labour movement and its allies must involve themselves actively in the process and participate in developing solutions, rather than just responding to measures taken by industry and government in defence of existing jobs.

What are the implications for solidarity across the Global South and North in the context of this local/global dimension? Under these conditions it is difficult for unions in the Global South and Global North to develop joint and sympathetic strategies and actions that will benefit all workers. We know that there has been resistance in both the Global South and Global North to accepting climate change as a threat because of the potential impact on jobs and unemployment[2] (Sikwebu/Aroun, Chap. 3). It is hardly surprising then that the workers in India for instance say: 'I will die quicker from not having a job than from climate change'. Under such dire conditions it is difficult for progressive unionists to put forward just transition climate change strategies on their union's agenda. This is not only a problem in the Global South. There are examples of unions in the Global North which have advocated protectionist policies such as border adjustments. This is where imported goods such as steel from low-cost economies are heavily taxed in order to protect jobs rather than create an equitable market, what Dimitris Stevis has referred to as 'green nationalism' (Stevis 2013).

If the climate crisis is not to be a union crisis and traditional core union beliefs in solidarity are not to be thrown aside, then globalisation has to be interpreted and used by unions in constructive ways to their advantage. This is a key challenge for unions in the coming years.

---

[2] For example, in Brazil pre-COVID unemployment was at about 13% (and approximately 27% in respect of youth unemployment); in South Africa, pre-COVID unemployment was about 30% (and 58% youth unemployment); and in Spain, pre-COVID unemployment was about 14% (and 31% youth unemployment) (Trading Economics 2020).

## Environmentalism of the Poor and Environmental Justice

Environmental labour studies have so far covered predominantly the environmental engagements of trade unions in industry, predominantly in the Global North. Environmental engagements of Indigenous communities, peasants, poor farmers and fisherfolk have been researched for a longer period of time and predominantly in the Global South. This research has covered environmental resistance in many places, as well as a variety of struggles, strategies and concepts used to define and justify environmental demands.

In his book, *Environmentalism of the Poor*, Joan Martínez-Alier (2002) describes the environmentalism of rural communities before the term 'environmentalism' existed. Examples are the struggles of workers against a contaminating copper mine in Japan in 1907 and of peasants against the contamination of water, soil and air threatening their health caused by copper mines and cement production (for China today see Liu, Chap. 24). In 1880, at the Rio Tinto in Andalucía workers and peasants fought together against the contamination of their lands and waters by the Rio Tinto Zinc company.

Such struggles, aiming to prevent environmental destruction caused by mines, oil extraction, deforestation, tree plantations for industrial use, damns and other practices destroying the working and living conditions of rural and Indigenous communities, were termed the 'environmentalism of the poor' by academics and activists in the late 1980s in India (Centre for Science and Environment, https://www.cseindia.org) and Peru (see Joan Martínez-Alier et al. 2016). Later on the term was established through studies conducted in Latin America, Asia and Africa (Guha and Martínez-Alier 1997). As Martínez-Alier explains, the

> 'environmentalism of the poor' does not assert that as a rule poor people feel, think and behave as environmentalists. This is not so. The thesis is that in the many resource extraction and waste disposal conflicts in history and today, the poor are often on the side of the preservation of nature against business firms and the state. This behaviour is consistent with their interests and with their values. (Martínez-Alier 2014)

The term is also meant to defy the notion that environmentalism is something of a luxury conviction and practice that is only possible to develop when people's basic needs are already satisfied. While this may be true for some consumer practices like buying ecologically (Bell, Chap. 19), the 'environmentalism of the poor' is not oriented towards consumption but towards production. What Indigenous and peasant communities are fighting for is to preserve the foundations of their livelihoods: uncontaminated soils, water, air, forests and the oceans. In this way they are conservationists but not in the conventional sense of preserving a pristine nature which human beings only enter for relaxation, but in the sense of preserving nature as an indispensable source for the production of the means of life.

Another concept that is used to describe these forms of resistance is 'environmental justice' (but see Snell, Chap. 7, who promotes the usage of this term for trade union environmentalism). Labour environmentalists and their allies played a critical role in developing environmental justice during the 1970s (Gordon 2004, 214–229; Rector 2018). During the 1980s environmental racism gave more impetus to questions of environmental justice, as evident in North Carolina and other mobilisations during which people fought against waste dumping in their neighbourhoods. This involved not only opposition against waste dumping but also against the siting of contaminating industries in areas of poor citizens, predominantly people of colour are described as fights for environmental justice or against 'environmental racism' (for a history of environmental justice see Taylor 1997).[3]

While such forms of resistance occur in urban areas, the 'environmentalism of the poor' has traditionally been researched in rural communities. Anguelovski and Martínez-Alier (2014) analyse the need to extend the concept further to forms of resistance in urban areas. They mention gentrification, contamination, the loss of liveable spaces in cities through the invasion of private traffic, the lack of green spaces and poor environmental services as troubles against which neighbourhood communities fight. Arguing for the need to connect different kinds of environmentalism in rural and urban areas—articulated in concepts like land grabbing, water justice, food sovereignty, health and safety at work, urban squatters and many more—they suggest understanding environmental justice as ultimately 'the defence of the right to place and territory and the right to stay without being displaced' (ibid., 173).

Martínez-Alier et al. (2016, 748) contend that there is a global environmental justice movement even if most struggles are local and organised by local movements. However, they are often supported by international NGOs, Environmental Justice Organisations (EJOs), academics and other institutions. Also, there are often connections and networks, creating global alliances and the actors against which these struggles are oriented are often the same, namely transnational corporations. Accounts of such struggles are offered in

---

[3] The term 'eco-justice' was first used by religious activists during the early 1970s and combined social and ecological justice in ways that today is called ecological justice.

Parts I, II and III of this volume. A formidable resource to learn more about these struggles on a global scale is offered by the Environmental Justice Atlas (https://ejatlas.org).

What the different forms of environmental justice struggles have in common is that communities aim to prevent changes in their living and working conditions which threaten their survival, their needs to live in affordable and healthy urban environments and their identities. In this sense they bear a similarity to trade union struggles against forms of production threatening them and their communities' health and safety. However, if we think about environmental practices and policies against global environmental destruction in terms of the climate crisis, biodiversity, acidification of the sea, deforestation, environmental toxins and other harms, the nature of environmental justice struggles and trade union environmental practices and politics are quite different. While the former are directly connected to people's livelihoods and living conditions and tend to *preserve* (or improve) them, the latter need to engage in *changing* their working and living conditions substantially, depending on the production process of which they are part. In the case of fossil fuel extraction and production it implies leaving behind their jobs, skills and capabilities, which also shape their identities and leap into the unknown with little security (as history has shown) and assurance that they will land in a secure place. It is therefore not surprising that trade union environmentalism is often defensive rather than pro-active. In a sense, like rural and urban communities they want to defend their livelihoods. The problem is that while in the case of many rural struggles this includes fighting for a better environment, in the case of many industrial jobs it implies fighting for the preservation of jobs, which are responsible for the destruction of the environment. Thus, while alliances between industrial workers and rural workers are necessary, since neither is strong enough to bring about transformative change on their own, such alliances are extremely difficult. Rural and industrial workers (to use a shorthand) are bound together in mutual dependency. But in the short run, what connects them are mutually exclusive interests. For instance, while communities in mining areas fight against the destruction of their living and working areas, workers in industrialised processes depend for the jobs on such mining. Their common interests exist on a more abstract level namely on the fact that, as the international trade union movement states, there are no jobs on a dead planet, neither rural nor urban ones.

# Environmental Labour Studies: An International and Interdisciplinary Research Area Bridging the Academic/Practitioner Divide

One of the most stimulating features of working in the area of environmental labour studies is that it has at least three key characteristics. First, it is *interdisciplinary*. If this area of work we have called environmental labour studies is to provide critical analyses of the ways in which production in the broadest sense creates and destroys environments and how working people can act towards these processes in order to identify and promote solutions to some of the most pressing crises the planet faces, then it has to call upon the theories, perspectives and methodologies of academic disciplines across the natural, physical, and biological sciences, the social sciences and the arts and humanities. The authors in this handbook work in the area of social sciences, focusing on working people as actors. Within this general parameter authors have referred to a variety of disciplines. A number of studies related in this book are based on research projects with interdisciplinary teams (Leandro, Trópia Vieira, and Räthzel, Bottazzi/Boillat, Uzzell/Räthzel, Thomas/Pulignano). Interdisciplinary work has much to offer our understanding of how workers as mediators of the society-nature relationship cannot only be part of the solution to environmental crises but also have the potential to connect environmental solutions with solutions to inequality and injustices.

A second feature of environmental labour studies is that it is *international*. Most research on the environment and labour has an international dimension, but the interaction of nature and labour is particularly acute when dealing with globalised economies and globalised labour relations, and the divisions of labour that produce them. Transnational corporations cause environmental damage in countries of the Global South where environmental regulations are weaker because governments are desperate for investments. Workers and their communities in these countries are at the sharp end of environmentally, socially and economically damaging policies and modes of production and, as Sikwebu and Aroun (Chap. 3) show, can therefore only engage in environmental policies when they can benefit from them. Given the ways in which workers of the Global South and of the Global North are pitched against each other as competitors, trade unions of the Global North often fail to overcome these divides by treating unions of the Global South as receivers of 'help' rather than allies in a global battle (Uzzell and Räthzel 2013). Thus, there is a lack of joint and solidaristic strategies to create a global-level playing field for workers and unionists in the Global South and North.

In addition, there is a lack of solidarity by unions in the Global North but also in the Global South for the needs of subsistence workers and frontline communities. The authors in Parts I–V demonstrate the need for alliances between trade unions and other workers' organisations. We have made an effort to invite contributions from a variety of countries across the world and hope that with the authors coming from 18 countries of the Global North and South the handbook can contribute to understanding the need for such broader alliances.

Thirdly, ELS furnishes *engaged* research (i.e., research that connects research, policy and practice) that can be useful not only for social actors, especially for working people and their representatives, but also for organisations and institutions striving for transformative change towards environmentally and socially just societies. Of course, this should not mean that research must be directly applicable to specific issues. As we know, theoretical work that does not seem to have any effect on solving problems can be crucial to changing the way we think and thus ultimately the way we treat 'real-life' problems. What is necessary though is to bridge another gap, namely the one between 'practitioners' and academics. While they should not replace each other, they can and should learn from each other. In this volume we have taken some steps towards this process. One is to listen to the stories of workers engaged in putting the environment on the trade union agenda. In the fifth section of the handbook which presents and analyses the life histories of individuals as promoters of change this has been our intention. In addition, we have been able to invite some researchers outside of academia to analyse the work of their organisations, namely of the ILO (Saget, Luu and Karimova), the ETUI (Galgóczi), civil society (Dobrusin) and NUMSA (Sikwebu and Aroun). In addition, a number of contributors hold or have held both academic and activist positions (Azzi, Bell, Elliott, Snell, Pillay, Rajan, Vachon and Ytterstad). This makes the political and institutional context of environmental strategies and their opportunities and constraints more visible, and is a welcome correction to suggestions from academics who are mostly unfamiliar with the nitty-gritty of everyday struggles.

## Structure of the Book

The handbook is divided into six parts, which reflect the main themes and simultaneously the future challenges that environmental labour studies face. We avoided a structure that would divide the book geographically, wanting to highlight that even if the conditions are different, working people across the globe are facing common threats. Of course, boxing chapters into specific thematic parts is always an ambiguous endeavour since every chapter has multiple dimensions and can fit into several boxes at once. Nevertheless, arranging the chapters thematically can offer an idea of some of the main issues and challenges discussed in environmental labour studies to date.

## Part I: Histories

While ELS is an emerging discipline, the political practices which it wants to explore and promote as central for an environmentally and socially just transition have a much longer history. We wanted to begin the book with some aspects of its diverse histories to create a context for the present and the future. The histories presented here cover sometimes countries (Rajan, Pillay, Vachon and Snell), and sometimes specific unions and the ways in which their environmental policies have changed over the years (Sikwebu and Aroun, Loomis).

In India, the relationship between environmental movements, the environmentalism of the poor and industrial workers is scant. As Ravi Rajan in Chap. 2 argues, state policies towards transforming the environment have been disenfranchising the poor, while industrial trade unions have yet to address the crucial significance of environmental destruction for workers. He presents a wide variety of literature on connections between environmental transformations and labour, which can serve as a groundwork for environmental labour studies.

Dinga Sikwebu and Woodrajh Aroun in Chap. 3 present a detailed analysis of the environmental policies of the National Union of Metalworkers of South Africa (NUMSA), of which they are both members. They show how NUMSA rose to the challenge presenting a programme for social ownership of renewable energy and developing a far-reaching programme educating their members on the climate crisis and strategies for its solution. However, for the time being the environmental policies of NUMSA have broken down as a result of an unjust transition as the authors explain, arguing that only if workers can benefit from a transition will they be ready to fight for it.

In Chap. 4 Devan Pillay takes a different look at NUMSA's history, a union in the vanguard environmentally among metal worker unions globally from the perspective of a 'decent life' in addition to 'decent work'. He argues that the working class is in need of a strategy that overcomes the fixation on GDP and thus economic growth as the measurement of well-being, discussing conflicts within NUMSA between 'ecosocialist' and 'Marxist-Leninists' worldviews.

Drawing on his experience in the labour movement as a union carpenter, an organiser with a large manufacturing union, and as the founding president of a large education union, Todd E. Vachon in Chap. 5 introduces us into the history of the 'just transition' concept in the USA to explain today's opposition by the majority of US labour unions to the Green New Deal presented by congresswoman Alexandria Ocasio-Cortez and Senator Ed Markey. An opposition that is difficult to understand since their roadmap towards zero carbon emissions focuses on workers, workers' rights and a just transition.

In the USA there is a long history of connections between environmentalists and unionists, which tends to be forgotten amidst present-day conflicts between the two movements. In Chap. 6 Erik Loomis tells the story of the close collaboration between timber workers and ecologists in the USA from the early twentieth century until its demise in the 1970s.

Darryn Snell in Chap. 7 argues that it is largely ignored that trade unions worldwide have a history as agents of environmental justice. He recounts this history focussing on Australia but including also trade unions' environmental justice practices on the international level. He asks environmental justice researchers to take the role of unions in this movement more seriously.

## Part II: Seeking Common Ground

A challenge for the success of workers' environmental engagements is alliances. It cannot be stressed often enough that neither environmentalists in the North and South, nor industrial workers, Indigenous peoples, food producers or miners can turn around our destructive mode of production on their own. More positively put, all these and other social agents have much to learn from each other. Thus, the theme of alliances, their success, prospects and losses are a central concern. While the theme surfaces in many chapters, those whose main concern is with alliances comprise the second part of the book.

In Chap. 8 Jacklyn Cock brings the strained relationships between labour and environmental movements in South Africa into focus. Moreover, based on her research in coal mining communities she argues that frontline communities need to be integrated into the unions' environmental policies. Despite, or even because of, their contradictory position of being the victims of the destructive effects of mining but also benefitting from it, they can play a powerful role in the process of environmental transformation.

Silpa Satheesh in Chap. 9 explores a conflict that is less visible in environmental labour studies, namely between different groups of working people. This is a place where industrial workers and agricultural workers and fisherfolk are in direct contact with each other. However, instead of understanding each other's needs their immediate needs of survival pitch them against each other in struggles against and in favour of industrial production, which destroys the livelihoods of farmers and fisherfolk.

Chapter 10 by Diego Azzi brings us back to industrial workers in the Global North and South. He analyses to what degree different just transition strategies of unions remain within given societal structures (consensus) or go beyond them (dissensus) particularly in their relations to business, the state and to other social movements and NGOs.

An issue on which unionists and environmentalists have been able to come together are the campaigns for climate jobs. In Chap. 11 Andreas Ytterstad, co-founder of this campaign in Norway, presents it as a strategy that involves a Climate Jobs Plan framed as a *just transition* away from fossil fuels and with a cosmopolitan perspective. He explains the difficulties of keeping the allies together while maintaining a radical transformative focus.

From Norway we move to Ecuador where Sara Latorre in Chap. 12 questions the conceptual separation of struggles around agricultural issues as 'social movements' and labour studies. Arguing that they are both workers' struggles, she recounts alliances and divisions between struggles of Indigenous peoples, agricultural workers and oil workers, representing the most important Ecuadorian economic sectors.

A hopeful story seems to be unfolding in Canada, from which Bruno Dobrusin in Chap. 13 presents just transition movements like the 15&Fairness in Ontario, which comprises Indigenous groups, community groups and workers in extractive and other industries putting forward a 'societal shift'. He argues that the Canadian Pact for a Green New Deal combines a worker as well as a societal focus.

## Part III: Farmers, Commoners, Communities

This part is dedicated to struggles that have been described as the environmentalism of the poor. Most but not all of them take place in the Global South. There are also Indigenous communities struggling against industrial projects that destroy their livelihoods in the Global North and engagements for liveable urban space belong to such struggles as well.

Stefania Barca and Felipe Milanez have visited the extractivist movement in Brazil, a coalition of waged workers, Indigenous people and peasants known as Aliança dos Povos da Floresta (Alliance of Forest Peoples), who fight for the preservation of the forest as a commons and develop ways of usages that simultaneously feed people and protect nature. They show in Chap. 14 the extent of violence that such promising movements face from corporations and politics.

With Beatriz Leandro, Patrícia Vieira Trópia and Nora Räthzel in Chap. 15 we remain in Brazil and learn more about the environmental struggles of peoples in the Brazilian Hinterland. An analysis of their issues is further deepened by the presentation of influential environmentalists in trade unions and other organisations of peasants in general and women agricultural workers in particular.

In Chap. 16 Erik Kojola examines the famous conflict over the North Dakota Access Pipeline that includes Indigenous people, environmental organisations and trade unions acting against and in favour of the construction of the pipeline. He weaves together the different dimensions of power relations that shape these conflictual relationships: colonialism, racism and capitalist exploitation.

Gustavo García-López in Chap. 17 argues for the importance of the concept and practices of the commons and commoning for environmental labour studies. He broadens the study of labour environmentalism to include 'labourers of the earth', who he sees at the forefront of resistance against environmental degradation, creating also alternatives for non-exploitative, cooperative work, jointly owning and carefully using lands and forests in solidarity. His overview over the literature in this area is complemented with examples of commoning in Puerto Rico.

In Chap. 18 Patrick Bottazzi and Sébastien Boillat provide an insight into farmers' movements in Sub-Saharan Africa, focusing on Senegal which is different from similar movements in South America. They show the difficulties that agroecology face in the region, where competition with conventional agriculture is strong and farmers are taught agroecological methods by organisations coming from outside the country.

This part of the books ends with another connection that workers and thus environmental labour studies has to make, namely the connection between workers as citizens and workers as producers. Karen Bell in Chap. 19 shows how working-class communities fight the environmental degradation in their neighbourhoods, sometimes as a result of polluting production in the vicinity. She suggests that there is space for an alliance between trade unions and working-class citizens to create a 'community environmentalism'.

## Part IV: Trade Unions and the State

Workers and their organisations not only need to create alliances with other social movements, but also need to influence or respond to state policies which may aim to address environmental degradation but pay no or insufficient attention to the effect these have on workers. Equally, they need to create alliances among themselves, since as they are dependent on national context, level of internationalisation and economic sector their interests differ considerably.

In Chap. 20 Catherine Saget, Trang Luu and Tahmina Karimova working for the ILO present an exemplary overview of state policies aiming at a reduction of $CO_2$ emissions, while protecting the rights of workers. Taking their point of departure from the guidelines for a just transition of the ILO they select three different countries in terms of their economic position on a world scale to illustrate the effect of the diversity of such policies: Spain, South Africa and Ethiopia. While states declare their policies as progressive, unions in these countries do not necessarily share this view.

Camilla Houeland in Chap. 21 analyses trade union fights against fuel subsidy removals in Nigeria, Sudan, Ghana and Zimbabwe. This is an issue also hotly discussed in Europe, especially in France where the removal of such subsidies gave life to the protests of the 'gilets jaune'. As Houeland shows, fuel subsidies removal does not only hurt middle-class people and SUV owners, as is often argued, but even more seriously poor people. This is why trade unions resist them and call for alternative strategies to reduce $CO_2$ emissions.

In Chap. 22 Adrien Thomas and Valeria Pulignano give us an overview of different just transition policies of trade unions in different sectors, different countries and on different levels of union organisation. International unions are more open to transformational change, while unions organising workers in the fossil fuel area and at the factory level tend to prioritise the immediate interests of their members to retain their jobs.

Nevertheless, as Béla Galgóczi demonstrates in Chap. 23, more or less wide-reaching just transition policies are implemented by states, and trade unions play a role in shaping them. Discussing the phasing out of coal in some European countries, the transformation of the car industry and environmental changes at the company level, Galgóczi argues that these policies can only become '*just* transitions' when they are complemented by a transformation of the existing structure of European welfare states into eco-social states.

A different relationship between state policies and workers' rights plays out in China as Juan Liu explains in Chap. 24. While there is legislation in place to protect workers' rights to healthy and safe working conditions, employers and local administrations find ways to deprive workers of these rights. This has led to such a high number of pneumoconiosis cases among rural workers who migrated to the mines and construction sites that she speaks about pneumoconiosis villages. Liu concludes that these production processes are not only damaging to workers' health but need also to be understood as environmentally destructive.

## Part V: Organic Intellectuals

Part V collects chapters that are similar in terms of their methodological approach. Here the authors have aimed to write the stories of environmental unionists from their point of view using the life histories of influential unionists in the USA (Dimitris Stevis, Chap. 26), the UK (David Uzzell, Chap. 27) and Spain (Nora Räthzel, Chap. 28). Starting from the notion that movements can make history, but that in order to do this, not only specific societal conditions are needed but also individuals with the capacity to seize the moment, take their comrades with them and thereby cooperatively realise the societal possibilities for change. The authors analyse the intersections between individual life histories, organisational and national histories.

## Part VI: Rethinking and Broadening Concepts

The last part of the book unites chapters in which authors contribute to the theoretical frameworks that are needed in environmental labour studies by rethinking some of its central concepts. Alf Hornborg in Chap. 29 examines the categories of labour and energy, reviewing physicalist and constructivist notions to develop a third perspective that avoids the naturalisation of societal processes of asymmetric exchange and accumulation to which the first two are prone.

In Chap. 30 Markus Wissen and Ulrich Brand have a comparable goal, namely, to show how the 'imperial mode of living', based on unequal divisions of labour globally, results in unequal power and economic relations between the Global North and the Global South, where the former's more favourable living conditions are made possible by the constant exploitation of the work and lives of the latter. This largely hidden unequal dependency is one of the causes for the reluctance of workers in the Global North to address the ways in which economic growth causes environmental destruction.

Reviving the work of André Gorz, Emanuele Leonardi and Maura Benegiamo in Chap. 31 set out to make his work fruitful for environmental labour studies. Gorz's labour-based political ecology, they claim, allows us to assess how the link between capital and nature has changed over the past and how these changes affect union politics.

In Chap. 32 Anna Saave and Barbara Muraca scrutinise the ways in which degrowth scholars have understood the concept of work and workers' agency. Including ecofeminist scholars, they suggest ways in which degrowth scholars and movements can develop a more inclusive, richer concept of work and workers.

From a different point of departure but with a similar goal, Thomas Barth and Beate Littig in Chap. 33 argue for the need to centre environmental policies and practices around the concept of work. This requires that the concept of work is liberated from its reductionist meaning that includes only waged employment. Including unpaid, mostly women's care work opens up the perspective to another kind of environmentally and socially just society that can also overcome unequal gender relations.

Picking up the threads of gender relations as relations of production and connecting them with conceptions of the society-nature and labour-nature relationships, Nora Räthzel in Chap. 34 claims that it is possible to envisage a theoretical framework that can analyse the connections between different categories of workers in industries, subsistence production, extraction and services (particularly care work), the North-South divide and the divide between paid and unpaid work.

Coming from the area of design, Damian White in Chap. 35 takes us through 100 years of history as well as prospective connections between the work of design, the working classes and ecological and feminist perspectives. He argues that environmental labour studies may well accommodate approaches to an inclusive just transition through labour and ecologically friendly designs.

David Elliott closes the volume (Chap. 36) with a wide-ranging discussion of the future of work that includes the question of automation and its perspectives and threats, new forms of low-scale production 'at home' with 3D printing methods, and the prospects of a new technology in the wake of Mike Cooley's ideas that technology should enhance not replace workers' skills. He concludes with the notion that the future is open and it is for workers to make it.

## Possible Futures for Environmental Labour Studies

With this handbook we have broadened the spectrum of the labour-nature relationships that environmental labour studies could explore to include not only industrial workers and their alliances with Indigenous peoples and ecologists, but also food and forest workers and workers in the commons. We have argued that environmental labour studies should also address marginalised workers in all sectors, including the industrial sector, as they are subject to the most serious environmental harm. It should expand the universe of workers to include non-waged workers of all types.

There is a wealth of research on the 'environmentalism of the poor', especially in the area of political ecology. So why is there a need to create yet another label to include this work? There are two reasons, in our view: firstly, the specificity of environmental labour studies lies in including working people across the globe as agents into our analyses. Secondly, there is a need for a dialogue not only between researchers working in the Global North and the Global South but between workers in these different areas. To be clear, focusing on working people as agnets and not only as victims of exploitation, appropriation and dispossession does not mean that we see the 'working classes of the world' as a homogeneous entity of subordinated people, who only have to 'unite and fight' to save the planet. As some authors in this volume have demonstrated, relations between different categories of workers are wrought with unequal gender relations, unequal power relations benefitting workers of the Global North over those of the South, as well as relations of power and subordination between workers within the Global North and South, including racism and the effects of colonialism. To face these contradictions by analysing their history, their present-day effects and the ways in which they impede solidarity not only between workers but also between workers and the nature they are transforming is only one, but a necessary step towards overcoming them. For environmental labour studies, thus, exploring the agency of workers towards environmental destruction and social inequality does not only mean exploring their struggles against these processes, but also the limitations of these struggles, the ways in which they might be themselves implicated in and part of the forms of exploitation, oppression and subordination they aim to overcome. Environmental labour studies has the potential to provide evidence-based support for policies and practices to protect and support working people in the struggle for environmentally and socially just societies, but only if it engages in the creation of a critical knowledge base.

To become more inclusive in terms of geographical spaces and the varieties of work, environmental labour studies needs to further develop its concepts and scientific methods. This includes among others the concepts of work and workers, the understanding of the society-nature and the labour-nature relationships and the relationship between dispossession, appropriation and exploitation in a world now controlled by various forms of the capitalist mode of production. In other words, environmental labour studies not only needs to expand its investigation of alliances between workers but also create alliances between researchers of different disciplines with a similar concern: what is needed to propel the world out of its trajectory towards the destruction of humans and non-human nature? Environmental labour studies has the capacity to learn from other areas of scholarly endeavour, for example, environmental history, environmental sciences, political ecology and Marxist ecology. It needs to learn from ecofeminists since gender relations and the relations between labour and nature are entangled with each other and with all other relations of power.

## References

Adkin, Laurie E. 1998. *The Politics of Sustainable Development: Citizens, Unions and the Corporations*. Montréal; Buffalo, NY: Black Rose Books.

Anguelovski, Isabelle, and Joan Martínez-Alier. 2014. The 'Environmentalism of the Poor' Revisited: Territory and Place in Disconnected Glocal Struggles. *Ecological Economics* 102 (June): 167–176. https://doi.org/10.1016/j.ecolecon.2014.04.005.

Barca, Stefania. 2014. Laboring the Earth: Transnational Reflections on the Environmental History of Work. *Environmental History* 19 (1): 3–27.

Blake, William. 1810. Preface. In *Milton: A Poem in Two Books*. London.

———. 1993. *Milton a Poem, and the Final Illuminated Works: The Ghost of Abel, On Homers Poetry, [and] On Virgil Laocoön*. Blake's Illuminated Books 5. Princeton, NJ: William Blake Trust/Princeton University Press.

Bloch, Ernst. 1973. *Das Prinzip Hoffnung 3 3*. Frankfurt am Main: Suhrkamp.

Castree, N., and B. Braun. 2001. *Social Nature. Theory, Practice and Politics*. London: Wiley-Blackwell.

Chakrabarty, Dipesh. 2009. The Climate of History: Four Theses. *Critical Inquiry* 35 (2): 197–222. https://doi.org/10.1086/596640.

Dewey, Scott. 1998. Working for the Environment: Organized Labor and the Origins of Environmentalism in the United States, 1948–1970. *Environmental History* 3 (1): 45–63.

———. 2019. Working-Class Environmentalism in America. In *American History*. https://oxfordre.com/americanhistory/view/10.1093/acrefore/9780199329175.001.0001/acrefore-9780199329175-e-690.

Farnhill, Thomas. 2014. Environmental Policy-Making at the British Trades Union Congress 1967–2011. *Capitalism Nature Socialism* 25 (1): 72–95. https://doi.org/10.1080/10455752.2013.879196.

Federici, Silvia. 1975. *Wages against Housework*. Bristol: Falling Wall Press.

Gereluk, Winston, and Lucien Royer. 2001. *Sustainable Development of the Global Economy: A Trade Union Perspective*. Geneva: International Labour Office. https://citeseerx.ist.psu.edu/viewdoc/download?doi=10.1.1.455.9836&rep=rep1&type=pdf.

Gordon, Robert W. 2004. *Environmental Blues: Working-Class Environmentalism and the Labor-Environmental Alliance, 1968–1985*. Dissertation, Wayne State University.

Gottlieb, Robert. 1993. *Forcing the Spring: The Transformation of the American Environmental Movement*. Washington, DC: Island Press.

Guha, Ramachandra, and Juan Martínez-Alier. 1997. *Varieties of Environmentalism: Essays North and South*. London: Earthscan Publications.

Haraway, Donna Jeanne. 2016. *Staying with the Trouble: Making Kin in the Chthulucene*. Durham: Duke University Press.

Hartig, Terry, Agnes E. van den Berg, Caroline M. Hagerhall, Marek Tomalak, Nicole Bauer, Ralf Hansmann, Ann Ojala, Efi Syngollitou, Giuseppe Carrus, and Ann van Herzele. 2011. Health Benefits of Nature Experience: Psychological, Social and Cultural Processes. In *Forests, Trees and Human Health*, 127–168. Springer.

International Labour Office. 2018. *Care Work and Care Jobs for the Future of Decent Work*. Geneva: ILO.

Korpela, Kalevi, and Ulla Kinnunen. 2010. How Is Leisure Time Interacting with Nature Related to the Need for Recovery from Work Demands? Testing Multiple Mediators. *Leisure Sciences* 33 (1): 1–14.

Laclau, Ernesto, and Chantal Mouffe. 2014. *Hegemony and Socialist Strategy: Towards a Radical Democratic Politics*. London: Verso Trade.

Lévi-Strauss, Claude. 1992. *The Raw and the Cooked: Introduction to a Science of Mythology*. Harmondsworth, Middlesex, England: Penguin Books.

Martínez-Alier, Joan. 2002. *The Environmentalism of the Poor: A Study of Ecological Conflicts and Valuation*. Cheltenham: Elgar.

———. 2014. The Environmentalism of the Poor. *Geoforum* 54 (July): 239–241. https://doi.org/10.1016/j.geoforum.2013.04.019.

Martínez-Alier, Joan, Leah Temper, Daniela Del Bene, and Arnim Scheidel. 2016. Is There a Global Environmental Justice Movement? *The Journal of Peasant Studies* 43 (3): 731–755. https://doi.org/10.1080/03066150.2016.1141198.

Obach, Brian K. 2004. *Labor and the Environmental Movement: The Quest for Common Ground. Urban and Industrial Environments.* Cambridge, MA: MIT Press.

Olsen, Lene, and Dorit Kemter. 2013. The International Labour Organization and the Environment: The Way to a Socially Just Transition for Workers. In *Trade Unions in the Green Economy. Working for the Environment*, eds. Nora Räthzel and David Uzzell, 57–73. London and New York, NY: Earthscan Routledge.

Räthzel, Nora, and David Uzzell, eds. 2013. *Trade Unions in the Green Economy: Working for the Environment.* London and New York, NY: Earthscan Routledge.

———. 2019. The Future of Work Defines the Future of Humanity and All Living Species. *International Journal of Labour Research. ILO* 9 (1–2): 145–171.

Rector, Josiah. 2018. The Spirit of Black Lake: Full Employment, Civil Rights, and the Forgotten Early History of Environmental Justice. *Modern American History* 1 (1): 45–66. https://doi.org/10.1017/mah.2017.18.

Rome, Adam. 2003. "Give Earth a Chance": The Environmental Movement and the Sixties. *The Journal of American History* 90 (2): 525–554.

Silverman, Victor. 2004. Sustainable Alliances: The Origins of International Labor Environmentalism. *International Labor and Working-Class History* 66: 118–135. https://doi.org/10.1017/S0147547904000201.

———. 2006. Green Unions in a Grey World: Labor Environmentalism and International Institutions. *Revue belge de philologie et d'histoire* 84 (4): 1123–1139. https://doi.org/10.3406/rbph.2006.5063.

Stevis, Dimitris. 2013. Green Jobs? Good Jobs? Just Jobs? US Labour Unions Confront Climate Change. In *Trade Unions in the Green Economy. Working for the Environment*, eds. Nora Räthzel and David Uzzell, 179–195. London and New York, NY: Earthscan Routledge.

Stevis, Dimitris, Edouard Morena, and Dunja Krause. 2020. Introduction: The Genealogy and Contemporary Politics of Just Transitions. In *Just Transitions: Social Justice in the Shift Towards a Low-Carbon World*, eds. Edouard Morena, Dunja Krause, and Dimitris Stevis, 1–31. London: Pluto Press.

Stevis, Dimitris, Edouard Morena, and Dunja Krause. 2020. Introduction: The Genealogy and Contemporary Politics of Just Transitions. In *Just Transitions: Social Justice in the Shift Towards a Low-Carbon World*, ed. Edouard Morena, Dunja Krause, and Dimitris Stevis, 1–31. London: Pluto Press.

Taylor, Dorceta. 1997. American Environmentalism: The Role of Race, Class and Gender in Shaping Activism 1820–1995. *Race, Gender & Class*, 5(1), 16–62. Accessed January 16, 2021. http://www.jstor.org/stable/41674848.

Trading Economics. 2020. Trading Economics. 2020. https://tradingeconomics.com/.

UN Department of Economic and Social Affairs. 1992. United Nations Sustainable Development: Chapter 29 Strengthening the Role of Workers and Their Trade Unions. Rio de Janeiro, Brazil. https://sustainabledevelopment.un.org/content/documents/Agenda21.pdf.

Uzzell, David, and Nora Räthzel. 2013. Local Place and Global Space: Solidarity across Borders and the Question of the Environment. In *Trade Unions in the Green Economy. Working for the Environment*, 241–256. London and New York, NY: Earthscan Routledge.

Walton, John. 2013. The Northern Rambler: Recreational Walking and the Popular Politics of Industrial England, from Peterloo to the 1930s. *Labour History Review* 78 (3): 243–268. https://doi.org/10.3828/lhr.2013.14.

World Bank. 2021. 'Employment in Industry (% of Total Employment) (Modeled ILO Estimate)'. The World Bank. Accessed June 20, 2021. https://data.worldbank.org/indicator/SL.IND.EMPL.ZS.

World Commission on Environment and Development, ed. 1987. *Our Common Future*. Oxford Paperbacks. Oxford; New York: Oxford University Press.

# Part I

## Histories

# 2

# Labour and the Environment in India

S. Ravi Rajan

## Introduction

Indian environmental movements, and by extension, scholarship about the environment in India, have largely focused on how state policies have disenfranchised the poor (Guha 1988; Guha and Martínez Alier 2000). They have also addressed the conditions of work and livelihood of the labouring classes in a wide range of sectors. Although labour movements and organizations have tended to focus on more traditional issues, they have addressed topics such as workplace safety, pollution and risk. There is also a significant body of scholarship from Marxist and kindred traditions that has specifically addressed environmental issues (Qadeer and Roy 1989; Roy 2000). Therefore, the groundwork for a field of study explicitly addressing the interface between labour and the environment in India exists. The purpose of this chapter is to analyse this wide-ranging literature with the view of fleshing out some of the broad insights that the Indian case contributes to the emerging field of environment-labour studies. The chapter also attempts to identify emergent threads and opportunities for future research and public engagement in this field. It has four parts, reflecting the broad sectoral ranges of topics that are at the forefront of concerns. Section 2 begins with a discussion of natural

S. R. Rajan (✉)
University of California, Santa Cruz, CA, USA
e-mail: srrajan@ucsc.edu

© The Author(s), under exclusive license to Springer Nature Switzerland AG 2021
N. Räthzel et al. (eds.), *The Palgrave Handbook of Environmental Labour Studies*,
https://doi.org/10.1007/978-3-030-71909-8_2

resource regimes, from forestry and agriculture to irrigation, which have institutional histories stemming from the colonial era. Section 3 addresses the confluence between industry, labour and the environment. Next, the chapter explores a host of issues concerning the urban sector, where labour and the environment are increasingly juxtaposed in tangible, material ways. Last, but by no means least, Sect. 5 focuses on the topic of displacement of labour across the sectors.

## Natural Resource Regimes

Studies of the environment and society in India have long addressed the consequences of state policies about natural resource regimes such as forestry and irrigation, and in particular, the conflicts between state policies and the needs of local resource users. While the topic of labour in this vast literature is present, it is not foregrounded. This section will describe the main arc of the scholarship, and on that basis, attempt to identify some of the key dimensions of the environment-labour dynamics therein.

Conflicts over natural resources date back to the nineteenth century, during which the British colonial state dramatically altered the manner in which forests and water resources were harnessed, controlled and managed. The systems, processes and infrastructures that they established served imperial goals—of extracting as much revenue as possible (Rajan 2006). In the case of forestry, although there were negotiations in some parts of India between local and traditional stakeholders and the colonial government, the broad trend in state policy during this period was to diminish or even abrogate traditional community rights to access and use forest produce (Guha 2000). At the same time, as trade in forest products increased, albeit differentially in different parts of India, new social groupings, often caste based, emerged, whose primary business was in timber and other forest-related commodities. These business elites developed at the cost of livelihoods of local people denied local people their traditional access rights to forests (Tucker 2011).

This state of affairs continued into the post-colonial era, with the passions and interests that won out under colonialism consolidating under the new political reality of independent India. Over the decades, however, there were many local protests against forest department policies, and by the early 1970s, full-blown social movements that addressed the interface between environment and human rights emerged (Hardiman 1999; Miśra and Tripathi 1978). With such movements gaining steam in different parts of the country, new discourses about resource use emerged, along with myriad experiments in alternative resource management strategies that emphasized inclusiveness and explored the economic viability of devolving forest management to local communities and firms (Khare et al. 1999, 33–41). Among them are burgeoning social movements on gender and the environment, addressing the relationship between deforestation, increasing female workload and labour, and decreasing access for girls and women to education, nutrition and community (Agarwal 2000, 2001; Mehta 2009; Rao 2012).

Similar to forestry, the colonial establishment sought to transform the landscape of agriculture, emphasizing cash crops and other revenue enhancing approaches geared to increase the colonial bottom line. Irrigation infrastructures such as canal building were a central part of this strategy. The investments largely had their desired impacts during the colonial era: yields expanded and per capita output of crops increased by nearly 45 per cent between 1891 and 1921 (Whitcombe 1970). However, local resource users suffered. Water, which previously had been largely a common property resource with some constraints of access and fees, was now a commodity, to access which, farmers had to pay fees. The colonial state undertook ruthless water fee recovery on all lands deemed to be irrigated, and thereby forced farmers to grow cash crops to generate cash. This exacerbated economic inequities by privileging those who could afford the service payments—water rates for the use of canals—and also engendered widespread corruption. Moreover, the system caused a number of ecological problems. Many areas of Punjab, once known for its good, well-watered soils, are today saline deserts, with alkaline and unproductive lands. The system also neglected, and in some cases, destroyed, traditional water systems where they did not serve the revenue needs of the colonial state (Whitcombe 1970; D'Souza 2006).

After independence in 1947, a major rethink of state policy began, with water works being seen as public goods (Shah 2011). This rethink reflected the nature of the coalition of interests, including landowners who profited from the colonial regime on one end of the spectrum, and the movements for land reform and redistribution, on the other. These tensions resulted in a number of local contestations. At the broad, policy level, though, irrigation charges were drastically reduced, and significant public funds were expended on irrigation. At the same time, the canal systems were poorly maintained, partly due to the reduced collection of maintenance fees, resulting in cycles of maintenance and decline, and further tensions and contestations on the ground. The result was the emergence of what has been called a "scavenging irrigation economy" with a proliferation of wells (initially, traditional and artisanal, but soon, tube wells, riding on subsidized electricity) (Dubash 2002; Iyer 2013). Gradually, this resulted in the depletion and breakdown of traditional lakes, and indeed, medium and major systems that were maintained by communal work (Shah 2010). Moreover, with the rapid spread of tube wells, groundwater began to be depleted, posing severe threats to hydrological systems (Dubash 2002; Iyer 2013). Compounding these problems has been the huge human cost of large dams, hundreds of which were built in independent India, the world's third largest builder of dams (2000) (Khagram 2004). As in the case of forestry, the irrigation economies also witnessed a number of localized peoples movements focused on redressing the injustices related to water access. There has also been a vibrant debate about equity and rights in access to water resources; democracy and participation; inter-state conflict management; groundwater regulation; and pollution, amongst a host of issues (Batliwala and Brown 2006).

Although the literature on forestry and irrigation does not explicitly address the interface of the environment and labour, it does so implicitly. At least five key issues can be identified that connect labour and the environment in the cases of both forestry and irrigation. First, both these resource management regimes required significant human power to carry out projects that involved significant transformations of landscapes. In the case of forestry, human labour was critical to artificial regeneration projects and a range of other infrastructure needs—from road building to harvesting. In the case of irrigation, the bulk of the work to carve out canal streams and maintain them could only be done by human labour in an era in which mechanization was not widespread, and this, needless to say, often meant poor working conditions and wages. Second, the regimes of natural resource management such as forestry and canal irrigation created by the colonial state ended up displacing more traditional systems of resource management. In both instances, there were strong traditions of communal natural resource management, embedded deeply in grounded socioeconomic and cultural systems (Mosse and Sivan 2003). The advent of state regimes, along with their legal and property regimes, eroded them, and as a consequence, traditional skills and organizational systems of management dissipated in many areas.

A third dimension of the labour-environment connect is migration. In the case of the canal projects, for example, labour was imported from other states. This created tensions between local resource users and outsiders, and this engendered tensions that were exploited by the interests and castes who benefitted most from the canal economy. Next, in the case of both forestry and irrigation, new forms of labour emerged—with expertise in civil engineering, public administration and forest management—whose agendas often ran up against the needs of local communities (Whitcombe 1970). Last, but no means least, is the topic of gender. In the case of forestry, with the extinguishing of traditional rights, and subsequently with increased deforestation, the brunt of fuel wood collection fell on women, which meant a lack of time for education, lack of resources for health and the breakdown of social networks, amongst others (Agarwal 1992). In conclusion, the advent of modern forestry and irrigation completely transformed social, cultural and economic relations, upending the community-based traditions of labour to those that served wider interests—initially imperialism, and subsequently, of those who emerged as the primary executors and beneficiaries of the imperial era. This dynamic is a critical component of the politics of natural resource regimes until today.

## Industry, Labour and the Environment

As India embarked on an industrial policy as part of its five-year plan driven developmental agenda starting with the second five-year plan during 1956–1961, environmental problems arose both within workplaces as well as in the world outside. Given that there were a wide range of industries, a wide range of hazards emerged—in and around industrial townships, industrial estates and the myriad unorganized and sometimes illegal clusters of factories that began to dot the landscape across the country starting the late 1960s. In response, a number of small organizations associated with labour—mainly outside the formal trades unions—started to document cases reflecting adverse work and living conditions—that impacted them since the mid-1970s. Broadly speaking, there were three types of issues and concerns—the acute, the chronic and the emergent.

Acute events are those that cause immediate harm. For example, I estimate, based on newspaper reports about gas leaks, boiler blasts or a catastrophic mechanical breakdown leading to a worker losing a limb, or even his or her life, that at least a couple thousand lives are lost each year. Sometimes, the acute events can be catastrophic, as was the case with the infamous gas leak disaster in Bhopal, which happened on the night of December 2–3, 1984, immediately killing between 2000 and 6000 people (the actual number of fatalities remains disputed), and maimed about half a million others and their progeny (Hazarika 1989). This case demonstrates that events that are seemingly accidental are embedded within broader political and economic dynamics that define them. It also speaks to the interface between power and poverty, and between the environment and labour.

Although the city of Bhopal has a long and history, it was, in 1984, a classic company town. Its political economy was dominated by Union Carbide India

Limited, the Indian subsidiary of Union Carbide Corporation (Chouhan et al. 1994). Union Carbide came to Bhopal as part of a complex of technologies that came to India in the wake of the Green Revolution, the agro-industrial complex that ushered in "high-yielding varieties" of cereals, along with chemical fertilizers and pesticides. Heralded as the panacea to address food shortages and poverty, the Green Revolution was also an opportunity for Western multinationals to rake in significant profits in the developing world (Shiva 1991). Union Carbide had an operational culture characterized by chronic lax maintenance, lower number of supervisory employees and inadequate safety training. This was true both with the US parent in the United States and in subsidiaries the world over (Morehouse et al. 1986). In Bhopal, many of the key design decisions made by the company, both during the planning and subsequent phases of the plant, meant that it was constantly plagued by accidents and mishaps throughout its history. The accident itself was not a surprise to any of the workers or operators in the plant or to the city's journalists who had been writing about it over a period of time (Chouhan et al. 1994).

Almost all the people killed and maimed lived near the plant and were part of the labour that had been recruited to the city from elsewhere in the hinterland and the country at large to work there. Yet, instead of helping their workers, the corporate response after the disaster was to look after its own economic interests. The company hired a top public relations firms, lobbied governments, denied that the accident was a result of its actions, advanced an unproven theory of sabotage and embarked upon strong divestment strategies culminating first in record profits, and then in the sale of the company to Dow Chemicals. The victims were neither rehabilitated nor adequately compensated, while the perpetrators of the world's worst industrial accident got away with little penalty or consequence (Rajan 1999). Another facet of the Bhopal story is the failure of both the state and the national governments. They did not ensure that the company was adequately regulated and failed to respond immediately and effectively when the accident happened. Nor did they respond adequately to the need to create the capacity for the long-term rehabilitation of the victims (Rajan 2001, 2002).

The state also failed to address the intrinsic problems in its own processes and systems. For example, in Bhopal, eligibility for rehabilitation required the certification of victims, for which certain sets of documents were needed, but not always present. The result was the emergence of a parallel economy of false documents and bribery (Rajan 1999). The state machinery also did not have the ability or the expertise to recognize or respond to such tragic choices (Calabresi and Bobbitt 1978). While it might be argued that Bhopal was an extreme case, accidents are a part of the industrial landscape in India (Indiastat

2020). On the other end of the spectrum, despite localized protests by workers, including at the Union Carbide factory in Bhopal, the labour movement at large in India did not, until Bhopal and even later, highlight or organize around either environmental hazards in the factory or toxic spills in poor and working communities that were increasingly becoming ubiquitous around India. International labour organizations also did not raise safety issues prominently in India until after the Bhopal disaster.

Chronic events, in contrast to acute, kill and maim over a long period of time. An example in point is air pollution which, by the end of the 1980s, had taken a huge toll on the environment and society in India. According to the World Health Organization's *Global Burden of Disease Report for 2010*, air pollution became the fifth leading cause of death in India, with 620,000 premature deaths in 2010, representing a six-fold increase since 2000 (Singh 2018; Narain and Bell 2006). Moreover, as the case of attempts at regulating air pollution in the capital city of New Delhi shows, the problem of pollution was only addressed partially, with the populations exposed to a range of other toxins that were in the ambient environment due to other sources besides those stemming from vehicular fuel (Véron 2006; Sahu 2008; Mathur 2005; Bell et al. 2004; Gauri 2009). Moreover, the poorest of the road users, notably, the drivers of auto-rickshaws, who had to face the brunt, as the infrastructure for the provision of CNG took a long time to establish. It was quite common, for months after the Supreme Court ruling, for drivers to wait many hours in line for their fill, thereby foregoing their income, whilst continuing to be exposed to pollution. In the meantime, those that had not converted to CNG or had not invested in a host of other devices were subject to fines, and harassment by the police.

A third category of issues might be termed emergent, referring to topics that, while not new, raise a host of emerging issues. A case in point is that of genetically modified organisms (GMOs). The regulatory structure for GMOs adopted by the Government of India mandated a regime of the safeguards that include pre-release testing following a US EPA precedent; and prescribed punitive action for any violation and non-compliance. Critics, however, rejected governmental claims about the environmental and health safety of GM crops and foods, and raised a host of questions concerning biosafety, consequent to the import of GM foods into India (Lianchawii 2005). A particular focus in the Indian GM debate was Bt (*Bacillus thuringiensis*) Cotton, which was formally released in 2002. The regulatory approval process for Bt Cotton in India was highly contested, and the struggles encompassed science, values, world views, political economic gradients and institutional processes and procedures. However, many farmers adopted the technology illicitly, thereby subverting the regulatory process and short-circuiting public debate (Lianchawii 2005; Subramanian and Qaim 2010; Qaim et al. 2006; Raju 2007; Scoones and Studies 2003). This case reflects differences between middle-class perceptions and institutions, along with internationalist environmentalist norms, on the one hand, and the needs of other classes. Moreover, it raises the question of how democracies decide between competing perceptions and needs. What matters more, the livelihoods of farmers, against a backdrop of indebtedness and farmer suicides, or purported risks to consumers?

Another type of emergent issue addresses a different set of trade-offs. A case in point is the Tehri Dam—a massive multi-purpose power project located in the state of Uttarakhand in India, the first phase of which was completed in 2006. Tehri is a component of a multi-purpose river valley project, and the main dam in Tehri town is the eighth tallest in the world. By mid-2006, when the dam was commissioned, more than US $1 billion had been spent on its construction (Fink 2000). The Tehri Dam has been controversial because it is sited in the Central Himalayan Seismic Gap, a major geologic fault zone, and barely 50 km from the epicentre of a magnitude 6.8 earthquake in October 1991 (Brune 1993; Govardhan 1993). Many stakeholders have been embroiled in these debates, ranging from the Indian central government to local NGOs (Paranjpye 1988). The result has been that the authority of a sequence of expert committees, constituted by the central and state governments as well as courts, was questioned, and their scientific validity publicly disputed. Where the Tehri Dam issue related to labour, specifically, is in the impact of the dam on those who lived there, as the dam submerged large areas of land and forced approximately 85,600 families to relocate against their will (Dogra 1992). This issue has raised another big question—about the fairness of forcing some people to sacrifice their livelihoods and even cultural contexts in order to supply something that others need—in this case energy. This reflects topic, addressed in more detail later, that raises broader ideological arguments about development priorities and trade-offs.

Perhaps an extreme example of such trade-offs is the case of nuclear energy. Significantly, due to the secrecy regime under which the nuclear programme functions in India, safety statistics are not easily forthcoming. Journalists and scholars who study Indian nuclear programme however argue that there have been too many safety lapses, including several instances of radiation leakage, to sustain the government's claim that it is safe and clean (Ramana 2006; Ramana and Reddy 2003; Mathai 2015). Moreover, the industry is not unionized, making it difficult for the workforce to demand accountability. The problem is rendered more acute because of the close association of the nuclear programme with the national security infrastructure, as a result of which protests by workers, and even journalistic and academic work is significantly curtailed (Bidwai 2011). In addition to the issue of safety at the workplace, the nuclear establishment poses significant questions about the safety of the places in which the labour force, and indeed, a wider community, lives. Closely related to safety is transparency and accountability, for honesty about the actual record can help forge institutional practices that can help foster a better record (LaPorte and Consolini 1991; Roberts et al. 2001). In India, the atomic energy establishment has, from its inception, taken decisions without parliamentary or public scrutiny. In recent times, a raft of public controversies has opened up a public debate. However, it is not however evident that public protest has enabled the establishment to think differently or create newer, more democratic and accountable systems of operation. On the contrary, it appears that the agency has instead hired firms to advise it on how better to communicate to the general public, on grounds that what is needed is not a dialogue, but a top-down communication.

Controversies about nuclear energy raise serious questions about the capacity of the state, and especially, of the atomic energy establishment to protect the labour force and respond adequately to the threat to the community at large in the case of nuclear accidents (Kaur 2020). The issue of state capacity, in turn, raises two further questions. Foremost amongst these is the question of whether hierarchical and technocratic establishments such as the atomic agency have the know how or the methodologies to understand the concerns of lay peoples, and indeed its own labour force (Leach et al. 2005; Wynne 1992; Collins and Evans 2008; Irwin and Wynne 2009; Smith and Wynne 1989; Wynne 2010; Gopalakrishnan 1999). Consequently, the communications and interactions between atomic agencies and the communities they are located in have not engendered the trust needed for local peoples to understand and abide by safety protocols, such as warning alarms or follow evacuation procedures. It is noteworthy here that the history of the interactions between atomic energy experts and lay people worldwide is full of examples of mistrust and miscommunications (Wynne 1992; Sands 1988; Mackay and Thompson 1988; Mould 1988). The second question concerns the capacity of the institutions that constitute the nuclear regulatory establishment—with management structures based on established hierarchies and decision-making structures—to respond to rapidly changing scenarios far away from decision-making centres. Where there are high public and environmental safety stakes, the design of management systems that enable groups within organizations to interact and collaborate becomes crucial. Here, the literature on high reliability organizations is particularly important, especially in its emphasis on the need to pay particular heed to the expertise or the tacit knowledge that labour further down the value chain possesses—expertise that can often help prevent an emerging threat from cascading into a catastrophe. This literature is also important in that it draws attention to the need to incorporate recalcitrant information, and integrate whistle blowing and alternative ideas and analytical perspectives into their decision-making iterative churn. Owing to the culture of secrecy, little is known on these issues about the Indian case. What is obvious is that there remains the prospect of significant collateral damage should an adverse event happen.

The case studies mentioned earlier are amongst the more prominent concerning industry, labour and the environment. For the most part, they paint a rather bleak picture of the labour organization, and their capacity to be agents of change. There are many other sectors in which Indian labour movements have made a difference, albeit at smaller or more localized scales. Extractive industries such as mining are important sites in this regard, with labour organizations increasingly adopting environmental issues in their mobilization agendas (Krishnan 2016; Lahiri-Dutt 2016). Another example lies in the ship-breaking yards such as those in Alang, which involve toxic substances such as asbestos, on the one hand, and child labour, on the other. Here, there have been increasing attempts in India, in the sub-continent as a whole, and more globally, to organize and mobilize for better working conditions (Ghosh et al. 2016). There are some who argue that labour movements and environmental movements in India have not worked productively together. For example, the Indian Supreme Court Advocate S. Muralidhar writes:

> Similarly, the trade unions do not see the problem of child labour as a labour problem. This brings us back to the question of a decisive approach. In case of the labour movement, we see they have adopted the decisive approach within the movement, that is not recognising unorganised labour. I, for instance, being involved in the Bhopal gas litigation, was always surprised how out of that litigation one concern was not highlighted. I mean the voice from the workers themselves. They were dealing with a very hazardous process. But no one in the unorganised labour force protested about working in those conditions. I think we haven't had that kind of response from within the labour force about environment, about the conditions of work about the health of workers. (CSE 2015)

Others have pointed out that the formal Left in India have traditionally failed to recognize the significance of environmental issues, focusing instead on their traditional concerns with class. The noted writer, Praful Bidwai, for example, observes:

> Until fairly recently, it (the Left) saw environmentalists fighting against destructive "development" projects such as high dams, river diversion schemes, gigantic mining ventures, highways cutting through rainforests, and highly polluting chemical and metallurgical industries or coal-fired power stations, in largely adversarial terms, as enemies of "development." (Bidwai 2015, 285)

He adds that Left in India has been deeply suspicious of non-party political formations, especially those that address issues such as the environment, and that they have rarely shown any recognition of ecological processes, the long-term consequences of the destruction of biodiversity, or the risks involved with nuclear technology—choosing in most instances to oppose people's movements in these sectors, even when they were led by their own cadres. He remarks, rather scathingly:

> The Left's dogma-driven approach to ecological questions dispensed with the need to engage in any concrete, issue-specific, well-reasoned arguments about the appropriateness or relevance of nuclear power and other high-risk technologies for India, their economic costs and generic safety problems, or India's embarrassingly poor record on assessing risks and regulating them. (Bidwai 2015, 288)

Moreover, he observes,

> The Left has paid little attention to renewables, to energy conservation and efficiency, and to other "green" measures. Nor does it have a position on adaptation to climate change, based on innovative ways of promoting low intensity agriculture, afforestation, water and coastline conservation, and so on. (Bidwai 2015, 289)

The contrast between the Indian formal Left, comprising the Communist parties, and their counterparts in China, from which at least one thread has historically drawn inspiration, as the environment has been on the radar of theory and ideology in China since at least the early 1970s.

# Urbanization

Industrialization in India after independence from the British resulted quite rapidly in the proliferation of urban and peri-urban areas. With the poor in the rural hinterlands pushed out of their homelands due to underemployment, displacement and the opportunities afforded by cities, a large number flooded India's city. According to one report, during the 40 years from 1971 to 2008, India's urban population rose by nearly 230 million. Significantly, though, the report states that it will take half that time to add the next 250 million. The report states further that the expansion will affect almost every state and points out that five of its largest states will have more of their population living in cities than in villages (McKinsey 2020). India's cities therefore face significant challenges—to employ, house, transport and provide education and health care for millions of the urban poor, as well as lower- and middle-class people who struggle to make a living. Against the backdrop of this demographic, social and environmental reality, there has been a vigorous debate about who urban areas are for and what are the rights of citizens and the duties of states.

The question assumed particular salience for the first time during the so-called Emergency in 1976, when then Prime Minister Indira Gandhi, in a show of force against her political opponents, misused state power during the 21 months in which the proclamation was in effect. Amongst them were two events that directly concerned the urban poor and the urban environment. The first of these involved population control. Indira Gandhi had long nurtured an interest in the environment and had built up a public image as a patron of conservation. She was particularly prone to inviting leading international experts and had long lent an ear to their advice. On the advice of Neo-Malthusian overseas experts and her son, Sanjay Gandhi, a technocratic environmentalist with an authoritarian bent, her government decided to act to enforce population control policies in urban areas. In the capital city of New Delhi, the government ran sterilization camps, and in many instances, police arrived in peri-urban parts of the city early in the morning, rounded up the men and forcefully sterilized them. According to official statistics during 1976–1977, the programme led to 8.3 million sterilizations. At the same time, her government embarked on a radical slum clearance project, with bulldozers razing the ostensibly illegal dwellings of the urban poor. These two acts of self-proclaimed environmentalist intent subsequently became a lightning rod for introspection about urban policy.

A host of issues arise as the topic of the urban environment gains traction in public consciousness. Among them is transportation. A significant number of the working-class labour force cannot afford public transportation systems and have to either walk or cycle to work. They therefore face the constant threat of being struck by vehicular traffic. India has 1 per cent of the world's vehicles, but accounts for 6 per cent of road accidents. Of the total number of deaths, 36 per cent impact those who ride two wheelers such as cycles, and 15 per cent, pedestrians (Salve 2019). This means that the poor face the brunt of accidents. There is also a gender dimension, as women pedestrians have had, on top of the regular hazards, also to bear the burden of sexual harassment by passers-by. Yet another group of people challenged by pedestrian transportation are hawkers on wheeled trolleys, who, in addition to having to face injuries on the road itself, are regularly threatened by policemen and other municipal officers who impounded their trolleys. Street vendors who transport perishable goods are threatened similarly by officials. The reason for the harassment is that the city laws often rendered street hawking illegal—on grounds that it impedes traffic (Roy 2000). The only way out for these people is to pay a bribe, and thereby add to their income burden. As Dunu Roy wrote, reflecting on this syndrome:

> All these road users, therefore, had much in common: they were not given licences or were on contract; their "illegality" enabled more value to be extracted out of them; the "environment" was a weapon to deepen their insecurity; and their unity was disrupted. They compared themselves with the cars, malls, retail centres, and waste compactors to argue that they created little congestion, used less energy, provided cheap services. ... Yet, they did not receive just value in return. They also found it curious that policymakers wished to convert a car-free society into a car-worshipping one. For them, it was illogical to tackle the problems created by motorised traffic after motorisation had taken place. Rather, one should build upon what is of existing economic and ecological value. There was a natural opposition between collectives of workers and those who profit from the products of their work. But, the opposition was often buried by preventing collectivisation of labour through division, contracting, outsourcing, and automation of work. (Roy 2000, 61)

Another dimension of the labour and the environment in the urban context is waste. Cities everywhere produce an enormous amount of waste, and Indian cities are no exception. India generates more than 60 million tonnes of waste every year. Less than 60 per cent is collected and about 15 per cent is processed (Swaminathan 2019). The task of cleaning up this waste is left to labourers who are largely in the unorganized sector. These people, including children, toil under difficult conditions, for poor wages. They have little by way of protective equipment, even whilst working with highly toxic substances as in the case of e-waste. Given their liminality in the social context, they are also unrecognized and invisible as a labour force. If anything, they are subject to harassment, both by the government and by criminal cartels who exploit them. Against this backdrop, there have been a number of organizations that have arisen in the civic sector. A well-known early one was the Self Employed Women's Association (SEWA) in the city of Ahmedabad, which began organizing female waste pickers from the 1970s onwards (Jhabvala and Bali 1990). These women were provided with personal protective environment, and infrastructures that helped them sort, store and re-sell waste products, and thereby increase their income. SEWA's approach has been adopted by and taken forward by other organizations. For example, in the capital city of New Delhi, an environmental research and action group called Chintan organized waste workers, helping them rise up the value chain so that they transform from scavengers to managers, collecting and recycling more than 30 tonnes of solid and electronic waste every day (Chintan n.d.). Further, they work to ensure that the children of waste pickers enter and stay in school. They also conduct research to help cast light on the conditions of work of Delhi's waste pickers, and advocate on their behalf. Organizations like Chintan have now proliferated across the country. One example is Safai Sena (cleaning army), which is a registered association of over 12,000 waste pickers across 3 states in North India. Safai Sena advocates for the recognition of waste pickers and small traders as a key aspect of waste management in India and against their stigma and exclusion. It works on the basis of local leadership and collective advocacy. Even as income from waste increases through such efforts and the conditions of labour improves, with concomitant societal benefits, such initiatives are threatened by new macro-economic trends, stemming from the privatization of waste management and the increasing adoption of mechanization and new logistical technologies.

The evolving urban context in India is a barometer of the environmentalism of the poor. Rather than sustain the binary of environment versus development, it raises broader questions—especially about the quality of life, and especially, the role of civic spaces, laws and institutions, in creating, nurturing and regenerating people's spaces. The debate over the last two decades has been productive, and has yielded a number of exciting initiatives, including those of sustainable cities, especially, urban transportation tuned to the needs of the labouring poor. Crucially, in some contexts, the plight of labour has also been politicized in a manner that has resulted in easing the burden of the labouring urban poor. A case in point is the Aam Aadmi Party (Common Person's Party; AAP) that started as a people's movement in 2012 and evolved into a political party that got elected to run the Delhi state government, first as a minority government in 2013, and subsequently as the majority in 2015. The AAP government has been a good example of how progressive policies can begin to tackle some of the chronic problems mentioned in this section and do so in a manner that is both sustainable and equitable.

## Displacement

As briefly alluded to in the previous sections, displacement from traditional homelands is one of the central reasons for the proliferation of urbanization in India. According to some statistical accounts, about 50 million people have been uprooted from the rural hinterland—and the agro and forestry based livelihoods that sustained them (Internal Displacement Monitoring n.d.). They made way for "development," an euphemism to describe a host of projects, from hydroelectric dams to coal-fired plants to pesticides factories such as the one that killed at least 5000 people in Bhopal in December 1984. Of the various sources of development-induced displacement globally, dams are amongst the worst—accounting for more than 60 per cent of people displaced by World Bank funded projects, although they constituted about 26 per cent of all of the Bank projects causing displacement (Dams 2000).

Since independence, India has built more than 3500 large dams, and several hundreds are in the works. Amongst the many criticisms of dam projects is that governments and lending institutions focus primarily on end-use benefits, such as electricity, and fail to systematically count the numbers of people that any given project would displace. This often results in incorrect estimates of displacement, and thereby skew cost-benefit analyses in favour of the projects in question. According to one study by the World Bank, more than half a million people across 192 projects had not been accounted for. A number of concrete examples spell out the implication of this statistic. In the Hirakud Dam, while the government figures state that 1.1 lakh (1 lakh = 100,000) people were displaced, unofficial figures are almost double that. Again, in the case of the Sardar Sarovar Dam, it was estimated at the outset of the project that about 6000 families would be displaced. According to the Narmada Bachao Andolan, a people's movement organized around opposition to the many dams on the river Narmada, the figure today is 85,000 families, or half a million people. Many more such examples can be mentioned, and one statistic puts this trend in relief: the number of people displaced in India since independence is more than three times the number displaced by India's partition, during with large masses of humanity were forced to migrate to or from Pakistan. The vast majority of the displaced were the rural poor, small and marginal farmers, landless labourers and tribal people (Mathur and Marsden 1998; Fernandes and Thukral 1989; Mehta 2009).

Another facet of displacement is that some populations have been displaced multiple times during their lives. A case in point is in Singrauli, where more than 200,000 people were displaced when the Rihand Dam was constructed in 1984, and then again when thermal power plants, coal mines and other industries were established in the area. According to one estimate, some of the Rihand oustees had been displaced four times. This is by no means an isolated instance of multiple displacement—the Soliga tribals in the state of Karnataka were first displaced in the 1970s by the construction of a dam, and again a couple of decades later with the establishment of the Rajiv Gandhi National Park. Again, fishing families were first displaced with the establishment of the Mangalore Port in the 1960s and again by the Konkan Railway in the 1980s—after they had made a transition from fisherfolk to farmers. For the millions of people who have been displaced, the process is dehumanizing. Many of them start out highly skilled in their places of origin, adept at agriculture, agro-forestry, animal husbandry and a host of other jobs that form the backbone of the rural landscape. However, when they move to peri-urban areas, these skills are less relevant, and as a consequence, they are seen as unskilled. This process of de-skilling, in addition to the trauma of migration itself, is deeply disempowering, and has had severe traumatic psychological and socio-economic consequences. The process of displacement involves a disconnect with the lands and production systems of origin, and along with it, the destruction or desecration of symbols and institutions, such as temples and sacred groves, that gave their life meaning, and which made up their cultural identities. Displacement also results in families being scattered; kinship groups disrupted, often irreversibly, and informal social networks being dismantled, which, in turn, means the lack of mutual support systems. Such processes are often gendered, with women finding themselves doing both domestic and quasi-industrial work—as informal construction labourers, for example, whilst having no time or space to build sustaining relationships (Mehta 2009).

Strikingly, despite the issue of displacement front and centre in Indian environmentalism for more than three decades, India has a poorly defined rehabilitation policy for those who are displaced. Thus far, less than a quarter of displaced people have been resettled. Rehabilitation is still poorly conceived, even where it is, at the outset of any so-called developmental project. In most instances, the process of rehabilitation begins years after people were first evicted. For example, those displaced by the Machkund Dam began to be resettled after a decade. Again, only 730 of the 2108 families displaced in the early 1950s by the Bhakra-Nangal Dam had been displaced 25 years later. Yet another poignant case is that of the Pong Dam. Of the 30,000 families displaced in the 1960s, the state determined that only 16,000 were eligible for compensation. Of them, only a fraction, that is 3756, were moved hundreds of miles. However, they were taken to a geography and a culture that was totally different to what they had hitherto known—from the Himalayan valleys to the desert of Rajasthan, where they found that some of the land allocated to them had been already occupied, whilst other lands were uncultivable. Ultimately, three quarters of these people moved back to their native state of Himachal state, only to find no support from the state government. Significantly, displaced people have no compensation rights, and demands for compensation often end up nowhere (Somayaji and Talwar 2011).

Given the sheer marginality of the people displaced by development related projects such as dams, and the absence of a rights regime, the saga of displacement can be understood as a callous, if not conscious, way of acquiring cheap labour. In many respects, displaced peoples find themselves ontologically not very different from those under indentured labour, an economic system under which Indian agricultural labour was enticed to work and serve in far flung plantations of Asia, South Africa and the Caribbean under British colonialism. If anything, the indenture system for all its duplicitousness, at least made a nod to free-will. With development-induced displacement, we have a system in which people are forced to migrate against their will. Ironically, they are all free citizens in a democracy—but with no real rights on account of them being either illegible or invisible to the economic planners, industrialists or urban dwellers, or being considered expendable for the greater cause of progress.

Not all migrations are due to development-induced displacement. People have moved due to natural disasters, political conflict, communal violence and a host of other reasons. Many also moved in search of productive work in the context of severe underemployment in their places of origin. The growing urban and peri-urban areas in India absorbed them by the millions, and this migrant labour formed the basis of the economy of urban India. One could argue that the process of urbanization was predicated on the migration of unskilled labour to cities, where they provided free labour that was otherwise hard to obtain. There is however a coda to the story of displacement. When COVID 19 struck, Indian cities could not support these people. Many were left destitute, with no way to sustain a living, and no easy way to return to their villages. While the novelty of COVID-19 could be adduced as a reason for the failure, the fact of their sheer liminality—which is the underlying cause of the broader saga of displacement—is perhaps a greater causative factor (Pal and Siddiqui 2020).

## Conclusion

A common theme across each of the four aspects of the environment-labour interfaces in India is that of estrangement. In each instance, human beings were forced in to doing the physical labour that was intended quite consciously to transform nature. In various ways, these processes of social reproduction involved alienation of people from themselves and nature. In the case of natural resources, the denial of traditional rights and the transformation of entire landscapes meant that local people, formerly very adept at using their natural resources, found themselves alienated from the newly emergent landscapes, which emphasized very different end users and goals. In the case of industry, the estrangement went further, as the labour force therein had within generations been forced into these new modes of working and found themselves in extremely toxic and unsafe contexts. Again, the urban context involved estrangement at even more levels of complexity—from the agrarian contexts from which much of the labour came from, from their families, and from other sections, classes and strata of society itself. This was particularly evident in the case of displacement, which represents the pinnacle of alienation, and results in de-skilling and the conversion of highly knowledgeable workers into liminal labour at the margins of economy and society. The reasons for the estrangement, alienation and liminality are quite evident—the changing political economies of the state, and the passions and interests that captured it from the colonial time onward. These political economies were abetted by discourses of development and nation building that were based on some clear normative choices. To these we can even discourses about the environment which sometimes also preferred some similar outcomes—such as population control, rational urbanization and auto-mobility—over the needs of the poor.

None of this is surprising. What makes India interesting, when compared with other developing country contexts, is the framework of democracy, and with it, civil debates and people movements that have helped re-frame discourses, expose the political economies and dream of alternatives (Shrivastava and Kothari 2012; Kothari and Joy 2017). Also interesting is the willingness and ability of public institutions, especially the courts, to consider these pleas for relief, although their rulings have also had their contradictions, and have had many collateral impacts. Looking ahead, the role for the social movements is only larger, for three reasons. First, the percentage of formally organized labour is decreasing, and India, in which labour has long been unorganized or semi-organized, is mired, like many other countries in gig economies which bring in their wake a further diminishing of rights. Second, the Indian state today is in transition, like perhaps everywhere in the world, and one can perhaps make the argument that it is neither a developmental state nor a protective one. For all the rhetoric, there also seems to be the diminishing of ideologies, be they conservative, socialist, liberal, Gandhian or a different variant. Instead, there is the rise of a culture of opportunism and vulture capitalism, and the framing of politics in terms of either entrepreneurialism or technocracy, neither of which is adequate to address the political economies underlying complex social problems. Third, the formal Left parties, and for the most part, informal labour unions and groups, have largely avoided grappling with the sheer materiality and importance of environmental issues. Thus, although both the environment and labour are threatened in ways that are historically unprecedented, all that remains in the short run is the prospect of Polanyian double movements seeking to salvage small spaces of habitation for the rapid liminality of the multitude (Rajan and Duncan 2013).

**Acknowledgements** I am extremely grateful to Dr Rohan d'Souza for his many comments and suggestions, and to the editors of this volume for their extremely useful peer review and suggestions for revision.

# References

Agarwal, Bina. 1992. The Gender and Environment Debate: Lessons from India. *Feminist Studies* 18 (1): 119–158.

———. 2000. Conceptualising Environmental Collective Action: Why Gender Matters. *Cambridge Journal of Economics* 24 (3): 283–310.

———. 2001. Participatory Exclusions, Community Forestry, and Gender: An Analysis for South Asia and a Conceptual Framework. *World Development* 29 (10): 1623–1648.

Batliwala, Srilatha, and L. David Brown. 2006. *Transnational Civil Society: An Introduction*. Bloomfield, CT: Kumarian Press.

Bell, Ruth Greenspan, Kuldeep Mathur, and Urvashi Narain. 2004. Clearing the Air: How Delhi Broke the Logjam on Air Quality Reforms. *Environment: Science and Policy for Sustainable Development* 46 (3): 22–39.

Bidwai, Praful. 2011. People Vs Nuclear Power in Jaitapur, Maharashtra. *Economic and Political Weekly* 46, no. 8: 10–14.

———. 2015. *Phoenix Moment: Challenges Confronting the Indian Left*. New Delhi: Harper Collins.

Brune, James N. 1993. The Seismic Hazard at Tehri Dam. *Tectonophysics* 218 (1–3): 281–286.

Calabresi, Guido, and Philip Bobbitt. 1978. *Tragic Choices*. New York: Norton.

Chintan. n.d. Chintan Environmental Research and Action Group. https://www.chintan-india.org.

Chouhan, T.R., et al. 1994. *Bhopal: The Inside Story – Carbide Workers Speak*. Panjim: The Other India Press.

Collins, Harry, and Robert Evans. 2008. *Rethinking Expertise*. Chicago: University of Chicago Press.

CSE. 2015. Labour vs Pollution. https://www.downtoearth.org.in/coverage/labour-vs-pollution-20536.

D'Souza, Rohan. 2006. *Drowned and Dammed*. New York: Oxford University Press.

Dams, World Commission on. 2000. *Dams and Development: A New Framework for Decision-Making: The Report of the World Commission on Dams*. London: Earthscan.

Dogra, Bharat. 1992. *Forests, Dams, and Survival in Tehri Garhwal*. New Delhi: B. Dogra.

Dubash, Navroz K. 2002. *Tubewell Capitalism*. New Delhi: Oxford University Press.

Fernandes, Walter, and Enakshi Ganguly Thukral. 1989. *Development, Displacement and Rehabilitation: Issues for a National Debate*. New Delhi: Indian Social Institute.

Fink, A.K. 2000. Tehri Hydro Power Complex on the Bhagirathi River in India. *Hydrotechnical Construction* 34 (8/9): 479–484.

Gauri, V. 2009. *Public Interest Litigation in India: Overreaching or Underachieving?* Policy Research Working Paper 5109, Development Research Group, World Bank.

Ghosh, Nilanjan, Pranab Mukhopadhyay, Amita Shah, and Manoj Panda. 2016. *Nature, Economy and Society: Understanding the Linkages*. New Delhi: Springer.

Gopalakrishnan, A. 1999. Issues of Nuclear Safety. *Frontline* 16 (6): n.p.

Govardhan, Veerlapani. 1993. *Environmental Impact Assessment of Tehri Dam*. New Delhi: Ashish Pub. House.

Guha, Ramachandra. 1988. Ideological Trends in Indian Environmentalism. *Economic and Political Weekly* 23 (49): 2578–2581.

———. 2000. *The Unquiet Woods: Ecological Change and Peasant Resistance in the Himalaya*. Berkeley: University of California Press.

Guha, Ramachandra, and Juan Martínez Alier. 2000. *Varieties of Environmentalism: Essays North and South*. New Delhi: Oxford University Press.

Hardiman, David. 1999. Of Forests and Its Plunderers. *Economic and Political Weekly* 34 (27): 1771–1772.

Hazarika, S. 1989. Reporting Bhopal. *Economic and Political Weekly* 24 (29): 1614–1614.

Indiastat. 2020. Industrial Accidents in India. https://www.indiastat.com/crime-and-law-data/6/accidents/35/industrial-accidents/18115/stats.aspx. Accessed November 2020.

Internal Displacement Monitoring. n.d. https://www.internal-displacement.org/countries/india. Accessed November 2020.

Irwin, Alan, and Brian Wynne, eds. 2009. *Misunderstanding Science?: The Public Reconstruction of Science and Technology*. Cambridge: Cambridge University Press.
Iyer, Ramaswamy R. 2013. *Water: Perspectives, Issues, Concerns*. New Delhi: SAGE.
Jhabvala, Renana, and Namrata Bali. 1990. *My Life, My Work: A Sociological Study of SEWA's Urban Members*. Ahmedabad: SEWA Academy.
Kaur, Raminder. 2020. *Kudankulam: The Story of an Indo-Russian Nuclear Power Plant*. Oxford: Oxford University Press.
Khagram, Sanjeev. 2004. *Dams and Development: Transnational Struggles for Water and Power*. Ithaca; London: Cornell University Press.
Khare, A., M. Sarin, N.C. Saxena, and S. Palit. 1999. Joint Forest Management: Policy, Practice and Prospects. In *Policy that Works for Forests and People*, ed. James Mayers and Stephen Bass, 33–41. Policy that Works Series No. 7: Series Overview. London: International Institute for Environment and Development.
Kothari, Ashish, and K.J. Joy, eds. 2017. *Alternative Futures: India Unshackled*. New Delhi: Independent.
Krishnan, Radhika. 2016. Red in the Green: Forests, Farms, Factories and the Many Legacies of Shankar Guha Niyogi (1943–91). *South Asia: Journal of South Asian Studies* 39 (4): 1–15.
Lahiri-Dutt, Kuntala. 2016. *The Coal Nation: Histories, Ecologies and Politics of Coal in India*. London: Routledge.
LaPorte, Todd R., and Paula M. Consolini. 1991. Working in Practice But Not in Theory. *Journal of Public Administration Research and Theory: J-PART* 1 (1, January): 19–48.
Leach, Melissa, Ian Scoones, and Brian Wynne, eds. 2005. *Science and Citizens*. London: Zed Books.
Lianchawii. 2005. Biosafety in India: Rethinking GMO Regulation. *Economic and Political Weekly* 40 (39): 4284–4289.
Mackay, Louis, and Mark Thompson. 1988. *Something in the Wind: Politics After Chernobyl*. London: Pluto.
Mathai, Manu V. 2015. *Nuclear Power, Economic Development Discourse and the Environment: The Case of India*. London: Routledge.
Mathur, Kuldeep. 2005. Battling for Clean Environment: Technocrats and Populist Politics in Delhi. In *Administrative Reforms: Towards Sustainable Practices*, ed. Amita Singh, Chapter 6. New Delhi: Sage.
Mathur, Hari Mohan, and David Marsden. 1998. *Development Projects and Impoverishment Risks: Resettling Project-Affected People in India*. Delhi: Oxford University Press.
McKinsey. 2020. *India's Urbanization: A Closer Look*. McKinsey & Company. https://www.mckinsey.com/featured-insights/urbanization/indias-urbanization-a-closer-look.
Mehta, Lyla. 2009. *Displaced by Development: Confronting Marginalisation and Gender Injustice*. New Delhi, India; Thousand Oaks, CA: Sage.

Miśra, Anupama, and Satyendra Tripathi. 1978. *Chipko Movement: Uttarakhand Women's Bid to Save Forest Wealth*. New Delhi: People's Action.

Morehouse, Ward, M. Arun Subramaniam, and Citizens Commission on Bhopal. 1986. *The Bhopal Tragedy: What Really Happened and What It Means for American Workers and Communities at Risk*. New York: Council on International and Public Affairs.

Mosse, David, and M. Sivan. 2003. *The Rule of Water*. Oxford University Press.

Mould, Richard F. 1988. *Chernobyl: The Real Story*. Oxford: Pergamon.

Narain, Urvashi, and Ruth Greenspan Bell. 2006. Who Changed Delhi's Air? *Economic and Political Weekly* 41 (16): 1584–1588.

Pal, Alasdair, and Danish Siddiqui. 2020. The Long Road Home: A Coronavirus Journey in India (Report). *Reuters*. https://www.reuters.com/investigates/special-report/health-coronavirus-india-migrants/. Accessed November 2020.

Paranjpye, Vijay. 1988. *Evaluating the Tehri Dam: An Extended Cost Benefit Appraisal*. New Delhi: Indian National Trust for Art and Cultural Heritage.

Qadeer, Imrana, and Dunu Roy. 1989. Work, Wealth and Health: Sociology of Workers' Health in India. *Social Scientist* 17 (5/6, May–June): 45–92.

Qaim, M., A. Subramanian, and G. Naik. 2006. Adoption of Bt Cotton and Impact Variability: Insights from India. *Review of Agricultural Economics* 28 (1, Spring): 48–58.

Rajan, S. Ravi. 1999. Bhopal: Vulnerability, Routinization, and the Chronic Disaster. In *The Angry Earth: Disaster in Anthropological Perspective*, ed. Anthony Oliver-Smith and Susanna Hoffman, 257–277. Routledge: New York.

———. 2001. Toward a Metaphysic of Environmental Violence: The Case of the Bhopal Gas Disaster. In *Violent Environments*, ed. Michael Watts et al., 380–398. Ithaca: Cornell University Press.

———. 2002. Disaster, Development and Governance: Reflections on the 'Lessons' of Bhopal. *Environmental Values* 11 (3): 369–394.

———. 2006. *Modernizing Nature*. Oxford: Oxford University Press.

Rajan, S. Ravi, and Colin A.M. Duncan. 2013. Ecologies of Hope: Environment, Technology and Habitation-Case Studies from the Intervenient Middle. *Journal of Political Ecology* 20 (1): 70–79. https://doi.org/10.2458/v20i1.21758.

Raju, K.D. 2007. *Genetically Modified Organisms Emerging Law and Policy in India*. New Delhi: TERI.

Ramana, M.V. 2006. Nehru, Science and Secrecy. http://www.geocities.ws/m_v_ramana/nucleararticles/Nehru.pdf. Accessed November 2020.

Ramana, M.V., and C. Rammanohar Reddy, eds. 2003. *Prisoners of the Nuclear Dream*. New Delhi: Orient Blackswan.

Rao, Nitya. 2012. *"Good Women Do Not Inherit Land" Politics of Land and Gender in India*. New Delhi: Social Science Press: Distributed by Orient Blackswan.

Roberts, Karlene H., Robert Bea, and Dean L. Bartles. 2001. Must Accidents Happen? Lessons from High-Reliability Organizations [and Executive

Commentary]. *The Academy of Management Executive (1993–2005)* 15 (3, Themes: Insights from Sports, Disasters, and Innovation, August 2001): 70–79.

Roy, Dunu. 2000. Organising for Safe Livelihoods: Feasible Options. *Economic and Political Weekly* 35 (52/53): 4603–4607.

Sahu, Geetanjoy. 2008. Implications of Indian Supreme Court's Innovations for Environmental Jurisprudence. *Law, Environment and Development Journal* 4 (1): 1. http://www.lead-journal.org/content/08001.pdf. Accessed November 2020.

Salve, Prachi. 2019. Poor Enforcement, Training: The Reasons Why There Are So Many Road Accidents in India. https://www.thesangaiexpress.com/Encyc/2019/11/21/India-s-traffic-laws-are-stricter-than-those-of-other-countries-but-these-laws-are-not-enforced-Road-accidents-cost-India-3-5-of-gross-domestic-product-every-year-and-are-avoidable-if-India-.html. Accessed November 2020.

Sands, Philippe. 1988. *Chernobyl: Law and Communication: Transboundary Nuclear Air Pollution – The Legal Materials*. Cambridge: Grotius.

Scoones, Ian, and University of Sussex, Institute of Development Studies. 2003. *Regulatory Manoeuvres: The Bt Cotton Controversy in India*. IDS Working Papers 197, 1–43.

Shah, Tushaar. 2010. "Past, Present and the Future of Canal Irrigation in India." Working Paper 1–46.

———. 2011. Past, Present and the Future of Canal Irrigation in India. In *India Infrastructure Report 2011 – Water: Policy and Performance for Sustainable Development*, ed. Infrastructure Development Finance Company Limited (IDFC), 69–89. New Delhi, India: Oxford University Press.

Shiva, Vandana. 1991. *The Violence of the Green Revolution: Third World Agriculture, Ecology and Politics*. London: Zed.

Shrivastava, Aseem, and Ashish Kothari. 2012. *Churning the Earth: The Making of Global*. New Delhi: Penguin.

Singh, Siddharth. 2018. *The Great Smog of India*. New Delhi: Penguin Random House India.

Smith, Roger, and Brian Wynne. 1989. *Expert Evidence: Interpreting Science in the Law*. London: Routledge.

Somayaji, Sakarama, and Smrithi Talwar. 2011. *Development-Induced Displacement, Rehabilitation and Resettlement in India: Current Issues and Challenges*. Abingdon, Oxon; New York: Routledge.

Subramanian, Arjunan, and Matin Qaim. 2010. The Impact of Bt Cotton on Poor Households in Rural India. *The Journal of Development Studies* 46 (2): 295–311.

Swaminathan, Mathangi. 2019. How Can India's Waste Problem See a Systemic Change? *Economic and Political Weekly* 53 (16): 4284–4289.

Tucker, Richard P. 2011. *A Forest History of India*. New Delhi; Thousand Oaks: SAGE.

Véron, René. 2006. Remaking Urban Environments: The Political Ecology of Air Pollution in Delhi. *Environment and Planning A* 38 (11): 2093–2109.

Whitcombe, Elizabeth. 1970. *Agrarian Conditions in Northern India*. Oakland: University of California Press.

Wynne, Brian. 1992. Misunderstood Misunderstanding: Social Identities and Public Uptake of Science. *Public Understanding of Science* 1 (3): 281–304.
———. 2010. *Rationality and Ritual: Participation and Exclusion in Nuclear Decision-Making*. London: Routledge.

# 3

# Energy Transitions in the Global South: The Precarious Location of Unions

Dinga Sikwebu and Woodrajh Aroun

## Introduction

In February 2013, the Congress of South African Trade Unions (COSATU) which is the country's biggest union federation adopted a range of resolutions on climate change. Sponsored mainly by the National Union of Metalworkers of South Africa (NUMSA) and the federation's affiliate at the time, the resolutions called for a broad approach to environmental sustainability informed by the principle of a just transition where the right of developing countries to industrialise is recognised as well as their duty to mitigate climate change. COSATU's resolutions on global warming and climate change followed in the footsteps of a Green Economy Accord concluded between the South African government, business and labour on the eve of the United Nations Framework Convention on Climate Change (UNFCCC) Conference of the Parties (COP17) held in the country in 2011. From around 2009, South African trade unions began to identify climate change as one of the threats that faced the planet and took steps to formulate mitigation policies (Cock and Lambert 2013, 92).

The policy positions on climate change adopted by South Africa's labour movement have been widely celebrated and described as progressive (Räthzel

D. Sikwebu (✉) • W. Aroun
National Union of Metalworkers of South Africa, Johannesburg, South Africa
e-mail: dingas@numsa.org.za

et al. 2018). As a result of its declaration that the trade-off between jobs and the environment as false, COSATU's climate change policy framework was lauded as 'an important organisational step towards strengthening the linkages between labour and environmental activists' (Cock 2018, 222). The work of its affiliate NUMSA, which combined building union capacities on climate change, policy formulation and campaigns, was described as novel and strategic (Satgar 2015, 278). In an introduction to the Green Economy Accord, the then President of South Africa Jacob Zuma stated that by signing the accord South Africans demonstrated that social partners were 'committed to reduce our dependence on coal-based energy' (EDD 2011, 2).

A decade after the praiseworthy policies and resolutions to shift to a low-carbon economy were adopted, it appeared that the convergence to wean South Africa off its dependence on fossil fuels had become unstuck. The labour movement that championed the transition to low-carbon energy was fighting the government and appeared to be now acting as defenders of fossil fuels. In March 2018, NUMSA together with a coal transporters association called Transform SA rushed to the High Court of South Africa for an interdict to stop the Minister of Energy Jeff Radebe from signing renewable energy (RE) contracts with 27 Independent Power Producers (IPPs). NUMSA argued that renewable energy by its nature was intermittent and incapable of providing base-load electricity that is required to ensure supply along the grid. In its bid to stop the signing of the contracts, the union also warned of massive job losses along the coal supply chain.

Although the High Court rejected the union's application (Omarjee and Cronje 2018), the National Union of Mineworkers (NUM) joined NUMSA in its condemnation of the signing of the contracts with Independent Power Producers. NUM threatened not to support the ruling African National Congress (ANC) in the general elections scheduled for 2019 'if government continues with its renewable energy programme, saying clean power will destroy jobs and create ghost towns in coal mining areas' (Niselow 2018, para. 1). In a joint statement the two unions described the signing of the contracts as a form of state capture by private interests, and an act against state-owned electricity utility Eskom and the broader public.

A few months after the court challenge and in negotiations to update the Integrated Resource Plan (IRP) which is the country's plan to procure electricity, labour emerged again as a defender of coal. At the time, there were reports that in the consultation amongst social partners within the social dialogue forum known as the National Economic Development and Labour Council (NEDLAC), labour and community representatives raised 'concerns over the exclusion of new nuclear, as well as the diminishing role of coal in the IRP' (Creamer 2019, para. 9).

In this chapter, we explain the pendulum swing on the part of South Africa's labour movement, from an initial endorsement of a low-carbon economy transition back to a defence of fossil fuels. Our argument is that there exists overlapping technological, institutional, historical and behavioural forces that lead to policy indeterminacy, slow the diffusion of low-carbon technologies, entrench dependence on fossil fuels and thwart processes towards decarbonisation. Unruh (2000, 817) describes the interaction between these interlocking forces as 'carbon lock-in'. Using South Africa as a case study, our argument is that the existence of carbon lock-in creates difficulties of transitioning to a different energy system. This is particularly the case, in countries in the Global South. To date, scholarly work on the just transition has focussed in the United States and Australia (Pai et al. 2020, 11).

Whereas we use the notion of a carbon lock-in to explain the inertia in moving to a system where cleaner energy sources are dominant, in this chapter we stretch the concept and show how trade unions themselves can either be locked in or act as facilitators of carbon lock-ins. While unions may want to move beyond fossil fuel-based energy systems, the fact is that they draw their members from the sub-sectors that constitute or support the fossil fuel industry.

With the international division of labour privileging the Global North with technological know-how, it becomes extremely challenging for countries in the Global South to develop sustainable sub-sectors that manufacture renewal energy technologies and that can absorb workers from traditional energy sectors. The contradictory location of the labour movement creates inconsistencies, leads to flip-flops and makes unions susceptible to pressures about potential job losses. In the case of South Africa, the inability of the energy transition to deliver on the promises of jobs, training and economic empowerment has made workers and their organisation sceptical or has turned them against the proposed shift to a low-carbon economy.

In making this argument, we will first explain how the existence of a carbon-intensive economy acts as a lock-in. Secondly, we will outline government initiatives to decarbonise South Africa's economy. Thirdly, we will sketch out labour's responses to government's initiative to move the country towards a low-carbon economy. Fourthly, we will put forward our explanation of why the convergence on the transition unravelled. As a conclusion, we will paint sketches of what can be done to move beyond a carbon lock-in and ensure a shift to a low-carbon economy.

## South Africa's Carbon-Intensive Economy

Historically, the abundance of cheap coal drove South Africa's economic development. On the basis of its coal endowment, the country built a network of energy intensive industries, bringing the energy sector together with mining and associated sub-sectors of manufacturing. The existence of this network as the core site of accumulation made Fine and Rustomjee (1996, 71) to characterise South Africa's economic development as the 'Minerals-Energy Complex' (MEC). The interlocking sub-sectors of the MEC were the energy industry, mining, basic metals, smelting, synthetic resins, basic chemicals, quarries, fertilisers and pesticides. Important MEC companies were the state-owned electricity utility Eskom and the South African Synthetic Oil Limited (SASOL) that uses both coal-to-liquids (CTL) and gas-to-liquids (GTL) technologies to produce petroleum products. Fine and Rustomjee (1996, 5) argued that the MEC 'lies at the core of the South African economy, not only by virtue of its weight in economic activity but also through its determining role throughout the rest of the economy'.

More than a hundred years later, the MEC remains intact as fossil fuels and coal in particular continue to power the country's mining, manufacturing and agricultural development. According to a government report released in 2019, fossil fuel continues to dominate the country's energy supply, with coal constituting 69%, followed by crude oil with 14% and natural gas contributing 3% in 2016. Renewables make up 11% of the remainder of total primary supply, while nuclear contributes 3% during the same period. The country also produced its petroleum products from coal (46.8%), crude oil (46.6%) and approximately 6.6% from gas (DoE 2019, 8). The dominance of coal is more acute in the electricity supply industry. Eskom generates more than 90% of South Africa's electricity, with more than 80% coming from coal-based power stations (Eskom 2019b, 34).

In 2016, the mining and industrial sectors consumed more than half of the energy generated in the country. The transport sector consumed 19%, commerce and public services 14%, residential 8% and agriculture 6% (Department of Energy 2019, 23). The national utility Eskom dominates the electricity sector in South Africa. Eskom which is a vertically integrated utility generates, transmits and distributes electricity to industrial, mining, commercial, agricultural and residential customers in South Africa. It also sells the electricity to municipalities, who in turn distribute power to businesses and households in certain areas. Through power purchase agreements, the utility also purchases electricity from Independent Power Producers (IPPs) and exports electricity to the region through Southern African Power Pool (SAPP).

Since 1999, just more than two dozen companies in mining, materials beneficiation and materials manufacturing who together consume more than 40% electrical energy in South Africa have constituted a voluntary and non-profit association of energy intensive consumers called the Energy Intensive Users Group (EIUG). They collectively contribute over 20% to the country's Gross Domestic Product (GDP) and active in energy policy discussions.

The weight of the MEC in terms of people it employs in South Africa is enormous. According to EIUG, its members employ 657,984 workers across sectors such as mining and quarrying, manufacturing, electricity, gas, transport, agriculture, retail and construction (EIUG 2018). Drawing on employment statistics from the Minerals Council of South Africa (formerly Chamber of Mines), mining with a contribution of 8.1% (R360.9 billion) to GDP in 2019 employed 454,861 workers of which 92,230 were coal miners. Table 3.1 shows the spread of employment across key sectors of the South African mining industry:

According to SASOL's South Africa (SSA) fact-book (SASOL 2018) the company employs 17,000 people. Eskom's headcount at the end of March 2019 shows that there were 39,292 employees in the public utility (Eskom 2019a). Employment in other MEC-related sub-sectors was significant: 13,100 employees worked in foundries servicing construction and general engineering (33%), automotive (30%), mining (25%), energy (5%), agriculture (5%) and other making up the remaining 2% (Davies 2015).

But South Africa's dependence on fossil fuel has made the country one of the highest greenhouse gas (GHG) emitters in the world. According to the Greenhouse Gas National Inventory Report 2000–2015, South Africa's gross greenhouse gas (GHG) emissions were estimated at 540,854 Gg $CO_2$e in 2015 (Gg = gigagram), with energy being the largest contributor 'responsible

Table 3.1 Employment in key sectors of the South African mining industry

| Commodity | 2010 | 2019 |
|---|---|---|
| Coal | 73,817 | 92,230 |
| Platinum/PGM | 181,969 | 164,513 |
| Iron Ore | 18,216 | 19,092 |
| Manganese | 5879 | 10,846 |
| Chrome | 13,971 | 19,693 |
| Gold | 157,019 | 95,130 |
| Diamonds | 11,143 | 15,728 |
| Total employment (all mining) | 498,141 | 454,861 |

Source: Own table based on figures from DMR in Chamber of Mines (2010) and Minerals Council of SA (2019)

for 84.8% of the increase over the 15 year period' (GHG National Inventory Report 2000–2015, viii).

Based on 2015 figures the report further highlights trends for the following sectors (GHG National Inventory Report 2000–2015, x–xi):

- GHG emissions from the energy sector were estimated to be 429,907 Gg $CO_2e$ representing 79.5% of South Africa's total gross emissions. Energy industries accounted for 60.4%; transport 12.6%; manufacturing industries and construction 8.6% and other sectors 11.4%
- Industrial processes and product use (IPPU) produced 41,882 Gg $CO_2e$. Overall the metal industry category contributes 73.9% to the total IPPU sector emissions; sub-sectors include iron and steel production (14,093 Gg $CO_2e$) and ferroalloys production (13,420 Gg $CO_2e$)
- Agriculture, forestry and land use change (AFOLU) GHG emissions were 49,531 Gg $CO_2e$ or 8.8% of South Africa's gross GHG emissions in 2015
- In 2015 the waste sector accounted for 19,533 Gg $CO_2e$ or 3.5% of South Africa's gross GHG emissions.

## South Africa's Winding Energy Transition

Attempts to diversify the country's energy mix away from fossil fuels go back to the first broad energy framework adopted after the demise of apartheid, the 1998 *White Paper on the Energy Policy of the Republic of South Africa*. Acknowledging that 'South Africa has neglected the development and implementation of renewable energy applications' (DME 1998, 79), the new policy framework targeted what it described as '6 000 MW of non-utility generation' that could be exploited. Forming part of this non-utility generation was electricity to be produced through technologies such as hydro, wind, solar thermal and waste incineration.

The next policy intervention was the unveiling of the 2003 *White Paper on Renewable Energy*. The policy re-affirmed government's support for renewable energy (RE) for non-grid rural communities and large-scale utilisation of renewables to reduce the country's carbon emissions. Government set as its target 10,000 GWh of renewables to final energy consumption by 2013. This was to be produced mainly from biomass, wind, solar and small-scale hydro (DME 2003, i).

But 2011 was the year of action and when things began to move. In March 2011, the Department of Energy (DoE) announced the Integrated Resource Plan (IRP) 2010–2030 (DoE 2011a) outlining the country's energy mix by

### 3 Energy Transitions in the Global South: The Precarious Location... 65

Table 3.2 Policy adjusted IRP 2010–2030 and IRP2019 electricity energy share

|  | 2010 (%) | 2030 (%) | 2030 (Adjusted in IRP2019) (%) |
|---|---|---|---|
| Coal | 90 | 65 | 59 |
| Nuclear | 5 | 20 | 5 |
| Hydro | 5 | 5 | 8 |
| Gas & Storage | 0 | 1 | 2 |
| Renewables | 0 | 9 | 18 (wind) |
|  |  |  | 6 (PV) |

Source: Own table based on figures from DoE Policy Adjusted IRP 2011, 7; IRP2019, 42

2030. The IRP set a 9% target for renewable energy as share of electricity to be generated.

This target was further changed when the Department of Mineral Resources and Energy released its revised IRP2019 in October 2019 (Department of Mineral Resources and Energy 2019). While coal remained the primary supplier, the promise to increase renewable energy (RE) from nil in 2010 to 24.7% in 2030 is a significant step towards decarbonisation. Table 3.2 illustrates the approved energy mix.

Another significant development in 2011 was the launch of the Renewable Energy Independent Power Producer Procurement Programme (REIPPPP). Since then, South Africa has embarked on a multi-billion-dollar programme to introduce renewable energy onto the country's energy system. The REIPPPP was launched to acquire 3275 megawatts (MW) of renewable energy (DoE 2011b). Through successive bid windows (BW) the Department of Energy plans to increase this by an additional 3200 MW of renewable energy capacity by 2020 (Reuters 2012).

The government's plan to accelerate the entry of renewable energy onto the national grid has been heralded as a major development. The Office of the Independent Power Producers (IPP Quarterly Report 2019) reported in its fourth quarterly report that sizeable investments totalling some R209.7 billion had been injected into the programme and 40,134 job years were created for South African citizens from the first bid window (BW1) to the fourth bid window (BW4).

In 2012 the South African government adopted the National Development Plan 2030 (NDP). The plan has what it describes as an overarching vision for environmental sustainability, protection of the ecosystem and a managed transition towards a low-carbon economy. The NDP targets the procurement of about 20,000 MW of renewable electricity by 2030 and the importing of electricity from the rest of the Southern African region. The plan outlines the intention to decommission 11,000 MW of ageing coal-fired power stations (NDP 2030, 15).

## Workers' Anxieties

As the government introduced policies and took measures to further diversify the country's energy mix, the labour movement began to build its capacity to engage with the new situation. For instance, in 2011, NUMSA constituted an energy research and development group (RDG) made up of workers drawn from energy intensive companies, the electricity utility Eskom and a few renewable energy companies. Using participatory research methods and intense training, the energy RDG developed policies for adoption by the rest of the union on climate change, building a renewable energy sector in South Africa, energy efficiency and carbon tax (Satgar 2015).

Although energetic about their work and endeavours to develop policies for the union, members of NUMSA's energy group were anxious about the initiatives to decarbonise South Africa's energy system. This emerged in interviews conducted in December 2011, seven months after the start of the group's intense programme. Uppermost in the concerns raised was the issue of potential job losses. When asked what they thought of the just transition to renewable energy sources, the majority were adamant that the shift will lead to job losses.

> The challenges that I see with renewables is that if they are implemented there will be a shortage or I can say the cutting of jobs. (*Alfred Sibiya, quality analyst at steel smelter Arcelor Mittal*)

> It is going to lead to job losses if they have to close down some of the fossil plants and energy intensive plants. There is going to be job losses. (*Nomfundo Ndlovu, a worker at Kestrel Wind Turbines*)

> I am afraid of a just transition. I am thinking about the employees. How many employees are going to lose their jobs? You can do just transition but I guarantee that some people would be crying. I can see we all want to move from fossil fuels to renewables. But the question is what about jobs? What about families that would remain foodless … nothing on the table. (*Michael Nene, a worker at Ellies Electronics*)

> I think we need to be slow on the transition because now we have to consider things pertaining to job losses because if you go about it quickly then there will be more job losses. (*Patrick Sefolo, a spot welder at a solar geyser manufacturing company, Kwikot*)

## 3 Energy Transitions in the Global South: The Precarious Location…

The workers also questioned the quality and sustainability of jobs in the renewable energy industry.

> I am sceptical in terms of the timing of the move. How fast can we move or how slow can we be? One is not entirely sure. It is almost impossible to give a very objective answer except to say that it would look like the renewable energy that is being punted is very expensive and without jobs that are really sustainable. The other thing is that it has never been clear how reliable renewable energy is to sustain the jobs, especially when people work in intensive users companies at the moment. (*Mbulaheni Mbodi, employed by the state-owned electricity utility Eskom*)

> In the renewable sectors, I am not sure about the guarantee of the jobs. We might produce the turbines and things like that and install them. But after that, what happens to the jobs? (*Nomfundo Ndlovu*)

> The jobs that are likely to be created will be temporary and not so much permanent ones. Permanent ones will be the technicians and engineers mostly. (*Alfred Sibiya*)

Although there was general support for the transition to a low-carbon economy, members of the RDG were clear that developing countries should not bear the cost of the shift. There was caution and pragmatism.

> I think we need to be very much strategic and cautious. We need to have a smooth transition at the right time and at a reasonable pace because I anticipate other countries robustly exploiting the process if you are not strategic enough. (*Christopher Chonco, a worker and safety representative at manganese smelter Assmang*)

> Because we have coal as a resource, the transition for us would mean that we still continue with fossil fuels while in the meantime we vigorously invest in renewable energy technologies. We need at the same time to invest in technologies that are cleaning the emissions from coal plants so that gradually we are able to have this transition without affecting in a big negative way our whole economy. (*Thanduxolo Ndyenge, engineer at Eskom*)

> I would say let us not rush into this transition yet. Let's first deeply look into the advantages of renewable energy and disadvantages of it. Because we might rush into doing this and 3 to 4 years down the line, it comes back to give us problems. (*Nomthandazo Nkosi, an operator at Assmang Chrome*)

A number of conditions that had to be met before commencing with the transition were identified. These conditions included training and reskilling of workers, technology transfers and financing mechanism.

> All these things should be quantified. We should be able to see them before we can say; 'let's just jump into the lake' (*Mbulaheni Mbodi*).

## Labour Contestations

Restructuring South Africa's coal-dependent electricity to meet government's commitments to climate change mitigation and introduction of renewable energy sources has been protracted and characterised by intense contestation. Not long after the celebrations on the signing of the Green Economy Accord in 2011, differences on the restructuring of the electricity industry began to emerge. The anxieties that workers had propelled the contestation that occurred in the aftermath of COP17 that was held in the country.

A fundamental difference emerged right at the time the government introduced the REIPPPP programme. The trade union, particularly NUMSA, criticised the programme for being private sector driven. NUMSA (2012, 4) in a statement said:

> Capital as is always the case, views the introduction of renewables as a new site of accumulation. Unfortunately, government and other policymakers see the new renewable energy sector as being privately owned. Thus far it appears that the state is content to outsource the country's renewable energy sector to the private sector. We are currently facing a situation in which the expansion of the renewable energy sector is rapidly developing along capitalist lines.

The union called for the building of a socially owned renewable energy sector, which NUMSA defined as a mix of different forms of collective ownership including energy parastatals, cooperatives, municipal-owned entities and other forms of community energy enterprises. Unless renewables projects were under public ownership and democratic control, and there was use of local content instruments in manufacturing new energy technologies, the union felt that foreign multinational corporations (MNCs) will drive the shift to renewables as they flood the country with their products. COSATU and NUMSA's views fell in the 'transformative environmental justice' category instead of the 'affirmative environmental justice' parameters that Stevis and Felli (2015) refer to in their schema on varieties of just transitions.

The second fissure was around the Independent System and Market Operator (ISMO) Bill (2012). In line with the objective of restructuring the electricity industry and after consultation with the electricity utility Eskom, the Department of Energy (DoE) introduced the ISMO Bill in March 2012. The Bill aimed to create a new state-owned company outside of electricity utility, responsible for system operations and the purchase of electricity from generators. Talking at a power summit organised by Investec Bank in Johannesburg two years before the introduction of the ISMO Bill, the then Minister of Energy Dipuo Peters stated that the intention of establishing an independent system and market operator was 'to level the playing field and eliminate conflict of interest between the buyer and the seller of electricity in a manner that protects all players from potential market abuse' (SA Government 2010, para. 31).

At the National Economic Development and Labour Council (NEDLAC) the labour constituency representing three of the country's trade union federations expressed reservations about the Bill. They felt that the proposed legislation would lead to the restructuring of the electricity sector and the unbundling of Eskom. They expressed concerns about the impact of the establishment of the market operator on jobs and electricity pricing. COSATU opposed the proposed legislation and argued that 'the establishment of ISMO would weaken Eskom's balance sheet and strengthen the calls for its privatization' (COSATU 2014, para. 8).

The third reflection of doubts amongst some sections of the labour movement about the manner in which transition away from fossil-based energy system was proceeding became apparent after the South African government announced in May 2015 plans to roll out a nuclear procurement process to build a fleet of new nuclear power stations. For a while after the announcement, COSATU held to its historical position on nuclear and issued a statement in which it categorically stated that 'the introduction of nuclear energy will have negative socio-economic effects in the country' and called for the expansion of investment in renewable energy which is 'safer, cheaper and more conducive for low-skill job creation' (COSATU 2015, para. 16–17).

But it was not long before COSATU's biggest public sector affiliate—the National Education, Health and Allied Workers Union (NEHAWU) challenged the federation's long-held position on nuclear energy. In June 2016, NEHAWU issued a discussion paper that commended 'the government of South Africa, especially the Minister and Department of Energy for the positive steps taken to grow the South African nuclear sector' (NEHAWU 2016, 23).

More surprising was the swipe that the union took at renewable energy sources. As a union that organises workers at various nuclear companies such as the National Radioactive Waste Disposal Institute and the Nuclear Energy Corporation of South Africa (NECSA) NEHAWU argued that there were safe ways of disposing nuclear waste and that nuclear power was the lowest emitter of greenhouse gasses, the cheapest form of energy, a great contributor to job creation and safer than other energy sources. NEHAWU called on COSATU not to embrace the move to renewables. The public sector union argued that renewable energy sources were intermittent and unreliable. It further argued that there is an inverse relationship between the renewable energy sector growth and the growth of the labour movement. 'By moving towards these [renewable] energy sources, we are actually reducing the workforce subscriptions to the labour federation and the federation should therefore see this as threat to its future, and the socialist movement as a whole' (NEHAWU 2016, 19).

The fourth and publicly visible disagreement on how a just transition would occur in South Africa emerged after Eskom announced in March 2017 the company's intention to go ahead with the mothballing and decommissioning of five coal-fired power stations. COSATU (2017a, para. 1) described Eskom's decision 'not just an arrogant decision but a hostile act of provocation directed at workers and their unions'. The federation called for a meeting with Eskom and government on the issue. Together with the National Union of Mineworkers (NUM), the federation was at pains to argue that the introduction of renewables should not be at the expense of workers and that 'climate change obligations to introduce renewable energy into the electricity grid should not result in back door privatisation and further commercialisation of Eskom' (COSATU 2017b, para. 6).

More strident was NUMSA's reaction. The union gave notice to strike and threatened to embark on a 'mother of all strikes' that would bring the economy on its knees. 'For a government just to say we are going to deal with emissions and we impose a carbon tax, we are going green, we are willing to flush 30,000 jobs so that we look good in G20 and all international platforms where they've made these commitments, is rubbish', said NUMSA's general secretary Irvin Jim (ENCA 2017, para. 9–10).

## The Vice-Grip of a Carbon Lock-In

How does one explain the shift on the part of South Africa's labour movement from being signatories to the Green Economy Accord to a position where unions were resorting to the courts to block the introduction of renewable energy? How do we account for what appears to be a pendulum swing?

At a conceptual level, the stubbornness of fossil fuels has led to the deployment of the concept of carbon lock-in and its related sub-concepts to understand the continued dominance of carbon-intensive industries in the face of scientific evidence that shows how greenhouse gasses contribute to global warming. Unruh (2000), Scholvin (2014), Seto et al. (2016) and more recently Stein (2017) use the concept to explain the inertia in the transition from fossil-based energy systems to clean sources of energy such as renewables. They argue that path-dependent carbon lock-in technologically, behaviourally and through a range of institutions thwarts the shift to cleaner energy forms.

Unruh (2000, 817) argues that carbon lock-in occurs when various forces such as 'technological, institutional and social' act together and in unison to preserve fossil fuel-based energy systems despite the negative impact that such systems may have on the environment. Seto et al. (2016, 19.22) identify three types of carbon lock-in: lock-in associated with technologies and infrastructure that emit $CO_2$, institutional lock-in associated with governance and decision-making in energy production and consumption; and behavioural lock-in related to habits and norms associated with the demand for energy-related goods and services. A broader 'legal and regulatory framework governing energy derived from fossil fuels' has the potential to block alternative energy sources like renewables from developing its full potential (Stein 2017, 559).

According to the concepts of technological feedback, sunk costs and institutional cultures, within carbon industries entrench reliance on fossil fuels despite known negative effects of greenhouse gasses and the existence of least-cost alternatives such as renewable energy. Increasing returns to scale and specialised knowledge also skew investment choices towards known and traditional energy sources. For Seto et al. (2016), the fossil fuel industry must be understood as a network of infrastructure, technologies, institutions, policies and consumer practices developed over two centuries. Non-emitting infrastructure such as pipelines and metal fabrication are dependent and support carbon-emitting industries. So do energy- demanding sectors. The usefulness of the concept of a carbon lock-in is that in its definition of a fossil fuel economy, similar to Stevis (2018, para. 4), it 'includes a wide range of related activities—infrastructure, transportation, housing and so on—that employ vast numbers of people across global production networks'.

When one looks at the REIPPPP overall programme in terms of scale, there can be no argument that the steps introduced represent a break from the carbon lock-in created by the minerals-energy complex (MEC). The programme does not constitute periods of 'critical junctures' that Scholvin (2014, 189) and Stein (2017, 562) refer to. These are phases where structural conditions and limitations open up a range of options for change and breaks from the constraints imposed on the energy trajectory by coal, gas and oil. For the break to happen, South Africa's energy sector requires a break from the infrastructure and technological lock-in and a switch over by the carbon-emitting infrastructure industry. The break also requires reconfiguration of institutional and behavioural arrangements.

The REIPPPP programme is very miniscule and does not threaten the fossil fuel industry. It also does not pose itself as an alternative to workers who work in carbon-intensive sectors. While the programme received wide acclaim, concerns have been expressed over ownership and control, the dominance of international corporations in engineering procurement and construction and capital flows. Baker and Wlokas (2015, i) also cite 'emerging tensions between "bankability" required by banks and investors and the economic benefits and community ownership criteria' of the programme.

The South African Renewables Energy Initiative argues that in spite of the many partnerships that South Africa has in the global energy field, none of these partnerships are on a 'scale that can deliver a critical mass of renewables' (SARi 2011, 7) to facilitate the country's entry into renewable energy markets. Although attempts have been made to develop a renewable energy industry in the country through policies of localisation and the use of instruments such as local content requirements much more clarity is needed on value and supply chains for the different renewable energy technologies. According to a joint report released by the International Renewable Energy Agency (IRENA) and the Global Wind Energy Council (GWEC) the establishment of a local supply chain for the wind industry can only be successful provided the domestic demand and growth opportunities are of a scale that ensures 'a sustained uptake of locally produced parts and machines' (IRENA-GWEC 2012, 24).

An overall evaluation of the REIPPPP programme through a series of Bid Windows (BW1–BW4) reveals that the scheme has been unable to build a sustainable renewable energy industry that can be the basis of a critical juncture. Although the South African government developed detailed local content requirements, winning bidders in BW1 conceded that during the initial stages of development technology and skilled technicians would have to be imported.

To illustrate the limitation imposed by this approach the case of Isivunguvungu Wind Energy Converter (I-WEC) and its principal partner DCD-Dorbyl deserves some comment. In its submission to the Department of Energy, the company felt that the 'local content threshold for onshore wind technologies' was surprisingly low and that they were 'capable of manufacturing complete wind energy converters for onshore wind energy technologies, with a local content of more than 60%' (Hancock 2011, para. 1–3). This was double the requirement for local content specified in the department's request for proposal document.

While the local service industry stood to benefit from the planned projects most developers indicated their preference for imported wind turbines and components: 'the norms of project finance still favour contractors and technology suppliers with extensive experience that to date tend to be European' (Baker and Wlokas 2015, 12). Local companies have also been disadvantaged by technological certification and standards required by international industries with a sizeable portion of investments spent on acquiring technology hardware from abroad (Moldvay et al. 2013 cited in Baker and Wlokas 2015, 25).

According to Baker and Wlokas (2015) European MNCs dominate the list of developers. In wind the main players are Globeleq (UK), Mainstream (Ireland) Gestamp, ACCIONA (Spain), EDF & GDF (France) and Enel Green Power (Italy). In solar PV the dominant players were Scatec (Norway); Mainstream (Ireland); Soitec (France), Gestamp, Tenesol, AE-AMD, Acciona (Spain), Solaire Directe (France) and Enel Green Power (Italy). Abengoa (Spain) and GDF Suez (France) lead in concentrated solar power (CSP).

Apart from tariffs and non-tariff barriers the use of Intellectual Property Rights (IPRs) in trade and trade-related agreements has proved to be a tricky one for developing countries and emerging markets. A report released by the United Nations Environment Programme (UNEP) and European Patent Organisation (EPO) identifies four factors that are responsible for obstructing technology transfer (UNEP–EPO 2013, 14):

- Access to the real know-how from the source companies (including access to trade secrets)
- Availability of suitably skilled staff
- Scientific infrastructure
- Favourable market conditions

## Explaining Labour's Pendulum Swing

Although job creation was a mandatory requirement in the REIPPPP programme and all bidders were required to meet certain minimum thresholds for the different renewable energy technologies, most of the jobs created were in the construction phases. Baker and Wlokas (2015, 24) argue that although

25% of the economic development criteria in the REIPPPP programme was allocated to job creation, 'the greatest opportunity for job creation occurs during construction'. To illustrate the limited nature of job creation, in their submission to the Department of Energy, a wind and solar project CDM Africa Climate Solutions indicated that a total of 189 jobs (of which 128 were low skilled) would be created in construction over 10–14 months duration. Only a total of ten operational jobs would be created for the project (DoE RSA PDD 2012, 7). Although there is some reference to transfer of skills and technology, the company admitted that it would fill high skill positions with imported technicians.

A second example to show the paltry nature of job creation is from the Cookhouse Wind project. In its application to register for Clean Development Mechanism (CDM) carbon credits, Cookhouse indicated that the use of imported skills for both construction and operation is a necessity and that imported equipment will be used given the urgency of project. In its application, the company submitted that in the construction phase it was to employ approximately 75 people over a 24-month period and '20 of the construction phase employment opportunities will be for low skilled job categories' while the remaining jobs will be filled by skilled employees. Operational and maintenance skills will initially be recruited from overseas (CDM PDD 2012, 4).

According to Baker and Wlokas (2015), MNCs proved to be also dominant in the engineering procurement and construction phases with Germany's Nordex and Siemen and Denmark's Vestas playing the leading part. In solar PV, the following companies dominated the space: Enertronica, Terni Energia, Moncada and Enel (Italy); ABB (Swiss); Juwi, Siemens (Germany); Iberdrola, Gestamp Solar and ACCIONA (Spain); Scatec Solar Solutions (Norway); Kentz (Ireland); and Solaire Directe (France).

GreenCape (2020) reported that several local manufacturing companies bore the brunt of the stop-start nature of the REIPPPP programme. Local solar PV company Jinko Solar closed down a 120 MW manufacturing facility, Sun Power (60MW facility) is currently dormant while Art Solar (75MW facility) scaled down operations. DCD wind tower manufacturing and construction company (capacity of 200 towers a year) is currently dormant while GRI towers (capacity of 150 towers a year) is not running at full capacity due to the slow pace of the REIPPPP programme.

A more spectacular illustration of how the country's renewable energy initiative failed to create *en masse* scale is around the solar water heaters programme. After much fanfare when it was launched, the programme failed to reach its target of one million solar water heaters by 2014. Kretzmann (2018) reported on the closure of several manufacturing solar plants after the

Department of Energy's abrupt announcement of the slashing of the budget for the National Solar Water Heater (NSWH) programme.

It is therefore not surprising that despite all the grandeur surrounding the REIPPPP programme, workers in traditional energy sectors across the Energy Intensive Users Group (EIUG) were not entirely convinced of this new energy paradigm and reluctant to embrace renewables as an alternative energy source. It is also not surprising that the labour movement has swung back to a position where calls for a just transition are replaced with a strong defence for fossil fuels.

## Precarious Position of Labour

While the concept of carbon lock-in has its usefulness in explaining the interaction of economic, technological, institutional and regulatory factors that block the rapid diffusion of cleaner energy sources, it is our contention that the concept needs to be further stretched to include the possibility of trade unions being part or facilitating carbon lock-ins. Through a footnote and in a passing reference (Unruh 2000, 823–824) cites John Kenneth Galbraith on how automobile unions in the United States were part of institutions that created 'non-market forces of lock-in'.

The story in this chapter shows how union positions can change; at one point be in support of a transition to a low-carbon economy and to later emerge as opponents of the introduction of renewables. When everything is said and done, it is union members who were faced with the loss of jobs. As the promised jobs did not materialise with REIPPPP programme, the initial anxieties morphed into opposition. Under the circumstances and for the workers who were to lose jobs, the slogans of some of the environmental groups—'to keep oil in the soil and leave coal in the hole'—made little sense.

Again, this chapter shows, as Stevis (2018) insists, that context matters. Although there was support at a central level for a shift to renewable energy, the impact was to be varied for different workers as well different sections of the labour movement. The dominance of MNCs in the REIPPPP programme, the failure of the localisation programme and the dismal building of a local sector producing renewable energy technologies meant that there were no new battalions of workers that were joining unions and that had a stake in the emergence of a clean energy sector.

In a recent study conducted in four communities surrounded by coal plants in South Africa, Cock (2019, 1) 'dependence on coal creates socially complex ambiguous patterns of resistance, which sometimes undermine the possibilities of a just transition'. The material dependence on the mines for coal as a source of energy, direct and indirect employment and the provision of a market for the informal sector activities, meant that there was little imagination in the struggles that these communities waged that was about moving beyond coal. The research also found out that the understanding of climate change and just transition was skin-deep in these communities. In many respects the more recent study challenges earlier reports (Cock 2014) about a burgeoning climate justice movement and prospects of 'red-green alliances' in South Africa.

## Conclusion

In this chapter we have tried to demonstrate what happens when an energy transition is unjust and when it happens at the expense of workers in a developing country. From a general support for the transition to a low-carbon economy, South Africa's labour movement is in the forefront of resistance to the introduction of renewable energy. A revolt against attempts to diversify South Africa's energy mix away from fossil fuel is brewing.

This need not be the path of the transition and a different path could have been traversed. Failure to address workers' concern and to build a socially owned renewable sector with a strong manufacturing base that will absorb workers who are displaced has led to a policy cul-de-sac. To unlock the transition and ensure a carbon lock-out, a number of steps need to be considered.

Carbon lock-in theorists do not dismiss the idea of the emergence of a new, more environmentally friendly energy paradigm. As Seto et al. (2016) suggest, the concept of a carbon lock-in does not suggest inevitability and pessimism. A carbon lock-in can burst asunder as a result of exogenous shocks that create windows of opportunity for change. Stein (2017, 562) and Scholvin (2014, 189) call the moments 'critical junctures'. They represent a break from a particular path of development; when structural conditions and limitations open up a range of options for change. According to Seto et al. (2016, 21) it is at these 'moments of plasticity' that it is possible to alter the system structure and its equilibrium. The impending environmental and socio-economic catastrophe associated with climate change and global warming may be the 'critical juncture' to move us forward.

But one does not have to wait for propitious circumstances, as intentional efforts and policy interventions are required for us to move towards a low-carbon economy. These steps are necessary to galvanise social forces that stand to benefit from the transition. Seto et al. (2016) identify incentives for carbon-reducing industries, government support research and development (R&D) in the renewable energy innovation, incentives for diversification away from fossil fuels, funding mechanism for transition costs, enactment of pro-decarbonisation legislation and abolishment of subsidies that favour carbon-intensive industries in the raft of measures that can be taken to facilitate the move towards renewables.

These policies to loosen the vice-grip of a carbon lock-in require multi-dimensional and meta-institutional interventions. It is not only the fossil fuel industry that structures the carbon lock-in. The vice-grip involves allied industries such as infrastructure, commerce, transportation, housing and agriculture. Policies for institutional and behavioural shake-up are necessary.

To support the transition, workers, their communities and organisations must be material beneficiaries of these changes. For the people of the Global South, obstacles that hinder their development and the possibilities of leapfrogging into a low-carbon era must be removed. Unless all these take place, the talk about an energy transition will remain rhetorical.

# References

Baker, Lucy, and Holle Linnea Wlokas. 2015. South Africa's Renewable Energy Procurement: A New Frontier? Energy Research Centre, University of Cape Town, South Africa. http://www.erc.uct.ac.za/sites/default/files/image_tool/images/119/Papers-2015/15-Baker-Wlokas-RE_frontier.pdf. Accessed 21 June 2020.

CDM PDD. 2012. *Cookhouse Wind Farm in South Africa Version 04 16 April 2012.* http://www.energy.gov.za/files/esources/kyoto/2012/Clean%20version_PDD_Cookhouse_Version%2004_16%20April_2012.pdf. Accessed 6 April 2012.

Chamber of Mines. 2010. *Facts and Figures 2010.* https://www.mineralscouncil.org.za/industry-news/publications/facts-and-figures. Accessed 18 May 2020.

Cock, Jacklyn. 2014. The 'Green Economy': A Just and Sustainable Development Path or a 'Wolf in Sheep's Clothing'? *Global Labour Journal* 5 (1): 23–44.

———. 2018. The Climate Crisis and a 'Just Transition' in South Africa: An Eco-Feminist-Socialist Perspective. In *The Climate Crisis: South African and Global Democratic Eco-Socialist Alternatives*, 210–230. Johannesburg: Vishwas Satgar, Wits University Press.

———. 2019. Resistance to Coal Inequalities and the Possibilities of a Just Transition in South Africa. *Development Southern Africa* 36 (6): 860–873. https://doi.org/10.1080/0376835X.2019.1660859.

Cock, Jacklyn, and Rob Lambert. 2013. The Neo-liberal Global Economy and Nature: Redefining the Trade Union Role. In *Trade Unions in the Green Economy: Working for the Environment*, ed. Nora Räthzel and David Uzzell, 89–100. London: Routledge.

COSATU. 2014. *COSATU Will Fight Attempts to Privatise Eskom to the Bitter End.* https://www.politicsweb.co.za/politicsweb/page/en/about/well-fight-attempts-to-privatise-eskom-to-the-bitt. Accessed 30 September 2019.

———. 2015. *COSATU Statement on the Introduction of Nuclear Energy 31 July 2015.* https://www.politicsweb.co.za/politics/our-concerns-over-govts-nuclear-build-plans%2D%2Dcosat. Accessed 23 April 2020.

———. 2017a. *COSATU: Decision by Eskom Board to Unilaterally Close Five Power Stations is a Hostile Act of Provocation Against Workers and Their Unions.* http://salabournews.co.za/press-releases/36398-cosatu-decision-by-eskom-board-to-unilaterally-close-five-power-stations-is-a-hostile-act-of-provocation-against-workers-and-their-unions. Accessed 23 April 2020.

———. 2017b. *COSATU Accuses Eskom of 'Hostile Act of Provocation'. Sowetan Live*, March 8. https://www.sowetanlive.co.za/business/2017-03-08-cosatu-accuses-eskom-of-hostile-act-of-provocation/. Accessed 23 April 2020.

Creamer, Terrence. 2019. Cabinet Considers, but Fails to Approve, IRP Update. *Engineering News*, September 19. http://www.engineeringnews.co.za/article/cabinet-considers-but-fails-to-approve-irp-update-2019-09-19/rep_id:4136. Accessed 30 September 2019.

Davies, John. 2015. *South African Foundry Industry.* [PowerPoint Presentation]. http://www.foundries.org.za/wp-content/uploads/2015/02/2-John-Davies_Vamcosa_Workshop_20022015.pdf. Accessed 18 May 2020.

Department of Energy. 2011a. *Integrated Resource Plan for Electricity 2010–2030, Revision 2, Final Report.* Republic of South Africa. http://www.energy.gov.za/IRP/irp%20files/IRP2010_2030_Final_Report_20110325.pdf. Accessed 21 September 2019.

———. 2011b. *Fact Sheet for the Media Briefing Session on 31 August 2011re the Renewable Energy Independent Power Producer (IPP) Programme.* Republic of South Africa. http://www.energy.gov.za/files/media/pr/2011/Fact%20Sheet%20for%20the%20Media%20Briefing%20Session%20on%2031%20August%202011%20re%20the%20REIPPPP.pdf. Accessed 26 September 2019.

———. 2019. *The South African Energy Sector Report.*

Department of Energy Republic of South Africa. 2012. *Clean Development Mechanism Designated National Authority PDD: CDM Africa Wind and Solar Programme of Activities for South Africa.* http://www.energy.gov.za/files/esources/kyoto/2012/20120405_PDDApplicationFormCDMAPoA_wind_solar.pdf. Accessed 29 September 2019.

Department of Environmental Affairs. 2015. *GHG National Inventory Report South Africa 2000–2015.*

Department of Mineral Resources and Energy. 2019. *Integrated Resource Plan 2019.* Republic of South Africa. http://www.energy.gov.za/IRP/2019/IRP-2019.pdf. Accessed 25 April 2020.

Department of Minerals and Energy. 1998. *White Paper on the Energy Policy of the Republic of South Africa December 1998.* Republic of South Africa. http://www.

energy.gov.za/files/policies/whitepaper_energypolicy_1998.pdf. Accessed 2 September 2019.

———. 2003. *White Paper on Renewable Energy November 2003*. Republic of South Africa. https://unfccc.int/files/meetings/seminar/application/pdf/sem_sup1_south_africa.pdf. Accessed 21 September 2019.

Economic Development Department. 2011. *New Growth Path: Accord 4 Green Economy Accord*. Republic of South Africa.

EIUG. 2018. *EIUG By Numbers*. Energy Intensive Users Group. https://eiug.org.za/. Accessed 18 May 2020.

ENCA. 2017. *NUMSA Fumes over Eskom Plant Closure Plans*. ENCA, March 17. https://www.enca.com/south-africa/NUMSA-fumes-over-eskom-plant-plant-closure-plans. Accessed 1 October 2019.

Eskom. 2019a. *Eskom Company Information Overview*. http://www.eskom.co.za/OurCompany/CompanyInformation/Pages/Company_Information.aspx. Accessed 19 May 2020.

———. 2019b. *Integrated Report: 31 March 2019*. https://www.eskom.co.za/IR2019/Pages/default.aspx. Accessed 1 December 2020.

Fine, Ben, and Zavareh Rustomjee. 1996. *The Political Economy of South Africa from Minerals-Energy Complex to Industrialisation*. Johannesburg: Witwatersrand University Press.

GHG National Inventory Report South Africa 2000–2015, Department of Environmental Affairs, 2015.

GreenCape. 2020. *Utility-Scale Renewable Energy 2020 Market Intelligence Report 2020*. https://www.greencape.co.za/assets/ENERGY_SERVICES_MARKET_INTELLIGENCE_REPORT_20_3_20_WEB.pdf. Accessed 1 December 2020.

Hancock, Tracy. 2011. Wind Turbine Maker Wants DoE to Shed Light on Local Content Threshold for Onshore Wind Technologies. *Engineering News*, September 23. https://www.engineeringnews.co.za/article/call-for-clarification-on-wind-energy-local-content-target-2011-09-23/rep_id:4136. Accessed 21 May 2020.

IPP Quarterly Report, March 2019.

IRENA-GWEC. 2012. *30 Years of Policies for Wind Energy Lessons from 12 Wind Energy Markets*. http://www.gwec.net/wpcontent/uploads/2012/06/IRENA_GWEC_WindReport_Full.pdf. Accessed 23 April 2020.

Kretzmann, Steve. 2018. Companies Face Closure as Rebate Scheme Ditched. *City Press*, May 20. https://www.news24.com/fin24/Economy/companies-face-closure-as-rebate-scheme-ditched-20180518. Accessed 12 June 2020.

Minerals Council of South Africa. 2019. *Facts and Figures Pocketbook 2019*. https://www.mineralscouncil.org.za/industry-news/publications/facts-and-figures. Accessed 18 May 2020.

National development Plan. 2030. Our Future-make it work, National Planning Commission, the Presiodency Republic of South Africa.

National Union of Metalworkers of SA (NUMSA). 2012. *Towards a Socially-Owned Renewable Energy Sector: A National Executive Committee (NEC) Position Paper*.

NEHAWU. 2016. *NEHAWU Discussion Paper on South African Nuclear Programme June 2016.*

Niselow, Tehillah. 2018. NUM Hits Out at Radebe on IPP Deal, Threatens to End ANC Support. *Fin24*, April 5. https://www.fin24.com/Economy/num-hits-out-at-radebe-on-ipp-deal-threatens-to-end-anc-support-20180405. Accessed 23 April 2020.

Omarjee, Lameez, and Jan Cronje. 2018. Court Rejects Numsa's Urgent Bid to Block IPP Signing. *Mail & Guardian*, March 29. https://mg.co.za/article/2018-03-29-court-rejects-numsas-urgent-bid-to-block-ipp-signing/. Accessed 2 May 2020.

Pai, Sandeep, Kathryn Harrison, and Hisham Zerriffi. 2020. *A Systematic Review of the Key Elements of a Just Transition for Fossil Fuel Workers.* Clean Economy Working Paper Series: WP 20-04. Smart Prosperity Institute.

Räthzel, Nora, Jacklyn Cock, and David Uzzell. 2018. Beyond the Nature–Labour Divide: Trade Union Responses to Climate Change in South Africa. *Globalizations* 15 (4): 504–519. https://doi.org/10.1080/14747731.2018.1454678.

Reuters. 2012. South Africa Okays R47billion in Clean Energy Projects. *Sowetan Live*, October 30. https://www.sowetanlive.co.za/business/2012-10-30-south-africa-okays-r47-billion-in-clean-energy-projects/. Accessed 24 September 2019.

SARi. 2011. *Partnering for Green Growth Summary.* https://sarenewablesinitiative.files.wordpress.com/2011/03/sari-partnering-for-green-growth-summary-november-2011.pdf. Accessed 1 October 2019.

SASOL. 2018. *SASOL SOUTH AFRICA (SSA) Fact Book.* https://www.sasolkhanyisa.com/wp-content/uploads/2018/10/7.-sasol-south-africa-limited-factbook.pdf. Accessed 19 May 2020.

Satgar, Vishwas. 2015. A Trade Union Approach to Climate Justice: The Campaign Strategy of the National Union of Metalworkers of South Africa. *Global Labour Journal* 6 (3): 267–282. https://doi.org/10.15173/glj.v6i3.2325. Accessed 20 May 2020.

Scholvin, Sören. 2014. South Africa's Energy Policy: Constrained by Nature and Path Dependency. *Journal of Southern African Studies* 40 (1): 85–202. https://doi.org/10.1080/03057070.2014.889361. Accessed 8 August 2019.

Seto, Karen C., Stephen J. Davis, Ronald B. Mitchell, Eleanor C. Stokes, Gregory Unruh, and Diana Ürge-Vorsatz. 2016. Carbon Lock-In: Types, Causes and Policy Implications. *Annual Review of Environment and Resources* 41: 19.1–19.28. https://www.annualreviews.org/doi/10.1146/annurev-environ-110615-085934. Accessed 18 December 2019.

South African Government. 2010. *Remarks by Ms Dipuo Peters, Minister of Energy, at Investec Power Summit, Investec Bank headquarters, Sandton 6 September 2010 South African Government.* https://www.gov.za/remarks-ms-dipuo-peters-minister-energy-investec-power-summit-investec-bank-headquarters-sandton. Accessed 1 October 2019.

Stein, Amy L. 2017. Breaking Energy Path Dependencies. *Brooklyn Law Review* 82 (2): 559–604. https://brooklynworks.brooklaw.edu/blr/vol82/iss2/7. Accessed 9 August 2019.

Stevis, Dimitris. 2018. (Re)claiming Just Transition. *Just Transition(s) Online Forum.* https://medium.com/just-transitions/stevis-e147a9ec189a. Accessed 11 June 2020.

Stevis, Dimitris, and Romain Felli. 2015. Global Labour Unions and Just Transition to a Green Economy. *International Environmental Agreements* 15 (1): 29–43.

UNEP–EPO. 2013. *Patents and Clean Energy Technologies in Africa.* http://documents.epo.org/projects/babylon/eponet.nsf/0/f87537c7cbb85344c1257b24005e7119/$FILE/patents_clean_energy_technologies_in_Africa_en.pdf. Accessed 22 May 2020.

Unruh, Gregory. 2000. Understanding Carbon Lock-In. *Energy Policy* 28 (12): 817–830. https://doi.org/10.1016/S0301-4215(00)00070-7. Accessed 18 December 2019.

# 4

# The New Struggles to Be Born: The Difficult Birth of a Democratic Ecosocialist Working-Class Politics

Devan Pillay

## Introduction

When the gross domestic product (GDP)[1] became the talismanic metric that claimed to measure economic development (for many a proxy for societal wellbeing) after the Second World War, trade unions by and large fell in line behind economists, investors and employers, in bowing before its apparent explanatory power. More GDP growth meant more economic activity, which ostensibly meant more jobs and better overall societal wellbeing—as long as the surplus was equitably redistributed. Of course, under 'neoliberal' capitalism since the 1980s, where global market liberalisation brought increased global competition, the problem of net 'jobless' growth emerged. Labour-saving technology either generated high unemployment, or the increased informalisation of labour, generally under lower wages and poorer working conditions (see Bieler et al. 2008).

---

[1] Earlier, the Gross National Product (GNP) was more commonly used, as it refers to goods and services of residents, no matter where production occurred, whereas GDP only counts goods and services, at market prices, produced within a country. GDP became the preferred method of accounting from the 1990s, as economic globalisation became more widespread (Fioramonti 2013).

---

D. Pillay (✉)
Department of Sociology, University of the Witwatersrand, Johannesburg, South Africa
e-mail: Devan.Pillay@wits.ac.za

In the 1980s attempts to produce a more accurate measure of societal wellbeing resulted in the United Nations Development Programme (UNDP) producing the annual Human Development Index (HDI). While a better metric than GDP, as it includes health and education, it remains subordinated to GDP growth, given the compromises that had to be made during its formulation through the UN process. In any case, HDI is marginal to public policy discourse, whilst GDP growth remains central. The GDP metric does not measure environmental and social impacts—the externalised costs of production and consumption. It does not measure societal wellbeing (or gross national happiness [GNH] as Bhutan puts it), which in holistic development terms includes both material and socio-psychological goods which accrue from socio-economic development (e.g. health, education, living standards, psychological wellbeing, work-life balance, community vitality, ecological diversity and political participation), minus the bads (e.g. pollution, waste, climate change, social disharmony, unfulfilling work, atomisation, deforestation and bad governance) (see Boulding 1966; Fioramonti 2013; Stiglitz et al. 2009; Pillay 2020).

There was a moment during the 1960s–1980s when hours of work became an issue within the left, and amongst some trade unions (especially in France, where real reductions in hours were achieved). In a sea of unemployment, unions have not made this a campaign issue, such that Decent Life instead of Decent Work[2] became the rallying cry. A refocusing of priorities around Decent Life requires the search for the full development of human potential as envisaged by Karl Marx. This includes the restoration of the 'metabolic rift' between humans and non-human nature (Foster 2009), and in that process also addressing the alienation amongst human beings, as well as within human beings. This is central to a holistic development vision that goes well beyond GDP growth as the main measure of societal wellbeing (Fioramonti 2017).

This chapter focuses on this challenge to the labour movement, through a case study of the National Union of Metalworkers of SA (NUMSA), formerly the second-largest affiliate of the Congress of South African Trade Unions (COSATU) (see Sikwebu and Aroun, Chap. 3 and Cock, Chap. 8). In recent years the union began a journey towards a democratic 'ecosocialist' working-class politics, which excited the imagination of environmental and labour

---

[2] Decent Work (focused around better working conditions and pay) is a campaign of the International Labour Organisation (ILO) and has been aimed mainly at workers in the formal sector, where trade unions traditionally organise. While efforts have been made to broaden the focus to include workers in the informal sector (e.g. street traders), the campaign has been unable to meaningfully address rising unemployment and informalisation of work, a systemic problem of global capitalism today (see Bieler et al. 2008).

activists yearning for a rupture with the suffocating alliance politics of the ruling African National Congress (ANC), South African Communist Party (SACP) and COSATU. It was an unprecedented moment within the labour movement—for the first time globally, a manufacturing trade union adopting policies around socially owned renewable energy—but did it prove too good to be true?

## Ecosocialism and Working-Class Politics

The global labour movement has, since its inception, been locked into a particular logic of 'development', namely that of economic growth with a substantive redistribution of the social surplus to the working class (Barry 2013). As argued in Pillay (2013), economistic unions tended to support the dominance of the market economy, and primarily favoured redistribution through the bargaining relationship with employers. Political unionism sought redistribution primarily through political parties and the state mechanism, where gains made in collective bargaining are supplemented by a substantial social wage (as in free or heavily subsidised housing, public transport, education, health care etc.). The dominance of the state (statism) took two main forms: social democratic and 'Marxist-Leninist' state socialism (or if you prefer, state capitalism).[3] In the process unions lose their independence to political parties and the state.

Social movement unionism (SMU) has been more robust, relying on class alliances and social mobilisation to achieve both workplace and broader societal redistribution of wealth, through substantive participation in the developmental process. In other words, social movement unionism, as an ideal-type, has sought to subordinate both the state and market to a democratically organised society, to achieve equitable social outcomes—a form of democratic-socialist working-class politics, where unions retain their independence from parties and the state.[4]

The South African Congress of Trade Unions (COSATU) during the 1980s is held up as one of the best examples of SMU (Baskin 1991) along with the Central Unica dos Trabalhadores (CUT) in Brazil during the same period (Seidman 1994). They both combined strong independence with alliances

---

[3] 'National-democratic' regimes that emerged after national liberation struggles often deployed the rhetoric of social-democracy or Marxism-Leninism, and practised a mix of the two (i.e. until neoliberal globalisation from the 1980s onwards severely constrained the state and increased the power of the private market, in particular the corporate sector—as is the case in South Africa after 1994).

[4] There is a vast literature on SMUs, in relation to other types of unionism. See Pillay (2013) for a full discussion on the different types and sub-types of unionism, as well as Scipes (2014).

with political actors, in the struggle for democracy and socialism. COSATU during the late 1980s, when political organisations were banned, played a leading role in the anti-apartheid struggle, giving that political campaign a strong working-class imprint. Once political democracy was achieved, however, these unions ceased to display strong SMU characteristics, often lapsing into a narrower form of political unionism, which severely undermined their independence. They also served to legitimise a democratic order constrained by neoliberal economic globalisation (with diminished prospects of achieving democratic socialism). This has led some authors to question whether the SMU concept is only applicable to authoritarian contexts (Von Holdt 2002).[5]

Studies of SMU in the USA have drawn the criticism that since these unions were not anti-systemic, in that their activities were not directed at the seizure of state power, they can at best be regarded as reformist[6] 'social justice' unionism (Scipes 2014). Pillay (2013) seeks to rescue the concept of SMU within democratic contexts, by creating two sub-types, namely anti-systemic SMUs and social justice SMUs, as ideal-types which display the characteristics of internal democracy and operational independence, and class alliances in support of both workplace and broader social struggles.[7]

Extending the SMU approach to recognise the crises of ecological destruction, and the manner in which it is intertwined with social crises of inequality, poverty and dispossession through the historical process of what Altvatar (2007) called 'fossil capitalism', takes the labour movement into the realm of democratic *ecosocialist* working-class politics—the building of red-green coalitions[8] (Pepper 1993; Lowy 2018). Such a politics combines radical utopian visions that avoid the pitfalls of dreamy utopianism, on the one hand, and the politics of strategic engagement that avoids the narrow pitfalls of 'possibilism', on the other (Boron 2012; Olin-Wright 2010). While in recent years the

---

[5] The strong participation of the Korean Confederation of Trade Unions (KCTU) in the struggle for democracy in South Korea during the 1980s has also been held up as an example of SMU (see Suzuki 2012).

[6] The debate on reform versus revolution (or transformation) is addressed later.

[7] As argued in Pillay (2013) anti-systemic SMUs can be further sub-divided into unions that supported 'popular-democratic' struggles, including alliances with political parties and a future democratic state, and anarcho-syndicalist unions that rejected alliances with political parties (and any future democratic state), on the grounds that they will inevitably be compromised by such alliances, and lose their independence, and militant movement character (see Van der Walt (2019) for a syndicalist perspective that avoids the SMU association). COSATU and CUT both had predominantly popular-democratic orientations, although there were 'syndicalist' impulses within both movements at different periods of their history and pre-history.

[8] This can of course be extended to red-green-brown-purple coalitions, to differentiate between rural conservation issues (green) and urban pollution and waste issues (brown), as well as gender issues of various kinds (purple). For purposes of brevity, the red represents the social in all its manifestations (including class, race and gender issues, amongst others), and the green non-human nature.

ecological left has been dismayed by what seemed to be the retreat into 'neo-extractivism' in Bolivia and elsewhere (Solon 2018), the *idea* of *buen vivir* (living well), and the granting to the earth constitutional rights, remains inspirational (Acosta and Abarca 2018). In the words of now-deposed Bolivian president Evo Morales:

> For us, what has failed is the model of "living better" (than others), of unlimited development, industrialisation without frontiers, of modernity that deprecates history, of increasing accumulation of goods at the expense of others and nature. For that reason we promote the idea of Living Well, in harmony with other human beings and with our Mother Earth. (Morales 2009)

These sentiments have encouraged a growing movement within the rich (or 'over'-developed) countries around the concept of 'degrowth'. It counterposes incessant economic growth to the Buddhist concept of 'enoughness' or *sufficiency*—what is needed to live a healthy, meaningful and comfortable life, without harming your neighbour, other sentient beings or the earth. These ideas were promoted in the twentieth century by amongst others Mahatma Gandhi, and informed the localist Buddhist economics of Ernst Schumacher (1973) as well as the more radical ecosocialism of Joel Kovel (2001).[9]

The sufficiency principle resonates with the ideas of the French Marxist Andre Gorz (1962), who made a forceful argument about the need for reduced working time, if we are to address the problem of unemployment, and reduce unnecessary consumption. In recent years this movement has accepted that there needs to be economic growth in the south—but balanced, ecologically sensitive growth that does not 'carbon copy' the dead end of western development trajectories (D'Alisa et al. 2015).

The degrowth paradigm explicitly embraces the 'utopian' thinking of *buen vivir* and *ubuntu* (Terreblanche 2018) and some variants also include ecological Marxist thinking (Magdoff and Foster 2011)—all of which feed into thinking about ecosocialism and more recently a Green New Deal (The Green New Deal for Europe 2019; Pettifor 2019; Aronoff et al. 2019). While ecosocialism is explicitly anti-capitalist, the Green New Deal is a broader concept which keeps open the debate around no growth, low growth and/or higher growth in developing countries (Cassidy 2020), although there seems to be a separate discourse around wellbeing and happiness perspectives, despite scope for a fruitful engagement between them (Pillay 2020).

---

[9] The concept of sufficiency is close to that of 'conviviality', as conceptualised by Ivan Illich (see Deriu 2015).

It is, however, one thing to have democratic ecosocialist aspirations, and another to generate forms of political struggle that uphold those values—both in the policies pursued, as well as in the conduct of leaders. Bolivia's Evo Morales came to power in 2006 through a coalition between indigenous movements and urban trade unions fighting, in the first instance, water privatisation, and building a movement for socialism that embodied the red-green alliance (Webber 2011). However, by reneging on many promises around ecological sustainability, pursuing extractivist pathways that boosted revenues for social development, he alienated large swathes of his support base. When he sought a third term in office as president, which went against the original constitution, he betrayed the democratic ecosocialist ethos that ushered him into power in the first place (leading to a tragic US-supported right-wing coup in November 2019).[10] This is a malady that has affected many left-wing leaders and parties over the past century, and is a lament expressed more recently by disillusioned activists in Nepal in relation to communist party leaders who came to power after the ousting of the monarchy in 2008 (Sharma 2019; Patel 2018).[11]

In other words, movements and parties have to directly confront George Orwell's *Animal Farm* (Orwell 1945) effect—the tendency of leaders to behave like the elites they replaced. This often includes becoming co-opted into the dominant paradigm, but retaining the revolutionary discourse that brought them to power. This tendency is magnified by the adage that 'power corrupts, but absolute power corrupts absolutely'—which is why the lessons of twentieth-century 'dictatorships of the proletariat' common to Marxist-Leninist regimes need to be learnt, as the South African Communist Party (SACP)'s Joe Slovo himself warned in his seminal 1990 pamphlet *Has Socialism Failed?* This is a fate that has befallen most liberation movements throughout the past century, whether nationalist or 'Marxist-Leninist'. Notable exceptions, arguably, include the Communist Party of India (Marxist) which, while retaining its 'Marxist-Leninist' discourse, has periodically taken power within a multi-party democracy in the state of Kerala, India, since the

---

[10] Morales maintains huge popular support, and under his leadership much progress was made in improving social development. The critique is that his movement and party did not create sufficient space for new leadership to emerge after his term ended, thus giving ammunition to the right-wing opposition.

[11] During a visit to Nepal in November 2019, the author spoke to Marxist intellectuals who expressed disillusionment with the Maoist leaders who enriched themselves since coming into office during various periods over the past decade. They had no faith in renewed armed struggle in parts of Nepal, as they felt that peasants who placed their trust in Maoist leaders during the 1990s, only to see their trust betrayed, were unlikely to again trust a new generation of Maoist leaders claiming to represent their interests. The reunited communist party achieved a huge majority in the 2018 elections, but their programme for socio-economic transformation is a modest one.

late 1950s, and in recent decades implemented deep participatory-democratic reforms to achieve substantial social redistribution, in partnership with allied organisations such as trade unions (Heller 1999; Williams 2008).[12]

Amongst others Gandhi and the feminist movement warned that activists must *be* the change they want to see, if true radical transformation is to be achieved. Drawing on the thinking of the ancients (Armstrong 2006), this involves *personal transformation* and continuous introspection, as well as a *deep participatory politics*, where leaders are always held accountable to their organisations, members and communities. In South Africa this was embodied in the thinking and practice of Rick Turner, who was an inspiration for a new generation of trade unionists in the 1970s (Keniston 2010).

Turner was a highly popular and influential Political Science lecturer at the University of Natal in the late 1960s and early 1970s, before he was banned by the apartheid regime and later assassinated. He promoted 'workers' control' of both unions and industry, as a stepping stone towards maximum participatory democracy in society as a whole—a *society-focussed* socialist vision, as opposed to the traditional *statist* emphasis of much of the 'socialist' world at the time. His brand of Humanist Marxism, primarily influenced by Jean-Paul Sartre's *Critique of Dialectical Reason* (Sartre 1960), was infused with a 'transcendent' (or if you like 'spiritual') essence that believed in non-violence, universal love and the unity between inner (introspective) and outer (structural) transformations.

Turner's notion of workers' democracy is captured in this quote from his seminal work *Eye of the Needle: Participatory Democracy in South Africa* (the title is a biblical reference to the alleged saying of Jesus that a rich person had as much chance of going to heaven as the camel had of going through the eye of the needle):

> 'Workers' control is not only a means whereby I can control a specific area of my life. It is an educational process in which I can learn better to control all areas of my life and can develop both psychological and interpersonal skills in a situation of co-operation with my fellows in a common task … participation in decision-making, whether in family, in the school, in voluntary organisations,

---

[12] It remains, of course, a matter of debate regarding the extent to which the male leadership of the party *fully* practises a democratic non-patriarchal politics, and to what extent the party maintains a vanguardist internal political hierarchy. In addition, the 'grassroots' faction of the party was contested by the more orthodox Marxist-Leninist faction, with the latter receiving support from amongst the trade unions (Williams 2008). The depth of participatory democracy implemented by the CPI (M) whilst in power may be put into question by the fact that they were always voted out of power after five years (alternating with the Congress Party). Nevertheless, they played a major role in establishing a social democratic consensus within the state, no matter which party coalition assumed power. In May 2021 CPI(M) led coalition broke the cycle and were re-elected for a second consecutive term.

or at work, increases the ability to participate and increases the competence on the part of the individual that is vital for balanced and autonomous development. Participation through workers' control lays the basis for love as a constant rather than as a fleeting relationship between people. (Turner 1972/1980, 39)

In many ways he embraced the concept of revolutionary love articulated by the Cuban-Argentinian revolutionary Che Guevara:

At the risk of seeming ridiculous, let me say that the true revolutionary is guided by a great feeling of love. It is impossible to think of a genuine revolutionary lacking this quality. (Guevara 1965)

Indeed, if socialist 'love' means rising above ourselves as individuals, and embracing the whole of humanity, then it is no different from the 'spiritual' essence conveyed by religious philosophers, some of whom imagine an external god or gods as the embodiment of the totality of Love, whilst others, like the Samkhya school and the Buddha, find that capacity *within* all of us. Turner easily made connections between his humanist Marxism and the 'spiritual' (without necessarily embracing a theism or belief in an external god).

Like the Buddha, Turner easily extended this humanism to the realm of non-human nature (i.e. the 'environment'). He understood the limits to growth argument, arguing that capitalism is 'intrinsically growth-oriented', but there are 'limits to the resources of our planet' as well as 'limits to our ability to dispose of our own rubbish' (Turner 1972/1980, 98). Unless this obsession with growth ended, and resources re-allocated to vital material needs (food, shelter and health), Turner argued, 'we can only look forward to a future of famine, growing inequality, social conflict, and universal hate and fear in the struggle for survival' (Turner 1972/1980, 97). This 'ecosocialist' prescience was, unfortunately, not taken up by his trade union colleagues at the time.

As mentioned earlier, Eastern and other traditional beliefs, such as that of the Native American (*buen vivir*) and African (*ubuntu*), inform much of the ecosocialist perspectives that converge around the degrowth movement, and considerations of happiness, wellbeing and localised Buddhist economics. While ecological Marxism has some differences with these perspectives, there is no huge barrier between them. Marx as a caring humanist (as opposed to the cold 'scientist' imagined by twentieth-century Leninism), whose theory of alienation had strong 'spiritual' meaning (Fromm 1961), desired a form of social equality and human flourishing that connected strongly with the yearnings of ancient philosophers seeking the end of human (and often animal) suffering (Armstrong 2006; Sardesai 1982; Chattopadhyaya 1970; Bensaada n.d.; Duchrow and Hinkelammert 2012).

Turner's utopian vision is different from a utopian politics that underestimates power relations, and the need to navigate choppy waters that involve both struggle and negotiation, and inevitably compromises. It seeks short-term tactical victories that are embedded in longer-term strategic visions that can only be guaranteed by a fundamentally democratic project, where power truly resides with the people. This was, indeed, the *modus operandi* of the re-emerged South African trade union movement in the 1970s and 1980s, of which the Metal and Allied Workers Union (MAWU), which went on to become the National Union of Metalworkers of SA (NUMSA), was a core component (Forrest 2011).

These early unions grappled with twentieth-century debates around 'reform or revolution' (Luxembourg 1908). Reformism seeks gradual reforms within the context of the dominant order, while revolution ordinarily seeks the overthrow of the state, through mass action and force, if necessary. Andre Gorz (1962) reformulated this to distinguish between 'reformist reforms', which are a dead end, and 'revolutionary reforms' that have short-term tactical value and longer-term revolutionary or transformative potential. The South African unionists were particularly interested in Antonio Gramsci's (1982) 'war of position', a military metaphor about holding your position in the face of a stronger enemy, and 'war of manoeuvre', which was the moment of all-out assault on the enemy when weaknesses appeared. MAWU and its allies, who went on to form the Federation of South African Trade Unions (FOSATU) in 1979, were careful to stress the need to build solid shopfloor structures, with built-in consultative and accountability mechanisms. Through this a strong worker leadership was nurtured, and a resilient organisation was built which could withstand state repression. When conditions allowed this, they argued for the need to entrench their power by participating in official industrial relations forums, thus displacing bureaucratic unions of the old order, whilst simultaneously engaging in mass mobilisation, including strike action. More militant unions allied to the ANC and SACP tended to reject this approach, and sought to directly confront the apartheid state—a strategy that invited massive state repression, which had a severe impact on the durability of their organisations (and tended to encourage more hierarchical or vanguardist forms of leadership, with weak consultation of their members).

The more cautious approach of the FOSATU unions was labelled 'workerist' and 'abstentionist', while they in turn labelled the ANC/SACP-allied unions 'adventurist' and 'populist' (because in their deference to their political principals, they did not emphasis working-class issues and the anti-capitalist nature of the struggle for democracy). Eventually these two streams came together to form COSATU in 1985, combining the shopfloor, independent and internal-democratic emphasis of FOSATU and allied unions, with the more militant, politically charged emphasis of the ANC/SACP-allied unions (Friedman 1987).[13] By 1990 COSATU became part of the triple alliance with the ANC and SACP, after the latter two were unbanned. While it maintained its operational independence, many of the fears of the 'workerist' unions were realised, in that COSATU gradually lost much of its SMU character (Pillay 2011).

## Case Study: NUMSA's Lost Moment

NUMSA was formed in 1987 through the merger of MAWU and a range of other unions, and became the second-largest affiliate of COSATU (Baskin 1991). Although MAWU/NUMSA for a long time argued for the formation of an independent workers' party, it succumbed to the gradual dominance of the SACP within the union movement during the late 1980s (see later), and became an important part of the triple alliance from 1990. In 2013 it made the momentous decision to stop being part of the alliance with the ANC and SACP, and hoped to convince its federation COSATU to follow suit (NUMSA 2013). However, this decision instead led to its expulsion from COSATU in 2014, along with COSATU's general secretary, Zwelinzima Vavi, who supported NUMSA (Pillay 2015). NUMSA went on to form the SA Federation of Trade Unions (SAFTU) in 2017, with Vavi as general secretary (NUMSA comprises about half the federation's 700.000 members).

The optimistic view was that the NUMSA split represented a return to the 'social movement union' roots of the union, where in the 1970s and 1980s, as MAWU, it led the argument for an independent but politically engaged labour movement, uncontaminated by the nationalist politics of the dominant liberation movements. This was influenced by the participatory-democratic, humanist Marxism of Rick Turner and Antonio Gramsci (Forrest 2011). The optimism was reinforced by NUMSA's path-breaking focus on climate change, alternative energy and green jobs since 2011, which did not receive as much media attention as its resolve to leave the alliance (see below). This had the potential of moving the union out of its traditional

---

[13] These unions were affiliated to the umbrella body, the United Democratic Front (UDF), which was formed in 1983 as the legal expression of the ANC/SACP (even though it by and large operated with a high degree of independence inside the country).

concentration on workplace bargaining issues and its assumptions about fossil-fuel growth paths, towards a broader focus on arguably *the* major issue facing capitalism: the natural limits to growth. Indeed, this occurred during a period when COSATU itself adopted a far-reaching policy focus on climate change and renewable energy in 2011, although it remained at the level of policy, rather than meaningful action (COSATU 2011).

For many on the independent left, the December 2013 NUMSA breakaway from the Alliance was a hopeful sign that at last the scales were falling from the eyes of large sections of the working class, as they saw that the ANC/SACP emperor had no clothes. The promise of NUMSA's broad United Front, and its 'movement for socialism', came on the heels of an ostensible break with 'fossilised' development paradigms, and a serious consideration of alternatives that suggested possibilities of forging a new, ecosocialist working-class politics.

These developments occurred within a context of growing concern about climate change and the need for renewable energy. The now moribund Democratic Left Front (DLF), at its launch in 2011, envisioned bold ecosocialist alternatives (DLF 2011), but failed to establish itself within the labour movement. The Climate Jobs Campaign, launched in 2011 by the Alternative Information and Development Centre (AIDC), produced research findings indicating that jobs in renewable energy sectors (including the building of wind, wave tide and solar power, the renovation and insulation of homes and offices, and the provision of public transport) could create 3.7 million decent jobs based on the principles of ecological sustainability, social justice and state intervention (One Million Climate Jobs Campaign 2011, 13).[14]

The Climate Jobs Campaign sought to show how shifted priorities and political will can generate the ideas and resources necessary to create meaningful alternatives (see Ytterstad, Chap. 11). The campaign however did not take root in South Africa, partly because of the turmoil within the labour movement. Nevertheless, the 2011 COSATU policy paper on the environment raised critical issues regarding a just transition from the current economic paradigm to that of a low carbon economy, which was a major step forward. However, as Jacky Cock (2013) pointed out, labour was caught between a *reformist* position, which sought accommodation within the logic of green capitalism, market-based solutions such as carbon trading, and technologies such as carbon capture and storage, and a *transformative* position, which

---

[14] The Campaign was an alliance of relatively small labour and social movements mainly located in Cape Town, and pulled together by the AIDC. It was updated in 2016 (see One Million Climate Jobs Campaign 2016). It was inspired by a similar campaign in the UK, which attracted mere attention from organised labour (see https://campaignncc.org/climatejobs).

stressed the need for a class analysis and the recognition that the capitalist system is at the heart of the crisis of climate change (see Cock, Chap. 8).

The most encouraging development occurred within NUMSA, which put forward an innovative transformative policy perspective around climate change in 2012. Back then it criticised the government's market-based proposals around renewable energy which give private companies (independent power producers) the lead in providing alternatives such as 'onshore wind, concentrated solar thermal, solar photovoltaic, biomass, biomass, landfill gas and small hydro' (NUMSA 2012: 1). NUMSA's 'socially owned' alternative involved:

- public, community and collective ownership of land sites which can produce renewable energy;
- social ownership of utilities that generate, transmit, and distribute energy;
- social ownership and control of the fossil fuel industry such as coal and synthetic fuel to harness their revenues and fund renewable alternatives;
- local content requirements in the building of a renewable energy manufacturing base, in order to create local jobs;
- the creation of municipal solar and wind parks;
- the use of workers' pension funds to finance socially-owned renewable companies;
- the promotion of gender equity at all levels of the occupational ladder in such companies; and
- the setting up of a network, in collaboration with local and international friends of NUMSA, to monitor the bidding process around government tenders for the provision of renewable energy. (NUMSA 2012)

In these proposals NUMSA made an implicit distinction between *social* ownership, which involves maximum democratic participation from below (by workers and citizens), and *state* ownership, which is bureaucratic control over public resources, increasingly within a framework of market principles, where workers are exploited and domestic consumers fleeced in the interests of large corporations—as is the case of the state-owned power utility Eskom, and the Central Energy Fund (CEF). NUMSA's proposals gave substance to its more general views on nationalisation where, in contrast to state-controlled 'nationalisation', it called for *worker-controlled* nationalisation of the commanding heights of the economy (NUMSA 2013).

In addition, there was growing convergence between the United Front and the South African Food Sovereignty Campaign. The UF actively supported the Hunger Tribunal in 2015 and the Drought Speak Out and Bread March in 2016, connecting the dots between the energy, climate and food crisis (www.safsc.org.za).

The political crisis and the expulsion of NUMSA in 2014, however, derailed this focus, putting it on the back burner in both NUMSA and COSATU. They subsequently lost or sidelined key union activists leading these processes, and it remains to be seen whether these seeds of a twenty-first-century ecosocialist politics can be re-planted. The declaration of the Workers' Summit in May 2016 to announce the launch of the process of a new federation surprisingly made no mention of ecological and gender issues—indicating the extent to which such issues were uppermost in the minds of unionists. However, buried in NUMSA's December 2016 congress declaration were references to ecological alternatives. This included its advice to the electricity state monopoly Eskom to 'open up to alternative energy', its demand that the country 'explore alternative sources of renewable energy (solar, wind, water), decrease reliance on coal, and reaffirm that the manufacturing and servicing of solar systems and power stations must be done locally', and a firm commitment to join NGOs like Section 27 to 'campaign against nuclear power stations'. The congress resolved to link struggles around climate change to the struggle against global capitalism, and to 'find allies in that effort across the globe'. It also re-affirmed its social movement union aspirations by stressing that it is 'a union that links shopfloor struggles with community struggles' (NUMSA 2016).

These resolutions, although not prominently featured, seemed *at face value* to re-affirm previous commitments to social green economic objectives. However, according to a NUMSA insider, the social forces behind the 2012 ecological moment included a worker-based research and development group, a research and development programme, and an alliance building approach at local and international levels. The latter led to the establishment of a broad-based Electricity Crisis Campaign in South Africa, and participation in an international union coalition, Trade Unions for Energy Democracy, and energy-related campaigns. These were initiatives in their infancy, with a strong potential to grow into something powerful. According to this insider, 'all these are gone. So how are the 2012 resolutions on climate change and renewable energy going to be taken forward, including the lines that pepper the terrible national congress declaration?'

Clearly, a 'socially owned' and 'worker controlled' orientation, which is more in accordance with a bottom-up ecosocialist (or eco-Marxist) approach, is contradicted by a 'Marxist-Leninism' that is normally associated with bureaucratic statism and productivism (or economic growth at all costs, including unsustainable environmental costs). Productivism is the orientation of many of NUMSA's top leaders and key advisers, and became union policy at the 2016 congress (culminating in the formation of the Socialist Revolutionary Workers' Party [SRWP] in 2018).

The SACP-derived 'Marxist-Leninist' ideological discourse—within both COSATU and NUMSA—is a major departure from the heritage of the 1970s, which embraced the more flexible, participatory-democratic and humanist Marxism of Rick Turner. After 1990 key intellectuals of the independent Marxist tradition went on to join the ANC and/or the SACP, and some became wealthy businessmen—thus contributing to the retreat of independent Marxism into the academy, and as minority strands within the union movement.

The SACP's Marxist-Leninism is of the mechanical Stalinist lineage where, throughout its history, the party followed all the twists and turns of the Soviet Union. With the fall of the Berlin Wall in 1989, and the publication of Joe Slovo's *Has Socialism Failed* in 1990, the SACP began to shed some of this baggage (Williams 2008). The party rapidly influenced COSATU affiliates after its unbanning in 1990, including NUMSA, whose top leadership picked up some of this mechanical baggage.[15] With the exception of the more flexible ecosocialist or democratic Marxist current (found within some social movements and unions), the dominant discourse and practice within the Left remains mired in a narrow vanguardist interpretation of Lenin's notion of 'democratic-centralist' politics (Satgar and Roger Southall 2015). Nevertheless, despite its 'Marxist-Leninist' discourse, does NUMSA have the potential to revive its participatory-democratic ethos and play a significant role in reinvigorating working-class politics in South Africa?

Indeed, some may argue that, in principle, there is not a huge barrier between a 'Marxist-Leninist' vanguard (as opposed to vanguard*ist*) approach, and participatory democracy, as the example of the Communist Party of India (Marxist) in Kerala alluded to earlier indicates. The Kerala example shows how a 'vanguard' can provide leadership from the centre, and be a catalyst for bottom-up democratic processes; in contrast to the vanguard*ist* approach, which pays *lip service* to genuine participatory democracy, and leads *exclusively* from the centre. In NUMSA the vanguardist faction sidelined the democratic Marxist current during 2015–2016, and remains dominant (but not without internal contestation).

The unprecedented attack on so-called middle-class Marxism at NUMSA's 2016 congress, and the side-lining of key activists who championed that more open, democratic ethos, severely undermined prospects of union and Left

---

[15] According to a NUMSA insider, those behind the assertion of Marxist-Leninism are to the 'right of the SACP', in terms of their adherence to the older, Stalinist orthodoxy.

revitalisation. Those who led the participatory-democratic, ecological moments were not given an opportunity to debate these ideological differences with the congress representatives. The leadership gave the line in its secretariat report, and the congress duly adopted it—thus severely undermining the union's own impressive history as a democratic union that respected diversity of opinion, and robust internal debate. This set the stage for the complete marginalisation of the United Front, which accommodated a diverse range of left opinion, and the formation of the dogmatic 'Marxist-Leninist' SRWP.

Despite these set-backs, the future of left revitalisation remains an open project globally and locally, given the crisis of fossil capitalism. A narrow, moribund twentieth-century 'Marxist-Leninism' has little appeal outside the Alliance and its off-shoots. For optimists who attended the 2016 NUMSA congress, there was a 'catchy twenty-first century melody somewhere within the raucous nineteenth/twentieth-century M-L music'.[16] This captured the hopes of many on the Left who are desperate for an alternative to the unpalatable choices of neoliberal capitalism, crony capitalism or authoritarian racial-statism (or 'third world nationalism'). The catchy melody, however, may reside more in a hopeful imagination, than in the post-congress reality of NUMSA's politics.

More recently, as Cock (2019) optimistically points out, both NUMSA and the National Union of Mineworkers (NUM) began to work with climate justice activists (although the individuals involved may not necessarily have the full endorsement of the respective unions). Encouragingly, SAFTU general secretary Vavi, in a *Sunday Times* article penned with climate justice campaigner Alex Lenferna, added his voice to growing concerns about climate justice (Vavi and Lenferna 2019). SAFTU participated in the Johannesburg climate strike on 20 September 2019—one of the largest environmental marches ever seen in the country (alongside others around the country, and the world). Vavi, as COSATU general secretary, played a critical role in ensuring the labour movement defended democratic rights and freedoms during Zuma's reign, and now combined these sensibilities with an explicit identification of the need for a just transition to a post-carbon future.

However, the unions remain highly compromised by relations of patronage, which inhibit their ability to become pioneering visionaries. NUMSA in 2018 and 2019 sided with BEE[17] coal interests (leaders of which had in the

---

[16] This is a view expressed by the Left scholar Patrick Bond in an email exchange on the Debate list, 17/12/16.
[17] Black Economic Empowerment (BEE) was called Black Economic Enrichment by the SACP, as it empowered a few connected individuals. It broadened to Broad-Based Black Economic Empowerment,

past been closely aligned with the state capture faction of the ANC), to oppose the extension of renewable energy by power utility Eskom. When challenged, they dusted off their socially owned renewable energy policy to act as a fig leaf, to cover up their coal interest, in a court case they eventually lost (Cloete 2018; Bloom 2020). Journalist Ferial Haffajee warned in early 2020 that power utility Eskom remains a key site of state capture, with unionists directly implicated:

> It's procurement budget is still large and trade union members are said to have cornered coal and coal transport contracts for family members. The fixes on the Medupi and Khusile power stations are going to need big budgets and political patronage networks are lining up for these. These networks are believed to have their lines of support on the party's NEC which explain why Eskom is subject to such a tussle.

## Conclusion

As the history of socialist struggles reveal, there is no necessary relationship between socialism (or indeed Marxism) and democracy—despite assertions by Marxist-Leninists and others that socialism IS democracy. Hence the re-assertion in recent times of *democratic* socialism (which implies the existence of undemocratic forms of socialism). Similarly, there is no necessary connection between environmentalism, and by extension, ecosocialism (or ecological Marxism) and democracy. All types of struggle lend themselves to either bureaucratic-oligarchic forms of organisation, or democratic forms. Democracy has to be continuously nurtured and protected, both within organisations and within the state and broader society.

A democratic ecosocialist working-class politics recognises the social and natural limits[18] to fossil-fuel growth pathways, and the need to forge a new form of counter-hegemonic politics, based on red-green alliances. This is a major challenge to trade unions, rooted as they are in defensive bargaining relationships with employers, and/or relations of patronage with political parties and the state. There are signs of hope that organised labour around the world is beginning to take issues of climate change and green jobs seriously. Whether they can move further and question the GDP growth paradigm, and

---

but this merely served to open up relations of patronage to trade union leaders and others from hitherto marginalised sectors, such as women's and rural organisations.

[18] The Marxist writer David Harvey (2014) cautions against the view that there are absolute limits to growth, as capital has always found ways to overcome barriers to growth, the latest being green capitalism.

move from the narrow concerns of Decent Work to a more holistic Decent Life agenda, remains a difficult challenge.

In South Africa, the green shoots of an ecosocialist working-class politics, planted within NUMSA and extended to COSATU, was a sign of a new awakening. NUMSA was one of the few unions worldwide to take up the ecological challenge, through its 2012 proposals for a socially owned renewable energy pathway. This opened up the possibilities of a new kind of working-class politics. However, this innovative ecological and democratic Marxist thrust was not shared by the key leaders and their advisors in the union. They clung to the twentieth-century paradigm, under the banner of an unreconstructed 'Marxist-Leninism' derived from the SACP (in its more dogmatic form). This culminated in an extraordinary attack on independent thinkers at NUMSA's December 2016 congress, publicly labelling them 'middle class Marxists'. While some still see a sweet twenty-first-century tune within the discordant twentieth-century music, others, like former NUMSA spokesperson Castro Ngobese, declared that 'the NUMSA Moment is lost'.[19]

While it, along with COSATU, has allowed these initiatives to wither, it can still be brought to life again, given the enduring crisis of fossil capitalism. The union movement still has within it worker impulses that want to assert a democratic, ecosocialist working-class politics—in conjunction with environmental justice and other social movements. While it may be a long way off, the seeds of broader worker unity, and the possibility of an imaginative ecosocialist working-class politics, have been planted. As the crisis of fossil capitalism deepens, it will take both workers' pressure from below and a visionary leadership from above to see these seeds sprout into green shoots, both within South Africa and across the world. As the ITUC's Sharon Barrows said a few years back, 'there are no jobs on a dead planet'. The environmental crisis in general and the climate crisis in particular are working-class issues of paramount importance.

## References

Acosta, A., and M.M. Abarca. 2018. Buen Vivir: An Alternative Perspective from the Peoples of the Global South to the Crisis of Capitalist Modernity. In *The Climate Crisis. South African and Global Democratic Eco-Socialist Alternatives*, ed. Vishwas Satgar, 131–147. Johannesburg: Wits University Press.

---

[19] Ngobese, who was dismissed by the union in September 2016, made this claim in a tweet before the NUMSA congress.

Altvatar, Elmar. 2007. The Social and Natural Environment of Fossil Capitalism. *Socialist Register* 43.

Armstrong, Karen. 2006. *The Great Transformation: The World in the time of Buddha, Socrates, Confucius and Jeremiah*. London: Atlantic Books.

Aronoff, Kate, Alyssa Battistoni, Daniel Aldana Cohen, and Thea Riofrancos. 2019. *A Planet to Win. Why We Need a Green New Deal*. London: Verso.

Barry, John. 2013. Trade Unions and the Transition Away From 'Actually Existing Unsustainability'. From Economic Crisis to a New Political Economy Beyond Growth. In *Trade Unions in the Green Economy. Working for the Environment*, ed. Nora Räthzel and David Uzzell, 227–240. London and New York: Earthscan.

Baskin, Jeremy. 1991. *Striking Back. A History of COSATU*. Johannesburg: Ravan Press.

Bensaada, M.T. n.d. The Islamic Liberation Theology of Ali Shariati. www.decolonialtranslation.org. Accessed 10 Feb 19.

Bieler, Andreas, Ingemar Lindberg, and Devan Pillay. 2008. *Labour and the Challenges of Globalisation: What Prospects for Transnational Solidarity*. London: Pluto.

Bloom, K. 2020. 'Renewable vs coal: the timeline of a fatally rigged fight', Daily Maverick 16 January.

Boron, Atilio. 2012. *Buen Vivir and the Dilemmas of the Latin American Left*. Trans. Richard Feder. http://climateandcapitalism.com/2015/08/31/buen-vivir-and-dilemmas-of-latin-american-left. Accessed 28 Sept 2015.

Boulding, Kenneth E. 1966. The Economics of the Coming Spaceship Earth. In *Environmental Quality in a Growing Economy*, ed. H. Jarrett, 3–14. Baltimore, MD: Resources for the Future/Johns Hopkins University Press.

Cassidy, John. 2020. Can We Have Prosperity Without Growth? *The New Yorker*, February 3.

Chattopadhyaya, Debiprasad. 1970. Some problems of early Buddhism. In *Buddhism: The Marxist Approach*, ed. Rahul Sankrityayan, Debiprasad Chattopadhyaya, Y. Balaramamoorty, Ram Bilaas Sharma, and Mulk Raj Anand, 2–15. New Delhi: People's Publishing House.

Cloete, Karl. 2018. Op-Ed: Numsa Supports a Transition from Dirty Energy to Clean Renewable Energy. *Daily Maverick*, March 15.

Cock, Jacklyn. 2013. Ask for a Camel When You Expect to Get a Coat: Contentious Politics and the Climate Justice Movement. In *New South African Review 3: The Second Phase: Tragedy or Farce?* ed. John Daniel, Prishani Naidoo, Devan Pillay, and Roger Southall, 154–172. Johannesburg: Wits University Press.

———. 2019. *Resistance to Coal and the Possibilities of a Just Transition in South Africa*. SWOP Institute, University of the Witwatersrand.

Congress of SA Trade Unions (COSATU). 2011. *A Just Transition to a Low-Carbon and Climate Resilient Economy. COSATU Policy on Climate Change. A Call to Action*. Booklet based on COSATU'S Policy Framework on Climate Change, adopted by the Central Executive Committee in November 2011.

D'Alisa, Giacomo, Frederico Demaria, and Giorgos Kallis, eds. 2015. *Degrowth: A Vocabulary for a New Era*. New York and London: Routledge.
Democratic Left Front (DLF). 2011. *Another South Africa and World is Possible! 1st Democratic Left Conference Report*, January 20–23, University of the Witwatersrand, South Africa.
Deriu, Marco. 2015. Conviviality. In *Degrowth: A Vocabulary for a New Era*, ed. D'Alisa Giacomo, Frederico Demaria, and Giorgos Kallis, 79–82. New York and London: Routledge.
Duchrow, Ulrich, and Franz J. Hinkelammert. 2012. *Transcending Greedy Money: Interreligious Solidarity for Just Relations*. New York: Palgrave Macmillan.
Fioramonti, Lorenzo. 2013. *Gross Domestic Problem. The Politics Behind the World's Most Powerful Number*. London: Zed Books.
———. 2017. *Wellbeing Economy: Success in a World Without Growth*. Johannesburg: MacMillan.
Forrest, Kally. 2011. *Metal that will not Bend: National Union of Metalworkers of South Africa 1980–1995*. Johannesburg: Wits University Press.
Foster, John Bellamy. 2009. *The Ecological Revolution: Making Peace with the Planet*. New York: Monthly Review.
Friedman, Steven. 1987. *Building Tomorrow Today*. Johannesburg: Ravan.
Fromm, Erich. 1961. *Alienation*. Extract. https://www.marxists.org/archive/fromm/works/1961/man/ch05.htm. Accessed 28 Sept 2015.
Gorz, Andre. 1962. *Strategy for Labor. A Radical Proposal*. Boston: Beacon Press.
Gramsci, Antonio. 1982. *Selections from the Prison Books*. London: Lawrence and Wishart.
Guevara, Ernesto Che. 1965. *Man and Socialism in Cuba*. https://www.marxists.org/archive/guevara/1965/03/man-socialism-alt.htm. Accessed 28 Sept 2015.
Harvey, David. 2014. *Seventeen Contradictions and the End of Capitalism*. Oxford: Oxford University Press.
Heller, Patrick. 1999. *The Labor of Development: Workers and the Transformation of Capitalism in Kerala, India*. Cornell University Press.
Keniston, William H. 2010. *Richard Turner's Contribution to a Socialist Political Culture, 1968–1978*. Unpublished MA mini-thesis, University of the Western Cape.
Kovel, Joel. 2001. *The Enemy of Nature. The End of Capitalism or the End of the World?* 2nd ed. London: Zed Books.
Lowy, Michael. 2018. Why Ecosocialism: For a Red-Green Future. *Great Transition Initiative*. October. https://greattransition.org/publication/why-ecosocialism-red-green-future.
Luxembourg, Rosa. 1908. *Social Reform or Revolution?* London: Militant Publications.
Magdoff, Fred, and John Bellamy Foster. 2011. *What Every Environmentalist Needs to Know about Capitalism*. New York: Monthly Review Press.
Morales, Evo. 2009. How to Save the World, Life and Humanity. In *People First Economics*, ed. David Ransom and Vanessa Baird. Oxford: New Internationalist Publications.

National Union of Metalworkers of South Africa (NUMSA). 2012. *Building a Socially—Owned Renewable Energy Sector in SA*. Resolution of the Numsa 9th National Congress, June/

NUMSA. 2013. *Numsa Special National Congress December 17 to 20, 2013 Declaration*.

———. 2016. *Secretariat Report*. 10th National Congress, December 12–15, Cape Town.

Olin-Wright, Eric. 2010. *Envisioning Real Utopias*. London: Verso.

One Million Climate Jobs Campaign. 2011. *One Million Climate Jobs: A just Transition to a Low Carbon Economy to Combat Unemployment and Climate Change*. Cape Town: One Million Climate Jobs.

———. 2016. *One Million Climate Jobs: Moving South Africa Forward on a Low-carbon, Wage-led, and Sustainable Path*. Cape Town: AIDC.

Orwell, George. 1945/2008. *Animal Farm*. London: Penguin Classics.

Patel, Pawan. 2018. *The Making of 'Cash Maoism in Nepal: a Thabangi Perspective*. New Delhi: Adarsh Books.

Pepper, David. 1993. *Eco-Socialism. From Deep Ecology to Social Justice*. London: Routledge.

Pettifor, Ann. 2019. *The Case for the Green New Deal*. London: Verso.

Pillay, Devan. 2011. The Enduring Embrace: COSATU and the Tripartite Alliance during the Zuma era. *Labour, Capital and Society* 44 (2): 56–79.

———. 2013. Between Social Movement and Political Unionism: COSATU and democratic politics in South Africa. *Rethinking Development and Inequality* 2: 10–27.

———. 2015. COSATU and the Alliance: Falling Apart at the Seams. In *COSATU in Crisis: The Fragmentation of an African Trade Union Federation*, ed. Vishwas Satgar and Roger Southall, 115–133. Johannesburg: KMM/FES.

———. 2020. Happiness, Wellbeing and Ecosocialism—A Radical Humanist Perspective. *Globalizations* 17 (2): 380–396.

Sardesai, S.G. 1982. The Riddle of the Geeta. In *Marxism and the Bhagvat Geeta*, ed. S.G. Desai and Dilip Bose. New Delhi: People's Publishing House.

Sartre, Jean-Paul. 1960/1976. *Critique of Dialectical Reason*. Vol 1 (English translation). Paris: Gallimard.

Satgar, Vishwas, and R. Roger Southall, eds. 2015. *Cosatu in Crisis: The Fragmentation of an African Trade Union Federation*. Johannesburg: FES/KMM Review Publishing.

Schumacher, Ernst Friederich. 1973. *Small is Beautiful. A Study of Economics as if People Mattered.* https://web.archive.org/web/20141014171926/http://www.ditext.com/schumacher/small/small.html.

Scipes, K. 2014. Social Movement Unionism or Social Justice Unionism? Disentangling Theoretical Confusion Within the Global Labor Movement. *Class, Race and*

*Corporate Power* 2 (3), article 9. https://doi.org/10.25148/CRCP.2.3.16092119. http://digitalcommons.fiu.edu/classracecorporatepower/vol2/iss3/9.

Seidman, Gay. 1994. *Manufacturing Militance: Workers' Movements in Brazil and South Africa, 1970–1985*. Berkeley and Los Angeles: University of California Press.

Sharma, Sudheer. 2019. *The Nepal Nexus. An Insider Account of the Maoists, the Durbar and New Delhi*. Gurgaon: Viking/Penguin.

Solon, Pablo. 2018. The Rights of Mother Earth. In *The Climate Crisis. South African and Global Democratic Eco-Socialist Alternatives*, ed. Vishwas Satgar, 107–130. Johannesburg: Wits University Press.

Stiglitz, Joseph, Amartya Sen and Jean-Paul Fitoussi. 2009. *Report by the Commission on the Measurement of Economic Performance and Social Progress* www.stiglitz-sen.fitoussi.fr. Accessed 7 Feb 2019.

Suzuki, A. ed. 2012. Cross-National Comparisons of Social Movement Unionism. Oxford: Peter Lang.

Terreblanche, Christelle. 2018. Ubuntu and the Struggle for an African Eco-Socialist Alternative. In *The Climate Crisis. South African and Global Democratic Eco-Socialist Alternatives*, ed. Vishwas Satgar, 168–189. Johannesburg: Wits University Press.

The Green New Deal for Europe. 2019. *A Blueprint For Europe's Just Transition. Edition II*. https://report.gndforeurope.com/cms/wp-content/uploads/2020/01/Blueprint-for-Europes-Just-Transition-2nd-Ed.pdf.

Turner, Richard. 1972/1980. *Eye of the Needle: Toward Participatory Democracy in South Africa*. New York: Orbis Books.

Van der Walt, Lucien 2019. *Beyond Decent Work. Fighting for Unions and Equality in Africa Friedrich Ebert Stiftung*. Occasional Paper.

Vavi, Z. and Lenferna, A. 2019. 'It's time to fight for climate justice', Sunday Times, 15 September.

Von Holdt, Karl. 2002. Social Movement Unionism: The Case of South Africa. *Work, Employment and Society* 16 (2): 283–304.

Webber, Jeffrey R. 2011. From Rebellion to Reform in Bolivia. In *Class Struggle, Indigenous Liberation and the Politics of Evo Morales*. Chicago: Haymarket.

Williams, M. 2008. *The Roots of Participatory Democracy: Democratic Communists in South Africa and Kerala, India*. New York: Palgrave.

# 5

# The Green New Deal and Just Transition Frames within the American Labour Movement

Todd E. Vachon

## Introduction

On November 13, 2018, more than 200 youth activists flooded House Minority Leader Nancy Pelosi's office demanding a 'Green New Deal' to address the growing crises of climate change and runaway inequality in America. Three months later, on February 7, 2019, Representative Alexandria Ocasio-Cortez and Senator Ed Markey introduced House Resolution 109 'Recognizing the duty of the Federal Government to create a Green New Deal.' The resolution called for a ten-year national mobilization to 'achieve net-zero greenhouse gas emissions through a fair and just transition for all communities and workers' (Ocasio-Cortez and Markey 2019). While the resolution was a roadmap rather than an actual piece of legislation, it nonetheless stands out from previous climate mitigation plans brought before the U.S. Congress because of its focus on jobs, workers' rights, unions, and the need for a just transition.

On the day the resolution was released, Local 32BJ of the Service Employees International Union (SEIU), representing 175,000 office cleaners, security officers, doormen, porters, maintenance workers, engineers, window cleaners,

T. E. Vachon (✉)
School of Management and Labor Relations, Rutgers University, New Brunswick, NJ, USA
e-mail: todd.vachon@rutgers.edu

and food service workers, became the first union to endorse the Green New Deal and urge Congress to swiftly adopt the resolution. However, despite the great deal of attention given to workers, unions, and worker voice, the resolution was not met with resounding support from the rest of the U.S. labour movement. One month later, on March 8, the powerful Energy Committee within the national union confederation, the American Federation of Labor and Congress of Industrial Organizations (AFL-CIO), issued an open letter to Rep. Ocasio-Cortez and Senator Markey which acknowledged the problem of climate change but went on to state: 'We will not accept proposals that will cause immediate harm to millions of our members and their families. We will not stand by and allow threats to our members' jobs and their families' standard of living go unanswered' (AFL-CIO Energy Committee 2019). In the weeks and months that followed, unions and labour federations took turns praising or disparaging the Green New Deal.

The detractors were typically from energy unions whose members earn a living in the fossil fuel industry, and the supporters were typically from other unions, particularly service and public sector unions whose members were already suffering the ill effects of climate change on the job and in their communities (see Stevis 2019 for a more nuanced categorization of unions). Supporters were also more likely to adopt a reimagination of unions as social movements (Räthzel and Uzzell 2011; see also Scipes 2014 for a discussion of social movement vs. social justice unionism). However, as with any other issue within U.S. labour, there was a great deal of local variation as well, with some building trades locals, including some locals of the International Brotherhood of Electrical Workers (IBEW) in California, New York, and New England taking a more progressive stance than their national leadership. But in sum, the major distinction was that the unions whose members would suffer job losses as a result of climate change mitigation were less likely to support aggressive plans to decarbonize the economy. Anticipating this problem, the authors of the Green New Deal resolution placed a very strong emphasis on the need for providing a just transition for displaced workers. Despite this focus on justice for workers, many unions are still resistant to or at least sceptical of the Green New Deal. Why? And among the unions that do support the Green New Deal, is there a consensus on what a just transition would actually entail? These are the major questions that motivate this chapter.

Drawing from my personal experiences in the labour movement as a union carpenter, an organizer with a large manufacturing union, and as the founding president of a large education union local, I describe what I see to be the major factors contributing to opposition to climate protection measures by some U.S. unions. I also draw from my participant observation within several labour-climate social movement organizations and 34 in-depth interviews with union leaders and activists to describe the variation in just transition frames put forth by labour activists who support a just transition. The interviews were conducted between 2014 and 2018 with members of what I will refer to broadly as 'The Labor-Climate Movement,' a movement of climate activists within the U.S. labour movement that seeks to move labour as a whole to a more progressive stance on climate change.

This chapter will begin with a broad overview of the concept of just transition to situate the concept within the nascent field of environmental-labour studies. It will then provide a grounded analysis to identify the major factors contributing to opposition and support for just transition within the U.S. labour movement with particular attention devoted to the variation in usage of the term by supporters. The final section explores the Green New Deal as an organizing space for this contestation to play out within labour and between labour and other social movements. The chapter concludes by considering what these findings at the granular level within the context of the U.S. labour movement can tell us more generally about efforts to provide a just transition for workers as we confront the existential crisis that is human-caused climate change.

## Just Transition: A Brief Overview

The term 'just transition' appears to have originated from, or at least been popularized by, the late American labour and environmental leader, Tony Mazzocchi of the Oil, Chemical, and Atomic Workers Union (now merged with the United Steelworkers) in the 1990s. The term combines the often-conflicting projects of economic transition and the pursuit of social justice into one combined endeavour (Cunniah 2010; Smith 2017; Stevis and Felli 2015; Stevis 2019). According to Meadowcroft (2011), the idea of a 'transition' provides a convenient political and policy frame to approach the societal movement towards sustainability. By talking about transition, aspirations for sustainability can be channelled into efforts to achieve long-term goals such as a zero-emissions economy (Scoones et al. 2015).

Looking at the 'just' side of the just transition concept, we find social movement actors raising the key political and economic questions of who wins, who loses, how, and why, as they relate to the existing energy system and any proposed transition to an alternative system. In determining whether a transition will be 'just,' the following questions must be raised: Who currently suffers from the effects of fossil fuel extraction, production, and energy generation?

Who benefits from it? Who will bear the social costs and reap the greatest benefits of decarbonizing our economy? For any transition to be just, there must be an analysis of the distributional outcomes and a corresponding plan to mitigate the ill effects, particularly for the most vulnerable elements of society, including marginalized communities and workers (Evans and Phelan 2016; Newell and Mulvaney 2013).

From the outset, just transition has been conceptualized as a political project. Any economic transition towards sustainability, even without pursuing social justice, is immediately political as our current social institutions do not naturally put us on a course of sustainable development. That is, the market system alone will not produce a carbon-free energy system, at least not in time to avoid climate catastrophe. This means that some form of state intervention is inevitably required to initiate a transition (Meadowcroft 2011). A transition to a sustainable economy will almost certainly require changes to law and regulations, such as the imposition of a carbon tax or a greenhouse gas (GHG) emissions cap and trade system. It will likely also require a significant expenditure of public revenue to accelerate the development and deployment of new technologies and to ease societal adjustment to new patterns of production and consumption. Such changes can only be engineered through political processes and enforced through institutions of the state.

Incorporating demands of social justice further politicizes the process (Healy and Barry 2017; McCauley and Heffron 2018). In a pluralist society such as the U.S., with countless competing interest groups, the government is inevitably preoccupied with negotiating the distributional fallout of any attempts to address social or environmental injustices (Hyde and Vachon 2019). When transition policies are adopted, the state must address the consequences of rising or declining industries, impacts on regions, workers, community members, business owners, and so on (Scruggs 1999). Often the most politically organized and resource-rich elements of society are able to win concessions, as has been the case with many not in my backyard movements. Over time, this unequal access to power and resources has created an environmentally racist configuration of the U.S. energy production and waste disposal systems (Bullard 1993; Newell 2005; Parks and Roberts 2010; Schlosberg 2013). To alleviate these forms of injustice requires a degree of economic and social planning that is anathema to the current neoliberal institutional arrangements embedded in the U.S. state (Hampson and Reppy 1996; Harvey 1996, 2005).

Despite the inherently political nature of just transition, many people from across the political spectrum agree that the world is on an unsustainable path and that business-as-usual is not an option (Leiserowitz et al. 2005). However, as Scoones et al. (2007) describe, there still remain many hard and soft disagreements about the most desirable form of transition. Hard disagreements arise from those fundamentally opposed to change, such as institutions and businesses whose profits and power are fundamentally interlocked with the status quo. Yet even among those sharing a consensus that change is needed lies a range of hotly contested visions of sustainability that define the framing of and approach to transition (see: Labor Network for Sustainability 2016; Snell 2018; Stevis and Felli 2015). These soft disagreements are equally important, and they too implicate material questions of economy, interest, and resource allocation. They also lead to varied conceptions of what just transition means and what it should entail. That is, there arises a set of distinct, contested, but inter-related, frames for the concept just transition. The following two sections will in turn explore the hard and the soft disagreements over just transition within the U.S. labour movement.

## Jobs Versus the Environment: Labour Resistance to Climate Protection and Just Transition

When asked about a just transition, Cecil Roberts, President of the United Mineworkers of America, said: I've never seen one. At the heart of labour opposition to just transition measures is the less-than-stellar experience with previous government programmes for workers who have been displaced from jobs as a result of public policy. A prime example is the Trade Adjustment Assistance programme which was designed to assist workers harmed by international trade deals such as the North American Free Trade Agreement. The programme is generally seen as inadequate for a number of reasons, including the limited number of workers that actually qualify for assistance, the level of assistance, and the inability of the job retraining programme to guarantee a job at the end. In fact, most workers who received training with the programme did not end up with jobs in the fields they had trained for (Barret 2001). In effect, most workers who lost good-paying blue-collar jobs went through an inadequate job placement programme and ended up with low-paying jobs they could have landed without the programme. This has led many unions to be highly sceptical of government transition programmes. AFL-CIO president, Richard Trumka, even went so far as to refer to just transition as an invitation to a fancy funeral.

Additionally, unlike most other rich capitalist democracies, the U.S. has a very weak social safety net. When workers lose their jobs, they not only lose their income, but they also lose health insurance for their families. Retraining and education are expensive endeavours that place nearly all of the risk, usually in the form of student debt, on the individual worker, with no guarantee that the investment will lead to a job in the end. The transition for workers from one occupation to another is made easier when education is free or highly subsidized and when health insurance and other benefits exist during the interim period of unemployment—social protections that are largely nonexistent in America. The inadequacy of existing transition programmes and the social safety net in general is a result of the neoliberal governing ideology that dominates American politics and which loathes taxation, social spending, and most forms of state intervention into markets (Harvey 2005).

In the absence of reasonable social protections and given the history of insufficient transition programmes, American energy workers are rightfully sceptical of government plans to help them and many instead adopt an oppositional stance towards climate protection overall. In the remainder of this section, I draw from my personal experiences, fieldwork, and interviews with labour leaders to identify the major structural explanations for labour support for the fossil fuel industry and resistance to strong climate change mitigation efforts, such as the Green New Deal. In sum, I identify 4 broad categories of explanations: fear of job loss, decline in bargaining power, promise of job creation, and the structure of the U.S. labour movement.

## Fear of Job Loss

Since the establishment of environmental and workplace protections in the early 1970s, private employers in the U.S. have resisted further curbs on corporate conduct by threatening job destruction. The refrain has been that environmental standards wipe out existing jobs and make new ones impossible. Kazis and Grossman (1982) showed in detail the use of this job blackmail to peel off trade unionists from environmentalists, making unnatural enemies of those who should be allies. The interest of workers in protecting their jobs is used by employers to achieve their own policy objectives and thus workers are often presented as the public face of opposition to environmental protection. This tactic works to the advantage of employers who are opposed to both workers' rights and environmental protections, allowing them to maximally exploit labour and nature for profit.

However, not all threats of job loss are mere blackmail. Addressing the causes of climate change will not only radically alter the way we generate electricity, travel, heat our homes, and consume goods and services, but it will also disrupt labour markets, employment, and work in whole industries. Replacing fossil fuels with renewable energy sources will reduce the number of jobs in the fossil fuel industry while simultaneously increasing the number of jobs in renewable power generation. There is no guarantee the new jobs will be in the same geographic location as the old or require the same skills—in fact, given the different resources and technologies associated with the two forms of energy generation, it is likely they will not. Unlike the *threat* of job loss by employers described earlier, decarbonizing the economy is guaranteed to eliminate a certain number of jobs in the fossil fuel industry. Unions have been responding to these changes in varied ways (Felli 2014; Hampton 2015; Räthzel and Uzzell 2011). However, the promise of green jobs elsewhere in the economy is generally of little comfort to the individual worker losing *their* job. In addition to the material consequences of unemployment, job loss also threatens people's identities (Breakwell 1986; Cha et. al 2021) . As the old saying goes, a bird in the hand is worth more than two in a bush. For fossil fuel workers, a job today is worth two jobs promised for tomorrow—especially when they are promised by a politician.

## Decline in Bargaining Power

Fossil fuel jobs in the energy sector, ranging from extraction to transportation to power plant operations, are very high-paying jobs that do not require a college degree. These jobs were not always good jobs but became increasingly better over the course of the twentieth century due to the collective efforts of workers bargaining for increases in wages, benefits, and workplace safety. Many of the new green energy jobs, particularly in the residential solar industry—the largest employer in the renewable energy sector—do not offer wages or benefits comparable to jobs in the fossil fuel industry. For example, in 2018 the mean annual salary for a fossil fuel power plant operator was $78,030 while the mean annual salary for rooftop solar installer was $46,010 (Bureau of Labour Statistics 2018).

This difference in pay is due in large part to the disparity in worker power between these two sectors. In the fossil fuel sector, workers have a long history of unionization and bargaining over wages, hours, and working conditions. The solar sector is characterized by at-will employment and anti-union employers that invest heavily in union avoidance campaigns to prevent workers from bargaining collectively for higher wages. Through concerted action by workers, residential solar jobs could become as high paying as fossil fuel jobs over time; however, the initial experience for a worker shifting from one occupation to the other is an immediate reduction in salary and benefits. For unions, organizing new workers in the private sector has become increasingly difficult. The steady weakening of labour protections over the past 50 years coupled with extreme anti-union sentiment among employers makes new union organizing considerably more challenging (Bennett and Kaufman 2016; Bronfenbrenner and Juravich 1995; Wallace et al. 2009).

Related to the disparities in bargaining power between the fossil fuel and renewable energy industries is the general lack of good union job alternatives for blue-collar workers in any sector. The ongoing history of globalization and deindustrialization—the shipping of jobs overseas in order to pay lower wages and pollute more in order to maximize profits in a competitive global market—has hallowed out the middle-class labour market (Bluestone and Harrison 1982; Bronfenbrenner and Luce 2004). The decline in well-paying, blue-collar manufacturing jobs has left few good alternatives for displaced workers without a college degree. Those who still hold one of the few union jobs left in the private sector, just 6% of workers in 2018, understand very well how fortunate they are in the contemporary American economy and are not prepared to give up those jobs. They also know that the majority of alternatives readily available are low-paying jobs without union representation or fringe benefits.

These disparities are amplified further in particular geographic regions, especially in rural or isolated areas with less diversified economies. There are countless cities and towns scattered across the map that are reliant on just one fossil fuel employer for the majority of wages and local tax revenue. For these communities, the economic impact of decarbonization can result in deep and widespread economic hardships. For workers and unions in these regions, such as Appalachia, the consequences of job loss are amplified considerably (Estep 2016). These geographic factors can contribute to resistance to climate protection measures by unions.

## The Promise of Job Creation in Fossil Fuels

Commercial construction workers are responsible for building fossil fuel power plants and pipelines as well as wind farms and commercial solar plants. Like fossil fuel jobs in the energy sector, unionized commercial construction jobs are highly paid blue-collar occupations that do not require a college degree. However, unlike most jobs, employment in construction is sporadic and comes in fits and starts depending on the state of the economy and level of investment in construction projects. An average construction worker may work for two or three different employers in a given year and dozens more over the course of their career. They may also spend several months each year unemployed, waiting for a construction project to begin.

Unlike other sectors of the labour movement, construction unions (known as the building trades) serve as an employment service, connecting workers looking for jobs to construction companies looking to hire large numbers of skilled workers on short notice for temporary jobs. The workers in return are guaranteed the same negotiated pay rate and benefits regardless of the employer and carry their benefits with them seamlessly throughout their career from one employer to another, even across durations of unemployment.[1] Unlike a typical factory worker who becomes a union member by way of employment at a particular workplace and only maintains membership so long as they are employed by that company, construction workers are union members first and employees of a particular company second. In other words, they are connected to the labour market via the union hiring hall (Weil 2005).

The sporadic nature of construction work combined with the hiring hall structure creates an incentive for building trades unions to offer political support for any and all construction projects in order to secure adequate employment opportunities for their members. All of the aforementioned approaches often lead building trades unions to support projects that might otherwise be harmful to their members' communities such as waste incinerators, casinos, and coal-fired power plants.[2] The problem is further amplified when companies in polluting industries actively seek labour support for their unpopular projects by signing project labour agreements (PLAs). A PLA, also referred to as a community workforce agreement, is a pre-hire collective bargaining agreement between construction companies and labour organizations that establishes the terms and conditions of employment for a specific construction project before it begins (Mayer 2010). For construction companies, it creates a broad base of support for their projects, for unions it offers assurance the jobs will be unionized once the project is approved.

As the threat of climate mitigation has increasingly loomed over the fossil fuel industry, oil and gas infrastructure companies have become increasingly supportive of signing PLAs with building trades unions to garner their

---

[1] The hiring hall structure successfully addresses a very serious structural problem in the construction industry—sporadic unemployment. Without this system or a very strong social safety net, construction workers would be in perpetual poverty or very few people would pursue a career in the trades long enough to develop the skills required to safely build large-scale infrastructure projects.
[2] The "all of the above" approach has also led some building trades unions to support renewable energy projects. An example of this was the construction of the Block Island Offshore Windfarm—America's first offshore windfarm.

powerful political support for projects that would face another form of organized political opposition—the environmental movement (see e.g. Cox 2016; Oil&Gas360.com 2019). While PLAs themselves are not inherently a bad thing—they certainly help to ensure that labour gets a fair share of the pie—the increased weaponization of them by employers to effectively hire unions as an anti-environmental political army has led to several jobs versus the environment clashes, such as those associated with various pipeline projects in recent years. If renewable energy companies were not so anti-union, they too could sign PLAs with building trades unions which could build a powerful pro-green political alliance and turn the tide of labour support away from fossil fuel projects.

## The Structure of the U.S. Labour Movement

The U.S. labour movement comprises over 100 unions with 15 million union members (Bureau of Labor Statistics 2020). Local unions in the U.S. are typically affiliates of international unions and may be part of union confederations with other local or international unions. Union confederations are umbrella organizations with which unions in a geographically defined area may choose to affiliate in order to build cross-sector solidarity and increase workers' power for collective actions and legislative efforts. Examples of union confederations include: central labour councils, which operate at the county or metropolitan level and typically comprise union locals; state federations operate at the state level and typically comprise local- and state-level unions; and the AFL-CIO is a national confederation of unions comprising 55 national unions, representing some 12.5 million workers in almost all industries. Most major unions are affiliated with the AFL-CIO, with a handful of notable exceptions, such as the SEIU, the Teamsters, the United Brotherhood of Carpenters, and the International Longshore and Warehouse Union. The building trades unions have their own national-level confederation in the U.S. called North America's Building Trades Unions which, as a block, has significant influence within the AFL-CIO.

Union confederations often take stances on political and economic issues based on the interests of their member unions. However, since confederations rely on voluntary membership dues in order to operate, they are often at the mercy of any one union or group of unions that can effectively veto an issue by threatening to disaffiliate from the confederation. This structural feature of union confederations is critically important in shaping labour's position on climate change mitigation policies. In fact, the AFL-CIO has a long history of supporting coal and other fossil fuels because of the influence of member unions with individual members who work for mining, drilling, transportation, and power generation companies. As a result, the confederation has generally taken a very cautious, conservative approach to climate policy (Vachon and Brecher 2016). Unlike the International Trade Union Confederation, and national confederations in other countries, the AFL-CIO has never supported either the Kyoto Protocol or other science-based emissions reduction targets (Brecher 2013). Historically, the AFL-CIO's climate and energy policy has been shaped predominantly by a small number of unions in extraction and construction who form the very powerful Energy Committee, and the rest of the confederation—unions representing the vast majority of union members—have mostly steered clear of what is, or could be, a divisive issue.

While distinct, these four explanations for labour opposition to just transition and climate protection more broadly are deeply intertwined, overlapping, and reinforcing. All are underpinned by the powerful free market ideology that dominates American political-economic discourse—neoliberalism. This belief system, rooted in what Adam Smith referred to as the unseen hand of the market, dictates what can and cannot be on the table for political discussion (Harvey 2005). In general, government intervention in the market in order to solve social problems—including climate change, lack of health care, student debt, job loss—is akin to heresy from this perspective. For workers, this constraint on democracy often leads to a false choice between having good jobs or having a healthy environment in which to live and work (however, for examples of unions promoting environmental protections, see Burgmann and Burgmann 1998; Dewey 1998; Farnhill 2016; Mayer 2009; Obach 2004; Snell and Fairbrother 2011; Savage and Soron 2011; Vachon and Brecher 2016). Workers in the fossil fuel industry become acutely aware of this dichotomy when federal, state, and local governments discuss solutions to the crisis of climate change that involve decarbonizing the economy. The hardships associated with job loss that result from the lack of a strong social safety net and the lack of good job alternatives for displaced workers are two examples of the ways in which the hegemonic neoliberal ideology can fuel anti-environmental activity on the part of unions. Each of the other explanations outlined earlier is underpinned by this same ideological commitment to markets as the fundamental guidepost for all economic decisions. Demanding a just transition involves challenging this ideology in exactly the sorts of ways that are imagined in the Green New Deal resolution.

# Clean Air and Good Jobs: The Labour-Climate Movement for a Fair and Just Transition

Counter to the resistance to climate protection by some unions (as described earlier), there is a movement of climate activists within the U.S. labour movement seeking to address the climate crisis from the perspective of the working class; they are the Labour-Climate Movement. Comprising climate-conscious union leaders and rank-and-file union members, the Labour-Climate Movement (LCM) seeks to create better jobs for workers now and a better environment for their communities and the planet's future. The LCM is the organized voice within the labour movement pushing for ecological sustainability as well as protections for workers and communities hurt by the effects of climate change and/or the changes that must be made in order to transition to renewable energy. The LCM is the wing of U.S. labour that is promoting the Green New Deal and fighting for a just transition.

Despite their overall support for the Green New Deal, I have in the course of my fieldwork distinguished three uniquely identifiable, but not entirely mutually exclusive frames for the usage of the term just transition by LCM participants. I borrow the three category names from an unpublished presentation to global union leaders by Sweeney (2016) in which he referred to just transition as being protective, proactive, and transformative. Each frame will be explored next.

## Protective Just Transition

Tony Mazzocchi once said: There is a Superfund for dirt. There ought to be one for workers (Mazzocchi 1993, 41). Summarizing Mazzocchi's definition of the superfund for workers, Brecher (2015, 34) says: It is a basic principle of fairness that the burden of policies that are necessary for society—like protecting the environment—shouldn't be borne by a small minority, who through no fault of their own happen to be victimized by their side effects. The idea of a superfund for workers is perhaps the best example of what I will refer to as a protective just transition, the most conservative of the three frames I have identified. In this frame, just transition, first and foremost, must take appropriate measures to protect jobs in vulnerable industries, but where job losses are unavoidable, it should provide adequate support by creating a safety net for those workers and communities that stand to lose out as a result of decarbonizing the economy (Pollin and Callaci 2016; Young 1998).

At the core of the protective vision of just transition is a set of protections for workers who are displaced as a result of government policy that is deemed necessary to protect the common good. This would primarily affect workers in the extraction industries, but also some building trades workers as well as some in manufacturing. Again, quoting Mazzocchi: Paying people to make the transition from one kind of economy—from one kind of job—to another is not welfare. Those who work with toxic materials on a daily basis in order to provide the world with the energy and the materials it needs deserve a helping hand to make a new start in life (Mazzocchi 1993, 40).

While it is true that, on balance, environmental policies tend to create more jobs than they eliminate (Bezdek et al. 2008; Goodstein 1999; Morgenstern et al. 2002), this fact is of little comfort to the workers in fossil fuel producing and using industries who lose their jobs as a result of climate protection policies, including coal miners, power plant workers, and oil refinery workers. Similar to existing trade adjustment programmes, proponents of protective just transition argue that a better programme can be developed for workers affected by energy transition policies. Specifically, people who lose their jobs because of the transition to a climate-safe economy should be eligible for full wages and benefits for a period of time and receive education or training, including free college tuition, or decent pensions with healthcare for those close to retirement age.

The protective vision of just transition typically focuses on the site of employment and the surrounding community. Proponents of the protective just transition frame argue that without a clear programme to protect working people from the effects of climate protection-related policies such as plant closures and drilling bans, the struggle for clean energy can easily come to be perceived as an environmentalist struggle against American workers—even though climate protection will benefit rather than harm American workers. A protective just transition programme could provide a critical element for drawing together workers, unions, and allies around a broader programme for protecting jobs by protecting the climate. Making a protective just transition programme for workers a central feature of climate change mitigation plans, argue proponents, could make the difference between united support for a sustainable and equitable economy and a never-ending battle over jobs versus the environment.

## Proactive Just Transition

The second frame of the term just transition, which I refer to as proactive, incorporates many aspects of the protective frame but adds two key elements: (1) a forward-looking plan, usually involving large-scale public investment to transition the economy to a more sustainable model, and (2) a seat at the table for labour in the development and administration of the transition programme. The first component was often referred to as a Second World War-style mobilization of the economy or a Green New Deal by many interviewees when they described their vision of a massive infrastructure programme to replace fossil fuels with renewable sources and increase energy efficiency. The second component is summarized in the following statement from a publication by the International Labour Organization (ILO): The notion of Just Transition is in line with the longstanding philosophy that has inspired the creation and the history of the International Labour Organization: the idea that social concerns have to be part and parcel of economic decision-making, that the costs of economic transition should be socialized as much as possible, and that the economic management of the economy is best achieved when there is genuine social dialogue between social partners (Cunniah 2010, 122). The proactive framing goes beyond the mere creation of a superfund or safety net to protect workers who are displaced locally, it envisions a large, economy-wide public programme developed with input from labour that seeks to proactively address climate change and minimize the social costs of transition rather than merely reacting to closures and job loss on a case-by-case basis.

The phrase Green New Deal is a play on President Franklin D. Roosevelt's New Deal programmes which comprised government spending and economic planning to address the economic hardships of the Great Depression in the 1930s. Seeing climate change as a similarly universal, if not an existential problem, the supporters of proactive just transition promote massive government investment in renewable energy projects that create good jobs while simultaneously reducing dependence on fossil fuels (Brecher 2019). The ideas of social dialogue and industrial planning as envisioned by the ILO provide a roadmap for what a proactive just transition might look like. Beyond merely protecting workers when disaster strikes, proactive just transition demands labour leaders be at the table to help craft policies that will protect both the environment and workers simultaneously.

Proponents of the proactive just transition frame argue that efforts to steer society towards a lower carbon future must be underpinned by attention to issues of equity and justice. The transition away from fossil fuels must not only protect the livelihoods of those that are affected by and dependent on the fossil fuel economy, but also proactive in addressing the social injustices faced by those currently excluded from the benefits of the energy economy or disproportionately suffering the consequences of the fossil fuel economy (Newell and Mulvaney 2013). A proactive just transition creates support structures for people and sectors that stand to lose out as a result of decarbonizing the economy through compensation and retraining for new employment opportunities but also ensures that new jobs created in low-carbon sectors are good jobs that pay a living wage, provide decent working conditions, are accessible to people with a range of skills, and offer opportunities for career growth. For proponents of this frame, the only reasonable way to ensure that any transition plan serves the interests of workers is to have workers and their organizations at the table to help craft the plan. As various participants told me in the course of my interviews, 'If you're not at the table, you're probably on the menu.'

## Transformative Just Transition

The third and most radical framing of just transition is what I refer to as a transformative just transition. Transformative just transition is not just about protecting workers or planning ahead for sustainability, but the central goal is transitioning to an entirely different socio-economic model which addresses the root causes of environmental degradation, worker exploitation, and social injustices (Vachon and Sweeney 2018). The two key elements of the transformative transition frame I identified in my data are: (1) enacting different models of ownership and economic decision-making, and (2) eliminating existing social and economic inequalities and injustices.

Transformative just transition is built on alliances between the climate justice, environmental justice and labour movements in what some participants have referred to as a process of de-silofication, or recombination of activists that have been sorted into specific issue-based silos that focused exclusively on one issue such as climate change, labour rights, environmental justice, or economic, racial or gender equality (Vachon et al. 2019). The transformative just transition frame unites activists at the intersection of jobs, justice, and the environment. For most promoters of this frame, just transition is a means to address the multiple and overlapping concerns of many individual social movements if they would just come out of their silos and see the interconnectedness of their issues.

Transformative just transition also embodies a radical reorganization of society and envisions a post-capitalist world order. Where the protective frame makes minor demands on the state to participate in the economy and the proactive vision makes a stronger challenge on free market ideology by demanding some limited economic planning and a voice for labour in policy making, the transformative vision blows the entire free market ideology out of the water by calling for a radical reorganization of the political economy of the world to eliminate the distinction between the owners of capital and labour. It also understands the historical use of racism and colonialism to sew division among the working classes of the world and seeks to directly remedy that history of injustices through its solution to the climate crisis.

In *This Changes Everything*, Naomi Klein (2014) makes a similar case for linking a variety of social, economic, and environmental struggles into programmes designed to address the climate crisis. Referencing Al Gore's documentary film (*An Inconvenient Truth*) about climate change, Klein (2014, 8) argues that the real inconvenient truth is that it's not about carbon emissions—it's about capitalism. She argues that the convenient truth, however, is that citizens can seize upon the existential crisis posed by climate change to transform the failed capitalist system and build something radically better. The real solutions to the climate crisis, she argues are also our best hope of building a much more stable and equitable economic system, one that strengthens and transforms the public sphere, generates plentiful, dignified work, and radically reins in corporate greed (Klein 2014, 125). This argument mirrors the frame put forward by supporters of transformative just transition.

## The Green New Deal and Just Transition

The tensions between unions supporting climate protection measures and those fighting to save their existing jobs in the fossil fuel energy sector predate the Green New Deal resolution (Vachon and Brecher 2016). We need not look any further than the battles over the Keystone XL Pipeline or the Dakota Access Pipeline to see this divide within the house of labour (Sweeney 2015). As I have outlined in this chapter, American energy workers have very real and well-justified fears and concerns about job loss. The lack of a strong social safety net, the dearth of good job alternatives, and the loss of bargaining power all make job loss the equivalent of an economic death penalty.

The arrival of the Green New Deal has brought some of these tensions into sharper focus (see e.g. AFL-CIO Energy Committee 2019). Despite its very strong demands for worker protections, some in labour have shown resistance, citing the long history of inadequate worker transition plans and the lack of consultation with the leaders of energy unions when drafting the resolution. The lack of faith in government programmes to address the economic concerns of workers has led some in the mainstream news media to jump on the opportunity to replay one of their beloved refrains, jobs versus the environment! This analysis of course fails to look at the primary reason for the scepticism among workers—a powerful neoliberal governing ideology that eschews state interventions into markets, either to protect the environment or to help workers.

For those in the LCM, the Green New Deal has opened countless doors for activists to challenge the gloves off, neoliberal form of capitalism that has defined the American economy for decades (Aronoff et al. 2019; Klein 2019). It has created a space for constructive conversations about labour and climate change within labour and between labour and other social movements. And it has created considerable opportunities for organizing at the intersection of climate change and work. Groups like the Labor Network for Sustainability have been advocating support for the Green New Deal within unions at the local, state, and national level in the U.S. Trade Unions for Energy Democracy has been building an international network of unions fighting for public ownership and democratic control over the energy sector. The Bargaining for the Common Good network, a growing network of predominantly public sector and education unions within the U.S. that are incorporating community demands into their contract campaigns, has begun working on ways to incorporate climate justice into local union collective bargaining (Vachon et al. 2019). Countless state and local coalitions are shaping their own green new deal plans at the subnational level.

But even among these unions and labour activists that are supportive of the Green New Deal, there is not a shared vision for just transition. As I have identified in this chapter, there are multiple, distinct uses of the term just transition among labour-climate activists, ranging from less to more radical in their prognosis, with each challenging free market ideology to varying degrees. The Green New Deal has created a space where these competing visions for a just transition can be discussed and debated. These debates can serve as a vehicle for deep organizing and shifting elements of labour further along the spectrum towards the transformative vision of just transition; what Stevis and Felli (2015, 32) have viewed as a path from reaction to proaction. However, overcoming the lingering resistance and mistrust of government programmes to protect workers will require a massive educational campaign and organizing drive coupled with real, tangible commitments by the government to support a truly robust and adequate programme for displaced workers. Without such assurances in place, asking workers to give up their jobs on the promise of a just transition sounds like asking them to do a trust fall with a group of people that have never played catch. Only the most foolish or naïve would do it. It is in this struggle for a genuine just transition programme that lies the greatest challenge but also the greatest potential of the Green New Deal to achieve its stated goals and live up to or possibly exceed the legacy of its historical namesake.

## Conclusion

This chapter has taken a grounded approach to examine, at the granular level, some of the debates around just transition that are unfolding within the U.S. labour movement. These discussions illuminate several key points which can be informative for more macro-level discussions about just transition in general. First, for any transition to be truly just, it must have the full support and faith of the workers that will be dependent on it—this requires the participation of workers and their organizations in the design and implementation of the plan. Second, building that trust will also require a combination of education and organizing among workers and leveraging political power to force governments to commit adequate resources to such programmes. Third, there can be no just transition without first challenging the neoliberal governing ideology which seeks private, market-based solutions to social problems like job displacement; state intervention, market regulation, taxation, and redistribution are all fundamental ingredients in the recipe for any flavour of just transition. Finally, the emerging field of environmental-labour studies is uniquely equipped to conduct cutting edge research at the intersection of jobs and the climate crisis and distil and disseminate this knowledge to fellow scholars in other disciplines, policy makers, and the public at large. In other words, we too have a role to play in ensuring that there is a truly just transition to a more equitable and sustainable world.

## References

AFL-CIO Energy Committee. 2019. *Letter to Senator Markey and Representative Ocasio Cortez.* March 8. https://ecology.iww.org/node/3229?bot_test=1. Accessed June 2020.

Aronoff, Kate, Alyssa Battistoni, Aldana Cohen Damiel, and Thea Riofrancos. 2019. *A Planet to Win: Why We Need a Green New Deal.* New York: Verso.

Barret, J. 2001. *Worker Transition and Global Climate Change.* Pew Center on Global Climate Change.

Bennett, James T., and Bruce E. Kaufman. 2016. *The Future of Private Sector Unionism in the United States.* New York: Routledge.

Bezdek, Roger H., Robert M. Wendling, and Paula Diperna. 2008. Environmental Protection, the Economy, and Jobs: National and Regional Analyses. *Journal of Environmental Management* 86: 63–79.

Bluestone, Barry, and Bennett Harrison. 1982. *The Deindustrialization of America: Plant Closings, Community Abandonment, and the Dismantling of Basic Industry.* New York: Basic Books.

Breakwell, Glynis M. 1986. *Coping with Threatened Identities.* London: Methuen.

Brecher, Jeremy. 2013. Stormy Weather: Climate Change and a Divided Labor Movement. *New Labor Forum* 22: 75–81.

———. 2015. A Superfund for Workers: How to Promote a Just Transition and Break Out of the Jobs vs. Environment Trap. *Dollars and Sense.* November/December. http://www.dollarsandsense.org/archives/2019/1119toc.html. Accessed June 2020

———. 2019. Making the Green New Deal Work for Workers. *In These Times.* April 22.

Bronfenbrenner, Kate, & Tom Juravich. 1995. *The Impact of Employer Opposition on Union Certification Win Rates: A Private/Public Sector Comparison.* Working Paper #113, Economic Policy Institute, Washington, DC

Bronfenbrenner, Kate, & Stephanie Luce. 2004. *The Changing Nature of Corporate Global Restructuring: The Impact of Production Shifts on Jobs in the US, China, and Around the Globe.* Research Paper, The US-China Economic and Security Review Commission.

Bullard, Robert D. 1993. *Confronting Environmental Racism: Voices from the Grassroots.* Boston: South End Press.

Bureau of Labor Statistics. 2018. *Occupational Employment Statistics.* https://www.bls.gov/oes/current/oes_stru.htm#00-0000. Accessed 22 May 2019.

———. 2020. *Union Members Summary.* https://www.bls.gov/news.release/union2.nr0.htm. Accessed June 2020

Burgmann, Meredith, and Verity Burgmann. 1998. *Green Bans, Red Union: Environmental Activism and the New South Wales Builders Labourers' Federation.* Sydney: University of NSW Press.

Cha, J. Mijin, Dimitris Stevis, Vivian Price, and Todd E. Vachon. 2021. *Workers and Communities in Transition: Report of the Just Transition Listening Project.* Labor Network for Sustainability https://www.labor4sustainability.org/jtlp-2021/jtlp-report/

Cox, John. 2016. Labor Agreement Divides Kern Oil Fields. Bakersfield.com. Accessed June 2020.

Cunniah, Dan. 2010. Climate Change and Labour: The Need for a Just Transition. *International Journal of Labour Research* 2: 121–123.

Dewey, Scott. 1998. Working for the Environment: Organized Labor and the Origins of Environmentalism in the United States, 1948-1970. *Environmental History* 3: 45–63.

Estep, Bill. 2016. Coal Jobs in Kentucky Fall to Lowest Level in 118 Years. *Lexington Herald-Leader*. May 2.

Evans, Geoff, and Liam Phelan. 2016. Transition to a Post-Carbon Society: Linking Environmental Justice and Just Transition Discourses. *Energy Policy* 99: 329–339.

Farnhill, Tom. 2016. Characteristics of Environmentally Active Trade Unions in the United Kingdom. *Global Labor Journal* 7: 257–278.

Felli, Roman. 2014. An Alternative Socio-Ecological Strategy? International Trade Unions' Engagement with Climate Change. *Review of International Political Economy* 21: 372–398.

Goodstein, Eban. 1999. *The Trade-Off Myth: Fact and Fiction about Jobs and the Environment*. Washington, DC: Island Press.

Hampson, Fen Osler, and Judith Reppy, eds. 1996. *Earthly Goods: Environmental Change and Social Justice*. Ithaca: Cornell University Press.

Hampton, Paul. 2015. *Workers and Trade Unions for Climate Solidarity: Tackling Climate Change in a Neoliberal World*. New York: Routledge.

Harvey, David. 1996. *Justice, Nature, and the Geography of Difference*. Cambridge, MA: Blackwell.

———. 2005. *A Brief History of Neoliberalism*. Oxford: Oxford University Press.

Healy, Noel, and John Barry. 2017. Politicizing Energy Justice and Energy System Transitions: Fossil Fuel Divestment and a Just Transition. *Energy Policy* 108: 451–459.

Hyde, Allen and Todd E. Vachon. 2019. Running with or against the Treadmill? Unions, Institutional Contexts, and Greenhouse Gas Emissions in a Comparative Perspective. *Environmental Sociology* 5: 269–282.

Kazis, Richard, and Richard Lee Grossman. 1982. *Fear at Work: Job Blackmail, Labor and the Environment*. New York: Pilgrim Press.

Klein, Naomi. 2014. *This Changes Everything: Capitalism vs. the Climate*. New York: Simon and Schuster.

———. 2019. *On Fire: The (Burning) Case for a Green New Deal. 2019*. New York: Simon & Schuster.

Labor Network for Sustainability. 2016. *Just Transition: Just What Is It?* Working Paper, Labor Network for Sustainability and Grassroots Policy Project.

Leiserowitz, Anthony, Robert Kates, and Thomas Parris. 2005. Do Global Attitudes and Behaviors Support Sustainable Development? *Environment* 47: 22–38.

Mayer, Brian. 2009. *Blue-Green Coalitions: Fighting for Safe Workplaces and Healthy Communities*. Ithaca, NY: Cornell/ILR Press.

Mayer, Gerald. 2010. *Project Labor Agreements*. Washington, DC: Congressional Research Service.

Mazzocchi, Tony. 1993. A Superfund for Workers. *Earth Island Journal* 9: 40–41.

McCauley, Darren, and Raphael Heffron. 2018. Just Transition: Integrating Climate, Energy and Environmental Justice. *Energy Policy* 119: 1–7.

Meadowcroft, James. 2011. Engaging with the Politics of Sustainability Transitions. *Environmental Innovation and Societal Transitions* 1: 70–75.

Morgenstern, R.D., W.A. Pizer, and J.-S. Shih. 2002. Jobs Versus the Environment: An Industry-Level Perspective. *Journal of Environmental Economics and Management* 43: 412–436.

Newell, Peter. 2005. Race, Class and the Global Politics of Environmental Inequality. *Global Environmental Politics* 5: 70–94.

Newell, Peter, and Dustin Mulvaney. 2013. The Political Economy of the Just Transition. *The Geographical Journal* 179: 132–140.

Obach, Brian K. 2004. *Labor and the Environmental Movement: The Quest for Common Ground*. Cambridge, MA: MIT Press.

Ocasio-Cortez, Alexandria, & Edward Markey. 2019. *Recognizing the Duty of the Federal Government to Create a Green New Deal*.

Oil&Gas360.com. 2019. *Labor Unions, Enbridge Sign Project Labor Agreement*. https://www.oilandgas360.com/labor-unions-enbridge-sign-project-labor-agreement/. Accessed June 2020.

Parks, B., & Roberts, T. 2010. Climate change: Social theory and justice. *Theory, Culture & Society*, 27(2–3): 134–166.

Pollin, Robert, & Brian Callaci. 2016. *The Economics of Just Transition: A Framework for Supporting Fossil Fuel-Dependent Workers and Communities in the United States*. Working Paper, Political Economy Research Institute, UMass Amherst.

Räthzel, Nora, and David Uzzell. 2011. Trade Unions and Climate Change: The Jobs versus Environment Dilemma. *Global Environmental Change* 21: 1215–1223.

Savage, Larry, and Dennis Soron. 2011. Organized Labor, Nuclear Power, and Environmental Justice: A Comparative Analysis of the Canadian and U.S. Labor Movements. *Labor Studies Journal* 36: 37–57.

Schlosberg, D. 2013. Theorising Environmental Justice: The Expanding Universe of a Discourse. *Environmental Politics* 22: 37–55.

Scipes, Kim. 2014. Social Movement Unionism or Social Justice Unionism? Disentangling Theoretical Confusion within the Global Labor Movement. *Class, Race and Corporate Power* 2: 1–43.

Scoones, Ian, Melissa Leach, Adrian Smith, Sigrid Stagl, Andy Stirling, & John Thompson. 2007. *Dynamic Systems and the Challenge of Sustainability*. STEPS Working Paper 1. Brighton: STEPS Centre.

Scoones, Ian, Peter Newell, and Melissa Leach, eds. 2015. *The Politics of Green Transformations*. New York: Routledge.

Scruggs, Lyle. 1999. Institutions and Environmental Performance in Seventeen Western Democracies. *British Journal of Political Science* 29: 1–31.

Smith, Samantha. 2017. *Just Transition: A Report for the OECD*. Brussels: ITUC-Just Transition Centre.

Snell, Darryn. 2018. 'Just Transition'? Conceptual Challenges meet Stark Reality in a 'Transitioning' Coal Region in Australia. *Globalizations* 15: 550–564.

Snell, Darryn, and Peter Fairbrother. 2011. Toward a Theory of Union Environmental Politics: Unions and Climate Change in Australia. *Labor Studies Journal* 36: 83–103.

Stevis, Dimitris. 2019. *Labour Unions and Green Transitions in the USA: Contestations and Expectations.* Adapting Canadian Work and Workplaces to Respond to Climate Change, Working Paper #108.

Stevis, Dimitris, and Romain Felli. 2015. Global Labour Unions and Just Transition to a Green Economy. *International Environmental Agreements* 15: 29–43.

Sweeney, Sean. 2015. Standing Rock Solid with the Frackers: Are the Trades Putting Labor's Head in the Gas Oven? *New Labor Forum* 26: 94–99.

———. 2016. *What Do We Mean by Just Transition?* Unpublished Presentation to Labour Leaders at the International Program for Labor, Climate, and Environment. September 15. Provided to the author by request.

Vachon, Todd E., and Jeremy Brecher. 2016. Are Union Members More or Less Likely to Be Environmentalists? Some Evidence from Surveys. *Labor Studies Journal* 41: 185–203.

Vachon, Todd E., and Sean Sweeney. 2018. Energy Democracy: A Just Transition for Social, Economic, and Climate Justice. In *Agenda for Social Justice: Global Solutions*, ed. Glen Muschert et al., 63–72. Bristol: Policy Press.

Vachon, Todd E., Gerry Hudson, Judith Lebanc, & Saket Soni. 2019. How Workers can Demand Climate Justice. *American Prospect*. September 2.

Wallace, Michael, Andrew S. Fullerton, and Mustafa E. Gurbuz. 2009. Union Organizing Effort and Success in the US, 1948–2004. *Research in Social Stratification and Mobility* 27 (13): 34.

Weil, David. 2005. The Contemporary Industrial Relations System in Construction: Analysis, Observations and Speculations. *Labor History* 46: 447–471.

Young, Jim. 1998. Just Transition: A New Approach to Jobs vs. Environment. *WorkingUSA* 2: 42–48.

# 6

# Working-Class Environmentalism: The Case of Northwest Timber Workers

## Erik Loomis

## Introduction

American media headlines often highlight tensions between organized labour and the environmental community. In 2019, the Illinois state AFL-CIO supported a bill in the state legislature called the 'Critical Infrastructure Protection Bill.' This would have criminalized climate change activists' actions such as climbing on oil pipelines, making many actions that worked effectively in protests against the Keystone XL Pipeline and at the Standing Rock Reservation in North Dakota a felony. To date, there have been no serious protests of this kind in Illinois, yet organized labour allied with employers to pre-emptively criminalize the same kind of direct action protest the labour movement once engaged in effectively. This bill, although shelved for the 2019 legislative session, was the latest in an increasingly long list of public provocations from organized labour towards environmentalists, most notably from Laborers International Union of North America President Terry O'Sullivan (Loomis 2017; Stoner 2019).

This is all quite damaging to the labour movement's relationships with other social movements, allies they desperately need to survive the attacks that have withered that movement from the robust years of the 1950s to a mere

E. Loomis (✉)
University of Rhode Island, Kingston, RI, USA
e-mail: eloomis@uri.edu

fraction of that membership and power today. After all, the AFL-CIO and Chamber of Commerce might agree on a given bill to criminalize environmental protest, but the Chamber still wants to eliminate unionized workplaces entirely. Yet, O'Sullivan and his allies have attacked the Green New Deal programmes coming from congressional liberals such as New York congresswoman Alexandria Ocasio-Cortez and Massachusetts senator Ed Markey and even have dismissed green energy development entirely as less useful for union jobs than dirty energy. This is not only outright false, but it also elides the point that, like we have seen with COVID-19, climate change will impact the working class with far more intensity and mortality than the middle and upper classes (Bonn 2019; Street 2019).

Yet the reality of labour-environmental interactions is far more complex than Sullivan's political stances. In fact, unions and environmentalists have a long and sometimes fruitful history that is little understood today. Scholars have noted the significant interactions between these movements in the 1970s and 1980s (Brucher 2011; Gordon 1998; Kazis and Grossman 1982; Szasz 1994; Rose 2000; Obach 2004; Leopold 2007). In order to understand the wariness of the labour movement towards environmentalists today, we must place recent hostility within historical context of a vastly transformed economic reality for the working class since 1973. Yet, while tenuous alliances still exist, the robust alliances of that period declined not because of inherent contradictions between the two movements but because of structural changes in both the American economy and politics that drove them apart. Specifically, economic stagnation beginning with the 1973 recession combined with deindustrialization and capital mobility to threaten working-class jobs, which employers cynically used to cleave labour and environmentalists apart by threatening to close factories if new environmental regulations passed (Loomis 2015). Moreover, the growth of the conservative movement and the election of Ronald Reagan in 1980 created a new politics that required environmentalists to force the government to comply with environmental laws through lawsuits. This strategy was necessary but also required an emphasis on fundraising over building popular support, moving popular environmentalism towards an increased elitism that had little to say to workers. Ultimately, the story of labour-environmental interactions over the past century reflects a broader narrative of the building of the New Deal state and the stability of post-war liberalism undermined by the devolution of the state fuelled by growing economic uncertainty and resentment that go far to define our politics today (McCartin 2011; Rosenblum 1995; Kingsolver 1989; Cowie 2010; Short 1989; Bevington 2009; Klyza and Sousa 2007; Klyza 1996; Turner 2012).

What scholars have examined in less depth are the roots of labour-environmental connections into the early twentieth century. This chapter looks at timber workers in the Pacific Northwest in the twentieth century as a window into the changing relationships between workers and environmentalism in the context of the broader economic transformations of the twentieth century. Arguing that timber workers used their unions for their own environmental agendas, it demonstrates the nearly unknown history of working-class environmentalism, the potential of alliances between the labour and environmental movements, and the structural transformations in capitalism that explain much about the decline of those relationships since the 1970s.

This discussion will borrow material from my 2016 book *Empire of Timber: Labor Unions and the Pacific Northwest Forests* (Loomis 2016). Beginning with the Industrial Workers of the World, the radical syndicalist union seeking to unite all industrial workers to overthrow capitalism, organizing around sanitation in the 1910s and ending with union responses to the ancient forests campaigns of the 1980s, working-class organizations consistently fought for environmental positions they believed would protect their members' health, safety, and recreational opportunities while also preserving jobs. These stories reflect transformations in American capitalism and political economy. They begin in the Gilded Age era of uncontrolled capitalism that exploited both labour and nature. Even though workers lacked the vocabulary to talk about their exposure to toxicity and the elements in modern environmental language, they still responded much as workers have in later decades. The 1930s through the 1950s was a period where both workers and government responded to labour exploitation through a significant expansion of federal power that brought workers unions and placed those unions in a government framework. That same government responded to the resource depletion that accompanied worker exploitation through the same expansion of federal power. In both the unions and the federal resource agencies, a vigorous debate took place over how to manage the environment. Ultimately, over the objections of more left-leaning unions, an industrial resource model emerged in the post-war period that maximized production for short-term gains. Unions won contracts that included higher wages and benefits, but also suffered a loss of influence over resource management. The 1970s saw a new era of protest, but it ran up against the shifting economic model of deindustrialization, automation, and neoliberalism that made the lives of workers increasingly tenuous and cleaved the nascent alliances between the two movements. Tensions between greens and labour today need to be understood within this context, denaturalizing their battles and placing them within the context of the decline of unions and well-paying blue-collar work.

The early twentieth-century Northwest timber industry was a maelstrom of massive environmental change. Deforestation, erosion, and wildlife depletion are well-known results of this industry. It also enacted a huge environmental toll upon workers. Loggers smelled, saw, heard, and tasted the products of industrializing nature created to maximize profit of both nature and workers. Industrial logging created a filthy and dangerous landscape in early twentieth-century Northwestern timber camps. Transforming the sun's energy into homes, mine timbers, railroad ties, and fuel did not require creating a landscape of illness and death, but timber operators indifferent to workplace safety or sanitation produced a deadly environment. Loggers suffered from a slow violence of disease, filth, insect bites, bodies ground down through hard work, draughty housing, and soaked clothing. Carrying their waterlogged and flea-infested bedrolls on their backs, loggers suffered through respiratory illnesses, filth, and miserable discomfort. Operators constructed draughty housing with hay for a mattress, refused to install modern sanitary facilities, and provided adulterated food that made workers sick. New technologies lacking safety precautions killed and maimed workers daily. From the venereal disease racking their bodies to the tree limbs crushing their heads, loggers may have known nature through working in a beautiful forest, but they suffered the consequences of an industrialized, polluted landscape.

The hay, unwashed bedding, and moisture from rain-soaked bedrolls combined to create a perfect environment for flea infestations. Torger Birkeland remembered his first night in a logging camp, at the age of eleven. He looked around for leftover hay to create his bed. He described his sleep as, 'Not too good. The fleas, having found their way out, were having a grand time, jumping around and feeding on brand-new, tender skin. Sleep came after my little friends had all been fed and had crawled back under the bottom blanket again' (Birkeland 1960). George Davidson's underwear became so full of vermin, he had to throw them out and sleep outside, exposed to the elements (*Industrial Worker* June 1917a). Kitchen facilities were as shoddily built as bunkhouses. Companies stored meat in the open air, allowing flies unlimited access. A lack of garbage disposal meant mounds of trash that attracted even more flies (U.S. Commission on Industrial Relations 1916). A logger writing in 1910 described the butter served in camps as 'white as wax, and as rotten as a putrid carcass, if smell goes for anything' (*Industrial Worker* 1910). A man sitting next to Egbert Oliver in one camp opened three successive eggs, each inedible. In the third was a half-formed chick. The logger ran outside and vomited (Oliver n.d.).

Timber employers were unconcerned with these issues until workers started organizing with the Industrial Workers of the World (IWW). The IWW entered the Northwest timber industry with a brief strike in Portland mills in 1907. A series of free speech battles in the region brought national attention to the organization and created brief mass movements that included loggers, but more substantive organizing lagged until about 1912 when their appeals as to why loggers should join the union focused on the working and living conditions in the camps. Effectively, though they lacked the language to describe it this way, loggers began to use the IWW as an environmental justice organization demanding clean and healthy bodies in the forest.

IWW newspapers began running stories about food, sanitation, bunkhouses, human waste, vermin, and other environmental issues. As Donald Worster has said, 'environmental history begins in the belly,' a sentiment with which loggers with their internal organs rumbling from adulterated and poor quality food might well have agreed (Mink 2009). 'A. Rebel' wrote of his experiences working in the Fordney Lumber Company camp cookhouse. Rebel described how the company focused on 'efficiency' in the cookhouse, with the head cook paid $25 extra per month for saving money. This gave the cook incentive to buy low-quality food. He created tapioca cooked in a dirty pot with food colouring to give it an appealing look after 'the dirt and flies were skimmed off.' Rebel noted that if workers knew what they ate, 'there would be a lot of vomiting' (*Industrial Worker* 1916).

The rise of IWW organizing began to motivate reformist elements in the timber industry to improve the camps. Some employers began inviting the Young Men's Christian Association into the camps to provide after-hours entertainment, sanitary facilities, and educational opportunities. In some camps, food improved and so did the overall quality of life. But this was strictly voluntarily and by 1917, little had changed for most loggers. That spring, IWW-led strikes plagued the Northwest and primarily revolved around these issues of environmental injustice. Strikes had very specific demands over the creation of space in the camps. One IWW strike poster listed requirements for housing, with '[s]anitary sleeping quarters with not more than 12 men in each bunkhouse' that included good lighting and reading tables. It also demanded laundry rooms and bathrooms with showers. Finally, workers wanted 'wholesome food on porcelain dishes with no overcrowding at dining tables,' and well-staffed cookhouses 'to keep them sanitary' (*West Coast Lumberman* 1917). Employers were outraged and called for government intervention to crush the strikes, but they also started fixing up the camps. A Wobbly named Dowling called 'the change between now and three months ago in grub, steam and hours, makes it seem as though we had jumped a century ahead' (*Industrial Worker* 1917b).

The next year, the IWW planned another round of strikes over sanitation. But American entry into World War I changed the context of what activism could accomplish. With wood from the Northwest's trees needed to build airplanes, the strikes became a national security issue. The Wilson administration ordered the Army to get out the needed wood. The Army turned the Northwest woods into a militarized camp. The military demanded companies grant it authority over labour standards in the industry. It organized the loggers into an industry-wide company union called the Loyal Legion of Loggers and Lumbermen. It banned the IWW from the forests, forcing workers to renounce membership to get work. But it also provided the changes to the camp and personal environments that loggers had struck for. It created modern bunkhouses with running water, bedding, and quality food, sending inspectors to mess halls and to renovate 'the entire physical conditions under which the lumber workers lived' (United States War Department 1919).

Recalcitrant owners were reprimanded. Paul E. Page protested an order that he clean up his company's water supplies. He used its log pond for its camp drinking water and an inspector worried that the men would defecate in the water, leading to a 'serious epidemic.' Complaining that an inspection of his camp was 'unfair and unreasonable' he placed the water system in his camp within the context of the entire industry, arguing '[i]f the entire water supply west of the Cascades is to be condemned for the reason that some log may be pulled through human excreta and come in contact with some brook that supplies drinking water, we are certainly in a bad way' (Spruce Production Division 1918).

While military repression absolutely had a lot to do with the collapse of the IWW in the forests, the environmental revolutions in the camps made a huge difference. IWW documents show growing frustration that loggers were now ignoring Wobbly calls for action and refusing to join. Even committed Wobbly organizers such as Ern Hanson recognized this fact, remembering, 'The bulk of the membership were just card carriers and would fight for higher wages and better conditions but had a pretty vague idea about syndicalist organizations and revolution' (Russell 1978). After the war ended, the federal government pulled out of regulating labour conditions in the forest but the Loyal Legion continued as a company union for many firms until the National Labor Relations Act banned them in the 1930s.

In that decade, working-class environmentalism in the forest took on a new tone, one that serves as more of a direct antecedent to contemporary environmentalism. Beginning in 1935, Northwest loggers successfully unionized. Most timber workers ended up in the International Woodworkers of America (IWA). Initially a communist-led union, though that ended after 1940, the

IWA organized around the despoliation of the forests by the timber company. The IWA was the first American union to make natural resource planning a central policy position.

In doing so, the IWA built upon pre-existing divides within the forestry profession that had developed in the 1930s. Although the United States Forest Service's first chief Gifford Pinchot had pressed a conservationist agenda, in the years since, the agency had developed a cosy relationship with timber barons. But in the 1930s, progressive forces within the field began to challenge this relationship and its environmentally unsustainable practices such as clearcutting. This movement received a boost when Franklin Roosevelt named Ferdinand Silcox, one of these reformers, as head of the Forest Service. Silcox roiled the timber industry by pressing for public regulation of private forests, the expansion of national forest lands, and limiting clearcutting. At the 1935 Society of American Foresters conference, Silcox told the timber industry to choose between its current exploitative path and federal regulation of private forestry to stabilize communities (Ficken 1987; Clary 1984).

In March 1938, Roosevelt toured Washington. During his trip, he called the nation's disappearing timber resources 'a matter of vital national concern' (Roosevelt 1941). The IWA used Roosevelt's speech to launch a forestry programme based upon its vision of an activist government creating working-class security through full employment, income redistribution, and the end of corporate control over the landscape. Don Hamerquist, a logger and organizer, wrote in the IWA newspaper *The Timber Worker* that he had given up hope that capitalist timber companies would manage the forests sustainably, stressing 'The workers must drive hard for conservation and reforestation before the state is turned into a Gobi desert.' He concluded that timber executives should face charges for 'conspiracy against posterity' (Hamerquist 1938).

For the next decade, the IWA tied sustainable forestry with good wages guaranteed by union contracts to create the long-term basis for working-class security in the Northwest forests. Union leadership told members in 1938, 'We shall unite in every honest effort to save the forests. Real conservation, selective logging, sustained yield, reforestation, fire preventions—coupled with union recognition, union wage scales, means sustained prosperity in the lumber industry for all!' (*The Timber Worker* 1938a). Viewing the exploitation of both labour and nature as the result of corporate greed, IWA officials saw the New Deal state as the last best hope to reroute forestry's trajectory in order to prevent a future of deforestation and poverty, erosion and community instability.

The IWA thus began working with progressive foresters to stabilize timber policy. Rain and snowmelt washed soil from deforested mountains into the

Puget Sound in the spring of 1938, causing it to turn brown near the shoreline. A *Timber Worker* editorial blamed the erosion and brown water on private forestry, calling corporate leaders 'criminally guilty of causing floods. Their reckless depletion of the Northwest's greatest natural resource has recently been scathingly criticized by President Roosevelt' (*The Timber Worker* 1938b). As fires consumed the forest that summer, union writers accused companies of allowing them 'to burn a little, or much, rather than take precautionary steps' (*The Timber Worker* 1938c). It was an important ally in the creation of Olympic National Park in 1938, over the strong objections of the timber companies.

By 1944, the IWA convinced the Congress of Industrial Organizations to pass a resolution to regulate private forestry. The next year, it hired Ellery Foster, a professional forester, as its research director to turn the IWA's environmental programme into policy. Foster articulated a detailed union forestry agenda at the 1945 IWA Convention. It included federal regulation over all private forests on fire control, reforestation, conservation, and logging methods and federal acquisition of non-producing lands that could be reforested. It called for government assistance to small landowners to develop their timber, employment programmes in conservation, and wood efficiency programmes to maximize resource use. To accomplish these goals, the IWA needed to harness federal power to shift forest control away from corporations. Foster had a Jeffersonian vision of 'dirt foresters,' assisted by government forestry experts, producing sustainable wood products for the American working class (Foster 1946).

At the core of this critique was an attack on clearcutting, the process which strips all trees off a tract of land for maximum efficiency. Logging every tree was efficient, profitable, and destructive, leading to erosion, fires, and watershed degradation. The IWA claimed that clearcutting destabilized communities by wiping out local timber harvests for decades and that avoiding manual reforestation doomed the forests to wasteland. Companies supported the practice as the only way they could fulfil the nation's wood needs. For the IWA, this missed the point. Denuded forests threw lives into turmoil. George Prokopovich, an IWA millworker at a Bend plant, killed himself after his mill shut down. Tim Sullivan, his district president, used the suicide as an example of clearcutting's toll on workers (*International Woodworker* 1951).

Foster urged government leadership to develop the forests responsibly. First, it should invest in forest roads to both provide timber and democratize access. Earl Mason, an Oregon forester and IWA supporter, told union members that private roads incentivized clearcutting and encouraged 'a type of logging that customarily is incompatible with sustained yield forestry'

(*International Woodworker* 1949). Public forest roads would only serve sustained yield forestry if the federal government managed logging operations, was Foster's second demand. Government foresters would visit all logging operations and tag trees the companies could take. This would make selective logging national policy. As Foster said in a 1946 talk, 'piddling regulation calling merely for a few seed trees to be left,' can never 'safeguard our forests and bring security to communities which depend on wood industries for job and income.' Instead, foresters would implement a strict construction of sustained yield forestry, making sure that trees could not be cut at any rate faster than the forest grew. When industry officials dismissed these ideas as utopian, Foster replied, 'we don't believe sustained yield and competent marking is as tough a job as organizing the mass production industries was,' making clear connections between the current forestry campaign and the larger trajectory of union victories over the past decade (International Woodworkers of America [hereafter 'IWA'] 1946).

The IWA never believed that the timber industry would accede to labour's demands on forestry. Foster told union members that they should 'keep right on getting more and more people to understand and fight for the Hook Forestry Bill' (Foster 1946). It had found a champion in Representative Frank Hook of Michigan. A former logger, Hook was a committed conservationist motivated by the timber industry's deforestation of his home state. Hook introduced the bill that created Isle Royale National Park in Michigan in 1940. He was also one of the CIO's strongest congressional supporters. An enemy of Dixiecrats, Hook publicly accused Martin Dies of fascism while John Rankin physically attacked him during a shouting match over the CIO. In Hook, the IWA found a dedicated, if controversial, defender of sustainable forestry (Allen 2004).

On April 30, 1946, Hook introduced the Forestry Conservation and Development Act. The Hook Bill attempted to codify much of Foster's forestry programme. Small foresters would receive conservation payments and credit aids similar to those paid to farmers under New Deal agricultural legislation. The government would provide technical assistance in the managing and marketing of forest crops and build logging roads to help farmers get the trees to market. Government foresters would choose the individual trees to log on federal land. Finally, a flat rate of 2 per cent of the appraised value of national forest land would fund county governments reliant upon timber taxes for their budgets. Hook believed his bill would create long-term security for both the forest and the communities that relied upon it (*International Woodworker* 1946).

But Hook lost his seat in the Republican landslide of the 1946 congressional elections. The IWA called for the bill's reintroduction as the next step in its forestry programme. But the passage of the Taft-Hartley Act in 1947, which sharply limited union tactics and forced communist unions out of the labour movement, changed the playing field for labour politics. Repealing it became the IWA's primary legislative goal and a forestry bill a secondary concern. Foster also represented an increasingly unfashionable position in a forestry profession that had long cooperated with the timber industry before the strained relations of the Roosevelt Administration. He resigned from the IWA in 1948.

Yet the IWA continued promoting an environmentalism that served its members. It passed resolutions to revive its forestry programmes and showed frustration with the AFL-CIO for its tepid support. In 1956, union president Al Hartung expressed disappointment with the AFL-CIO for not taking conservation seriously. He decried that 'we are so lax in protecting the one natural resource that's going to keep your lakes full of water and your homes supplied with water … a place for recreation where you can get away from the city streets and get out in the shade and into the green of your timbers. These are some of the things that for some reason seem to steel and auto unimportant' (IWA 1956). The IWA spoke publicly in opposition to a proposal to reduce the size of Olympic National Park in the early 1950s. The IWA also gave support to protections for what became the Three Sisters Wilderness (Marsh 2009). When Hartung testified before the Senate Committee on Interior and Insular Affairs in support of a 1958 wilderness bill, he articulated an expansive vision of post-war prosperity. Envisioning a future where growing population combined with technological innovation and a federal commitment to full employment, he saw no alternative but a shorter workweek. Wilderness gave workers a place to enjoy their expanding free time. Hartung argued that a failure to act would cause the nation's children to 'hold us responsible for having cheated them of part of their birthright as Americans' (Hartung 1958).

By the 1970s, the future of the timber industry seemed less rosy than in 1958. Jobs began to disappear. Increasingly efficient mills automated and laid off unnecessary workers. In many areas, the gigantic old-growth trees were logged out, leaving less profitable second-growth stands in their place. Moreover, beginning in 1962, American export policy prioritized sending unprocessed logs to Japan instead of employing American loggers to turn them into saleable products. By 1970, over 2.5 billion board feet of timber was exported from west coast ports, a number up 16.6 per cent from the previous year. About 96.2 per cent of that timber went to Japan. Exports exploded during the Reagan years, peaking in 1989 at 1.944 billion board feet of timber, twice of their peak during the Carter Administration. Between 1979 and 1989, lumber production in the Northwest increased by 11 per cent while employment dropped by 24,500 jobs. This devastated the IWA. In October 1978, IWA District 3, covering the Pacific Northwest, had 36,300 members. By February 1981, that had declined to 22,273 (IWA 1981).

These economic shifts were replicated in many ways throughout the American economy by the 1970s, often leading unions to become fearful of supporting environmentalism. Yet the IWA not only continued its tradition of working with environmentalists, it made new environmental demands of their employers. Specifically, under the leadership of union president Keith Johnson and research director Denny Scott, it used the Occupational Safety and Health Administration to empower workers on the shop floor to protect themselves from chemical exposure. At the same time, with the increasingly powerful environmental movement seeking to save the last ancient forests and the northern spotted owls that live in them, the IWA continued to work towards alliances to save both forests and jobs. This continued despite a decade of timber industry efforts to blame environmentalists for the job losses it itself created.

The timber industry remained one of the nation's most dangerous industries throughout the twentieth century. Between 1970 and 1979, 1372 timber workers died in California, Oregon, and Washington (Rose 61). To fight this death toll, the IWA drew from the language of environmentalism to promote health and safety for workers. Conceptualizing the workplace as a corporate-created environment helped the IWA to become a national leader among unions on workplace health. In doing so, the IWA encouraged workers to demand safe conditions, stop work if they felt unsafe, and use the nation's new workplace safety regulatory structure as a way to channel the unrest that affected much of the nation's working class in the 1970s into productive action.

The IWA's strategy was what Denny Scott called the 'total work environment.' Borrowing this language from the nascent environmental movement, Scott meant thinking about the relationship between worker and workplace in a holistic context that considered how the nature of work affected the human body. This could empower workers to see themselves as part of a work ecosystem that could be transformed. It meant workers and their unions should have the full knowledge of the plant environment. Scott wanted companies to open their books about 'accidents, toxic materials, environmental monitoring and the cost of safety materials and safety improvements' (IWA 1977a).

This meant a variety of strategies to protect workers from unsafe working environments. First, it meant regulations on noise exposure. This was especially important with chainsaws, which had become dominant in the timber industry by the 1940s. OSHA began working on a noise standard in 1972, but progress was glacial. Second, it meant protecting loggers' hands from vibration white finger, a form of Raynaud's syndrome that creates discoloration in the fingers and toes, with workers' fingers turning white because of the constant hours of vibration they suffered in cold weather. Numbness and pain were other symptoms and the long-term effects included the loss of hand function, arthritis, and permanent nerve damage. Extreme cases led to amputation. A 1979 IWA survey showed 20 per cent of timber fallers suffering permanent disability due to chainsaw injuries (Johnson 1981).

But winning these battles was difficult for a small union. Logger Jim Menefee supported an anti-noise resolution the IWA passed in 1977, which called timber executives 'callous' and 'inhumane' because they opposed reducing the noise standard in mills to 85 dba from the common dba of well over 105. But Menefee could not figure out how to put teeth into such a resolution when companies controlled the testing and monitoring process. Workers took hearing tests and received no information from the companies who would not share noise levels with workers, saying they would have to take the company's word (Scott 1978). Creating federal standards on white finger proved no more successful. Scott noted that the lack of a successful workers' compensation claims for white finger meant that employers had no financial incentive to prevent the condition (Scott 1977). The IWA teamed up with researchers at Portland State University in 1979 to submit a grant for federal funding to develop safeguards for upon workers using chainsaws. But the government rejected the application. The IWA lacked the power to force changes to the workplace (Johnson 1979).

Equally intractable was chemical exposure. In 1985, the IWA, in coalition with several partners that included the Oregon Environmental Council, lobbied the Oregon state legislature to pass House Bill 2254, strengthening so-called right-to-know regulations, obligating employers to inform workers of the chemicals they handled on the job. Eight months after Morris Sweet started work in 1969, he woke up with the skin on his nose split open. A test revealed exposure to pentachlorophenol, a dioxin-based chemical frequently used in wood preservation after World War II. Sweet continued at the job for another twelve years, suffering rashes and cramps. He did not know what was happening to him, but '[t]he thing I noticed is that I always felt sick'). Although the right-to-know bill failed, the testimony of workers like Sweet provided powerful first-hand narratives that made unsafe working environments a central point of contention between IWA members and timber employers in the 1970s and 1980s.

Scott built connections with medical researchers, international timber health experts, government officials, national health and safety specialists, and environmental organizations in an attempt to build the knowledge base necessary to craft new regulations and provide workers information they needed to ensure they could recognize an unsafe workplace and act upon it. The union began holding workplace safety seminars around the Northwest. Calling these workshops, 'our front-line defence in any attempt at which we have gained,' they focused on workers' rights under OSHA, how to document violations, and engaging in an actual inspection at an employer who agreed to participate, all with the larger purpose of empowering workers to use federal and state regulations to make their own lives cleaner and safer on the timber mill shop floor (*International Woodworker* 1976). But only timber companies had access to the injury and illness logs companies had to turn into OSHA. Ultimately, the IWA's total work environment programme had its limits, but is still a window into the possibility of how unions could create their own environmental programmes with demands that were specific to working-class Americans.

The IWA also used environmental language to make connections between poor forest policy and other regional environmental problems. A 1977 union resolution criticized President Gerald Ford for cutting reforestation funding, noting that it would depress timber industry employment and 'result in floods and erosion because lands are bare but should be forested.' Instead, it urged Ford to crack down on pollution that 'kills trees and animals and is injurious to the health of all living beings including humans' and noted 'a cleaner, more healthy environment' would create jobs (IWA 1977b). Tim Skaggs, business agent for an Arcata, California local, excoriated Georgia-Pacific (G-P) in 1982 testimony over its water pollution, urging stricter controls over pesticide usage. Skaggs talked about how environmentalism had come about as a result of 'outrageous forest practices' that 'led to the adoption of regulations to protect water, wildlife, and the multiple uses of public and private lands.' But more was needed because G-P and other companies relied on 'massive clearcutting' that 'caused substantial erosion and stream siltation, resulting in a loss of water quality' (*Hard Times* 1983).

The union challenged timber companies directly on their environmental footprint. In 1978, Weyerhaeuser decided to build a new export facility at Dupont, Washington, to centralize its export of raw logs at one large facility. Environmentalists opposed it because it bordered Nisqually National Wildlife Refuge, one of the last wild spots on the heavily industrialized south Puget Sound. The Washington Environmental Council sued to block the facility. The IWA briefly joined in the opposition before local leaders asked the leadership to withdraw. The facility's environmental impact made union leaders uncomfortable. Denny Scott noted that while the export facility itself was relatively small, Weyerhaeuser owned 3000 acres around the site and he did not believe the company would leave it 'in a natural, undeveloped state' (Loomis 2016).

This was not the only time the IWA considered job losses an acceptable cost for environmental protection. The Oregon Wilderness Coalition sent Scott a Wilderness Economic Impact study it conducted in 1977, finding that significantly increased wilderness protection would have only minor impact on employment. The study noted that increased wilderness would reduce forest employment by 1.1 per cent. Scott corrected the OWC's numbers to 2.9 per cent, but he also called it 'a minimal reduction' in jobs and offered his support (Scott 1977).

On the other hand, sacrificing any jobs to expand wilderness was a difficult position for a timber union in the late 1970s. There were limits to union environmentalism, with growing grassroots pressure in the union by scared rank-and-file workers to stand up to environmental protection, especially as greens sought to list the northern spotted owl under the Endangered Species Act. This bird requires old-growth timber to live and although there was not enough harvestable old-growth to maintain the industry for many more years, the owl served as a symbol of outsiders—environmentalists, hippies, urban dwellers, liberals—who sought to change what it meant to be a Northwesterner. Still, the IWA maintained lines of communication with environmental organizations. As jobs disappeared, companies shifted blame from their own responsibilities overcutting the resource and exporting logs to Asia to environmental restrictions. IWA president Keith Johnson urged workers to ignore their employers blaming environmental protection for job losses and instead listen to the Sierra Club: 'We cannot permit industry their motion alone to determine when the natural resources of your country can be depleted and you, in turn, will be out on the bricks looking for a job' (Johnson 1974). For Johnson, employers, not environmentalists, had destroyed jobs through unsustainable forestry.

The IWA saw itself as holding a middle ground on environmental issues between corporate rapacity and greens' overreaching. It framed its support for the Environmental Protection Agency (EPA) by respecting members' discomfort with environmentalists. It worried about 'unrealistic EPA actions' that could cost workers in other industries jobs but also hoped greens would force corporations to clean up their polluted factories instead of relocating to the developing world. IWA leaders saw themselves as mediators between 'those who never cut down another tree and those who feel a full stomach is a fair exchange for a few more particles of crud in their lungs.' But ultimately, they believed the EPA could provide workers 'a paycheck <u>and</u> a decent world to live in' if workers' voices played a central role in environmental debates (Spohn 1974).

By 1987, though growing anger among rank-and-file timber workers susceptible to anti-environmentalist messages around the northern spotted owl and ancient forest protection and continuing employment declines led to the end of IWA environmentalism. As early as 1978, when the union supported the expansion of Redwood National Park, grassroots protest within the union had created a significant ruckus at the IWA's annual convention. The union's ageing leadership was replaced that year with new officers who rejected the past and united with companies to protest against environmental protection. The IWA moved towards the 'Wise Use' stance so many stakeholders in western public lands had taken in the wake of the Sagebrush Rebellion in the late 1970s (Echeverria and Eby 1995). The union newspaper began publishing bitter anti-environmentalist screeds and encouraged members to buy 'Save a Logger, Eat an Owl' bumper stickers. IWA Local 3-2 placed a wood statue of an owl pierced by an arrow on its roof, which the newspaper proudly reproduced in a picture (*The Woodworker* 1989).

This new stance did not go unchallenged from the rank and file. Retired logger Crawdad Nelson wrote to the newspaper befuddled at the new anti-environmental stance. Nelson lambasted the IWA's participation in 'anti-preservation rallies' and asked union leadership whether 'it is in the interest of workers to continue working as if old-growth were an infinite resource?' He called supporting bosses, 'the practical equivalent of suicide.' The official union response to this disgruntled member stated the Labor Timber Coalition 'was not formed to further the position of our unscrupulous employers, it was formed to bring some reason to the question of timber supply.' But the new IWA position could not answer the longer-term questions Nelson asked about the effects of disappearing old growth on jobs (Loomis 2016).

A half-century of union environmentalism ended as a declining union threw its lot in with employers. It's hard to blame them. Complicated economic transformations are not something easily communicated to workers with limited time and patience, as IWA leadership discovered. American workers and American unions have too often identified with their employers, both culturally and politically. For timber workers who saw their way of life threatened, it was easier to blame outsiders than their employers who claimed they wanted the industry to continue as it had for the last half-century. Companies found it easy to place blame for their own actions on environmentalists. And too many greens were in fact indifferent to the future of timber workers, particularly younger radicals in such organizations as EarthFirst! From 1987 until the IWA's merger with the International Association of Machinists in 1994, they joined other unions in lambasting environmentalists over the spotted owl. That's the story that is known today. But equally important is the decades of union environmentalism that came before this.

## Conclusions

There are as many lessons for us in that history as there is in the contemporary tensions between the two movements. First is that environmental protection does not often cause job loss. The many predictions that spotted owl protection would destroy the timber industry was belied by employment numbers. Between 1983 and 1988, average timber employment in Oregon and Washington was 105,000. By 1994, after the vast majority of the Northwest's national forest land was closed to logging, it was still 91,000 (Goodstein 1999). This was a significant decline, but much less so than what had decimated timber employment since the 1960s, processes that also still continued during the 1990s. Yet we still see environmentalists blamed for job losses. From the coal mines of Appalachia to the fishing industry of New England, industry has convinced workers that environmentalists are at fault for the job losses they themselves caused. Sadly, in those industries that are unionized, unions have often gone along with this messaging.

The second is that the experience of Northwest loggers was repeated in various ways through the American workforce. As early as the 1920s, the American Federation of Labor built connections with conservationists for big planning projects that would come to fruition in the 1930s and 1940s. The United Auto Workers had a full-time staffer working on nuclear issues, eventually leading to him becoming one of the pioneering anti-nuclear activists by the late 1950s. The Congress of Industrial Organizations had an assistant general counsel working heavily on a variety of environmental projects that included national parks and other recreational efforts; he eventually became the head of the National Parks and Conservation Association. Moreover, many unions, especially the United Steelworkers of America, United Farm Workers, and the Oil, Chemical, and Atomic Workers took leadership roles in building connections with environmentalists in the 1970s and 1980s (Lipin 2007; Fine 2012; Dewey 1998; Brucher 2011; Gordon 1998; Rector 2018).

Third, many of the tensions between the two movements today can also be traced back to changes in the composition of the American labour movement. It was the industrial unions emerging from the CIO that took the lead in building alliances with environmentalists. The building trades, including the United Brotherhood of Carpenters in the timber industry, expressed contempt for these connections going back to the UBC attacking the IWA for it in the 1930s. The trades, more politically conservative and less interested in wide-scale social change, have often eschewed alliances with other social movements. Yet, while the trades have also suffered from the unionbusting of the past half-century, it's those industrial unions who have been most decimated. Those mass-based politics are lost. The building trades again became dominant in the labour movement and they had rarely worked with environmentalists and often held politically conservative positions. The rise of public sector unionism and the new organizing of service worker unions such as the Service Employees International Union did bring a new set of political actors into the labour movement that helped fill the gap in left-leaning social policy, but they have focused on issues that matter more to their membership than the environment—health care, immigration rights, fighting for public education. If meaningful alliances between labour and greens are to again develop, it will probably require unions of government workers, health care providers, grocery store clerks, and hotel workers to lead the way, fighting to protect their own workplaces and the lives of low-wage workers from communicable disease, climate change, and other environmental disasters while fighting to create a world that is both economically and environmentally sustainable.

Finally, there is nothing inherently oppositional about workers and greens. Rather, the current tensions between the labour and environmental movements can ultimately be traced to the structural economic and political shifts defining the United States since 1973. Deindustrialization and capital mobility happening at the same time as the inflation and oil crises gave employers a tremendous tool to cleave unions from supporting environmental regulations, claiming they would close the plants and move overseas. The rise of the conservative movement took over the Republican Party and turned the bipartisan nature of environmentalism in the 1960s and 1970s into something very partisan. Moreover, after Ronald Reagan won the presidency in 1980s on an explicitly anti-environmentalism platform, environmental groups responded by turning away from grassroots organizing and towards fundraising for court cases to protect previous victories. While this made sense, it also gave the movement a public perception of elitism, one not helped by focusing more on charismatic animals than the toxicity killing working-class people. By the late 1980s, with protection of the northern spotted owl grabbing headlines when timber workers protested against it, environmentalists' indifference towards the fate of loggers demonstrated this elitism for all to see. In many ways, alliances between unions and environmentalists have never recovered.

Today, we live in a world filled with precarity, both ecological and economic. The spectre of climate change threatens to destroy the planet's ecosystem. The poor, both within the United States and globally, will bear the brunt of the human suffering. As greens fight with increasingly desperate measures to save the planet, workers find their futures increasingly unstable. Any answer to these problems has to create both environmental and economic stability. While we seek creative ways to bridge the contemporary gaps between these two movements, understanding that workers have long histories of environmentalism can help generate understanding, dialogue, and a sense of hope about the future among all of us.

# References

Allen, Mary Louise Hook. 2004. *Fightin' Frank: The Biography of Upper Peninsula's 12th District Democratic Congressman*. M.L.H. Allen.
Bevington, Douglas. 2009. *The Rebirth of Environmentalism: Grassroots Activism from the Spotted Owl to the Polar Bear*. Washington, DC: Island Press.
Birkeland, Torger. 1960. *Echoes of Puget Sound: Fifty Years of Logging and Steamboating*, 19–21. Caldwell, ID: The Claxton Printers, Ltd.
Bonn, Tess. 2019. Union Leader Says Green New Deal Would Make Infrastructure Bill Absolutely Impossible. *The Hill*. https://thehill.com/hilltv/rising/431145-union-leader-says-green-new-deal-would-make-infrastructure-impossible.
Brucher, William. 2011. From the Picket Line to the Playground: Labor, Environmental Activism, and the International Paper Strike in Jay, Maine. *Labor History* 52 (1): 95–116.
Clary, David A. 1984. *Timber and the Forest Service*. Lawrence: University Press of Kansas.
Cowie, Jefferson. 2010. *Stayin' Alive: The 1970s and the Last Days of the Working Class*. New York: The New Press.
Dewey, Scott. 1998. Working for the Environment: Organized Labor and the Origins of Environmentalism in the United States, 1948–1970. *Environmental History* 3 (1): 45–63.
Echeverria, John D., and Raymond Booth Eby, eds. 1995. *Let the People Judge: Wise Use and the Private Property Rights Movement*. Washington, DC: Island Press.
Ficken, Robert. 1987. *The Forested Land: A History of Lumbering in Western Washington*. Seattle: University of Washington Press.
Fine, Lisa M. 2012. Workers and the Land in US History: Pointe Mouillée and the Downriver Detroit Working Class in the Twentieth Century. *Labor History* 53 (3): 409–434.
Foster, Ellery. 1946. The American Forestry Congress. *International Woodworker*.

———. 1947. Trees Good Crops. *International Woodworker*.
Goodstein, Eban. 1999. *The Trade-Off Myth: Fact and Fiction about Jobs and the Environment*. Washington, DC: Island Press.
Gordon, Robert. 1998. 'Shell No!' OCAW and the Labor-Environmental Alliance. *Environmental History* 3 (4): 460–487.
Hamerquist, Don. 1938. Timber Is a Crop?
*Hard Times*. 1983. IWA Demands Safe Jobs and Clean Water, 7.
Hartung, Al. 1958. Statement in Support of the Wilderness Bill. IWA, UO, Box 280, Folder 47.
*Industrial Worker*. 1910. A Logger, 'Who Says a Logger Lives?'
———. 1916. Stomach-Robbing the Lumberjacks, November.
———. 1917a. Poor, Often Rotten, June.
———. 1917b. Timber! Timber! Hours Coming Down, November.
*International Woodworker*. 1946. IWA-CIO Forestry Program Is Answer to Hook Opponents.
———. 1949. Letters.
———. 1951. IWAer's Suicide May Be Related To 'Slash and Get Out Policy'.
———. 1976. OSHA Seminar Teaches Worker Rights.
International Woodworkers of America. 1945. IWA Executive Board Meeting. International Woodworkers of America Papers, University of Oregon Special Collections, Box 14, Folder 12.
———. 1946. Brief for Presentation at American Forestry Congress. IWA, UO, Box 344, Folder 6.
———. 1956. Fifteenth Annual Convention Proceedings.
———. 1977a. IWA Convention Minutes, IWA, UO. Box 26, Folder 2.
———. 1977b. R-61, Revival of the American Economy. IWA, UO, Box 54, Folder 10.
———. 1981. IWA, UO, Box 54, Folder 12.
———. n.d. International Woodworkers of America Papers. University of Washington Special Collections, Box 10.
Johnson, Keith. 1974. Address to IWA Western States Regional Council, IWA, UO, Box 31, Folder 22.
Johnson, Keith to Dr. Anthony Robbins. 1979, IWA, UO, Box 291, Folder 24.
Johnson, Keith to Dr. Pah Chen. 1981. IWA, UO Special Collections, Box 291, Folder 24.
Kazis, Richard, and Richard L. Grossman. 1982. *Fear at Work: Job Blackmail, Labor, and the Environment*. New York: The Pilgrim Press.
Kingsolver, Barbara. 1989. *Holding the Line: Women in the Great Arizona Miners' Strike of 1983*. Ithaca: Cornell University Press.
Klyza, Christopher. 1996. *Who Controls Public Lands?: Mining, Forestry, and Grazing Policies, 1870–1990*. Chapel Hill: University of North Carolina Press.
Klyza, Christopher McGrory, and David J. Sousa. 2007. *American Environmental Policy, 1990–2006: Beyond Gridlock*. Cambridge, MA: The MIT Press.

Leopold, Les. 2007. *The Man Who Hated Work and Loved Labor: The Life and Times of Tony Mazzocchi*. White River Junction, VT: Chelsea Green Publishing.
Lipin, Lawrence. 2007. *Workers and the Wild: Conservation, Consumerism, and Labor in Oregon, 1910–1930*. Urbana: University of Illinois Press.
Loomis, Erik. 2015. *Out of Sight: The Long and Disturbing Story of Corporations Outsourcing Catastrophe*. New York: The New Press.
———. 2016. *Empire of Timber: Labor Unions and the Pacific Northwest Forests*. New York: Cambridge University Press.
———. 2017. *The Unions Betraying the Left*. *The New Republic*. https://newrepublic.com/article/140423/unions-betraying-left.
Marsh, Kevin. 2009. *Lines in the Forest: Creating Wilderness Areas in the Pacific Northwest*, 28–29. Seattle: University of Washington Press.
McCartin, Joseph. 2011. *Collision Course: Ronald Reagan, the Air Traffic Controllers, and the Strike that Changed America*. New York: Oxford University Press.
Mink, Nicolaas. 2009. It Begins in the Belly. *Environmental History* 14 (2): 312.
Obach, Brian K. 2004. *Labor and the Environmental Movement: The Quest for Common Ground*. Cambridge, MA: The MIT Press.
Oliver, Egbert S. n.d. Sawmilling on Grays Harbor in the Twenties. IWA, UO, Box 275, Folder 6.
Rector, Josiah. 2018. The Spirit of Black Lake: Full Employment, Civil Rights, and the Forgotten Early Spirit of Environmental Justice. *Modern American History* 1 (1): 45–66.
Roosevelt, Franklin D. 1941. *The Public Papers and Addresses of Franklin D. Roosevelt 1938*, 144–150. New York: The Macmillan Company.
Rose, Fred. 2000. *Coalitions across the Class Divide: Lessons from the Labor, Peace, and Environmental Movements*. Ithaca: Cornell University Press.
Rosenblum, Jonathan D. 1995. *Copper Crucible: How the Arizona Miners' Strike of 1983 Recast Labor-Management Relations in America*. Ithaca: Cornell University Press.
Russell, Bert, ed. 1978. *Hardships and Happy Times in Idaho's St. Joe's Wilderness*, 108. Caldwell, ID: Lacon Publishers.
Scott, Denny to Kurt Kutay. 1977. IWA, UO, Box 300, Folder 25.
Scott, Denny to Ted Bryant. 1978. IWA, UO, Box 106, Folder 7.
Short, C. Brant. 1989. *Ronald Reagan and the Public Lands: America's Conservation Debate, 1979–1984*. College Station: Texas A&M Press.
Spohn, Dick to Ladd. 1974. IWA, UO, Box 300, Folder 9.
Spruce Production Division Correspondence. 1918. University of Washington Special Collections. Box 88, Folder Spruce Production Division, Correspondence—Surgeon.
Stoner, Rebecca. 2019. Why Are Unions Joining Conservative Groups to Protect Pipelines. *Pacific Standard*. https://psmag.com/environment/why-are-unions-joining-conservative-groups-to-protect-pipelines.

Street, Paul. 2019. Slandering the Not-So Radical Green New Deal: A Bipartisan Operation. *Counterpunch*. https://www.counterpunch.org/2019/02/25/slandering-the-not-so-radical-green-new-deal-a-bipartisan-operation/.

Szasz, Andrew. 1994. *EcoPopulism: Toxic Waste and the Movement for Environmental Justice*. Minneapolis: University of Minnesota Press.

*The Timber Worker*. 1938a. Who Will Stop the Fires. *The Timber Worker*.

———. 1938b. When Puget Sound Is Muddy. *The Timber Worker*.

———. 1938c. Support Grows for Selective Logging. *The Timber Worker*.

*The Woodworker*. 1989. Untitled Image.

Turner, James Morton. 2012. *The Promise of Wilderness: American Environmental Politics since 1964*. Seattle: University of Washington Press.

United States Commission on Industrial Relations. 1916. *Industrial Relations: Final Report and Testimony, Vol. V*, 4211–4212. Washington, DC: Government Printing Office.

United States War Department, Office of the Secretary. 1919. *A Report of the Activities of the War Department in the Field of Industrial Relations During the War*, 45–46. Washington, DC: Government Printing Office.

*West Coast Lumberman*. 1917. Inland Empire I.W.W. Demands.

# 7

# Trade Unions and Environmental Justice

Darryn Snell

## Introduction

At the AFL-CIO[1]'s Illinois State Convention in October 1965, Martin Luther King Jr. was invited to give the open address to union affiliates and delegates. In his speech he spoke about the union movement's many achievements. He stated:

> The labor movement was the principal force that transformed misery and despair into hope and progress. Out of its bold struggles, economic and social reform gave birth to unemployment insurance, old-age pensions, government relief for the destitute and, above all, new wage levels that meant not mere survival but a tolerable life. The captains of industry did not lead this transformation; they resisted it until they were overcome. When in the thirties the wave of union organization crested over the nation, it carried to secure shores not only itself but the whole society. (King 1965)

In his speech that day, Martin Luther King Jr. also made the important observation that unions have not been sufficiently recognised for their struggle for social and economic accomplishments: 'It is a mark of our intellectual

---

[1] American Federation of Labor and Congress of Industrial Organizations.

---

D. Snell (✉)
School of Management, RMIT University, Melbourne, VIC, Australia
e-mail: darryn.snell@rmit.edu.au

backwardness that these monumental achievements of labor are still only dimly seen' (King 1965). This chapter argues that the union contributions to social and economic advancement that Martin Luther King Jr. spoke about in 1965 continue to be overlooked some 50 years later including by the environmental justice (EJ) community and EJ scholars. While Martin Luther King Jr. did not speak explicitly about EJ in his many speeches, he was adamant about the need to address the inequities that working people and people of colour in particular were confronting including those associated with health, wellbeing and environmental harm. The union movement, working in coalition with other social movements particularly the civil rights movement, was considered by Martin Luther King Jr. as a primary force of progressive change including change related to environmental injustices. On 3 April 1968, the day before Martin Luther King Jr. was assassinated, he was marching with Memphis's striking sanitation workers. Some 1300 African American members of the American Federation of State, County, and Municipal Employees (AFSCME) were demanding improvements in pay and the dangerous working conditions that exposed them to health and safety risks such as the malfunctioning garbage truck which had crushed to death two of their members (Estes 2000). As recently noted by Eddie Bautista, executive director of the New York Environmental Justice Alliance, 'You don't get more environmental justice than that' (cited in Rosa-Aquino 2019) of the striking Memphis sanitation workers of 1968. This history and contribution of unions to advancing EJ, however, continues to be overlooked.

While there is some debate among EJ scholars about the origins of the movement, one common view is that the EJ movement began in 1982 in Warren County, North Carolina, when a coalition of residents, civil rights groups, environmental organisations and clergy mobilised to stop the building of a hazardous waste landfill which would disproportionately impact the low-income people living in the area, the majority of whom were also nonwhite (Cutter 1995). However, the roots of the EJ movement are much earlier and involved trade unions. For example, as early as 1976 American unionists organised the 'Working for Environmental and Economic Justice and Jobs: A National Action Conference' in Black Lake Michigan involving activists from many different spheres including labour unions, civil rights, women's rights and environmental organisations to discuss strategies for achieving EJ for working people (Rector 2014, 2018; Gordon 2004).

This chapter does not seek to do a comprehensive review of the EJ literature (others are better placed to carry out this task) but rather to make the theoretical and empirical case that trade unions are important EJ actors with a long history of leading and participating in EJ struggles in the workplace, among local communities and globally. In presenting this argument, however, there are two important caveats presented. First, trade unions' role and capacity as an EJ actor is not straightforward and often more complex than it is for other 'traditional' community-based EJ actors. Principally, this is due to their membership base and representation among workers who are involved in transforming the material world in ways which may have harmful environmental impacts. The second caveat is that unions often have a difficult relationship with other EJ actors and may at times be reluctant to address environmental harms in ways advocated by other EJ actors. The complexities surrounding their relationship to firms, communities and most importantly workers are critical to understanding the tensions and differences which emerge between unions and other EJ actors. Union differences with other EJ actors, however, should not discredit them as an EJ actor or justify them being overlooked. As will be presented, the voice of unions has at times influenced EJ activists to reconsider their community concerns and campaigns in alternative more progressive ways which aid in the strengthening coalitions between unions and others EJ actors (union positions and views have no doubt also been influenced by those of other EJ actors but this is not the focus of this chapter). The chapter begins with a brief discussion of EJ followed by a theoretical consideration of union capabilities as an EJ actor. A series of EJ case studies involving unions is then presented to illustrate the chapter's key arguments followed by a brief conclusion.

## What Is Environmental Justice?

Environmental justice is diverse and has multiple meanings but tends to refer to the notion that environmental burdens should not be disproportionately experienced by groups of people such as those who are socio-economically disadvantaged, people of colour and regional communities (see Schlosberg 2007; Walker 2012; Holifield et al. 2017), whilst environmental benefits are enjoyed solely by the privileged (Cutter 1995). What constitutes an environmental justice concern varies with context and situation but, overall, where groups of people have been identified as unfairly and overwhelmingly impacted by environmental harms, EJ takes the standpoint that this situation should be rectified. Environmental harms often relate to air and water quality or the location and treatment of waste and toxins all of which have implications for human health. Hazards associated with the built environment are also frequent environmental justice concerns when, for example, marginalised groups are forced to live in older houses containing asbestos, lead paints or other harmful substances which place them at higher risk of acquiring health problems (Jacobs et al. 2009; National Academies of Sciences, Engineering, and Medicine 2017). EJ maintains the position that the development, implementation and enforcement of environmental laws, regulations and policies must ensure that groups of people are not unfairly treated and burdened. EJ campaigners, thus, work to ensure a sustainable and healthy environment for all individuals and communities and to protect the rights of victims of environmental injustices. Who are the 'victims' can include those in the immediate area in which people are living and working, a broader geographical region or globally as in the case of the EJ concerns emerging from climate change impacts (Schlosberg 2013). As this chapter highlights, unions have been active at all levels (i.e. the workplace, the local community and the broader global environment) and often in ways distinct from other EJ actors.

## Unions and EJ: Theoretical Considerations

While EJ scholars have not widely considered the role of trade unions in securing EJ outcomes, there is some recognition of their contributions (Loomis 2015; Montrie 2008; Rector 2018). Often this is recognised in terms of their role as an EJ coalition partner with others or as a 'stakeholder' with EJ concerns. Consequently, the theoretical analyses underpinning such studies tend to assume that unions are not dissimilar to a wide range of other 'stakeholders' (e.g. environmental organisations, religious groups and community-based organisations) with similar capabilities and capacities. Unions, however, have very distinctive organisational structures and characteristics which differentiate them from other EJ actors and condition their EJ capabilities and responses. First, it is important to note that trade unions are democratic organisations accountable to their members. Unions are collective agents of workers who pursue self-declared goals on behalf of workers. They are reliant upon fee-paying members to provide the financial resources for their ongoing reproduction and the pursuit of union goals. Solidarity among members, and worker interests more generally, therefore, underpins the organisational structure of unions and their success in challenging forms of exclusion and oppression. While it is recognised that solidarity is not exclusive to trade unions and has historically been 'a central practice of the political left' (Featherstone 2012, 5), research suggests there is a much stronger mutually reinforcing interplay between unions and solidarity as a consequence of their particular structure and location in the capitalist economic system (see Dixon et al. 2004; Hyman 2007).

The interplay between unions, solidarity and individual member interests, however, is complex and has implications for union capacities as an EJ actor which brings us to the second point about the distinctive characteristics of trade unions. As collective agents, unions shape the context for all members—albeit not necessarily in the way any particular member desires on every matter. As democratic organisations, unions are accountable to members and, as individual members, workers are active agents in their own right with the ability to respond to and influence the agency of their unions. Unions are caught in a tension between certain vested interests, typically defined in terms of 'bread and butter' industrial matters (i.e. wages and conditions), and matters pertaining to social or environmental justice (e.g. equality, human rights, pollution and environmental harm) (Flanders 1970; Hyman 2007). Pursuing goals that are not necessarily perceived by members as in their collective vested interest in the short term, such as some EJ issues, requires sustained member support over the longer term. When EJ concerns emerge from the same industries in which workers make their living, the relationship between vested interests and 'the sword of justice' can often become polarised in the interplay between unions, solidarity and the interests of individual members.

A third distinctive aspect of trade unions which separates them from other EJ actors relates to their particular relationship to industry which is often the culprit of EJ concerns. As noted by Harvey et al. (2017):

> Trade unions occupy a distinctive position in this landscape of stakeholders in that they are at once and the same time internal stakeholders (a significant stake in the success, continuity and growth of businesses, formalised channels of voice and interaction) and external stakeholders (formal/structural independence from the firm, contrasting interests with the firm in some areas, somewhat narrower issue/advocacy interests). (45)

Being independent from the 'firm' but also internal to it provides unions a range of unique opportunities to influence EJ concerns within, emerging from and beyond the firm.

One final point about unions as an EJ actor is to acknowledge that labour agency and union engagement, industrially or in some other capacity, is embedded within an institutional, spatial and temporal context (Zeitlin 1987, 168). This context is both globally and locally determined and its 'landscape' conditions union agency and action (Coe and Jordhus-Lier 2010). Union engagement in sustainability and 'green' agendas which consider 'just transition' aspects, for example, have become an international phenomenon (see Räthzel and Uzzell 2011). Unions, however, are also embedded in locally defined institutional arrangements and social, economic and environmental problems which enable and constrain union agency. In many Coordinated Market Economies (CMEs) in Northern Europe, for example, unions are included as social partners in company consultations, including issues related to environmental and social responsibility (see Harvey et al. 2017; Preuss et al. 2015). As noted by Delautre and Abriata (2018), these corporatist institutional arrangements—which vary from information to consultation to joint-decision making—can empower unions and employees and 'help to promote higher labour standards and corporate responsibility' (5). However, such arrangements can also leave unions open to accusations of complicity in company strategies which can prejudice their effectiveness when it comes to broader issues such as EJ campaigns. In Liberal Market Economies (LMEs), union capacity to influence environmental policies, programmes and strategies of firms internally is more limited (Hofman et al. 2017; Preuss et al. 2015; Sobczak and Havard 2015). Their role as a collective bargaining agent, however, is one of the ways unions have sought to address environmental concerns within such contexts (Markey and McIvor 2019). Industrial contexts also condition the way specific unions respond to environmental questions. Education and public sector unions, for example, find it much easier to take principled positions on regulating agricultural and industrial pollutants (e.g. carbon emissions and asbestos) than unions whose membership are employed in polluting industries. Even within the same union, differences may emerge between locals or branches on the best ways to address environmental harms due to local industrial particularities and implications as well as union leadership stances and membership positions. Drawing upon case study evidence, these key conceptual arguments are now explored in greater detail.

## Unions and Environmental Justice in the Workplace

For workers, the environment is typically defined in terms of the workplace. Unions, as collective organisations, draw upon worker solidarity to influence power relations within workplace settings (Loomis 2015). The forging of solidarities among workers has enabled workers to escape dangerous, oppressive and alienating conditions since the early industrial revolution (see Featherstone 2012; Loomis 2015; Thompson 1991). In E.P. Thompson's *The Making of the English Working Class* (1991, original version published in 1963) the hazardous working conditions that workers confronted throughout the industrial revolution are laid bare. He documents how workers, including young children, were expected to work among unprotected machinery and hazardous chemicals without proper safety attire and equipment for long hours resulting in a significant physical toll on workers' lives (Thompson 1991). To illustrate, Thompson presents a diary entry from a labourer in the 1840s:

> [A] factory labourer can be very easily known as he is going along the streets; some of his joints are almost sure to be wrong. Either the knees are in, the ankles swelled, one shoulder lower than the other, or he is round-shouldered, pigeon-breasted, or in some way deformed. (1991, 328)

Drawing upon other texts of the period, Thompson shows how 'the rich lose sight of the poor, or only recognise them when attention is forced to their existence by their appearance as vagrants, mendicants, or delinquents' (1991, 322).

Contemporary EJ scholars widely acknowledge that it is working-class people who have overwhelmingly confronted environmental injustices. Daniel Faber (2017), for example, argues 'environmental injustices are rooted in processes of capital accumulation and power structures' (62) which exploit and disempower labour in the workplace. Like E.P. Thompson argued many years earlier (1991), Faber points out the ways the working class are more at risk of environmental harm with the expectation they perform dangerous jobs (2017). In this assessment, the treatment of the 'working class' can be considered an environmental injustice. What Faber (2017), however, fails to acknowledge are examples of how unions have fought to address this environmental injustice through industrial action and collective bargaining campaigns aimed at improving the environmental conditions under which work is expected to be performed and how work and working conditions are regulated. The emergence of trade unions in the early 1800s was a direct response to growing awareness of environmental injustices (Perry 2019; Webb and Webb 1920).

As noted by Roelofs et al. (2017):

> Low-income people and people of color are more likely to encounter chemical, physical, and biological hazards and psychosocial stressors in their communities and at work. ... Disparities in work-related exposures arise from disproportionate employment in hazardous jobs—compounded by workplace discrimination, ineffective prevention and training, and restructuring of the workplace, creating less secure jobs. (23–24)

Addressing health inequity of working-class people constitutes a core activity of unions (Barca 2012) and no other social movement can claim greater achievements in this domain (Bennett 2011a; Gordon 2004). Unions have fought to improve occupational health and safety legislation and the rights of its occupational health and safety officers to educate and train workers about workplace safety and monitor safe practices in the workplace. As noted by Malinowski et al. (2015), as unions win increases in wages and benefits, make improvements in workloads, secure safety provisions and standards, 'they are raising their collective voice to challenge the accepted forms of resource distribution, hierarchical power dynamics, and traditional levers of societal health' (267) creating healthier workplaces but also healthier communities. Environmental safety is part of the trade union's role in raising class consciousness and the transformative political practice underpinning union solidarity (Bennett 2011b). This class and environmental safety awareness extends beyond the workplace and into the union home and broader community. In addition, the achievement of unions in one workplace often typically extends to non-union workplaces as employee and social expectations more generally, become elevated.

The role of unions as an EJ actor in the workplace, nevertheless, is not without its challenges. Unions have typically relied on collective bargaining or industrial action to address workplace environmental concerns. Employers have often resisted union efforts to bring about improvements in the workplace due to cost implications and question the harm associated with a workplace practice or substance used in the workplace. This can result in expensive and lengthy union campaigns whereby unions work with medical researchers and industrial hygienists to gather the evidence needed to demonstrate workplace harm and mount legal cases against firms (e.g. see Malinowski et al. 2015; Rector 2014, 2018). Union resources dedicated to these efforts can be quite extensive but are often critical for advancing workplace EJ outcomes. Their close association with workers also enables them to build trust among the workforce for employee examinations and quality data (Rector 2014).

While the overarching argument of this chapter is that unions are important, albeit largely overlooked, EJ actor, two important caveats were presented. One of those caveats was that that unions' role and capacity as an EJ actor is not straightforward and more complex than it is for other 'traditional' community-based EJ actors. Principally, this is due to their membership base and representation among workers who are involved in and whose lives depend upon the continued activities which result in harmful environmental impacts. Addressing environmental harms in the workplace, therefore, is rarely easy or straightforward for unions. Depending upon costs and viable alternatives, employers can be resistant to introducing changes which would reduce environmental harm. In addition to discrediting the medical evidence, employers may threaten to close facilities and use 'job blackmail' to undermine solidarity and union efforts to address environmental concerns. Workers and unions also have their own traditions which may result in members not viewing identified environmental harms as priorities, particularly if jobs are threatened. The following case of Australian campaigns to rid the workplace and broader society of harmful asbestos are illustrative of such challenges for unions.

## Solidarity Divided and Reunited: Australian Unions and the Case of Asbestos

Concerns about the negative health impacts of asbestos exposure had been raised by workers since the late 1890s (Australian Asbestos Network 2020b; McCulloch and Tweedale 2008; Ruff 2012; Ruff and Calvert 2014). By the 1930s medical research emerged connecting lung cancer to asbestos exposure (McCulloch and Tweedale 2008). Soon after, unions across a range of countries began campaigning for better protection for workers exposed to asbestos dust. They fought for additional medical studies; reductions in dust levels at mine sites and factories manufacturing asbestos products; special 'asbestos-dust' payments to cover workers expected to work with asbestos; and legal compensation for those acquiring asbestos-related diseases.

In Australia, where asbestos resources were some of the largest in the world, unions identified workplaces failing to reduce asbestos exposure and warned workers not to take up positions there (Australian Asbestos Network 2019). At some worksites, however, unions struggled to convince workers of the dangers of asbestos particularly in localities where there was a long tradition of accepting occupational health risks (Australian Asbestos Network 2020b). In asbestos mining communities, where there were few other job opportunities, unions were often conflicted with needing to protect the health of workers and their economic livelihoods (Ruff and Calvert 2014). In Australia, when one of the largest crocidolite (blue) asbestos mines was closed in the 1960s, for example, the Australian Workers' Union described it as 'callous treatment' despite the growing health concerns surrounding the mine (Australian Workers Union 1966, cited in Haigh 2006, 115).

By the late 1960s, however, cases of asbestosis and mesothelioma were being recorded in every Australian state and unions were becoming more united in campaigns to ban asbestos from Australian workplaces and Australian products (Australian Asbestos Network 2020b). By the late 1960s crocidolite (blue) asbestos was banned, followed by amosite (brown) asbestos in the mid-1980s and chrysotile (white) asbestos in 2003 (Australian Asbestos Network 2019). Beginning in the 1970s, Australian unions increased their industrial struggles around asbestos exposure through protests, strikes and worker stoppages which continue through to today (Australian Asbestos Network 2019). A significant aspect of more recent struggles has been to ban the import of asbestos products which was finally secured in 2003. Asbestos-related diseases, however, continue to plague many communities and ageing workers globally (McCulloch and Tweedale 2008). It is estimated that asbestos related-diseases account for some 4000 deaths per year in Australia (Asbestos Council of Victoria 2019). Unions continue to fight for compensation for these workers and support asbestos-related disease support and advocacy groups in impacted local communities (Australian Asbestos Network 2019).

One of the largest compensation campaigns centred around the James Hardie Company which had previously dominated Australia's asbestos-product industry (Australian Asbestos Network 2020a). James Hardie fought an extensive battle to distance itself from its asbestos liabilities including relocating its company to the Netherlands in an attempt to avoid compensation payments (Australian Asbestos Network 2020a). Bernie Banton, a trade unionist who worked for James Hardie in the 1960s and 1970s where he was exposed to asbestos daily, became the public face of the union-led campaign 'Make James Hardie Pay' (Australian Asbestos Network 2020a). Bernie Banton lived with asbestosis for many years before he acquired mesothelioma which eventually took his life in 2007 at the age of 61. In 2005, however, he was well enough to witness the culmination of the Campaign when the Australian Council of Trade Unions, the New South Wales Government, James Hardie and victims signed off on an agreement that secured compensation for Hardie's victims from the company for the next 40 years (Peacock 2011). The environmental and health improvements achieved for poor and working-class people by removing asbestos from workplaces and its continued use in low-cost housing, public schools and roofing materials cannot be underestimated and extend well beyond the workplace.

## Trade Unions and EJ Beyond the Workplace

As noted by Schlosberg (2013), EJ has focused primarily on addressing environmental harms experienced by local communities. This community focus partially explains why trade unions have tended to be overlooked. However, as highlighted in the case of asbestos, health equity and EJ issues emerging within the workplace often extend into the local community. Periodically, this type of issue energises unions to become involved in broader EJ campaigns. In 1948, for example, the US Steel Corporation released a toxic cloud that killed twenty people and left hundreds sick or dying in Donora Pennsylvania (Loomis 2015). The 'Donora Death Fog' led the United Steelworkers Union (USW) to recognise the close relationship between health and safety issues in the plant and environmental justice issues in surrounding communities. The USW became a strong voice for the Donora community and began to advocate for Federal environmental protection as an extension of the union's responsibilities for its members' health and safety (Vachon and Brecher 2016). The USW continued to demonstrate their commitment to community-based EJ campaigns throughout the early 1960s, for example, in their strong support for the first Federal Clean Air Act of 1963. In 1973, the alignment of union, community and environmental interests underpin industrial action towards petrochemical company Shell in the United States. The Oil, Chemical and Atomic Workers International Union (OCAW) organised action for better health and safety provisions for workers on the premise that the dangerous substances they were being exposed to by the company also had disastrous environmental repercussions for the surrounding community and environment (Loomis 2015). Although ultimately unsuccessful in its original aim, the anti-Shell strikes served to strengthen the links between unions and the environmental movement and contributed to later health and safety legislation (Loomis 2015).

When and why trade unions choose to become involved in EJ community campaigns is not well explored by either union or EJ scholars. EJ scholars have been strongly influenced by the new social movement literature which assumes EJ movements emerge from community activism or environmentalists' activities but may involve the 'old' trade union movement as a coalition member (Sicotte and Brulle 2017). Trade union literature, on the other hand, tends to explain the activities of trade unions beyond the workplace in terms of 'community unionism' or 'social movement unionism' involving a broadening of union purpose with prospects for union revitalisation (Diani 2018). In this analysis, trade unions are considered a central actor in addressing community-based issues as opposed to a coalition member. Depending upon the particular EJ campaign, its history and evolution, there is likely to be varying levels of evidence to support each theoretical position. The factors that motivate unions to move beyond industrial issues into community campaigns are often complex as are often the reasons that they join in coalition with other EJ actors. Union involvement in an EJ coalition to improve the air quality in poor neighbourhoods surrounding Los Angeles's ports illustrates these complexities.

## Los Angeles and the Coalition for Clean and Safe Ports

Low-income and minority communities located in Los Angeles have a long history of campaigning around air pollution hazards that contribute to heightened levels of asthma and other respiratory problems for residents (Motavvef 2020). Central to many of these campaigns are concerns about diesel smoke from trucks and buses driving through their neighbourhoods which are often located near highway exit ramps or major thoroughfares. Addressing environmental health disparities for these communities is not always easy given they result from poor urban planning not easily reversed. The focus of campaigns, subsequently, tends to be on reducing the level of air pollutants (e.g. carbon monoxide, sulphur dioxide and other particulate matter) originating from passing vehicles. The Coalition for Clean and Safe Ports exemplified such an approach. The campaign centred primarily around improving the air quality of working-class neighbourhoods located along the major truck routes to LA's ports of Los Angeles and Long Beach; the two largest container ports in the United States. The campaign has been described as a 'Labour-Green Alliance' (Greenhouse 2010) in which unions, environmentalist and community groups become 'unusual allies' (Larrubia 2008) through 'interest convergence' (Cummings 2014). The campaign began in 2006 when the Los Angeles Alliance for a New Economy (LAANE), a nonprofit closely tied to the Los Angeles County Federation of Labor, brought together the Teamsters and Unite Here Unions with the National Resources Defense Council, the Sierra Club and the Coalition for Clean Air. During this year, LAANE and the unions also pulled together neighbourhood and community groups, advocates for asthma sufferers as well as the National Association for the Advancement of Colored People (NAACP) and other social justice organisations committed to fight for cleaner air. The Teamsters argued that the air quality was so poor around port neighbourhoods because port truck drivers, most of whom were owner-drivers, earned so little that they could not afford to buy new less polluting trucks (Cummings 2018). Working with LAANE, they convinced environmentalist and other coalition members that improving the air quality required ameliorating the working conditions and wages for port truck drivers. They advocated that trucking firms be held accountable for the air pollution, not the owner-drivers. They maintained that drivers needed to be more appropriately categorised as employees of trucking firms rather than independent contractors and, as such, it was the responsibility of the firms to buy and maintain cleaner trucks (LAANE 2008). Owner-operators had always presented challenges for the Teamsters Union in improving wages and conditions for truck drivers. Firms used owner-operators precisely to de-unionise their workforce, reduce their risks and drive down costs. If the ports could force trucking companies to employ drivers directly, the Teamsters argued, they would have a better chance of improving wages and conditions for drivers while at the same time improving air quality.

The Clean Trucks campaign achieved a major success in 2008 when campaign organisers convinced the mayor of Los Angeles, Antonio Villaraigosa (a former union organiser), and the city's ports to require trucking firms to employ their drivers directly and bear the costs of buying new trucks (Greenhouse 2010). Subsequently, the Clean Air Action Plan and the Clean Trucks Program were launched resulting in the progressive banning of older, heavily polluting trucks from the ports and the conversion of owner-operators into employees (Greenhouse 2010). The latter programme, however, was challenged in a long legal battle that involved the American Trucking Association suing the Port of Los Angeles over the requirement that they employ workers directly (Greenhouse 2010). In 2011, the US 9th Circuit Court of Appeals sided with the trucking industry which meant the Clean Trucks Program was allowed to proceed but without the mandate that drivers had to be employed by trucking firms (Cummings 2014). The Teamsters hailed it a disaster for drivers as they would unjustly shoulder the costs of cleaning up the air. Many truck drivers went on strike or sued the City of Los Angeles claiming the arrangements violated labour laws by misclassifying them as contractors when they were employees. Many trucking firms responded to these concerns not by employing them directly but by agreeing to purchase new trucks on their behalf and then leasing them back to the drivers while deducting lease payments from their pay cheques (Greenhouse 2010).

From an environmental perspective, the Clean Truck Program was heralded a success by environmentalists and community organisations. By 2012, the Clean Truck Program had banished the last remaining older, more polluting trucks from the port terminals reducing emissions by 90 per cent (Port of Long Beach 2019). The neighbouring communities benefitted from these environmental justice outcomes which set the stage for other similar campaigns throughout the country. Unions were strategically important to this success. Unfortunately, however, the campaign did not significantly improve the wages and working conditions for truck drivers as was the hope of unions and their allies (Greenhouse 2010). Union efforts to bring about the end of the independent, contractor-based system of port trucking were ultimately defeated by the Federal Government law which regulated trucking and supported such a system. As noted by Cummings (2014), while the 'Labour-Green Alliance' aimed to deliver a win for the environment, the disadvantaged local communities suffering from poor air quality and the working arrangements for drivers, it ultimately failed workers:

> [W]hile the ports are now on track to achieve green growth, it is still on the backs of their most vulnerable workers. (1165)

The Coalition for Clean and Safe Ports Campaign nonetheless illustrates how unions can perform important EJ roles beyond the workplace and work with other EJ actors. Certainly, in this instance, the unions were not simply EJ 'coalition partners' but the major initiators and drivers of the campaign. In addition, it was through their political connections with the Los Angeles mayor and other council members that some environmental and community outcomes were ultimately secured. Analytically, the motivation for unions initiating and leading this campaign could be presented as simply 'vested interests' with the unions' primary aim being to convert owner-operators to trucking firm employees whom they could represent and organise. Such an analysis, however, would underestimate the unions' commitment to improving the air quality in neighbourhoods along the port arterial thoroughfares. During the time of the campaign, the Teamsters' national leadership were beginning to articulate clear policy positions on environmental issues including support for 'green jobs' and abandoning support for oil drilling in the Arctic (Larrubia 2008). Understandings of unions' roles and motivations in community-based EJ campaigns, therefore, must also consider union commitments to the broader 'sword' of social and environmental justice.

## Global EJ and Trade Unions

Within the EJ literature discussion has emerged about EJ's historical concerns about local community issues and how they might relate to broader national and international environmental justice concerns (Martinez-Alier et al. 2016;

Schlosberg 2004; Schlosberg and Collins 2014; Gleeson and Low 1998; Osler Hampson and Reppy 1996). There is growing recognition that a number of environmental inequities at the local level are global in nature. The global trade in waste (including plastic, electronic waste, decommissioned ships, etc.), for example, often results in poorer nations, and poorer communities within these nations, left dealing with environmental hazards. Biopiracy has meant the appropriation of genetic resources from indigenous peoples who do not fully understand the implications of property rights (Rose 2016). EJ researchers are starting to chart these global environmental inequities and the environmental justice movements challenging these inequities (see Martinez-Alier et al. 2016; Pellow 2007). Others are investigating the interaction between local and global environmental justice with some suggesting that 'global movements for environmental justice are spreading geographically, globalizing their claims, sharing resources and becoming increasingly networked amongst themselves' (Martinez-Alier et al. 2016, 748). However, it is unclear to what degree many local EJ concerns become connected to global EJ movement action for change with similar priorities. Trade unions, as EJ actors, are often better positioned to connect local to global EJ campaigns due to their relationship between local union branches, their national unions and global union federations which are confederations of national and regional trade unions associated with specific industry sectors.

International solidarity and struggle has a long and well-documented history in the labour movement (Burgmann 1995; Featherstone 2012; Kang 2012). International efforts by unions to improve health and safety on farms and in the workplace for working people in the Global North and South including the global eradication of toxic chemicals and substances like asbestos are illustrative of global environmental justice concerns.

International Workers Memorial Day, held annually on 28th April, is the union movement's international day of remembrance and action for workers killed, injured or harmed by work. April 28th commemorates the anniversary that the US Occupational Safety and Health Act of 1970 was enacted. This date has come to be a day of international solidarity and campaigning with the slogan 'Remember the Dead, Fight for the Living!'. Unions often join with other organisations on this day to campaign around environmental justice concerns. In 2019, for example, the International Trade Union Confederation (ITUC) joined with the Alliance for Cancer Prevention on International Workers' Memorial Day to campaign for 'No More' exposure to harmful chemicals in the workplace (Alliance for Cancer Prevention 2019). These types of activities involve local branches, national unions and global union federations working together to make local concerns global and vice versa. Aligning local EJ concerns of unions with those of global union campaigns and priorities, however, is rarely straightforward. Union concerns about climate change and the policies to address these concerns are used to further highlight these challenges.

## Trade Unions and Climate Justice

One of the caveats made at the beginning of this chapter was that unions often have a difficult relationship with other EJ actors and may at times be reluctant to address environmental harms in ways advocated by other EJ actors. Climate change has been one of the most divisive and challenging issues for unions and other EJ actors. Climate change, however, has also tended to divide unions locally, national and globally. These differences have principally emerged over the future of fossil fuels and those that make their living from fossil fuel-based industries and those that do not. Despite the divisive aspect of climate change for unions and their relationship with other EJ actors, this issue also illustrates how union activism at the local level has influenced EJ activists and global unions to re-evaluate their positions and advocate more socially justice ways to address climate change.

Environmentalists have led the charge for climate action through rapid reductions of carbon emissions. Unions associated with fossil-fuel industries have found it difficult to find common ground with environmentalists and have often accused them of being reckless and callous for the policies they advocate. Rapid closure of coal mines and coal-fired power stations, they argue, will drive up power prices and cost of living for poorer people and devastate the lives of workers and communities that depend on these industries. The emergence of the Climate Justice (CJ) movement is often presented as resolving these tensions by advocating that climate change is both an environmental and social, ethical and political issue. As noted by Schlosberg and Collins (2014), EJ and CJ are different in orientation but share many principles including respect for indigenous peoples and the importance of self-determination of people regardless of socio-economic status, ethnicity, geographic location and so forth. Both EJ and CJ have raised concerns about how disadvantaged groups will be disproportionately impacted by the destruction caused by climate change (e.g. poor communities impacted by Hurricane Katrina in the United States, small island states experiencing rising ocean levels in the South Pacific). More recently EJ and CJ have also come to appreciate that climate action also has ethical and social implications and have come to embrace the notion of 'just transition' developed by trade unions (Routledge et al. 2018; Morena et al. 2020).

Unions had long maintained that efforts to address environmental harms may also result in unintentional burdens being placed on groups of workers and that local communities and workers who have suffered a disproportionately high cost for environmental improvements should not be left behind (Stevis et al. 2020). Unions, working through global union federations and the ITUC have sought to advance these views through UN conventions and initiatives aimed at addressing environmental concerns. At the Rio Earth Summit in 1992, for example, unions advocated for union involvement in sustainable development discussions and policy development through the Agenda 21 proposals (Räthzel and Uzzell 2011; Silverman 2006). This enabled the Internationl Confederation of Free Trade Unions (ICFTU) and the Trade Union Advisory Committee (TUAC) to become influential players in the UN Commission on Sustainable Development where they could raise concerns about social and economic development and the implications for workers (Gereluk and Royer 2003). This early involvement in the UN's sustainable development discussions meant the union movement was well positioned to present a union perspective about climate change within UN meetings.

Much has been written about union positions on climate change particularly its advocacy for 'just transition' for communities and workers adversely impacted by climate change policies (Morena et al. 2020). As noted by Stevis et al. (2020) the concept of 'just transition' has trade union roots dating back to the 1970s in the United States where unions sought to reconcile environmental and social concerns. Just transition became an important component to labour environmentalism both at the local level where workers and communities confronted significant industrial and structural adjustment challenges as environmental priorities changed and at the international level as decarbonising the global economy became a major goal. Where climate justice campaigners sought to draw attention to how certain communities would be disproportionately impacted by the destruction caused by climate change the union movement reminded us that certain communities and workers dependent on polluting industries would be unfairly disadvantaged by efforts to address global warming. In 2015, after years of campaigning and lobbying of government representatives at UN Climate conventions, union succeeded in getting 'Just Transition' included in the Preamble to the Paris Climate Agreement noting that climate change policies needed to 'take into account the imperatives of a just transition of the workforce and the creation of decent work and quality jobs in accordance with nationally defined development priorities' (UN 2015, 21). The international trade union movement intends to advance these aims at future UN Climate Change Conferences.

Union pursuits of just transition at the international level have also meant unions have come into contact with other climate justice and environmental justice actors. This interaction may explain why climate justice and environmental justice actors have come to appropriate the concept of justice transition in much of their policy advocacy (Stevis et al. 2020). The Climate Justice Alliance, for example, which formed in 2013 and claims to be 'an alliance of 70 urban and rural frontline communities, organizations and supporting networks in the climate justice movement' makes just transition a defining principle of their mission (Climate Justice Alliance 2020). Unfortunately, however, it is not common for CJ or EJ actors to acknowledge the trade union origins and contributions to advancing just transition principles.

Debates about just transition, whether at the local or international level, have included extensive debate among unions and other CJ and EJ actors about energy production technologies, electricity generation, supply and pricing and the future of energy systems. Energy unions often express frustration with environmentalists and climate justice activities for not fully appreciating the challenges of ensuring stable electricity supply using renewable technologies and at prices which working people can afford in the transition to renewable energy (see Maher 2016). Reaching a common understanding about the technological, social and political challenges of energy system transitions and consensus on how best to proceed to meet socio-environmental concerns has been a central aim of Trade Unions for Energy Democracy (TUED). TUED is an international union-led climate justice initiative—epitomising the synergy between the union movement and climate justice movement that has been encouraged at local, national and international levels by the ITUC since its inception in 2006 (Rosemberg 2013). TUED emerged from a global trade union roundtable discussion on energy and climate change held in New York in 2012. Today, TUED involves some 72 unions across 24 countries including four Global Union Federations, three regional trade union organisations and eight national union confederations. In addition, TUED involves academics, energy and social policy advocates and environmental organisations. TUED's aim is to build a global trade union community for energy democracy which 'promote[s] solutions to the climate crisis, energy poverty, the degradation of both land and people, and responds to … attacks on workers' rights and protections' (TUED 2019). Environmental justice is at the centre of TUED's activities, including in the organisation's promotion of solidarity between workers in the Global North and South and between unions that represent energy workers and those that are concerned about energy-related issues. TUED advocates for a progressive 'transformative' political position (see Sweeney and Treat 2018) in which just solutions to the climate crisis, energy poverty and environmental degradation are best achieved through 'democratic control and social ownership of energy'. They support 'reclaiming' the power sector from private enterprise who continue to profit from carbon pollution while also advancing the public ownership of new energy assets. These policy positions reflect the challenges experienced in securing just transition in local contexts where electricity privatisation has proven a major barrier (van Niekerk 2020; Snell 2020). Unions aligned with TUED have used their political associations with progressive parties and politicians in their home countries to advance these 'transformative' positions at a national level. In this regard, TUED's activities further illustrate the ways unions have sought to align local, national and international positions related to just transition, the transitioning of energy systems and the importance of public ownership in achieving climate change mitigation outcomes that balance the types of social and environmental priorities central to EJ.

## Conclusion

EJ researchers have taken a perfunctory interest in the role of trade unions with a common view being unions' involvement in environmental justice issues is 'limited and episodic' (Bailey and Gwyther 2010, 2). This chapter has challenged this misconception. Unions fought for improvements in working people's lives long before what is widely considered the birth of the EJ movement. If one accepts that working-class people suffer disproportionately from environmental harms and that this situation needs to be addressed (Loomis 2015; Faber 2017; Holifield et al. 2017; Rector 2018), then unions should surely be considered one of the most successful EJ movements the world has ever witnessed. However, this chapter has also argued that unions are sui generis from other EJ actors. Principally, this is due to their membership base and representation among workers who are involved in transforming the material world which may have harmful environmental impacts. While it is generally recognised that achieving EJ involves struggle, for unions this struggle is often as much internal as it is external as they balance environmental and job protection concerns. This dynamic is often shaped by industrial and local contexts resulting in the prospects of differences emerging between unions and within the same union at the local, national and global levels.

At times this also results in differences between unions and other EJ actors as unions balance justice for the environment, members and local communities. Union differences with other EJ actors, however, should not discredit them as an EJ actor or justify them being overlooked. As demonstrated by LA's Coalition for Clean and Safe Ports and the global Climate Justice movement the voice of unions has had positive influences on the positions and campaign strategies of EJ and CJ activists in not dissimilar ways that Martin Luther King, Jr. acknowledged in relation to the civil rights movement some 50 years ago. As EJ research advances into the future, EJ scholarship would also benefit from greater acknowledgement of union contributions to advancing EJ struggles.

# References

Alliance for Cancer Prevention. 2019. Supporting International Workers' Memorial Day 2019. Accessed October 3, 2019. https://allianceforcancerprevention.org.uk/2019/04/28/supporting-international-workers-memorial-day-2019/

Asbestos Council of Victoria. 2019. Asbestos Related Disease Facts and Figures Australia 2018. Accessed August 26, 2019. https://gards.org/asbestos-related-disease-facts-and-figures-australia-2018/.

Australian Asbestos Network. 2019. Role of Unions. Accessed August 26, 2019. https://www.australianasbestosnetwork.org.au/asbestos-history/battles-2/campaigners/role-unions/.

———. 2020a. The Battles: Battling James Hardie. Accessed April 8, 2020. https://www.australianasbestosnetwork.org.au/asbestos-history/battles-2/battling-james-hardie/.

———. 2020b. The Health Disaster: Growing Medical Awareness. Accessed April 8, 2020. https://www.australianasbestosnetwork.org.au/asbestos-history/health-disaster-2/growing-medical-awareness/.

Bailey, Janis, and Ross Gwyther. 2010. Red and Green: Towards a Cross-Fertilisation of Labour and Environmental History. *Labour History* 99 (November): 1–16.

Barca, Stefania. 2012. Bread and Poison: Stories of Labor Environmentalism in Italy, 1968–1998. In *Dangerous Trade: Histories of Industrial Hazard Across a Globalizing World*, ed. Christopher Sellers and Joseph Melling, 126–139. Philadelphia, PA: Temple University Press.

Bennett, David. 2011a. Health, Safety and Environmental Education at the Canadian Labour Congress. *New Solutions: A Journal of Environmental and Occupational Health Policy* 21 (2): 283–290.

———. 2011b. Labour and the Environment at the Canadian Labour Congress—The Story of the Convergence. In *Northern Exposures: A Canadian Perspective on Occupational Health and Environment*, ed. David Bennett, Charles Levenstein, Robert Forrant, and John Wooding, 29–35. Abingdon, UK: Routledge.

Burgmann, Verity. 1995. *Revolutionary Industrial Unionism: The Industrial Workers of the World in Australia*. Cambridge, UK: Cambridge University Press.
Climate Justice Alliance. 2020. About Climate Justice Alliance. https://climatejusticealliance.org/about/.
Coe, Neil, and David Jordhus-Lier. 2010. Constrained Agency: Re-evaluating the Geographies of Labour. *Progress in Human Geography* 35 (2): 211–233.
Cummings, Scott L. 2014. Preemptive Strike: Law in the Campaign for Clean Trucks. *UC Irvine Law Review* 4 (3): 938–1165.
———. 2018. *Blue and Green: The Drive for Justice at America's Port*. Cambridge, MA: MIT Press.
Cutter, Susan L. 1995. Race, Class and Environmental Justice. *Progress in Human Geography* 19 (1): 111–122.
Delautre, Guillaume, and Bruno Dante Abriata. 2018. *Corporate Social Responsibility: Exploring Determinants and Complementarities*. ILO Research Department Working Paper No. 38. Geneva: ILO. Accessed March 30, 2020. https://www.ilo.org/global/research/publications/working-papers/WCMS_654735/lang%2D%2Den/index.htm.
Diani, Mario. 2018. Unions as Social Movements or Unions in Social Movements? In *Social Movements and Organized Labour: Passions and Interests*, ed. Jürgen R. Grote and Claudius Wagemann, 43–65. London, UK: Routledge.
Dixon, Marc, Vincent J. Roscigno, and Randy Hodson. 2004. Unions, Solidarity, and Striking. *Social Forces* 83 (1): 3–33.
Estes, Steve. 2000. 'I Am a Man!': Race, Masculinity, and the 1968 Memphis Sanitation Strike. *Labor History* 41 (2): 153–170.
Faber, Daniel. 2017. The Political Economy of Environmental Justice. In *The Routledge Handbook of Environmental Justice*, ed. Ryan Holifield, Jayajit Chakraborty, and Gordon Walker, 61–73. Abingdon, UK: Routledge.
Featherstone, David. 2012. *Solidarity: Hidden Histories and Geographies of Internationalism*. London, UK: Zed Books.
Flanders, Allan D. 1970. *Management and Unions: The Theory and Reform of Industrial Relations*. London, UK: Faber.
Gereluk, Winston, and Lucien Royer. 2003. Sustainable Development: A Trade Union Perspective. *New Solutions: A Journal of Environmental and Occupational Health Policy* 13 (1): 35–41.
Gleeson, Brendan, and Nicholas Low. 1998. *Justice, Society and Nature: An Exploration of Political Ecology*. Melbourne, Australia: Routledge.
Gordon, Robert. 2004. *Environmental Blues: Working-Class Environmentalism and the Labor-Environmental Alliance, 1968–1985* (Ph.D. Dissertation) Wayne State University.
Greenhouse, Steven. 2010. Cleaning the Air at American Ports. *The New York Times*, February 25, 2010. https://www.nytimes.com/2010/02/26/business/26ports.html.

Haigh, Gideon. 2006. *Asbestos House: The Secret History of James Hardie Industries*. Melbourne, Australia: Scribe Books.

Harvey, Geraint, Andy Hodder, and Stephen Brammer. 2017. Trade Union Participation in CSR Deliberation: An Evaluation. *Industrial Relations Journal* 48 (1): 42–55.

Hofman, Peter S., Jeremy Moon, and Bin Wu. 2017. Corporate Social Responsibility under Authoritarian Capitalism: Dynamics and Prospects of State-Led and Society-Driven CSR. *Business and Society* 56 (5): 651–671.

Holifield, Ryan, Jayajit Chakraborty, and Gordon Walker. 2017. Introduction: The Worlds of Environmental Justice. In *The Routledge Handbook of Environmental Justice*, ed. Ryan Holifield, Jayajit Chakraborty, and Gordon Walker, 1–11. Abingdon, UK: Routledge.

Hyman, Richard. 2007. How Can Trade Unions Act Strategically? *Transfer: European Review of Labour and Research* 13 (2): 193–210.

Jacobs, David E., Jonathan Wilson, Sherry L. Dixon, Janet Smith, and Anne Evens. 2009. The Relationship of Housing and Population Health: A 30-Year Retrospective Analysis. *Environmental Health Perspectives* 117 (4): 597–604.

Kang, Susan L. 2012. *Human Rights and Labor Solidarity: Trade Unions in the Global Economy*. Philadelphia, PA: University of Pennsylvania Press.

King, Martin Luther, Jr. 1965. Speech to the Illinois AFL-CIO State Convention, October 7: Springfield, IL.

LAANE. 2008. Clean Trucks Program. Accessed March 23, 2016. https://laane.org/blog/timeline/2008-port-los-angeles-clean-truck-program/support-ctp-1/.

Larrubia, Evelyn. 2008. Labor, Environmentalists Unusual Allies. *Los Angeles Times*, November 27, 2008. https://www.latimes.com/archives/la-xpm-2008-nov-27-me-green27-story.html.

Loomis, Erik. 2015. *Out of Sight: The Long and Disturbing Story of Corporations Outsourcing Catastrophe*. New York, NY: The New Press.

Maher, Tony. 2016. Why Australia Needs a Just Transition. https://me.cfmeu.org.au/leadership-message/why-australia-needs-just-transition.

Malinowski, Beth, Meredith Minkler, and Laura Stock. 2015. Labor Unions: A Public Health Institution. *American Journal of Public Health* 105 (2): 261–271.

Markey, Raymond, and Joseph McIvor. 2019. Environmental Bargaining in Australia. *Journal of Industrial Relations* 61 (1): 79–104.

Martinez-Alier, Joan, Leah Temper, Daniela Del Bene, and Arnim Scheidel. 2016. Is There a Global Environmental Justice Movement? *The Journal of Peasant Studies* 43 (3): 731–755.

McCulloch, Jock, and Geoffrey Tweedale. 2008. *Defending the Indefensible: The Global Asbestos Industry and Its Fight for Survival*. Oxford, UK: Oxford University Press.

Montrie, Chad. 2008. *Making a Living: Work and Environment in the United States*. Chapel Hill, NC: University of North Carolina Press.

Morena, Edouard, Dunja Krause, and Dimitris Stevis, eds. 2020. *Just Transitions: Social Justice in the Shift Towards a Low-Carbon Economy*. London: Pluto Press.
Motavvef, Athena. 2020. 7 Reasons Why Asthma Is an Environmental Justice Crisis. Accessed April 13, 2020. https://www.weact.org/2017/05/7-reasons-asthma-environmental-justice-crisis/.
National Academies of Sciences, Engineering, and Medicine. 2017. *Communities in Action: Pathways to Health Equity*. Washington, DC: The National Academies Press. https://doi.org/10.17226/24624.
Osler Hampson, Fen, and Judith Reppy. 1996. *Earthly Goods: Environmental Change and Social Justice*. Ithaca, NY: Cornell University Press.
Peacock, Matt. 2011. *Killer Company: James Hardie Exposed*. Melbourne, Australia: Harper Collins.
Pellow, David N. 2007. *Resisting Global Toxics: Transnational Movements for Environmental Justice*. Cambridge, MA: MIT Press.
Perry, Matt. 2019. Unions Can—And Will—Play a Leading Role in Tackling the Climate Crisis. *The Conversation*, October 9. Accessed April 10, 2020. https://theconversation.com/unions-can-and-will-play-a-leading-role-in-tackling-the-climate-crisis-113226.
Port of Long Beach. 2019. The Port of Long Beach Clean Trucks Program. Accessed September 30, 2019. https://www.polb.com/environment/clean-trucks/#program-details.
Preuss, Lutz, Michael Gold, and Chris Rees, eds. 2015. *Corporate Social Responsibility and Trade Unions: Perspectives Across Europe*. Oxford, UK: Routledge.
Räthzel, Nora, and David Uzzell. 2011. Trade Unions and Climate Change: The Jobs versus Environment Dilemma. *Global Environmental Change* 21 (4): 1215–1223.
Rector, Josiah. 2014. Environmental Justice at Work: The UAW, the War on Cancer, and the Right to Equal Protection from Toxic Hazards in Postwar America. *The Journal of American History* 101 (2): 480–502.
———. 2018. The Spirit of Black Lake: Full Employment, Civil Rights, and the Forgotten Early History of Environmental Justice. *Modern American History* 1: 45–66.
Roelofs, Cora, Sherry L. Baron, Sacoby Wilson, and Aaron Aber. 2017. Occupational and Environmental Health Equity and Social Justice. In *Occupational and Environmental Health*, ed. Barry S. Levy, David H. Wegman, Sherry L. Baron, and Rosemary K. Sokas, 23–39. Oxford, UK: Oxford University Press.
Rosa-Aquino, Paola. 2019. What the Environmental Justice Movement Owes Martin Luther King, Jr. https://grist.org/article/what-the-environmental-justice-movement-owes-martin-luther-king-jr/.
Rose, Janna. 2016. Biopiracy: When Indigenous Knowledge Is Patented for Profit. *The Conversation*, March 8. Accessed April 13, 2020. https://theconversation.com/biopiracy-when-indigenous-knowledge-is-patented-for-profit-55589.
Rosemberg, Anabella. 2013. Developing Global Environmental Union Policies through the International Trade Union Confederation. In *Trade Unions in the*

*Green Economy: Working for the Environment*, ed. Nora Räthzel and David Uzzell, 15–28. Abingdon, UK: Routledge.

Routledge, Paul, Andrew Cumbers, and Kate Driscoll Derickson. 2018. States of Just Transition: Realising Climate Justice through and Against the State. *Geoforum* 88: 78–86.

Ruff, Kathleen. 2012. Canada's Role as Producer, Exporter and Defender of Asbestos. *Women & Environments International Magazine Spring* 2012 (90/91): 20–22.

Ruff, Kathleen, and John Calvert. 2014. Rejecting Science-Based Evidence and International Co-Operation: Canada's Foreign Policy on Asbestos under the Harper Government. *Canadian Foreign Policy Journal* 20 (2): 131–145.

Schlosberg, David. 2004. Reconceiving Environmental Justice: Global Movements and Political Theories. *Environmental Politics* 13 (3): 517–540.

———. 2007. *Defining Environmental Justice: Theories, Movements, and Nature.* Oxford, UK: Oxford University Press.

———. 2013. Theorising Environmental Justice: The Expanding Sphere of a Discourse. *Environmental Politics* 22 (1): 37–55.

Schlosberg, David, and Lisette B. Collins. 2014. From Environmental to Climate Justice: Climate Change and the Discourse of Environmental Justice. *WIREs Clim Change* 5: 359–374.

Sicotte, Diane M., and Robert J. Brulle. 2017. Social Movements for Environmental Justice through the Lens of Social Movement Theory. In *The Routledge Handbook of Environmental Justice*, ed. Ryan Holifield, Jayajit Chakraborty, and Gordon Walker, 25–36. Abingdon, UK: Routledge.

Silverman, Victor. 2006. 'Green Unions in a Grey World'—Labor Environmentalism and International Institutions. *Organization & Environment* 19 (2): 191–213.

Snell, Darryn. 2020. Just Transition Solutions and Challenges in a Neoliberal and Carbon-Intensive Economy. In *Just Transitions: Social Justice in the Shift Towards a Low-Carbon Economy*, ed. Edouard Morena, Dunja Krause, and Dimitris Stevis, 198–218. London: Pluto Press.

Sobczak, André, and Christelle Havard. 2015. Stakeholders' Influence on French Unions' CSR Strategies. *Journal of Business Ethics* 129 (2): 311–324.

Stevis, Dimitris, Edouard Morena, and Dunja Krause. 2020. Introduction: The Genealogy and Contemporary Politics of Just Transition. In *Just Transitions: Social Justice in the Shift Towards a Low-Carbon Economy*, ed. Edouard Morena, Dunja Krause, and Dimitris Stevis, 1–32. London: Pluto Press.

Sweeney, Sean, and John Treat. 2018. *Trade Unions and Just Transition: The Search for a Transformative Politics*. TUED Working Paper #11. Accessed October 3, 2019. http://unionsforenergydemocracy.org/resources/tued-publications/tued-working-paper-11-trade-unions-and-just-transition/.

Thompson, Edward Palmer. 1991. *The Making of the English Working Class*. Toronto, Canada: Penguin Books.

TUED. 2019. About Trade Unions for Energy Democracy. Accessed September 15, 2019. http://unionsforenergydemocracy.org/.

UN. 2015. *Paris Agreement*. United Nations Treaty Collection.

Vachon, Todd E., and Jeremy Brecher. 2016. Are Union Members More or Less Likely to Be Environmentalists? Some Evidence from Two National Surveys. *Labor Studies Journal* 41 (2): 185–203.

Van Niekerk, Sandra. 2020. Resource Rich and Access Poor: Securing a Just Transition to Renewables in South Africa. In *Just Transitions: Social Justice in the Shift Towards a Low-Carbon Economy*, ed. Edouard Morena, Dunja Krause, and Dimitris Stevis, 132–150. London: Pluto Press.

Walker, Gordon. 2012. *Environmental Justice: Concepts, Evidence and Politics*. Abingdon, UK: Routledge.

Webb, Sidney, and Beatrice Webb. 1920. *The History of Trade Unionism*. London, UK: Longmans and Company.

Zeitlin, Jonathan. 1987. From Labour History to the History of Industrial Relations. *The Economic History Review* 40 (2): 159–184.

# Part II

Seeking Common Ground

# 8

# 'Beware of the Crocodile's Smile': Labour-Environmentalism in the Struggle to Achieve a Just Transition in South Africa

Jacklyn Cock

## Introduction

The chapter focuses on the potential of cooperation between trade unions, environmental organisations and community groupings to drive a transformative concept of a 'just transition' from fossil fuels. It shows how each social force has different priorities: mainly preventing job losses for a fractured and diverse labour movement, resisting the dispossession of land and livelihoods for mining-affected communities and the protection of nature for environmentalists. In contrast to those in the Global North who see the climate crisis as driving the imminent collapse of industrial civilisation, the chapter describes how, for some in the Global South, it is the closure of the coal mines and coal-fired power stations which is regarded as potentially catastrophic. The chapter points to four areas of innovative worker agency in South Africa, but emphasises the difficulties involved in promoting a 'labour environmentalism', particularly the continuing power of the 'minerals-energy complex' as well as fractures within both the labour and environmental movements and tensions between them. It concludes that the challenge to labour and environmental justice activists is to support building an alternative development path opposed to extractivism in coal mining-affected communities. These

J. Cock (✉)
University of the Witwatersrand, Johannesburg, South Africa
e-mail: Jacklyn.Cock@wits.ac.za

communities are the worst affected and the least considered in the current debates and could form the bedrock of a mass movement for a just transition. The chapter prioritises their experiences and understandings.

Recently a warning of 'the crocodile's smile' was made by a powerful trade union leader about the response of the environmental movement to the climate crisis.[1] It signals the difficulties in building alliances between trade unions and environmental justice organisations. The concept of 'labour-environmentalism' is a container of competing interests. All over the world energy transitions are highly contested and complex processes, but in South Africa, as in much of the Global South and parts of the Global North, the difficulties are compounded by the continued dominance of the minerals-energy complex.

## The Power of the Minerals-Energy Complex in the South African Energy Landscape

Contrary to the massive protest movement gathering momentum in the Global North, the current post-apartheid state is heavily involved in promoting extractivism. The coal-dominated electricity sector is an important component of the minerals-energy complex (MEC) system of accumulation which continues to dominate the economy and which has historically relied on cheap coal and cheap labour (Fine and Rustomjee 1996). It 'encompasses critical links and networks of power between the financial sector, government, the private sector and parastatals such as the Industrial Development Corporation and Eskom' (Baker et al. 2015: 8). With coal providing 95% of its electricity, South Africa is the largest carbon emitter in Africa and the fourteenth largest in the world.

A powerful body, the Mining Council (earlier called the Chamber of Mines), promotes the notion of coal mining and burning as the essential road to economic growth and stability and is seldom challenged. A recent report is deliberately aimed at countering 'negative public opinion about coal' and emphasises the creation of jobs, procurement spending and the 'adverse' effects of strict environmental laws (COM 2018: 13).

---

[1] The term 'movement' is highly contested. In this chapter the term 'labour movement' is used to refer to trade unions of a great variety of organisational forms and ideological persuasion, not all of which are connected through labour federations such as COSATU, FEDUSA, SAFTU and NACTU. The same usage applies to the 'environmental movement'. Neither has a coherent centre or tidy margins. Both are inchoate sums of multiple, diverse, uncoordinated struggles and organisations.

Eskom, the state-owned energy provider, is at the centre of the tensions between closing the coal mines to reduce carbon emissions and protecting existing jobs. This is a dilemma for a government of 'fragile stability' which is under pressure to deliver a commitment made in 2009 to reduce carbon emissions drastically. Eskom, which has always been committed to coal, is in crisis. Mismanagement, large-scale corruption and cost overruns in building two of the largest coal power stations in the world, Kusile and Medupi, have pushed Eskom to the point of collapse with heavy debt (R450 billion at present) and frequent black outs (euphemistically termed 'load-shedding') throughout the country. The 'unbundling' of Eskom into three separate units—generation, transmission and distribution—is currently underway and is strongly opposed by the unions, particularly National Union of Mineworkers (NUM) and National Union of Metalworkers South Africa (NUMSA), acting in unity on this issue to prevent job losses. The closure of older coal mines could be accelerated if the government agrees to a proposal to access concessional climate change finance to address Eskom's liquidity crisis (Paton 2019: 2). In several such proposals loans to finance Eskom's debt are conditional on the closure of coal plants to make way for renewable energy.

At the same time 'There is a new and promiscuous intimacy between governments and mining companies globally' (Marshall 2015: 64). This intimacy is at the centre of the National Development Plan, which is also promoting a 'social dialogue' approach to the climate crisis in a series of workshops convened by the National Planning Commission (NPC) over the past four years including some key actors connected to the MEC but has failed to attract labour. Several trade unionists commented that this NPC process 'had nothing for labour'. The emphasis in four national workshops was on building a shared vision and agenda between 'social partners' and on managing 'an orderly transition to avoid social and economic disruption', as one participant expressed it. Writing of the global scene, Sean Sweeney maintains that 'The political success of this (social dialogue) approach is largely due to the just transition being defined in a benign and non-confrontational way which poses little or no challenge to the mainstream, pro-growth, business-dominated narrative, a narrative that was largely created by the liberal wing of the global corporate elite' (Sweeney and Treat 2018: 27). In their view the social dialogue approach is simply not up to the task of bringing about the kind of revolutionary, transformative change that the climate crisis requires.

The continuing power of the MEC is a major obstacle. Power is concentrated in the Energy Intensive Users group of three dozen transnational corporations in the extractives and smelting sector. They consume over 40% of energy generated by Eskom and 'continue to dominate the drafting of energy policy in their favour, including pricing' (Swilling 2015: 13). 'The basic problem is national-monopolistic Eskom for closing all routes to a genuinely just transition and greening of industry, since they keep insisting on extreme base load for extremely subsidized power to extract and remove South African minerals resources for transnational corporations' (Interview, Patrick Bond, Johannesburg 14.3.2018). As Bond argues, 'attempts to reform the MEC from the inside are only as strong as the climate justice revolt that grows from the outside' (Interview 9.9.2019). This is the context in which the president of the South African Federation of Trade Unions (SAFTU) has referred to 'a highly indebted, dysfunctional coal power system'. He maintains that, 'the lack of climate action from our government means we must take to the streets, build a broad and diverse movement and demand action for a just transition from fossil fuels. … It is time for labour and the climate justice movement to stand together for a just and sustainable world' (Vavi 2019: 5). However, such 'standing together' presents formidable challenges.

## Fault Lines in the Labour and Environmental Movements

During the apartheid regime environmentalism operated effectively as a conservation strategy which was mainly concerned with the protection of threatened plants, animals and wilderness areas, and neglected social issues (Cock and Koch 1991; Agyeman 2005). Furthermore, for many black South Africans environmentalism was linked to dispossession as thousands were forcibly removed to create national parks as well as 'protected areas' and in the process they lost their land and livelihoods (Walker 2008).

The notion of environmental justice represents a dramatic shift away from this traditional authoritarian concept of environmentalism. It is linked to social justice as 'an all-encompassing notion that affirms the value of all forms of life against the interests of wealth, power and technology' (Castells 1997: 132). It provides a radical alternative to the discourse of ecological modernisation, which has been criticised for its reformism and overemphasis on consensual politics (Warner 2010). Environmental justice puts the needs of the poor and excluded at the centre of its concerns, particularly in relation to the impacts of the climate crisis. It rejects the market's ability to bring about social justice or environmental sustainability, and thus represents a powerful challenge to the economic growth model and the increasing commodification and financialisaton of nature packaged as 'the green economy' (Martinez-Alier 2014). The outcome is that the relationship between environmental justice and mainstream environmentalism has 'traditionally been uneasy' (Agyeman

2005: 1). Operating as a network, the key organisation of the environmental justice movement, Earthlife, was launched 30 years ago and is mobilising diverse grassroots communities on a variety of rights and claims, some of which have a constitutional grounding such as the Bill of Rights, section 24, which states that 'everyone has the right to an environment which is not harmful to their health and wellbeing' (Constitution of the Republic of South Africa 1996: Chapter 2). As a movement it is fuelled by a growing tension between the discourse of rights and the experience of unmet needs among the 60% of South Africans living in poverty. The possibility of realising a radical and transformative just transition from fossil fuels lies in an alliance between the environmental justice and labour movements, especially at the grassroots level.

## Fractures Within the Labour Movement and Just Transition

There are two broad approaches towards the nature of change in the concept of a just transition: a minimalist position that is primarily defensive, emphasising shallow, reformist change, the social protection of vulnerable workers, with green jobs and 'green growth', and an alternative notion involving deep, transformative change to produce a more just and equal society.

Beginning in 2010 the labour movement in South Africa, particularly the labour federation of 22 different unions, Congress of South African Trade Unions (COSATU), played a key role in introducing and promoting a transformative understanding of a 'just transition'. The concept was both grounded in peoples' lived experience and aspirational; it was at the heart of a powerful narrative of hope for a more just and sustainable world, a compass for alternative forms of producing, consuming and relating to nature. However, since the expulsion of the National Union of Metalworkers of South Africa (NUMSA) from COSATU and the formation of a new labour federation, SAFTU divisions within the movement have deepened. In the past few years they have retreated into a defensive position, focusing on protecting existing jobs. This protective stance is wholly comprehensible in light of South Africa's unemployment rate of 40% (one of the highest in the world), massive job losses in the mining industry (particularly gold and platinum) and high levels of poverty.

However, the power resources approach emphasises worker agency. For example, Schmalz et al. (2018) distinguish between structural power referring to the position of wage earners in the economic system; associational power derived from trade unions or collective political organisations which potentially promote solidarity and coordination; societal or symbolic power meaning cooperation with other social groupings and includes coalition power which involves shared networks which can be mobilised for collective action and make appeals to public consciousness; and institutional power which resides in the state regulating bodies, such as labour laws and tripartite institutions.

A crucial question now is whether, drawing from these power resources (structural, associational, institutional and societal) and through forging closer connections with two other social forces—the environmental justice movement and coal-affected communities—labour could reclaim the concept and drive a just transition. In doing so labour could manage the tensions between promoting workers' long-term interests through supporting the closure of coal mines to reduce the carbon emissions which are driving climate change, and protecting their short-term interests in the form of support for retrenched workers and coal-dependent communities.

Between 2010 and 2016 labour driven by the Congress of South African Trade Unions (COSATU) developed strategic initiatives in four key areas:

## Areas of Innovative Worker Agency

### Engaging the Binary Between Jobs or Environmental Protection

A major obstacle to an alliance between 'red' and 'green' social forces to drive a just transition is the conventional binary which poses the issue as a choice between protecting jobs or nature. Drawing from the British trade union experience, COSATU contributed substantially to the initiation of a Climate Jobs Campaign in 2011 which engaged in extensive research and has identified over one million new, alternative 'climate jobs', meaning specifically 'those that help to reduce the emissions of greenhouse gases and build the resilience of communities to withstand the impact of climate change' (Ashley 2018: 27). Examples include developing renewable energy plants, public transport and small-scale organic agriculture. However, one trade unionist insisted that the climate jobs campaign 'to have any traction … should be driven by labour and housed within the new labour federation South African Trade Union Federation of Trade Unions (SAFTU)', which was formed in opposition to COSATU (key informant interview, Johannesburg 14.3.2018).

In challenging the binary between jobs and environmental protection through the Climate Jobs Campaign the emphasis was on the synergies as captured in the current slogan: 'No jobs on a dead planet'.

In 2011 the single biggest trade union in South Africa, the National Union of Metal Workers of South Africa (NUMSA), with more than 338,000 members passed a resolution in favour of social ownership of renewable energy as part of developing 'a revolutionary and class approach to climate change' (NUMSA Press Statement 19.8.2011). It has been the most vocal proponent of the social ownership and control of energy and consistently opposed the state programme of privatised renewable energy. Instead, it has strongly promoted the notion of energy democracy as a building block towards socialism.

## Attempts to Unify the Labour Movement on the Relation Between Capitalism and Climate Change

The labour movement recognised that the climate crisis was an opportunity to demonstrate that the expansionist logic of capitalism means it is not only unjust but also unsustainable. The first principle of the 2011 COSATU climate change policy framework stated that 'Capitalist accumulation is the underlying cause of excessive greenhouse gas emissions' (COSATU 2011: 53). Endorsed by the COSATU Central Committee this document was based on 15 principles linking sustainability and justice and was widely distributed. It has been pointed out that, 'While this framework has not been abolished, it did not feed into any specific union policies' (Räthzel et al. 2018: 24). However, labour's focus on the unjust and unsustainable nature of capitalism opened up an intense debate on alternatives, particularly a new form of ecological, ethical and democratic socialism.

This is significant because for many people socialism is discredited because its claims have been marred by a history of authoritarianism, productivism, human rights abuses and environmental damage. But the labour movement promoted debates of an alternative socialist vision, as elaborated by writers such as Michael Löwy who wrote of 'a new eco-socialist civilization, beyond the reach of money, beyond consumption habits artificially produced by advertising, and beyond the unlimited production of commodities that are useless and or harmful to the environment' (Lowry 2006: 302). Other authors such as Kelly and Malone (2006) have stressed that eco-socialism is necessary because capitalist expansion threatens human survival.

The argument that capitalism is at the root of the current climate crisis was reflected in a joint (COSATU, National Council of Trade Unions [NACTU] and Federation of Unions of South Africa [FEDUSA]) submission in 2012 in response to the government's Green Paper on Climate Change. The submission stated that

> We are confident that any efforts to address the problems of climate change that does not fundamentally challenge the system of global capitalism is bound to fail, and to generate new, larger and more dangerous threats to human beings and our planet. Climate change is caused by the global private profit system of capitalism. Tackling greenhouse gas emissions is not just a technical problem. It requires a fundamental economic and social transformation to substantially change current patterns of production and consumption. (National Council of Trade Unions (NACTU) 2012: 2)

## Establishing Connections Between Labour and the Environmental Justice Movement

Given that the climate crisis is deepening and that workers in the extractive sector are most vulnerable in a shift away from coal to a new energy regime, it could be expected that the most powerful driver of a transformative just transition would be a cooperation between the labour and environmental movements, a 'red-green' alliance. Furthermore, South Africa has a tradition of 'social movement unionism', meaning a form of unionism that goes beyond workplace issues and engages actively in political and community struggles. These unions challenged minority rule and the lack of social infrastructure in the working-class communities of countries such as Brazil and South Africa. This new wave of worker militancy was labelled social movement unionism as it blurred the demarcation between unions as formal organisations and social movements as 'loosely structured networks of action' (Schmalz et al. 2018: 113). In the Global South this involved struggles not only over wages and working conditions, but also over living conditions and social services in working-class households (Seidman 1994: 2–3; Webster 2018: 174–196).

Besides social movement unionism the labour moment could draw from several iconic moments of 'red-green collaboration'. One such moment was the exposure of pollution by Thor Chemicals which imported toxic waste into South Africa. The environmental justice organisation, Earthlife Africa, worked closely with affected communities and local workers particularly with the Legal Resources Centre and the Chemical Workers Union in conducting research and public education. The Chloorkop campaign against the siting of a toxic waste dump between 1993 and 1994 was another watershed 'rainbow' event which involved collaboration (Cock and Koch 1991: 209).

However, today there are no formal coalitions or alliances between labour and environmental justice organisations. The best example of a formal relationship took the form of a reference group established by the research arm of COSATU in 2011 on which all 22 affiliate unions were represented, as well as activists from key environmental justice organisations. Drawing on 'coalition power', this formed the embryo of a red-green coalition to drive a just transition.

Over a six-year period this very active grouping engaged in

1. Workshops of popular education on climate change and the necessity of a just transition with all the COSATU affiliate unions as well as public seminars on concrete issues affecting the everyday experience of the working class such as rising food and energy prices and water shortages. Many discussions involved an understanding of the capitalist system as the cause of climate change, and a concern that the labour movement should not be seduced by the notion of a 'green economy' with its promise of 'green growth' and 'green jobs' (Cock 2014).
2. Research on developing policy responses to climate change in a range of industries and on energy policy alternatives culminated in the COSATU climate change policy framework supporting a just transition.

In hindsight, perhaps this group focused too much on questions of principle rather than on strategy. The focus was on understanding the nature, cause and effects of climate change rather than on the content and the modalities of a just transition. Furthermore, there were differences between the National Union of Mineworkers (NUM) and the National Union of Metalworkers (NUMSA) which should have been addressed more directly. The NUM was increasingly defensive of the interests of coal miners, in the face of the threats of job losses from mine closures, falling coal prices, mechanisation, absolutist demands from environmental activists like 'Keep the coal in the hole' and the divestment movement. Then (and now) the NUM continues to argue for 'clean coal' from expensive and untested technological innovations such as Carbon Capture and Storage.

## Labour and Coal Workers

In opposing closing of the coal mines the NUM is resuscitating the old 'jobs versus environment' binary. The NUM, formerly South Africa's largest trade union, is key to achieving a just transition, but is struggling with a loss of credibility and support, as well as competition from the formation of the Association of Mining and Construction Workers Union (AMCU), with membership dropping to 187,000. However, since 2018 NUM has been active in marches and picketing against the possibility of job losses for its members at Eskom. In numerous interactions NUM officials have supported the expansion of coal mining, protested job losses from mine closures, promoted controversial 'clean coal' technologies such as Carbon Capture and Storage, and even opposed the closure of old coal-fired power stations on the grounds that they 'can be rebuilt to extend their lives' (NUM President 19.6.2019). On at least one occasion a NUM spokesman opposed the notion of a just transition on the grounds that it is a 'northern notion' and inappropriate as 'coal is part of our African culture'. Labour has played a minimalist part in local anti-coal initiatives, and NUM is often described as 'pro-mining' and 'uncaring about the people'.

Clearly, a just transition requires extensive planning both regionally and nationally. But a lack of preparation is also evident in the labour movement and is in strong contrast to the 1987 National Union of Mineworkers (NUM) strike which led to 40,000 workers losing their jobs. On that occasion the NUM set up a job-creating programme to establish cooperatives and a Mobile Job Creation Unit with a 20-tonne truck that delivered training to mineworkers during the retrenchment processes that followed the strike. The union organised contact groups for the miners facing retrenchment; they discussed future livelihood strategies and how they would invest their retrenchment packages (Philip 2018).

Since those initiatives the notion of a red-green coalition to drive a just transition has faded. There is no single environmental justice movement and no single, coherent 'labour environmentalism'. Some elements of the labour movement remain sceptical of a just transition. Some unions, such as the Federation of Democratic Unions of South Africa (FEDUSA) affiliates and Solidarity, are largely silent on climate change. The Climate Jobs Campaign is now driven largely from outside the labour movement, by the Alternative Information and Development Centre, and has evoked some scepticism from labour activists. Despite extensive debate within intellectuals in the labour movement there is little shared understanding of a just transition today. This emerged clearly at the March 2018 'National Labour Climate Change Conference' convened by National Labour and Economic Development Institute (NALEDI) and attended by representatives of the three main labour federations: Confederation of South African Trade Unions (COSATU), Federation of Unions of South Africa (FEDUSA) and National Council of Trade Unions (NACTU). There was no expression of a clear vision of a world without coal, or of a world without full employment in wage labour. However, both the National Union of Metalworkers of South Africa (NUMSA) and the South African Federation of Trade Unions (SAFTU) were absent on this occasion and both are closer to a 'social power' approach. As Sweeney and Treat write, 'Unions must develop transitional strategies that are anchored in a paradigm of sharing solidarity and sufficiency. A just transition is possible but it will have to be demanded and driven forward by a broad democratic moment, with unions playing a key role' (Sweeney and Treat 2018: 43).

The labour movement is not against renewables but increasingly defensive and adamant that the state's privatised renewable energy policy is a threat because it will involve job losses and increased energy prices. At the same time the environmental movement is increasingly unwavering about the immediate closure of coal mines and coal-fired power stations and a shift to renewable energy (in whatever form) as essential to a just transition. It does not acknowledge that a deep just transition requires changes not only in the sources of energy, but also in who owns and controls various components of the energy system (Overy 2018: 8). Recently 'strong red-green' tensions and animosities have surfaced, expressing the 'rancour and open hostility between labour and environmental activists at worst' (Uzzell and Räthzel 2013: 1).

## Red-Green Tensions

Conflict first surfaced in 2017 when coal truck owners sent a convoy of about 100 coal trucks to blockade Pretoria. The Coal Transporters Forum announced they were against the new privatised renewable energy programme because it would create surplus energy capacity forcing Eskom to close five power stations as a result of which 30,000 jobs would be lost. In March 2018 NUMSA obtained an urgent court interdict to block Eskom from signing renewable energy contracts with 27 independent power producers. NUMSA argued that the contracts would be detrimental to the working class because electricity prices would rise. As a result, Eskom would need less coal to produce electricity, which would lead to the closure of coal-fired power plants and the loss of thousands of jobs. NUMSA maintained that 'our starting point should be to protect, expand and democratise the current capacity for manufacturing. (…) Whatever measures we are taking should champion a jobs-led industrial strategy which is needed to build a modern economy. In the energy sector, the goal should be a just transition from fossil fuels without destroying jobs' (Irvin 2017). Deputy General Secretary Karl Cloete said the NUMSA position is that 'renewable energy has great potential to give communities greater control of their resources and to satisfy their energy needs on a decentralised basis. We were and remain absolutely clear that renewable energy is essential to mitigate climate change' and should be socially owned and democratically controlled (Cloete 2018). This view has been endorsed by the rival labour federation COSATU and the South African Federation of Trade Unions (SAFTU) with 29 affiliated unions with a combined membership of nearly 800,000 (Declaration of the Inaugural Central Committee of SAFTU November 2019).

SAFTU vowed to 'mobilise workers to oppose Eskom's planned closure of five coal-powered power stations which could produce 30–40,000 job losses, fight the partial privatisation of Eskom by involving independent power producers, step up the campaign against nuclear energy and develop a position on transition to socially owned renewable energy'. NUM Eskom employees in the Free State region marched in Bloemfontein to demand that Eskom stops doing business with Independent Power Producers (NUM Press Statement 26.3.2018). COSATU in its application on the closure of coal-mined power stations, as well as the signing of the renewable energy IPP programmes, 'has emphasised the need for the economy to reduce reliance on fossil fuels through the use of alternative green energy sources such as water, wind and sunlight. However, the transition to a low carbon economy must address the issues of jobs, ownership and control as well as localisation' (Interview COSATU official 30.4.2018). NUM threatened to mobilise society 'and embark on a programme of destabilizing society should the government continue with the jobs bloodbath, in Eskom and among government employees' (NUM statement 24.10.2018).

Different groupings in the environmental movement reacted very differently to the NUMSA court interdict, variously emphasising the costs of Medupi and Kusile (the two largest coal power stations in the world currently under construction) and the environmental and health impacts of coal (2000 deaths a year are attributed to Eskom's coal-fired power stations). Earlier in the month the campaign, Life After coal (an alliance composed of Earthlife, the Centre for Environmental Rights and Groundwork) had written to the minister requesting a meeting and calling for 'urgent steps to ensure a rapid but just transition away from coal and towards publicly-owned renewable energy' (Bobby Peek, Director of Groundwork 2.3.2018).

Some comments from individuals were hostile and accusatory, describing the court application as 'mad', 'as abusing NUMSA', alleging that 'the trade union leadership in this country have lost the plot completely', as meant 'to sabotage renewable energy in favour of coal', 'stand in the way of progress' and other scattered observations. One publication referred to NUMSA's court action as 'an attempt to suppress the growth of renewable energy generation' and 'short sighted and futile' (*Business Day*, 14.3.2018). Such comments along with disinvestment campaigns from 350.org and slogans such as 'keep the coal in the hole' do not reflect workers' interests. NUMSA later clarified its position. Irwin Jim said on national television news on 13 March 2018: 'We are not against renewable energy but it must be socially owned. We need an energy mix but must have a just transition.' He later referred to the need to 'defend NUMSA from enemies with a crocodile smile' (Facebook post 18.3.2018). Clearly these tensions undermine the possibility of establishing a strong red-green coalition to drive a just transition. Furthermore, these tensions are surfacing in local community struggles, often fuelled by mining corporations with (often false) promises of employment which is equated with development. Jobs were at the centre of the conflict when the Mfolozi Community Environmental Justice Organisation held a peaceful and legal demonstration outside the Pietermaritzburg High Court in 2018 to draw attention to their strong opposition to the extension of the Tendele coal mine. The mining company bussed in pro-mining protestors and claimed resisters were anti-development.

## Illustration of Tensions Between Mining and Environmental Protection: The Mabola Protected Environment

This is an area of wetlands, pans and grasslands and the source of three major rivers. It is a key water resource area with high biodiversity. It is one of the few areas in the world which contains three different varieties of cranes (crowned, wattled and blue). To prevent coal mining, it was declared a protected area in 2014 but in 2016 the Ministers of Environmental Affairs and of Mineral Resources quietly granted permission for an Indian mining company Atha Africa's proposal to develop an underground coal mine within the protected area. Two of the trustees were related to the then President Zuma. Atha claimed that the mine would generate 500 jobs. A coalition of 8 different environmental organisations was formed to resist the decision on the grounds that the impact of mining activities would be 'environmentally catastrophic'.

A key environmental justice organisation, the Centre for Environmental Rights (CER), was appointed to act for the coalition and in January 2019 the North Gauteng High Court refused an application for leave for Atha Africa to appeal its decision to overturn the government's approval for the coal mine. The senior vice-president of Atha Africa has posted tweets challenging opposition to his coal project on the grounds that it frustrated development opportunities. He accused the CER of 'economic sabotage and treason', calling the head of CER a 'Liar in chief' (https://www.groundup.org.za) (Groundup 29.10.2018). Most seriously Atha Africa helped to organise a group of pro-mine activists and bussed a group 300 km from Volksrus to protest at the CER offices in Johannesburg. 'The community is becoming a battlefield' (key informant interview, Wakkerstroom 3.8.2018).

The case points to several themes that have surfaced in other coal struggles at the community level such as

- The absence of organised labour
- The increasing power of judicial activism by the emerging environmental justice movement
- The need to challenge a corrupt and dysfunctional state
- Ignorance of the environmental impact of coal mining and the importance of water resources
- The desperation for jobs especially in areas of high unemployment among young black men
- The tendency for mining companies to promote jobs as a signifier of development.

The most important challenge to labour and environmental justice activists is to build an anti-coal social movement which means focusing on mining-affected communities. In numerous elite debates and acrimonious exchanges, the voices of the coal workers and those living in coal-affected communities were not being heard.

The Mabola case stimulated the need to allow grassroots voices in threatened areas to be heard. As a consequence, a 2019 pilot project of the Institute of Society Work and Politics (SWOP) at the University of Witwatersrand, Johannesburg, involved eight exchange workshops and interviews with key informants in 3 Mpumalanga communities where most of the coal-fired power stations and coal mines are situated. The exchange workshops were organised by grassroots organisations and involved an exchange of two types of knowledge, *experiential*, meaning community participants sharing the problems of living in a mining-affected community, and *empirical*, meaning information from SWOP activist-researchers about other struggles, new policy developments and various understandings of a just transition. The aim was on empowering community members with the information and confidence to formulate demands and participate in the struggle to ensure that the transition from coal is just and transformative. This approach sought to replace what Mazibuko Jara has termed 'extractivist' research methods with cooperative processes involving community members at every stage. This involves the co-production of knowledge and is in tune with decolonial (and feminist) approaches to research which emphasise reflexivity and dialogic learning, validating lived experience, sharing and reciprocity (Otto and Terhorst 2011).

## Bringing in Coal Mining-Affected Communities

Most of the poor, black communities living close to the operational coal-fired power stations and open-pit working or abandoned mines are experiencing the direct damage to their health due to air pollution. In addition, they are dealing with forced removals; social dislocation and dispossession; loss of their land-based livelihoods, ranging from producing food, keeping livestock such as cattle, goats and chickens, threats to food security and limited access to clean water; violation of their ancestral graves; and inadequate consultation in the awarding of mining licences.

Coal dominates many of the communities especially in Mpumalanga where the majority of Eskom coal plants and mines are located and which has been described as 'the place where most of the challenges of a just transition are situated' (Interview Tasneem Essop Middleburg, 13.2.2019).

But while there is extensive collection action in these little towns it is not generally directed against coal per se as a form of extractivism. Especially in Mpumalanga, during the exchange workshops, there was a frequent rejection of the central charge implied in the notion of a just transition, namely the closure of the coal mines and coal-fired power stations. Most collective actions such as protest marches and roadblocks and burning down of municipal buildings (as in the case of the Phola township) by protestors angry about land allocated to mining, rather than to housing, are protests about how mining corporations or the local state or Eskom operate. These generate claims and demands regarding issues such as employment practices (particularly the neglect of local labour), compensation for relocation of ancestral graves, damage to homes from blasting, the loss of land and land-based livelihoods, water pollution and the lack of consultation about how mining licences are allocated. There is little generic critique of coal as a means of accumulation at the grassroots level and limited understanding of the climate crisis.

Furthermore, these mining-affected communities are not homogeneous socially or ideologically and frequent factionalised struggles occur, which mining corporations often manipulate to further divide them and promote their own interests. Further research is needed on the extent to which coal workers are integrated in these communities. Nationally, the majority of coal miners are migrants, living far from their homes (Burton 2018). They clearly benefit from their wages but some have expressed an uneasiness about the negative impacts of coal pollution, particularly air quality.

Residents know that the air pollution from coal makes them sick but there is a material dependence on coal which takes two main forms: the coal mines and power stations contain the possibility of jobs, and most of the women in these communities survive by informal sector activities and coal miners provide most of the customers.

This local dependence on coal generates a kind of socially complex, ambivalent resistance. For example, X is active in the grassroots organisation, Mining Affected Communities United in Action (MACUA), and participated in the mass protest of 5000 people organised by MACUA at the giant coal power station, Kusile, but she sells cooked foodstuffs outside the mine and her husband is employed there. Others rent backyard rooms to migrant coal miners, wash clothes, repair cars, run small roadside shops and shebeens, do cleaning work and rely on coal workers as the main consumers of these informal sector activities. These forms of dependence create a captive imaginary which makes it difficult to conceptualise a just transition to a world without coal. For some the notion of a 'just transition' is simply declarative, empty of substantive content with no relation to everyday life. For many it has had catastrophic implications.

In answer to that precise question—What would a world without coal look like?—most residents answered in catastrophic terms. For example, 'it would be a dark and dangerous world full of crime and hunger'. The majority of informants do not want the coal mines to close, for a range of reasons. One woman said, 'if the mine closes how will I get compensation for the damage to my house from blasting?'

To date, trade unions have largely failed to engage with these mining-affected communities, but despite this, resistance to coal is increasing and many new grassroots organisations are being formed and taking the form of self-organising horizontal networks of 'coalition power'. Several were established because of empowering work by environmental justice organisations such as Groundwork, Action Aid, WoMin and the Centre for Environmental

Rights. Writing of a different context, Nilsen described environmental justice activists in fenceline communities as 'catalytic agents', working in partnerships which involve informing and mobilising resistance to coal (Nilsen 2010: 76). In the South African context acquiring knowledge of citizenship rights and constitutional protections is empowering, as is connecting local awareness of loss and destruction from coal to the larger issues of environmental justice.

The environmental justice movement bridges ecological and social justice issues by putting the needs and rights of the poor and excluded at the centre of its concerns. It has initiated several alliances to protect local communities from the expansion of coal mining. An example is the Save our iMfolozi Wilderness Alliance which connected the Mpukongoni Environmental Justice Organisation to the Global Environmental Trust and was triggered by an ongoing threat of a new coal mine to be built at Fulani at the edge of the iMfolozi Reserve. This was established in 1879 and is the oldest nature reserve in Africa and the largest concentration of southern white rhino in the world (Interview with Sheila Berry of the Global Environmental Trust, Pietermaritzburg, 4.9.2018). The Labour movement has not been part of these alliances.

Educational workshops and engagements between organised labour and these local groupings could be a potentially powerful force driving a just transition. However, according to Matthews Hlabane, the founder of Mining Communities United in Action (MACUA), 'COSATU affiliates understand about the just transition but they also have reservations because it has not been demonstrated [through alternatives forms of energy production or different sources of employment, for example] and they are mainly active only at the national level' (Interview, Witbank, 21.6.2018).

Overall there is a serious disconnect regarding the role of labour at the local and national levels. Nor is Labour connecting with the environmental justice activists working in these communities. An environmental justice activist who has some thin connections with NUMSA but has not been able to engage with NUM stressed that 'there is a big gap between NUMSA at the national and local levels. (…) Most NUMSA officials at the local level know nothing about a just transition or about socially owned energy (…). For example, at the Hendrina power station working hours have been reduced which means that wages are lower. (…) Eskom has no decommissioning plan and most

Eskom workers are contract workers and have no knowledge of debates about a just transition' (key informant interview, 10.6.2018 Middleburg).[2]

Another environmental justice activist working with mining-affected communities recently met with NUMSA in Middleburg but said, 'I felt they were trying to intimidate me. They said the job losses from the closure of coal mines was because we insist on environmental compliance. At present the debates on a just transition are not connected to our experience and our struggle' (Interview, Middleburg 13.2.2019). These comments indicate that far more popular education needs to be conducted by the trade unions, especially in relation to the inevitability of a transition from coal. Moreover, it is debatable how successful the popular education workshops in urban areas were, or how widely the COSATU climate change policy document has filtered through the labour movement.

## Conclusion

'We are entering the declining decades of the fossil fuel era, a brief episode of human time' (Mitchell 2013: 231). It is a time of both increasing mobilisation for climate justice in the North and deepening pessimism. In the Global South for many, daily survival is a challenge. This is particularly true of South Africa which has one of the highest rates of unemployment in the world. In many Mpumalanga communities where the majority of coal mines and coal-fired power stations are situated, the notion of a just transition is associated with job losses and an uncertain future. The concept of a just transition could be at the heart of a powerful narrative of hope, justice, sustainability and radical change. But the labour movement is caught in the crossfire between the challenging task of protecting workers' long-term interests by addressing the climate crisis and closing the coal plants to reduce carbon emissions but also protecting workers' short-term interests by mobilising to secure a just transition which includes income support, reskilling and job replacement. SAFTU describes this dual role clearly: It is 'mobilising for a deep transformation of the current economic system of production and consumption while at the same time including protecting workers shop floor concerns' (SAFTU discussion paper on environment, Working Class Summit 21–22 July, 2018). In

---

[2] COSATU only takes responsibility for the permanent workers. However, the bulk of the workforce at Eskom's Hendrina coal-fired power station includes 2300 workers hired on short-term contracts by labour brokers for whom neither Eskom nor COSATU is taking any responsibility. These contract workers are the most precarious and vulnerable category (Interview Eskom official, Pullenhope, 8. 2018). NUM representing some 14,000 workers is the largest union at Eskom (Marrian 2019: 12).

another statement it stressed that it is 'in favour of a just transition (…) in a way that protects the livelihoods of mining and energy workers and the lives of communities most affected by environmental pollution' (SAFTU statement 2.6.2018).

What is needed is a deeper understanding of the climate and unemployment crises, and specifically the links between climate change, carbon emissions, the necessary closure of coal mines and power stations and a shift to a just transition with meaningful content relating to people's needs and aspirations. Unless labour reclaims its power and is able to do this and establish closer connections with the environmental justice movement, coal workers and mining-affected communities, the case of South Africa could demonstrate what an 'unjust transition' looks like.

## References

Agyeman, Julian. 2005. *Sustainable Communities and the Challenge of Environmental Justice*. New York: New York University Press.

Ashley, Brian. 2018. Climate Jobs and Two Minutes to Midnight. In *The Climate Crisis*, ed. Vishwas Satgar, 272–292. Johannesburg: Wits University Press.

Baker, Lucy, Jessie Burton, Catrina Godinho, and Brian Trollip. 2015. *The Political Economy of Decarbonisation: Exploring the Dynamics of South Africa's Energy Sector*. Energy Research Centre, University of Cape Town, South Africa.

Burton, Jesse. 2018. *Coal Transitions in South Africa*. Energy Research Centre, University of Cape Town, South Africa.

Castells, Manuel. 1997. *The Power of Identity*. Oxford: Blackwell.

Chamber of Mines. 2018. *Annual Report*. Johannesburg: Chamber of Mines.

Cloete, K. 2018. NUMSA Demands a Socially Owned Renewable Energy Programme. *The Daily Maverick*, April 3.

Cock, Jacklyn. 2014. The 'Green Economy': A Just and Sustainable Development Path or a 'Wolf in Sheep's Clothing. *Global Labour Journal* 5 (1): 23–44.

Cock, Jacklyn, and Eddie Koch. 1991. *Going Green. People, Politics and the Environment in South Africa*. Cape Town: Oxford University Press.

Constitution of the Republic of South Africa. 1996. https://www.gov.za/documents/constitution/chapter-2-bill-rights#24. Accessed 5 June 2020.

COSATU. 2011. *A Just Transition to a Low-Carbon and Resilient Economy*. Johannesburg: COSATU.

Fine, Ben, and Zavareh Rustomjee. 1996. *The Political Economy of South Africa: From Minerals-Energy Complex to Industrialisation*. Cape Town: Hurst Publications.

Irvin, Jim. 2017. Global Capitalism Is in Crisis and the Alliance Is in Irreversible Decline. *The Daily Maverick*, May 5.

Kelly, J., and S. Malone. 2006. *Ecosocialism or Barbarism*. London: Socialist Resistance.

Lowry, Michael. 2006. *Ecosocialism*. London: Haymarket Books.

Marrian, Natasha. 2019. Sparks fly at Eskom. *Mail and Guardian*, August 8.

Marshall, Judith. 2015. *Contesting Big Mining: From Canada to Mozambique*. Amsterdam: Transnational Institute.

Martinez-Alier, Joan. 2014. Between Activism and Science: Grassroots Concepts for Sustainability Coined by Environmental Justice Organisations. *Journal of Political Ecology* vik 21m: 19–60.

Mitchell, Timothy. 2013. *Carbon Democracy: Political Power in the Age of Oil*. London: Verso.

National Council of Trade Unions (NACTU). 2012. *Labour's initial response to the National Climate Change Green Paper 2010*. http://ilr.cornell.edu/globallabourresearchinstitute/research/upload.

Nilsen, Alf. 2010. *Dispossession and Resistance in India. The River and the Rage*. London: Routledge.

Otto, B., and P. Terhorst. 2011. Beyond Differences? Exploring Methodological Dilemmas of Activist Research in the Global South. In *Social Movements in the Global South. Dispossession, Development and Resistance*, ed. S. Mott and Nilsen, 299–223. Palgrave Macmillan.

Overy, Neil. 2018. *The Role of Ownership in a Just Energy Transition*. Research Report Project 90.

Paton, Carol. 2019. Bigger Bailout for Eskom Now on the Table. *Business Day*, May 20.

Philip, Kate. 2018. *Markets on the Margins*. London: James Curry.

Räthzel, Nora, Jacklyn Cock, and David Uzzell. 2018. Beyond the Nature–Labour Divide: Trade Union Responses to Climate Change in South Africa. *Globalizations* 15 (4): 504–519.

Schmalz, Stephan, Carmen Ludwig, and Webster Edward. 2018. The Power Resources Approach: Developments and Challenges, and Global Capitalism. *Global Labour Journal* 9 (2): 113–134.

Seidman, Gay. 1994. *Manufacturing Militancy: Workers' Movements in Brazil and South Africa. 1970-1985*. Berkeley: University of California Press.

Sweeney, Sean, and John Treat. 2018. *Trade Unions and Just Transitions The Search for a Transformative Politics*. New York: Trade Unions for Energy Democracy.

Swilling, Mark. 2015. *Greening the South African Economy*. Cape Town: Juta.

Uzzell, David and Nora Räthzel. 2013. Local Place and Global Space: Solidarity across borders and the question of the environment. In *Trade Unions in the Green Economy: Working for the Environment* eds. Nora Räthzel and David Uzzell. London: Earthscan/Routledge.

Vavi, Zwelinzima. 2019. South Africa's Energy Transition. *The Sunday Times*, September 13.

Walker, Cheryl. 2008. *Landmarked: Land Claims and Land Restitution in South Africa*. Johannesburg: Jacana.

Warner, Rose. 2010. Ecological Modernization Theory: Towards a Critical Ecopolitics of Change. *Environmental Politics* 19 (4): 538–556.

Webster, Edward. 2018. The Rise of Social Movement Unionism: The Two Faces of the Black Trade Union Movement in South Africa. In *Resistance and Change in South Africa*, ed. Phillip Frankel, Noam Pines, and Mark Swilling, 174–385. London: Croom Helm.

# 9

# Fighting in the Name of Workers: Exploring the Dynamics of Labour-Environmental Conflicts in Kerala

Silpa Satheesh

## Introduction

Existing literature on labour-environmental relations explains conflicts as a square-off between working-class trade unions and middle-class environmental movements (Foster 1993; Rose 2000; Obach 2002; Estabrook et al. 2018). Such conceptions of environmentalism as a middle-class phenomenon fall short in explaining the conflicts between labour movements and working-class environmental movements, particularly in the context of the Global South (Satheesh 2020a). Using a case of labour-environmental conflict surrounding industrial pollution in Kerala, a South Indian state, this chapter demonstrates how tensions can develop even when people from similar class backgrounds constitute both movement groups.

The chapter explores the conflicts between trade unions and green movements in the Eloor-Edayar industrial belt in Kerala and uncovers how the stand-off between the industrial workers union and the local green movement, constituted by poor and working-class members, challenges current theories. A careful exploration of the history of environmentalism in this industrial belt shows that the local green movement draws from a left ideology while employing Gandhian modes of non-violent protest/direct action

S. Satheesh (✉)
Azim Premji University, Bangalore, India
e-mail: silpa.satheesh@apu.edu.in

(Satheesh 2020b). Moreover, the long history of working-class movements and left politics combined with the red shades of green politics specifically in Kerala complicates pre-existing readings and interpretations of labour-environmental conflicts. It also shifts the discussion of labour-environmental conflicts from the realm of 'red-green conflicts' to that of 'red-red conflicts', where both movements declare their ideological affiliations to Marxism.

Against this backdrop, the chapter seeks to find answers to the following question: How do we explain conflicts between labour and environmentalists when working-class members constitute both movements? Relying on qualitative methods and frame analysis, it examines the collective action frames of unions and green movements as they strive to legitimize their goals in terms of the interest of workers, the role of class, and the complexities of left politics. The findings presented here expose the strong presence of workers' interest on either side of the conflict and highlight how resource dependence and sectoral location produce opposing economic interests and identities among workers. Most importantly, the analysis presented here reaffirms the importance of expanding the conceptions surrounding class to include the environmental inequalities and burdens created by industrial capitalism. In doing so, it reiterates the need to define class not just in terms of the economic exploitation stemming from the lack of access to the means of production but to also include environmental exploitation (Pellow 2000; Newell 2005) resulting from the destruction of the conditions of production and the unequal distribution of environmental burdens.

## Context: Labour-Environmental Tensions in the Eloor-Edayar Industrial Belt

Eloor-Edayar region is the industrial hub of Kerala, a South Indian state with an acclaimed model of development and public action (Franke and Chasin 1994; Jeffrey 1992; Issac and Harilal 1997). The history of working-class struggles, communist governments, and the shift of its class politics from class struggle to class compromise during the 1980s (Heller 1995, 1999) makes Kerala an excellent setting to understand the dynamics of labour-environmental relations. Moreover, the proliferation of grassroots environmental movements against state-sponsored development projects in Kerala complicates the relationship between labour and environmental movements in the state (Satheesh 2017, 2020b). Despite the accolades received for the Kerala model, studies have pointed out how it excludes the socially disadvantaged and historically marginalized sections from enjoying the benefits of Kerala's supposedly egalitarian development (Kurien 1995; Raman 2010). Against this background the chapter explores the interface between trade unions and green movements in Kerala surrounding industrial development and pollution.

The history of local industrialization in the region can be traced back to the 1940s when the princely state of Travancore, then the postcolonial state, decided to transform the Eloor-Edayar region into a centre of modernization and industrial development. The region soon expanded to house more than 280 chemical industries, and this expansion of industries was followed by the issue of industrial pollution. Some of the major industries located in the hub include the factories of public and private sector companies, including Fertilisers and Chemicals Travancore (FACT), Hindustan Insecticides Limited (HIL), Indian Rare Earths Limited (IRE), Travancore Chemical Corporation (TCC), bone-mill industries, and leather industries among many others (LAEC 2005). Pollution started affecting people's everyday lives in the region as early as 1970 and an organized environmental movement against pollution emerged during the late 1990s. Greenpeace declared the region as a toxic hotspot and the industrial belt is marked on the EJ Atlas as one of the critically polluted areas (Patra 2014; Joseph 2020). There have been many scientific studies that confirm the presence of heavy metal pollutants in Periyar, the river surrounding the industrial belt (Greenpeace 2003; NGT 2018).

The visible and discernible effects of pollution included river discoloration and rampant fish kills (The New Indian Express 2019; The Hindu 2020; Times of India 2020). According to local environmentalists, the constant release of untreated industrial effluents into Periyar has derailed the economic livelihoods of fisherfolk and *Pokkali* farmers and created adverse health consequences (Satheesh 2020c). Members of the community came together to launch a fight against fallouts of industrial pollution, which became more organized in 1998 when *Periyar Malineekarana Virudha Samithi* (PMVS hereafter, Periyar Anti-Pollution Campaign) was formed to coordinate and mobilize the environmental grievances. PMVS stands as an exemplar of the environmentalism of the poor (Martinez-Alier 2003; Guha and Martínez-Alier 2013), drawing its participant base from poor and working-class members (Satheesh 2020b). The constant struggles from the local greens resulted in the formation of the Local Area Environment Committee (LAEC) under the Supreme Court Monitoring Committee in 2004 (Mohan et al. 2010; Dwivedi 2001). However, the proactive role played by the LAEC in regulating industrial pollution in the region, particularly the issuance of temporary closure notice to non-compliant industries, was met with strong opposition from the industrial workers' unions in the region (Shrivastava 2007). To counter the greens, trade unions revamped the Standing Council of Trade Unions (SCTU hereafter), a collective of unions representing the interests of industrial workers in the Eloor-Edayar region. SCTU includes all major unions including Centre of Indian Trade Unions (CITU) affiliated to the Communist Party of India (Marxist) (CPIM), All India Trade Union Congress (AITUC), affiliated to Communist Party of India (CPI), Indian National Trade Union Congress (INTUC), affiliated to the Indian National Congress (INC) and BMS (*Bharathiya Mazdoor Sangh*), affiliated to the ruling Bharatiya Janatha Party (BJP). The unions affiliated to the left political parties (CITU, AITUC) dominate and hold the important offices at the union collective.

In short, SCTU is the labour movement and PMVS and *Janajagratha* (People's Vigilante) the green movement explored here. The labour and green movements in the region share an antagonistic relationship with each other on matters related to industrial pollution. The chapter details how both the unions and green movements seek to legitimize their movement's demands with worker's grievances. In doing so, I explore how class manifests itself in the tensions between SCTU and PMVS.

# Conceptualizations of Labour-Environmental Conflicts

The literature on labour-environmental conflicts can be broadly classified into two groups: (1) economic analyses and (2) class-based analyses. Economic analyses focus on explaining labour-environmental conflicts in terms of the jobs versus the environment trade-off between trade unions and green movements (Siegmann 1985; Dewey 1998; Gordon 1998; Jones and Dunlap 1992; Gottlieb 1993; Kazis and Grossman 1991; Bonanno and Blome 2001). The assumptions underlying this conception posit how environmental regulations lead to job loss, thereby creating an adverse consequence for labour (Cooper 1992; Obach 2002, 2004). Other studies contest the argument that environmental regulations lead to job loss (Porter and van der Linde 1995; Bezdek et al. 2008; Räthzel and Uzzell 2011) arguing that transforming the economy also creates new jobs. Class-based studies explain how different class-based interests and cultures engender conflicts between working-class trade unions and middle-class environmental movements (Buttel et al. 1984; Foster 1993; Gottlieb 1993; Gould et al. 2004; Rose 2000).

Additionally, it is important to notice that inquiries into the relationship between class and environmentalism are dominated by studies that reduce environmental movements as entities that 'ignore class and other social inequalities' (Foster 1993, 12). Exploring the limits of environmentalism without class, Foster notes that environmentalism is a middle-class organization that maintains disdain against the interest of workers. This chapter seeks to challenge this argument using a case of grassroots environmental movement, where members articulate their grievances in terms of class. In other words, critiques that singularly characterize environmental movements as being devoid of a class perspective overlook the long history of poor and working-class environmental movements in the Global South (Gadgil and Guha 1994; Martinez-Alier 2003; Nilsen 2008). The combination of material (red) and environmental grievances (green) in the protest lexicons of such struggles calls out the blanket depiction of environmentalism as a postmaterialist movement devoid of a structural and political-economic outlook (Dwivedi 2001). To show the complex operation of structural and individual factors, studies exploring the environmental movements in the Global South have often adopted a political economy approach (Bandyopadhyay and Shiva 1988). For example, Nilsen's (2008) Marxian analysis of the conflict over dam-building on the Narmada River in western India illustrates how the Narmada movement influenced the trajectory of capitalist development in postcolonial India. Nonetheless, critiques problematize the 'red and green' categorization of Indian environmental movements (Baviskar 2005).[1]

However, such one-sided conceptions of environmentalism as a middle-class phenomenon[2] guided by post-material values overlook the poor and working-class environmental movements in the Global South[3] (Gadgil and Guha 1994, 2013; Martinez-Alier 2003; Guha and Martínez-Alier 2013). Considering these debates surrounding the red shades of green politics in the Global South, the chapter revisits the 'dilemma of class vs. ecology' (Foster 1993) by tracing the class orientations and Marxian politics underlying PMVS. Much of the literature[4] on labour-environmental conflicts focuses on Western cases and contexts, thereby creating a near-complete absence of cases from the Global South (one exception being Räthzel and Uzzell 2012, 2013),

---

[1] Baviskar calls out the position that Indian environmental movements represent 'environmentalism of the poor' by positing how collaborations with middle-class actors and audience has been a defining feature of movements in India (Baviskar 2005, 161).

[2] This proposition is guided by the new social movement theories that infer the presence of a "New-Class" operating beyond class-based grievances or identities (Buechler 1995).

[3] Similarly, the environmental justice movements and labour environmentalism in the Global North challenge the mainstream conceptualizations of environmentalism as a middle-class phenomenon (Taylor 2002; Stevis et al. 2018; Banzhaf et al. 2019; Bell 2000).

[4] This is not to overlook the vast literature on working-class and labour environmentalism from both Global North and South (Barca 2012; Cock 2004; Satgar 2015; Stevis et al. 2018).

leaving no room for North-South comparisons of labour-environmental relations. Additionally, the focus on class within the literature on labour-environmental relations (LER hereafter) as an explanatory factor is also notable considering the declining popularity of class within mainstream social movement research, which is replete with claims about the demise of class in social movement mobilization (Cohen 1985; Buechler 1995; Rootes 2004). Much of this divergence between social movement studies and those on labour-environmental relations can be attributed to the little interactions between these two fields of knowledge (except for Obach 2004 and Mayer 2009). In this respect, this chapter is also an attempt to bridge these two fields by exploring labour-environmental conflicts using the analytic lens of framing and grievance interpretation (Benford and Snow 2000).

To demonstrate the limitations of generalized class-based explanations of labour-environmental conflicts, the chapter uses the case of ongoing tensions between trade unions and green movements in the Eloor-Edayar industrial belt in Kerala, surrounding industrial pollution. The working-class base of the two movement groups and the presence of workers' interests on either side of the conflict contradict existing theoretical explanations of such conflicts. Moreover, the long history of working-class movements and its unique model of development make Kerala an excellent postcolonial setting to understand the dynamics of labour-environmental conflicts and the role of class.

## Exploring the Intersection of Class and Environmental Inequality

The analysis presented in this chapter is grounded in a structural view of inequality that explicates how ecological degradation is an inherent part of capitalist society and class struggle (Foster 1993, 2000; Wright 2016). Traditional Marxist approaches interpret class *relationally* where class is defined in terms of the extraction of economic surplus based on the private ownership of the means of production and the social relations of production this constitutes (Wright 1997, 2000, 2005). Studies have underscored the importance of tracing environmental problems and consequent inequalities to the workings of the capitalist economic system (Magdoff and Foster 2011; Burkett 2006). Expounding an 'ecological Marxist theory', O'Connor (1991) distinguishes between the 'first and second contradictions of capitalism' where the second contradiction interprets environmental movements as a response to the capitalist destruction of the conditions[5] of production. These works highlight how capitalism not only creates inequality through the exploitation of workers, but also exacerbates environmental inequalities by destroying the conditions of production. Building on these approaches, the chapter explores how the destructions of the conditions of production under industrial capitalism engender tensions among labour and working-class environmental movements in Kerala.

The focus on industrial pollution and concomitant environmental burdens makes it imperative to tease out the interface between class, exploitation, and environmental inequalities. According to Burnham (2002, 117), 'exploitation (not consciousness or common awareness)' is the hallmark of class. He notes that it is important to consider the control of the conditions of production as much as the mode of ownership of the means of production when identifying class positions. Talking about class and environmental inequality, Newell (2005) explains how the latter reflects and reinforces other forms of hierarchy and exploitation along the lines of class, race, and gender. Elaborating on the class dimension of environmental exploitation, Newell notes, 'in environmental terms, an understanding of the operations of the ruling class reveals the ways in which decisions get made that systematically distribute risk and hazard to the poor while at the same time preserving the privilege and property of the bourgeoisie' (Newell 2005, 81).

Focusing on the intersection between environmental quality and social hierarchies, Pellow argues that environmental inequality 'addresses more structural questions that focus on social inequality (the unequal distribution of power and resources in society) and environmental burdens' (Pellow 2000, 582). Drawing from these approaches, the chapter conceives class based on both economic and environmental exploitation created and perpetuated by the capitalist system of production. Since the attempt here is also to look at the intersection of structural and individual factors, the chapter also uses subjective interpretations of class, where the individuals self-identify their location as well as the location of members of the opponent movement within the class schema.

The structural location within the matrix of capitalist domination and low economic status render members of the green movement extremely vulnerable to the fallouts of industrial pollution in the Eloor-Edayar belt. For the

---

[5] Ecological critiques of Marx argue that the productivist/Promethean nature of Marxian critique of the political economy overshadows the possibility to understand the environmental externalities created as a result of production. Other scholars have challenged this argument by demonstrating how an ecological critique of capitalism is inherent in the writings of Marx and Engels (Foster 1999; Burkett 1999, 2006; Foster 2000, 2002; Pepper 2002).

purpose of this chapter, I use Basu's classification of working class in India, where the author divides the working class into (1) unorganized workers and (2) workers in the organized sector (Basu 2009). The classification is meaningful here as it recognizes the internal heterogeneity of the working class in India. More so, by illustrating the diversity of economic interests prevailing among the workers on either side of the conflict, the chapter expands on research highlighting the fragmented nature of class formation and how that challenges the conception of the working class as a homogeneous and static feature of capitalism (Chandavarkar 1994, 1998; De Neve 2019). In other words, the square-off between the two movements in the Eloor-Edayar region demonstrates the diversity of overlapping and intersecting working-class sections marked in the sociological literature of Indian labour (Chandavarkar 1998, 9). In that respect, this chapter relies on an interpretation of class that considers both economic and environmental exploitation by capitalism (Harvey 1976; Stevis and Assetto 2001; Magdoff and Foster 2011).

## Methods and Analytical Approach

Relying on a combination of ethnographic methods (in-depth interviews, theory-driven participant observation) and document analysis, the chapter traces the ways in which the two movements frame their grievances surrounding workers. Framing, as used in the social movement literature, focuses on the processes of grievance interpretation used by social movements to convince, recruit, and motivate adherents, bystanders, and audiences about the movement's demands and goals (Benford and Snow 2000; Benford 1993). The collective action frames, the products of framing activities, are analysed here to find how the movements legitimize their demands (Satheesh and Benford forthcoming). The data used in this project was collected as part of the extensive field research conducted in the region from April to July 2018. In-depth interviews were conducted with 38 participants belonging to the trade unions and green movements, and pseudonyms are used to maintain anonymity. More than 1000 pages of documents (including campaign materials and media reports) and interview transcripts were coded and analysed to identify the dominant themes. Since the focus here is on the intersection of structure and agency, the analyses of the movement frames were guided by a combination of the extended case method (Burawoy 1998) and constructionist grounded theory (Charmaz 2014). The use of class as an analytic tool here is grounded in data, where the participants construct their narratives using the lexicons of class.[6] The attempt here is to decolonize and expand the theoretical understanding pertaining to labour-environmental conflicts using a postcolonial empirical case (Bhambra 2016) and ethnographic data. The focus on grounded narratives is important in understanding how theoretical constructs such as class, work, exploitation, and solidarity are interpreted by movement participants, thereby venturing theory construction from below. The chapter uncovers the influence of structural and political-economic factors on the frames constructed by labour and green movements.

## Tracing the Working-Class Origins of the Local Environmental Movement

In this section, I reaffirm the social roots of the local environmental movement constituted by different green groups, including PMVS, and *Janajagratha* (People's Vigilance) and how that problematizes existing conceptions of labour-environmental conflicts. The local environmental movement in the Eloor-Edayar industrial belt, led by PMVS, is constituted by a diverse group of poor and working-class members. It is a resource-poor movement constituted by factory workers, informal workers, contract and casual labour, members of the in-land fishing community, and farm workers.

Joining a long list of environmental movements formed by the poor and members of the working class in the Global South, the movement embodies a variant of 'empty-belly' environmentalism (Guha and Martínez-Alier 2013), fighting against the environmental burdens of industrial pollution with limited resource infrastructure. Detailed ethnographic profiling, subjective identification, and identification by members of SCTU reaffirm the working-class base of the local environmental movement. When inquired about the social class composition of the local environmental movement, Padmanabhan, the convener of SCTU, the union collective, said, 'They are all working-class people … people who thrive on daily wage or contract jobs. I would never say that they are rich or middle-class, even. These are people who struggle to eke out a living.' This working-class base of the green movement and the strong presence of unions affiliated to the communist parties within the Standing Council dub the conflicts as 'red-green' conflicts as opposed to the more prominent label, *blue-green conflicts*.

---

[6] The communist movement in Kerala played a huge role in popularizing Marxist theory and concepts such as class, class struggle, and class consciousness in the protest lexicons. The study classes (teach-ins) enabled the translation and diffusion of these terminologies in Kerala's sociopolitical landscape.

Moreover, the green movement clearly specifies how they perceive environmental action as part of the class struggle against capitalist exploitation. In a pamphlet that announces a discussion forum on *Marxism, Environment, and Development*, Kumar, the leader of PMVS, explains the relevance of Marx in understanding the environmental crisis. The pamphlet reads:

> The capitalist forces that plunder soil, water, and all the natural resources to reap profits threaten the survival of human beings and our nature. … Though such blatant conquests of nature or the state-capital nexus were not prominent during the period in which Marx formulated his philosophy, Marx and Engels actively questioned the invasion of nature by human beings. Conceiving the class struggle as the conflicts between workers and the capitalists, or to believe that environmental conservation does not form part of the class struggle … such assumptions are only going to embolden and aid the capitalist forces. (PMVS 2008,[7] PMVS campaign material archives)

In addition to establishing the continued relevance of Marx's writings in understanding environmental exploitation, the excerpt identifies how the state facilitates the capitalist onslaught on nature by creating policies that help accumulation. During an interview,[8] Kumar, the frontline leader of PMVS, highlighted how the state hands over the ownership of natural resources to the forces of capitalism and therefore is complicit in creating environmental damages. More importantly, the narrative underscores the need to consider environmental grievances as a working-class struggle against industrial capitalism. The excerpt urges to break away from reductive conceptions of class struggle as one between only workers and capitalists. Kumar reaffirms how the environmental movement is part of the class struggle, highlighting the intersection of environmental grievances and working-class issues. The excerpt explicates the working-class origins and anti-capitalist orientations of the local environmental movement in the region.

Among the green groups, PMVS is pronounced in its affiliations to Marxian ideology, and this mainly traces back to the members' strong ties to the communist movement, Communist Party, KSSP (People's Science Movement in Kerala), and other organizations affiliated to the political left. Perhaps due to the long history and experience with mobilization, PMVS is the movement group that actively invests and engages in the interpretive and framing tasks.

---

[7] The pamphlet, printed in Malayalam, was retrieved from the archive of campaign materials maintained by the frontline leaders of PMVS. The campaign materials were retrieved during field research in July 2018.

[8] Personal interview with Kumar, May 26, 2018.

Further, the narratives constructed by the PMVS argue that attempts to negate the working-class roots of the environmental movement would only help the capitalists to create a schism between the workers and the environmental activists.

The presence of movements that have participants from working-class backgrounds on either side of the conflict complicates the existing explanations discussed earlier, for this is not a conflict between two competing class interests or cultures. The ongoing tensions indicate the need to move beyond class-based explanations to understand the tensions between labour and green movements. The data presented in the following section question pre-existing assumptions about the presence of shared economic interests and cultures that unify people located in the same class positions.

## Fighting in the Name of Workers

One of the interesting aspects of the conflicts between the unions and green movements in the Eloor-Edayar industrial belt is that they both invoke the interests of workers to legitimize their goals and tactics. According to the collective action frames, the conflict between labour and environmental movements is one *for and among* workers who hold opposing interests. The different types of workers including the proletariat, contract workers, and self-employed workers and their varied relationship to natural resources play a decisive role in engendering the conflicts.

The thematic analysis of the frames constructed by the environmental movement presented later in the chapter explains how the green activists interpret the issue of pollution by focusing on its negative implications for the livelihoods of resource-dependent workers such as the inland fisherfolk and the farmers, whereas the unions focus on the income and job security of the factory workers. The divergent interpretations of workers' interests expose how singular and homogeneous class categorizations (Chandavarkar 1998) fail to explain schisms within movements constituted by members belonging to the same social class. The data presented here explicate how sectoral location, dependence on natural resources, and vulnerability to environmental damages produce distinctly different interests and outcomes for 'workers'.

## In the Name of Fisherfolk and Farm Workers

According to Kumar, the fight against pollution is also a fight to secure the livelihoods of the inland fishworkers and *Pokkali* farmers. Even though the local environmental movement is not constituted entirely by the fishworkers[9] and farmers, their economic livelihoods prevail as one of the important grievances used in the collective action frames of the green movement. The fishworkers include artisanal fisherfolk who eke out a living by catching fish from the inland waters using traditional tools and techniques (nets and country boats). In that sense, the subsistence fishing community is self-employed with low socio-economic standing. Of late, all major unions have branches among the fishworkers; however, independent organizations such as the National Fishworkers' Forum (NFF) are prominent in fighting for the interest of their communities in Kerala. The *Pokkali* farmers undertake a unique model of rice cultivation that combines paddy and prawn farming. The resource-dependent nature of these forms of work makes them more vulnerable to the negative fallouts of industrial pollution.

During an interview session, Kumar elaborated on how the fight against industrial pollution in the Eloor-Edayar region is also a fight for securing the livelihoods of fishworkers and *Pokkali* farmers. In his words:

> The present circumstances prevent these workers from selling their labour power. A fish worker going to the river to catch fish is returning barehanded. He is not getting enough fish to catch. ... If you go to the harbour area, you can see most of the boats returning empty-handed. The situation is pretty much the same for in-land fish-workers. Pollution has brought us here. ... It has also adversely affected the organic agriculture and aquaculture. Pollution has severely affected *Pokkali* culture [a unique saline tolerant rice variety that is cultivated using extensive aquaculture in an organic way]. The intrusion of chemical contaminants prevents these agrarian rice varieties from growing to its full potential. (Kumar, interview, May 26, 2018)

Heightened levels of pollutants in the river have resulted in a dramatic decline in the availability and diversity of fish in the River Periyar. Constant fish kills and dwindling supply have resulted in poor catchment across the local fisheries sector, thereby jeopardizing their economic livelihoods. The excerpt shows the extent of damage inflicted by pollution on agriculture and

---

[9] Considered as petty commodity producers according to traditional Marxist definitions, the artisanal fisherfolk are generally referred to as fishworkers (*Matsyathozhilaalikal*) in Kerala (Sinha 2012; Dietrich and Nayak 2006).

aquaculture and its effects on agrarian workers engaging in *Pokkali* farming. This illuminates how industrial pollution destroys the conditions of production, thereby derailing the livelihoods of the fishworkers and farmworkers.

Muneer, a local environmentalist, member of PMVS, recollected his interaction with a local fish vendor and how that motivated him to organize collective action. Muneer remarked:

> So, one day as we were standing here and chatting by the side of this village road right in front of Abdu's home, a woman in our neighbourhood walked past us. … She used to sell fish to make a living and carried her woven basket on top of her head … she walked back to us and said the supply of fish is really low in the market. She said the supply is low because of the company water. Company water, that's what we used to call the toxic effluents from the industry. And that struck me. (Interview, May 13, 2018)

This brief interaction, according to Muneer, helped him see the adverse effects of pollution on people's livelihoods. The workers who lost their daily catchment were some of the initial victims of industrial pollution. During the initial phase of the anti-pollution campaign, the increased vulnerability experienced by these workers played a crucial role in motivating them to participate in the movement events. Reconstructing the emergence of the movement, Abdu talked about the movement's strong support base among the inland fishworkers as follows:

> The majority of participants in our first protest were fish workers. Fishermen from Varappuzha, Kadamakkudy from all these regions … they need no convincing of the issue, you know … they were struggling with it and were looking out for options … openings, to express their dissent … and perhaps, a leadership to fight against this. (Abdu, member of PMVS, interview, May 30, 2018)

As Abdu points out, the fishworkers were aggrieved by the fallouts of pollution that it required little effort on the part of the green activists to recruit them to the movement against pollution. The PMVS mobilized the collective grievances of these workers and the people in the industrial belt to form organized collective action against the polluting industries. The first protest organized by the PMVS was also an ode to the struggles of the fishworkers where members formed a *human chain* by standing on an array of fishing boats parked across the River Periyar. Though these inland fishworkers' unions, CITU in particular, wholeheartedly backed PMVS in the initial phase, their support weakened in the later stages, especially after the industrial workers' trade unions intensified their attack on the local greens by forming SCTU to lead the countermovement. According to PMVS, the fishworkers and farmworkers offer solidarity to the green movement, even though their unions are careful not to antagonize the industrial workers' union while offering support.

Comparing the number of fishworkers and factory workers in the formal system, the greens argue that the loss of livelihoods from continued pollution is much higher than the number of jobs lost from the closure of polluting factories. During interviews, the greens also questioned the dominant model of development, whereby the mainstream logic would associate 'job' only with employment opportunities in the formal sector. Explaining how these 'jobs' are available with zero investment, the excerpt below highlights the importance of preserving them. Contrasting the number of factory workers to the workers who make a living from the River Periyar, Kumar stated:

> I mean, all these reveal the failure associated with development as a concept, that we have tried implementing. I mean a river is a place where many thousands of people are engaging in work with zero investment. We might have to spend 1000–2000 crores to produce 100 job opportunities, but here you don't even spend 100 paisa (lowest denomination of Indian currency). In one of the studies we had conducted, we found 22,000 in-land fish-workers; this is in comparison to the 8800 workers in the industries operating in Eloor. Such comparisons could vary in terms of value addition or contribution to GDP, but still, the difference in numbers is stark. But still, the primary productive side of our society includes agriculture and fishing. The present model of development is eliminating these two sectors of primary production, and that's when things become problematic. (Interview, May 26, 2018)

Using the economic rationales of cost and benefits, Kumar articulates how the conversations about securing jobs are always skewed in favour of employment within the organized manufacturing sector. He points out how pollution derails the lives and livelihoods of more than 22,000 inland fishworkers, livelihoods that do not require huge capital investments or infrastructure, adding how the model of modern development conceives progress in a scheme of valuation that prioritizes the secondary sector over the primary sector. However, the local environmentalists point out how the SCTU is focused solely on representing the interests of permanent workers in the factories to the complete exclusion of the tens of thousands of resources-intensive livelihoods derailed as a result of industrial pollution.

Despite the actual number of people employed, industry is considered more desirable because it signifies progress, defined in the modern economic sense, namely the transition from primary (agriculture) to secondary (industry) production. According to the local environmentalists, this model of development is responsible for the ongoing pollution in the Eloor-Edayar industrial belt. As Kumar's statement clarifies, the attribution of blame transcends the level of individuals and organizations when the politics of modern development is held responsible for the negative environmental burdens (Bandyopadhyay and Shiva 1988; Bell 2000).

## Defending the Interests of Factory Workers

Quite similar to the green movements, the unions attempt to legitimize their opposition by highlighting the interests of another set of workers, the factory workers. Focusing on the grievances of the workers employed in the industries, SCTU argues that any measure to regulate pollution will adversely affect the jobs and income of factory workers. According to the union leaders, the collective action against pollution organized by the local environmental movement hurts the economic interests of the factory workers as well as the political economy of industrial development. In Padmanabhan's words, 'An industry closing down would first affect its workers'. Often, they used stories of industrial closures to explain the effects of environmental activism on factory workers. Expressing anguish over an industry ordered to stop operation due to non-compliance to pollution regulation, Kabeer, another union leader, remarked:

> A set of so-called environmentalists have taken a stand to stop the operation of even the industries that comply with all the guidelines put forward by the PCB and they manage to use such pseudo activism in the name of the environment. The most recent victim of such activism was Sree Sakthi Paper Mills. … So, what outcome did this action produce? The company is closed, all its machinery left to rusting and more than 700 workers … the factory workers that includes allied and contract workers lost their jobs. An enterprise has disappeared altogether. But if these people, these pseudo environmentalists had approached this issue with some equivalence, then this wouldn't have happened. (Kabeer, CITU union leader, Interview, July 3, 2018)

Kabeer portrays workers of closed industrial units as victims of local environmentalism. By focusing on the plight of 700 workers and their families, he explicates the issue of job loss associated with closing non-compliant industries. He argues that the industry was compliant and followed the guidelines issued by the PCB and attributes the closure to what he describes as 'so-called pseudo-activism' of environmentalists. Citing the story of Sree Sakthi Paper Mills, the narrative brings out the resentment against the closure through the images of 'rusting' machinery as a sign of an enterprise disappearing altogether. It harshly critiques the model of local environmentalism for disregarding the interests of the 700 workers who lost their means to livelihood. The statement uncovers the diagnostic frames of the unions that pin the blame for job loss on what they call pseudo-environmentalism. In other words, the unions legitimize their resentment against greens citing the interests of factory workers.

Throughout the interviews, SCTU members reaffirmed the centrality of factory workers and industries to the larger economy. When inquired about the presence of working class or workers on either side of these conflicts, Padmanabhan, a CITU leader and convener of SCTU, who is also a member of the Communist Party, tried to put things into perspective:

> In India, the majority of workers are in the unorganised sector. The economy is not run by a few industries on the banks of the river, but the economy is run by the innumerable number of workers who work and later spend their wages in the market, thereby creating demand for products … the market moves as the purchasing power of the people changes. See, if the workers get no bonus or salary, then that will get reflected in the market. This affects all the workers who make a living from the river such as the fish workers … even the fish workers who catch fish from the river sell this fish to the families of the industrial worker … so that's a symbiotic relationship. The labour of the factory workers also affects other workers who largely remain in the unorganised sector. It affects all. To preserve the means of livelihood for all these workers … we need to secure the industries … the river and the environment in which the industry operates … it is a joint responsibility of all these people. We need to move in that direction. (Interview, May 21, 2018)

Padmanabhan succinctly presents the union's argument for taking the side of industrial workers. Even though he starts by reaffirming the importance of unorganized workers, he soon shifts to expounding the importance of factory workers for other workers and the rest of the economy, underlining the centrality of industries and factory workers for the creation of market demands. By describing how the wages of factory workers create demand for the product of the fishworkers he defines their interests and relations as symbiotic. However, he did not consider how pollution leads to fish kills and adversely affects the supply of fish in the local market. It is important to trace these narratives to the situatedness of factory workers and unions within the system of industrial capitalism. Such embeddedness poses unique impediments for trade unions to join the fight against pollution, since their economic interests are tied to the profits made by the factory owners.

## Jobs as a Mechanism for Manufacturing Consent

The frames adopted by the unions should also be considered in the backdrop of the political economy of industrial development in the region and the use of industrial workers' jobs as a mechanism manufacturing consent for capitalist accumulation and thus for the externalization of environmental destruction among industrial workers and the larger community. As he talked about the strategies industries should use to curb local activism, Madhu, a veteran leader of the BMS, explained how jobs could provide the much-needed local support for industries. Explaining the social psychology of people and its linkage to the political economy of industrial development, he explained the closure of an industry in terms of its lack of native workers:

> Let me offer a psychological explanation. When a private enterprise comes here, they employ local people … in the case of Mercum, they had 139 permanent employees … among the 139 only 9 were workers from this region, all the rest came from outside. Instead, if 50 among this 139 had been from Eloor then these workers and their families would fight for the industries … they would act as protectors for Mercum because I as a person won't prefer the closing down of a company that employs my son. He would not prefer that, nor would his wife, my wife, or our entire family. So they will get the support of that whole family and if there were 50 workers, they will get the support of 50 families. The blunder Mercum committed is this … they had employed workers from outside of this region. (Interview, June 7, 2018)

The excerpt demonstrates how workers can act as 'protectors' of industries as families will find it hard to fight against an enterprise that sustains them. Using the example of Mercum, a company where only 6% of the workforce was from the region, Madhu illustrates how the absence of local employees would imply a lack of support from families in the region. In fact, the narrative seeks to explain how the social psychology of workers' non-participation is deeply ingrained in their situatedness within the industrial system of production (Norgaard 2011), making their income contingent upon the financial performance and smooth operation of the industries.[10] Understanding

---

[10] Norgaard (2006) observes that the non-response to climate change in Norway can be explained as a matter of socially organized denial, where the economic prosperity of the community is tied to the oil industry.

the potential of jobs as a mechanism to win the support of the local people and curb anti-pollution activism, the industries actively highlighted the potential job loss of factory workers in their counterframes. Having a stronger worker base in the locality often helped industries escape the culpability of pollution. Citing one such example, Krishnan a member of AITUC continued:

> In the case of HIL (Hindustan Insecticides Limited), there was an incident where Toluene leaked from the plant into the river … the leaked Toluene (chemical) spread across the river as a film on top … it caught fire when someone threw a cigarette butt into the river … and many houses burned down in the fire … the chemical is highly inflammable and many of these houses were thatched with coconut leaves, so that accentuated the extent of this fire … but none of the people marched to HIL. And that happened because all the people affected by this fire worked in HIL. … It is true that the leaked chemical might have burned their homes, but if they demand closure of the industry then that would affect their daily bread and butter. They cannot survive without a job; therefore, they did not protest. (Interview, May 30, 2018)

The excerpt shows how jobs operated as a form of blackmail that deterred resistance against pollution even in the wake of visible pollution and associated damages. It explains how the local base of employees in HIL contained any direct action against the company, despite the company causing environmental damages, as many of the employees lived in and around the factory.

Additionally, the industries in collaboration with the trade unions instituted a system of control called 'token-system' which mandates that workers should have a daily token issued by their respective unions to be able to work in the factories. Such an arrangement enabled industries to issue tokens only to those workers who maintain silence and refrain from local environmental activism. This implies that the ongoing tensions between labour and green movements can also be traced back to the hegemony of local capital, which successfully convinces unions to establish a system that further disempowers the factory workers and their ability to sell their labour power. In doing so, the industries use jobs as a blackmail mechanism to pit workers against the local environmentalists, a trend well documented in existing studies on labour-environmental conflicts (Kazis and Grossman 1991; Bonanno and Blome 2001; Johnstone and Mando 2015). This warrants extensive inquiries to examine how the nexus between SCTU and industries manifests itself in labour-environmental relations.

## Discussion

The chapter set out to find an answer to the question: why do trade unions and green movements fail to form an alliance despite belonging to similar class backgrounds? The findings presented here point out that the antagonism between unions and green movements in the Eloor-Edayar region can be traced back to the opposing economic interests held by the workers who align on either side of the conflict. Highlighting the competing economic interests, the chapter examines why the attribution of generalized economic interests over the whole working class is problematic as the interests of workers are contingent upon their sectoral location and resource dependence. In doing so, the chapter demonstrates the simultaneous presence of competing and overlapping interests among the members of the same class and uncovers the importance of understanding class as a heterogeneous and dynamic category as opposed to a homogeneous and static feature of capitalism. Thus, the chapter underscores the need to consider the intersection of class and environmental inequalities and defines class based on both economic and environmental inequality.

Moreover, it highlights how the economic situatedness of unions within the system of capitalist production and the hegemony of capitalism constrain the ability of industrial workers to engage in environmental activism. Coalitions between labour and green movements are imperative to build safe workplaces and healthy communities as shown by examples from across the world (Obach 1999, 2004; Mayer 2011; Chomsky and Striffler 2014). More so, the burgeoning literature on 'just transition' and exemplars of labour environmentalisms uncover the possibilities of tackling environmental issues without hurting the interests of workers employed in extractive sectors (Snell and Fairbrother 2011, 2013; Snell 2018; Stevis et al. 2018). Nonetheless, the findings presented here invite fresh inquiries into the heterogeneity of working-class interests and whether the institutionalization of trade unions have influenced unions' relationship with other progressive social movements.

Additionally, it is important to consider structural factors that extend beyond this particular setting such as the shift in class politics of labour in Kerala (Heller 1995, 1999) to better understand the dynamics of labour-environmental interface in this area. The chapter establishes the need to further examine if and how this shift in class politics affects the labour movements' relationship with other progressive movements in Kerala. It also shows that the proliferation of poor and working-class environmental movements that fight against the unequal distribution of development burdens questions the much-acclaimed Kerala Model from the standpoints of environmental justice and sustainability (Satheesh 2020b). Furthermore, the Marxian orientation of many of these environmental struggles and the domination of trade unions affiliated with left-wing political parties problematize the idea that there is one homogeneous left that is either on the side of unions or on the side of environmentalists. Movements hold competing ideas on what it means to be on the 'left' of the political spectrum.

This analysis has shown that for future research to understand how conflicts between environmental movements and workers movements occur, it is important to undertake an analysis of the concrete contexts in which they occur. This includes avoiding imposing perceived class interests on either of the movements' members and instead investigating their specific social positions in the local contexts in which they operate and their relationships to the powers that be, locally as well as nationally and internationally.

# References

Ameerudheen, TA. 2018. RSS Game Plan: In Kerala, Land Acquisition Is a New Front in the BJP's Battle with the Left. *The Scroll*, August 8. Accessed December 3, 2018. https://scroll.in/article/889464/rss-game-plan-in-kerala-land-acquisition-is-a-new-front-in-the-bjp-s-battle-with-the-left.

Bandyopadhyay, J., and Shiva, V. (1988). Political Economy of Ecology Movements. *Economic and Political Weekly*, 1223–123.

Banzhaf, Spencer, Lala Ma, and Christopher Timmins. 2019. Environmental Justice: The Economics of Race, Place, and Pollution. *Journal of Economic Perspectives* 33 (1): 185–208.

Barca, Stefania. 2012. On Working-Class Environmentalism: A Historical and Transnational Overview. *Interface: A Journal For and About Social Movements* 4 (2): 61–80.

Basu, Deepankar. 2009. Analysis of Classes in India: A Preliminary Note on the Industrial Bourgeoisie and the Middle Class. Sanhati, November 4. http://sanhati.com/excerpted/1919/.

Baviskar, Amita. 2005. Red in Tooth and Claw? In *Social Movements in India: Poverty, Power, and Politics*, ed. R. Ray and F. Katzenstein, 161–178. New York: Rowman and Littlefield Publishers.

Bell, Karen. 2000. *Working-Class Environmentalism: An Agenda for a Just and Fair Transition to Sustainability*. Cham: Palgrave Macmillan.

Benford, Robert D. 1993. Frame Disputes within the Nuclear Disarmament Movement. *Social Forces* 71 (3): 677–701.

Benford, R.D., and D.A. Snow. 2000. Framing Processes and Social Movements: An Overview and Assessment. *Annual Review of Sociology* 26 (1): 611–639.

Bezdek, Roger H., Robert M. Wendling, and Paula DiPerna. 2008. Environmental Protection, the Economy, and Jobs: National and Regional Analyses. *Journal of Environmental Management* 86 (1): 63–79.

Bhambra, Gurminder K. 2016. Postcolonial Reflections on Sociology. *Sociology* 50 (5): 960–966.

Bonanno, Alessandro, and Bill Blome. 2001. The Environmental Movement and Labor in Global Capitalism: Lessons from the Case of the Headwaters Forest. *Agriculture and Human Values* 18 (4): 365–381.

Buechler, Steven M. 1995. New Social Movement Theories. *Sociological Quarterly* 36 (3): 441–464.

Burawoy, Michael. 1998. The Extended Case Method. *Sociological Theory* 16 (1): 4–33.

Burkett, Paul. 1999. *Marx and Nature: A Red and Green Perspective*. New York: Springer.

———. 2006. *Marxism and Ecological Economics: Toward a Red and Green Political Economy*. London: Brill.

Burnham, Peter. 2002. Class Struggle, States and Circuits of Capital. In *Historical Materialism and Globalisation*, ed. M. Rupert and H. Smith, 113–129. London: Routledge.

Buttel, Frederick, Charles Geisler, and Irving Wiswall. 1984. *Labor and the Environment*. Westport, CT: Greenwood.

Chandavarkar, Rajnarayan. 1994. *The Origins of Industrial Capitalism in India: Business Strategies and the Working Classes in Bombay, 1900–1940*. Cambridge: Cambridge University Press.

———. 1998. *Imperial Power and Popular Politics: Class, Resistance and the State in India, 1850–1950*. Cambridge University Press.

Charmaz, Kathy. 2014. *Constructing Grounded Theory*. California: Sage Publications.

Chomsky, Aviva, and Steve Striffler. 2014. Labor Environmentalism in Colombia and Latin America. *Working USA: The Journal of Labor and Society* 17 (4): 491–508.

Cock, Jacklyn. 2004. Connecting the Red, Brown and Green: The Environmental Justice Movement in South Africa. A Case Study for the UKZN Project: Globalisation, Marginalisation and New Social Movements in Post-Apartheid South Africa, Centre for Civil Society and School of Development Studies.

Cohen, Jean L. 1985. Strategy or Identity: New Theoretical Paradigms and Contemporary Social Movements. *Social Research* 52 (4): 663–716.

Cooper, Mary. 1992. Jobs vs. the Environment. *CQ Researcher* 2 (18): 411–431.

Dewey, Scott. 1998. Working for the Environment: Organised Labor and the Origins of Environmentalism in the United States: 1948–1970. *Environmental History* 1: 45–63.

Dietrich, Gabriele, and Nalini Nayak. 2006. Exploring the Possibilities of Counter-Hegemonic Globalization of the Fishworkers' Movement in India and Its Global

Interactions. In *Another Production Is Possible: Beyond the Capitalist Canon*, ed. B. de Sousa Santos, 381–341. London: Verso.

Dwivedi, Ranjit. 2001. Environmental Movements in the Global South: Issues of Livelihood and Beyond. *International Sociology* 16 (1): 11–31.

Estabrook, Thomas, Charles Levenstein, and John Wooding. 2018. *Labor-Environmental Coalitions: Lessons from a Louisiana Petrochemical Region*. Routledge.

Foster, John Bellamy. 1993. The Limits of Environmentalism without Class: Lessons from the Ancient Forest Struggle of the Pacific Northwest. *Capitalism Nature Socialism* 4 (1): 11–41.

———. 1999. Marx's Theory of Metabolic Rift: Classical Foundations for Environmental Sociology. *American Journal of Sociology* 105 (2): 366–405.

———. 2000. *Marx's Ecology: Materialism and Nature*. New York: NYU Press.

———. 2002. *Ecology Against Capitalism*. New York: NYU Press.

Franke, Richard W., and Barbara H. Chasin. 1994. *Kerala: Radical Reform as Development in an Indian State*. Monroe: Food First Books.

Gadgil, Madhav, and Ramachandra Guha. 1994. Ecological Conflicts and the Environmental Movement in India. *Development and Change* 25 (1): 101–136.

———. 2013. *Ecology and Equity: The Use and Abuse of Nature in Contemporary India*. Routledge.

Gordon, Robert. 1998. "Shell no!" OCAW and the Labor-Environmental Alliance. *Environmental History* 3: 460–488.

Gottlieb, Robert. 1993. *Forcing the Spring*. Washington, DC: Island Press.

Gould, Kenneth A., Tammy L. Lewis, and J. Timmons Roberts. 2004. Blue-Green Coalitions: Constraints and Possibilities in the post 9-11 Political Environment. *Journal of World-Systems Research* 10 (1): 91–116.

Greenpeace. 2003. *Status of Periyar's Health at the Eloor Industrial Estate, Kerala India*. Greenpeace Research Laboratories, UK: University of Exeter.

Guha, Ramachandra, and Joan Martínez-Alier. 2013 [1997]. *Varieties of Environmentalism: Essays North and South*. London: Routledge.

Harvey, David. 1976. Labor, Capital, and Class Struggle Around the Built Environment in Advanced Capitalist Societies. *Politics & Society* 6 (3): 265–295.

Heller, Patrick. 1995. From Class Struggle to Class Compromise: Redistribution and Growth in a South Indian State. *The Journal of Development Studies* 31 (5): 645–672.

———. 1999. *The Labor of Development: Workers and the Transformation of Capitalism in Kerala, India*. Ithaca, NY: Cornell University Press.

Isaac, T.M. Thomas, and K.N. Harilal. 1997. Planning for Empowerment: People's Campaign for Decentralised Planning in Kerala. *Economic and Political Weekly* 32 (1): 53–58.

Jeffrey, Robin. 1992. *Politics, Women and Well-being: How Kerala Became 'A Model'*. New York: Palgrave Macmillan.

Johnstone, Barbara, and Justin Mando. 2015. Proximity and Journalistic Practice in Environmental Discourse: Experiencing 'Job Blackmail' in the News. *Discourse & Communication* 9 (1): 81–101.

Jones, Robert Emmet, and Riley E. Dunlap. 1992. The Social Bases of Environmental Concern: Have They Changed Over Time? *Rural Sociology* 57 (1): 28–47.

Joseph, Neethu. 2020. How Pristine Kerala Island Transformed into World's Toxic Hotspot. *The News Minute*, January 28. Accessed January 30, 2020. https://www.thenewsminute.com/article/how-pristine-kerala-island-transformed-world-s-top-toxic-spot-116980#:~:text=Local%20body%20could%20have%20saved,to%20the%20mainland%20by%20bridges.

Kazis, Richard, and Richard Lee Grossman. 1991. *Fear at Work: Job Blackmail, Labor, and the Environment*. New York: Pilgrim Press.

Kurien, John. 1995. The Kerala Model: Its Central Tendency and the Outlier. *Social Scientist* 23 (1): 70–90.

LAEC. 2005. *Environment Impact Assessment Report on Eloor-Edayar 2004–2005*. Kochi: Local Area Environment Committee.

Magdoff, Fred, and John Bellamy Foster. 2011. *What every environmentalist needs to know about Capitalism: A citizen's guide to capitalism and the environment*. NYU Press.

Martinez-Alier, Joan. 2003. *The Environmentalism of the Poor: A Study of Ecological Conflicts and Valuation*. Northampton, MA: Edward Elgar Publishing.

Mayer, Brian. 2009. Cross-Movement Coalition Formation: Bridging the Labor-Environment Divide. *Sociological Inquiry* 79 (2): 219–239.

———. 2011. *Blue-Green Coalitions: Fighting for Safe Workplaces and Healthy Communities*. Ithaca: Cornell University Press.

Mohan, N. Shantha, Sailen Routray, and Kishor G. Bhat. 2010. *National Consultation on Water Conflicts in India: The State, the People and the Future*. National Institute of Advanced Research, Report No. R2-2010.

Neve, De, and Geert. 2019. The Sociology of Labour in India. In *Critical Themes in Indian Sociology*, ed. S. Srivastava, Y. Arif, and J. Abraham, 165–181. London: Sage.

Newell, Peter. 2005. Race, Class and the Global Politics of Environmental Inequality. *Global Environmental Politics* 5 (3): 70–94.

NGT (National Green Tribunal). 2018. Shibu Manuel Vs. The Government of India (OA. No. 396/2013(SZ), M.A. No.14/2018 (SZ)).

Nilsen, Alf Gunvald. 2008. Political Economy, Social Movements and State Power: A Marxian Perspective on Two Decades of Resistance to the Narmada Dam Projects. *Journal of Historical Sociology* 21 (2–3): 303–330.

Norgaard, Kari Marie. 2011. *Living in Denial: Climate Change, Emotions, and Everyday Life*. Cambridge, MA: MIT Press.

O'Connor, James. 1991. On the Two Contradictions of Capitalism. *Capitalism, Nature, Socialism* 1 (1): 11–38.

Obach, Brian. 1999. The Wisconsin Labor-Environmental Network: A Case Study of Coalition Formation among Organized Labor and the Environmental Movement. *Organization & Environment* 12 (1): 45–74.

Obach, Brian K. 2002. Labor-Environmental Relations: An Analysis of the Relationship between Labor Unions and Environmentalists. *Social Science Quarterly* 83 (1): 82–100.

———. 2004. *Labor and the Environmental Movement: The Quest for Common Ground.* Cambridge, MA: MIT Press.

Patra, Swapna K. 2014. Chronic Pollution in Eloor, Kerala, India. *Environmental Justice Atlas*, March 5. Accessed June 20, 2014. https://ejatlas.org/conflict/chronic-pollution-in-eloor-kerala-india.

Pellow, David N. 2000. Environmental Inequality Formation: Toward a Theory of Environmental Injustice. *American Behavioral Scientist* 43 (4): 581–601.

Pepper, David. 2002. *Eco-Socialism: From Deep Ecology to Social Justice.* Routledge.

Porter, Michael E., and Claas Van der Linde. 1995. Toward a New Conception of the Environment-Competitiveness Relationship. *Journal of Economic Perspectives* 9 (4): 97–118.

Raman, K. Ravi, ed. 2010. *Development, Democracy and the State: Critiquing the Kerala Model of Development.* Routledge.

Räthzel, Nora, and David Uzzell. 2011. Trade Unions and Climate Change: The Jobs versus Environment Dilemma. *Global Environmental Change* 21 (4): 1215–1223.

———. 2012. Mending the Breach between Labour and Nature: Environmental Engagements of Trade Unions and the North-South Divide. *Interface: A Journal for and about Social Movements* 4 (2): 81–100.

———, eds. 2013. *Trade Unions in the Green Economy: Working for the Environment.* Routledge.

Rootes, Christopher. 2004. Environmental Movements. In *The Blackwell Companion to Social Movements*, ed. D.A. Snow, S. Soule and H. Kriesi, 608–640. Malden, MA: Blackwell Publishing Ltd.

Rose, Fred. 2000. *Coalitions across the Class Divide: Lessons from the Labor, Peace, and Environmental Movements.* New York: Cornell University Press.

Satgar, Vishwas. 2015. A Trade Union Approach to Climate Justice: The Campaign Strategy of the National Union of Metalworkers of South Africa. *Global Labour Journal* 6 (3): 267–282.

Satheesh, Silpa. 2017. Development as Recolonization: The Political Ecology of the Endosulphan Disaster in Kasargod, India. *Critical Asian Studies* 49 (4): 587–596.

———. 2020a. Moving Beyond Class: A Critical Review of Labor-Environmental Conflicts from the Global South. *Sociology Compass* 14 (7). https://doi.org/10.1111/soc4.12797.

———. 2020b. The Pandemic Does Not Stop the Pollution in River Periyar. *Interface: A Journal For and About Social Movements* 12 (1): 250–257.

———. 2020c. Red-Green Rows: Exploring the Conflict between Labor and Environmental Movements in Kerala, India. PhD diss., University of South Florida.

Shrivastava, Arun Kumar. 2007. *Environment Trafficking.* New Delhi: APH Publishing Corporation.

Siegmann, Heinrich. 1985. *The Conflict between Labor and Environmentalism in the Federal Republic of Germany and the United States.* New York: St. Martin's.

Sinha, Subir. 2012. Transnationality and the Indian Fishworkers' Movement, 1960s–2000. *Journal of Agrarian Change* 12 (2–3): 364–389.

Snell, Darryn. 2018. 'Just Transition'? Conceptual Challenges Meet Stark Reality in a 'Transitioning' Coal Region in Australia. *Globalizations* 15 (4): 550–564.

Snell, Darryn, and Peter Fairbrother. 2011. Toward a Theory of Union Environmental Politics: Unions and Climate Action in Australia. *Labor Studies Journal* 36 (1): 83–103.

———. 2013. Just Transition and Labour Environmentalism in Australia. In *Trade Unions in the Green Economy: Working for the Environment*, ed. N. Räthzel and D. Uzzell, 146–161. New York: Routledge.

Stevis, Dimitris, and Valerie Assetto. 2001. History and Purpose in the International Political Economy of the Environment. In *The International Political Economy of the Environment: Critical Perspectives*, ed. D. Stevis and V. Assetto, 239–255. Boulder, CO: Lynne Rienner Publishers.

Stevis, Dimitris, David Uzzell, and Nora Räthzel. 2018. The Labour–Nature Relationship: Varieties of Labour Environmentalism. *Globalizations* 15 (4): 439–453.

Taylor, Dorceta E. 2002. Race, Class, Gender, and American Environmentalism. Vol. 534. US Department of Agriculture, Forest Service, Pacific Northwest Research Station.

The Hindu. 2020. NGT Flays Government Apathy to Periyar. September 29, 2020. https://www.thehindu.com/news/cities/Kochi/ngt-flays-govt-apathy-to-periyar-pollution/article32726342.ece.

The New Indian Express. 2019. Activists See Red Over Pollution of Periyar by Industries: Pollution Control Board Vows Action. January 12, 2019. https://www.newindianexpress.com/cities/kochi/2019/jan/12/activists-see-red-over-pollution-of-periyar-by-industries-pcb-vows-action-1923969.html.

Times of India. 2020. Periyar Remains Polluted: Report. *Times of India*, May 24, 2020. https://timesofindia.indiatimes.com/city/kochi/periyar-remains-polluted-report/articleshow/75925493.cms.

Wright, Erik Olin. 1997. *Class Counts: Comparative Studies in Class Analysis.* Cambridge University Press.

———. 2000. Working-Class Power, Capitalist-Class Interests, and Class Compromise. *American Journal of Sociology* 105 (4): 957–1002.

———. 2005. Social Class. In *Encyclopedia of Social Theory, 2*, ed. George Ritzer, 717–724. Sage Publications.

———. 2016. Two Approaches to Inequality and their Normative Implications. Items: Insights from the Social Sciences. https://items.ssrc.org/what-is-inequality/two-approaches-to-inequality-and-their-normative-implications/.

# 10

# Trade Union Politics for a Just Transition: Towards Consensus or Dissensus?

Diego A. Azzi

## Introduction

The present chapter discusses whether the main drivers of labour politics for a *just transition* are ones promoting a consensual or a *dissensus-driven* politics to the linkages between the world of labour, environment/climate change and the economy. Just transition visions state that solutions for saving the planet should not leave behind workers and communities whose existence has been linked to the polluting industries. Recently, the climate crisis has resulted in the adoption of the just transition concept—with varying degrees of radicalism—by a series of other actors, be them social movements, NGOs, the private sector or even states. Throughout the chapter, these different approaches will be framed in terms of their capacity of promoting political changes that not only contribute to solving environmental problems, but also that are capable of promoting a more just world.

In this chapter I analyse whether trade unions' strategies promote consensual or dissensual politics by examining three mutually constituted tiers: (1) *intra-union* (members; organising; awareness raising, education; strategy; youth); (2) *extra-union* (relations with the state, businesses, the UN system and financial institutions); and (3) *union-plus* (engagement with

D. A. Azzi (✉)
Federal University of ABC (UFABC), São Bernardo do Campo, Brazil
e-mail: diego.azzi@ufabc.edu.br

environmental movements and organisations, alliance building and concerns towards society as whole).

My approach here also shares the power resources analysis, which highlights the primary sources of workers' power (*structural power*) based on the ability to resolve conflicts, and the vitality of labour organisations (*associative power*) and their possibilities of cooperation with social movements (*social power*), taking into account institutional configurations (*institutional power*) (Schmalz et al. 2019, 88; Fichter et al. 2018).

In a short first part I outline the consensus-dissensus analytical scheme employed in this chapter and connect it to similar analytical schemes used to differentiate just transitions. The subsequent three parts explore each tier and I conclude with a discussion on the *dissensus-driven* just transition perspective and its relation to the *union-plus* tier and the *social power* of unions.

## Analytical Considerations: A Concept Prone to Both *Consensus* and *Dissensus*

Jacques Rancière provides the philosophical framework on which we stand to define the political meanings of consensus and dissensus. While I do not intend to present Rancière's thought here, readers can refer to *Disagreement* (1999) and *Dissensus* (2010) as well as May's works (2008, 2010) as introductory writings to his political concepts. Fundamentally, what matters to our discussion here are Rancière's definitions of *police and politics* and *consensus and dissensus*.

First and foremost, following the French philosopher's conceptualisation, *police* is referred to here not only as the coercive forces of the state, but also to any given social order with its well-established political hierarchies and inequalities, in which *the people* are denied real democratic power in favour of a liberal system of periodic elections managed by a political elite. That is what is normally called western democracy. In this consensual order, *politics* is often confronted and overcome by administrative controls and technical decision-making that justify and reproduce the *police order* as the natural, inevitable way one must live in.

In that sense then, Rancière argues that true *politics* is one that confronts the *police* order through the manifestation of a political *disagreement*—whenever a whole social world that is not been taken into account by the current order makes itself seen and heard. This can only be done through acts of *dissensus* that give visibility to new narratives contesting the existing unjust distribution of material and symbolic goods, the existing *distribution of the sensible* (Rancière 2004, 7–46)—the world as we share and perceive it through our senses and common language intellect.

The essence of politics is, then, this manifestation of *dissensus*, as the tense emergence of subaltern worlds that reclaim the equality of anyone and everyone, contesting the 'natural' hierarchies of a given *police* order, including the written and unwritten social norms that define who is allowed to take part in the political affairs of a community and the ones who are not to be heard at all. Reframing the terms and possibilities of bringing about equality among human beings is therefore paramount in Rancière's thinking of political action. This is why for Rancière several activities that we would usually interpret as political, are not really politics if they do not challenge political inequality. For instance, a labour strike over whether a wage raise will be of 3% or 5%, but that does not question the broader police order to which workers are submitted, is not *politics* in terms of producing dissensus, because it unfolds within the police order's established consensual hierarchies and social parts are playing their expected consensual role.

I argue that environmental-labour politics such as just transition and climate justice struggles have a real potential to promote a *politics of dissensus*, as they comprise a set of practices that operationalise an alternative, possible and necessary, society-nature relationship, confronting the current police order as well the forms of state and economic power that reinforce it. "They involve confrontational political action against the state; placed practices on the ground; the articulation of alternative global imaginaries through the fashioning of transversal collaborations between diverse constituencies beyond the state, and at the same time an engagement with the state" (Routledge et al. 2018, 7) and towards international organisations and regimes.

In this brief introduction, I seek to frame the discussion highlighting that, since its origins, two competing dynamics or visions have been present within the concept of just transition: consensual and dissensual. Stevis and Felli established similar categories to point out that one vision aims at the just transition for the labour force and may be called "affirmative" (*consensual*) because it seeks more equity within the parameters of existing political economy. Within the affirmative view, authors highlight the willingness of unions to work on *shared solutions* to promote the greening of the economy, in mutual understanding with international organisations and other stakeholders (2015, 34, 36). This solution often prioritises addressing those sectors of the workforce considered to be the main losers in the transition to a low-carbon economy.

In what follows I intend to discuss the proposition that trade unions' approaches to just transition are *consensus-driven* when relating to businesses, governments and international organisations—in a kind of action that traditionally emphasises unions' *institutional power*. This consensual approach seeks to collaborate with the United Nations and the private sector to find solutions to curb climate change while at the same time promoting green jobs. On the other hand, unions' approaches when relating to, or allying with, environmental and social movements are more prone to contentious, *dissensus-driven* political action and emphasise unions' *social power*. Both views coexist in the *intra-union* tier, with varying configurations depending on each specific union, national centre, federation and confederation.

Dissensual approaches are frequently disenchanted with the potential of the UN system or even national administrations to promote the transformation needed to save the planet and, therefore, these approaches "often challenge the post-political consensus of the UNCOP process through direct political contestation that critiques the green-washed capitalist solutions to climate change and sometimes offer alternative visions" (Routledge et al. 2018, 3).

The idea of just transition plays a role, thus, not only as a response to the transformation of the climate and the economy, but also as a pedagogical tool of workers education (strengthening *structural* and *associative power* in the *intra-union* tier) and a concrete framework for building bridges of dialogue and rapprochement with social movements also working around the social consequences of environment degradation and climate change (therefore strengthening the *social power* of trade unions and their political presence in the *union-plus* tier) (Schmalz et al. 2019; Fichter et al. 2018).

Thus, visions that aim to promote a completely different relationship between industries and the communities that live in their surroundings and between the economy and society in general—which may be called "transformative" (*dissensual*) political perspectives of a low-carbon society. The transformative vision puts forward a social ecological/eco-socialist agenda that relates just transition to the imperatives of rebalancing power in society. These two broad dimensions are linked to the potential, as well as the limits for a systemic transition towards a low-carbon economy.

Different transition approaches in the consensus-dissensus gradient relate to different notions of justice as perceived by trade unions as well as other social actors. There are also different understandings and realities of injustice, different views on the *distribution of the sensible*—in Rancière's terms—which in turn lead to a range of just transition understandings, as is often the case with social movements and trade unions.

Just transition is not exactly a new concept in the *intra-union* tier. What is new about just transition nowadays is that this narrative has reached the wider public and several relevant political as well as corporate actors in the *extra-union* tier. In the process, the trade union movement has managed to take part in a new political structure of opportunities provided by continuous just transition negotiations, events and initiatives at *all* three levels of our analysis: the *intra-union, extra-union and union-plus tiers*.

It has gradually built a more complex approach to the jobs-nature relationship than the original visions of labour and environment as contradictories and worked to educate rank and file members in this sense, as captured in the slogan *"there are no jobs on a dead planet"*, used by the ITUC from the 2010s onwards. As it will be shown throughout this chapter, just transition politics is directly linked not only to the move to a low-carbon economy but also to the reshaping of the trade union movement in the near future.

## The Intra-Union Tier: Overcoming the Labour-Environment Contradiction?

During a large part of the twentieth century the classic labour-environment contention (Räthzel and Uzzell 2011; Slatin and Scammell 2011; Silverman 2004) has been taken as hardly reconcilable, meaning that for unions, the protection and promotion of jobs would always come ahead of environmental concerns, often seen as long-term threats about which action can be postponed while short-term job needs are fulfilled. Industry regulation campaigns by environmentalists have been frequently faced with all kinds "of 'job blackmail' from employers, wielded to secure loyalty from workers who might otherwise oppose the company's greedy or polluting practices" and eventually end up allying with environmental and community activists (Young 1998, 42).

Building an "environmental-labour" vision for a politically progressive, economically just and environmentally sustainable future remains a major challenge to our days. It requires both an economic shift in mindset and a cultural shift in trade unionists political *modes of subjectivation* (Rancière 2004), that is, the ways through which they perceive and create themselves as political actors and not merely as workers.

In the *intra-union* tier in particular, the environment-labour vision may take different shapes when comparing international union bodies, national centres or local unions. These differences in view and the varying knowledge on the subject seem to be more related to the degree of engagement in international activities and bodies than to one's geographical base. This is due to the fact that at the international level union organisations are in closer relationship to the *extra-union* official spaces shaping the environment and climate negotiations as well as to other social actors at play. In this regard, Rosemberg points out to the need and challenges involved in "*better aligning national union positions to the international union message*" on climate change and just transition (JTRC 2020, 46–47).

If, on the one hand, pushing forward a coherent just transition agenda among its affiliates is a legitimate concern for the ITUC, across the Global South, on the other hand, national centres and local unions also have their own concerns, primarily if the just transition agenda is a Global North issue that will benefit mainly northern workers, or, if it indeed can become a tool for struggle in the Global South unionism as well. Diverse trade union cultures in the Global North and South play a role in shaping the way in which unions relate to environmental and climate issues, as well as on the possibilities of forging alliances with social movements and NGOs. This is especially true with regard to the ITUC regional bodies, TUCA in the Americas, ITUC Africa, ITUC Asia, plus the European Trade Union Confederation—each with its own set of political relations, social alliances and framing of the environmental-labour politics.

Overcoming the background of labour-environmentalism contradiction is a political need that would begin to be addressed with the rise of the *sustainable development* concept within the UN system from 1987 onwards. This opened a new window of opportunity for trade unions to meaningfully engage in the environment and climate negotiations with their own legitimate, specific agenda (Silverman 2004).

Within the *intra-union* tier, the idea behind what was to be called "just transition" from the 1990s onwards was born in the United States during the 1970s. Throughout the early 2000s, the concept was incorporated into documents and speeches of the International Confederation of Free Trade Unions (ICFTU) and of Global Trade Union Federations (GUFs), such as ITF (Transport) and ICEM (Chemicals) (Rosemberg 2010, 156–157). In 2006, during its founding Congress, the International Trade Union Confederation (ITUC) added climate change among the new issues of growing international importance to its priories. The potential millions of "green jobs" envisaged as an outcome in the move to a low-carbon industrial paradigm were drivers of this vision (UNEP et al. 2008; ITUC and TUAC 2009).

According to our use of Rancière's conceptualisation, this ITUC view on promoting green jobs through just transition is taken here as—mainly—a consensual one. It has been also classified as one that pushes for a mere *managerial reform* of industrial relations, meaning that greater equity and justice are sought within the existing economic system and without questioning the existing political and economic power relations.

By putting forward concrete sectoral plans and budget requirements, as seen in Pollin and Callaci (2019, 1–46), AFL and CTC (2015, 1–32), NZCTU (2017, 1–23) and Cooling et al. (2015, 1–28), in these approaches workers and their unions are at the centre of the just transition process as beneficiaries of support and as engines of change towards a world with low-carbon emissions. Special emphasis is placed on the *institutional power* of unions, on social dialogue and tripartite negotiations between governments, unions and employers as the process by which rights and benefits can be guaranteed (Just Transition Research Collaborative 2018, 13) in the relationship with the *extra-union* tier.

In a move to reach out to interact with other points of view and hear inputs from more dissensual perspectives, the ITUC itself has engaged in seminars with *union-plus* social forces such as women's movements in order to further develop the intersections with gender, race and class struggles. A feminist perspective on just transition highlights gender disparities within industrial branches (namely heavy manufacturing, energy and utility sectors), as well as persisting gender pay gap and women's massive presence in the informal and care economies, advocating for "gender just solutions" to women who, after all, are the most affected by climate change disruptions (see Acha 2016).

Dissensual approaches are more often led by social movements and some NGOs in the *union-plus* tier, but they are also present in strategic trade union political documents, such as the Labour Development Platform of the Americas (PLADA), a hemispheric platform for union action by the Trade Union Confederation of the Americas (a regional branch of the ITUC).

Beyond the dominant perspective centred on jobs creation and reconversion to clean industries, some other *intra-union* spaces such as the Trade Unions for Energy Democracy (TUED) initiative engage with the more dissensual approach to just transition, advocating that contestation through social empowerment of unions is a better way to achieve just transition policies than strategies of win-win negotiation with the private sector. Here, a central component of their strategy is the ideas of *public ownership and democratic control* over the energy production and distribution system as a condition for energy justice and just transition (TUED 2012).

According to the TUED initiative, the international trade union movement is in a process of searching for a bottom-up dissensual politics of just transition, but its dissensual profile is not to be taken for granted at all (Sweeney and Treat 2018). Although several ITUC affiliates from the global north and south take part in this initiative, the ITUC itself does not, given that public ownership over the energy sector agenda is not a top priority in its strategy, consensus building among unions at world level is hard to reach and that the Just Transition Centre (JTC) strategy prioritises building union-company partnerships.

Scholars have pointed out that while just transition has gained strength in the *extra-union* space of international politics and in the Global North as well, it is rarely mentioned in the Global South (JTRC 2018, 10), apart from some notable exceptions in the *intra-union* tier, such as in South African debates around energy security and the state-owned energy giant Eskom with NUMSA (National Union of Metalworkers) and SAFTU (South Africa Federation of Trade Unions).

Particularly in the Americas, through the Trade Union Confederation of the Americas (ITUC's regional branch, TUCA), from 2008 onwards, the issue begins to be developed with more intensity alongside national centres in the region, featuring CLC Canada, CUT Brazil, CGT RA and CTA-A from Argentina. Specific labour-environmental conferences are held, a stronger coordination of the Latin American delegation in the COPs is crafted and, finally, the concept of just transition is included in the aforementioned Labour Development Platform of the Americas (PLADA) strategy document in 2014 (Medeiros 2016, 257–306; CSA 2014).

Written in collaboration with social movements and allied NGOs, the PLADA platform represents a dissensual inflection within trade union political standards, as it addresses not only labour-driven aspects of sustainable development and just transition, but also takes into account 'environmental, social, economic and policy dimensions of sustainable development'. Particularly in the 'environmental dimension' section, the Trade Union Confederation of the Americas stands for radical ideas such as (a) 'environmental justice'; (b) 'defence and preservation of the commons'; (c) 'water as a human right'; (d) 'energy sovereignty and democratisation with a sustainable matrix'; (e) 'a just transition'; and (f) 'a new production, distribution and consumption paradigm with present and future environmental sustainability' (CSA 2014, 43–48).

TUCA has led the development of a Latin American perspective on the environmental-labour relationship (Anigstein and Wyczykier 2019) that, despite never becoming the mainstream position put forward by the ITUC, managed to push the organisation to take into account these more radical perspectives when acting (still very much driven by the AFL-CIO (USA)/ DGB (GER)/ RENGO (JPN) triad). Besides, it gave political capital to the region's national centres involved, such as CUT Brazil, UGT Brazil, CGT

Argentina, CTA-A Argentina, CNUS Dominican Republic, CGTP Peru and CUT Chile, as, through their participation in the TUCA Environment Working Group, they were also key Latin American national centres mobilising and lobbying during the COPs alongside ITUC.

With CUT Brazil in particular—TUCA's largest Latin-American affiliate and promoter of *both* consensual and dissensual just transition politics—the ITUC Just Transition Centre has started a closer collaboration about concrete just transition efforts from 2019 onwards. Following a two-day policy seminar held in Sao Paulo in August 2019, cooperation has evolved further, leading CUT Brazil to develop a concrete plan of action for just transition initiatives in two different northeastern states, Bahia and Rio Grande do Norte.[1]

In Rio Grande do Norte, Petrobrás state-owned oil company is currently dismantling onshore oil facilities and drills and selling others to private companies in the state. The project aims to help formulate transition policies for those workers to move into the renewable energy sector, mainly wind, solar and biogas, which is also growing in the state for the past years. In Bahia, the project will develop just transition education for workers in the Camaçarí industrial pole, particularly with unions in oil, chemicals and electricity workers. Projects are expected to be running until late 2021.

It would be far too normative to try and foresee whether consensual visions will gain influence and prevail over dissensual ones in the interplay across the three tiers of our analysis. In Rancière's terms, ITUC's perspective would be more concerned with the distribution of goods under the current *police order* and its established hierarchies and inequalities, while dissensual approaches would be more prone to question whether workers and communities are taking part as equals in the making of their own lives or not, if their political view is being heard or ignored (May 2008, 149).

## The Extra-Union Tier: Public and Private Interests on Just Transition

The various *consensual* notions of just transition usually have in common placing demands towards the *extra-union* tier, be it to states, to the international system or to corporations, in order *for them* to promote transition policies, safeguard workers and their communities and curb climate change.

---

[1] Information about the mentioned projects has been obtained through interview with Daniel Gaio, CUT Brazil Environment Secretary, in June 2020.

Mainstream union strategies of social dialogue and partnership with governments and businesses for putting forward a just transition agenda, on the other hand, focus on *institutional power* of trade unions when engaging with the *extra-union* tier, that is, their legitimacy and ability to negotiate with employers on behalf of their member workforce, and 'also their capacity to meaningfully engage in governmental talks', be it at domestic or international levels.

The *institutional power* strategy is often faced with contradictions of interest that arise amongst very heterogeneous trade union centres as well as within business groups, the latter involving a diversity of actors (so-called partners) that do not have the same approach towards labour, climate change and just transition (Tómasdóttir 2019). Those partners include business groups such as the B Team, We Mean Business, the ILO employers' representatives, Climate Action 100+ financial funds, the UN Global Compact and the World Economic Forum (WEF) businessmen annually gathered in Davos. While some see that green capitalism can succeed through the intensification of technology in production and the decrease in labour cost, others understand that the human aspect of production must, somehow, be protected from the innovation disruptions of the jobs-workers relationship.

The inclusion of just transition in the preamble of the official Paris Agreement (UNFCCC, 2015) opens possibilities of broadening its political content and application practices in the coming years, as the international trade unionism of the next decades faces the challenge of organising and mobilising labour branches in the broad and heterogeneous services sector, for which just transition policies also can and should be designed. Some initiatives are already underway with union participation in them, such as the campaign for the creation of 'One Million Climate Jobs', which presents concrete proposals on how governments can generate new decent jobs and promote emissions cuts at once (CECP 2017, 16–26, 51–61).

It is already clear that there is no single set of policies for a just transition (ILO 2015) and no 'one size fits all' recipe, but, rather, multiple different national transition plans that must take into account the specificities of local economies, labour markets and environmental impacts. Up to the Paris 2015 climate summit, the just transition concept had remained more focused on the union reality of the Global North than that of the Global South. The strong commitment to social dialogue and partnership with the private sector is one of those northern trade unionism features, taking into consideration that throughout the Global South, labour unions are being marginalised and workers' rights are seen purely as costs to be reduced in order to (supposedly) attract foreign investment (from the Global North).

In a short period of time, the just transition agenda has been interpreted and adapted more or less freely according to the interests at stake. The multiplicity of interpretations and presentations of this concept does not always assume a progressive contour, as can be seen in the classification proposed by the JTRC (2018) which shows a typology of some narratives of the just transition—status quo, *managerial reform, structural reform and transformation*, with varying degrees of radicalism and reformism along the consensual-dissensual continuum.

*Structural reform approaches* to just transition are those in which both distributive justice and procedural justice are ensured—taking into account workers and local communities' views and demands. Procedural justice implies an inclusive and equitable decision-making process that guides the transition—a democratising of the political inequalities among people in a given *police* order—and collective ownership and management of the new decarbonised energy system by different stakeholders, rather than just one interested actor.

From the mid-2000s onwards, trade union politics for a just transition became increasingly attached to lobbying governments in climate change negotiations and pushing for industry-union relations to take into account just transition policies, thus emphasising the *extra-union* strategy. At the UN COP 15 in Copenhagen 2009, for the first time, the ITUC openly championed the idea of "a just transition to a low-carbon economy, that integrates the decent work agenda and the interests of working people" (Hennbert and Bourque 2011, 154–156). Using the just transition language, unions managed to make themselves heard and to be recognised and counted as a legitimate stakeholder taking part at strategic *extra-union* decision-making negotiations, thus entering "*a close-knit community of individuals, [diplomats and international organisations' staff] who are traditionally wary of newcomers*" (JTRC 2020, 39).

As a result of the emphasis on the *extra-union* space, the international trade union movement became more technical, policy-driven and lobbyist about just transition. When one looks at ITUC's position papers (ITUC and TUAC 2009; ITUC et al. 2015; ITUC 2017, among others), this has led to a strategy that privileges formal pledges during ILO and UN COPs summits, pledges directed to the private sector and to governments in order to move forward a just transition agenda through national collective bargaining agreements.

Since then, in the period leading to the Paris UN COP 21 (2015), the international trade union movement was successful in achieving that "just transition" be explicitly mentioned in the *Green Jobs Initiative* (2009–2014)—a joint initiative of UNEP, the ILO, the ITUC and the International

Organisation of Employers (JTRC 2018, 8–9, 13). The substitution of 'old' for 'new' jobs is a feature in this transition approach, in which green job creation represents 'justice'.

By December 2015, a diverse range of public and private actors had gotten involved in the just transition debate, to the point that it was no longer just a jobs-centred approach only, but had rather been inserted into businesses, that is, environmental and local communities' narratives concerning both justice and transition (Friends of the Earth 2011; ITUC et al. 2015; ILO 2010, 2015).

In 2016, the ITUC started a new phase in its strategy for just transition with the creation of the Just Transition Centre (JTC) (a partnership with European unions around ETUC and the B Team, a business coalition), a sort of NGO think tank to help unions figure out concrete plans for just transition and promote collaborative industrial relations with companies, states and investors (Stevis et al. 2020, 19). The Just Transition Centre plays a role of promoting the linkages between the *intra-union* tier at the local level and the *extra-union* tier of transnational corporations' decision-making on just transition policies.

Companies themselves, then, have begun to understand that a just transition may also provide "associated opportunities" (B Team 2018) for their business (not just for their workforce), and fossil fuel–producing states also started to pose the needs for a just transition for countries such as Poland, according to a sort of "right to transition" vision (UNFCCC 2018; Kurtzman 2017). For the companies gathered in the B Team, "businesses who want to grasp the commercial opportunities [in the transition to a low-carbon economy] need to be able to change rapidly too. (…) From a commercial perspective, implementing a just transition allows companies to plan for, manage and optimise the operational and reputational effects of cutting emissions and increasing resource productivity" (B Team 2018, 2).

The 2018 COP 24 in Katowice, Poland, is perhaps the latest relevant moment in terms of a just transition strategy at the *extra-union* level. The worker movement's lobbying actions persuaded the COP secretariat and a number of national governments to endorse the *Solidarity and Just Transition Silesia Declaration*, which fundamentally reaffirms the recognition of the need of a just transition for the workforce, the importance of proper social dialogue and the necessity for governments' voluntary Nationally Defined Contributions (NDCs) to take into account just transition policies (UNFCCC 2018).

These sorts of demands ask for a transition that is planned and coordinated, with emphasis on cooperation and social dialogue, managing to promote social security policies and plans, labour rights and workforce retraining. However, 'private corporations and investors are unlikely to be moved by appeals for cooperation and sharing, even when such appeals come from scientists' (Sweeney and Treat 2018, 43).

As for governments, visions vary a lot from developing to developed countries in relation to supporting or opposing just transition policy agendas. Amongst international organisations (including the ILO) and governments prevails a realist perspective on the outcomes of climate talks that hardly conceives possibilities of a *just* transition for *all* nations, but rather foresees inevitable winners and losers that will arise on the path towards a low-carbon economy (ECLAC 2019a). Most of the losers are expected to be nations in the Global South, as climate change will hit them harder and the costs of the needed response measures will be higher (ECLAC 2019b).

Traditional financing mechanisms are being challenged by civil society groups that accuse them of being biased towards funding large projects and institutions, leaving other initiatives behind (Acha 2016, 2–3). Until now, the fact is that investors and financial markets more broadly have been far from being the engines for accelerating transition (Robins et al. 2018a, 22–23; 2018b) and, in fact, remain short in comparison to the financial needs stressed out in the Paris Agreement Article 2.1(c) (Zadek 2018, 1–8; Singh and Bose 2018, 1–68).

For Sweeney and Treat, it should be well recognised that "private, for-profit interests invariably fail to make investment decisions on the basis of long-term, collective benefit. And short-term thinking and time frames simply do not lend themselves to a Just Transition" (2018, 42–43). The fact that market players will not be interested in funding non-profitable transition measures (except, of course, for small-scale corporate responsibility projects run by foundations and NGOs) puts into question the effectiveness of the ITUC's and Just Transition Centre's *institutional power* strategy of social dialogue and partnership with the private sector and highlights the importance of a multi-tier strategy, coherently coordinating positions and alliances throughout the *intra-union*, *extra-union* and *union-plus* tiers.

Governments face the challenge of providing financial, industrial, social and economic policies, norms and regulations for the transition to a low-carbon future. The expected results, though, are very unlikely to be achieved through a weak state or through a political programme centred on austerity—especially in the Global South where unemployment, inequality and the distribution of wealth are at the centre of social problems, in small as well as in large countries. Moreover, many of the jobs in 'clean' industries are today frequently more precarious than those in traditional 'dirty' industries, the latter being perhaps still more committed to ideals of social protection and collective bargaining, the remains of the twentieth century's union institutional power (UNEP et al. 2008, 275–313).

## The Union-Plus Tier: Social Movements Space and Social Power of Unions

Taking advantage of the continuous structure of opportunities provided by the process of monitoring the United Nations COPs, the ITUC's just transition narrative expanded to reach other social actors also trying to influence the climate talks, both at the *extra-union* and at the *union-plus* tiers, with differing emphasis though. The years separating Rio+20 (2012) and the Paris Climate Conference (2015) were crucial in ensuring the inclusion of ITUC's labour-centred vision into other stakeholders' views on the need to better integrate the social justice dimension of climate action (JTRC 2020, 46–47).

In the *union plus* tier, the process of mutual recognition, trust building and working together must not be taken as a homogeneous one, for it is always dependent on the history, culture, interests and politics within each union organisation, as well as within each environmental movement. The relations within this tier are also subject to different geographies across north-south regions and across national-international engagement, sometimes leading to labour-environment alliances at the international level that are difficult to reproduce in some national contexts.

Political action by civil society organisations and social movements will be required to accelerate the ending of the fossil fuel era and to promote sustainable development. Evans and Phelan point out that a synergy of environmental justice movements and trade union's just transition campaigns may have the potential to challenge *extra-union* tier actors and the "economic, social, political injustices that cause oppression and insecurity, including social and economic inequity based on class, gender, race and other oppressions" (2016, 5). As Healy and Barry remind us tough, "the political space for civil society mobilisation is country dependent and normative interventions to stigmatize/delegitimize the fossil fuel industry may alienate coalitions (communities, unions) in fossil fuel dependent regions" (2017, 6).

In this regard, the global trade union movement is faced with tough strategic choices to be made. Neoliberal globalisation's industrial restructuring, growing precariousness and outsourcing have been eroding traditional trade union representation and their institutional power throughout the world—while also opening windows of opportunity for a more movement-like trade unionism, much more grounded in local communities through social power, instead of the traditional political *institutional power* strategy vis-à-vis the state and international organisations.

Just transition narratives have been making their way back to community-based approaches as well as in broad campaigns such as the Grassroots Global Justice Alliance (GGJ), Friends of the Earth (FoE) or the World March of Women (WMW). It acquires meanings that go beyond unionised workers to the 'underemployed' and the social movements space, where broader goals such as zero waste, the promotion of regionalised food systems, community-based public renewable energy (energy democracy), public transportation, affordable energy and ecosystem restoration are pursued. In other cases brought up by Healy and Barry (2017, 5), the 'just transition strategy overlaps with debates around 'green new deals', 'Green Keynesianism', and movements towards a 'circular economy''.

In a process of long-term mutual learning, environmentalist actors also began to modify their vision on the political *distribution of the sensible* and pay greater attention to the differentiated social implications of climate change in the world of work. This change in the environmentalist positioning towards workers and jobs increased the possibilities of fruitful dialogue and cooperation between the *social movements' space* and the *trade unions' field*,[2] as can be seen, for instance but not only, in the ITUC approximation towards Greenpeace, Friends of the Earth and the World Wildlife Fund (WWF), a process that had started back in the Rio 92 Earth Summit, but that was renewed and strengthened from 2010 onwards about climate change and financing for a just transition (Huxtable 2016, 251).

During the COP 21 negotiations the *union-plus* perspective among civil society organisations was deepened with an alliance forged among the ITUC, Friends of the Earth International, the International Alliance for Catholic Development Agencies, Action Aid International, Greenpeace International, Christian Aid, WWF International and Oxfam International. These organisations partnered alongside with the B Team and We Mean Business,[3] co-signing a *Call for Dialogue: climate action requires just transition* (ITUC et al. 2015, 1–2), in which pledges are made towards governments to 'show leadership', to

---

[2] Following Pierre Bourdieu, Lilian Mathieu's (2012) characterisation of social *fields* and *spaces* distinguishes trade unions from social movements. Whereas trade unions have much more formalised political relations, social norms, barriers to entry and exit, distinction marks, etc. that enable them to constitute a proper social *field*, social movements—in general—operate in a more loose, flexible, often informal and open social *space* of political relations.

[3] As civil society groups representing businesses' interests, these coalitions act as NGOs and are classified here as pertaining to the *union-plus* tier. Companies themselves are located in the *extra-union* tier.

provide investment and to set 'ambitious climate goals' based on 'social dialogue with all relevant parties'.

When unions interact with the social movements space they are subjected to different leadership selection processes, different logics of 'activist capital', identities, political legitimacy decision-making processes and so on. They are confronted with different political cultures and *modes of subjectivation* that are more or less hard to deal with for unionists, also depending on the personal skills involved (Mathieu 2012). In these interactions, unionists influence others' views on the *distribution of the sensible* at the same time that they have their own perspective transformed in order to surpass the labour-environment traditional contradiction.

While social movements are usually driven by the beliefs, interests and the immediate needs of grass-roots members and their leadership is not always elected, in the case of labour unions this is often codified in their structures: members select the leaders and determine the interests and the campaigns of the union through a democratic system (Transnational Institute 2020, 15). It means that political legitimacy, decision-making power, hierarchies, symbols and so on have to be represented according to negotiated manners among social actors, and in this process mutual political culture influences take place and shape visions.

Indeed, Anigstein and Wyczykier (2019, 13) show that in the process of addressing climate change, TUCA leaders and staff reformulated their stances as their interaction with other regional social movements increased, culminating with the third regional conference on energy, environment and labour (CSA, 2018). "The rapprochement with Latin American peasant, feminist, and environmental movements revealed problems and challenges not included in the United Nations agenda. The production of agrofuels and large hydroelectric projects, for example, forced [trade unions] to rethink the sustainability of some renewable energy sources in the region, given the way they affect local communities." Nevertheless, the challenge of building an environmental-labour vision that puts sustainable development and just transition as policy priorities for unions and governments still remains strong.

It's important to highlight that the movements which are today beginning to consolidate proposals under the name 'just transition' also have a rich history themselves and fight for their freedom to construct and realise their own emancipatory visions. In Latin America specifically, these are movements against free trade agreements and neoliberalism; the alter-globalisation movement; movements for energy sovereignty and democracy struggles; environmental justice movements; decolonisation and independence struggles; feminist and women's movements; movements against racism; and fights for agrarian reform, peasant rights and food sovereignty. "This diversity of backgrounds, political traditions, and strategic goals means that the dialogue that is creating a radical concept of just transition that is not free from tensions or contradictions [between trade unions and social movements in the region]" (Transnational Institute 2020, 5).

## Conclusion: Framing Just Transition as a Politics of Dissensus

Based on Jacques Rancière's concepts of consensus and dissensus, this chapter has analysed international trade union movement politics for a just transition, demonstrating that as of today, its strategy is primarily *consensual*, based on *institutional power* of unions, and targets the *extra-union* tier of labour action—states, corporations and the international system. In this concluding remarks, I highlight that framing an environmental-labour perspective alongside the *union-plus* tier progressive forces provides unions with opportunities to develop a more *dissensual* just transition politics, enhancing their *social power* and appeal towards younger generations.

The problem with the mainstream consensual approach is that it cannot be ignored that the neoliberal hegemony reigning almost worldwide since the 1990s has deepened inequality, intensified attacks on organised labour's *institutional power* and exacerbated the financialisation of the economy, harming formal jobs (Stiglitz 2002; Duménil and Lévy 2015). It seems as if the horizon for a *politics of dissensus* has been drastically reduced, and this has immersed left-wing forces in general in a terminal crisis. Trade union density has been falling all over the world, both in developing and developed nations (Rosenfeld 2014, 1–30). The world seems ever more hostile to labour protection and labour rights, let alone trade union organisations.

In this hostile context, the panorama of the world of work in the second decade of the twenty-first century is not a positive one and just transition narratives cannot afford to be naive about it. Growing precariousness has meant that jobs have become increasingly uncertain, unstable and insecure due to policies and power relations that—under the cover of entrepreneurship discourses—transfer risks form companies, employers and governments to common workers with limited or no rights.

However, the climate crisis—a common challenge to all peoples by definition—has provided a continuous political structure of opportunities for progressive social forces, including progressive trade unions, to promote a renewed dissensual critique of business-as-usual capitalist exploitation of human labour and nature alike. Dissensual meaning that has the potential to reconnect and address the problems inherent in our current mode of production and consumption; gender, class and race inequality; the asymmetries in the international political economy of nations; and the untamed power acquired by private corporations, threatening democratic decision-making processes. Importantly, forging renewed class consciousness among different generations of environmental and labour movements through a critique of the capitalist environmental crisis is a challenge not only for unions but for left-wing forces as a whole.

The development of coherent views and policies about just transition, climate change and sustainable development can bring unionism closer to younger generations than the (today weakened) formal relationship of rank-and-file representation in collective bargaining. If the lobbying-bargaining strategy is not accompanied by trade union active participation in other, more *dissensual* political spaces, the labour movement risks recycling the old issues of contention between trade union interests and those of environmental and social actors. One might argue that it's precisely this different institutional bargaining approach that differentiates trade unions from social movements, which is correct. However, this *consensual* approach to just transition emphasises action through trade union *institutional power* (as we have seen, in decline) while placing *social power* in a secondary level gives margin to the old accusations of trade union self-interested corporatism.

The path towards just transition will require a reorganisation of the relations between labour and their counterparts in the *extra-union* tier: namely state and capital. It will cause not only an industrial transformation, but, moreover, it will require that labour itself re-imagines its place in the political economy (Stevis and Felli 2015, 39). Where trade unions have taken up the issues of informal workers in the *union-plus* tier, such as in Argentina's CTA-A, unions have also undergone internal changes in the *intra-union* tier, being somewhat more movement-like in terms of the dynamics of representation and relation towards rank-and-file members. For unions, a strategic objective would be to move, educate and inspire union members, progressive organisations and the broader public, tying this to the struggle to rejuvenate the trade unions and the political left (Sweeney and Treat 2018, 43).

Dealing with just transition policies leads to consider not only the future of work and the work of the future, but also on the future of the trade union movement as an institutional form of organisation and collective action on behalf of workers' rights. In many parts of the world, the institutionally strongest unions remaining today are often the ones attached to polluting industries or else unions that are attached to the public sector. They are also those of greater *associative power*. Transitioning rank-and-file workforce in those industries implies the challenge of not destroying a union by moving its own members away to other branches in the new, sustainable productive structure needed to curb emissions. That would be union suicide. Thus, to demand and guarantee the implementation of just transition policies that maintain and expand trade union representation, the right to association and collective bargaining remains an indispensable aspect in opposing the current dynamics of regressive labour reforms both in the north as well as in the Global South. If transitioned green jobs are in fact going to be decent jobs is still to be proved, but the prospects are not optimistic.

Making use of ITUC's concept (2019) that was the driver of its action in the run up to COP 21 in Paris, just transition can indeed be seen as 'a tool' the international trade union movement shares with the international community. It has the potential to be a tool for promoting a broader politics of dissensus, and not simply one for lobby and partnership building with businesses and the state. It is also a tool that may position unions "as an anchor and as an engine for social justice and equity in and beyond the workplace", reconnecting the workers' movement with its social roots, as suggested by Rosemberg (JTRC 2020, 53).

As Räthzel and Uzzell state (2011, 7), in the process of finding common ground with social actors, the traditional trade union goals "are broadened to include working and living conditions, and the way in which societies are organised to provide workers (and all citizens) with social and political rights. […] Consequently, a union defining itself in this way, needs to go beyond defending solely the rights of its members and include the interests of all those who want a just society."

In its dissensual and anti-systemic versions, ideas of just transition get closer to a utopia, to a holist resetting of societal relations towards sustainability and justice. However, if taken not in the sense of a naive and intangible goal but, rather, a real utopia (Wright 2010) that can empower labour forces to bargain and make their voice heard both by the state and market forces, just transition does have a place as tool for struggle and strengthening of class consciousness. This is one of the reasons why just transition is not to be taken as a 'false solution'—a term frequently used by those critical to the climate negotiations and green capitalism approaches.

Nevertheless, it may indeed turn into a false solution if limited to its consensual version based on unions' *institutional power*, as the risk of being captured by the *police order* is always lurking. As argued by Evans and Phelan (2016, 9), the risk is that unions' strategies for just transition remain attached to the *extra-union* tier logic of hierarchies and inequalities and that "the labour movement is likely to rely heavily on government-led initiatives focused on influencing energy markets and subsidies, with some public investment in new job creation".

A *politics of dissensus* for a just transition is expected to reside in the interaction of unions with the *union-plus* tier social forces—many of which are now moved forward by young generations (Svampa 2020)—targeting coordinated demands and challenges towards the *extra-union* tier, be it state or market forces. For the relationship with other social actors that have more radical or holistic approaches themselves provides a mutually reinforcing dynamics in which trade unions can formulate ideas and practices that go beyond traditional institutional power. In the process of enhancing unions' social power strategy, they ultimately gain class legitimacy and leverage to contest the established *police order,* its hierarchies and inequalities sustaining the global political economy.

# References

Acha, Majandra. 2016. *Gender Equality and Just Transition*. WEDO Discussion Paper.

AFL-CTC (Alberta Federation of Labour—Coal Transition Coalition). 2015. *Getting It Right: A Just Transition Strategy for Alberta's Coal Workers*. Alberta Federation of Labour—Coal Transition Coalition.

Anigstein, Cecilia, and Gabriela Wyczykier. 2019. Union Actors and Socio-environmental Problems. The Trade Union Confederation of the Americas. *Latin American Perspectives*, 46 (6): 109–124.

B Team. 2018. *Just Transition. A Business Guide*. B Team/Just Transition Centre.

CECP. 2017. 100.000 Empregos para o Clima. *Climate Jobs Campaign Portugal*, 2a edição, Lisboa.

Cooling, Karen, Marc Lee, Shannon Daub, Jessie Singer. 2015. *Just Transition: Creating a Green Social Contract for British Columbia's Resource Workers*. Canadian Center for Policy Alternatives.

CSA. 2014. *PLADA. Plataforma Laboral de Desarrollo de las Américas*. Confederación Sindical de Trabajadores/as de las Américas. São Paulo.

———. 2018. Declaración Final de la 3er. Conferencia Regional de Energía, Ambiente y Trabajo. Accessed http://csa-csi.org/NormalMultiItem.asp?pageid=12399. San José, Costa Rica, 11 de Octubre de 2018.

Duménil, Gérard, and Dominique Lévy. 2015. Neoliberal Managerial Capitalism. Another Reading of the Piketty, Saez, and Zucman Data. *International Journal of Political Economy*, 44: 71–89.

ECLAC. 2019a. *Foro sobre transición Justa, empleos verdes y acción climática: intercambio de experiencias para Latinoamérica y el Caribe*. United Nations Economic

Commission for Latin America and the Caribbean (ECLAC). Accessed https://www.cepal.org/es/notas/expertos-dialogaron-la-cepal-transicion-justa-empleos-verdes-accion-climatica-america-latina.

———. 2019b. *Cambio climático: impactos y transición justa*. United Nations Economic Commission for Latin America and the Caribbean (ECLAC). Accessed https://www.cepal.org/sites/default/files/c_de_miguel_2019_10_cc_transicionjustalimpiafinal.pdf.

Evans, G., and L. Phelan. 2016. Transition to a Post-Carbon Society: Linking Environmental Justice and Just Transition Discourses. *Energy Policy* 99: 329–339.

Fichter, Michael, Stefan Schmalz, Carmen Ludwig, Bastian Schulz, and Hannah Steinfeldt. 2018. *The Transformation of Organised Labour. Mobilising Power Resources to Confront 21st Century Capitalism*. Berlin: Friedrich Ebert Stiftung.

Friends of The Earth. 2011. *Just Transition. Is a Just Transition to a Low-Carbon Economy Possible within Safe Global Carbon Limits?* London: FoE England, Wales and Northern Ireland.

Healy, Noel, Barry, John. 2017. Politicizing Energy Justice and Energy System Transitions: Fossil Fuel Divestment and a 'Just Transition'. *Energy Policy* 108: 451–459.

Hennbert, Marc-Antonin, and Reynald Bourque. 2011. The International Trade Union Confederation (ITUC): Insights from the Second World Congress. *Global Labour Journal* 2 (2): 154–159.

Huxtable, David. 2016. *The International Trade Union Confederation and Global Civil Society: ITUC Collaborations and Their Impact on Transnational Class Formation*. Dissertation. Department of Sociology. University of Victoria, Canada.

ILO. 2010. Climate Change and Labour: The Need for a 'Just Transition'. *International Journal on Labour Research* 2 (2).

———. 2015. *Guidelines for a Just Transition Towards Environmentally Sustainable Economies and Societies for All*. Geneva: International Labour Organization.

ITUC (International Trade Union Confederation). 2017. *Just Transition—Where We Are Now and What Is Next? A Guide to National Policies and International Climate Governance*. Brussels: ITUC.

———. 2018. Building Workers' Power: Change the Rules. 4th ITUC World Congress Statement (Final). Accessed https://www.ituc-csi.org/4co-e-5-building-workers-power?lang=en.

———. 2019. Índice Global de los Derechos de la CSI 2018: Reducción del espacio democrático y codicia corporativa sin freno. International Trade Union Confederation. Accessed https://www.ituc-csi.org/indice-global-de-los-derechos-de-20302?lang=en. Brussels, 2019.

ITUC and TUAC (Trade Union Advisory Committee). 2009. *A Just Transition: A Fair Pathway to Protect the Climate*. Brussels: ITUC and TUAC.

ITUC et al. 2015. *Call for Dialogue: Climate Action Requires a Just Transition*. Paris: ITUC, Friends of the Earth International, the International Alliance for Catholic Development Agencies, Action Aid International, Greenpeace

International Christian Aid, WWF International, Oxfam International, the B Team and We Mean Business.
Just Transition Research Collaborative (JTRC). 2018. *Mapping Just Transition(s) to a Low Carbon World*. Geneva: United Nations Research Institute for Social Development (UNRISD), Rosa Luxemburg-Stiftung (RLS) and University of London in Paris.
Morena, Edouard, Dunja Krause and Dimitris Stevis, eds. 2020. *Just Transitions. Social Justice in the Shift Towards a Low-Carbon World*. London: Pluto Press.
Kurtzman, Joel. 2017. The Low-Carbon Diet. How the Market Can Curb Climate Change. *Foreign Affairs* (September/October).
Mathieu, Lilian. 2012. *L'espace des mouvements sociaux*. Paris: Éditions du Croquant.
May, Todd. 2008. *The Political Thought of Jacques Rancière. Creating Equality*. Edinburgh: Edinburgh University Press.
———. 2010. *Contemporary Political Movements and the Thought of Jacques Rancière. Equality in Action*. Edinburg: Edinburgh University Press.
Medeiros, Josué. 2016. A constituição de um sindicalismo sociopolítico: o caso da CSA. In *Nuevos estilos sindicales en América Latina y el Caribe*, Emilce Cuda, ed., 257–305. Buenos Aires: CLACSO.
NZCTU (New Zealand Council of Trade Unions) 2017. *Just Transition. A Working Peoples' Response to Climate Change*. At https://union.org.nz/wp-content/uploads/2019/02/JustTransition.pdf
Pollin, Robert, and Brian Callaci. 2019. The Economics of Just Transition: A Framework for Supporting Fossil Fuel–Dependent Workers and Communities in the United States. *Labor Studies Journal* 44(2): 93–108.
Rancière, Jacques. 1999. *Disagreement. Politics and Philosophy*. Minneapolis: University of Minnesota Press.
———. 2004. *The Politics of Aesthetics. The Distribution of the Sensible*. Edited and translated by Gabriel Rockhill. London and New York: Continuum Publishing.
———. 2010. *Dissensus. On Politics and Aesthetics*. Edited and translated by Steven Corcoran. London and New York: Continuum Publishing.
Räthzel, Nora, and David Uzzell. 2011. Trade Unions and Climate Change: The Jobs versus Environment Dilemma. *Global Environmental Change*, 21: 1215–1223.
Robins, Nick, Vonda Brunsting, and David Wood. 2018a. *Investing in a Just Transition. Why Investors Need to Integrate a Social Dimension into Their Climate Strategies and How They Could Take Action*. LSE Grantham Research Institute on Climate and the Environment; Center for Climate Change Economics and Policy; The Harvard Hausser Institute for Civil Society. In Partnership with PRI and ITUC.
———. 2018b. *Climate Change and the Just Transition: A Guide for Investor Action*. LSE Grantham Research Institute on Climate and the Environment; Center for Climate Change Economics and Policy; The Harvard Hausser Institute for Civil Society. In Partnership with PRI and ITUC. September 2018.

Rosemberg, Anabella. 2010. Building a Just Transition: The Linkages between Climate Change and Employment. *International Journal of Labour Research*, 2(2): 125–161.

Rosenfeld, Jake. 2014. *What Unions No Longer Do*. Cambridge, MA: Harvard University Press.

Routledge, Paul, Andrew Cumbers and Kate Driscoll Derickson. 2018. States of Just Transition: Realising Climate Justice through and Against the State. *Geoforum*, 88: 78–86.

Schmalz, Stefan, Carmen Ludwig, and Edward Webster. 2019. Power Resources and Global Capitalism. *Global Labour Journal* 10 (1): 84–90.

Silverman, Victor. 2004. Sustainable Alliances: The Origins of International Labour Environmentalism. *International Labor and Working Class History*, 66: 118–135.

Singh, Harjeet, and Indrajit Bose. 2018. *History and Politics of Climate Change Adaptation at the United Nations Framework Convention on Climate Change*. Research Paper 89. Geneva: South Centre.

Slatin, Craig, and Madeleine Scammell. 2011. Environmental Justice and Just Transition. *New Solutions* 21 (1): 1–4.

Stevis, Dimitris, and Romain Felli. 2015. Global Labour Unions and Just Transition to a Green Economy. *International Environmental Agreements* 15 (1): 29–43.

Stevis, Dimitris, Edouard Morena and Dunja Krause, eds. 2020. *Just Transitions: Social Justice in the Shift Towards a Low-Carbon World*. London: Pluto Press.

Stiglitz, Joseph. 2002. *Globalization and Its Discontents*. New York: W. W. Norton & Co.

Svampa, Maristella. 2020. ¿Hacia dónde van los movimientos por la justicia climática? *Nueva Sociedad* n. 286. https://nuso.org/articulo/hacia-donde-van-los-movimientos-por-la-justicia-climatica/

Sweeney, Sean, and John Treat. 2018. *Trade Unions and Just Transition. The Search for a Transformative Politics*. Working Paper 11. New York: TUED.

Tómasdóttir, Halla. 2019. To Be or Not to Be in Davos. B Team, January 30, 2019. Accessed http://www.bteam.org/announcements/to-b-or-not-to-b-in-davos/.

TNI (Transnational Institute). 2020. *Just Transition: How Environmental Justice Organisations and Trade Unions Are Coming Together for Social and Environmental Transformation*. Workshop Report. Amsterdam: Transnational Institute.

TUED (Trade Unions for Energy Democracy). 2012. *Resist, Reclaim, Restructure. Unions and the Struggle for Energy Democracy*. Working Paper 1. New York: Trade Unions for Energy Democracy. Accessed http://unionsforenergydemocracy.org/wp-content/uploads/2013/12/Resist-Reclaim-Restructure.pdf.

UNEP (United Nations Environment Program), International Labour Organisation, International Organisation of Employers and International Trade Union Confederation. 2008. *Green Jobs: Towards Decent Work in a Sustainable, Low-Carbon World*. Nairobi: United Nations Environment Program.

UNFCCC. 2015. *Paris Agreement*. Accessed https://unfccc.int/sites/default/files/english_paris_agreement.pdf. United Nations, Paris.

———. 2018. *The Solidarity and Just Transition Silesia Declaration.* UN COP 24, Katowice, Poland.

Wright, Eric Olin. 2010. *Envisioning Real Utopias.* London: Verso Books.

Young, Jim. 1998. Just Transition: A New Approach to Jobs vs. Environment. *Working USA,* 2(2): 42–48.

Zadek, Simon. 2018. Financing a Just Transition. *Organization & Environment,* 32(1): 18–25.

# 11

# Climate Jobs Plans: A Mobilizing Strategy in Search of Agency

Andreas Ytterstad

# Introduction

'Climate jobs' sounds like a good idea. Jobs 'that directly contribute to preventing climate change and global warming' (One Million) speak directly to, and hold the promise to solve the 'jobs vs the environment dilemma' (Räthzel and Uzzell 2011), a key problematic within environmental labour studies (e.g., Hampton 2015; Stevis and Felli 2015). If the climate crisis can be credibly framed and accepted by trade unions as an opportunity for rather than a threat to good jobs, that would enhance the possibility of solving the crisis.

Climate jobs in this chapter are presented not just as an idea, but as a strategy. This strategy involves and begins with a Climate Jobs Plan (CJP), one that is framed simultaneously as a *just transition* away from fossil fuels. The main impulse of the Climate Jobs strategy is global—not national. Romain Felli observed how the very first Climate Jobs Campaign launched in the UK in 2009 had 'a cosmopolitan perspective in mind':

> Of course cuts in the UK on their own will make little difference to global climate change. But if we campaign for a million new jobs, and win them, people all over the world will see what we have done. They will know it is possible. And then they can do the same. And that will save the planet. (Felli 2011, 100)

A. Ytterstad (✉)
Department of Journalism and Media Studies, Oslo Metropolitan University, Oslo, Norway
e-mail: andreasy@oslomet.no

© The Author(s), under exclusive license to Springer Nature Switzerland AG 2021
N. Räthzel et al. (eds.), *The Palgrave Handbook of Environmental Labour Studies*,
https://doi.org/10.1007/978-3-030-71909-8_11

The most illuminating way of conceiving the general and global character of climate jobs is to present the CJPs as *a mobilizing strategy in search of agency*. After charting the origin and spread of CJPs, I spell out both the 'climate' and the 'jobs' part of the plans in more detail, noting some of the differences of emphasis in the way different coalitions have developed their plans. Subsequently, I elaborate on why all such plans amount to a just transition away from fossil fuels. I then exemplify how Climate Jobs Campaigns have included and attempted to integrate broader environmental and job concerns. Whereas the inclusiveness of the climate jobs coalitions is a key strength, it may also harbour weaknesses. For several of the campaigns it is hard, despite their intentions, to exclude the widespread and established perspectives of a 'green economy' from their climate jobs demands. Conversely, it may be hard but necessary for the campaigns to *include* and summon the state, and argue that climate jobs must become public sector jobs in every country (Ytterstad 2020b). The largest strategic challenge for the Climate Jobs Campaigns is therefore to mobilize agency powerful enough to transform nation-states and make this unlikely scenario happen.

Several scholars within environmental labour studies have included a Climate Jobs strategy in their analysis or typologies. Dimitris Stevis and Romain Felli locate climate jobs within a fairly deep and transformative 'social ecological' (2015, 138–139) version of just transition. Paul Hampton (2015) argues that "climate job" implies a class-based strategy, different from the state-centred, eco-modernist approach of 'green jobs'. Gunderson (2019, 35), by contrast, is dismissive of parts of the climate jobs approach, because they 'cater to the empty promise of "green growth"'. His assessment builds in part on the work of Stefania Barca, but Barca's (2017) own analysis is more nuanced. She finds the UK campaign for One Million Climate Jobs (OMCJ hereafter) Keynesian and state-centred, whereas the South African campaign is more firmly rooted in the multiple social movements of that country. In his book on South Africa's energy transition, Lawrence (2020), however, aligns the OMCJ campaign quite explicitly to classical Keynesianism.

Scholars have different theoretical and political positions when they evaluate social phenomena. But the variations in the assessments alluded to above also reflect broader tensions within the different coalitions working to implement the Climate Jobs strategy. Kenfack (2019), for example, finds a tension between the unionists and the non-unionists in the Portuguese campaign; the former are more prone to a 'softer' affirmative version of just transition and the latter support a more radical and transformative version. These tensions are not, he contends, insurmountable (Kenfack 2019, 136).

To develop a Climate Jobs strategy, and to search for a broader and more powerful agency for Climate Jobs Plans, is therefore a bridge building exercise. In this Handbook, I present and describe the Climate Jobs strategy as a scholar and bridge builder, including identifying some of the tensions within and between the different coalitions. My own theoretical position is inspired by Antonio Gramsci's concept of good sense. Solutions to climate change must emerge and develop from already existing aspects of good sense, amongst workers, environmentalists, school strikers, indigenous communities and others (Ytterstad 2020a). One important aspect of good sense consists of an interest in the truths of global warming among such subaltern groups. Several international surveys indicate that the demographic segments least invested in the prevailing order are also the most appreciative of climate science (Malm 2018, 136). The climate strikes of 2019, despite media attempts to infantilize the protests (Jacobsson 2020), provided further evidence of widespread interest in solutions that challenge current systems. In the conclusion, I place climate jobs within a contemporary global context of emergency and crisis, arguing that the Green New Deal and the coronavirus both signal that climate jobs are an idea whose time has come.

## The Origin and Spread of CJPs

This section relies on a formal interview undertaken with the 'grandfather' of the Climate Jobs strategy, Jonathan Neale. His book *Stop Global Warming— Change the World* (Neale 2008) contains much of the thinking behind the Climate Jobs strategy, but 'climate jobs' do not appear explicitly in the book. The phrase must have originated in conversations with some of the key individuals who read his book. One of them was Chris Baugh, assistant general secretary of the Public and Commercial Services Union (PCS). Baugh was instrumental in the launch of the Climate Jobs Campaign and 'could phone up general secretaries in the union movement' (Neale 2019). Further foothold in the trade unions came with the establishment of the Trade Union Group of the Campaign against Climate Change (CaCC) in the UK. Between 2006 and 2009 Neale was international secretary of the CaCC. I believe that his international travels in that period and after were an important causal factor for the spread of Climate Jobs Campaigns in South Africa, Canada, Norway, Portugal and New York State.

CJPs have emerged in the intersection of climate coalitions and trade unions. All initiatives have gained varied but broad support. The Canadian Labour Congress (cf. CBC News 2016) and the General Confederation of the Portuguese Workers (CGTP) have supported CJPs. Parts of the union movement have endorsed the Climate Jobs Campaigns in the UK, South Africa, New York State and Norway. Environmental organizations have been part of every initiative. The Protestant (and former state) Church of Norway is also part of the Norwegian alliance. Sharan Burrow, General Secretary of ITUC, and Kumi Naidoo, (former) leader of Greenpeace International, were impressed by the inclusion of the Norwegian Church after having visited the 2015 Bridge to the Future conference in Oslo (Burrow and Naidoo 2015).

When Environmental Labour Studies emerged as an academic field, common theoretical and historical ground was primarily sought between environmentalists and workers (Räthzel and Uzzell 2012a; Räthzel and Uzzell 2011). The inclusion of the Norwegian Church indicates the possibility of even broader alliances and coalitions. This inclusiveness underscores the challenges of bridge building for CJPs.

The CJPs are invariably designed as a response to both a climate change need and an employment need. Simultaneously, the coalitions behind the CJPs frame their strategy as a way of making a just transition away from fossil fuels more concrete. The particular configuration of the climate need, the employment need and just transition differ somewhat according to national contexts. For analytical purposes, I have chosen to first present the climate part of the CJPs—where the climate need is most salient—and then the jobs part of the CJPs—where the jobs need is most salient, before I show how the CJPs fit with (various versions of) just transition.

## The Climate Part of the CJPs

The most important distinctive and general feature of all CJPs lies in their close relationship to climate science. Climate science informs both the needs to mitigate and the need to adapt to the effects of a warming world (Malm 2018). All countries must cut emissions rapidly if we are to have a fighting chance of keeping the global temperature within safe limits, but rich countries, particularly rich fossil fuel producing countries, must cut the most (Bond 2012).

To mitigate global climate change, therefore, means to locate the sources of emissions in every given country and region and to find the kinds of jobs that will reduce the overall emissions. The British campaign estimated that a million climate jobs (710,000 of which should be in electricity and transport) will provide a 83 per cent cut in all greenhouse gas emissions over 20 years (Campaign Against Climate Change 2014). The Canadian campaign started by identifying the 'three sectors of the Canadian economy that were collectively responsible for up to 81 per cent of Canada's greenhouse gas emissions—energy, transportation and construction'. The 'Big Shift' of getting Canadian emissions down in line with climate science is therefore broken down into jobs that would provide the 'energy shift', the 'transportation shift' and the 'building shift' (Clarke 2018, 10–11).

CJPs thus prioritize jobs where the emission reduction effect is highest and most certain. Climate jobs have different timeframes. Buses and bus workers can be put to work quickly, especially if you close the main arteries of megacities to nothing but buses. Building renewable energy, and the new (super) grids needed to support it, takes longer. But the workhorses of renewable energy—wind and solar power—can still be assembled within a shorter time frame than carbon sequestration projects (often still in a pilot stage) or nuclear power plants (Jacobsen and Delucci 2011).

There are transition plans of 'climate neutrality' which make the targets of climate jobs campaigns pale in comparison. Such plans, however, usually rely on market schemes (Lohman 2011) or wishful thinking of future geoengineering capable of sucking out greenhouse gases from the atmosphere (Clarke 2018, 30). The estimates of the CJPs, by contrast, rely on current engineering and knowledge of what works within the decade that matters most: this one.

Trade unions sometimes complain about emission targets that are 'not rooted in an engineering-based approach' (AFL-CIO 2019). A climate jobs strategy should therefore aim to link climate science with the engineering knowledge of skilled workers, as well as the more practical side of good sense which unions, Indigenous communities and people in general possess (Ytterstad 2014, 2020a).

Finally, there is also room for adaptation jobs within the CJPs. The South African *One Million Jobs Campaign* puts numbers on many adaptation jobs, especially jobs that may help farmers adapt to climate change (Alternative Information and Development Centre 2016). João Camargo from Climáximo in Portugal argues that CJPs in the rich countries of the North should consist solely of mitigation jobs. In Portugal, the CJP also includes jobs needed for reforestation (Camargo 2020). Arguably, reforestation is an example of jobs that work for both mitigation and adaptation purposes, what O'Brien (2012) calls adaptation as 'deliberate transformation'.

## The Job Part of CJPs

The job part of the CJPs is no less central. Where the economic situation is most desperate, we find the highest salience of unemployment in the framing of CJPs. This is how Kenfack (2018, 58) summarizes the methodology used in the design of the South African Campaign:

> Four immediate research objectives were set for each Research Group and research area, namely to focus on 1) maximizing job creation; 2) minimizing carbon emission; 3) Identifying the primary agencies involved in achieving the first two objectives and 4) Indicating how the first two objectives are to be met.

Ashley confirms that the first 'fundamental point of departure' was unemployment, the second being 'to stop the advance of climate change' (Ashley 2018, 272). Using an eco-feminist perspective Barca finds the South African campaign particularly inclusive of 'community caregivers' by 'foreseeing up to 1,300, 000 jobs to be created in domestic/health care, land restoration and urban farming' (2017, 397). For Lawrence, climate jobs appear as a secondary vehicle for a different goal altogether. 'Guaranteed Full Employment: Achieving Economic Recovery with Climate Jobs' is a telling subtitle of his book (Lawrence 2020, 132).

The Norwegian plan for climate jobs is at the opposite side of the spectrum. Unemployment only briefly rose above 5 per cent in 2016 after a fall in the price of oil in 2014/2015. The CJP in Norway outlined how 50,000 jobs in offshore wind could give oil and gas workers jobs to transition into (Ytterstad 2013). But the mere suggestion of scaling down oil and gas tends to prompt a unanimous counter charge by the oil industry: you are prompting a downfall of the Norwegian economy, pulling the rug from under the welfare state (Sæther 2017). The Bridge to the Future can thus hardly make the sort of bold claims Lawrence (2020) makes for South Africa, that climate jobs is *the* avenue towards economic success. Differences between Norway and South Africa aside, all of the CJPs include suggestions or demands as to how to finance climate jobs.

All CJPs, moreover, demand that climate jobs should be *good* jobs. The Climate Jobs Campaign in New York State illustrates this well. In June 2017, Governor Andrew Cuomo announced a partnership with the Climate Jobs Campaign (Cuomo 2017). This joint initiative focused mostly on jobs in offshore wind (Climate Jobs n.d.). Cha provides a fascinating analysis of the preparatory discussions between unions, the researchers of the Workers Institute of Cornell University in New York and Cuomos own climate initiative. She finds that even where the climate goals of Cuomo have been high, 'it is important to note that there is no commitment to job quality under these goals' (Cha 2017, 460). This, of course, is a serious fault line for the possibility of increasing union support for CJPs, and would—if not countered—reintroduce the jobs versus the environment dilemma (Räthzel and Uzzell 2011) with a vengeance. Well-paid, unionized workers in the oil industry will not transition into low-paid, unorganized part-time work in offshore wind.

In a recent book chapter Skinner (2020) draws important lessons from the cooperation with the Cuomo administration. One of the key reasons why she thinks they were successful was that Climate Jobs New York was spearheaded by the unions in the building trades, energy and transport sectors:

> They launched the campaign and put the concerns of their existing members in the energy sector, as well as concerns about the quality of jobs in new clean-energy sectors, at the heart of their proposal. This meant that, at its core, the work was grounded in a vision of how New York could tackle the climate crisis but also create good jobs and build more equitable communities. (Skinner 2020, 139)

If and when authorities start to implement CJPs, the emphasis of the Climate Jobs Campaign may shift towards precisely this task of pressurizing those same authorities on the need for good jobs and more equitable communities.

## A Just Transition Away from Fossil Fuels

Taken together, the climate part and the job part of CJPs almost seamlessly become an integral part of a just transition away from fossil fuels. The Bridge to the Future Alliance, for example, 'have come together behind one specific aim: a democratic, planned, just transition that creates 100,000 climate jobs, and allows for slowing down the country's rapacious oil and gas extraction' (Broen til framtiden n.d.).

As Price succinctly puts it in her analysis of trade union resistance to fracking in the UK, 'the question of what employment will look like in a fossil-free economy lurks behind…all energy conflicts' (2019, 174). Her study gives a rare insight into how climate jobs demands are inserted into the life of specific and local campaigns purportedly about ways in which to stop fossil fuel emissions. Indeed, there are several other campaigns on just transitions away from fossil fuels that have incorporated climate jobs into their demands and reports. The reports produced by Washington D.C.-based Oil Change International, on how both Norway (Oil Change International 2017) and the UK (Oil Change International 2019) can manage a decline in its production of oil and gas, deserve special mention. They explicitly reference the Bridge to the Future in Norway, just transition and the importance of union rights for British workers. Inspired by the UK Million Climate Jobs Campaign, the Scottish Greens commissioned an innovative report on the jobs potential in decommissioning the oil platforms in the North Sea (Minio-Paluello 2015).

Last but not least, it is not hard to trace the core content of the climate jobs strategy amongst the school children who went on global climate strike on 20 September 2019. The 17-year-old protest organizer Daisy spoke to the Guardian's blog that day, of the demand of the Australian school strike, to fund a just transition and create jobs for all fossil-fuel workers and communities:

> Climate justice is not about jobs versus the environment. Just as climate change hurts people, unemployment hurts people (…). If our government cares about all of us then they need to get on with the job of stopping any new coal, oil and gas projects, powering Australia with 100% renewable energy by no later than 2030, and doing all this while funding just transition and jobs for all fossil fuel workers and their communities so that no one is left behind. (Live Blog 2019)

What is the most precise description of the CJPs, as versions of a just transition away from fossil fuels? Are CJPs an instance of a 'social dialogue' or a 'social power' approach (Sweeney and Treat 2018) to just transition? Are climate jobs 'affirmative' or 'transformational' demands? (Kenfack 2019). On the whole, Stevis and Felli (2015, 38) place the UK OMCJ within a 'social ecological' version of just transition.

Kenfack, however, quotes a representative of the CGTP in Portugal during a national climate justice meeting, which 'reveals the flexible position of the worker's confederation' and suggests a tension present within all climate jobs coalitions:

> We are all here struggling for climate justice and for climate jobs. We are all demanding divestment from fossil fuels and investment in clean energy, and we are all convinced that this is the way to go. Nonetheless, I would like to point out that we, as a trade union confederation have some characteristics that none of the movements in the hall has. We must defend the environment, yes! But we also have to defend workers and their jobs, and that is where it becomes com-

plex for us. If a fossil-fuel-dependent company asks us whether to continue operating or to divest from fossil fuels and stop its activities, for all of you the choice will be obvious. For us that will not be so easy to decide…. (Kenfack 2019, 235)

This statement is a useful cue for the rest of the analysis, and this quote also illustrates why the bridge-building efforts needed to invest Climate Jobs strategies with political agency involve both dialogue and power.

## The Inclusiveness of the Climate Jobs Strategy

The Green Economy Network in Canada was formed 'after consultations with various Indigenous communities, environmental organizations, labour unions and other public-interest organizations' (Clarke 2018, 10). Such consultations illustrate how the CJPs are a mobilizing strategy in search of *broader* agency. In this section, I will first elaborate on the inclusiveness of the Climate Jobs strategy. Then I briefly analyse how the Climate Jobs strategy navigates broader environmental/ecological and broader jobs concerns.

The CJPs are inclusive out of necessity. The sheer scale of halting climate change, providing drastic cuts of greenhouse gases and keeping most of the existing reserves of fossil fuels in the grounds, is so huge that it makes little sense to put forth climate jobs as the sole solution. Before spelling out the Climate Jobs strategy, the UK campaign emphasizes that 'There are thousands of things we need to do' (Campaign Against Climate Change 2014, 4). A crucial part of what Climate Jobs Campaigns do is to educate, inform and train. The main activity of the Bridge to the Future is its yearly conference, where 600 people hear climate scientists and others explain the gravity of the climate crisis. In Portugal:

> climate justice training is often organized for labour and trade unions that are more and more expressing their need to be trained on the issue. Primary and secondary schools also often solicit campaigners to come and give talks to pupils on climate issues and possible solutions. (Kenfack 2019)

One who gives many such talks is João Camargo from Climáximo, a climate scientist who has written a book on how to save the forests of Portugal, increasingly destroyed by severe forest fires (Camargo 2018b). He has also written, drawn and illustrated a Climate Change Combat Manual (Camargo

2018a), with climate jobs as an integral part, but where the main point is to help young people organize.

There is also a practical, movement side of the need to be inclusive. The impulse to unite, to join forces, is a natural and long-standing one. Climate change has contributed strongly to such unification processes. In her seminal *This Changes Everything—Capitalism vs The Climate* Naomi Klein praises the 'brave coalition' behind the South African OMCJ (Klein 2014, 127).

There is also a process of cross-fertilization between coalitions, where climate jobs become part of the demands of other organizations or coalitions. Satgar and Cherry chart the South African Food Sovereignty Campaign (SAFSC) and locate the OMCJ as part of the same post-apartheid protest cycle. The 'development of a domestic renewables industry as part of creating climate jobs' is on the list of SAFSC demands (Satgar and Cherry 2020, 328).

Although the CJPs build primarily on climate science, they are not oblivious to other environmental concerns. The subtitle of the latest UK report, for example, is 'Tackling the Environmental and Economic Crisis' (Campaign Against Climate Change 2014). There are tensions between some of the climate jobs demanded by the campaigns and broader ecological concerns. The Friends of the Earth in Norway had a big debate internally about wind turbines before joining the Bridge to the Future alliance (Skjellum Aas 2015). They would probably not have joined the alliance if it had not restricted the demand for climate jobs associated with wind turbines to *offshore* wind turbines. Although eagles will die in the rotor blades of wind turbine farms at sea, that was less painful for conservationists than the prospect of wind turbines onshore, which would take an even greater toll on Norwegian nature. Generally, environmentalists tend to accept that climate change is a 'threat multiplier', as Clarke puts it (2018, 32), for many other ecological issues, and that some destruction of local nature may be necessary to keep global temperature at levels nature and society alike can endure.

Another important issue that CJPs may find challenging to address is the question of job creation in general. What is the relationship between the general demand for jobs by workers and unions, and the specific climate jobs targeted towards reducing emissions? More jobs in renewable energy require mostly skilled workers, and this is likely to expand the membership of unions of electrical workers (Snell and Fairbrother 2011, 86; Angell 2018). But 'while there is a need for growth in unionized, high-quality jobs in renewable energy industries, there is also a need to dramatically improve working conditions for the most precarious in the workforce' (Karim 2019, 26).

Alia Karim tells the story of how the Toronto-based Good Jobs for All (GJFA) coalition has tried to link the struggles for high-quality jobs with social equity and climate justice. She has interviewed Carolyn Egan, president of the Steelworker Toronto Area Council, who sees a future for GJFA in 'continuing to work with racialized communities prioritizing their needs in the fight for climate jobs for all'. Karim also cites Naomi Klein's speech to the founding convention of UNIFOR in 2013:

> It's not just boilermakers, pipefitters, construction workers and assembly line workers who get new jobs and purpose in this great transition. There are big parts of our economy that are already low-carbon. They're the parts facing the most disrespect, demeaning attacks and cuts. They happen to be jobs dominated by women, new Canadians, and people of colour. (Karim 2019, 24, 27)

Klein's argument here is similar to the one we saw Barca make above in respect of the inclusion of 'community caregivers' in the South African 2011 OMCJ jobs report.

While this demonstrates the capability of CJPs to build bridges to workers outside of the 'big three'—shifts in energy, transport and buildings (Clarke 2018, 10)—a strategic dilemma results if the definition of what a climate job is becomes *too* inclusive. Brecher's notion of 'climate jobs for all' (Brecher 2018), as a key building block for the Green New Deal in the US, is not part of a Climate Jobs Strategy at all, but a project to alleviate poverty.

It may therefore be worth being stricter in the definition of climate jobs. It may be wiser to cross-fertilize the fight for jobs with the fight for climate jobs, than to remove the demarcation lines between the two types of struggles altogether. In recognition of this point, the latest version of the South African OMCJ (Alternative Information and Development Centre 2016) has a tighter focus on climate (mitigation) jobs than the 2011 one analysed by Barca (2017).

## Hard to Exclude: The 'green economy'

When some of the prime movers of the CJPs spell out their own strategy and positions, they draw up quite sharp demarcation lines against the 'green economy' (Ytterstad 2015; Ashley 2018). Clarke (2018, 56), for example, provides a scathing critique of the official Rio + 20 document from 2012 'The Future we Want'—a green economy strategy of commodification, privatization and financialization—'to generate new sources of profit or capital accumulation'. Leonardi finds that the South African OMCJ has been very successful in the ways in which they have 'inverted' the 'carbon trading dogma' of the green economy. 'Just like the green economists, OMCJ activists recognize the climate crisis as a terrain for development—as a job creation rather than a job killer—but do so by privileging the working classes' interest instead of the financial sector's needs' (Leonardi 2018, 111, 118).

The breadth and inclusiveness of the Climate Jobs Campaigns makes it hard to uphold sharp demarcations lines. The name of the coalition behind the Canadian OMCJ—the Green Economy Network—illustrates the difficulty of excluding all notions of "green economy" from Climate Jobs strategies altogether. When broad coalitions act together, the lines between 'climate jobs' and 'green jobs' are often blurred.

Climate jobs aim for emission cuts set by the global limits established by climate science. Most Trade Unions refer to green jobs. The International Trade Union Confederation contributed to an often quoted report by the United Nations Environmental Programme in 2008, where the definition of green jobs is centred on resource efficiency (Hampton 2015, 64–68). That may be a fitting definition in some cases like waste and farming, but it also allows the fossil fuel industry to present itself as 'green' or 'clean' if and when they increase their 'carbon efficiency' (Ihlen 2007; Ytterstad 2020b). Resource efficiency measures in the carbon intensive sectors will not achieve enough of an aggregate global emission cut and, as Stevis and Felli argue (2015, 39–40), risk 'displacing environmental costs across time and space'. Resource efficiency often forms the bedrock of strategies for national, corporate and industrial competitiveness, but pays scant attention to a limited global carbon budget (Ytterstad 2016). In the onshore based industry of Norway—aluminium production is a key case of this—resource efficiency measures taken over the last 20 years have helped reduce emissions substantially. Because of this history, the Industry and Energy Union of Norway proudly refers to their jobs as green jobs (Angell 2018).

Green jobs can mean more and many different things for unions (LRD Booklets 2019). The LO in Norway included demands for green union representatives in their 2009 book on green workplaces (Bjartnes et al. 2009). I pragmatically define green jobs as all measures that would assist the development of environmental concerns in existing workplaces (Ytterstad 2013). Such inclusion and pragmatism may explain why the Bridge to the Future conferences now have a *dual* focus, both on how to make existing work greener (green jobs) and on how to demand and win the fight for 100,000 new climate jobs.

The Norwegian case suggests a paradox in the search for broader agency for climate jobs: to broaden a Climate Jobs Campaign, you cannot exclude unions or other agents drawn towards the narrative of the green economy. But the dogmas of the green economy strategy tend to transfer agency back again—from the people to the market (Ytterstad 2016). Bridge building between climate jobs and green jobs is therefore a balancing act.

## Hard to Include: The State and the Public Sector

The CJPs demand *public* investment in climate jobs from the state. Some scholars within environmental labour studies have expressed concerns about the Climate Jobs strategy being *too* focused on state intervention. Räthzel and Uzzell (2012b, 9) think this state focus comes at the detriment of involving workers and unionists 'as makers of their own future'. Leonardi (2018, 111) even suggested that because states are so embedded in the attempt to enhance capital accumulation at the expense of the climate, the climate jobs strategy 'would benefit from a non-state based perspective'. The South African OMCJ is acutely aware of how hard it may be to include the state and the public sector as the key driving force for and employer of climate jobs, with the electricity public utility Eskom mired in crisis (Ashley 2018, 275–277).

In a neoliberal era, the state and the public sector are neither a trustworthy guarantor of good jobs nor of the public interest. That era is by no means over, but Klein's latest book on the burning need for a Green New Deal (Klein 2019) suggests a renaissance for the viability of state intervention (Felli 2019), where a Climate Jobs mobilization strategy might have more fertile ground in which to grow. The Trade Unions for Energy Democracy (TUED) were not alone in rejoicing (TUED 2019) over the ambitious and costly version of the Green New Deal put forward by the Bernie Sanders US presidential campaign (2019–2020), which said that 'renewable energy generated by the Green New deal will be publicly owned' (Bernie n.d.)

Such developments might assist the climate jobs mobilization in transforming the role of the state. From the perspective of acute climate crisis and with the enormous task of ensuring a rapid just transition away from fossil fuels, such a transformation of the purposes of the state *has* to happen (Roy et al. 2020). In such a vein, Sean Sweeney argues explicitly against a non-state, local and 'horizontalist' perspective, widespread in eco socialist writings and within climate justice activism more generally:

> while it is not difficult to imagine how local initiatives and community-level control can fit into an eco-socialist vision, it is much harder to see how the speed and scale of change required can be achieved without states playing a leading role in mobilizing the resources needed to create space for local efforts to be effective. (Sweeney 2020, 20)

Unlike Sweeney, the coalitions behind the CJPs do not have a socialist position. They have no agreed-upon theoretical or ideological foundation from which they could nationalize or expropriate parts of the economy for the purposes of rolling out and employing millions of well-paid unionized climate jobs within this decade. But like Sweeney, the coalitions tend to support the pragmatic reasons for turning attention '*toward* the state, not *away* from it' (ibid.). In a joint call for climate jobs published just before this chapter was finalized, this pragmatic reasoning for public sector jobs is apparent:

> Corporations and the market have had decades to solve the problem. They have not done so. We could argue about whether they can eventually do it. But it is clear that they will not act in time. Only governments can raise the amounts of money needed for climate jobs to replace almost all the fossil fuels we burn now. And only governments will do the many essential things which make no profit. So most of the jobs will have to be in the public sector. (Call 2020)

Behind such pragmatism, important and unresolved questions loom for the CJPs. How can states finance the climate jobs? Is classical Keynesianism an option (Lawrence 2020) or can modern monetary theory (MMT)—in vogue amongst proponents of a Green New Deal in the US—provide the 'policy space to purchase idle resources (e.g., unemployed labor power) and carry out policy programs' (Sweeney 2020, 22). According to MMT thinkers, governments that are 'monetarily sovereign' have an unlimited capacity 'to spend in their own currencies' (Mitchell and Fazi 2017, 263). If they are right, that would greatly increase the capacity of state-funded climate jobs.

The CJPs have no formal position, neither on Keynesianism nor on MMT, but the UK CJP would suggest that climate jobs are seen as public sector jobs, by demanding a National Climate Service on par with the National Health Service. The Norwegian CJP is less strident about demanding climate jobs from the state as the direct employer, perhaps because—as we have seen—they are more inclusive of green economy narratives. The main thrust remains the same for all CJPs, however. They demand that their governments must take the lead and roll out the various concrete projects necessary for the creation of climate jobs.

## In Search of an Agency with the Power to Win

The CJPs do not expect governments to do this without public pressure. The joint call for climate jobs cited above continues:

# 11 Climate Jobs Plans: A Mobilizing Strategy in Search of Agency 263

Climate jobs, and wider Green New Deals, are a necessity. They are also a strategy for mobilizing a mass climate movement…Public sector climate jobs will also mean we can promise retraining and new jobs to miners, oil workers and other carbon workers. That is morally right. It is also politically important. (Call 2020)

In practice, public sector unions, or unions of electrical workers, are most likely to become early supporters of climate jobs campaigns (Angell 2018; Hackett and Adams 2018, 29). Unions more entrenched in carbon intensive sectors, by contrast, may perceive climate jobs as a threat (cf Snell and Fairbrother 2011, 98; Stevis 2018). The Canadian and the Portuguese cases, however, suggest that it is possible to obtain the support of entire trade union confederations for CJPs.

Union support for CJPs is not the same as action though. Jacklyn Cock has interviewed trade unionists in South Africa who want unions to be much more actively involved in the OMCJ: one trade unionist insisted that the climate jobs campaign 'to have any traction should be driven by labour and housed within SAFTU' (Cock 2019, 863). Workers sometimes take action even without active union involvement. This was the case with the occupation of the Vestas wind turbine factory on the Isle of Wright in July-August 2009. The closing of this factory 'left the reputation of the new green economy—never mind the prospects for the 600 workers at the Vestas plant—in tatters' (Hampton 2015, 170–171). For Hampton, the Vestas occupation 'provides further evidence that workers and their trade unions have the potential to develop into swords of justice' (ibid., 177). It was the link between Vestas and the birth of the Climate Jobs mobilization strategy that prompted him to place climate jobs as an alternative *class*-based strategy to the state or market-centred versions of green jobs criticized in his *Workers and Trade Unions for Climate Solidarity* (2015).

The Vestas occupation lost, but pictures of this struggle are still on the front page of the UK climate jobs report. In Portugal, the activists of Climáximo creatively engage unions and groups of workers that might use the CJPs in their concrete struggles for jobs. With workers of the Sines coal power plant, they have tried to concretize the just transition process in this port city. Moreover, with the workers of the Moura Fábrica Solar, a solar panel factory being moved to China in search of higher profits, they have tried to mount demands for renationalization (Camargo 2020; Hansen 2020). None of these struggles have been successful so far, but these cases may serve to be inspirational.

Climáximo has probably been the most successful CJP in fusing the Climate Jobs strategy with the school strikes. As of the second school strike in Portugal, climate jobs were an integral and explicit part of the demands of Fridays For Future (FFF). FFF Portugal is also active in the Climate Jobs Campaign itself, which recently launched '10 measures to win in 4 years project'. FFF is the lead organization of three of these measures, 'national railway expansion', 'public buildings efficiency' and 'fighting for things of long lasting quality' (Empregos n.d.). Whether CJPs themselves are things of long-lasting quality remains to be seen, but the Portuguese campaign at least suggests that 'climate jobs' sounds like a good idea for young people too.

## Conclusion: Climate Jobs as Crisis and Emergency Jobs

'I want a concrete plan, not just nice words' (Milman 2019). With those words, Greta Thunberg sailed into New York for the Climate Action Summit in September 2019. The CJPs presented in this chapter are attempts to make a just transition away from fossil fuels more concrete. Plans in the UK, South Africa, Canada, Norway, Portugal and New York State all take climate science and sources of emissions as the methodological starting point for their estimates of jobs, first and foremost within energy, transport and buildings (Cha 2017). The coalitions behind these plans, moreover, emphasize that climate jobs must be unionized, well-paid, good jobs. Skilled jobs within renewable energy are particularly important as replacement jobs for fossil fuel workers. The climate coalitions built to realize these plans are very inclusive in nature.

This chapter has surveyed the limits of inclusiveness in the CJPs. It has been fairly easy for CJPs to include broader environmental and job concerns, although some demarcations lines have proven important to uphold. In South Africa, unemployment is such a desperate problem in its own right that it has been tempting to call all jobs outside the fossil fuel sector climate jobs. If that happens, CJPs risk becoming confused with a Keynesian recipe for full employment and economic recovery (Lawrence 2020), thus losing credibility as a plan geared to solve the global climate crisis. The South African OMCJ appears to have corrected itself on this score. Their 2011 report was too inclusive in its outline of climate jobs. Their latest report has a tighter focus on jobs that directly cut emissions (Alternative Information and Development Centre 2016, see Pillay, ch. 4). More concrete research on the climate effects of the CJPs is needed, but the biggest challenge for the Climate Jobs strategy is to mobilize for an agency which ensures that the bus drivers, construction workers and electricians we so desperately need are hired everywhere in time (Ytterstad 2020a).

In its search for powerful agency, the Climate Jobs strategy harbours some very real difficulties. The CJP in Canada emerged from a coalition called 'The Green Economy Network' (GEN). In theory, this is a people-driven version of the green economy at complete odds with market-driven versions. Tony Clarke, the founder of GEN, pits his Canadian plan for 'Getting to Zero' explicitly against the Rio + 20 summit version, which is about generating new and green sources of profit (Clarke 2018). Yet many trade unions take their ideas of what a green job is from such UN driven frameworks, centred on how to make green investments 'resource efficient' (UNEP n.d.). In practice, it has proven quite difficult for the Climate Jobs Coalitions to eschew 'green' jobs completely from their plans (Ytterstad 2013). But if and when CJPs blur the difference between 'climate jobs' and 'green jobs', they also risk blurring agency. Whereas climate coalitions, unions or workers can fight and even strike for climate jobs (Hampton 2015), investors are the most important agents for green jobs.

A related difficulty for the Climate Jobs strategy, noted by several scholars within environmental labour studies (Räthzel and Uzzell 2012b; Leonardi 2018), is their demand that climate jobs should be public sector, state jobs. The emergence of the Green New Deal may, however, be a sign that popular agency may become reinforced rather than eclipsed by the state. A strong Green New Deal organized around socio-ecological principles (Stevis 2018) may help create a solidarity strong enough to help groups of workers who stand to lose out in a climate jobs-driven transition. The CJP of New York State has succeeded in getting its governor on board, and this collaboration might represent a step in this direction, having secured some good climate jobs in offshore wind as well as building more equitable communities (Skinner 2020).

In combination, the climate crisis and the coronavirus crisis force nation-states, often against the will of their leaders, to see both our health and our climate as global public goods (Ytterstad 2020a). For the purposes of future research, the development of the Climate Jobs strategy and just transitions alike, the handling of the coronavirus crisis may prove decisive. On the one hand, several countries started opening up new coal power plants in the midst of the pandemic, something that prompted Greta Thunberg and Naomi Klein to tweet about 'recovery' as suicide (Klein 2020). I echoed this sentiment in an op-ed in April 2020: After the failures of the GND attempts by Obama after the financial crisis in 2008, the climate movement developed the slogan 'if the climate was a bank, we would have saved it already'. Rescue funds for fossil fuel companies suggest this slogan after COVID-19 will be updated to 'if the climate was an airline or fossil fuel company' (Ytterstad 2020c).

Yet on the other hand, Green New Deals were a prominent part of discussions on how to restart the economy and provide employment in the aftermath of COVID-19. In such a context, the CJPs could increase its sense of relevancy. A joint statement, 'Climate Jobs are an Idea whose Time has Come' (Call 2020), was published in May 2020, gathering signatures from the organizations and/or leaders of organizations from the climate jobs coalitions surveyed in this chapter, with the aim of increasing the size of their audience. At the same time, a new coalition called The Cry of the Xcluded was launched in South Africa, with a radical demand for three million jobs, one million of which were climate jobs (Cry of the Xcluded 2020).

If workers and Climate Jobs Coalitions find their way to action and victories in the struggle for climate jobs, that might still inspire and salvage a global and just transition away from fossil fuels, into a socially and ecological benign future, worthy of the Greta Thunberg generation.

## References

AFL-CIO. 2019. Unions Reject the Green New Deal.
Alternative Information and Development Centre. 2016. *One Million Climate Jobs—Moving South Africa Forward on a Low-Carbon, Wage-Led, and Sustainable Path*. Cape Town.
Angell, Frida Hambro. 2018. Klimaperspektiver i norsk fagbevegelse: En analyse av klimapolitisk handlingsrom i LO og utvalgte forbund. Master's thesis, University of Oslo.
Ashley, Brian. 2018. Climate Jobs at Two Minutes to Midnight. In *The Climate Crisis: South African and Global Democratic Eco-Socialist Alternatives*. South Africa: Vits University Press.
Barca, Stefania. 2017. Greening the Job: Trade Unions, Climate Change and the Political Ecology of Labour. In *The International Handbook of Political Ecology*. Massachusetts: Edward Elgar.
Bernie. n.d. The Green New Deal.
Bjartnes, Jon, Jon Olav Bjergene, Torgny Hasås, and Anne Beth Skrede. 2009. *Klimavennlige handlinger på arbeidsplassene*. Oslo: Gyldendal Arbeidsliv.
Bond. 2012. *Politics of Climate Justice: Paralysis Above, Movement Below*. Durban: University of KwaZulu-Natal Press.
Brecher, Jeremy. 2018. Climate Jobs for All: Building Block for the Green New Deal. LNS discussion paper.

Broen til framtiden. n.d. About 'Bridge to the Future'.
Burrow, Sharan, and Kumi Naidoo. 2015. Civil Society Will Build a Bridge to a Safe Climate Future. *Equal Times*, August 5.
Call. 2020. Climate Jobs are an Idea whose Time has Come.
Camargo, João. 2018a. *Manual de Combate às Alterações Climáticas*. Lisboa: Parsifal PT.
———. 2018b. *Portugal em Chamas—Como Resgatar as Florestas*. Lisboa: Bertrand Editora.
———. 2020. Interview during Bridge to the Future Conference.
Campaign Against Climate Change. 2014. *One Million Climate Jobs—Tackling the Environmental and Economic Crisis*. UK.
CBC News. 2016. Canadian Labour Congress Proposes Plan to Create 1 Million 'Climate Jobs,' March 3.
Cha, J. 2017. Labor Leading on Climate: A Policy Platform to Address Rising Inequality and Rising Sea Levels in New York State. *Pace Environmental Law Review* 34: 423.
Clarke, Tony. 2018. *Getting to Zero: Canada Confronts Global Warming*. James Lorimer & Company.
Climate Jobs, NY. n.d. Offshore Wind—Press Releases.
Cock, Jacklyn. 2019. Resistance to Coal Inequalities and the Possibilities of a Just Transition in South Africa. *Development Southern Africa* 0: 1–14. https://doi.org/10.1080/0376835X.2019.1660859.
Cry of the Xcluded. 2020. The Cry of the Xcluded: We Want a Radical New Deal that Provides Three Million Jobs. *Maverick Citizen*, May 13.
Cuomo. 2017. Governor Cuomo Announces Major Climate and Jobs Initiative in Partnership with the Worker Institute at Cornell University ILR's School and Climate Jobs NY to Help Create 40,000 Clean Energy Jobs by 2020. Governor of New York.
Empregos. n.d. Empregos para o clima.
Felli, Romain. 2011. Cosmopolitan Solutions 'From Below': Climate Change, International Law, and the Capitalist Challenge | Romain Felli—Academia.edu. In *Ethics and Global Environmental Policy—Cosmopolitan Conceptions of Climate Change*. Massachusetts: Edward Elgar.
———. 2019. Beyond the Critique of Carbon Markets: The Real Utopia of a Democratic Climate Protection Agency. *Geoforum* 98: 236–243. https://doi.org/10.1016/j.geoforum.2018.02.031.
Gunderson, Ryan. 2019. Work Time Reduction and Economic Democracy as Climate Change Mitigation Strategies: Or Why the Climate Needs a Renewed Labor Movement. *Journal of Environmental Studies and Sciences* 9: 35–44. https://doi.org/10.1007/s13412-018-0507-4.
Hackett, Robert A., and Philippa R. Adams. 2018. *Jobs vs the Environment? Mainstream and Alternative Media Coverage of Pipeline Controversies*. Corporate Mapping Project. Canadian Centre for Policy Alternatives.

Hampton, Paul. 2015. *Workers and Trade Unions for Climate Solidarity: Tackling Climate Change in a Neoliberal World*. London: Routledge.

Hansen, Nina. 2020. Minst halvparten av de som skal jobbe i solenergiverkene skal komme fra kullkraftverkene, er målet for klimaaktivisten i Portugal. *Fri fagbevegelse*, February 3.

Ihlen, Øyvind. 2007. *Petroleumsparadiset: norsk oljeindustris strategiske kommunikasjon og omdømmebygging*. [Oslo]: Unipub.

Jacobsen, Mark Z., and Mark A. Delucci. 2011. Providing All Global Energy with Wind, Water, and Solar Power, Part I: Technologies, Energy Resources, Quantities and Areas of Infrastructure, and Materials. *Energy Policy* 39: 1154–1169. https://doi.org/10.1016/j.enpol.2010.11.040.

Jacobsson, Diana. 2020. Young vs Old? Truancy or New Radical Politics? Journalistic Discourses about Social Protests in Relation to the Climate Crisis. *Critical Discourse Studies* 0: 1–17. Routledge. https://doi.org/10.1080/17405904.2020.1752758.

Karim. 2019. Carbon Cuts, Not Jobs Cuts: Toward a Just Transition in Canada. In *Local Activism for Global Climate Justice: The Great Lakes Watershed*, ed. Patricia E. Perkins, 1st ed. Routledge.

Kenfack, Chrislain Eric. 2018. *Changing Environment, Just Transition and Job Creation: Perspectives from the South*. Consejo Latinoamericano de Ciencias Sociales. https://doi.org/10.2307/j.ctvn96f9v.

———. 2019. Just Transition at the Intersection of Labour and Climate Justice Movements: Lessons from the Portuguese Climate Jobs Campaign | Global Labour Journal 10.

Klein, Naomi. 2014. *This Changes Everything: Capitalism vs. The Climate*. New York: Simon & Schuster.

———. 2019. *On Fire: The (Burning) Case for a Green New Deal*. New York: Simon & Schuster.

———. 2020. Germany Is Opening New Coal Plants.

Lawrence, Andrew. 2020. *South Africa's Energy Transition*. Cham: Springer International Publishing. https://doi.org/10.1007/978-3-030-18903-7.

Leonardi, Emanuele. 2018. Carbon Trading, Climate Justice and Labour Resistance: Definition Power in the South Africa Campaign One Million Climate Jobs. In *Climate Justice and the Economy: Social Mobilization, Knowledge and the Political*. 1st ed. Routledge.

Live Blog. 2019. Guardian Live Blog. *The Guardian*, September 26.

Lohman, Larry. 2011. Finalization, Commodfication and Carbon: The Contradictions of Neoliberal Climate Policy. In *Socialist Register 2012: Crisis and the Left*, ed. Leo Panitch, Gregory Albo, and Vivek Chibber, 1st ed., 85–108. The Merlin Press Ltd.

LRD Booklets. 2019. *Union action on Climate Change—A Trade Union Guide*. Labour Research Department Publications.

Malm, Andreas. 2018. *The Progress of This Storm: Nature and Society in a Warming World*. London; New York: Verso.

## 11 Climate Jobs Plans: A Mobilizing Strategy in Search of Agency 269

Milman, Oliver. 2019. Greta Thunberg 'wants a concrete plan, not just nice words' to Fight Climate Crisis. *The Guardian*, August 29.

Minio-Paluello, Mika. 2015. *Jobs in Scotland's New Economy—A Report Commissioned by the Scottish Green MSPs*.

Mitchell, William, and Thomas Fazi. 2017. *Reclaiming the State: A Progressive Vision of Sovereignty for a Post-neoliberal World*. London: Pluto Press.

Neale, Jonathan. 2008. *Stop Global Warming: Change the World*. Bookmarks.

———. 2019. How Climate Jobs Started and Spread.

O'Brien, Karen. 2012. Global Environmental Change II From Adaptation to Deliberate Transformation. *Progress in Human Geography* 36: 667–676. https://doi.org/10.1177/0309132511425767.

Oil Change International. 2017. *The Sky's Limit Norway: Why Norway Should Lead the Way in a Managed Decline of Oil and Gas Extraction*. Washington, DC.

———. 2019. *Sea Change: Climate Emergency, Jobs and Managing the Phase-Out of UK Oil and Gas Extraction*. Washington, DC.

One Million, Climate Jobs! One Million Climate Jobs! Green Econonomy Network Campaign Narrative.

Price, Vivian. 2019. Labour Organizing against Climate Change: The Case of Fracking in the UK. In *The Role of Non-State Actors in the Green Transition: Building a Sustainable Future*, ed. Jens Hoff, Quentin Gausset, and Simon Lex, 1st ed. Abingdon, OX; New York, NY: Routledge.

Räthzel, Nora, and David Uzzell. 2011. Trade Unions and Climate Change: The Jobs Versus Environment Dilemma. *Global Environmental Change* 21: 1215–1223. https://doi.org/10.1016/j.gloenvcha.2011.07.010.

———. 2012a. Mending the Breach Between Labour and Nature: Environmental Engagements of Trade Unions and the North-South Divide. *Interface* 4: 81–100.

———, eds. 2012b. *Trade Unions in the Green Economy: Working for the Environment*. Routledge.

Roy, Ashim, Benny Kuruvilla, and Ankit Bhardway. 2020. Energy and Climate Change: A Just Transition for Indian Labour. In *India in a Warming World: Integrating Climate Change and Development*, ed. Navroz K. Dubash. Oxford University Press.

Sæther, Anne Karin. 2017. *De beste intensjoner: oljelandet i klimakampen*. Oslo: Cappelen Damm.

Satgar, Vishwas, and Jane Cherry. 2020. Climate and Food Inequality: The South African Food Sovereignty Campaign Response. *Globalizations* 17: 317–337. https://doi.org/10.1080/14747731.2019.1652467.

Skinner, Lara. 2020. Building a Pro-Worker, Pro-Union Climate Movement. In *Labor in the Time of Trump*, ed. Jasmine Kerrissey, Eve Weinbaum, Tom Juravich, and Dan Clawson. London: Cornell University Press.

Skjellum Aas, Kristian. 2015. Hva er de grønne jobbene.

Snell, Darryn, and Peter Fairbrother. 2011. Toward a Theory of Union Environmental Politics: Unions and Climate Action in Australia. *Labor Studies Journal* 36: 83–103. https://doi.org/10.1177/0160449X10392526.

Stevis, Dimitris. 2018. US Labour Unions and Green Transitions: Depth, Breadth, and Worker Agency. *Globalizations* 15: 454–469. https://doi.org/10.1080/14747731.2018.1454681.

Stevis, Dimitris, and Romain Felli. 2015. Global Labour Unions and Just Transition to a Green Economy. *International Environmental Agreements : Politics, Law and Economics; Dordrecht* 15: 29–43. http://dx.doi.org.colorado.idm.oclc.org/10.1007/s10784-014-9266-1.

Sweeney, Sean. 2020. The Final Conflict? Socialism and Climate Change. *New Labor Forum* 29: 16–24. https://doi.org/10.1177/1095796020914987.

Sweeney, Sean, and John Treat. 2018. *Trade Unions and Just Transition: The Search for a Transformative Politics*. TUED Working paper 11. New York.

TUED. 2019. *US Presidential Candidate Bernie Sanders' Call for Public Ownership of New Renewable Energy*. TUED Bulletin 89.

UNEP. n.d. Green Economy.

Ytterstad, Andreas. 2013. *100 000 klimajobber og grønne arbeidsplasser nå!* Gyldendal.

———. 2014. Good Sense on Global Warming. *International Socialism* 4: 141–165.

———. 2015. Climate Jobs as Tipping Point—And Challenge to Norwegian Oil and Climate Change Hegemony. Lessons learnt (so far). In *The Politics of Eco Socialism—Transforming Welfare*, ed. Kajsa Borgnäs, Teppo Eskelinen, Johanna Perkiö, and Richard Wardenius. Routledge.

———. 2016. Vinn, vind eller forsvinn? Klima, miljø og økonomijournalistikk. In *Økonomijournalistikk. Perspektiver og metoder*. Fagbokforlaget.

———. 2020a. Indigenous Good Sense on Global Warming. In *Indigenous Knowledges and the Sustainable Development Agenda*, ed. Roy Krøvel and Anders Breidlid. Routledge.

———. 2020b. Solving the Climate Crisis—Time to Mobilize for the Climate Jobs of the Future. *Open Democracy*, January 22.

———. 2020c. Klimakrisen kommer stadig nærmere oss. *Dagbladet*, April 14.

# 12

## The Role of Ecuadorian Working-Class Environmentalism in Promoting Environmental Justice: An Overview of the Hydrocarbon and Agricultural Sectors

Sara Latorre

## Introduction

Within Latin America, Ecuador is a major site of class conflict concerning the extraction and exploitation of natural resources, thus providing an ample scope for exploring the interrelated processes of environmental injustice, which include the environmental deterioration of working and living conditions, and collective actions of resistance against them.

The political subjects of those collective actions form part of what Martínez-Alier calls 'popular environmentalism or environmentalism of the poor' (2002), that is, subaltern classes that mobilise in defence of the environment in which they work and live, or from which they gain their livelihood (mainly through non-extractive activities compatible with the integrity of their environment). This means that their struggle mainly responds to their material interest in the environment as their source of living. As such, the nature of this type of conflict is characterised predominantly by a resource-led dispute rather than by an inherent environmental consciousness. Since there is no reason

S. Latorre (✉)
Department of Development, Environment and Territory, Latin American Faculty of Social Science, Quito, Ecuador

why resource-led dispute and environmental consciousness should be mutually exclusive, they are also not necessarily connected. However, because of their direct dependence on the environment to make a living, it is more likely that these people are interested in managing environmental resources in a sustainable manner. A similar expression to refer to these forms of environmental collective action is 'environmental justice movements' (Martinez-Alier et al. 2016), a type of environmentalism which focuses on social justice and serves utilitarian purposes, thus differentiating itself from the post-materialist environmentalism of the Global North. The aforementioned movements are commonly regarded as an expression of new social movements (Melucci 1980). In fact, environmental justice movements rarely express their demands in strictly environmental terms. As Robbins (2012) points out, these actors commonly assert their identities through the way they make a living. By doing this, their 'livelihood identity' contributes to connect disparate groups by blurring other dimensions of identity such as ethnicity, race, religion, caste or gender. However, he further contends, the opposite trend also exists. That is, these actors might also articulate those subject positions based on class, ethnicity, race, caste or gender as their own political identities, while at the same time rallying demands for better environmental, working and livelihood conditions (Robbins 2012; Latorre et al. 2015; Lucero 2008).

This is precisely the case in Ecuador, where ethnic identity politics have played a relevant role since the 1990s, when the indigenous movement and, to a lesser extent, the Afro-descendant movement abandoned their class-based mode of organisation to adopt an ethnic strategy aimed at advancing land claims and other demands. Despite the hegemony of ethnic identity politics enacted by the Ecuadorian indigenous movement, other rural social groups have emerged and coexisted since the 1960s, while adopting different political collective identities such as peasant and rural wageworkers. Interestingly, the social composition of these movements and organisations at times overlaps, as is shown by (indigenous) subsistence peasants that engage in different ways with the wage labour system, for example.

Another peculiar aspect to Ecuador and, generally speaking, to Latin America, is that only a small sector of the working class is unionised. As Chomsky and Striffler (2014) affirm, labour studies in Latin America must necessarily reach further to include rural workers, workers in the informal sector and subsistence peasants. This is not often the case as many scholars tend to study peasant and indigenous struggles as expressions of (new) agrarian social movements rather than labour issues (Henderson 2015; Veltmeyer 1997; Lucero 2008). In order to encompass all these social expressions revolving around the relation between environment and labour, and to better understand the Ecuadorian environmental/labour justice movement, I rely on the concept of 'environmentalism of the working class' developed by several scholars such as Barca (2012), Bennett (2007) and Gordon (2004). This expression refers to

> the day-to-day struggles that workers at the bottom of the agriculture, industry and service sectors lead, both individually and in organised form, to defend the integrity and safety of their working environment and of the environment where their families and communities live. (Barca, 66)

In this regard, the Ecuadorian environmental/labour justice movement can be understood as the totality of collective actions enacted to defend the living conditions (of production and reproduction) of the working class (which includes both wage and non-wage labour). This group should not be seen as a unified social movement, but rather as a plurality of movements with their own organisational structures and collective identities that are increasingly networked and brought into strategic alliances among them and with other environmental justice organisations. Among the variety of movements that are being mentioned, the social movements and organisations against large-scale agriculture and mineral extractivism are the ones that have stood out the most through the last four decades. This is far from surprising since the Ecuadorian economy has been largely based on the agro-export (principally banana, shrimp and flowers) and hydrocarbon sectors (Latorre et al. 2015) over this period of time.

This chapter will look at several organisations and actors related to the agriculture and oil sectors, regarding the way they understand both their goals for social justice and how they incorporate an environmental agenda. In this sense, following previous work on Latin American working-class environmentalism (Carruthers 2008; Acselrad 2010; Chomsky and Striffler 2014; Alimonda et al. 2017), this chapter will examine the relation between work and environmental justice (in terms of internalisation of environmental and equality priorities, as well as the potential alliance with environmental justice organisations (mainly NGOs)), in order to contribute to a political ecology of work in Ecuador as an illustrative example of Global South countries.

The chapter begins with a brief review of the political economy of Ecuador by highlighting the heterogeneous social landscape of the country and the social transformations associated with the agriculture and hydrocarbon sectors. Next, it presents the case study of the Ecuadorian environmental/labour justice movement associated to the agriculture sector, followed by the hydrocarbon case. It concludes with a comparison between the two cases in relation to their environmental politics and some insights that emerge from them regarding the relationship between workers' rights and the environment.

## The Ecuadorian Political Economy

Ecuador is a resource-rich/low-GDP/low-labour-income periphery country whose economy relies heavily on agri-food and oil exports. In 2019, exports of primary products accounted for 80% of total exports, and oil constituted 43% of primary exports (Central Bank of Ecuador 2020). Regarding agri-food primary exports, shrimp, flowers and palm oil commodities are among the primary agricultural exports along with traditional cacao, coffee and bananas. According to the Central Bank of Ecuador, in 2019, exports of shrimp, bananas and flowers accounted for 80% of non-oil primary exports. While cacao, coffee and banana production are produced by both powerful landlords and by small and medium producers, shrimp, flower and palm oil commodities are dominated by agribusinesses. Moreover, they are highly polluting activities with exploitative and hazardous labour conditions.

### The Agrarian Political Economy

Generally speaking, Ecuador has a very high unequal agrarian structure despite two agrarian reform processes (1964 and 1973). According to the last National Agricultural Census (2001), the country's Gini coefficient only decreased from 0.86 in 1954 to 0.80 in 2000. Unfortunately, as Ecuador lacks an updated census to analyse its evolution in more recent years, the coefficient seems to have remained more or less at similar levels. This census also shows that 75% of the farmers own agricultural production units (UPAs) of under 10 ha, which represent 11.8% of the total hectares used for agriculture. However, 6.4% of the farmers own agricultural production units from 50 ha upwards, which represent 61% of the total hectares used for agriculture.

Today, the agricultural sector is still a fundamental source of employment in the rural areas, with more or less two-thirds of the rural economically active population working in this sector (Egas et al. 2018). There are marked agro-productive and socio-demographic differences between the three main regions of mainland Ecuador nonetheless.

Ecuador's lowland coast is the main agricultural export-oriented region where the land concentration is the highest. In this region, along with a powerful agribusiness sector, areas of peasant production exist, which are associated mainly with cocoa, banana, maize and rice farming. The coast-based peasantry, generally, has more access to land (20 ha in average) and a better quality of soil than its Andean highlands counterpart (2–5 ha in average) (Henderson 2015). These disparities are due to uneven processes of capitalist development and the application of an agrarian reform. In general terms, coast-based peasants identify as land-holding, commercially oriented and self-identified as *mestizo* peasantry. In turn, the Andean region has traditionally been an area of domestic agriculture and livestock production. However, new agro-export commodities have been introduced such as broccoli and flowers since the 1980s. This is the region where the *Kichwa* indigenous ethno-linguistic group has traditionally lived. Until the 1960s, this indigenous group was tied to the hacienda system (large estates owned by a *mestizo* powerful elite) characterised by a semi-feudal socio-economic organisation called 'huasipungo' (Clark 2017). The hacienda economy was linked to communities of *huasipungueros* or *kichwa* indigenous peasants who contributed permanent quotas of labour in exchange for small subsistence plots and low wage supplements. This pre-capitalist form of social relations ended with the first agrarian reform in 1964, which was followed by another agrarian reform in 1973. While these two reforms created a larger class of landed peasants, they did not significantly alter the agrarian structure and land concentration in the country as it has been already described. In the Andean region, the indigenous *ex-huasipungueros* generally obtained little and marginal land, which explains why nowadays they suffer from a pronounced process of dividing the land to create smaller farmsteads. A large mass of landless or landed-poor peasants exists today in both regions whose livelihoods rely more and more on off-farm wage labour activities, such as in the (banana, flowers, broccoli and oil palm) agribusiness sectors.

## The Oil Political Economy

The Amazon region remained relatively isolated from domestic and international market integration until late 1960s (Fontaine 2003). Oil exploitation started simultaneously with state-led and spontaneous colonisation processes during this decade in the northern provinces (Widener 2011). In the subsequent decades, the oil frontier expanded to southern provinces. This altogether led to major transformations in the Amazon ecosystems and the ways of life of the indigenous population, who initiated a process of organisation to resist encroachment (from settlers and oil companies) on their traditional territories, as will be explained in the second case study.

Today, oil extraction is mainly concentrated in the northern areas, but it is also present in some parts of the central Pastaza province, even within protected areas (Latorre et al. 2015). Several different ethno-linguistic indigenous groups live in the Amazon region with large differences across space and ethnicity in terms of market integration, organisational strength and access to collective land. Likewise, there is a large population of *mestizo* settlers organised in communities and cooperatives with loose ties to regional and national peasant organisations (Fontaine 2003). Both indigenous and *mestizo* inhabitants base their livelihoods on a combination of peasant production (mainly for subsistence) and off-farm wage labour activities (mainly in the oil industry and agribusiness). Sometimes, they are complemented with revenues from forest-based products (non-timber forest products, timber, tourism, etc.). Finally, this is one of the Ecuadorian regions with lowest rates in socio-economic indicators such as basic services, level of education and health infrastructure. This will influence the class struggle dynamics, which will be discussed later.

## The Agrarian Labour Movement: An Environmental Agenda Framed as Both Food Sovereignty and Plurinational State[1]

Most of the rural labour organisations that internalised environmental priorities in the next decades emerged in the aftermath of the first agrarian reform in order to advance the land struggle. In 1965, the Federation of Agricultural Workers (FETAP) was constituted and quickly became the leading organisation in the advancement of workers' rights and land redistribution until the 1980s (Altmann 2017), which in 1997 adopted its current name: the National Confederation of Peasant, Indigenous, and Black Organisations (FENOCIN).[2] Its constituencies come from both northern-central highlands provinces and coastal ones. In 1969, the National Federation of Agricultural Workers (FENACLE) was founded. Its members come from areas of short-cycle

---

[1] CONAIE's (1994) definition of the plurinational state is: 'the organisation of government that represents the joint political, economic, and social power of the peoples and nationalities of a country; that is, the Plurinational State is formed when various peoples and nationalities unite under the same government, directed by a Constitution' (p. 52).

[2] FETAP has changed its name several times. In 1968, it changed its named to the National Federation of Peasants Organisations (FENOC), which referred to its peasant composition, and in 1988 again changed it to FENOC-I to make its indigenous social base visible. In 1997 changed again its acronym to FENOCIN (National Confederation of Peasant, Indigenous, and Black Organisations). See table at the end of this chapter.

monoculture (maize and rice) and plantations (bananas and sugar cane) on the coast and, to a lesser extent, in the highlands (floriculture). From its very beginning, FENACLE's action has focused on the working conditions of rural wage labourers in the agribusiness (mainly banana and sugar cane) sector, as there were already other important organisations representing the peasants' interests. This notwithstanding, rural wage labourers were only a minority among its constituencies (Negreiros 2009). In this sense, FENACLE has combined a union agenda centred on wage labourers' rights (including safety and healthy occupational environments) with a traditional pro-peasant agenda focused on land, water, credit and market access. Since the late 1990s up until now, FENACLE and FENOCIN have adopted the food sovereignty discourse as their organising principle, as we further develop later (see also Kojola, ch. 16).

It is important to mention that there are very few rural unions due to several reasons. First, since the 1960s, the agricultural sector has been predominantly shaped by subcontracted forms of employment which prevent direct hiring, making it much more difficult for workers to unionise. Second, the Ecuadorian Labour Code only allows for unions to be created at the company level, not at a sector level (e.g. banana sector). These factors have contributed to greatly slowing the growth of rural workers' unions—even those that have been created have very little bargaining power. Last, the strong peasant identity that exists among rural poor land or landless workers even if their main source of income comes from off-farm wage labour (Henderson 2015).

A last prominent organisation advancing environmental justice, named ECUARUNARI, emerged in 1972 with the support of progressive Catholicism and some socialist organisations. Unlike the other two, it centred its fight not only on the agrarian reform, but also on the full citizenship of indigenous people[3] from the very beginning (Altmann 2017). ECUARUNARI operates as a federation of regional organisations based on the Highlands that are themselves shaped by traditional organisational forms already existing in the indigenous communities. In 1986, ECUARUNARI and its Amazonian counterpart Confederation of Indigenous Nationalities of Ecuadorian Amazon (CONFENIAE) joined together to create a national-scale indigenous organisation named Confederation of Indigenous Nationalities of Ecuador (CONAIE), which since mid-1990s became the most important political actor to confront the neoliberal policies that had been implemented in Ecuador since the 1980s. These structural adjustment programmes favoured large-scale agriculture exports and oil exploration that had severe

---

[3] Its full Indian name means 'awakening of the Ecuadorian Indians'.

consequences for the (indigenous) peasantry. As a reaction, the indigenous movement (MIE), led by CONAIE, adopted an ethnic-class agenda, in which its central points were an integral agrarian reform (including not only land, but also water and other productive resources), legalisation of ancestral territories (mainly in the Amazon as a way to confront extractive activities and settlers), bilingual education and plurinationality of the state as a strategy against long-standing processes of domination, violence and invisibility of the existence of the indigenous population (Lucero 2008). These central demands have an environmental dimension in the sense that the plurinational state was conceived to be rooted in an alternative economic model based on the principles of sovereignty, interculturality, equity and sustainability. In this sense, the nationalisation of strategic sectors (water, energy and communications) was promoted along with an agrarian model that prioritises the cultivation of food over biofuels or luxury crops (flowers or shrimps). Furthermore, it promotes agro-productive models that respect nature and traditional environmental knowledge and the use of own and non-genetically modified (GM) seeds (CONAIE 1994).

The emergence of MIE led to the political displacement of FENOCIN and FENACLE as the leading rural political actors, which had already entered into crisis since the early 1980s in a post-agrarian reform context (Giunta 2014). This fact, in turn, forced them to revise their agendas and strategies of action, which would lead them to adopt the discourse of food sovereignty as a banner of struggle by mid-1990s. The international articulation process of (indigenous) peasants that led to the formation of Via Campesina (VC) in 1993 played an important role in this discursive adoption. VC is an international organisation made up of more than 150 organisations in 70 countries. It is estimated that around 200 million farmers belong to it (Via Campesina n.d.). In 1996, this organisation officially launched the proposal of food sovereignty on a global scale as an alternative to the neoliberal agro-food system. This proposal is defined as the right of peoples to healthy and culturally appropriate food, produced through sustainable methods, as well as their right to define their own agricultural and food systems. It is a model of sustainable peasant production that benefits communities and their environment. Therefore, it is a proposal to confront current global environmental injustices reflected in the food, rural poverty and climate crises experienced at the global level.

FENOCIN and FENACLE are founding members of VC and its regional branch named Latin American Coordinating Body of Rural Organisations (CLOC), the latter formed in 1994. In the late 1990s, FENOCIN and

ECUARUNARI along with CONFEUNASSCNC,[4] which had emerged in 1992, created a consensus-building space called Mesa Agraria in order to achieve a common understanding of the agrarian issue. It lasted until 2009 and had the financial and technical support of several Ecuadorian NGOs working on environmental and rural issues during its existence. FENACLE joined this coordinating platform in 2005, but ECUARUNARI left it in 2003. The alliance between these class-based organisations and MIE (CONAIE and its regional members) has occurred at various times, but always with tensions. The former has the perception that CONAIE has dominated spaces within the state apparatus and has controlled state resources at the expense of other organisations.[5] It is also believed that CONAIE's ethnic discourse prevents class-based alliances, by reinforcing ethno-racial differences (Henderson 2015).

As a result of this collaborating space, in 2003 an innovative Agrarian Agenda was developed and framed in terms of food sovereignty (Giunta 2014). In this sense, the food sovereignty discourse developed in Ecuador can be understood as a unitary discourse to develop multiple dimensions of the emerging ecological agrarian issue: land/territory demands, gender equality, agroecology, respect for indigenous and traditional knowledge and recognition of cultural diversity.

Furthermore, the Agrarian Agenda was recognised as a reference point in the elaboration of the Minga para el Agro plan launched by the CONAIE indigenous leader and Minister of Agriculture in 2003. That same year, Lucio Gutiérrez became president with support of the MIE alliance, which explains why an indigenous leader took office. However, a few months later, MIE withdrew its support for Gutierrez, as he followed neoliberal policies of previous governments and its representatives resigned from all government posts. Therefore, the institutionalisation of this agenda was very limited.

During 2002–2005, Mesa Agraria's members along with MIE successfully rallied against the regional Free Trade Area of the Americas. Afterwards, the bilateral Free Trade Agreement was planned to be signed between the US and the Ecuadorian government. They denounced the negative impacts that it would have on peasants and biodiversity.

---

[4] Single Confederation of Social Security Affiliates in Ecuador. In the mid-twenties this organisation separated, which led to the emergence of CNC-EA (National Coordinating Body of Peasants-Eloy Alfaro) and CONFEUNASSC.

[5] In 1989, CONAIE assumed responsibility for helping to manage a programme of intercultural bilingual education in all Indian areas of the country. The budget is provided by the state. Furthermore, in 1998, it took control of CODENPE (Development Council of the Nationalities and Peoples of Ecuador) excluding other indigenous organisations such as FENOCIN.

The arrival to power of the progressive government of Rafael Correa in 2006 was another turning point for these rural labour organisations and their demands around food sovereignty. This year was the beginning of the process of institutionalisation of food sovereignty and the national political prominence, once again, of the class-based rural labour organisations, especially FENOCIN. Between November 2007 and July 2008, there was a national constituent process to elaborate the new Ecuadorian constitution, which would recognise the right to food sovereignty. FENOCIN and FENACLE had a direct participation as two assembly members came from these two organisations. The former was the most influential organisation, shaping constitutional articles relating to food sovereignty. It is important to note that these federations had the technical support of several NGOs and national activist platforms (such as the Agrarian Collective[6]) in favour of agro-ecology as a productive model to achieve food sovereignty.

The constitutional recognition of the right to food sovereignty and the promulgation of its Organic Law on Food Sovereignty (2009 and reformed in 2010) were a great achievement of these social organisations. However, with its approval, a hard fight for the content of the food sovereignty legal framework through the elaboration of the supplementary laws (related to water, agro-biodiversity, productive promotion, lands, territories and communes) had started. The process of formulation and approval of these supplementary laws lasted until 2017. For this policy-making process, the Plurinational and Intercultural Conference of food sovereignty (COPISA) was created. It is a civil society space made up of nine members meritocratically selected by the Council for Citizen Participation and Social Control. It had the task to draft legislative proposals on these supplementary laws by undertaking a truly participative process with Ecuadorian civil society. However, it was the National Assembly (dominated by the government party) which had the competence to pass the laws. As a result, COPISA's role in the development of the food sovereignty framework was very limited (Giunta 2014).

The food sovereignty Ecuadorian legal framework pretty much reflects the government's own vision, which is aligned with increasing domestic production and productivity based on conventional agrarian technologies, stimulating the national agro-industrial processing sector, substituting imports and increasing the production of both new and traditional agro-export commodities to generate foreign exchange revenues. Despite this situation, Ecuadorian

---

[6] The Agrarian Collective was made up of several NGOs such as HEIFER, IEE, Intermon-Oxfam, CAFOLIS and the Agroecological collective, which is a network of both rural growers and urban-based consumers working on 'healthy food'.

VC's members have continued to support the government until now and have participated in the COPISA institutional sphere. CONAIE, unlike the former, distanced itself from the government in 2008 once the Ecuadorian constitution was approved. The Ecuadorian government has adopted a very critical stance towards this social movement, accusing its leaders of being '*ponchos dorados*' (indigenous elites). It has also closed down the state spaces that MIE had controlled since the late 1980s as previously mentioned.

Since 2016, MIE, along with several other rural labour organisations, supportive of the VC's food sovereignty principle and government critics, has converged in a National Agrarian Agreement. The central points of the ecological agrarian issue raised in 2003 remain fundamental. While FENOCIN and FENACLE have not supported it, some of its base organisations—mainly those that have agroecology as a key organising principle—have (Cumbre Agraria n.d.).

It is important to highlight that while government's food sovereignty productivist policies distanced themselves from demands for food justice as those raised by VC and its national counterparts, some of them actually benefitted smallholders and agricultural wage workers. As for public policies related to agribusiness, Correa's government structured a new model of contract farming which tied small producers to national agribusiness firms. PRONERI, as it was called, was aimed at fostering domestic food production and processing while, at the same time, providing stable prices and market access to smallholders. The PRONERI programme was accompanied by policy measures that increased the availability of public credits for agriculture, through the National Development Bank, and subsidised agricultural inputs. Additionally, the government carried out the Reactivation Program of Coffee and Cocoa, which are export crops cultivated by smallholders. As Henderson (2015) notes, it is exactly because of those programmes that FENOCIN's coastal members were supportive of the government.

For what concerns rural working conditions and the rural labour force, Correa's government doubled the minimum wage to 310$/month and, as previously stated, prohibited labour outsourcing from 2008 to 2016. It is worth mentioning that most of the Ecuadorian rural workers did not have an anti-agribusiness stance; on the contrary, they demanded the formalisation of labour contracts and improved working conditions including health and safety aspects (Latorre et al. 2015).

Lastly, in 2009, the government approved a modest land reform programme, with the goal of legalising the communal land of peasant and indigenous organisations. The original plan was to reduce the Gini index from 0.81 to 0.70 by distributing 2,500,000 ha of land to smallholders. An assessment issued by the government in 2015 stated that only 114,500 ha had been redistributed to peasant associations and 897,000 ha, mainly in the Amazonian provinces, had been titled in favour of indigenous communes, communities and nationalities. In 2016, the government also issued a new Land Law, heralded as an improvement by FENOCIN (Clark 2017).

Summarising, since the beginning of the twenty-first century, the Ecuadorian agrarian agenda raised by rural labour organisations has transcended the traditional classist claims to encompass new issues related to territory-based identity and ecological sustainability. This new ecological agrarian question, rather than claiming indigenous/peasant inclusion in the dominant agrarian model, has aimed to reshape it towards more locally embedded food systems on the basis of agro-ecology. Furthermore, it has been framed in both food sovereignty and plurinational state terms. However, this narrative under-represents the interests of full-time crop workers and commercially oriented peasants that have less radical stances towards export production and agribusiness. These demands have been taken up by the Ecuadorian government, which explains their support of the government.

## The Anti-Oil Labour Environmental Politics: Between Territorial Rights, Cosmopolitics and National-Developmentalist Claims

Amazonian indigenous inhabitants and settlers have traditionally led resistance to oil activities in Ecuador with the support of national and international NGOs. Whereas the former has become key to resist the expansion of the oil frontier in the Amazonian region, the latter has stood out for implementing environmental justice actions in areas where the oil frontier has become old and resources have been emptied. In turn, the role of environmentalist NGOs has been fundamental for the networking of protesters within Ecuador and beyond, and for the support of legal actions and public campaigns. Finally, another actor, the public oil union FETRAPEC, played a prominent role during the 1990s and beginning of the twentieth century, in a context where the oil industry was experiencing a neoliberal-led transformation process.

These four political actors emerged during the 1980s in a context where the Ecuadorian Amazon region (mainly northern provinces) became an area of intensive exploitation of timber, oil and agricultural resources. Oil exploration and exploitation along with spontaneous and state-led colonisation processes were already around thirty years old, and the socio-environmental impacts caused by them were already widely visible. Moreover, the country was entering a neoliberal model of development after a severe economic and debt crisis, which led to a denationalisation process of the oil industry.

The 'Amazon for life' international campaign (1989–1994), launched by various national and international environmental NGOs in alliance with MIE, emerged within this context. It is perhaps the most well-known campaign on the matter, supported by a very clear environmental justice agenda. It was developed around three main targets: first, to avoid oil exploitation in the Yasuní National Park, a mega-biodiverse area and the traditional territory of isolated and non-isolated Huaorani indigenous peoples; second, the territorial legalisation of Huaorani indigenous peoples as a strategy to halt oil exploitation in their territory; and finally, the denouncing of the environmental liabilities caused by the company Texaco-Gulf since mid-1960s. This company operated in Ecuador until 1992, used obsolete technology and took advantage of the fact that during that time Ecuador did not have environmental laws. As a result, wide pollution, associated health problems, and the undermining of both indigenous and settlers' livelihoods were common patterns in the north of the Ecuadorian Amazon. While this campaign failed to halt the expansion of the oil frontier into this protected area and Huaorani's territory, it had much more success internationalising Texaco-Gulf's environmental injustices and strengthening local action. In this respect, 30,000 inhabitants affected by this company's environmental mismanagement filed a lawsuit in 1993 against the company for causing environmental pollution before the Supreme Court of New York, headquarters of Texaco-Gulf.[7] In 1994, the Amazon Defence Coalition (FDA) was formed to bring together legal action against Texaco-Gulf. It was assembled by 12 *mestizo* social organisations in alliance with Secoya, Cofan, Huaorani and Quichua indigenous organisations (Yanza 2004). Beyond this international legal action, it also exercised up until now functions of legal advice to local communities in their fight to achieve a level of economic compensation and environmental protection from oil pollution in accordance with Ecuadorian legislation. National environmental activist NGOs also supported the FDA with resources and technical expertise. As a result of all this work, it has played a key role in the institutionalisation of local environmental conflicts related with oil pollution.

[7] In 2002 this case was returned to the Ecuadorian justice system, and in 2011, Chevron, which acquired Texaco in 2001, was ordered to pay almost $9 billion in damages. At the same time, Chevron responded with an appeal as well as a civil lawsuit of its own against the plaintiff's lawyers arguing fraud under a racketeering act. In 2018 the Permanent Court of Arbitration in The Hague agreed with the company. Now Ecuador faces a million-dollar fine still to be determined.

Furthermore, its legal and political activities contributed very much to the strengthening of environmental legislation on hydrocarbons and the use of less polluting technologies (Fontaine 2003). It is important to highlight that the first piece of environmental legislation tackling hydrocarbon activities dates back to 2001. Until then both national and transnational oil companies had been conducting their activities under virtually no, or very low at the best, environmental regulation over several decades. As a result, there are high rates of cancer incidence, abortions and cancer-related deaths among the population living in the vicinity of oil wells (Ministerio del Ambiente-MAE 2016; Widener 2011).

As mentioned before, in the 1990s, the Ecuadorian indigenous peoples' movement (led mainly by CONAIE) embraced the organising principles of identity and territory and became a powerful nationwide actor. This ethnic discourse initially emerged among Amazonian indigenous organisations and federations (CONFENIAE) that lobbied for its adoption by CONAIE and its Andean counterpart ECUARUNARI (Fontaine 2003). As a result, where indigenous peoples are involved, resistance to oil extraction has usually been made through multilevel indigenous movement's organisations, using an identity politics discourse that stresses the recognition of indigenous rights—mainly territory and prior consultation—(before the 1998 Constitution)[8] and later for compliance with law (from 1998 onwards). In this sense, indigenous identity politics in Ecuador is inseparable from environmental and resource claims. During the 1990s, indigenous resistance outcomes varied across the Amazon region, with exploitation proceeding unabated in the north and also in some parts of the central provinces, while temporary stops were forced in other parts of central Amazon and in the southern provinces (Latorre et al. 2015).

However, MIE had more success fighting against the neoliberal agenda implemented by several governments from the early 1980s until 2006. During the first oil bonanza (1972–1982), foreign debt grew significantly, as international lending institutions were eager to lend money to petro-dollar countries such as Ecuador. In 1982, Ecuador could no longer finance the repayments. A rescheduling of the debt was negotiated, subject to the direction and economic adjustment plans of international financial institutions (IFIs) such as the World Bank. These programmes prioritised fiscal solvency and inflation control, which was to be achieved through market liberalisation, privatisation, integration and specialisation within the world economy as a raw

---

[8] This Ecuadorian constitution recognised for the first time the indigenous collective right such as the right to collective territory and to prior and informed consultation in line with the 169 ILO Convention.

material producer. While neoliberalism reactivated economic growth after the country's debt crisis, this was accompanied by increased poverty and inequality (Larrea 2004), which eventually led to massive emigration and the dollarisation of the economy in 2000 during the 1998–1999 Ecuadorian financial crisis. However, these policies were highly contested by wide sectors of the Ecuadorian society. From 1995 to 2001, MIE weaved alliances with other popular social actors, through the Coordinating Body of Social Movements (CMS), and thus increased its capacity to confront the very meaning and character of the neoliberal state. Two presidents, Abdalá Bucaram (1997) and Jamil Mahuad (2000), were overthrown during the period within which CMS was leading the anti-neoliberal struggle (Marega 2015) as a result.

One of CONAIE's key allies in CMS was the public oil union. For the Ecuadorian oil industry, neoliberalism meant the replacement of previous national policies with new ones promoting the liberalisation and privatisation of the sector. This union was formed in 1980s under the name FETRACEPE as the public oil company was named CEPE.[9]

From its beginnings, it adopted a national and classist political project where the defence of national sovereignty and strategic resources were core elements (Marega 2015). In the 1990s, an anti-imperialist and anti-transnational oil companies' sentiment rose among Ecuadorian civil society as private oil companies operating in the country obtained around 80% of the oil royalties share while the Ecuadorian state was left with only 20% (Latorre et al. 2015). MIE, FETRAPEC and environmental activism NGOs articulated an anti-imperialist, environmentalist critique of companies that they regarded as looting their country. The public oil company's environmental mismanagement was made invisible by a narrative that pointed to foreign oil companies as the ones obtaining Ecuadorian oil benefits at the expense of the Ecuadorians, who were left with the environmental externalities of this highly polluting activity. In this regard, while most of MIE's Amazonian bases confronting the expansion of the oil frontier (including state-led exploitation) were adopting a direct opposition to it, at a national level, MIE converged with FETRAPEC as it considered that the defence of the national oil sovereignty was a priority in terms of it being one of the most important Ecuadorian strategic natural resources.

---

[9] In 1972, the first Ecuadorian public oil company called Ecuadorian State Oil Corporation (CEPE) was created. In 1989, it was restructured into a holding company, comprising three subsidiary companies, which was named Petroecuador. When the company changed name, the union FETRACEPE also changed its name to FETRAPEC. In 2010, two public companies were created, Petroamazonas EP and Petroecuador EP.

This anti-imperialist and environmental political discourse critique was also very present among northern Amazonian inhabitants, mainly *mestizo* settlers, who were the ones that most experienced the political and economic marginalisation of the Amazonian oil extractive areas. After many years of oil exploitation, this region showed high levels of poverty and lack of basic infrastructure and public services. As a reaction to this marginalisation, the people self-organised in a civil society platform called Bi-Provincial Committee, through which they organised several strikes during the first years of the twentieth century. Most of these strikes were responded to by the government with the militarisation of the region and the declaration of a state of emergency. The strikers were mainly claiming a greater oil rent distribution for the extractive areas and inhabitants, environmental restoration of the highly polluted oil areas and better environmental practices by the oil companies (Latorre et al. 2015). Despite the fact that both *mestizo* and indigenous inhabitants share the same polluted and unhealthy environments in the northern Amazonian provinces, perceived differences of ethnicity or 'race' prevented them from organising themselves together to face this situation. In this respect, the Amazonian branch of MIE was absent during the aforementioned environmental justice collective actions (Widener 2011).

Following the same trend as in the agricultural case, the arrival to power of the progressive government of Rafael Correa in 2006 was a turning point for these anti-oil environmental politics. In fact, the political context had changed a year before during the transitional government of Alfredo Palacio (2005–2006), in which Rafael Correa was its economy minister. During his short period in government, Correa implemented some progressive economic policies that partially addressed the national oil claims raised by the anti-oil movement. He amended the Hydrocarbon Law, so as to nationalise 50% of the extraordinary earnings by foreign oil companies, terminated the contract with Occidental (which represented 12% of total oil production) and created an environmental fund to enact measures of environmental and social restoration in the northern Ecuadorian Amazon. Later on, as presidential candidate in 2006 and as President between 2007 and 2011, he rallied for a moratorium of oil operations in the south-centre provinces of the Ecuadorian Amazon. Furthermore, his government launched the Yasuní-ITT Initiative to keep over a billion barrels of oil in the Parque Nacional Yasuní (Yasuni National Park) in return for payments of $3.6 billion from the international community (half of what Ecuador would have realised in revenue from exploiting the resources at 2007 prices) in 2007. Additionally, rights to nature were recognised within the innovative post-development legal framework of the Good Living in the 2008 Constitution, which aimed to depart from modern western ideologies, mainly those of nature-society dualism and Eurocentric universalism (Villalba-Eguiluz and Etxano 2017).

This institutional permeability towards a post-oil economy ended around year 2011 in the case of oil, when the advance of the oil frontier towards the central-southern Amazon was announced. The Yasuní-ITT Initiative (2013) was put to an end and the government radicalised its frontal opposition against environmental activists NGOs, MIE and oil unionism. Generally speaking, the government of Correa implemented a neo-extractivist agenda that directed a greater share of revenues from extractivism towards redistribution, especially to the areas where oil production takes place. By 2015 the Ecuadorian government had already invested US$ 518,582,852 in the Amazon region to implement various projects related to environmental sanitation, road and education infrastructure, electrification and the relocation of the population through housing projects (Ecuador Estrategico n.d.). It had also strengthened its two public oil companies and created the Environmental Remediation Program (PRAS) to deal with oil-led environmental liabilities. Up until 2016, 931 oil-polluting sources in the northern Amazonian provinces had been eliminated (MAE 2016). As a result, the environmental justice claims raised by the Bi-Provincial Committee were met by the Ecuadorian government, which in turn led to its disarticulation. However, more critical voices and local concerns towards oil extraction have been delegitimised and criminalised (Latorre et al. 2015).

Within this neo-extractivist context, indigenous actors have implemented novel resistance strategies, by the deployment of an indigenous cosmopolitics (Marisol de la Cadena 2015) in the context of climate change international governance. That means the positioning of indigenous ontologies that recognise nature as sentient and humans as inextricably connected to nature as alternatives to dominant political and economic models. In the case of Ecuador, two indigenous cosmopolitics initiatives have emerged: the first, called *Kawsak Sacha* (Living Forest) and developed by Amazonian *kichwa* organisations of the province of Pastaza, pursues the declaration of kichwa's territories as *Kawsak Sacha*, a new legal category based on the concept that they are sacred (Bravo 2019). The second, called 'Cuencas Sagradas' (Sacred basins), is promoted by CONFENIAE and its Peruvian counterpart with the support of various national and international environmental NGOs and COICA.[10] It aims to protect a bio-national rainforest region (Ecuador-Perú) of 30 million hectares from extractive activities. This area covers the indigenous territories of half a million of indigenous inhabitants of 20 different

---

[10] Indigenous Coordination of the Amazon Basin.

nationalities (Cuencas Sagradas n.d.). Both proposals have been positioned in various climate change Conference of the Parties (COPs) as climate change mitigation proposals. In this respect, Amazonian indigenous organisations are claiming an oil moratorium for the south-central Amazon region while the world moves towards a post-carbon economy.

In short, strict opposition to oil exploitation existed mainly among indigenous organisations since the late 1980s up until now. These collective actions of resistance have been framed from the very beginning within cultural identity politics that emphasise the indigenous right to prior consultation on extractive activities in their territories. In recent years, these indigenous politics have been intertwined with cosmopolitics that position indigenous ontologies and territories as radical solutions to climate change and biodiversity loss. These new political repertoires are giving them more international visibility and access to transnational allies, which in turn might strengthen their struggle at a national level. For their part, *mestizo* settlers and the oil public union played a key role during 1990–2000, demanding the nationalisation of oil and the payment of the ecological debt in the form of greater distribution of oil revenues at the sites of extraction. These social actors have been demobilised in the recent years as the Ecuadorian government has met their demands.

## Conclusions

This chapter has analysed the environmental politics of labour organisations in the agricultural and oil sectors in Ecuador. It has shown that among the social groups belonging to the category of working-class environmentalism, indigenous and peasant organisations are the ones leading the struggles for environmental justice. In turn, union organisations are playing a more modest role. In this regard, not only class, but also ethnicity/ 'race' is the key social category that interplays in the way in which the Ecuadorian working class organises itself and advances an environmental agenda.

In both cases, the internalisation of explicit environmental priorities since the 1990s by labour organisations has been interlinked with the recognition of demands for cultural diversity and autonomy. This was to be expected since the environment is a livelihood issue in Ecuador. Therefore, the development of an environmental awareness by these social groups is tied to the defence of particular cultural ways of living with nature. This has been illustrated, for example, by the ecological valuation of the traditional (indigenous) peasant model of agricultural production and its associated knowledge and practices in the first case study. In the oil case, the entanglement between indigenous rights claims and the defence of the environment has been more straightforward. As it has been presented, the Amazonian indigenous organisations positioned themselves from the very beginning as guardians of the environment for the defence of their territories and natural resources. While there might exist a rhetoric usage of the environmental discourse by these indigenous organisations, in the Ecuadorian case, oil resistance is increasingly associated with alternative proposals for sustainable management of the natural resources located in their territories.

For the environmental politics of Ecuadorian labour organisations, related to both the agricultural and oil sectors, international alliances with other labour organisations and advocacy groups have been fundamental. For instance, in the agriculture-related labour organisations, the frames (food sovereignty/plurinational state) used by them since the 1990s to advance an environmental agenda are very much influenced by the agenda of VC and/or the international indigenous rights movements. In turn, Amazonian indigenous politics against oil has had the support of the COICA and indigenous rights and environmental advocacy groups. In this second case, these organisations have mainly helped indigenous organisations to become visible and scale up of their struggles to international bodies with the goal of putting pressure on the Ecuadorian state from outside. The need to put pressure on the government from outside is explained by the antagonistic positions between the state and anti-oil groups due to the key role that oil production has for the state's revenues. In both case studies presented, domestic advocacy groups have also been important, by providing legal and technical expertise and helping activists to institutionalise their demands.

Some particularities can be highlighted in these two case studies mainly with regard to the type of environmental politics and the institutionalisation of the claims of social actors. Among the agriculture-related labour actors, all of them have converged on an ecological agrarian question that seeks more equitable and ecologically sustainable food systems despite using different political frameworks. Notwithstanding this apparent consensus, controversies exist around what it means and how to move towards this goal. In particular, the role played by the agribusiness sector and export-crops in imagining sustainable and fair food systems has been very contentious. Therefore, both a radical ecological agrarian position (seen as antagonistic towards peasant-based agro-ecological productive models) and a moderate one (seen as complementary, but with better environmental working conditions and fair markets) can be discerned among labour organisations. Furthermore, these two positions can be read in terms of class differences existing among the

Ecuadorian peasantry. In this regard, organisations representing full-time plantation workers and commercially export-oriented peasants are adopting the moderate stance, whereas poor land peasants or landless peasants' organisations tend to adopt the radical one. This moderate position among some labour organisations has allowed them to make alliances with the Ecuadorian government from 2006 onwards, which explains their achievements in institutional terms. In this regard, the Ecuadorian agrarian legal framework and the policy implementation in the last fifteen years reflect, to some extent, this moderate position.

Regarding labour organisations around oil exploitation, there are two main positions that can also be read along ethnic/ 'race' and geographical lines. The first claims the abandonment of oil exploitation in order to move towards a post-carbon economy and is adopted by indigenous organisations in the areas of oil frontier expansion. The second demands a state-led oil exploitation model with both better oil rent distribution in the sites of extraction and better environmental management practices. This is embraced mainly by *mestizo* labour organisations in old oil frontiers. The actors of these two positions momentarily converged during the neoliberalisation of the Ecuadorian oil industry, when an anti-imperialist and environmentalist position emerged. During this period of time, anti-imperialist oil extraction was put at the front of their struggles. It was also the only moment in which the public oil union converged with environmentalist movements. In this regard, it can be said that the Ecuadorian public oil union adopted an implicit environmental stance in the sense that it opposed the environmental destruction transnational oil companies produced in their quest for profits. However, during the neo-extractivism focus of the last decade, the union of oil workers in the public sector is no longer a key political actor advancing environmental concerns. Regarding the institutionalisation of these two positions (the abandonment of oil production vs. the nationalisation of oil production), the second one has permeated, to some degree, the Ecuadorian state since 2000. The Ecuadorian government has sought to gain oil legitimacy by strengthening its role in social and rural development at the sites of extraction while increasing the oil exploitation by the public oil company.

Finally, it is important to highlight that Ecuador's integration with global markets has relied on cheap labour and the externalisation of the ecological damage associated with agricultural and mineral commodities. Therefore, despite the fact that in the last decade there has been some progress in the internalisation of some of these socio-environmental externalities, there is still scope to move this ecological modernisation environmental agenda forward in both economic sectors. Ecuador's economy relies on oil. In this regard,

Table 12.1 Timeline of the emergence of Ecuadorian working-class environmental organisations

| Organisation name | Year of foundation | Acronym |
| --- | --- | --- |
| Federation of Agricultural Workers | 1965 | FETAP |
| National Federation of Peasants Organisations/ The National Confederation of Peasant, Indigenous, and Black Organisations | 1968/1997 | FENOC/ FENOCIN |
| National Federation of Agricultural Workers | 1969 | FENACLE |
| Awakening of the Ecuadorian Indians | 1972 | ECUARUNARI |
| Confederation of Indigenous Nationalities of Ecuadorian Amazon | 1980 | CONFENIAE |
| Confederation of Indigenous Nationalities of Ecuador | 1986 | CONAIE |
| Single Confederation of Social Security Affiliates in Ecuador | 1992 | CONFEUNASSCNC |
| Via Campesina | 1993 | VC |
| Latin American Coordinating Body of Rural Organisations | 1994 | CLOC |
| Plurinational and Intercultural Conference of food sovereignty | 2009 | COPISA |
| Amazon Defence Coalition | 1994 | FDA |
| Coordinating Body of Social Movements | 1995 | CMS |
| Federation of CEPE Workers /Federation of Petroecuador Workers | 1980/1989 | FETRACEPE/ FETRAPEC |

Ecuador needs to make a transition to a post-oil economy as soon as possible, but to do that it could use the existing oil production (without amplifying this frontier) to enable this transition. In El Serafy's term (2013), 'sowing the oil' to create a post-extractivist economy. However, the political conditions to advance with more radical pro-environmental and pro-worker structural changes do not exist in Ecuador at this moment (Table 12.1).

# References

Acselrad, Henri. 2010. Ambientalização Das Lutas Sociais—O Caso Do Movimento Por Justiça Ambiental. *Estudos Avançados* 24 (68): 103–119.

Alimonda, Hector, Catalina Toro Pérez, and Facundo Martín. 2017. *Ecología política latinoamericana: pensamiento crítico, diferencia latinoamericana y rearticulación epistémica. Volumen II.* Ciudad Autónoma de Buenos Aires: CLACSO; México: Universidad Autónoma Metropolitana. http://biblioteca.clacso.edu.ar/clacso/gt/20171030104749/GT_Ecologia_politica_Tomo_II.pdf.

Altmann, Philipp. 2017. Una breve historia de las organizaciones del Movimiento Indígena del Ecuador. *Antropología Cuadernos de investigación* 12: 105–121. ISSN: 1390-4256.

Barca, Stefania. 2012. On Working-Class Environmentalism: A Historical and Transnational Overview. *Interface: A Journal for and about Social Movements* 4 (2): 61–80. ISSN 2009-2431.

Bennett, D. 2007. Labour and the environment at the Canadian Labour Congress – the story of the convergence. *Just Labour: A Canadian Journal of Work and Society* 10: 1–7.

Bravo, Andrea. 2019. *Mujeres indígenas y cambio climático en el contexto de la ampliación de las fronteras extractivas.* FLACSO Ecuador: Tesis de maestría.

Carruthers, David V., ed. 2008. *Environmental Justice in Latin America: Problems, Promise, and Practice.* London: MIT Press.

Central Bank of Ecuador. 2020. Statistical database. Accessed May 4, 2020. https://contenido.bce.fin.ec/docs.php?path=/documentos/PublicacionesNotas/Catalogo/IEMensual/Indices/m1997072018.htm.

Chomsky, Aviva, and Steve Striffler. 2014. Labor Environmentalism in Colombia and Latin America. *Working USA* 17 (4): 491–508. https://doi.org/10.1111/wusa.12135.

Clark, Patrick. 2017. Neo-developmentalism and a 'vía campesina' for Rural Development: Unreconciled Projects in Ecuador's Citizen's Revolution. *Journal of Agrarian Change* 17 (2): 348–364. https://doi.org/10.1111/joac.12203.

CONAIE. 1994. *Proyecto político de la CONAIE.* https://www.yachana.org/earchivo/conaie/proyectopolitico.pdf.

Cuencas Sagradas. n.d. Iniciativa de las cuencas sagradas territorios para la vida. Accessed March 5, 2020. https://cuencasagradas.org/.

Cumbre Agraria. n.d. Documento Operativo Acuerdo Nacional Agrario. Accessed March 16, 2020. https://cumbreagrariaecuador.files.wordpress.com/2016/07/acuerdo-agrario-nacional-documento-operativo.pdf.

De la Cadena, Marisol. 2015. *Earth Beings. Ecologies of Practice across Andean Worlds.* Durham: Duke University Press.

Ecuador Estrategico. n.d. Inversión por región. Accessed May 10, 2015. https://www.ecuadorestrategicoep.gob.ec/programas-servicios/#.

Egas, Juan J., Olga Shik, Marisol Inurritegui, and Carmine P. De Salvo. 2018. Análisis de la política agropecuaria en Ecuador. BID. Accessed March 17, 2020. https://publications.iadb.org/publications/spanish/document/analisis-de-politicas-agropecuarias-en-ecuador.pdf.

El Serafy, Salah. 2013. *Macroeconomics and the Environment.* Cheltenham Gloss: Edward Elgar Publishing.

Fontaine, Guillaume. 2003. *El precio del petróleo Conflictos socio-ambientales y gobernabilidad en la Región Amazónica.* Quito: FLACSO/IFEA.

Giunta, Isabella. 2014. Food Sovereignty in Ecuador: Peasant Struggles and the Challenge of Institutionalization. *The Journal of Peasant Studies* 41 (6): 1201–1224. https://doi.org/10.1080/03066150.2014.938057.

Gobierno del Ecuador. 2001. III Censo Nacional Agropecuario. https://www.ecuadorencifras.gob.ec/censo-nacional-agropecuario/.

Gordon, Robert W. 2004. *Environmental Blues: Working-Class Environmentalism and the Labor-Environmental Alliance 1968–1985.* Dissertation, Detroit, MI: Wayne State University.

Henderson, Thomas Paul. 2015. Food Sovereignty Food sovereignty and the Via Campesina in Mexico and Ecuador: Class Dynamics, Struggles for Autonomy and the Politics of Resistance. PhD diss., University of London.

Larrea, Carlos. 2004. Dolarización, exportaciones y pobreza en Ecuador. In *Efectos Sociales de la Globalización: Petróleo, Banano y Flores en Ecuador*, ed. Tania Korovkin, 157–184. Quito: Abya-Yala/CEDIME.

Latorre, Sara, Katharine N. Farrell, and Joan Martínez-Alier. 2015. The Commodification of Nature and Socio-environmental Resistance in Ecuador: An Inventory of Accumulation by Dispossession Cases, 1980–2013. *Ecological Economics* 116: 58–69. https://doi.org/10.1016/j.ecolecon.2015.04.016.

Lucero, Jose A. 2008. *Struggles of Voice. The Politics of Indigenous Representation in the Andes.* University of Pittsburgh Press.

Marega, Magali. 2015. Reconfiguración de la relación estado-sindicalismo petrolero público en el Ecuador de la Revolución Ciudadana. *Ecuador Debate* 94: 31–42.

Martinez-Alier, Joan. 2002. *The Environmentalism of the Poor: A Study of Ecological Conflicts and Valuation.* Cheltenham: Edward Elgar.

Martinez-Alier, Joan, Leah Temper, Daniela Del Bene, and Arnim Scheidel. 2016. Is There a Global Environmental Justice Movement? *The Journal of Peasant Studies* 43 (3): 731–755. https://doi.org/10.1080/03066150.2016.1141198.

Melucci, Alberto. 1980. The New Social Movements: A Theoretical Approach. *Information (International Social Science Council)* 19 (2): 199–226. https://doi.org/10.1177/053901848001900201.

Ministerio del Ambiente-MAE. 2016. *Pasivos ambientales y reparación integral: experiencias de gestión en el Ecuador.* Quito: Ecuador.

Negreiros, Janaina. 2009. La FENACLE y la organización de los asalariados rurales en la Provincia del Guayas. *Ecuador Debate* 78: 125–140.

Robbins, Paul. 2012. *Political Ecology: A Critical Introduction.* Oxford: Blackwell.

Veltmeyer, Henry. 1997. New Social Movements in Latin America: The Dynamics of Class and Identity. *The Journal of Peasant Studies* 25 (1): 139–169. https://doi.org/10.1080/03066159708438661.

Via Campesina. n.d. La vía campesina: la voz de las campesinas y de los campesinos del mundo. Accessed March 15, 2020. https://viacampesina.org/es/la-via-campesina-la-voz-las-campesinas-los-campesinos-del-mundo/.

Villalba-Eguiluz, C. Unai, and Iker Etxano. 2017. Buen Vivir vs Development (II): The Limits of (Neo-)Extractivism. *Ecological Economics* 138: 1–11. https://doi.org/10.1016/j.ecolecon.2017.03.010.

Widener, Patricia. 2011. *Oil Injustice Resisting and Conceding a Pipeline in Ecuador*. Rowman and Littlefield Publishers, Inc.

Yanza, Luis. 2004. El juicio a Chevron Texaco. Las apuestas para el Ecuador. In *Petróleo y desarrollo sostenible en Ecuador*, ed. Guillaume Fontaine, 37–44. Quito: FLACSO.

# 13

# A Just Transition for All? A Debate on the Limits and Potentials of a Just Transition in Canada

Bruno Dobrusin

## Introduction

A Just Transition strategy to address the need to curb emissions while at the same time protecting workers has been a leading element of the intervention by the International Trade Union Confederation (ITUC) and other labour groups, at least since the Copenhagen Conference of the Parts, known by its' acronysm COP (ITUC 2009). Just Transition existed as a concept and policy proposal earlier than the COP 2009 conference (see the JTRC (Just Transition Research Collaborative) 2018 for a full chronology of the uses of the term and also Rosemberg 2020), but the international climate negotiations, the inclusion of the term Just Transition in the official text and the push by labour groups like the ITUC have created a 'proliferation of Just Transition' (JTRC 2018: 9–10). This proliferation has often led to an overuse of the term, without clear attached implications for a Just Transition-focused set of the policies and the intended consequences.

Different labour and community groups have used this concept to mean different things. In the more restrictive format, Just Transition refers to the workers (and sometimes the communities) directly affected by a shift in industry away from high-carbon intensity and towards low-carbon economies. In a broader format, Just Transition incorporates the realities of the

---

B. Dobrusin (✉)
Toronto, ON, Canada

© The Author(s), under exclusive license to Springer Nature Switzerland AG 2021
N. Räthzel et al. (eds.), *The Palgrave Handbook of Environmental Labour Studies*,
https://doi.org/10.1007/978-3-030-71909-8_13

peripheral communities affected by those industries (such as indigenous groups), as well as the conditions under which workers in the low-wage service industry around those economies survive. This chapter analyses the reasons behind using each of these arguments, and the political impacts it has had on policy and on alliance-making for labour and social movements, with a focus on the Canadian context.

## Framing a Just Transition

Just transition has been present in different trade union policy proposals and forums for decades. As Sweeney and Treat (2018) discuss, the two most used meanings of Just Transition focus on for whom the transition is required. They frame the difference in terms of a 'worker-focused' approach, and a societal shift approach. In the case of the 'worker-focused' Just Transition, they explain it as a term 'used to highlight concerns about the likely impacts of climate and environmental policies on specific categories of workers (say, in a coal-fired power station that faces closure)' (Sweeney and Treat 2018: 2). The second use of Just Transition, based on societal power shift, refers to the increasing use of the term to explain broader and deeper socio-economic transformations that are necessary to transform society and transition to a low-carbon economy. In their words,

> With this broader usage, it is acknowledged that in order to address climate instability and its consequences, serious social and economic changes will be necessary—changes that will need to be both rapid and radical, if there is to be any serious attempt both to mitigate the impact of emissions (to minimize further damage to the earth's climate systems), and to help communities adapt to the consequences of warming that are already 'locked in' (from emissions already released). This shift in usage reflects an increasingly clear and explicit recognition that transitioning to a sustainable future society will involve a deep transformation of the current one. (Sweeney and Treat 2018: 2)

These two approaches are also separated by strategies vis-à-vis business and other potential allies. In the context of the worker-centred approach to Just Transition, the strategy of social dialogue has been predominant, while a strategy of social power has dominated in the societal shift understanding of Just Transition. The social dialogue approach has been prominent among international trade union bodies like the ITUC, heavily influenced by the tradition of the European trade union movement (Sweeney and Treat 2018: 18–19).

While social dialogue was a large component of post-war recovery in Europe, and a leading element within the International Labour Organization, with trade unions losing power over the years, it has become more of an aspirational conversation than a discussion between similarly weighted players. As a result, the lighter version of social dialogue has dominated. In the words of Sweeney and Treat (2018: 24): 'Where once it was understood to be part of a "social contract" arrangement between roughly equal partners, social dialogue has since become more aspirational, centring around an appeal for governments and companies to embrace a different economic paradigm, one that is more socially and ecologically sustainable'. The predominance of this approach has meant that unions have limited scope for moving away from mainstream ideas that tweak the system without significantly challenging its composition, while intending to win small concessions from the partners.

In contrast, there are sections of the labour movement building a social power approach to climate action (Sweeney and Treat 2018: 30–41). The social power approach acknowledges the limitations of the system to address climate change and pushes for a transformative approach that goes beyond the needs of a specific group of workers. It considers: 'A programmatic shift is necessary—one that reflects the challenge posed by Just Transition and can point to real solutions. Such a program may be able to win the support of some climate-concerned businesses in certain countries, that is, if they are willing to acknowledge that their pro-market approach to decarbonization is not working, and that approaches grounded in cooperation and the common good—rather than competition and private profit—need to be pursued' (Sweeney and Treat 2018: 31).

The difference between these two approaches can be perceived in the international trade union movement as well as in the Canadian context. The International Trade Union Confederation has been focused on a limited use of Just Transition, creating a Just Transition Centre in 2016 and partnering with a business-led initiative called the 'B-Team' which promotes collaboration between employers, workers and governments in preparing for the transition. In the 2018 'Business Guide' produced by the ITUC's Just Transition Centre and the B-Team, the definition of Just Transition is clearly aligned with that of a 'worker-focused' approach that depends on social dialogue mechanisms: 'Simply put, a Just Transition is a process involving employers, unions, and sometimes governments and communities, planning and delivering the transition of economies, sectors, and companies to low carbon, socially just and environmentally sustainable activities' (Just Transition Centre/B--Team 2018). Put in these terms, it is clearly a policy rather than a 'politics' focus. Other union organizations, including Trade Unions for Energy

Democracy, have taken the expansionary, societal shift perspective on Just Transition, emphasizing the need for systemic change, including the balance of relations between the Global North and the Global South (Bertinat 2016; JTRC 2018; TUED 2018; TNI and Taller Ecologista 2019; TNI 2020; Morena et al. 2020).

In Canada, labour groups, led by the Canadian Labour Congress, have worked closely with the ITUC Just Transition Centre, and pushed the Canadian government under Justin Trudeau to create a Federal Task Force for Coal Power Workers and Communities (Canadian Labour Congress 2018). This Task Force toured the provinces where coal mining will be phased-out, and organized town halls with entire communities (Government of Canada 2019a, 2019b). The Task Force was a unique effort to negotiate a remediation for the workers and communities affected by the coal phase-out determined by the federal government. The Task Force, and the policies it will push for, can be perceived within the 'worker-focused' notion of a Just Transition, focusing on those 3000 estimated coal miners, their communities (estimated at 50,000 people country-wide) and their future (Meyer 2018).

A different perspective on Just Transition is expressed by indigenous groups, community movements and campaigns like the 15&Fairness in Ontario, who are pushing for a transition into a low-carbon economy that not only focuses on the workforce of the fossil fuel industry, but also considers the realities of the frontline communities affected by fossil fuels, as well as the low-carbon workers that work in precarious working conditions mainly within the service industry. These two approaches can complement each other, by bringing together groups that may already collaborate in other issues but have not created an alliance around the issue of climate justice. This chapter compares both positions in the context of Canada and the approaches they have taken (social dialogue vs. social power) and discusses the possibility of combining them into the vision of the Green New Deal (Klein 2019).

## Just Transition Debates in Canada: A Worker-Focused Approach

In Canada, the trade union movement has been engaged on the issue of Just Transition for many years. Unions like the Communications, Energy and Paperworkers Union, and its predecessor the Canadian section of the Oil, Chemical and Atomic Workers International Union (now merged with the Canadian Autoworkers into Unifor since 2013), were pioneers in putting forward policies for Just Transition (CEP 2000). Similarly, the United Steelworkers (USW) in Canada has also put forward policies on Just Transition and recreated the US-based Blue Green Alliance in 2008, called Blue Green Canada (Smith and Neumann 2019).

Many of the members from Unifor as well as from USW work in extractive industries. In the case of Unifor, the experiences of the forestry industry in BC during the 1990s have affected the views on Just Transition for many workers (Cooling et al. 2015). The Forestry Renewal Act in 1994 intended to put in place different Just Transition programmes for workers and communities in the forestry industry, putting money into retraining and diversifying the industry. The programme included successful initiatives, but overall has been perceived as insufficient for workers who had high-paying jobs in the forestry industry and could not be transitioned to jobs with those standards in other industries (Cooling et al. 2015: 20–21).

Experiences such as the one in British Columbia and the current transformation in the coal industry in Alberta, Nova Scotia and Saskatchewan affect the ways in which a majority of the labour movement (and workers affiliated with it) perceive Just Transition as either a positive or negative policy. The main labour organizations discuss Just Transition in terms of assisting workers directly affected by the transition away from fossil fuels. The coal phase-out by the federal Trudeau government and some provincial governments like the Notley administration in Alberta were focused exclusively on the transition for workers and their communities in coal-producing areas. After pressure from the trade union movement, the federal government announced in 2017 the establishment during 2018 of a Task Force on Just Transition for Canadian Coal Power Workers and Communities (CLC 2018). This initiative included representatives from the unions of affected workers (USW, UNIFOR), as well as the Canadian Labour Congress (CLC), the power workers' union (International Brotherhood of Electrical Workers, IBEW), local government representatives and a representative of the utilities industries. It was chaired by the president of the CLC and the executive director of the Conservation Council of New Brunswick (Government of Canada 2019a).

The Task Force backgrounder details that the group will provide recommendations for the federal government, and it will be a consultation process (Government of Canada 2019a), not a binding mechanism to be put into law. The Task Force was chaired jointly by the president of the CLC, Hassan Yussef, and the executive director of the Conservation Council of New Brunswick, Lois Corbett. The federal government provided a technical secretariat that provided logistical and administrative support (Government of Canada 2019a). The recommendations were published in March 2019. The

Task Force's initial documents provide a broad definition of Just Transition in the following way:

> Just Transition means that society shares the costs of transitioning to a low-carbon economy. It would be unjust for workers and communities in affected sectors to shoulder the full cost of transition. These workers and communities, like all Canadians, have earned a better future. (Government of Canada 2019a: 1)

However, at the outset of the Task Force, it is very clear that this is a 'worker-focused' approach, when it clarifies that the task force will deal with the transition of affected workers, meaning workers who have permanent jobs (full-time or part-time) at a coal mine or a coal-fired generating station, now and throughout the transition, including those workers who were laid off from these facilities starting in 2017. It will also deal with affected communities, meaning communities that depend on a coal mine or a coal-fired generating station for employment, tax or royalty revenue, services, impact benefit agreements or economic activity (Government of Canada 2019a: V). The shortfalls of the Task Force are multiple, including that from the outset it is not going to determine specific laws, but rather make recommendations for a broader strategy, eventually to be implemented by the federal government in collaboration with affected provinces. Mertins-Kirkwood (2019: 7) outlined a problem with restricting the work of the Task Force to coal workers and not adopting a broader approach to all fossil fuel workers. The Task Force also falls short of incorporating workers from the service industry, public service workers, who belong to those coal communities as well. There is no representative from indigenous communities, who have borne the brunt of extractive industries and have been leading the call to make Just Transition a concept and practice inclusive of their realities (Brown 2019; Indigenous Climate Action 2020).

The recommendations include embedding the concept and practice of Just Transition into legislation, creating Just Transition centres in the affected communities, providing workers a path to retirement by creating a bridge programme for pensions and providing funding for local communities directly affected (Government of Canada 2019a: ix).

Due to the fact that the unions and community groups are members of the Task Force, and therefore sign on to the recommendations, they have generally come out in support of its work and recommendations (CLC 2019a; Blue Green Canada 2019). Some of the criticism expressed publicly point to the fact that there needs to be a longer than five-year financial commitment from the federal government in order to provide a more stable transition in time (Meyer 2019), while the head of the Alberta Federation of Labour (a member of the Task Force) expressed: '[Workers are] happy to get training, but that's not actually what they want. They're happy to get pension bridging, but that's not actually what they want. They're happy to get relocation allowances, that's certainly not really what they want. What they really want is to go from one job to another. And that's a desire that went unmet with the Just Transition that we negotiated, and it's a desire that was not properly addressed with the federal Just Transition task force either. And one of the reasons it went [un]addressed here in Alberta and at the federal level is it would require a more active approach to labour market policies than we've traditionally had here in Canada' (Gray-Donald and Eaton 2019). In other words, a possible shortfall of the Task Force is that, while not being an ambitious project, it creates expectations among the workers and affected communities that it cannot later meet in the policy, because it would require a more radical transformation than the parties partaking in this social dialogue are willing to carry out. The federal government's response to the recommendations has been to focus on infrastructure funding for the affected communities, to the tune of 150 million Canadian dollars, to start in 2020. In the case of Alberta, the provincial government has provided benefits such as bridge to retirement for workers, tuition vouchers and re-employment opportunities (Mertins-Kirkwood and Deshpande 2019: 11). These responses are a step forward but are not sufficient to deal with the needs of those communities, workers, as well as the larger society in the affected provinces. One special element to consider is that for all the talk of partnership with employers, the Task Force and the policies put forward do not address the responsibilities of the industries themselves, leaving governments and communities to bear the brunt of the transition.

While the main labour bodies in Canada and the government Task Force for coal workers have focused on the first kind of Just Transition, a 'worker-focused approach' implemented through social dialogue, campaigns like 15&Fairness in Ontario and those pushing for a Canadian version of the Green New Deal have taken a societal shift approach, which includes a transformational platform that goes beyond shifting industries from fossil fuel to renewables, and attempts to build social power as a way to reach it. In the following section, this chapter presents a different perspective from the 15&Fairness campaign in Ontario on what a Just Transition means, who it represents and what kind of change it could lead to.

# An Expanded Just Transition: Fighting for 15&Fairness

One way to reconsider positions around Just Transition is to look at campaigns that do not rise from the core of the labour movement but rather from the margins, aiming to shape the ways in which the labour movement organizes and who it represents, while enhancing the capacity to change realities. The 15&Fairness campaign is an example of that. The campaign was born with the goal of fighting for increases in the minimum wage in Ontario and improving labour laws in the province, to especially protect workers in sectors affected by expanding precarious work. Its core organization is not a union, but the Workers' Action Centre (WAC) in Toronto, a workers-led organization that organizes informal and non-unionized workers, many of them from migrant communities. The union movement has supported and financed the campaign, but it runs as an autonomous organization. WAC has had the capacity to reach workers beyond the conventional labour circles, by borrowing from the tool kit of the workers' centres in the US: providing information on labour rights, immigration and how to access public services, while pushing for an organizing agenda that reaches those without experience in the labour movement in Canada (Kumar and Schenk 2006). The campaign organized thousands of people around the province, generally outside of the labour movement, although increasingly connecting with unionized workers that also saw the big picture and benefits of changing the legislation. After more than 4 years of pressure from the 15&Fairness campaign, the Ontario legislature passed in the fall of 2017 one of the most ambitious labour reforms in the country: increased the minimum wage to 14 dollars an hour (a 33% increase) and adopted a roadmap to 15 dollars in 2019, including cost of living annual adjustments from then on; equal pay for equal work (especially important for part-time, precarious workers); ten days of job-protected emergency leave, including two paid sick days; fairer scheduling for workers; and easier unionization for security guards, cleaners, home care and community workers (Bush 2017).

These changes to the labour laws resulted in direct improvements to the lives of an estimated 1.7 million workers in Ontario (Fight 15&Fairness 2018). The new labour legislation also benefitted unionized workers whose collective bargaining agreements were renegotiated to incorporate some of the floors new benefits that the law brought forward. Workers in some sectors like steel did not have two paid sick days, and through the general legislation were granted rights that went beyond what they had negotiated for their members earlier on (United Steel Workers 2017). The capacity of the 15&Fairness campaign to mobilize both unionized and non-unionized workers showed that the interests of these two sectors can be aligned and combined to put forward a potent force.

The successes of the campaign were challenged by the victory of Doug Ford and the Conservative Party in Ontario in June 2018, which backtracked on many of the labour law changes that had been won during the previous government, and froze the minimum wage at the 14 dollars per hour mark (Crawley and Janus 2018). This put 15&Fairness in a defensive position, trying to save as many of the acquired rights as possible. It also became clear that the campaign had to expand beyond labour rights, something that it had already done in the past but now gained renewed urgency. The attacks on labour rights were coupled with retrenchment on environmental rights and an increasing wave of anti-immigrant stances across the political spectrum in Canada (Lukacs 2019; Migrant Rights 2019).

15&Fairness had historically been organized in different caucuses, allowing people to participate in a sector closer to their workplaces, or places of residence. Following previous years of debate, the campaign decided to revitalize the climate caucus, with two objectives: to bring the issues of precarious workers, especially in the service sector, to climate discussions and to incorporate new allies into the campaign by linking climate and labour rights. A key element in the discussion within the campaign was to remodel the idea of 'Just Transition' to incorporate the majority of Canadian workers, who are not in the fossil fuel sector, but in the service industry which employed more than 70% of workers in 2019 (STATSCAN 2019a). Among the main sectors that 15&Fairness organizes in (accommodation and food services), unionization rates are low, compared to the 30% overall union density in Canada, with less than 6% for workers in the accommodation and food services sector and just over 12% for workers in the wholesale and retail services (STATSCAN 2019b). The need to incorporate those workers meant revisiting the idea of a Just Transition, to bring it closer to what Sweeney and Treat (2018) have called the 'social-power Just Transition' approach, which expands from the workers specifically affected in the fossil fuel sector. In the words of one of the organizers of the campaign:

> For some, the notion of 'Just Transition' includes related industries—so not just energy and polluting industries and affected communities, but also investments

in housing, retrofitting, waste management, and other environmentally sustainable sectors. While this is a leap forward in broadening the definition of Just Transition, it is still not enough. When we speak of Just Transition in work, we must include transition to decent work for all low-wage, precarious, migrant, undocumented, exploited, Indigenous, excluded and unemployed workers. Here we can look towards the 'Fight for $15' campaigns as well as struggles by unionized postal workers and migrant workers to lead the way. (Hussan 2019)

The climate caucus in 15&Fairness embraced the idea of an expansive Just Transition that included workers in the fossil fuel sector as well as those in the service industries, the rights of migrant workers and indigenous people, using the climate crisis as an opportunity to reorganize society and redistribute wealth beyond those specific sectors that need to be transitioned. As explained in the main tool used by the campaign's climate caucus, getting higher benefits and wages to workers in the service and care sector is a climate solution in itself:

> We need to keep organizing, especially among those who work in areas that are a huge part of a more planet-friendly future, like care workers, recycling workers, transit workers, food service workers, education workers, health workers, and other public service workers. (Fight for 15&Fairness 2019a)

Furthermore, making it easier to unionize workplaces can allow workers to have a voice in those spaces, push for greener workplaces and challenge existing production and services practices that affect workers, communities and consumers. One example of a union-led campaign that embraces this approach is the Canadian Union of Postal Worker's *Delivering Community Power* campaign that proposes making Canada Post a leading force in fighting climate change, by using electric vehicles, including manufacturing an entire new fleet made in Canada's GM Oshawa plant (Aquanno 2019), putting charging stations on its shops for electric vehicles, using renewable energy, promoting postal banking for remote communities and checking-in on seniors (CUPW 2019). This is the clearest example of a broad-based idea of a Just Transition that goes beyond the workers' specific locations and engages communities, social movements and consumers (Cox 2016; Hutt 2016, 2018; Keith 2017).

In the 15&Fairness campaign, a great deal of effort is put into this social power approach to Just Transition. The following statement from their climate caucus flyer encompasses a comprehensive approach to an alternative narrative of Just Transition in Canada so far.

## From the Tar-Ssands to Fossil-Free Power: A Just Transition for All of Us

> We can only stop the climate crisis by transforming the economy and empowering workers to be part of the solution: migrant workers must have permanent resident status; existing low-paid workers in low-carbon jobs need higher wages, fairness, and respect; and better- paid workers in high-carbon jobs need a Just Transition to low-carbon jobs. (Fight for 15&Fairness 2019a)

This statement synthesizes an ambitious Just Transition position that departs from the one promoted by trade unions in the Canadian context, as explained earlier in the chapter. It incorporates the traditional block fighting for a Just Transition (high-carbon workers, mainly in the fossil fuel sector, but not restricted to it) by outlining the need to transition them to low-carbon jobs (and keeping their union benefits); it talks about the need to turn low-carbon jobs into decent, unionized jobs; and it incorporates the plight of migrant workers, many of whom are victims of climate disasters in their places of origin and were pushed to migrate as a result.

By stating this view of a Just Transition, the 15&Fairness campaign in Ontario, and increasingly at the national level, has raised the bar for ambitious climate action that incorporates workers, while moving beyond the group of workers historically at the centre of the debate. The work on Just Transition is recent for the campaign, which has historically focused on decent work issues. However, there are achievements worth taking note of, even if the climate caucus initiative is still in its initial steps. The campaign managed to connect the dots with a climate movement that has historically had problems in talking to issues of class and decent work (McAlevey 2019) and include organizers within 15&Fairness that came from the climate justice movement, as well as being an active participant in the climate strikes that took place throughout 2019 (15&Fairness 2019b). While most unions were slow to react to the climate strikes and the potential to ally themselves with the growing movement, the 15&Fairness campaign through its climate caucus and a comprehensive vision of Just Transition was present throughout (15&Fairness 2019b). A second relevant achievement was the capacity of the campaign to push for an agenda of Just Transition within the Pact for a Green New Deal that included workers in low-wage sectors, and that made a '15 dollars' minimum wage a central theme of the Canadian version of a Green New Deal (Pact for a Green New Deal 2019b).

## Reaching a Green New Deal in Canada

In order for ambitious climate action to take place and leave no workers and their communities behind, the two narratives on Just Transition discussed in the previous sections need to be combined. Combining them implies creating an alliance between the labour groups and organizations behind those narratives, as well as engagement by the group of workers they claim to represent. The proposal for a Green New Deal in Canada intended to reach that common purpose and create an alliance between fossil fuel workers, low-wage workers in low-carbon jobs, indigenous communities and environmental groups (Lewis 2019; Camfield 2019; Smith and Neumann 2019). The efforts to form alliances are ongoing, despite contrasting views between environmental groups on how to address worker and union participation. The long-established distrust between unions and the environmental movement (McAlevey 2020a) is a major obstacle to overcome in Canada in order to build cross-movement coalitions that challenge the current status quo in battling climate change. Following the rise of the Sunrise Movement in the United States (Klein 2019) and the resolution for a Green New Deal put forward by US Congress Representative Ocasio-Cortez and Senator Markey, a group of Canadian environmental and community groups began putting forward the idea of a Green New Deal project that could be specific to Canada, and build from the momentum seen in the US. News from the US usually have a high impact on the Canadian media landscape, and therefore this provided an opening for ambitious climate action in line with previous projects like the Leap Manifesto (Lukacs 2019; Klein 2019). The Leap Manifesto was an initiative very similar to the Green New Deal launched in 2015 in Canada, which at its height created a coalition of unions and social, indigenous and environmental movements pushing for transformative action. Due to disputes within the coalition on whether to push for Canada's left-leaning New Democratic Party (NDP) to endorse the manifesto in its 2016 Alberta Convention, the Leap Manifesto became a source of tension between unions closely aligned with the NDP's mainstream politics and movements that were willing to push further to the left (Klein 2019: 169–190; Lukacs 2019).

From the beginning of the discussions, it was clear that the Canadian version of a Green New Deal needed to avoid some of the controversies that took place in the US, where a significant portion of trade unions have remained sceptical of the Green New Deal proposal (Foster 2019). The role of organized labour in the coalition as well as the discussion of the 'red lines' that the coalition was not willing to cross became central issues (Pact for a Green New Deal

2019b). The initial group of organizations pushing for the project in Canada was composed mainly by environmental groups and some community groups with a focus on climate, such as the Council of Canadians, a Canada-wide community organization that advocates for climate justice, trade justice, democratic reform and indigenous rights. The effort to bring labour groups to the table was hampered by the fact that labour organizations were hesitant to join a coalition that they perceived as outside of their control and problematic for many of their members (Dembicki 2019). As mentioned in the first section of this chapter, Just Transition and climate action projects from the Canadian labour movement have been heavily influenced by a focus on making the case for transitioning workers in directly affected sectors (Canadian Labour Congress 2017, 2018, 2019b). The 2016 Leap Manifesto attempted to broaden the scope of these proposals, with the support of Canada's largest union, the Canadian Union of Public Employees (CUPE, as well as other smaller unions), but eventually faced stiff opposition from sectors of labour that saw ambitious climate action a challenge to their members in the fossil fuel sector (Lukacs 2019).

The Canadian Green New Deal, however, benefitted from having the endorsement of the 15&Fairness campaign, putting the needs of migrant workers and precarious workers at the centre of the most ambitious climate programme civil society organizations had put together. Following its launch in May 2019, more than six months ahead of the Canadian federal election, the Pact for a Green New Deal (as a coalition) organized town halls around the country, putting together a platform that was debated extensively by a variety of communities and organizations (Dembicki 2019; Camfield 2019). The participation of the 15&Fairness campaign made issues like Status for All migrant workers and the push for expanding unionization, guaranteeing at least a $15 minimum wage, paid emergency leave and the right to organize and unionize in all sectors of the economy central planks of the Green New Deal in Canada. It also allowed for groups of workers that had been engaged in the campaign in Ontario to reach out to their workplaces and place climate discussions at the forefront. Incorporating the 15&Fairness demands into the Pact for a Green New Deal coalition meant putting an ambitious Just Transition position forward.

The coalition gained momentum ahead of the federal elections, to the point that different parties adopted policies similar to those put forward in the Pact for a Green New Deal, although none endorsed the platform as such (Cruickshank 2019). Few trade unions were willing to join the coalition, despite the efforts by the conveners to reach out to labour groups, engage them and welcome their contributions (Armstrong 2019). The Ontario regional affiliate of the Canadian Union of Public Employees (CUPE) was the largest union organization that joined the coalition (representing 270,000 workers in that province), while the Canadian Union of Postal Workers (54,000 members) was the only member with a federal reach (Pact for a Green New Deal 2019a). In Quebec, the Confédération des syndicats nationaux (CSN) also endorsed the call. In early debates, unions such as CUPE at the federal level, the Fédération des travailleurs et travailleuses du Québec, known as FTQ in Quebec and even Unifor (the largest union organizing fossil fuel workers) seemed willing to engage in the coalition, but eventually desisted from participation (Dembicki 2019). From the unions that endorsed the coalition, only the postal workers actively engaged in the town halls, while the rest remained involved in name only. The reasons behind non-involvement were never made publicly clear by union representatives. There were calls to work together (Smith and Neumann 2019), and also for 'labour specific engagement' (UNIFOR 2019), but there was no open, public opposition to it; neither was there an active participation beyond the initial stages. Part of the explanation might be that unions did not want to be perceived as pushing for radical proposals that question the positions of the government and of the opposition parties. Despite these tensions, there was a clear choice from the major unions to avoid open confrontation as it happened in the United States following the launch of the Green New Deal resolution (Cohen 2019).

## The Challenges of Building Blue-Green Coalitions

The lack of labour participation in the Pact for a Green New Deal symbolizes the gap between more ambitious approaches to climate transformation and more reform-oriented tendencies. An ambitious approach that encompasses what Sweeney and Treat refer to as a Power approach, can build on the social dialogue approach experience of the trade union movement, but to be more inclusive of marginalized groups of workers it also needs to go beyond that approach. The labour movement has participants in both of these approaches, but the vast majority of its organizations are siding with a version of Just Transition that restricts the policy to directly affected workers and their communities. There are certainly serious concerns on the effects of ambitious climate policy for the members of some unions, especially those in the fossil fuel sector. Workers in those sectors will have to transition to new occupations and move away from an industry that has provided high wages and good standard of living for them (McAlevey 2019). The climate movement needs to certainly do better to speak, and persuade, those workers to join in ambitious policies like the Green New Deal (McAlevey 2020a). Even those unions that could significantly benefit from a broad Just Transition, such as the construction sector, tend to be dismissive of ambitious climate action and usually are amongst the most active defenders of large-scale fossil fuel-related infrastructure projects (Cohen 2019). The distrust created by the experience of the Leap Manifesto in Canada also contributes to the dynamics of passive participation in the Green New Deal.

Part of the task for a campaign like 15&Fairness, in its efforts to fight climate change, is to reach out to those groups of workers, including unionized ones, that were supportive of the fight for a higher minimum wage and better working conditions, and with whom the campaign had built relationships and trust as a legitimate voice fighting for working people during the fight for decent work in Ontario. The challenge for the campaign is to bring the voices of those workers and the unions they belong to, to the climate justice movement more decisively and to function as a bridge between the two.

As Stevis (2018) asserts, Just Transition is needed, especially in those countries and locations where the welfare system is absent. This does not mean that there is no need for policy helping the workers directly affected by closures in their industries. Rather, it means that narrowing Just Transition to only that sector is a limited political outlook. A social power approach to Just Transition can actually create an alliance between those who are fighting for their survival in the fossil fuel industries, and those who are also struggling in precarious jobs in sectors of the economy that do not receive the same attention. The effectiveness of such an alliance is yet to be seen, but there are examples of where it could lead. The Delivering Community Power campaign is an example of that combination, bringing formal collective bargaining proposals together with an ambitious societal agenda (Hutt 2018). Another example may be taking place in the efforts by a group of trade unionists, community groups and environmental organizations around the platform to reconvert the Oshawa General Motors plant into an electric vehicle facility that produces for Canada Post (Aquanno 2019). Exploring these experiences could help in building a broad Just Transition alliance within the Green New Deal coalition, but requires further work to bring the labour movement inside the coalition, talk to workers in fossil fuel sectors in a language that includes them and also work with the broader communities of those workers to get them behind the vision (McAlevey 2019).

## Concluding Remarks

The debate around Just Transition in Canada is representative of the broader debate of the trade union movement on climate change. The majority of the unions in the fossil fuel sector, and the workers they represent, maintain a position anchored in the idea that broad policies that fight for everyone will not put the focus on those sectors most affected by transitions to a low-carbon economy. Different reasons explain these dynamics. The capacity of unions in the fossil fuel sector to shape debates around climate action within the labour movement is greater than that of other blocks of workers. Lack of direct engagement with fossil fuel workers in putting forward a specific alternative proposal by the environmental movement has also affected these positions. Fossil fuel workers are not inherently attached to the industry, but rather they are conscious of the lifestyle that working in the sector has provided them with and the fights it took the unions to get there (UNIFOR 2018). Asking workers and their unions to move away from their main source of income without a concrete vision that they can be part of and that maintains their livelihoods in the context of growing insecurity is a major challenge for the climate justice movement. As a USW union leader in the Albertan coal mines put it: 'Just to have lots of plans and say this will be all wonderful, doesn't work when you get your pink slip and you still have a mortgage and kids to feed' (Milne 2018). The challenge is to build a coalition in which the workers in the fossil fuel sector, and their unions, see that a broader Just Transition can lift everyone up. In the Canadian context, Just Transition policies have had a representation problem, by focusing almost exclusively on workers who are white, middle-aged men, leaving behind other group of workers and community members that should also be included in Just Transition policies (Mertins-Kirkwood and Desphande 2019).

This chapter discussed two different stances on Just Transition from sectors within the labour movement. The majority of the formal trade union movement has stood behind a 'worker-focused' approach that prioritizes workers and communities directly affected by the transition to a low-carbon economy. The 15&Fairness campaign has put forward a more ambitious, broader notion of Just Transition that takes the debate on climate action beyond the group of workers directly affected to those who should also be part of the discussion: precarious workers, migrant workers, indigenous communities. These two stances are not in conflict with each other, but they run on parallel lanes. While several organizations in the labour movement have been focused on influencing the government's Just Transition Task Force on Coal Power

Workers and Communities, the 15&Fairness campaign made a decision to actively engage with the Green New Deal platform. These two approaches can complement each other, by taking the specific policies of the worker-focused approach and making them part of the societal shift. The Fight for 15&Fairness managed to build alliances between formal and informal workers, unionized and non-unionized, in the struggle for better working conditions in Ontario, proving that unions can work beyond the immediate interest of their membership. In embracing a broader conceptualization of Just Transition, the campaign could play a role in bringing the labour movement into a broader coalition for ambitious climate action. In a sense, 15&Fairness has built into its proposal the language similar to the one been discussed by union organizers in the US context, of an 'an aggressively pro-union Green New Deal guaranteeing all workers in the current fossil fuel sector can keep their current union contracts, while enabling "essential" workers in low-carbon-emission sectors (home care and childcare workers, those that harvest and deliver our food, etc.)' (McAlevey 2020b). The challenge is to build that proposal into a credible and effective programme of action that workers in the fossil fuel sector can be a part of.

**Acknowledgement** I would like to thank Dimitris Stevis, David Uzzell and Nora Räthzel for the valuable comments to earlier drafts of this chapter.

# References

Aquanno, Scott. 2019. GM Workers Can Lead Us Into the Future. *Our Times Magazine*, May 1. https://ourtimes.ca/article/gm-workers-can-lead-us-into-the-future. Accessed 27 April 2020.

Armstrong, Helen. 2019. A Green New Deal for Canada Set to Launch. *Now Magazine*, April 29. https://nowtoronto.com/news/green-new-deal-climate-change-TD-RBC/. Accessed 27 April 2020.

Bertinat, Pablo. 2016. *Transición energética justa. Pensando la democratización energética*. FES SINDICAL: Montevideo. https://library.fes.de/pdf-files/bueros/uruguay/13599.pdf. Accessed 25 April 2020.

Blue Green Canada. 2019. Statement on Federal Just Transition Task Force Report. *Press Release*, March 13. http://bluegreencanada.ca/blog/statement-federal-just-transition-task-force-report. Accessed 24 April 2020.

Brown, Valine. 2019. This Is What Indigenous Energy Sovereignty Looks Like. A Just Transition Case Study. *Briarpatch Magazine*, April 29. https://briarpatch-magazine.com/articles/view/indigenous-climate-action.

Bush, David. 2017. 15 and Fairness Shakes Up Ontario. *Canadian Dimension*, June 13. https://canadiandimension.com/articles/view/15-and-fairness-shakes-up-ontario. Accessed 27 April 2020.

Camfield, David. 2019. What Will It Take to Win a Green New Deal? *Canadian Dimension*, July 2. https://canadiandimension.com/articles/view/what-will-it-take-to-win-a-green-new-deal. Accessed 28 April 2020.

Canadian Labour Congress. 2017. Unions Applaud Canada's Commitment to a Just Transition for Coal Workers. *Press Release*, November 16. https://canadianlabour.ca/news-news-archive-unions-applaud-canadas-commitment-just-transition-coal-workers/. Accessed 24 April 2020.

———. 2018. 'Canada's Unions Will Help Shape a Just Transition for Coal Workers. *Press Release*, April 24. https://canadianlabour.ca/news-news-archive-canadas-unions-will-help-shape-just-transition-coal-workers/. Accessed 24 April 2020.

———. 2019a. Just Transition Task Force Report Has Potential to Put People at the Heart of Climate Policy. *Press Release*, March 11. https://canadianlabour.ca/just-transition-task-force-report-has-potential-to-put-people-at-the-heart-of-climate-policy/. Accessed 24 April 2020.

———. 2019b. Canadian Labour Congress Stands in Solidarity with Student-Led Global Climate Strike. *Press Release*, September 20. https://canadianlabour.ca/clc-solidarity-student-led-global-climate-strike/. Accessed 28 April 2020.

Canadian Union of Postal Workers, CUPW. 2019. *Delivering Community Power*. https://d3n8a8pro7vhmx.cloudfront.net/themes/5b72fccf4445ea195a9818d0/attachments/original/1564754931/DeCoPo_2019_aut_en.pdf?1564754931. Accessed 28 April 2020.

Cohen, Rachel. 2019. Labor Unions Are Skeptical of the Green New Deal, and They Want Activists to Hear Them Out. *The Intercept*, February 28, https://theintercept.com/2019/02/28/green-new-deal-labor-unions/. Accessed 27 April 2020.

Communications, Energy and Paperworkers Union of Canada, CEP. 2000. *Just Transition to a Sustainable Economy in Energy*. Montreal Convention, Policy 915. https://wayback.archive-it.org/288/20120913110105/http://www.cep.ca/sites/cep.ca/files/docs/en/policy-915-e.pdf. Accessed 27 April 2020.

Cooling, Caren; Marc, Lee; Shannon, Daub and Jessie, Singer. 2015. Just transition. Creating a green social contract for BC's resource workers. Vancouver: CCPA.

Cox, Mojdeh. 2016. Powering Up Our Post Office. *Our Times Magazine*, August 26. https://ourtimes.ca/article/powering-up-our-post-office. Accessed 27 April 2020.

Crawley, Mike, and Janus, Andrea. 2018. Ford Government Freezing $14 Minimum Wage as Part of Labour Reform Rollbacks. *CBC*, October 23. https://www.cbc.ca/news/canada/toronto/doug-ford-open-for-business-bill-148-repeal-1.4874351. Accessed 28 April 2020.

Cruickshank, Ainslie. (2 August 2019). Poll Shows Majority of Canadians Say Economy Should Shift from Oil and Gas, But Most Don't Know the Green New Deal. *Toronto Star*, August 2. https://www.thestar.com/news/canada/2019/08/02/

poll-shows-majority-of-canadians-say-economy-should-shift-from-oil-and-gas-but-most-dont-know-the-green-new-deal.html. Accessed 29 April 2020.

Dembicki, Geoff. (5 June 2019). Inside the Race to Unify Progressives Behind a Canadian Green New Deal. *The Tyee*, June 5. https://thetyee.ca/Analysis/2019/06/05/Inside-Green-New-Deal/. Accessed 28 April 2020.

Fight for 15&Fairness. 2018. $15 & Fairness Victory for Ontario Workers as Bill 148 Passes Final Reading. https://www.15andfairness.org/_15_fairness_victory_for_ontario_workers_as_bill_148_passes_final_reading. Accessed 15 Sept 2019.

———. 2019a. *Let's Make Climate Work, Decent Work*. Toronto: Workers' Action Centre.

———. 2019b. Climate Justice = Economic Justice = Racial Justice. September 20. https://www.15andfairness.org/join_the_global_fight. Accessed 26 April 2020.

Foster, David. 2019. What We Learned from the First Green New Deal. *The Hill*, October 2. https://thehill.com/opinion/energy-environment/429319-what-we-learned-from-the-first-green-new-deal. Accessed 27 April 2020.

Government of Canada. 2019a. *A Just and Fair Transition for Canadian Coal Power Workers and Communities. Task Force on Just Transition for Canadian Coal Power Workers and Communities*. Ottawa: Environment and Climate Change Canada. http://publications.gc.ca/collections/collection_2019/eccc/En4-361-2019-eng.pdf. Accessed 20 Sept 2019.

———. 2019b. *What We Heard. From Canadian Coal Power Workers and Communities. Task Force on Just Transition for Canadian Coal Power Workers and Communities*. Ottawa: Environmental and Climate Change. http://publications.gc.ca/collections/collection_2019/eccc/En4-362-2019-eng.pdf. Accessed 20 Sept 2019.

Gray-Donald, David, and Eaton, Emily. 2019. A Just Transition Requires a Planned Economy. But Whose Plan? *Briarpatch Magazine*, October 10. November/December 2019 edition.

Hussan, Syed. 2019. Canada and a Green New Deal. *National Observer*, January 17. https://www.nationalobserver.com/2019/01/16/opinion/canada-and-green-new-deal. Accessed 20 Sept 2019.

Hutt, James. 2016. Recharging a Dying Province. *Our Times Magazine*, April 18. https://ourtimes.ca/article/climate-justice-recharging-a-dying-province. Accessed 27 April 2020.

———. 2018. Putting Climate on the Bargaining Table. *Our Times Magazine*, September 20. https://ourtimes.ca/article/putting-climate-on-the-bargaining-table. Accessed 27 April 2020.

Indigenous Climate Action. 2020. Indigenous Climate Action Responds to Teck Resources Pulling Frontier Application. *Press Release*, February 23. https://58e3a4e9-6eac-481c-a0c8-162cb51fe9ef.usrfiles.com/ugd/58e3a4_92819e1b417647ea946b503a35d3f408.pdf. Accessed 15 April 2020.

International Trade Union Confederation. 2009. *A Just Transition: A fair pathway to protect the climate*. Brussels: ITUC. https://www.ituc-csi.org/IMG/pdf/01-Depliant-Transition5.pdf. Accessed 25 April 2020.

Just Transition Centre/ The B-Team. 2018. *Just Transition: A Business Guide.* ITUC. https://www.ituc-csi.org/IMG/pdf/just_transition_-_a_business_guide.pdf.

Just Transition Research Collaborative. 2018. *Mapping Just Transition(s) to a Low-Carbon World.* Geneva: UNRSID, RLS Brussels, University of London Institute in Paris. http://www.rosalux-nyc.org/wp-content/files_mf/reportjtrc2018_1129.pdf.

Keith, Melissa. 2017. Connecting Canada. Postal Hubs and the Promise of Community Power. *Our Times Magazine*, February 24, https://ourtimes.ca/article/connecting-canada. Accessed 27 April 2020.

Klein, Naomi. 2019. *On Fire: The Burning Case for a Green New Deal.* Toronto: Knopf Canada.

Kumar, Pradeep, and Christopher Schenk. 2006. *Paths to Union Renewal: Canadian Experience.* Toronto: University of Toronto Press.

Lewis, Avi. 2019. Top 5 Reasons the Green New Deal is Workable, Winnable and the Idea We Need Right Now. *Canadian Dimension*, July 30. https://canadiandimension.com/articles/view/top-5-reasons-the-green-new-deal-is-workable-winnable-and-the-idea-we-need. Accessed 28 April 2020.

Lukacs, Martin. 2019. *The Trudeau Formula. Seduction and Betrayal in an Age of Discontent.* Montreal: Black Rose Books.

McAlevey, Jane. 2019. Organizing to Win a Green New Deal. *Jacobin Magazine*, March 26. https://jacobinmag.com/2019/03/green-new-deal-union-organizing-jobs. Accessed 27 April 2020.

———. 2020a. The Climate Movement Doesn't Know How to Talk with Union Members About Green Jobs. Interviewed by Alleen Brown, *The intercept*, March 9. https://theintercept.com/2020/03/09/climate-labor-movements-unions-green-new-deal/. Accessed 25 April 2020.

———. 2020b. Celebrating May Day Starts by Taking Workers Seriously. *The Nation Magazine*, May 1. https://www.thenation.com/article/society/trump-gop-coronavirus-immigration/. Accessed 1 May 2020.

Mertins-Kirkwood, Hadrian. 2019. *Heating Up, Backing Down: Evaluating Recent Climate Policy Progress in Canada.* Adapting Canadian Workplaces- Working Paper N.203. https://www.policyalternatives.ca/sites/default/files/uploads/publications/National%20Office/2019/06/Heating%20Up%2C%20Backing%20Down.pdf. Accessed 25 April 2020.

Mertins-Kirkwood, Hadrian, and Zaee Deshpande. 2019. *Who Is Included in a Just Transition? Considering Social Equity in Canada's Shift to a Zero-Carbon Economy.* Ottawa: CCPA/ACW. https://www.policyalternatives.ca/sites/default/files/uploads/publications/National%20Office/2019/08/Who%20is%20included%20in%20a%20just%20transition_final.pdf. Accessed 22 Sept 2019.

Meyer, Carl. 2018. Federal Panel Privately Urges Trudeau Government to Do More for Coal Workers. *National Observer*, November 5. https://www.nationalobserver.

com/2018/11/05/news/federal-panel-privately-urges-trudeau-government-do-more-coal-workers. Accessed 20 April 2020.

———. 2019. Trudeau-Appointed Task Force Proposes Solutions to Address Fear, Anxiety and Mistrust Among Coal Workers. *National Observer*, March 10. https://www.nationalobserver.com/2019/03/10/news/trudeau-appointed-task-force-proposes-solutions-address-fear-anxiety-and-mistrust. Accessed 20 April 2020.

Migrant Rights Network. 2019. *Truth & Lies About Immigration*. Information Factsheet. https://migrantrights.ca/wp-content/uploads/2019/08/MRN-Immigration-FAQ.pdf. Accessed 29 April 2020.

Milne, Roy. 2018. Some Jobs in New Energy Industries Come with a Pay Cut of $50K: Coal Miner. Interviewed by Ana Maria Tramonti, *CBC The Current*, December 13. https://www.cbc.ca/radio/thecurrent/the-current-for-december-13-2018-1.4943449/some-jobs-in-new-energy-industries-come-with-a-pay-cut-of-50k-coal-miner-1.4943040. Accessed 29 April 2020.

Morena, Edouard, Dunja Krause, and Dimitris Stevis, eds. 2020. *Just Transitions. Social Justice in the Shift Towards a Low-Carbon World*. London: Pluto Press.

Pact for a Green New Deal. 2019a. *Endorsements*. https://act.greennewdealcanada.ca/endorsements/. Accessed 15 Sept 2019.

———. 2019b. *What did we hear at the Pact for a Green New Deal Town Halls?* https://act.greennewdealcanada.ca/what-we-heard/. Accessed 20 Sept 2019.

Rosemberg, Anabella. 2020. No jobs on a dead planet': The international trade union movement and Just Transition. In *Just Transitions. Social Justice in the Shift Towards a Low-Carbon World*, ed. Edouard Morena, Dunja Krause, and Dimitris Stevis, 32–55. London: Pluto Press.

Smith, Rick, and Kevin Neumann. 2019. Labour a Key Partner in a Canadian Green New Deal. *Canadian Dimension*, May 23. https://canadiandimension.com/articles/view/labour-a-key-partner-in-a-canadian-green-new-deal. Accessed 28 April 2020.

Statistics Canada. 2019a. Table 14-10-0023-01 Labour Force Characteristics by Industry, Annual (x 1,000). Accessed 22 Sept 2019.

———. 2019b. Table 14-10-0070-01 Union Coverage by Industry, Annual (x 1,000). Accessed 25 April 2020.

Stevis, Dimitris. 2018. *Reclaiming Just Transition*. https://medium.com/just-transitions/stevis-e147a9ec189a. Accessed 20 Sept 2019.

Sweeney, Sean, and Treat, John. 2018. *Trade Unions and Just Transition. The Search for a Transformative Politics*. TUED Working Paper, No. 11. http://unionsforenergydemocracy.org/wp-content/uploads/2018/04/TUED-Working-Paper-11.pdf.

Trade Unions for Energy Democracy, TUED. 2018. *Just Transition: A revolutionary Idea—Bulletin 73*. http://unionsforenergydemocracy.org/just-transition-a-revolutionary-idea-tued-bulletin-73.

Transnational Institute. 2020. *Just Transition: How Environmental Justice Organisations and Trade Unions Are Coming Together for Social and Environmental Transformation*.

TNI, Amsterdam. https://www.tni.org/files/publication-downloads/web_just-transition.pdf. Accessed 26 April 2020.

Transnational Institute and Taller Ecologista. 2019. *Towards a Corporate or a People's Energy Transition?*. TNI, Amsterdam. https://www.tni.org/files/publication-downloads/02_energy_transition.pdf. Accessed 26 April 2020.

UNIFOR. 2018. *The International Climate Crisis and Just Transition*. Toronto: Unifor 2018 Lobby Document. https://www.unifor.org/sites/default/files/documents/document/unifor_justtransition_backgrounder_en_web.pdf. Accessed 5 Sept 2019.

———. 2019. *What Is the Deal with a Green New Deal?*. https://www.unifor.org/en/what-deal-green-new-deal/. Accessed 27 April 2020.

United Steel Workers. 2017. Ontario Labour Reforms a Start Toward Fairness in a Changing Economy. *Media Release*, November 23. https://www.usw.ca/news/media-centre/releases/2017/ontario-labour-reforms-a-start-toward-fairness-in-a-changing-economy-says-usw. Accessed 27 April 2020.

# Part III

Farmers, Commoners, Communities

# 14

# Labouring the Commons: Amazonia's 'Extractive Reserves' and the Legacy of Chico Mendes

Stefania Barca and Felipe Milanez

> *At first I thought I was fighting to save the rubber tappers, then I thought I was fighting to save the Amazon, but now I realize I am fighting for humanity*
> —Chico Mendes (Quoted in Weiss (2008: 331; our translation))

## Introduction

Three decades after the death of Chico Mendes (1944–1988), the Brazilian trade unionist who wanted to save the Amazon forest, his legacy continues to be disputed between different political constituencies. In fact, his work is now evoked as inspirational not only by environmental activists but also by advocates of green capitalism. His name is used to brand green-washing operations of retail multinationals, gas and sanitation companies, aluminium producers (which use energy and raw materials from Amazonia), hydroelectric and mining companies or the development of carbon markets in the state of Acre. The

---

S. Barca (✉)
Center for Social Studies, University of Coimbra, Coimbra, Portugal
e-mail: sbarca@ces.uc.pt

F. Milanez
Institute for Humanities, Arts and Sciences, Federal University of Bahia, Bahia, Brazil

union he founded in 1985, Conselho Nacional dos Seringueiros (National Rubber-Tappers' Council—CNS), has turned into an NGO and received financial support from companies that were denounced by forest communities for violating their rights. Such contradictory uses of Mendes' legacy, amplified by the media at each anniversary of his murder, not only obfuscate the true meaning of his work and that of the decades-long activism that led to the founding of the rubber-tappers' union, but also jeopardize the endurance of his most important material legacy, the Extractivist Reserves (Resex), that is, a path-breaking type of conservation unit that Mendes contributed to creating (Milanez 2017a, b).

This chapter locates the contested legacy of Chico Mendes against the backdrop of Brazil's 'extractivist' movement, namely a coalition of waged workers, Indigenous people and peasants known as Aliança dos Povos da Floresta (Alliance of Forest Peoples) that took shape in the state of Acre in the mid-1980s. Our aim is to assess the historical significance of this experience of environmental/labour activism as embedded within the political ecology of the commons (see also Garcia López, this volume). In short, the chapter makes two related points: first, it argues that one key potentiality of labour environmentalism is that of countering the degradation of non-human nature by resisting the annihilation of the commons. Second, it shows how this potentiality is dialectically produced through a historical process of violence and (r)existence (Porto-Gonçalves 2002), that is, peasant and Indigenous struggles for autonomous existence via resistance to the destruction of the commons.

## Labour and the Common/s

One of the most fundamental traits of capitalist modernity is the annihilation of the common/s. By this term, we mean a non-commodified relationship between work and nature, based upon non-alienated labour—namely autonomous, subsistence-oriented and eco-sufficient modes of production (Barca 2019a). Consequently, we see commoners as non-waged workers, that is, people who make their living through a direct interaction with non-human nature through their labour (Bennholdt-Thomsen and Mies 2000; Salleh 2009; Burkett 2009; Brownhill et al. 2012; De Angelis 2017). The term common/s is meant to signify both the work of *common*-ing and the material dimension of what is (re)produced through it—*the commons*. Since the common/s run counter to accumulation, capital needs to annihilate them via the enclosure of land and other natural resources, and by exercising violence against the commoners (Peluso and Watts 2001). As various authors have made clear, this is not a historically concluded phase but an ongoing process constantly re-enacted by capital in order to maintain and create new resources for accumulation (Luxemburg 2003; Mies 1986; Caffentzis and Federici 2014; De Angelis 2007). Dispossessions and expulsions, witch-hunting and violence against women (including anti-abortion laws), colonization, enslavement, systematic violation of the rights of peasants and workers, killing or incarceration of those who resist, debt peonage, are all ways for capital to sever people's direct relationship with nature (including their own bodies) so to increase their productivity and appropriate their surplus labour.

Political ecology sees the enclosure of the common/s as a fundamental cause of environmental degradation via the encroachment and cutting of forest areas, the reduction of biodiversity in monoculture and plantation regimes, habitat contamination and the extermination of wildlife (see Garcia López, this volume). This approach allows to see how, in the process of annihilating the common/s, violence against humans and natural habitats is inextricably linked (Peluso and Watts 2001) via 'environmental violence' (Barca 2014a). In our understanding, environmental violence is not only a material phenomenon, but entails a very important symbolic dimension. In classical political economy, the common/s were considered as waste, in the sense that they were seen as a missed opportunity for increasing production via private appropriation and capital investment—the improvement ideology (Barca 2010). The non-productivity (or low productivity) of both labour and natural resources in a common/s regime is thus a basic axiom of capitalist political economy. An important consequence of this ideology is that violence against the common/s is often legitimized as necessary in order to increase productivity as expressed by the hegemonic discourse of GDP growth (Milanez 2019).

The capitalist notion of productivity has had profound implications for environmental policies. On the one hand, it has led to conceiving conservation as the politics of setting parts of nature aside from production, putting them under state control (as in wilderness parks). On the other, the logic of capital productivity has led to formulate the politics of environmental management (as in the so-called scientific forestry), so that nature could be subjected to management techniques and investments that would increase its productivity—the principle of eco-efficiency (Martínez-Alier 2002; Barca and Bridge 2015). Both kinds of conservation politics have implied the dispossession and displacement of people from their ancestral lands and territories, or the biopolitical control of people within protected areas. Since the late twentieth century, once environmental degradation has become evident as a planetary process, the eco-efficiency model has become predominant, advocating for the extension of capital control upon nature at the global scale. As Stevis et al. (2018: 440) write: 'Nature has become, and arguably is increasingly becoming commodified, that is, part of Capital'. A fundamental claim of Environmental Labour Studies, in fact, is that the commodification of nature has been an important cause of opposition between environmental and labour movements, contributing to a false perception of their relationship on both sides. From the point of view of labour, once become capital, nature stands on the other side of the labour/capital conflict, and saving natural resources is seen as coming at the expense of saving labour (Räthzel and Uzzell 2011). Our point here, however, is that these two forms of capitalist conservation—setting aside and commodifying nature—respond to a dualist vision of access to it (public vs. private) that excludes a third possibility, that is, the common/s. In other words, capitalist conservation is alimented by and actively contributes to the annihilation of the common/s, both materially and discursively.

This chapter will show how the common/s constitute an alternative, labour-driven (rather than state or capital-driven), form of conservation. They do so, we argue, because commoning work is not oriented towards productivity, that is, increasing GDP, but towards re-productivity (Biesecker and Hofmeister 2010), that is, increasing human well-being by enhancing the reproductive capacity of nature. This apparently utopian work/nature relationship has been long practised and defended by different people in a multiplicity of places: among them are Brazil's 'traditionally occupied' territories (Berno de Almeida 2004), including Indigenous and Quilombo areas, and areas where subsistence farming predominates (Leroy 2017). Taking inspiration from these ancestral practices, the *seringueiros*—a community of waged workers from the state of Acre—broke the chains that enslaved them to the owners of the means of production (the rubber barons) and struggled for an autonomous relationship with the forest. This, we claim, was an act of insurgent commoning: it consisted in abolishing the private property and waged labour regimes that agrarian capitalism had imposed upon them, establishing a new institution—the 'extractivist reserve' (Resex). Resulting from working-class, Indigenous and peasant life-projects that disputed precisely the annihilation of the common/s, the National Rubber-Tappers Union (CNS) initiated a double counter-movement. On the one hand, it reaffirmed workers' control over both their labour and the land—countering original accumulation and the various forms of alienation associated with it. On the other, it aimed to safeguard the well-being of both human and non-human nature, by protecting it from exploitation.

Since 1990, Brazil has instituted 94 Resex, part of the national system of conservation units, joined by 381 agroecology projects (PAE), part of the country's agrarian reform programme, totalling 26 million hectares of land that are protected via a common/s regime (Almeida et al. 2018: 26). Extractive reserves are defined as 'forests (and other biomes) with high biodiversity inhabited by populations with low demographic density that use low intensity techniques' (ibid.: 25). The conservation benefits of this system are the object of heated debates, which are part and parcel of the struggle for (or against) the common/s. The most significant aspect of such dispute, from the perspective of work/nature relationship, is that concerning the productivity of the Resex: as Almeida et al. (2018) have shown, when measured in terms of quantity of rubber extracted per hectare, the Resex fares significantly lower than a monoculture plantation; however, if measured in terms of biodiversity per hectare, the Resex results between 100 and 200 times more productive. High biodiversity means a qualitatively richer form of productivity—one which allows for a variety of products to be extracted from the same area, allowing human communities to thrive in their territory, resisting dispossession and proletarianization. In short, this means that the common/s can be models of strong sustainability *as well as* human well-being and social equality due to their high capacity for (re)productivity.

By pointing to the ecological relevance of work in a commoning regime, the Brazilian Resex holds global significance for environmental justice struggles (Barca 2014a; Martínez-Alier 2002). We cannot understand the political ecology of the Resex, however, without considering the murders, land grabbing, death threats, public defamation and other forms of pressure deployed against them over time. All these must be read in a contest of structural politico-economic violence, namely the selective non-compliance of the Brazilian State with functions of law enforcement, legality and justice, and the protection of human rights, as well as the promotion of human development in the Resex via the provision of the basic infrastructures that grant people access to citizenship rights. This must be understood as a strategic impotence, due to the Brazilian state's commitment to capitalist development and to the colonial/patriarchal projects that underlie it (Milanez 2019).

In this chapter, we point to the importance of placing violence against the commoners, but also misrepresentations of the common/s as *unproductive*, high on the research agenda of anyone with an interest in the political ecology of labour, especially in the rural contexts that global capitalism considers as commodity frontiers. Like Amazonia's Resex areas, a number of other places where subsistence work is expended in agroecology practices, as well as in community fisheries and other forms of food sovereignty premised upon access to the commons, are experiencing similar attacks (Scheidel et al. 2020). This, we argue, must be seen as capital's response to labour's autonomous relationship with non-human nature, and to forms of non-waged work allowing for the production of income and well-being outside the circuit of capital.

## The Geo-historical Context: Amazonia, 1960s–1970s

Violence has a long history in Brazil's rubber plantations (Santos 1980; Taussig 1984). Since the beginning of the activity, in the nineteenth century, rubber barons devised means to impose a culture of ferocious individualism, with the aim of dividing workers from both the Indigenous populations and from each other, while lobbying to keep forests out of public control. While, in the majority of rubber plantations, Indians and rubber-tappers lived closely together, a system of racial differentiation was created by rubber barons, in which the Indians were often enslaved or assigned the lowest-paid jobs, and rubber-tappers were enrolled for anti-Indian expeditions aimed at hunting for slaves and women. On the other hand, with the pretence of preventing workers from selling rubber to their competitors, landowners hired some of them to spy on the others, and brutally punished any failure in delivering the requested crop, often with death.

Starting in 1975, Brazil's military dictatorship began to organize rural unions through CONTAG (National Confederation of Farmworkers). This had been created in 1963–1964 (right before the coup d'état) in response to rural workers' mobilizations for access to land, with the official mission of guaranteeing the implementation of an agrarian reform, which was issued with the Land Statute of 1964. Following the crisis of export-led rubber production due to changes in the international market, the agrarian reform aimed at regularizing the cultivation of (relatively) 'unproductive' lands, with a view to opening them up for agrarian capitalism, by encouraging settler colonization in areas of rubber forest. After a Maoist guerrilla group in the south of Pará had been massacred by the army, between 1968 and 1974, the regime came to see CONTAG as a means to prevent new outbreaks and acquire political control over the rural workers who migrated to the Amazon in search for land, ending up as waged labour in the region's big estates (Paula and Silva 2008). In this context, CONTAG opened the first sections of its farmworkers union *Movimento Sindical de Trabalhadores Rurais* (Rural Workers Trade-union Movement, MSTR) in the state of Acre (1975–1977).

Soon enough, however, the *ruralistas* (i.e. the organization of agrarian landowners), operating against both rubber-tappers in Acre and land squatters in southern Pará and Rondônia, organized violent reaction, and the assassination of union leaders—with the complacency of the regime—became a key component of their counter-strategy. In Acre, the murder of MSTR representative Wilson Pinheiro, in 1980, opened a period of extremely violent conflicts, which continued well after the murder of Chico Mendes in 1988—coming after six failed attempts—and has intensified in the last decade. Contradicting the Weberian concept of the state as exclusive holder of the legitimate monopoly of force, the Brazilian dictatorship shared this monopoly with the rural elites (Loureiro and Pinto 2005). This explains the practice, that spread across the Amazon region during the 1980s, of hiring gunmen to exercise social control over the countryside, and especially over land—a fundamental aspect of Brazil's agrarian problem. Hired gunmen, often police officers, served to protect the latifundia—large unoccupied areas, mostly left to cattle—against landless squatters, but they were also employed to kill religious, political and trade union representatives. Soon they evolved into private militias and security companies, and, after the end of the dictatorship, gunmen have become an integral part of the *grilagem* (land encroachment) process in Brazil's rural hinterlands; in other words, they form the armed core of capital's strategy for the enclosure and annihilation of the common/s. In Chico Mendes' words, violence against rural workers was genocidal politics consciously deployed in order to get hold of the Amazon forest: 'Its destruction, I think, implies the genocide of all of us who live in these forests, with dire consequences for the rest of the country and for humanity itself' (Smith 1989).

The *ruralistas*' plan to dominate workers by dividing them from each other and from the Indigenous people came to a halt in the early 1980s, and it underwent a historical defeat later in that decade, with the creation of the Alliance of Forest Peoples. To understand how this was possible, we need to look at the relationship of both Indians and rubber-tappers with the land. Since the early 1970s, the regime had been setting in motion a plan for the integration of Indigenous peoples through an Indian Emancipation statute (1973), which had grouped all Indigenous populations of Brazil into three categories of classification, depending on their relationship with white settler society: (1) isolated, (2) in the process of integration and (3) integrated. Through the prospected 'emancipation', Indians would acquire the right to sell their land on the market—which would have led to their final integration in Brazilian capitalism as a proletarian workforce. The Indians, however, were not interested in this project. This became clear when large mobilizations spread across the country, initiating the emergence of Brazil's Indigenous movement which, in the following years, began to expand strongly in the state of Acre. Its main aim was the demarcation of their ancestral territories—land where they would be allowed to live autonomously and that they would manage according to their traditional knowledge and techniques. In the early 1980s, the Indigenous movement in Amazonia started to organize direct actions to reclaim these territories, and expelled the rubber tapping companies. Witnessing these Indigenous struggles, based on their refusal of the 'Emancipation' statute, inspired rubber-tappers to do the same. As Ailton Krenak (in Cohn 2015: 57) recalled: the alliance between Indians and rubber-tappers in Acre emerged out of a common interest towards pushing capitalists out of the forest.

## Insurgent Commoning: The Extractivist Movement and the Resex Law, 1980s–1990s

At the time of the 1965 coup, Chico Mendes was a young worker in the Cachoeira rubber plantation, in the state of Acre. Unlike most of his co-workers, he had had the chance to get (self)educated, thanks to a former communist *guerrilla* man, Euclides Távora, who was hiding in the forest to escape the regime's persecution while also working for the politicization of rural migrant workers (Mendes 1998). Távora instructed Mendes to join a rural union as soon as this would be formed in his area—which he did, joining MSTR in 1975. Being one of the few literate workers in his region, Mendes soon was elected secretary. At the time, the union's strategy to improve the lives of peasants in Acre was based on three pillars: first, ending the debt peonage system; second, recognizing the rights of land squatters; third, guaranteeing the rights of rural workers employed on farms. Chico Mendes' leadership emerged from this movement, and was instrumental in struggles for the first and the second point, but differentiated on the third point. In Acre, the majority of workers affiliated with MSTR were rubber-tappers: inspired by the struggles of their fellow Indians, they did not consider the official union strategy appropriate for their context, because they were starting to see the possibility of reclaiming a new and better identity for themselves, that of autonomous subsistence workers and citizens of the forest—that is, commoners. This possibility was premised upon the primary aim of saving the forest itself—their home and source of livelihood—from destruction. In this political vision, the individual allotment system of the land reform would represent an absolute failure for rubber-tappers, whose livelihood depended on access to extended areas of forest and on multiple subsistence uses of non-timber forest resources. Thus, they felt the need to redefine the aims and strategy of the agrarian reform.

The result of this internal debate among rural workers was a union split: a section of the MSTR, connected to the Pastoral Land Commission, remained linked to the government and participated in institutional political actions, while representatives of the Xapuri Rural Workers Union, led by Chico Mendes, decided to organize an autonomous and oppositional rubber-tapper movement, called Conselho Nacional dos Seringueiros (CNS), founded in 1985. According to Mendes (1988), the legal expropriation of land for the creation of agrarian settlements would have the perverse effect of recognizing previous encroachments of forest on the part of big landowners. In fact, this was already happening in the South of Pará, where vast areas of rubber and Brazil-nut forest had been expropriated in accordance to the National Agrarian Reform Plan of 1985, with the scandalous overpricing of illegal properties to the benefit of local elites. In addition, the shift of existing rubber allotments into individual settlements would create yet another problem: in the attempt to making agrarian capitalists out of commoners, the way of life of rubber-tappers would be profoundly transformed—initiating an unavoidable process of deterritorialization (Paula 2004). To benefit rubber-tappers, agrarian reform should be different. In short, while MSTR remained faithful to the prevailing distributional idea of agrarian justice, the workers of Xapuri understood that the effective resolution of agrarian conflicts in their region necessarily involved the recognition of their collective rights and the protection of the forest as commons (Paula and Silva 2008).

It was from this analysis that the idea of *reservas extractivistas* (extractive reserves) arose, inspired by a dialogue with the Indigenous peoples who lived in what were then termed 'reserves' (before the concept of 'Indigenous lands' appeared in the 1988 Constitution). The concept materialized at a national meeting of rubber-tappers in 1985, together with the growing awareness of the need for a broader alliance with Indigenous people on one side, and with the environmental movement on the other. What rubber-tappers and Indigenous people shared, together with other 'traditional' Brazilian populations, like Quilombo communities, was the practice of commoning: collective access to the land they traditionally occupied, their territories, understood as sources of livelihood, freedom and cultural autonomy through consolidated practices of work-sharing. In Mendes' words:

> The Indians did not want to be colonists, they wanted to use the land as a community, and the rubber tappers shared this vision as well. We did not want ownership of the land, we wanted it to be held by the government, granting and regulating the rubber-tappers use. This thing caught on and started to attract the attention of the Indians, who started to articulate with us. (Mendes 1998: 78; our translation)

The Indigenous leader Ailton Krenak, who joined Mendes in building the Alliance of Forest People, recalls him saying:

> We learned our way of raising children from the Indians and from the forest itself. We attend to all our basic needs and have already created a culture of our own, which brings us much closer to the Indigenous tradition than to the 'civilized' tradition. (Krenak, in Diniz: 83; our translation)

The *seringueiros* movement can thus be considered a significant example of struggle for the de-alienation of work through the affirmation of an insurgent right to the common/s, or the natural conditions of production. While the agrarian reform freed rubber-tappers from their debt peonage towards the landowners, it threatened them with dispossession from their customary access to the forest, offering them monetary compensation in exchange. Accepting this offer would mean for the rubber-tappers to give way to ranchers and loggers—which was precisely the intent of the reform bill—with irreversible consequences for the extension of forest cover, and for the *seringueiros*' way of life. Here the notion of 'rubber-tappers territoriality' (Porto Gonçalves 1999) emerged as a strong identification with the place where they lived and made their living, and the will to fight for it. Rejecting the logic of universal monetary valuation and the consequent commodification of land and labour, the *seringueiros* understood their life-project as priceless, and organized to gain legal entitlement to it. Unlike other similar territorial struggles, however, this one was framed as a labour struggle—one led by a workers' movement to achieve autonomy from capital: in the words of a rubber-tapper: 'They have capital, we have the union' (ibid.: 74).

In a short time, the movement spread from Acre to the entire Brazilian Amazon, throughout the states of Rondônia, Amazonas, Amapá and Pará, mobilizing direct collective action as the main form of struggle for land and forest (Paula 2009), via an original form of non-violent protest called *empate*. Soon enough, it became clear that a more successful strategy would be that of uniting the *seringueiros* with collectors of Brazil-nut (*castanha*)—a plant widely diffused in the state of Pará, and with all other subsistence workers and forest dwellers of Amazonia that shared the same commoning life-project (Allegretti 2008; Milanez 2015). Through this process, the CNS acronym came to signify a union of all 'extractivist' workers reclaiming their right to the common/s. These different groups were united by two main struggles: (1) that for improving the quality of life of their communities through public policies that directly supported the sustainable use of non-timber forest products (such as subsidized rubber prices) and, (2) that for securing their collective access to land through a dedicated legislation. Articulating these two struggles, the Resex proposal emerged as a collective construction, a collaborative concept formulated via interaction between 'extractivists', environmental activists and natural scientists from national and international circles (Hochstetler and Keck 2007).

Mendes' leadership was instrumental in attracting the attention of national and international supporters. The Resex institution, he convincingly argued, represented an unprecedented opportunity to overcome a wicked problem of capitalist extractivism: that of the increasingly destructive interactions between human work and the natural world. International actors played a key role in contributing external resources and put pressure on international organisms, such as the World Bank, to guarantee territorial rights in the Amazon region, initiating a 'boomerang effect' (Hochstetler and Keck 2007). Their argument was that the protection of rubber-tappers would imply the protection of the forest, which was coming to the attention of international media as a vital global common. This strategy proved successful. With ordinance No. 627, of July 30, 1987, only two years after the founding of CNS, the first Extractive Settlement Project (PAE) was instituted in the state of Acre. Eighteen months later, on December 22, 1988, Chico Mendes was assassinated.

Initially, the PAE was intended for the exploration of areas with extractive rubber plantations through economically viable and ecologically sustainable activities, and represented an innovation in the settlement regime by granting collective use autonomously regulated by the concessionary community. With Law No. 7804, of July 18, 1989, the Resex institution—defined as 'territorial spaces designated to the autonomous and sustainable exploration and conservation of renewable natural resources by extractivist populations'—gained momentum within the scope of national environment policy. The body responsible for the creation of new PAEs became the IBAMA (Brazilian Institute of Environment and Natural Resources) and, later on, the 'Chico Mendes Institute for Conservation and Biodiversity', with a specific section named the National Center for Sustainable Development of Traditional Populations.

The political and economic sustainability of the Resex, according to the law, depended on mutual efforts of the state and the extractivists' unions. The state would be responsible for promoting a legal framework for collective property and guaranteeing territorial integrity amid areas of violent conflicts over land; meanwhile, residents would contribute to the conservation of the settlement with their re/productive work (Hochstetler and Keck 2007).

The story of how the Resex institution emerged is highly relevant to discussions of labour/environmental dialogue and collaboration. The Resex came out of a process of 'intercultural translation', that is 'the procedure that allows to create reciprocal intelligibility among different experiences of the world, both available and possible' (Santos 1980: 125). In this specific case, multiple intercultural translations emerged: between Indigenous and rubber-tapper knowledge about the forest; between working-class, peasant and Indigenous political cultures; between North American environmental activists and South American labour activists, and their respective organizational traditions; between academics and social movement activists; and between multiple languages (English, Portuguese and Indigenous languages) in their local/national, rural/urban differentiations. As a leading figure in this intercultural exchange, Chico Mendes was able to build bridges and harmonize political differences, circulating between different social groups to find common interests, opening new political opportunities and translating different cultural perspectives into a unity of vision and strategy. The extractive reserves were expressly thought of as an alternative form of conservation: as social subjects directly interested in preserving the country's forest cover, rubber-tappers came to be seen as the ideal strategic alliance for an environmental movement which shared with them the aim of protecting the Amazon from destruction (Allegretti 2008). This alliance, one of the most striking examples of labour-environmental coalitions in modern history, acquired worldwide significance by showing that working-class communities could design truly path-breaking environmental policies (Paula 2009, Paula e Silva 2008).

During the 1990s and early 2000s, a number of *seringueiro* leaders entered institutional politics, becoming governors, mayors, city councillors, deputies, senators and even ministers. The most important result of this political process was the emergence of the concept of *florestania*—literally, forest-zenship, that is, the citizenship of the forest dweller. This highly innovative political principle, which links human rights to the rights of nature, has long helped the *extrativistas* to persist in their life-project, seeing their existence recognized by the Constitution and the local laws; this has granted that the Resex remained relatively more protected than other rural areas, maintaining a forest coverage incomparably larger than the private properties that surround them (Almeida 2008; Almeida et al. 2018).

## A Disputed Legacy

Over the past two decades, a variety of practices in public discourse on conservation and development in Brazil have aimed at dismantling rubber-tappers' identity as commoners. The framework of *florestania* has been slowly eroded and replaced by 'green economy' discourses, deployed to the effect of green-washing the implementation of 'accelerated growth' programmes (called PAC) on the part of Lula da Silva's and Roussef's governments. Gradually but steadily, the Resex have shifted from common/s to sustainable development hotspots, with increasing external pressure to improve their combined eco-efficiency and commercial productivity. Rather than autonomous territorialities, they have been framed as production units competing with others on the international market. The capitalist logic of productivity, rejected by the movement in the 1980s, has come back in green clothes, threatening once again the logic of reproductivity put in place by the *seringueiros*. Green governmentality has made its power more pervasive than ever, convincing public opinion that a 'sustainable' commodification of forests represents the best interest of the country and the *extrativistas* themselves, who are now pressured to turn into entrepreneurs. Meanwhile, media discourses and academic debates have changed significantly, and even the big landowners have reconfigured their identity, adopting a strategic environmentalist posture with the aim of gaining support in sectors of academia and public opinion, and to attract investors. Promoting themselves as efficient managers of big estates, they claim that private property would grant better performances than the commons in terms of conservation of biodiversity (Peres and Schneider 2012; Michalski et al. 2010). In this context, the commercial extraction of timber, initially prohibited in the Resex, has become one of the main factors of internal division in the rubber-tapper movement and a source of violent conflict with farmers, loggers and ranchers surrounding the Resex (Paula and Silva 2008).

These shifts in public discourse have been co-constitutive with the shifts in international markets and in national economic policies that characterize what Maristella Svampa (2019) has called the 'commodity consensus' turn of Latin America. With it, the capitalist/developmentalist version of extractivism, a.k.a. commodification of nature—an ideology which dogmatically transits across the political spectrum (Danowski and Viveiros de Castro 2017)—has prevailed over the *extrativistas*' model. It has been embodied by economic plans and fiscal policies adopted by both neoliberal and progressive governments, assuming an extremely violent turn with the most recent far-right governments of Temer and Bolsonaro, which have put a halt to Indigenous and Quilombo land demarcations, while actively dismantling environmental protection programs (Araujo et al. 2019). The fact that timber, hydropower, soy and other commodities have a much higher market potential is one explanation for the tremendous pressure that surrounds Resex areas. Driven by such powerful pressures for 'accelerated growth', environmental violence has led many *extrativistas* to leave the Resex, provoking a rural exodus towards the urban areas—thus becoming proletarians. Others have tried to avoid this destiny by abandoning the traditional practice of rubber extraction, turning to livestock or more intensive farming (Salisbury and Schmink 2007), in an attempt at becoming entrepreneurs. As *extrativistas*' practices have diversified under the pressure of the Brazilian state and economic interests surrounding their territories, now including logging and cattle ranching, so their political identity as insurgent commoners, struggling for territorial rights, has been fractured (Vadjunec 2011).

Not unexpectedly, the 'commodity consensus' turn has spurred a new wave of environmental violence against the Resex, aimed at getting hold of their invaluable resources—timber, oil, water, minerals, biodiversity—or at turning them into monocrop plantations and pasture (Milanez 2019). Physical violence, death threats and assassination of unionists, peasants and Indigenous people have become the daily reality for the *extrativistas*, who live in a permanent state of insecurity. In this scenario, the discursive disqualification of the common/s as 'non-productive' has become a powerful tool for the *ruralistas* to legitimize violence. In the words of a prominent rancher from South Pará: 'these people (i.e. the *extrativistas*) do not contribute to lifting millions of people out of poverty. Sometimes they abuse their rights, and *they need to be excluded from Brazilian society*'.[1] In order to produce the non-existence of the commons, the commoners must be eliminated, both symbolically and materially.

---

[1] See the documentary movie *Toxic: Amazonia* (by Bernardo Loyola and Felipe Milanez), USA 2011: 32'05 to 32'15.

## Conclusion

More than three decades after Mendes' assassination, the memory of the *seringueiros* struggle does not seem to be reflected in labour's environmental politics as represented by the main trade union organizations worldwide. Why is it so? Our hypothesis is that the international labour movement has been dominated by an eco-modernist approach—the belief that the response to ecological problems would come from the greening of the 'forces of production', rather than from the common/s (Barca 2019b). This belief is being radically shaken by the climate justice movement today, and new possibilities are arising for labour environmentalism to converge with struggles for the commons towards a grass-roots Just Transition. With the aim of contributing to this transformation, we offer here some reflections on the lessons that labour environmentalists can draw from the Resex experience.

First and foremost, the *seringueiros* story tells us that defending labour rights and the rights of nature need not be contradictory political aims, but can in fact be one and the same struggle, insofar as this takes the common/s as its horizon. While the rubber-tappers could have taken advantage of the agrarian reform law to turn into land-owners, they demanded instead full emancipation and autonomy in a non-capitalist (commoning) relation to the forest. By struggling for their de-alienation, they also liberated non-human nature from commodification and from the logic of capitalist productivity. Rejecting the capitalist route to agrarian reform did not only imply for them to liberate their labour from the wage relation, but also to liberate non-human nature from the money-commodity-money circuit of capitalist valuation. They perfectly understood that the two processes were organically related. As a consequence, while the government's Land Statute aimed at turning them into a petty propertied class competing in the capitalist market—which would have led them to remove the forest cover to plant crops—they organized for an alternative reform, inspired by a vision of forest preservation and community access to it. Such a struggle for the right to an autonomous life-project, different from agrarian capitalism, was fundamental in the construction of rubber tapping territoriality. Their story illuminates the existence of a particular type of labour environmentalism—one in which labour and environment are organically connected by an anti-capitalist horizon of commoning. The latter, we argue, is a key precondition for what Räthzel and Uzzell (2011) consider a 'revolutionary reformist' approach to labour environmentalism, that is, one based on the active and creative involvement of workers in devising 'new forms of production that include a concern for nature and avoid its destruction' (ibid.: 88).

Second, the Resex experience tells us that the common/s are a crucial arena for convergence between labour and environmental organizations. In fact, CNS came to represent labour's visions and interests vis-à-vis decision-making processes with vast potential impact upon the Amazon region—thus establishing an unprecedented connection with the Brazilian environmental movement. This epistemic innovation of Resex must be seen as a highly original and vital contribution of Brazilian rubber-tappers to environmentalism worldwide, representing a concrete alternative to both the hegemonic theory and practice of conservation, and to the capitalist green economy. In fact, the struggle for the common/s is probably the most crucial point of contact for different forms of counter-hegemonic environmental mobilization in recent decades. Being an important part of this global struggle, the Resex is highly inspirational because it proves that autonomous, non-alienated work and the common/s are irreducible obstacles to capitalist degradation of both human and non-human nature—in other words, to the metabolic rift (Foster 2000). This is why, when allied to each other, anti-capitalist labour movements and commoning movements become a powerful driver of what Ariel Salleh (2010) calls 'metabolic value', that is, embodied and embedded (non-abstract) forms of value with the capacity to usher in truly sustainable human development.

Finally, the Resex story speaks to scholars in ELS as not mere observers, but as active members of a labour/environment alliance. As this chapter has shown, the struggle for the common/s is, at the same time, material and discursive, involving both the physical preservation and reproduction of people and the non-human world, and the symbolic defence of the political principle of the common, or non-alienation—and its collective memory. In this sense, the struggle for the common/s involves academics as knowledge-producing subjects in important ways: this certainly should not lead us to an attempt at freezing certain social identities, arising in given historical moments and spaces, with the intention of extending them into the future; rather, it means refusing to normalize the annihilation of the common/s by maintaining a constant effort at documenting and making sense of violent practices triggered by governments and corporations against both the commons and the commoners. The violence that abruptly took the life of Chico Mendes did not put an end to the struggle for the common/s. Keeping its memory alive serves the scope of reminding us that the commodification of land and labour is not inherent in human nature, but a historical product of unequal class relations.

The best way to illustrate this last point is by recalling the testimonies of four representatives of the Alliance of Forest People, given in a session on the legacy of Chico Mendes at a political ecology conference held in March of 2019.[2] Coming from different sections of the CNS movement, they all agreed that the most important and enduring legacy consisted in the political unity of different rural and coastline communities (Indigenous, Quilombos, Caboclos), representing a variety of biomas and modes of re/production from within and beyond Amazonia, into a single movement struggling for a common project of (r)existence and autonomy from capital. They evoked a new *empate* in response to the persistent violence against the forest and its people—a call that included social scientists from Latin America and beyond as members of that larger transnational alliance of nature defenders that Mendes had envisioned and brought to life. This chapter is our response to that call: we hope it will inspire environmental/labour scholars worldwide to mobilize in solidarity with Indigenous and *extrativistas* communities against the tremendous violence they are experiencing, and to embrace their struggle for the common/s as a path-breaking possibility for saving nature by de-alienating labour.

## References

Allegretti, Mary. 2008. A construção social de políticas públicas. *Chico Mendes e o movimento dos seringueiros. Desenvolvimento e Meio Ambiente* 18: 39–59.

Araújo, Roberto, Ima Vieira, Peter Mann de Toledo, Andréa dos Santos Coelho, Iveloi Dalla-Nora, and Felipe Milanez. 2019. Territórios e alianças políticas do pós-ambientalismo. *Estudos Avançados* 33 (95): 67–90.

Almeida, Barbosa de Mauro W. 2004. Direitos á floresta e ambientalismos: seringueiros e suas lutas. *Revista Brasileira de Ciências Sociais* 19 (55): 33–53.

———. 2008. A enciclopédia da floresta e a florestania. *Pagina 20*, January 3.

Almeida, Barbosa de Mauro W., Mary Helena Allegretti, and Augusto Postigo. 2018. O legado de Chico Mendes: êxitos e entraves das Reservas Extrativistas. *Desenvolvimento e meio ambiente* 48: 25–49.

Barca, Stefania. 2010. *Enclosing Water. Nature and Political Economy in a Mediterranean World*. Cambridge, UK: White Horse Press.

———. 2014a. Telling the Right Story. Environmental Violence and Liberation Narratives. *Environment and History* 20 (4): 535–546.

---

[2] They were: Sônia Guajajara, Indigenous leader and candidate to Brazil vice-presidency for the PSOL (Socialism and Freedom) party in 2018; Ângela Mendes, daughter of Chico, and secretary of women's office in the CNS; Edel Moraes, vice-president of CNS and Cludelice Santos, former vice-president of CNS. See III Congresso Latino-Americano de Ecologia Política, Universidade Federal da Bahia (Salvador, BR): roda de conversa "Chico Mendes vive", March 18, 2019 (https://www.congressoecologiapolitica.org/programacao).

———. 2019a. The Labour(s) of Degrowth. *Capitalism, Nature, Socialism* 30 (2): 207–216.

———. 2019b. Labour and the Ecological Crisis. The Eco-modernist Dilemma in Western Marxism(s), 1970s–2000s. *Geoforum* 98: 226–235.

Barca, Stefania, and Gavin Bridge. 2015. Industrialization and Environmental Change. In *The Routledge Handbook of Political Ecology*, ed. T. Perreault, G. Bridge, and J. McCarthy. London: Routledge.

Bennholdt-Thomsen, Veronika, and Maria Mies. 2000. *The Subsistence Perspective. Beyond the Globalized Economy*. London: Zed Books.

Berno de Almeida, Alfredo W. 2004. Terras tradicionalmente ocupadas: processos de territorialização e movimentos sociais. *Revista Brasileira de Estudos Urbanos e Regionais* 6 (1): 9–32.

Biesecker, Adelheid, and Sabine Hofmeister. 2010. Focus: (re)productivity. *Ecological Economics* 69 (8): 1703–1711.

Brownhill, Lee, Teresa E. Turner, and Wahu Kaara. 2012. Degrowth? How About Some 'De-alienation?'. *Capitalism Nature Socialism* 23 (1): 93–104.

Burkett, Paul. 2009. *Marxism and Ecological Economics. Toward a Red and Green Political Economy*. Chicago: Haymarket Books.

Caffentzis, George, and Silvia Federici. 2014. Commons Against and Beyond Capitalism. *Community Development Journal* 49 (S1, Jan.): i92–i105.

Cohn, Sérgio, ed. 2015. *Ailton Krenak*. Rio de Janeiro: Azougue editorial.

Danowski, D., and E. Viveiros de Castro. 2017. *The Ends of the World*. Cambridge: Polity Press.

De Angelis, Massimo. 2007. *The Beginning of History: Value Struggles and Global Capital*. London: Pluto.

———. 2017. *Omnia sunt communia. On the Commons and the Transformation to Postcapitalism*. London: Verso.

Diniz, Nilo. 2001. *Chico Mendes: um grito no mundo*. Master diss, University of Brasilia.

Foster, John B. 2000. *Marx's Ecology. Materialism and Nature*. New York: Monthly Review.

Hochstetler, Kathryn, and Margareth E. Keck. 2007. *Greening Brazil*. Durham: Duke University Press.

Leroy, Jean P. 2017. Markets or the Commons? The Role of Indigenous Peoples, Traditional Communities and Sectors of the Peasantry in the Environmental Crisis. In *Brazil in the Anthropocene: Conflicts between Predatory Development and Environmental Policies*, ed. Liz-Rejane Issberner and Philippe Lena, 104–124. London: Routledge.

Loureiro, Violeta, and Jax Pinto. 2005. A questão fundiária na Amazônia. *Estudos Avançados* 19 (54): 77–98.

Luxemburg, Rosa. 2003. *The Accumulation of Capital*. London: Routledge Classics.

Martínez-Alier, Joan. 2002. *The Environmentalism of the Poor: A Study of Ecological Conflicts and Valuation*. Cheltenham: Edward Elgar Publishing.

Mendes, Francisco. 1988. A Preservação da Floresta Amazônica. Filmed May 1988 at the department of Geography of the University of São Paulo, São Paulo, SP, Video, 109:45 min. Accessed June 29 2020. https://www.youtube.com/watch?v=oKS5JVTDmWU.

———. 1998 A defesa da Vida. In: Martins, Edilson. *Chico Mendes: um povo da floresta*. Rio de Janeiro: Garamond.

Michalski, Fernanda, Jean Metzger, and Carlos Peres. 2010. Rural Property Size Drives Patterns of Upland and Riparian Forest Retention in a Tropical Deforestation Frontier. *Global Environmental Change* 20 (4): 705–712.

Mies, Maria. 1986. *Patriarchy and Accumulation on the World Scale*. London: Zed Books.

Milanez, Felipe. 2015. 'A ousadia de conviver com a floresta': uma ecologia política do extrativismo na Amazônia. PhD diss., University of Coimbra.

———. 2017a. 25 anos sem Chico Mendes. *CartaCapital*, December 22. https://www.cartacapital.com.br/blogs/blog-do-milanez/25-anos-sem-chico-mendes-1140.html.

———. 2017b. 'Chico Mendes está pulando dentro do túmulo', diz amigo. *CartaCapital*, December 22. https://www.cartacapital.com.br/blogs/blog-do-milanez/osmarino-amancio-rodrigues-chico-mendes-era-libertario-e-esta-pulando-dentro-do-tumulo-3045.html.

———. 2019. Countering the Order of Progress: Colonialism, Extractivism and Re-existence in the Brazilian Amazon. In *Towards a Political Economy of Degrowth*, ed. Ekaterina Chertkovskaya, Alexander Paulsson, and Stefania Barca, 121–136. London: Rowman & Littlefield International.

Paula, Eder. 2004. O Movimento Sindical dos Trabalhadores Rurais e a Luta Pela Terra no Acre: conquistas e retrocessos. *Revista Nera* 7 (5): 86–101.

———. 2009. No limiar da resistência: luta pela terra e ambientalismo no Acre. In *Lutas camponesas contemporâneas: condições, dilemas e conquistas, v.1: o campesinato como sujeito político nas décadas de 1950 a 1980*, ed. Bernardo Fernandes, Leonilde Medeiros, and Maria Paulilo, 201–222. São Paulo: Editora UNESP.

Paula, Eder, and Silvio Silva. 2008. Movimentos sociais na Amazônia brasileira: vinte anos sem Chico Mendes. *Revista Nera* 11 (13): 102–117.

Peluso, Nancy, and M. Watts, eds. 2001. *Violent Environments*. Ithaca: Cornell U.P.

Peres, Carlos, and Maurício Schneider. 2012. Subsidized Agricultural Resettlements as Drivers of Tropical Deforestation. *Biological Conservation* 151: 65–68.

Porto Gonçalves, Carlos. W. 1999. "A territorialidade seringueira: geografia e movimiento" social. *GEOgraphia* 1 (2): 67–88.

———. 2002. O Latifúndio Genético e a R-existência Indígeno-Camponesa. *GeoGraphia* 4 (8) https://doi.org/10.22409/GEOgraphia2002.v4i8.a13431.

Räthzel, Nora, and David Uzzell. 2011. Mending the Breach between Labour and Nature: Environmental Engagements of Trade Unions and the North-South Divide. *Interface: A Journal for and about Social Movements* 4 (2): 81–100 (November 2012).

Salisbury, David, and Marianne Schmink. 2007. Cows Versus Rubber: Changing Livelihoods among Amazonian Extractivists. *Geoforum* 38 (6): 1233–1249.

Salleh, Ariel. 2009. *Eco-Sufficiency and Global Justice. Women Write Political Ecology.* London: Pluto Press.

———. 2010. From Metabolic Rift to 'Metabolic Value': Reflections on Environmental Sociology and the Alternative Globalization Movement. *Organization & Environment* 23 (2): 205–219.

Santos, Boaventura de Sousa. 2006. *A Gramática do tempo: para uma nova cultura política.* Porto: Ed. Aforamento.

Santos, Roberto. 1980. *História Econômica da Amazônia (1800–1920).* São Paulo: T. A. Queiroz.

Scheidel, Armin, Daniela Del Bene, Juan Liua, Grettel Navasa, Sara Mingorría, Federico Demaria, Sofía Avila, Brototi Roy, Irmak Ertör, Leah Temper, and Joan Martínez-Alier. 2020. Environmental Conflicts and Defenders: A Global Overview. *Global Environmental Change.* https://doi.org/10.1016/j.gloenvcha.2020.102104.

Smith, Marianne. 1989. *Chico Mendes: Voice of the Amazon.* Doc. 56 min.

Stevis, Dimitris, David Uzzell, and Nora Räthzel. 2018. The Labour–Nature Relationship: Varieties of Labour Environmentalism. *Globalizations* 15 (4): 439–453.

Svampa, Maristella. 2019. *Neo-extractivism in Latin America: Socio-Environmental Conflicts, the Territorial Turn, and New Political Narratives.* Cambridge: Cambridge University Press.

Taussig, Michael. 1984. Culture of Terror-Space of Death. Roger Casement's Putumayo Report and the Explanation of Torture. *Comparative Studies in Society and History* 26 (3): 467–497.

Vadjunec, Jacqueline. 2011. Extracting a Livelihood: Institutional and Social Dimensions of Deforestation in the Chico Mendes Extractive Reserve, Acre, Brazil. *Journal of Latin American Geography* 10 (1): 151–174.

# 15

## Connecting Individual Trajectories and Resistance Movements in Brazil

Beatriz Leandro, Patrícia Vieira Trópia, and Nora Räthzel

## Introduction

In 2020, the Women's Protest at Aracruz celebrated their 14th anniversary. On 8 March 2006, 1800 women from Via Campesina (International Peasant's Movement) carried out a major action against the monoculture of eucalyptus. They occupied the nursery tree farm of Aracruz Celulose in Barra do Ribeiro, Rio Grande do Sul, and destroyed greenhouses and eucalyptus trays.

They decided it was time to show the world the consequences of the large-scale planting of eucalyptus. Eucalyptus trees need much water for their development—an average of 30 litres per day. This causes a shortage of water for plants, animals and humans. The women identified capital as the enemy of the working class and while arguing that there is no women's liberation without the destruction of capital. Via Campesina aims to guarantee land for rural workers and create a positive relationship with nature through agroecological

---

B. Leandro (✉)
KOINONIA, Ecumenical and Interreligious Dialogue for the Amazon, São Paulo, Brazil

P. V. Trópia
Federal University of Ûberlandia, Uberlândia, Brazil

N. Räthzel
Department of Sociology, Umeå University, Umeå, Sweden

and other forms of sustainable production and ways of living. The Women's Protest at Aracruz is one of the internationally best-known marches organised by Via Campesina. It also reflected the activism of social movements in Brazil in general.

## Agrarian Protests

The large number of agrarian protests taking place in Brazil is a response to the growth of violence in the hinterlands. The country is considered, by Global Witness, as the most violent country in the world against environmentalists (Carta Capital 2017).[1]

Agrarian protests aim at defending agrarian reform, demarcation of indigenous land, resisting violence and environmental degradation (CPT-Comissão Pastoral da Terra 2018a, b).

Environmental conflicts expose the destructive power of corporations, agrobusiness or the State, and show how aggression towards nature affects life and the local and popular economy and culture. Martínez Alier (2008) calls environmental conflicts 'popular ecologism', 'ecologism of the poor' or 'environmental justice movements'. Martínez-Alier and O'Connor (1996) introduced the term 'distributive ecological conflict' to characterise the conflictual dynamics of economic and political power. Fernandes explains, 'for the peasantry their land is both where they produce and where they live, while for agribusiness the land is only a workplace' (Fernandes 2016, 54).

This chapter analyses resistance against environmental degradation in Brazil, in a context of agribusiness that is expanding capitalist relations of production in the countryside. Firstly, we describe the socio-economic conditions for rural protest; secondly, we analyse data of these protests, showing their aims and strategies and the ways in which they redefine 'development' and build an alternative model of agriculture based on the principles of social justice, ecological sustainability and respect for culture (Desmarais 2007). Thirdly, we analyse the life histories of organisers of these movements. The intersectional analysis of collective and individual subjects allows us to identify: (1) the context in which activists became leaders, (2) the collective experiences they were part of, (3) the demands at the core of their struggles and (4) the values influencing them.

---

[1] Between 1985 and 2019, the Pastoral Land Commission (CPT 2018a, b) recorded 50 massacres killing 247 people in ten Brazilian states. Between 1985 and 2017, it recorded 1438 cases of conflict including 1904 murder victims, mostly in the Amazon forest. Only 113 cases (8%) were judged in which 31 killers and 94 executors were convicted. This shows how impunity is still a pillar sustaining violence.

Our goal is to draw attention to the distributive character (Martínez-Alier and O'Connor 1996) of environmental conflicts and social protests. In the Global South ecologism is often agrarian, comprising movements for whom nature is an essential ally of their work and life, which is why they need to defend it.

## Part I: Social Movements and Their Roots and Conditions

### The Role of Brazilian Liberation Theology

The creation of the largest social movement, the Landless Workers' Movement (MST: Movimento dos Trabalhadores Rurais Sem Terra), was influenced by Liberation Theology. Most Brazilian militants are motivated by Christian beliefs: land for the landless, food for the hungry, literacy for the uneducated—not through charitable work, but by forcing the state to take its responsibilities for its poorest citizens seriously.

Shortly after the end of the Second Vatican Council in 1965, Brazilian bishops, priests and nuns began to practise Liberation Theology (French 2007). Important sectors of the Brazilian Catholic Church adopted a 'preferential option for the poor', 'at a time when economic development, modernization, and democracy were not considered meaningful during the repressive environment characterized by the Brazilian military dictatorship (1964–1985)' (French 2007, 409).

According to Löwy, of all the countries in Latin America, it is Brazil where 'Liberation Theology and its pastoral followers […] won a decisive influence' (French 2007, 410). During the dictatorship the 'Brazilian Church was practically the only effective space of liberty … the voice of the voiceless', according to Gómez de Souza (French 2007, 410). French claims that 'Liberation Theology played a critical role in the Church's changing attitudes toward indigenous rights and land struggles and served as the catalysing force for the creative use of the law to advance those goals' (2007, 410).

Bishops organised the impoverished rural and urban communities politically through the Basic Ecclesial Communities (CEBs: Comunidades Eclesiais de Base). By 1981 there were around 80,000 CEBs across Brazil comprising some two million people (Sader 1995, 155–6).

The CEBs were connected to the popular masses during the dictatorship (Frazão and Souza 2007), encouraging individuals to make sense of the world by linking their faith with the questioning of social inequality produced by the rapid expansion of capitalism. The CEBs resulted from a process initiated by the Medellín Conference of Latin American Bishops in 1968, who agreed that the church should take on the concerns of the poor. This movement was later officially named 'Liberation Theology' by the Peruvian bishop Gustavo Gutiérrez Merino (Gutiérrez Merino 1973). Based on Marxist philosophy, Gutiérrez Merino argued that 'people must build their own history rather than wait for divine intervention' (Menezes and Julio 2007, 332).

While unions were recognised as legitimate actors in dialogues with government and employers, social movements (precarious workers, housewives, unemployed young people, slum dwellers) had neither a formal position in the productive sphere (Sader 1995), nor a welfare state on which to rely. Individuals turned to communitarian structures such as the CEBs which supported their struggle and their discourse of solidarity and justice (Sader 1995; Löwy 2000).

Liberation Theology was developed 'under specific historical circumstances' in Latin America: dependent capitalism, mass poverty and popular religiosity, including the impact of the Cuban Revolution (Löwy 2000, 63). It can be comprehended as an 'elective affinity' between interpretations of Christian faith that views society through the eyes of the poor and oppressed and social utopia (Löwy 2000).

Liberation Theologists used educational methods based on Paulo Freire's 'Pedagogy of the Oppressed' (Freire 1996), which explained how capitalism works through the exploitation of the working class and how struggles emerged out of the worker's experiences. Freire's ideas were influenced by Gramsci arguing that ' "everyone" is a philosopher and it is not a question of introducing from scratch a scientific form of thought on everyone's individual life, but of renovating and making "critical" an already existing activity' (Gramsci 1999, 637–8). This would enable the 'oppressed' to intervene in reality in order to change it (Freire 1996, 64).

It was Leonardo Boff who in the mid-70s connected Liberation Theology with the ecological paradigm. His analysis gained support from 'The Limits to Growth' (Meadows et al. 1972), which predicted an environmental and economic collapse within a century if 'business as usual' continued. Boff (1993) recognised the co-responsibility of Christianity for the environmental crisis as the effects of climate change would be felt by the most vulnerable and miserable ones. He argued that Liberation Theology and the environmental discourse come from the same root: from two 'plagues that bleed': the plague of poverty that tears up the social tissue and the plague of systematic aggression against the Earth (Boff 1996, 163). Boff pointed out that the ecological, human, social and spiritual aspects of life need to be bound together. Today, during the totalitarian times in South America, the Catholic Church seems to be 'redeeming' itself, supporting the struggles of the 'Wretched of the Earth'.

## Agribusiness Still Calls the Shots in Agrarian Reform

Since 1945, peasants and rural workers organised as civil associations (known as Peasant Leagues) under the initiative and direction of the recently legalised Brazilian Communist Party (PCB: Partido Comunista Brasileiro). Rural leagues and associations were created in almost every state in the country, notably in the Northeast. However, with the banning of the PCB in 1948, these organisations collapsed.

Peasant Leagues resurged in 1955, with the creation of the Pernambuco Agricultural and Cattle Raising Society of Planters (Gaspar 2005). The movement's objective was to fight for agrarian reform and land ownership. The Military Coup of 1964 prohibited the movement and arrested and exiled its main leader, Francisco Julião. All rural social movements were violently persecuted and went underground, re-emerging in the 1960s and the second half of the 1970s, supported by the Catholic Church and left-wing parties, such as the Workers' Party (PT: Partido dos Trabalhadores).

During the 1970s, land reform and indigenous struggles were often intertwined.[2] The Liberation Theology–inspired CPT was created in 1975 by radical pastoral activists, who sought an end to the widespread violence of landowners and the state against the rural poor. It offered legal services, encouraged the creation of rural unions, denounced the use of violence against the rural poor and offered courses in faith and politics. It became a catalysing force for a series of land struggles attracting support from rural trade unions, academics, journalists and politicians, some of whom were to become the organisers of the Workers' Party in the state of Sergipe in 1980—the same year that the MST was established there, initially informally (French 2007, 431). It was formally founded in 1984, emerging from a process of occupying *latifundios* (large landed estates) and is currently the main rural social movement in Brazil. Its collective actions have three phases: occupation, camping and settlement.

The MST is more than a peasant's movement. It seeks to restructure Brazil socially, economically and politically by '(1) strengthening civil society

---

[2] In 1975, a series of meetings with indigenous leaders and trade-union federations resulted in a document entitled 'A Single Outcry of Indigenous and Peasants: The Land for Those Who Work It' (French 2007, 430).

through the organization and incorporation of marginalized sectors of the population; (2) highlighting the importance of public activism as a catalyst for social development; (3) facilitating the extension and exercise of basic citizenship rights among Brazil's poor; and, (4) engendering a sense of utopia and affirmation of ideals imbued in Brazil's long term, complex and open-ended democratization process' (Carter 2005). Struggles of the MST against the use of transgenics and pesticides in agriculture, or against deforestation, go beyond agrarian reform, with the MST denouncing the predatory character of capitalism as an annihilator of nature and lives.

The MST has led over 2500 land occupations, with about 370,000 families settling on 7.5 million hectares of land that they won as a result of the occupations (Land Struggle Database 2017). In rural communities organised by the MST, there are about 120 high schools, 200 complete elementary schools and more than 1000 elementary schools spread across 24 states (Lima 2019). In 2019 about 200,000 students were registered in these elementary schools (ibid.).

Between 1988 and 2016, the rural movements in Brazil have led to 9748 land occupations, 27% of those occurring being between 1996 and 1999 (Land Struggle Database 2017). In this period, the centre-right (Fernando Henrique Cardoso's governments 1994–2001) implemented neoliberal policies, converging with a new phase of spreading monopoly capitalism in the countryside. The problems experienced by the majority of the rural population were caused by agricultural modernisation leading to exploitation and marginalisation (Grzybowski 1994):

> Around 11.3 million rural inhabitants live on 3,775,826 farms and are left with just 4.03 percent of the wealth produced, and the families living on 2,014,567 farms have incomes of half the statutory minimum wage or less. Most farms with the smallest share of the wealth are family farms. (Fernandes 2016, 54)

Brazilians huge inequalities have arisen through the dominance of the agribusiness model. The concentration of land creates rural exodus, destruction of natural resources and common goods, and the devastation of the environment. It also empowers an elite associated with pesticide-intensive agribusiness producing commodities for export and reinforces gender inequality: Men own 87.32% of all farms which represent 94.5% of all agriculturally used land. Women own 12.7% of all farms and these represent only 5.5% of all used land. Also, 8.1% of rural workers without land are women, while only 4.5% of them are men (OXFAM 2016, 10). The exploitation of natural resources affects women disproportionately, since they lose essential elements of their livelihood (food, herbal medicines and access to water) and are exposed to violence.

Moreover, giant national and multinational corporations which control agribusiness are represented in Parliament by a group of deputies and senators, called 'rural bloc' or 'cattle bloc'. Their influence is evident in the permits they get for transgenic plants; the ways in which the government responds to conflicts between landowners, indigenous people and peasants; and in how they prevent governments from creating improvements for the poor (Sauer 2017). Furthermore, these corporations receive most of the rural credit and influence how technology is invested in agriculture and livestock. Therefore, commodity-oriented agribusiness overpowers family farming which produces most of the food consumed in Brazil (Fernandes 2016, 53).

According to Navarro, 'To maintain its dominance, agribusiness presents itself as the only possible model of development, questioning whether family or ecological farming is worthwhile' (Fernandes 2016, 53). Lula and Dilma Rousseff (PT) were unable to abate this dominance. They were elected by a broad class alliance, including sectors of the agrarian ruling class, such as the Brazilian Agriculture and Livestock Confederation (CNA). While maintaining an ambiguous relationship with the agrarian social movements, the PT governments did not earn the trust of the agrarian elite in spite of partly supporting their interests:

> Coupled with economic concessions and incentives granted to agribusiness, this alliance was critical in neutralizing and derailing structural land policies such as the expansion of land expropriation and settling landless families, resulting in an increase in monocropping and economic dependence on the export of raw materials. (Sauer 2017, 103)

The implementation of policies supporting family farming and of policies strengthening large agribusiness enterprises expresses the ambiguous nature of Brazilian's neo-developmentalism, formed during the PT's governments. It was directed by the Brazilian bourgeoisie, involving several sectors of the economy—mining, heavy construction, the agribusiness, the manufacturing industry and, to a certain extent, the large predominantly national private and state-owned banks (Boito Jr. 2012). All these sectors expected the state to protect them from competition by foreign capital. The same is true for the lower middle class, the working class, peasants and workers of the 'marginal mass'.[3]

Brazil has 'encouraged the dissemination of policies […] on rural territorial development and food acquisition programmes to other Latin American countries' (Sauer et al. 2018, 2).[4] On the other hand, PT governments 'implemented the "national champions policy" (BNDES 2017 quoted in Sauer et al., 2) […] and reinforced the presence of Brazilian agribusiness multinationals' (Sauer et al., 2). The state supported the interests of large business groups more than the networks supporting family farming. This had consequences for the environment.

Agribusinesses account for almost 70% of Brazil's greenhouse gas emissions (Nobre and Biderman 2017). The 'big four' agricultural commodities—palm oil, wood products, soy and cattle—also account for more than a third of tropical deforestation each year (McCarthy 2016). If deforestation continues at the current pace (some 7000 km$^2$ per year in the Amazon) in three to four decades there will be an accumulated loss intensifying global warming regardless of all efforts to reduce greenhouse gas emissions (Scott et al. 2018). Brazil is the only of the world's large economies to increase pollution without generating wealth for its society. Deforestation hampers Brazil's goals for the International Policy on Climate Change for 2020 (Estrada 2017) and compromises its target set during the Paris Agreement on zero illegal Amazon deforestation by 2030.

Far-right, agribusiness-friendly, President Bolsonaro, 'the exterminator of the future' according to the former environment minister Marina Silva, has initiated a significant change in Brazil's environmental policies and regarding the Amazon deforestation. Silva and seven other former environment ministers—who served governments across the political spectrum over 30 years—denounced the president's assault on rainforest protections, warning that he was systematically destroying Brazil's environmental protection policies (Kaiser 2019).

In December 2019, the Brazilian Bar Association for Human Rights and the Arns Commission[5] denounced Bolsonaro to the International Criminal Court for 'crimes against humanity' and inciting the genocide of Amazon's

---

[3] José Nun (1969) elaborated the concept of 'marginal mass' identifying the formation of a class fraction within the Latin American proletariat composed of (1) the reserve industrial army, a surplus population of workers at the disposal of capitalism, with the possibility of incorporation into the productive process; and (2) relative surplus population exceeding the limits of incorporation into the productive system.

[4] Positive experiences in the areas of family farming and social policy are mostly related to actions undertaken by the Brazilian Cooperation Agency (ABC), the now-dissolved Ministry of Agrarian Development, the Ministry of Social Development and the Brazilian Agricultural Research Corporation (EMBRAPA).

[5] The Arns Commission is a non-partisan group comprised of lawyers, former ministers and civil society activists. It was named after Cardinal Paulo Evaristo Arns, who protected hundreds of South American activists during dictatorships (Phillips 2019).

indigenous peoples. 'The living conditions and lifestyles of the indigenous peoples are being destroyed by river pollution and invasion of their lands by wildcat miners, loggers and land-grabbers' (Phillips 2019 in the Guardian).

Bolsonaro slashed funding for scientists and fired the head of the National Institute for Space Research (INPE), Ricardo Galvão, accusing him of lying about the data that proved the spike in deforestation and of being 'in cahoots' with environmentalists. Galvão was chosen as one of the ten most important scientists by the journal *Nature* later that year (Tollefson 2019).

During seven months in office, Bolsonaro approved 290 new pesticides, some of them banned in the EU, the United States and elsewhere. A total of 1942 registered pesticides were re-evaluated, with the number considered extremely toxic dropping from 702 to 43 (Gonzales 2019):

> Communities are being exposed regularly to pesticides by aerial spraying, including homes and schools, suffering symptoms of acute poisoning and undoubtedly chronic effects for many as well. (Dowler 2020)

In the next section we will examine the degree, contents and forms of resistance of rural movements.

## The Profile of the Agrarian Protests and Manifestations

The agrarian protests and land occupations by different social movements taking place in Brazil between 2000 and 2016 were a reaction against the agribusiness model, which 'controls the land in two ways: through monopoly ownership of peasants' holdings (Oliveira 1991 quoted in Fernandes, 53) and 'the imposition of production technologies on the peasant families whose income it capitalizes' (Fernandes 2016, pp. 52–3).

The Land Struggle Database has identified a broad repertoire of struggles by agrarian workers: public building sit-ins, occupations of public buildings, marches, encampment and road obstructions, among others. Tilly and Tarrow define such struggles as 'the ways in which people act together in pursuit of shared interests' (2007, 11). To identify particular characteristics of struggles, Comerford considers that protests use 'a certain 'style' of mobilization and demonstration, crystallizing a well-defined repertoire of collective action forms with public visibility' (1999, 127). The style of mobilisations and the way of giving visibility to demands and problems help to clarify their analysis.

In 2014, Brazil ranked third in the world concerning environmental conflicts (Environmental Justice Atlas, cf. Agência Envolverde Jornalismo 2014). They derived from three disasters (Carta Capital 2016).

Firstly, the construction of Belo Monte Hydroelectric Plant under Dilma Rousseff's government led to the outbreak of strikes and protests by workers, as well as demonstrations by social movements, environmentalists and indigenous groups. The plant was built on the Xingú River, in the municipality of Altamira, Pará, affects more than 20,000 people and threatens the survival of the Kayapó indigenous tribes and *quilombolas*[6] who depend on the river.

The Movement of People Affected by Dams (Movimento dos Atingidos por Barragens—MAB), created in the 1980s, is the most active movement. In February 2020, families affected by dams occupied the city hall building and the office of Norte Energia, a concessionaire of the Belo Monte plant in Altamira. They demand to be recognised as affected by the company and to receive compensation or resettlement (MAB 2020).

Secondly, the environmental crime committed by Samarco (Vale/Brazil and BHP Billiton/United Kingdom) in 2015, when the disruption of the Fundão Dam killed 19 people and left more than 600 families homeless in the city of Mariana, in the state of Minas Gerais.[7] In the fifth anniversary of the rupture of the dam the justice established that Samarco had until February 2021 to complete the resettlements of all victims, a late victory for MAB.

Finally, the construction of the Rio Petrochemical Complex de Janeiro (Comperj) has affected artisanal fishermen. The conflict intensified in 2009, when the fishermen occupied the construction site for 38 days. After this, fishermen from the *Men of the Sea* movement have been menaced and four of them were murdered.

Between 2000 and 2016, 12,554 agrarian protests took place involving 7,049,073 people (Land Struggle Database 2017, 51). In the same period, 6469 land occupations took place, carried out by 137 social movements. The MST is responsible for most land occupations, according to Land Struggle Database (2017).

Most of the protests have various causes, involving land tenure and environmental issues (CPT 2018a, b). To identify their claims, we analysed 10,621 agrarian protests occurring in Brazil between 2000 and 2014 (Land Struggle Database 2017), allocating them according to claims: land tenure and agrarian reform, environmental issues (water, pesticides, mining companies, dams,

---

[6] *Quilombolas* are descendants from black slaves who, in colonial times, fled seeking shelter in isolated regions. They remained there after abolition, maintaining their identity and culture.

[7] In January 2019, another disaster, caused by the same company (Vale), took place in Brumadinho, in the state of Minas Gerais that killed 270 people.

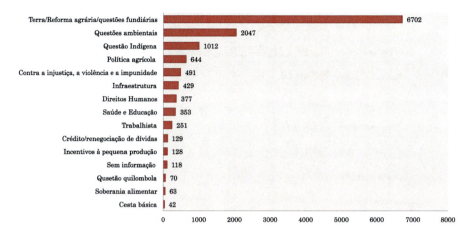

Fig. 15.1 Agrarian protests per type of claim, Brazil (2000–2014). (Source: DATALUTA 2015. Our elaboration)

monoculture), agricultural politics, indigenous and *quilombolas* issues, human rights, workers' issues, injustice and violence, and others. Most of the protests (51.2%) are claims for land, agrarian reform and land issues like expropriation, settlement and resettlement (Fig. 15.1). Environmental issues represent 15.9% of all claims. Indigenous movements fight for the demarcation of land, health, education, public policies concerning water and the environment, against hydroelectric plants and violence (7.9%).

Water scarcity—exacerbated by climate change and water-related disasters—can cause tensions leading to violent conflicts between people, communities and countries. As Fig. 15.2 shows, water is the main source of environmental conflicts in Brazil (50.1%). Half of these include defending rivers and aquifers against corporations or hydroelectric plants. These are fight for survival since peasants, rural workers, indigenous people and family farmers depend on water as their source of life.

As resources move through the commodity chain from extraction, processing to disposal, at each stage environmental impacts are externalised onto the most marginalised populations. Something consumers are not aware of.

Conflicts over water, mining, oil and agribusiness often overlap. The significance of commodity chains for the overlapping of environmental conflicts is made visible by a collaborative map (Environmental Justice Atlas 2020) that identifies them around the world, collecting stories of communities struggling for environmental justice. By increasing their visibility, it attempts

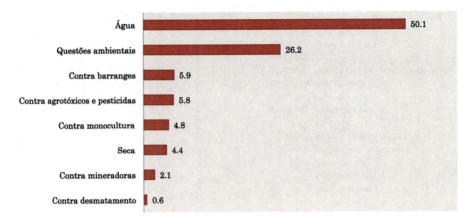

**Fig. 15.2** Type of claims from environmental protests (2000–2014) Brazil. (Source: DATALUTA 2015. Our elaboration)

to serve as a virtual space where those working on environmental justice issues can get information and find counterparts (Temper et al. 2015). In this atlas there are currently about 680 socio-environmental conflicts documented in South America—171 in Brazil (Environmental Justice Atlas 2020). Leal (2013, 84) talks about 'dormant' socio-environmental conflicts, in which competing interests and different social conditions of re-production generate inequalities concerning access to environmental resources, without manifesting themselves openly.

We can understand the development of rural social movements as a product of the way in which capital, labour and power move historically *through*, and not *upon*, nature in what Moore calls the 'web of life' (2013). Thus, capital produces new natures, including 'cheap workers' separated from their land and exploited like nature. Simultaneously, these very workers act against capital, forcing it to change and at times to withdraw.

Having presented the societal conditions under which social movements in Brazil developed and the issues for which they are fighting, we now shift to the micro-level, focussing on how individuals become members and leaders of such movements to understand the ways in which individual, organisational and national trajectories interrelate.

## Part II—Becoming Organic Intellectuals: Poverty, Inequality and the Politicisation of Experiences of Exploitation

The following interviews were conducted within the international research project 'Moments of danger, moments of opportunity: the role of individuals as change agents in organisations'. In Brazil, South Africa, India, Spain, Sweden and the UK we interviewed over 120 trade unionists responsible for environmental issues in industrial and rural unions (see also Chaps. 27 and 28 in this volume). In Brazil 20 interviews, lasting between one and three hours, were conducted with environmentally engaged union officials and members of rural workers' movements between 2012 and 2013. In the context of the rural struggles we outlined above, we focus here on members of Via Campesina, the Rural Women's Movement and the Rural Workers Union in the Amazon. We chose three environmentalists with comparable backgrounds: coming from poor families they have had to engage in precarious work at a young age. They grew up during the military dictatorship when Liberation Theology and Paulo Freire's ideas were building the pillars of resistance. They share an engagement with Liberation Theology, which made them aware of Capital's double exploitation of workers and nature: Isabel Castro, board of Directors of the Movement of Rural Women in Via Campesina (MMTR); Rosângela Piovizani, Director and National Coordinator of the Rural Women's Movement (MMC); Peixe, President of the Rural Workers Union, Santarém, Pará.[8]

Articulating macro- and micro-social analyses, we avoid explaining social processes solely based on the role of individuals (subjectivism) or solely in terms of structures, disregarding the importance of subjects (objectivism). Portelli writes about the connection between individual stories and social relations explaining that the stories people tell

> are both highly individual expressions and manifestations of social discourse, made up of socially defined and shared discursive structures […]. Through these structures, […] we can see how each individual […] negotiates the interplay between the individual and the social, of individual expression and social practice. This negotiation varies with each […] performance but is always carried out on the basis of recognizable, socially defined 'grammars'. (1997, 82)

---

[8] All interviewees agreed to their words and their real names being used.

Previously, we have laid out the societal 'grammars' with which our protagonists grew up: the Brazilian dictatorship, the limited power of progressive governments to tackle the unjust distribution of land, the poverty of small farmers and landless workers, including gendered relations of power in society at large, and Liberation Theology as a belief system that linked religiosity with a legitimation of resistance. Presenting exemplary (Flyvbjerg 2006) life stories of environmentally engaged leaders helps us to understand how individuals translate these societal 'grammars' into personal motivations for actions.

Gramsci's concept of 'organic intellectuals' is helpful here. He insisted that 'in any physical work, even the most degraded and mechanical, there exists a minimum of technical qualification, that is, a minimum of creative intellectual activity'. 'All men are intellectuals [...] but not all men have in society the function of intellectuals' (1999, 140). The function of an intellectual meant to organise consent *for* the specific class, which they represented and to create 'homogeneity' *within* that class. Organic intellectuals help to create a new hegemony challenging existing power relations. 'The mode of being of the new intellectual can no longer consist in eloquence, [...] but in active participation in practical life, as constructor, organiser, "permanent persuader"' (ibid., 141). How did our interviewees become such 'permanent persuaders'?

## They Do Not Call It Poison

At an early age, Isabel wanted to transcend the limitations of her birthplace:

> I was born in São José do Ouro, [...] in Rio Grande do Sul, [...] in the countryside. I stayed there until I was 13. From there I went to the city because I wanted to study. [...] at that time we didn't have transport [...] The 4th grade finished and [...] it wasn't possible to continue.
>
> I went to work as a housemaid and [...] I found that I was a house slave … because the family paid my studies. But studying, I came to discover, was in a public school which didn't cost anything. [...] I didn't earn anything; I got food, some used clothes, the books that they gave me were used, the only new thing [...] was a uniform. [...] At that time, it was the military dictatorship, [...] we had to wear a school uniform.

When she realised, she had been lied to, she 'went to a relative's house; [...] I was able to get another job, also as a housemaid, but I received a salary there. [...] a housemaid is a job which is exploited [...] there isn't a timetable'.

Around 1978, during the military regime, Isabel was asked to become a member of the Youth Pastoral; 'that was my entrance', she recalls: 'We began

## 15 Connecting Individual Trajectories and Resistance Movements...

to find out about the other struggles, the other organisations, [...] at the time, many organisations were born, including the women's movement which I am part of today.'

Isabel creates a connection between the experience of being exploited, her decision to leave her first housemaid job and her learning about and joining 'other struggles'. The Youth Pastoral as the 'entrance' into struggles exemplifies the role of the church as an organiser of resistance. It could overcome the individualised working conditions of housemaids, which were difficult to reach by unions or political organisations.

With her husband, whom she met in the Youth Pastoral, Isabel decided to move back to the countryside. There,

> the organisation of women began to appear, the rural women workers, [...] and I joined the women's movement there [...] the struggle for rights began there, [...] social security, maternity pay. All the social security rights that we have today, we won there.

In the midst of these struggles Isabel experiences motherhood as a difficult time because it took her out of the movement for over six years:

> But then came the children [...] I had two daughters very close to each other [...] And I stopped. So, now it's motherhood, [...] It was hard to deal with ... I wanted to be in the fight too. [...] It was a friend from the party who pushed me to return to the movement.

Isabel reflects how her personal situation, the difficulty deciding how to combine the political work she cares for with the care of her children, is

> part of our struggle, the construction of that equality [...]. It's the reason for the existence of the movement of women [...] the women's struggle is for the freedom of women. [...] There is so much in the [...] culture, of everyday life [...] that wouldn't let us speak. [...] And when I went my friend told me, be careful. It was not always peacefully discussed. [...] But it was worth it [...] to take part. From then on, I have always been present in the struggles of the movement, in the day to day.

Isabel experienced that exploitation was present not only in unequal power relations between employer and employee, not only in the lack of legislation protecting women's needs, but also in the everyday life among her comrades, where the right of women not to be reduced to their role of mothers was not respected. Sexism, she recognises, is present everywhere: 'the decisions within the families continue to be made by men, the rural way of life continues to be very sexist; it's a sexist, capitalist society.'

And capitalism, she realises, also dominates the state:

> For some time, our claim was with the state, the question of rights. But then we realised, that wasn't enough, [...] Today, the exploitation takes many forms [...] Today the enemy has another face [...] the state, it's dominated by big groups. [...] we all saw that it is big capital that is exploiting us. The argument for a different society, a just society, equal, that is a very old argument. From the beginning of the movement, and from the beginning of the women's struggle I have said it's not ours; it came many years before, all women have always been fighting in the way they were able to.

From linking the everyday with an analysis of societal structures and the history of resistance, Isabel proceeds to connect the exploitation of workers, specifically women, with the exploitation of nature by agrobusiness:

> The pesticides, that these days are on our table every single day, [...] come in a form that people are obliged to use them [...] there are a few companies dominating the seeds, the fertilisers, the remedies, food, marketing. [...] everything is within the same companies. And there are 4 or 5 companies in the whole world, dominating all of that.

Isabel regards the fertilisers provided by those big companies as poison, but:

> They don't call it poison. Because the people were also educated to say fertiliser, to say fungicide, to say that it isn't poison. [...] There was a period in history when we were working [...] on agroecology, it was then that transgenic seeds came. [...] It began illegally and then all the work of rescuing the seeds that we, principally the women, had done was lost [...]. In the meetings that's what we do, we exchange many seeds, also recipes, we also do a lot of work rescuing the past, that which the grandmothers, the mothers had done [...] And it's lost.

Everyday experiences include also the broader effects of the destructive nature of agribusiness: 'There is a very big change, each year, the climatic changes [...] are more accelerated. There are droughts [...] it has rained in one day what it had to in one month, then it takes it all.'

Isabel summarises her experiences and struggles in a central ecofeminist notion: 'Women have been treated as nature is. Violence against the Earth is the same as violence against women, exploited, everything that can be is taken from it.'

Workers' rights—women's rights—the state—its domination by big capital—violence against the earth: this is the chain of equivalences that Isabel draws as a result of a life-long learning process. Resistance against these power relations is put into a historical context. It is in the process of taking part in 'the everyday of the struggles' that Isabel becomes an organic intellectual in the Gramscian sense, 'from technique-as-work one proceeds to technique-as-science and to the humanistic conception of history' (Gramsci 1999, 142). To connect the practices of work with a theorisation of the work process and to situate this within an understanding of the history of societal-human development is what enables Isabel to become a leader within the working class.

Isabel's story conveys the ability of overcoming specialisation (the housemaid experience) by connecting hers with 'other struggles'. The intersections of individual experience and societal structures are mediated through individuals sharing the same experiences while being at a different stage in making sense of their experiences within the societal 'grammar': the people who invite her into the Youth Pastoral, the friend, who helps her to come back into the struggle and the older women, who connect present-day resistance with a historical knowledge that allows them to create alternatives. Through these individuals Isabel enters a struggle in which the personal and the political converge presenting her with the possibilities existing at a certain moment to overcome exploitation and individualisation.

## Those Who Devastate Are Not the Small Ones

Rosângela Piovizani is the national coordinator of the Rural Women Movement (MMC) and co-founder of the Rural Workers Union of Roraima. She was born in Paraná, her parents being of Spanish and Italian origin. There are similarities between her and Isabel's story, one of them being the conflict between motherhood and participation in political struggles. While Isabel went straight from the Youth Pastoral into the women's movement, Rosângela's trajectory was more diverse. She gets married at the age of 18, moves to Roraima, joins the Liberation Theology–oriented Ecclesial Catholic community, the small farmers' association, and becomes the co-founder of the PT in her region. The Land Pastoral Commission encouraged her and funded the creation of the Rural Women Workers' Movement in 1996. Rosângela details the role of Liberation Theology:

> The Catholic Church had a very important job […] with regard to Brazil, but especially in the Amazon. In our region, in our parish, it had the role of giving

awareness to all the leaders, all the people who were in the communities, for the formation of politics, the awareness of the system which had been put there, the importance of the fight, the importance of militancy, and to be not only Christians, but also [...] an agent of social transformation.

'Giving awareness to all the leaders' can be analysed in the light of the three steps of social transformation according to Liberation Theology. The first step is that of being aware, which implies a direct confrontation with anti-reality, an experience of 'existential shock' (Boff 1996, 170). The 'harsh reality wakes you up' can be related to the second step, the analytical questioning in search of a critical knowledge in the light of faith (Boff 1996). According to Boff, 'the causes of suffering must be scrutinised taking into consideration the relations of political, economic and ideological power. [...] Poverty is neither innocent nor natural; it is produced; and that's why the poor are exploited and impoverished beings' (1996, 171). Rosângela translated this perspective into the need to develop practical 'tools' for action:

> Then I, we [...] began discussing that we needed to move forward with constructing the tools for the class struggle. We saw that the rural workers' unions were a tool, the Small Farmers' Association was a tool, the Workers' Party was also a tool, so from when you become aware of reality, of suffering and of rights, you begin also to become more political, and you join in the fight, you organise yourself.

The creation of movements to press for environmental and gender rights relates to the most important moment of the process of social transformation: practices of change (Boff 1996, 171). When you 'join in the fight' 'Christian faith' materialises in alliance with other transformative forces. By legitimising the workers' and environmental movements with a coherent religious canon, Liberation Theology contributed to their expansion and strength (Löwy 2000).

Rosângela left the PT because at the time they did not acknowledge the need to fight sexism in their own organisation and in society at large (Räthzel et al. 2015). Above we discussed the statistics showing that movements fight for land and water. However, the fight for women's rights is not reflected in those statistics. It might be subordinated under the title of 'human rights'. While the Rural Women's Movement, which Rosângela co-founded, promotes the motto: 'Strengthening the fight in defence of life', their understanding of 'life' is broader than the defence of land and water and their notion of women is more diverse and inclusive than the term 'rural women' may suggest:

Lilac is the colour that represents feminism and the struggle of women. The straw hat, the diversity, so our banner, for example, has three women, one represents more the black, indigenous, migrant, immigrant. […] The children and production. Because for us, production and reproduction of life are very connected.

Both Isabel and Rosângela become leaders by connecting their everyday experiences and struggles with a perspective of a society based on equality, justice and self-determination, but their trajectories finding these connections differ. Isabel comes to a broader perspective through the women's movement. Rosângela comes to the women's movement out of frustration with the workers' movement. Both their trajectories exemplify the limitations of specific experiences becoming part of the general societal 'grammar'. While they connect women's rights with the rights of workers and the protection of the earth and human reproduction, political parties and societal analyses do not learn from them how to make these connections.

## Land, Source of Livelihood: Do Not Give Up Your Land

Manoel Edivaldo Santos Matos (Peixe) was born and raised in a small community by the river Arapiuns in Pará. His mother was a Portuguese descendent and his father descended from indigenous Brazilians and Africans. They had 13 children, of which only 7 survived. 'Because it was like this: one survived, another died, one survived, another died. […] I was in between the ones who died. But I didn't die, I stayed.'

His family worked in agriculture—in the production of cassava flour, rubber tapping and livestock. All the work was done within the community on the basis of an equal division of work. 'It was just one family.' If someone hunted, everyone got a piece. The families joined together in what was called 'muxirum', a collective effort.

At home Peixe cooked, washed the clothes, went fishing, cleared the land and cut the rubber trees. At the age of 16 he became the leader of the Catholic youth group. He travelled to other communities and promoted debates about 'the importance of the organisation, the struggle'. He was also elected to manage a football club in the community. Later he became involved with the Rural Workers Union and joined a left-wing group within the PT. Peixe created different 'tools for the class struggle': the Youth Pastoral in his region; he transformed the structure of the rural workers' union, and initiated political groups in other communities.

When the soya plantations arrived in the Amazon and small farmers began selling their land to the Cargill soya consortium, Peixe helped to launch a campaign titled 'Land, source of livelihood. Do not give up your land'. It was carried out by the Land Pastoral Commission in partnership with other environmental groups (e.g., the Amazon Defence Group, the Federation of Communitarian Organisations of Santarém, Greenpeace). A major limitation of these organisations is the lack of resources compared to those of the powers they are fighting against. This is one explanation for their limited success.

For Peixe, protecting the environment equals protecting the livelihood of the workers in the community: that is what the union is there for. After they had successfully fought for the rights of the communities to own the land they had been living on, people could decide between two modes of ownership: collective or private. While the former included working the land according to their needs but excluded the right to sell it, the latter allowed it. About half of the communities opted for the first, half for the second mode of ownership. As a result, about half of the privately owned land was sold with devastating results for the sellers, who lost their land and could live only a short time from the money they had received for it.

In a way the union is trying to prevent change not to provoke it: to prevent loggers and soya growers from appropriating the land. But in order for things to stay as they are, they had to change. The communities had to change and become owners of their land and then organise to create new means of livelihoods that were protective of the Amazon. To protect and guide the communities, the unionists needed to change and become lawyers, experts in understanding government documents and the workings of the bureaucracy.

Peixe described how deforestation and agrobusiness is causing climate change. There were more frequent floods, torrential rains and intense droughts. 'I've never seen it. Our river over there dried up … I've never seen that.' The experience of climate change also changed the course of the communities' struggles which are now focused not only on protecting land ownership but also on influencing general politics to confront the climate crisis. Being part of organisations creates a sense of collective action that encourages Peixe to believe 'we are sure that we are able to fight, but in the end the decisions are taken by those who have the political power'. He is aware of the need of individuals capable of leading the necessary collective struggles: 'People all over the region and my relatives all admire me. I think I have a certain leadership capability, it's the way I am.' Defining a learned competence as a character trait, Peixe simultaneously praises himself and tones down this praise by presenting his capability as not of his own making.

Peixe's sense of being part of a group, a political party, a community, relates to the principle of 'common destiny' that is central to Liberation Theology and the environmental movement. It implies that all living creatures have a common origin and a common destiny. Describing the struggle of the reserves against selling the land 'out of fourteen communities, seven resisted' this story resonated with the story of his childhood, where he was among half of the siblings who survived. His narrative can be read as 'an encounter of two stories' (Bourdieu 2001). It means that the principle of action lies 'in the complicity between two states of the social, between history in bodies and history in things' (Bourdieu 2001, 184). The agent 'does what is in his power to make possible the actualization of the potentialities inscribed in his body in the form of capacities and dispositions shaped by the conditions of existence' (2001, 183). Similarly, Portelli (1997) argues that an individual story represents the expression of possibilities available to people at a specific time and place. That Peixe was among half of the surviving children of his family and was able to empower half of the Amazon communities in his region to protect their land is a coincidence demonstrating that there is always a possibility for resistance and survival, but also that they are limited and their achievements fragile.

## Conclusion

We began our chapter outlining the socio-economic and cultural contexts within which rural social movements organise their resistance against the destruction of nature, the basis of their life and work. We found that international and national corporations, especially agribusiness, are dominating the Brazilian political space through the direct representation of their interests by political parties in parliament, while this is not the case concerning the interests of the poor, small farmers and agricultural workers. This has weakened the power of progressive governments (Lula and Rousseff) to realise meaningful land reform that could have put an end to the exploitation of people and nature in the Brazilian hinterland. Thus, the proliferation of direct confrontations over land has made Brazil one of the countries with the highest number of rural conflicts and the highest number of violent attacks against environmental activists in the world.

More than half of the rural conflicts we analysed concern access to land and water resources. Martínez-Alier and Martin O'Connor (1996) define these struggles as 'distributive ecological conflicts', meaning that environmental gains and losses are distributed unequally between corporations and landowners, on the one hand, and workers and small farmers, on the other, the latter bearing the brunt of ecological destruction caused by the former.

However, as an intersectional analysis of the societal 'grammar' of such unequal distribution and their translations in individual life stories revealed, each conflict is an 'overdetermination' (Althusser 1971) of numerous conflicts, which feed off each other: (1) the exploitation of women and their paid and unpaid work, (2) the exploitation of nature, (3) the individualisation of workers and small landowners, which seduces them to sell their land. Since this land is then used for monocultures (or for the wood industry) it destroys traditional forms of production which are protective of humans, disintegrating communities.

But our analyses have also suggested that the ways in which members of exploited rural communities become organic intellectuals of social resistance movements are precisely by connecting these arrays of conflicts and finding their common origins. Such capabilities are developed by taking part in everyday struggles, as Gramsci suggested, and developing their experiences and insights into a broader perspective of societal transformation. However, these abilities do not develop automatically out of their experiences. In the case of our protagonists, parts of the Catholic Church and its teaching based on the Theology of Liberation became the mediator of experiences, practices of resistance and theories of exploitation.

The intersections of individual trajectories, structures of exploitation and institutionalised political-religious guidance enabled members of the rural poor to become leaders of national and international movements by connecting the sources of the conflicts they experienced with concepts that included but also transcended the concept of ecological distributive conflicts. Describing their exploitation as the result of 'predatory capitalism' and connecting the liberation of women with the fight against capitalism grounds the notion of distribution in the relations of production. Recounting their life stories, the interviewees described succinctly that it was the mode of production (agribusiness with its pesticides) and the relations of production (private corporations and big landowners) which resulted in environmental and human exploitation. Unjust distribution is not the origin of exploitation but itself the result of the private appropriation of natural resources on a national and corporate scale. The exploitation of humans and nature in the Global South is no more reducible to a problem of distribution than the exploitation of humans and nature in the Global North. The source of both is the global profit-oriented system of production, of which unequal distribution is the consequence.

For Environmental Labour Studies (ELS) this means that in spite of the differences between rural and industrial workers in terms of historical economic, political, and cultural conditions and in spite of the power relations between workers of the Global South and North, they are, in terms of their everyday struggles and their broader societal perspectives, much closer to each other than they might be aware of themselves. It could therefore be a task for ELS scholars and activists to further explore the possibilities and necessities of South-North, country-city alliances based on the insights that working people in both hemispheres are constantly developing their capabilities through their practices of resistance.

**Acknowledgement** The research for this chapter was funded by Vetenskapsrådet (The Swedish Research Council: DNR 421-2010-1990).

# References

Agência Envolverde de Jornalismo. 2014. Atlas: Brasil é o terceiro país com mais conflitos ambientais. March 25. https://envolverde.com.br/atlas-brasil-e-o-terceiro-pais-com-mais-conflitos-ambientais/. Accessed 2 March 2020.

Althusser, Luis. 1971. *Lenin and Philosophy and other Essays*. New York: Monthly Review Press.

Boff, Leonardo. 1993. *Ecologia, Mundialização, Espiritualidade: A emergência de um novo paradigma*. São Paulo: Editora Ática.

———. 1996. *Ecologia: Grito da Terra, Grito dos Pobres*. São Paulo: Editora Ática.

Boito Jr. Armando. 2012. *As bases políticas do neodesenvolvimentismo*. São Paulo: FGV. https://bibliotecadigital.fgv.br/dspace/bitstream/handle/10438/16866/Painel%203%20-%20Novo%20Desenv%20BR%20-%20Boito%20-%20Bases%20Pol%20Neodesenv%20-%20PAPER.pdf. Accessed 2 Oct 2020.

Bourdieu, Pierre. 2001. *Meditações Pascalianas*. Rio de Janeiro: Bertrand Brasil. https://nepegeo.paginas.ufsc.br/files/2018/06/BOURDIEU-Pierre.-Meditações-pascalianas.pdf. Accessed 2 Oct 2020.

Carta Capital. 2016. *Os 10 conflitos ambientais mais explosivos do mundo*. https://ecoa.org.br/os-10-conflitos-ambientais-mais-explosivos-do-mundo/. Accessed 21 Oct 2020.

———. 2017. *Brasil é o país mais perigoso do mundo para ambientalistas*. https://www.cartacapital.com.br/sociedade/brasil-e-o-pais-mais-perigoso-do-mundo-para-ambientalistas/. Accessed 21 Oct 2020.

Carter, Miguel. 2005. *The Landless Rural Workers' Movement (MST) and Democracy in Brazil*. Centre for Brazilian Studies, University of Oxford, Working Paper 60. https://www.lac.ox.ac.uk/sites/default/files/lac/documents/media/miguel20carter2060.pdf. Accessed 21 Oct 2020.

Comerford, John Cunha. 1999. *Fazendo a luta: sociabilidade, falas, e rituais na construção de organizações camponesas*. Rio de Janeiro: Relume Dumará.
Comissão Pastoral da Terra (CPT)—Centro de Documentação Dom Tomás Balduíno. 2018a. https://www.cptnacional.org.br/publicacoes/cedoc-dom-tomas-balduino-da-cpt Accessed 21 Oct 2020.
———. 2018b. https://www.cptnacional.org.br/publicacoes-2/destaque/4683-sugestao-de-pauta-cpt-ira-lancar-o-relatorio-conflitos-no-campo-brasil-2018-na-proxima-semana. Accessed 21 Oct 2020.
Desmarais, Annette. 2007. *La Vía Campesina: Globalization and the Power of Peasants*. London: Pluto Press.
Dowler, Crispin. 2020. Soya, Corn and Cotton Make Brazil World Leader for Hazardous Pesticides. *Unearthed*. February 2. https://unearthed.greenpeace.org/2020/02/20/brazil-pesticides-soya-corn-cotton-hazardous-croplife/. Accessed 21 Oct 2020.
Environmental Justice Atlas. 2020. https://ejatlas.org. Accessed 21 Oct 2020.
Estrada, Rodrigo. 2017. Brazil Considering to Open Up Several Conservation Areas for Deforestation. https://www.greenpeace.org/usa/news/brazil-considering-open-several-conservation-areas-deforestation/. Accessed 21 Oct 2020.
Fernandes, Bernardo Mançano. 2016. Development Models for the Brazilian Countryside Paradigmatic and Territorial Disputes. *Latin American Perspectives* 43 (2): 48–59. https://doi.org/10.1177/0094582X15616117.
Flyvbjerg, B. 2006. Five Misunderstandings About Case-Study Research. *Qualitative Inquiry* 12: 219–245.
Frazão, Maria Goreti, and Ney Souza. 2007. Catolicismo No ABC Paulista: Da Doutrina Social Ao Movimento Popular. *Revista Eletrônica Espaço Teleológico* 2. http://tinyurl.com/obopreh. Accessed 21 Oct 2020.
Freire, Paulo. 1996. *Pedagogy of the Oppressed*. London; New York: Penguin Books.
French, Jan Hoffman. 2007. A Tale of Two Priests and Two Struggles: Liberation Theology from Dictatorship to Democracy in the Brazilian Northeast. *The Americas* 63 (3): 409–443.
Gaspar, Lúcia. 2005. Peasant Leagues (Ligas Camponesas). *Pesquisa Escolar On-Line*. Joaquim Nabuco Foundation, Recife. https://pesquisaescolar.fundaj.gov.br/pt-br/artigo/ligas-camponesas/. Accessed 21 Oct 2020.
Gonzales, Jenny. 2019. Bolsonaro Administration Approves 290 New Pesticide Products for Use. *Mongabay*. August 12. https://news.mongabay.com/2019/08/bolsonaro-administration-approves-290-new-pesticide-products-for-use/. Accessed 21 Oct 2020.
Gramsci, Antonio. 1999. *Selected Writings from the Prison Notebooks of Antonio Gramsci*. Ed. Quentin Hoare and Geoffrey Nowell Smith. London: ElecBooks.
Grzybowski, C. 1994. A Comissão Pastoral da Terra e os colonos do sul do Brasil. In: V. Paiva. (Org.) *Igreja e questão agrária*, 248–276. São Paulo: Loyola.
Gutiérrez Merino, Gustavo. 1973. *A Theology of Liberation: History, Politics, and Salvation*. Maryknoll, NY: Orbis Books.

Kaiser, Anna Jean. 2019. 'Exterminator of the Future': Brazil's Bolsonaro Denounced for Environmental Assault. *The Guardian*. May 9. https://www.theguardian.com/world/2019/may/09/jair-bolsonaro-brazil-amazon-rainforest-environment. Accessed 21 Oct 2020.

Land Struggle Database. 2017. http://www.lagea.ig.ufu.br/rededataluta/relatorios/brasil/dataluta_brasil_en_2017.pdf. Accessed 21 Oct 2020.

Leal, Giulana Franco. 2013. Justiça Ambiental, Conflitos Latentes e externalizados: Estudo de caso de Pescadores artesanais do norte fluminense. *Ambiente & Sociedade* 16 (4): 83–102. https://www.scielo.br/pdf/asoc/v16n4/06.pdf. Accessed 21 Oct 2020.

Lima, Wesley. 2019. *Educação do Campo: conquistas e resistência popular*. MST. May 13. https://mst.org.br/2019/05/13/educacao-do-campo-conquistas-e-resistencia-popular/. Accessed 21 Oct 2020.

Löwy, Michael. 2000. *A Guerra Dos Deuses: Religião e Política Na América Latina*. São Paulo: Editora Vozes.

MAB (Movimentos dos Atingidos por Barragens). 2020. Fundação Renova e a violação do direito à moradia. June. https://mab.org.br/2020/05/04/funda-renova-e-viola-do-direito-moradia/# Accessed 21 Oct 2020.

Martínez-Alier, Joan. 2008. Conflictos ecológicos y justicia ambiental. *Papeles* 103: 11–27. https://www.fuhem.es/papeles_articulo/conflictos-ecologicos-y-justicia-ambiental/. Accessed 21 Oct 2020.

Martínez-Alier, Joan, and M. O'Connor. 1996. Ecological and Economic Distribution Conflicts. In *Getting Down to Earth*, ed. R. Costanza, O. Segura, and J. Martínez-Alier. Washington, DC: Island Press.

McCarthy, Ben. 2016. *Supply Change: Tracking Corporate Commitments to Deforestation-free Supply Chains*. Washington, DC: Forest Trends. https://www.forest-trends.org/wp-content/uploads/2016/07/doc_5248.pdf. Accessed 21 Oct 2020.

Meadows, Donella H, and Club of Rome. 1972. *The Limits to Growth; a Report for the Club of Rome's Project on the Predicament of Mankind*. New York: Universe Books. https://collections.dartmouth.edu/teitexts/meadows/diplomatic/meadows_ltg-diplomatic.html. Accessed 21 Oct 2020.

Menezes, Neto, and Antonio Julio. 2007. A Igreja Católica e os Movimentos Sociais do Campo: a Teologia da Libertação e o Movimento dos Trabalhadores Rurais sem Terra. *Caderno CRH* 20 (50): 331–341.

Moore, Jason 2013. El auge de la ecología-mundo capitalista (i). *Laberinto, Málaga* 38: 9–26. https://jasonwmoore.com/wp-content/uploads/2017/08/Moore-El_Auge_de_la_ecologia-mundo_capitalista__Part_I__Laberinto__2013.pdf. Accessed 21 Oct 2020.

Nobre, Carlos, and Rachel Biderman. 2017. Brazilian Agribusiness Can Change Its Image with Carbon-Neutral Production. *Folha de S Paulo*. September 27. https://www1.folha.uol.com.br/opiniao/2017/09/1922037-agro-pode-mudar-imagem-com-producao-carbono-neutro.shtml?origin=uol. Accessed 21 Oct 2020.

Nun, José. 1969. Superpoblación relativa, ejército industrial de reserva y masa marginal. *Revista Latinoamericana de Sociologia* 5 (2): 178–236.

OXFAM. 2016. Terrenos da desigualdade: Terra, agricultura e desigualdade no Brasil rural. https://oxfam.org.br/wp-content/uploads/2019/08/relatorio-terrenos_desigualdade-brasil.pdf. Accessed 21 Oct 2020.

Phillips, Dom. 2019. Indict Jair Bolsonaro Over Indigenous Rights, International Court Is Urged. *The Guardian*. November 28. https://www.theguardian.com/world/2019/nov/27/jair-bolsonaro-international-criminal-court-indigenous-rights. Accessed 21 Oct 2020.

Portelli, Alessandro. 1997. *The Battle of Valle Giulia. Oral History and the Art of Dialogue*. London: University of Wisconsin Press.

Räthzel, Nora, David Uzzell, Ragnar Lundström, and Beatriz Leandro. 2015. Spaces of Civil Society and the Practices of Resistance and Subordination. *Journal of Civil Society* 11 (2): 154–169. https://doi.org/10.1080/17448689.2015.1045699.

Sader, Emir. 1995. *Quando Novos Personagens Entram Em Cena*. São Paulo: Editora Paz e Terra.

Sauer, Sérgio. 2017. Rural Brazil During the Lula Administrations: Agreements with Agribusiness and Disputes in Agrarian Policies. *Latin American Perspectives* 46 (4): 103–121.

Sauer, Sérgio, Moisés V. Balestro, and Sérgio Schneider. 2018. The Ambiguous Stance of Brazil as a Regional Power: Piloting a Course Between Commodity-Based Surpluses and National Development. *Globalizations* 15 (1): 32–55. https://doi.org/10.1080/14747731.2017.1400232. Accessed 21 Oct 2020.

Scott, C.E., S.A. Monks, D.V. Spracklen, et al. 2018. Impact on Short-Lived Climate Forcers Increases Projected Warming Due to Deforestation. *Nat Commun* 9 (157). https://doi.org/10.1038/s41467-017-02412-4. Accessed 21 Oct 2020.

Temper, Leah, Daniela del Bene and Joan Martinez-Alier. 2015. Mapping the Frontiers and Front Lines of Global Environmental Justice: The EJAtlas. *Journal of Political Ecology* 22: 255–278. http://jpe.library.arizona.edu/volume_22/Temper.pdf. Accessed 21 Oct 2020.

Tilly, C., and S.G. Tarrow. 2007. *Contentious Politics*. Boulder, CO: Paradigm Publishers.

Tollefson, Jeff. 2019. Ricardo Galvão: Science Defender. *Nature* 576 (19): 361–363. https://www.nature.com/immersive/d41586-019-03749-0/index.html. Accessed 21 Oct 2020.

# 16

## Whose Labour, Whose Land? Indigenous and Labour Conflicts and Alliances over Resource Extraction

Erik Kojola

## Introduction

In the spring of 2016 thousands of Indigenous peoples and climate activists assembled near the Standing Rock Sioux Tribe reservation in the U.S. state of North Dakota to protest the construction of the Dakota Access Pipeline (DAPL) that would transport oil from fracking wells to distribution centres and refineries. The protest grew into the national and international NoDAPL movement that connected opposition to the pipeline with demands for climate justice, protecting clean water, and defending Indigenous self-determination. The movement brought together a coalition of environmental, human rights, and Indigenous groups. Yet, the labour movement was divided. Segments of organized labour, particularly in construction, aligned with industry and anti-environmental forces to promote the pipeline. Other unions voiced opposition to DAPL and solidarity with Indigenous people as part of their broader climate and social justice activism.

What do the tensions and collaborations around the NoDAPL movement reveal about relationships between unions, environmentalists, and Indigenous peoples? Why does conflict between unions over an infrastructure project matter for addressing the climate crisis and promoting environmental justice?

E. Kojola (✉)
Department of Sociology and Anthropology, Texas Christian University, Fort Worth, TX, USA
e-mail: e.kojola@tcu.edu

And what can labour-Indigenous relationships show about the interconnections between capitalism and colonialism?

In this chapter, I argue that examining labour-Indigenous relationships, particularly in settler colonial countries, is important theoretically and politically for understanding the causes of environmental crises and the possibilities and challenges in forming alliances to advance socio-environmental justice. Indigenous and climate justice movements could be stronger with the resources and power of the labour movement while solidarity with Indigenous people and environmentalists could bolster unions (Dreiling 1998; Obach 2004; Mills 2011; Mills and McCreary 2012). While there is research on labour-environmental relationships (Obach 2004; Mayer 2009; Sweeney 2013; Loomis 2015) and environmental-Indigenous relationships (Gedicks 1993; Ali 2003; Grossman 2017), labour-Indigenous relationships are under-examined but require analysis of the distinct cultural, political, economic, and historical dynamics. Theories about cross-movement coalitions that are useful for analysing labour-environment relations (Rose 2000; Obach 2004; Mayer 2009; Hultgren and Stevis 2020) are inadequate for analysing union-Indigenous interactions. Indigenous Nations are distinct from other groups because they are nations, not interest groups or social movement organizations, while Indigenous-led movements differ from other movements because they are often engaged in decolonial struggles and defending territory and Indigenous notions of sovereignty.

This chapter focuses on settler colonial countries in North America (what many Indigenous people call Turtle Island) and issues surrounding mining and energy production. Resource extraction impacts Indigenous people because a large share of mineral and fossil fuel reserves is on or near Indigenous lands (Ali 2003). A recent resurgence in Indigenous mobilization in the U.S. and Canada has focused on resistance to oil and gas infrastructure and formed alliances with the climate justice movement to oppose fossil fuels. Meanwhile, mining and energy industries have significant levels of unionization in the U.S. and Canada, and some unions support new extractive projects, even mobilizing against Indigenous and climate justice movements.

This chapter first provides an overview of settler colonialism and capitalism to understand the context for labour-Indigenous relationships. The next section explores labour-Indigenous conflict and collaboration around resource extraction and climate change, using protests over DAPL as an emblematic case. The chapter concludes with reflections on pathways for new research and building labour-Indigenous-environment solidarity.

## Theoretical Perspectives: Settler Colonialism and Capitalism

Understanding labour-Indigenous relationships requires analysis of settler colonialism and the interconnections with capitalism (Mills and Clarke 2009). Settler colonialism is a particular form of colonialism in countries like the U.S. and Canada where new settler societies are created on the existing homelands of Indigenous people through genocide, removal, and displacement (Wolfe 2001). The expansion of European settler colonialism in North America was driven by capitalism and the desire to extract materials from nonhuman nature, acquire land, and generate profits—leading to Indigenous dispossession and a process of primitive accumulation (Coulthard 2014). Indigenous people lost territory as they were forced off their homelands and industrialization damaged ecosystems that were vital to Indigenous livelihoods and cosmologies (LaDuke 1999; Whyte 2018).

Ongoing appropriation of Indigenous lands and the destruction of ecosystems to extract natural resources under settler colonialism is what drives many environmental crises, threatens Indigenous livelihoods, and commits violence against Indigenous people (Clark 2002; Schlosberg and Carruthers 2010; Estes 2019). Colonialism dislocates human relationships to nonhuman nature and for Indigenous people, 'the disruption of relationships to land represents a profound epistemic, ontological, cosmological violence' (Tuck and Yang 2012, 5). Whyte (2018) describes settler colonialism as a form of social and ecological domination and violence. Indigenous communities face environmental injustices from greater exposure to industrial pollution and the economic, health, cultural, and spiritual impacts of ecological destruction and land dispossession (Schlosberg and Carruthers 2010).

In the U.S. and Canada, Indigenous tribes have retained limited legal rights as nations which creates a unique situation of nations within nations that have semi-sovereign status. In Canada, there are 634 recognized First Nations governments while in the U.S. there are 573 federally recognized Native American tribes.[1] Many tribes govern reservations that are lands created by colonial governments with the intent of containing and relocating Indigenous people (National Congress of American Indians 2015). Tribal members also have dual tribal and U.S. or Canadian citizenship. The legal status and powers of First Nations and Native American tribes differ but they have some degree of authority in both countries, such as the ability to elect tribal governments and

---

[1] There are also tribes not recognized by federal governments which means they do not have the same legal status, access to federal resources and programs, and control over territory.

control some public services. Control over land and the right to gather, hunt, and fish were established in various treaties signed by tribes with the U.S. and Canadian governments in the nineteenth century. However, treaties have been violated and changed to enable settlers to acquire Indigenous land and facilitate resource extraction (LaDuke 1999; Churchill 2002). In the U.S., court rulings have established the principles of federal 'trust responsibility' to protect tribal lands, resources, and self-government, and to provide federal assistance for the livelihoods of tribal communities (National Congress of American Indians 2015). However, these obligations have been underfunded and broken.

Varying claims to territory, self-determination, and sovereignty are often central issues for many tribes (Clark 2002; Pickerill 2018). Indigenous notions of sovereignty are not tantamount to ideas of Western liberal democracy as they draw from distinct Indigenous worldviews and social relations (Estes 2019). Bruyneel (2007) argues that Native Americans operate within a 'third space of sovereignty' that defies the boundaries of the U.S. colonial system and have resisted colonialism from the inside and outside. Indigenous nations have strategically used claims to sovereignty and legal tactics to advance Indigenous self-determination and protect livelihood practices (Deloria and Lytle 1984; Norman 2017; Pasternak 2017). For example, since the 1970s tribes have increasingly asserted treaty rights that protect their ability to harvest plants, fish, and hunt on and off reservation lands. These rights have been used to advance environmental justice by contesting polluting industries that would damage plants and animals and, therefore, the ability to exercise treaty rights (Norman 2017; Estes 2019). However, some tribes and Indigenous activists have opposed working within dominant settler colonial laws and politics, and have sought de-colonization outside of these frameworks (Bruyneel 2004).

Assessing labour-Indigenous relationships requires attention to these distinct cultural, historical, economic, and political dynamics of settler colonialism and Indigenous movements for sovereignty. Native American tribes and First Nations are nations, not interest groups or racial categories. Indigenous social movements are different from other social movements because they are not demanding rights and recognition but are engaged in decolonial struggles (Steinman 2012; Simpson 2017). Still, the views of Indigenous peoples are complex and vary between and within tribes. Tribal governments, as with any nation, do not represent all views as Indigenous social movements and activists, and elders and spiritual leaders might have different perspectives than tribal government leaders. There are also class, gender, age, and other intersectional differences within Indigenous communities.

Indigenous peoples and unions have a complex relationship. Indigenous workers have been active in organizing unions and radical class politics (Neufeld and Parnaby 2000; Parnaby 2006), although often with little support from white-led unions (O'Neill 2018). For example, Indigenous workers in Canada were active in early organizing of workers in extractive industries, like logging (Neufeld and Parnaby 2000; Parnaby 2006). Indigenous workers are currently members of unions in industries ranging from construction to food service and retail to the public sector, but there is not a large or organized Native American presence in the North American labour movement. There is also a history of tension between unions and Indigenous peoples. Unions have protected jobs for their members, who have historically been white and male, often through the maintenance of racial hierarchies (Roediger 1991). This has included outright exclusion of Indigenous workers and relegating Indigenous workers to marginalized positions (Parnaby 2006; Fernandez and Silver 2017). Indigenous nations may also see unions as a threat to their authority and control when they organize workers at tribally owned businesses, like casinos (O'Neill 2018). Thus, Indigenous people may view unions as a colonial institution that protects the interests of non-Indigenous workers (Mills 2007; Fernandez and Silver 2017; O'Neill 2018).

## Labour-Indigenous Relationships Around Mining and Energy

In this section I focus on labour-Indigenous relationships around environmental issues, particularly mining and fossil fuel production. I describe three different scenarios for labour-Indigenous relations. First, there can be conflict between pro-growth unions and Indigenous opponents to extractive development. Second, there is potential for pro-growth coalitions between unions and tribal governments who support extractive development. Finally, there is potential collaboration around environmental justice and resistance to the socio-environmental risks of resource extraction.

### Conflict over Sovereignty, Land, and Natural Resources

Debates over who controls land and resources and how, or if, resource extraction occurs, have generated conflict between pro-growth unions and tribal governments and Indigenous activists. Indigenous peoples have opposed extractive development for many reasons, including violations of sovereignty, disruptions to spiritual and cultural practices, economic exploitation, and environmental and human health risks (Kulchyski and Bernauer 2014; Fernandez and Silver 2017). Yet, self-determination is a central reason. Ali's (2003) comparative analysis of Indigenous reactions to potentially hazardous mining finds that questions of sovereignty, not simply the extent of environmental risks, account for why communities resisted.

On the other hand, support for expanding extractive industries by some unions and workers, who are predominantly white and male, is framed as a way to create jobs and revenue. Supporting extractive industries is not only motivated by material interests, but also produced by political, cultural, and ideological processes of settler colonialism and racism. Demand for extractive jobs is motivated and legitimized by workers' identities and relationships to land. In North America, the meaning and history of land and resource extraction is defined by colonialism and ideologies of white settlers bringing modernity to productively use the land which valorized masculine labour in extractive industries (Wolfe 2001).

## Pro-development Alliances

Pro-development labour-Indigenous alliances are also possible. While Indigenous groups have mobilized against extraction, there is no universal Indigenous position and there are internal political struggles (Pasternak 2017). Thus, there can be divisions between tribal governments and elites that support extraction as a path to economic development, and Indigenous environmental activists, elders, and spiritual leaders who more often oppose these projects as threats to Indigenous ways of life. In some contexts, tribes have supported extractive development, including projects opposed by non-Indigenous environmental groups, based on the jobs and revenue for Indigenous workers, tribal governments, and businesses (Ali 2003).

Labour-Indigenous alliances could push for more equitable extractive development. Some Indigenous leaders view active engagement with extractive industries, rather than outright opposition, as a practical approach that can reduce environmental impacts, provide for communities, and increase Indigenous sovereignty (Mills 2011). For example, First Nations in Canada are using community-benefit agreements with mining companies and provincial governments to secure royalties and jobs for Indigenous workers (Smith and Frehner 2010; Mills 2011; Thompson 2018). There is some union support for these efforts, such as the United Steelworkers (USW) in Canada using collective bargaining to strengthen community-benefit agreements including provisions for hiring Indigenous workers who then become union members (Mills and McCreary 2012). Efforts to create more equitable extractive development are complex and contested. They reflect attempts by industry to ameliorate and co-opt Indigenous opposition but are also efforts by Indigenous nations to secure resources for their communities and to retain profits from extraction on their lands. However, pro-growth alliances do not challenge an industry that is fundamentally socially and ecologically destructive.

## Coalitions for Environmental Justice

Another direction is also possible in which Indigenous people and labour movements align in solidarity to push for environmental justice and challenge the interconnected forces of settler colonialism and capitalism. Non-Indigenous workers and residents near mining may share concerns with Indigenous peoples about pollution, loss of agricultural and recreation land, and risks to human health. The same companies that destroy Indigenous lands may also expose workers to occupational hazards. There is also emergent labour-Indigenous solidarity around environmental justice struggles from unions led by immigrants, women, and workers of colour who are less privileged by settler colonialism and white supremacy. Estes (2019) traces the long history of Indigenous resistance and internationalist politics, particularly the 1960s–1970s Red Power movement that joined other global liberation struggles against colonialism and capitalism, which could align with radical working-class politics.

# Case Study of the Dakota Access Pipeline: Conflicts and Alliances over Fossil Fuels

The accelerating climate crisis has spurred the climate justice movement to address the unequal impacts of climate disruptions. The movement has created alliances between Indigenous and environmental groups to stop fossil fuel infrastructure and defend Indigenous rights which has generated collaboration and conflict with unions (Whyte 2017; Jenkins 2018). Some progressive unions have joined Indigenous peoples and tribes, and environmentalists to oppose fossil fuel projects and mobilize for climate justice (Sweeney 2012; Fernandez and Silver 2017). Meanwhile Indigenous environmental groups, like the Indigenous Environmental Network and Honor the Earth, have supported just transition policies and alternatives like community-run solar energy projects on reservations (Indigenous Environmental Network 2019). However, demands to stop fossil fuel extraction have also created conflict with other unions. Many mining and construction unions support new fossil fuel projects arguing they create jobs and provide necessary energy (Aronoff 2016a). I use conflicts over DAPL to explore these dynamics.

## The NoDAPL Movement

Native American tribes and activists, and environmentalists mobilized to block construction of DAPL in 2016 which was part of a broader movement to defend Indigenous sovereignty and keep fossil fuels in the ground. The proposal was a 1200-mile pipeline that would carry 570,000 barrels of crude oil daily from fracking wells in the Bakken Shale of North Dakota to distribution points and refineries in Illinois (Estes 2019). Opposition to DAPL arose in order to defend Indigenous self-determination, protect clean water, and advance climate justice. The route was initially planned to go near Bismarck, North Dakota—a predominantly white city—but was re-routed due to concerns about contaminating the city's water. The new route would go closer to tribal lands and underneath the Missouri River (*Mni Sose* for Lakota and Dakota peoples) which is the primary drinking water source for the Standing Rock Sioux reservation. This rerouting—which many environmental and Indigenous leaders called an act of environmental racism—sparked what became a high-profile struggle.

The Standing Rock Sioux Tribe and Indigenous activists argued that the pipeline would violate Indigenous sovereignty and treaty rights. The pipeline would be built on territory that the Standing Rock Sioux Tribe never consensually ceded to the U.S. government. The land was acquired through force, coercion, and deception as the U.S. government failed to uphold treaty obligations and used complicated legal manoeuvres to push tribal members into giving up their land (Estes 2019). The Standing Rock Sioux Tribe also argued that the pipeline violates Article II of the Fort Laramie Treaty of 1868 that guarantees 'undisturbed use and occupation' of reservation lands surrounding the proposed pipeline because construction and pipeline leaks would contaminate clean water and damage ecosystems that are integral to using and living on the land (Kimmel 2016). The Tribe also argued that they had not been adequately consulted about the project whose construction would destroy sacred religious and burial sites that are protected by federal and international law (Kimmel 2016; Whyte 2017). Despite high levels of poverty and unemployment on the Standing Rock reservation, promises of economic opportunity from the pipeline were flatly rejected.

Starting in April 2016, Lakota, Dakota, and Nakota people (who Europeans called the Sioux) and allies calling themselves water protectors formed a resistance and spiritual camp named Sacred Stone to block construction of DAPL. They targeted the site where the pipeline would cross a reservoir of the Missouri River called Lake Oahe near the Standing Rock reservation. They held ceremonies, prayer circles, and horseback rides to protect spiritual and cultural connections to sacred lands and water threatened by the pipeline (Whyte 2017). The initial camp was organized by Indigenous activists and elders, and eventually gained the support of Tribal governments. The movement grew as more people joined and formed other protest camps including non-Indigenous allies from environmental, racial justice, and other social movements. At the protests' peak in late 2016, thousands of people filled multiple camps in this rural part of North Dakota. Police and private security responded with force and violence including using attack dogs, water cannons in freezing winter weather, and tear gas against the nonviolent demonstrators (Prupis 2016). Images of violent repression in the media sparked public outrage (Bearak 2016).

The opposition grew into the NoDAPL movement with the active #NoDAPL social media campaign, protests in cities across the U.S., and boycott and divestment campaigns against the project's financial backers. Environmentalists framed the pipeline as a threat to clean water and climate change and a concrete target to stop the extraction of fossil fuels. In Minneapolis, Minnesota, multiple protests were held throughout 2016 and 2017, including a march led by Indigenous youth where a few thousand people marched through neighbourhood and downtown streets chanting the slogan of the NoDAPL movement, '*Mini Wiconi*—Water is Life.' Native dancers and drummers performed at intersections and speakers from Indigenous, climate justice, and racial justice organizations addressed the crowd when the march culminated at a Native American community centre for a fundraising meal and workshops. The public attention and lawsuits by the Standing Rock Sioux Tribe pushed the U.S. government under President Obama to delay construction and eventually deny use of a piece of federal land, putting construction on hold.

However, there were divisions between Tribal governments and Indigenous and non-Indigenous activists. In January 2017 as the camps dwindled to several hundred people, the Standing Rock Sioux Tribal Council passed a resolution asking the activists to disband due to concerns about health and safety, and social and economic disruptions (McKenna 2017). Some protestors, including Indigenous and non-Indigenous, refused because they felt maintaining the occupation was necessary. Meanwhile the Tribe shifted their focus to legal and political tactics. Yet, both strategies ultimately failed to stop DAPL. In January 2017, recently elected U.S. President Donald Trump reversed Obama-era decisions to allow the project to move forward (Cheree 2017). By the end of February 2017, the last camp inhabitants were removed by police.

Regardless of the final outcome, NoDAPL placed national attention on often invisible Indigenous issues and made connections between capitalism and settler colonialism as underlying causes of climate change. The NoDAPL movement was not simply an environmental protest. It was driven by Indigenous resistance to settler colonialism and assertions of sovereignty (Whyte 2017). The movement centred around Indigenous demands to uphold treaty rights, preserve sacred sites, protect clean water, as well as stopping climate change. Indigenous scholars and activists argue that the movement is part of a broader Indigenous political resurgence in North America (Simpson 2017).

Opposition to DAPL energized the growing climate justice movement that includes a coalition of Indigenous and environmental groups. NoDAPL built on previous alliances established between environmental groups, rural white landowners, and tribes, nicknamed the 'Cowboy and Indian Alliance,' to block the Keystone XL pipeline that began in 2011 (Grossman 2017). Support for actions like NoDAPL reflects a progressive shift in the U.S. environmental movement to engage social justice issues. Grassroots environmental activism led by youth, people of colour, and immigrants is challenging corporate and state power and advancing visions for socio-ecological justice (Mendez 2020).

## Labour Union Responses to NoDAPL

Labour unions had divergent responses to the NoDAPL movement which reflect two different potential scenarios for labour-Indigenous relations. One reaction led to conflict. Unions representing workers in the fossil fuel and construction industry largely supported the pipeline. This reaction is based on a business unionism perspective (Fantasia and Voss 2004; Moody 2007) focused narrowly on securing pay and jobs for members and working collaboratively with industry. The other reaction led to collaboration. Some unions in other sectors joined with Indigenous and environmental groups opposing DAPL to promote climate justice. This reaction is based on a social movement union approach (Clawson 2003; Milkman and Voss 2004) and forming

## 16 Whose Labour, Whose Land? Indigenous and Labour Conflicts... 375

alliances to transform the economy and advance social and environmental justice. However, the third scenario, a pro-development labour-Indigenous alliance, did not form as tribes largely rejected promises of potential economic growth from DAPL.

Unions in construction and energy production who have a material interest in DAPL and other fossil fuel projects, joined industry and conservative politicians in promoting the pipeline and mobilized against Indigenous and environmental groups. They claimed that DAPL would create jobs and bolster energy security while following environmental regulations to prevent pollution (Aronoff 2016a; Brecher 2016; Hand 2016). This pro-pipeline response needs to be situated in the context of construction unions. The building trades are among the most conservative unions in the North American labour movement, exemplifying business unionism (Fantasia and Voss 2004; Erlich and Grabelsky 2005; Moody 2007). They have a history of excluding people of colour and women while protecting predominantly white and male members (Paap 2008). They are organized around specific crafts or trades which they protect by managing the supply of skilled workers. Employment is also subject to the boom and busts of the construction industry and members regularly move between projects; thus, unions often collaborate with construction companies to secure contracts on new projects and compete against non-union companies (Erlich and Grabelsky 2005). Energy Transfer Partners, the company that owns the DAPL, strategically agreed to hire unionized construction workers through a project labour agreement which helped garner union backing.

Construction unions led by the Laborers International Union of North America (LIUNA) issued statements supporting the project using jobs versus environment and nationalist rhetoric. They also mobilized their members to attend government hearings and other events to show worker support (Laborers International Union of North America 2015). Discourses of resource nationalism and energy security are used by these unions to argue that expanding fossil fuel production is necessary to protect the national interest—equating the nation with industry and white male workers (Stevis 2012; Laborers International Union of North America 2014; Stevis and Felli 2014). For example, a group of unions representing construction workers and truck drivers stated that pipelines will benefit America and workers' families:

> Our members make careers out of jobs created by projects like Dakota Access, and our jobs depend on the investments of conscientious employers. If companies like Energy Transfer Partners cannot trust that the regulatory process outlined in federal law will be upheld, who will continue to invest in America? The

family-sustaining jobs and benefits that this project provides are in jeopardy. (IUOE, IBT, LIUNA, UA, and IBEW 2016)

Union support is shaped by economic conditions. Unionized construction jobs are among the few decent paying blue-collar jobs in a region impacted by deindustrialization and disinvestment. Construction workers also struggle to find consistent employment and face declining wages, dwindling union power and density, and dangerous working conditions (Erlich and Grabelsky 2005). Large industrial infrastructure projects, such as pipelines, are one of the few segments of the construction industry that still have modest unionization levels as other segments, like residential construction, have become largely non-union (Erlich and Grabelsky 2005). This situation increases the power of fossil fuel companies to leverage worker support and stifle opposition if they agree to hire union labour. Still, proactive defence of DAPL prioritizes jobs for male and white construction workers over the livelihoods and sovereignty of Indigenous peoples as well as the impacts of climate change on marginalized communities. Union support for Energy Transfer Partners also ignores how work in pipeline construction and the oil and gas industry is dangerous, and large segments of the industry are non-union (Juhasz 2018).

In addition to the typical job creation argument, leaders of pro-DAPL unions also dismissed the claims of Native American tribes and reproduced settler colonial and conservative rhetoric about radical and violent pipeline protesters. For example, LIUNA President O'Sullivan expressed 'frustration' that Native Americans 'have disregarded the evidence and the review process to vilify a project' (Hand 2016). These unions argued that the Standing Rock Tribe had been adequately consulted and did not provide credible evidence of harm to cultural sites. They rationalized arrests and violence against the non-violent protesters for supposedly violating the law and putting workers at risk. LIUNA claimed in a statement that their members were 'facing intimidation and dangerous confrontations as protests against the Dakota Access Pipeline have intensified' (Laborers International Union of North America 2016). Thus, it's the male and white construction workers who face physical harm, not the peaceful demonstrators, that included many elders, women, and children, who are subject to state violence. Framing protestors as a violent threat legitimized repression and violence that has been used against Indigenous peoples throughout U.S. history. Thus, pro-DAPL unions became complicit in state violence as they supported dispossession of Indigenous lands and the violation of treaty rights. Some unions even joined with conservative groups to promote laws punishing and criminalizing protests against pipelines and energy infrastructure, which civil liberties and environmental groups have condemned as an attack on free speech rights and peaceful protest (Stoner 2019).

Yet, these divisions are not inevitable or pre-determined. Construction and energy-sector unions do not necessarily align with industry. They can be critical of fossil fuel companies during contract negotiations and organizing campaigns. Work in the industry is hazardous and corporations push down wages and benefits. Thus, there is potential for collaboration with Indigenous and environmental groups to challenge the exploitative and ecologically destructive practices of fossil fuel companies. They could align around creating good jobs in safer and cleaner industries, and developing alternative energy. But, these efforts are hampered by the limited success of previous green jobs initiatives. The scale of unionized jobs in renewable energy has not surpassed the oil and gas sector that is growing in North America and been a source of construction projects in the midst of broader economic insecurity, lack of public infrastructure spending, and declining unionization in other construction sectors (National Association of State Energy Officials and Energy Futures Initiative 2019).

Other unions and segments of the labour movement opposed DAPL and aligned with Indigenous and environmental groups. Support for NoDAPL ranged from direct involvement in the protest camps to issuing statements of support. Five national unions, including the Service Employees International Union (SEIU), the Communication Workers of America (CWA), and the National Nurses Union (NNU), publicly supported the NoDAPL movement (Funes 2016). Importantly, these unions primarily represent workers in the service, healthcare, and public sectors without a direct material interest in the fossil fuel industry. Support for Indigenous rights and climate justice reflects how these unions have adopted social movement unionism, such as organizing new industries and different demographics of workers, largely women, people of colour, and immigrants, and engaging broader social justice issues (Clawson 2003; Milkman and Voss 2004). Unions framed their opposition to DAPL in terms of climate justice but also solidarity with Indigenous peoples. For example, SEIU said in a statement:

> The two million members of SEIU stand beside the Standing Rock Sioux Tribe in their fight to protect their sacred lands and burial grounds from being dug up if the construction of the Dakota Access Pipeline is allowed to continue as planned. The history, culture and lives of Tribal people, the first Americans, should be respected and protected. (Service Employees International Union 2019)

SEIU supported Standing Rock's claims, unlike pro-pipeline unions, and the need to protect treaty rights, in addition to climate and environmental concerns.

Solidarity with Indigenous peoples reflects the social locations and progressive politics of their members. SEIU, for example, is a leader in organizing immigrant workers for whom the ideologies of settler colonialism might be less engrained and who do not have the same privileges provided to white U.S. citizens by settler colonialism and extractive capitalism. SEIU framed DAPL as an environmental injustice indicative of the inequalities facing marginalized peoples, including their members:

> Historical disregard for low income communities and communities of color, including those where many SEIU members live and work, has subjected them to toxic air pollution and contaminated waterways for decades. In these communities, asthma and other respiratory ailments caused by toxic air and poisonous toxins such as lead in the water supply, affect our children's health and ability to thrive. (Service Employees International Union 2019)

Similarly, the Labor Coalition for Community Action, a group of union constituency groups representing people of colour, women, and LGBTQ workers, issued a statement supporting Indigenous sovereignty and the need to 'protect Native lands from exploitation by corporations and the U.S. government' (Labor Coalition for Community Action 2016). These groups represent workers who have experienced environmental injustices and will be among those most impacted by climate change. Many of them work in industries dealing with the health and social consequences of climate change, which helps foster solidarity with Indigenous people fighting polluting industries.

However, LIUNA also represents a large number of immigrants and Latinx workers—although the workers in North Dakota are more likely to be white—and represent the lowest paid and least skilled in the construction industry and are one of the more progressive construction unions. Still, they view pipeline jobs as acceptable, even necessary, given the lack of substantial policies to create other employment in green sectors.

Rank-and-file union members and broader working-class movements also supported the NoDAPL movement, often based on critiques of capitalism, colonialism, and racism. Aronoff (2016b) documents how grassroots union members and workers were active in NoDAPL, even members of pro-pipeline unions who broke from official union positions. Working-class and union activists formed the grassroots Labor For Standing Rock protest camp (Aronoff 2016b). A nurse and United Food and Commercial Workers Union member who helped organize the group explained that this action was part of a broader movement against the fossil fuel industry and energizing the labour movement: 'we can make the connections between all union members and begin to organize a new labor movement that fights for full employment and builds a sustainable world where working people, not CEOs, are the new leaders' (New York Nurses Association 2016).

Different responses to NoDAPL created conflicts within the labour movement. Pro-pipeline unions attacked other unions and environmentalists for blocking job creation. LIUNA left the Blue Green Alliance, a union-environmental coalition, due to disputes over the Keystone XL and DAPL pipelines. LIUNA's president, Terry O'Sullivan, issued fiery statements remarking, 'We're repulsed by some of our supposed brothers and sisters lining up with job killers like the Sierra Club and the Natural Resources Defense Council to destroy the lives of working men and women' (Restuccia 2012). Claims that pipeline opponents are hurting working-class families privilege a particular vision of the working class and ignore that opposition involved working-class Indigenous people, youth climate activists, and people of colour.

The divergent union responses to NoDAPL are indicative of a growing split within the North American labour movement over climate change and between more progressive and conservative unions (Bogado 2016). Sweeney (2016) argues that conflicts over pipelines represent a major struggle within organized labour among those who see oil and gas development as a path to prosperity and those who see climate change as a socio-environmental crisis that demands immediate action to transform and de-carbonize the economy. This conflict is about different material interests but also different political ideologies and strategies for strengthening the labour movement.

## Conclusion

This chapter argues that examining labour-Indigenous relationships around issues of resource extraction, energy production, and climate change in the U.S. and Canada is productive for understanding how processes of settler colonialism, industrialism, and capitalism are intertwined in producing environmental injustice. If unions and environmentalists want to resolve socio-environmental problems and create a more just and sustainable society, then settler colonialism and racism must be addressed since, as Indigenous scholar and activist Whyte (2018) argues, settler colonialism is a form of socio-ecological violence intertwined with capitalism and racism that produces environmental crisis. Economic growth and accelerating extraction of natural resources—drivers of climate disruption and exploitation—depend upon dispossession of Indigenous lands which is facilitated by settler colonial structures and legitimated by racist ideologies. Yet, the political challenge is that these same systems and ideologies create privileges for white, particularly male, workers, and creating alliances across multiple forms of difference is difficult. Extractive industry jobs are appealing for workers and unions in the context of declining wages, a lack of decent jobs, and anti-unionism. Conservative political movements in the U.S. have channelled this economic insecurity into racial resentments and anti-environmentalism, including defence of polluting and extractive industries (Hochschild 2016).

A new research agenda is needed exploring the complex relationships between unions and Indigenous peoples, which will advance scholarship on labour environmentalism and environmental and climate justice. Scholarship on labour environmentalism has focused on union-environmental relationships overlooking Indigenous groups and tribes (Obach 2004; Räthzel and Uzzell 2011; Stevis 2018) while scholarship on Indigenous-environmental relationships has overlooked labour issues (Clark 2002; Dove 2006; Grossman 2017; Pickerill 2018). The nascent research on union-Indigenous relationships has not focused on the environment (Mills and Clarke 2009; Mills and McCreary 2012; Fernandez and Silver 2017).

Labour-Indigenous relationships are important for understanding possibilities and challenges for climate justice movements and the complex intersections of class, Indigeneity, gender, and race. As seen in the NoDAPL movement, the climate justice movement is beginning to build alliances between Indigenous peoples and non-Indigenous environmentalists. Yet, having labour involvement is also vital for the climate justice movement. Unions and workers have some power to disrupt and transform the fossil fuel economy through strikes, workplace actions, and collective bargaining. On the other hand, vocal union support for resource extraction allows capital and the state to present workers as populist defenders of extractivism and block class solidarity (Kojola 2019). How, or if, unions join the climate justice movement will also depend on how they relate to Indigenous groups who are a significant and powerful part of the movement. The legal rights of Indigenous nations have been effective in blocking hazardous industries, such as asserting treaty-protected fishing rights to stop construction of new mines or using tribal authority to deny permits for fossil fuel infrastructure on reservations (Gedicks 1993; Clark 2002; Norman 2017). Collaboration with tribes and Indigenous activists requires that labour and environmental movements recognize the perspectives of Indigenous peoples and overcome legacies of conflict (Moore 1998; Dove 2006; Mills and McCreary 2012; Grossman 2017). Indigenous spiritual, cultural, and political connections to land need to be honoured in order to have meaningful and respectful alliances as well as recognition of different decision-making practices that impact the ability to sustain coalitions.

Labour-Indigenous relationships are also relevant for strengthening unions. Critiques of settler colonialism and support for Indigenous peoples are being incorporated into of range of contemporary social movements from racial to climate justice (Pellow 2017). If unions oppose these issues, they will be on the opposite side of vibrant social movements which will weaken their ability to build coalitions and attract younger and more diverse members. Construction unions vehemently supporting fossil fuel projects and opposing Indigenous groups matters for the entire labour movement because these actions often get media coverage that reproduces dominant perceptions of unions as anti-environmental and representing white men (Kojola 2017). Aligning with Indigenous and environmental justice movements can be part of union revitalization through activism and coalition building to advance social justice and challenge capital (Mills 2007; Fernandez and Silver 2017). While unions have been slow to recognize and support Indigenous struggles, there are emerging efforts among a few social movement–oriented public sector unions in Canada that have used popular education to develop members' knowledge and analysis of colonialism and collaborated with First Nations to promote social justice (Mills and Clarke 2009; Mills and McCreary 2012; Fernandez and Silver 2018).

Overcoming internal union divisions and external tensions with Indigenous peoples around environmental and climate justice is possible but challenging. It partially depends on open and respectful dialogue between different movements that recognizes difference while identifying shared forces of oppression to create solidarity. Coalition politics could help overcome cleavages of race, nationality, class, and gender to envision and demand substantive transformations in how nonhuman nature is valued and used, and how governance systems are organized. A just transition framework has potential to align unions and settler workers with Indigenous peoples around sustaining livelihoods and protecting workers and marginalized communities in moving towards a less resource and carbon-intensive economy. Yet creating good paying jobs in green industries alone will not be enough if this simply puts a green veneer on existing economic and social relations. A transformative approach could challenge capitalism and settler colonialism to de-commodify nonhuman nature and social reproduction, and foster democratic governance. Ultimately, the current model of global capitalism and settler colonialism is socially and ecologically unsustainable, and Indigenous peoples and non-Indigenous workers will suffer if the climate crisis is not addressed because, as the saying goes, there are no jobs on a dead planet.

# References

Ali, Saleem. 2003. *Mining, the Environment, and Indigenous Development Conflicts.* Tucson: University of Arizona Press.

Aronoff, Kate. 2016a. AFL-CIO Backs Dakota Access Pipeline and the 'Family Supporting Jobs' It Provides. *In These Times*, September 17, 2016. Accessed December 1, 2020. https://inthesetimes.com/working/entry/19475/afl-cio_backs_dakota_access_pipeline_and_the_family_supporting_jobs_it_prov.

———. 2016b. LIUNA's Rank-and-File is Challenging Union Leadership on Standing Rock—and Beyond. *In These Times*, November 7, 2016. Accessed December 1, 2020. https://inthesetimes.com/working/entry/19605/liunas_rank_and_file_is_challenging_union_leadership_on_standing_rockand_be.

Bearak, Max. 2016. U.N. Officials Denounce 'Inhuman' Treatment of Native American Pipeline Protesters. *The Washington Post*, November 15, 2016. Accessed December 1, 2020. https://www.washingtonpost.com/news/worldviews/wp/2016/11/15/u-n-officials-denounce-inhuman-treatment-of-north-dakota-pipeline-protesters/.

Bogado, Aura. 2016. Big Labor has an identity crisis, and its name is Dakota Access. *Grist*, September 28.

Brecher, Jeremy. 2016. Dakota Access Pipeline and the Future of American Labor. *Labor Network for Sustainability*.

Bruyneel, Kevin. 2004. Challenging American Boundaries: Indigenous People and the 'Gift' of U.S. Citizenship. *Studies in American Political Development* 18 (1): 30–43.

———. 2007. *The Third Space of Sovereignty: The Postcolonial Politics of U.S.-Indigenous Relations*. Minneapolis: University of Minnesota Press.

Cheree, Franco. 2017. The Final, Messy, Defiant Days of the Standing Rock Camps. *Vice*, February 21, 2017. Accessed December 1, 2020. https://www.vice.com/en_us/article/gvmypx/the-final-messy-defiant-days-of-the-standing-rock-camps.

Churchill, Ward. 2002. *Struggle for the Land: Native North American Resistance to Genocide, Ecocide, and Colonization*. San Francisco: City Lights.

Clark, Brett. 2002. The Indigenous Environmental Movement in the United States Transcending Borders in Struggles against Mining, Manufacturing, and the Capitalist State. *Organization & Environment* 15 (4): 410–442.

Clawson, Dan. 2003. *The Next Upsurge: Labor and the New Social Movements*. Ithaca: ILR Press.
Coulthard, Glen Sean. 2014. *Red Skin, White Masks: Rejecting the Colonial Politics of Recognition*. Minneapolis: University of Minnesota Press.
Deloria, Vine, and Clifford M. Lytle. 1984. *The Nations Within: The Past and Future of American Indian Sovereignty*. New York: Pantheon Books.
Dove, Michael R. 2006. Indigenous People and Environmental Politics. *Annual Review of Anthropology* 35 (1): 191–208.
Dreiling, Michael. 1998. From Margin to Center: Environmental Justice and Social Unionism as Sites for Intermovement Solidarity. *Race, Gender & Class* 6 (1): 51–69.
Erlich, Mark, and Jeff Grabelsky. 2005. Standing at a Crossroads: The Building Trades in the Twenty-First Century. *Labor History* 46 (4): 421–445.
Estes, Nick. 2019. *Our History is the Future: Standing Rock versus the Dakota Access Pipeline, and the Long Tradition of Indigenous Resistance*. London and New York: Verso.
Fantasia, Rick, and Kim Voss. 2004. *Hard Work: Remaking the American Labor Movement*. Berkeley: University of California Press.
Fernandez, Lynne, and Jim Silver. 2017. *Indigenous People, Wage Labour and Trade Unions: The Historical Experience in Canada*. Winnepeg: Canadian Centre for Policy Alternatives.
———. 2018. *Indigenous Workers and Unions: The Case of Winnipeg's CUPE 500*. Winnepeg: Canadian Centre for Policy Alternatives.
Funes, Yessenia. 2016. SEIU is Latest Union to Declare Support of the Standing Rock Sioux Tribe's #NoDAPL Fight. *Colorlines*, October 3. Accessed December 1, 2020. https://www.colorlines.com/articles/seiu-latest-union-declare-support-standing-rock-sioux-tribes-nodapl-fight.
Gedicks, Al. 1993. *The New Resource Wars: Native and Environmental Struggles against Multinational Corporations*. Boston: South End Press.
Grossman, Zoltán. 2017. *Unlikely Alliances: Native nations and White communities join to defend rural lands*. Seattle: University of Washington Press.
Hand, Mark. 2016. Dakota Access Foes Call on AFL-CIO to Retract Support of Pipeline. *Counter Punch*, September 20, 2016. Accessed December 1, 2020. https://www.counterpunch.org/2016/09/20/dakota-access-foes-call-on-afl-cio-to-retract-support-of-pipeline/.
Hochschild, Arlie Russell. 2016. *Strangers in Their Own Land Anger and Mourning on the American Right*. New York: The New Press.
Hultgren, John, and Dimitris Stevis. 2020. Interrogating Socio-ecological Coalitions: Environmentalist Engagements with Labor and Immigrants' Rights in the United States. *Environmental Politics*. 29 (3): 457–478.
Indigenous Environmental Network. 2019. *Indigenous Principles of Just Transition*. https://www.ienearth.org/justtransition/.
IUOE, IBT, LIUNA, UA, and IBEW. 2016. Joint Letter to President Obama: Dakota Access Pipeline Permit. October 3, 2016.

Jenkins, Kirsten. 2018. Setting Energy Justice Apart from the Crowd: Lessons from Environmental and Climate Justice. *Energy Research & Social Science* 39: 117–121.

Juhasz, Antonia. 2018. Death on the Dakota Access. *Pacific Standard*, October 19.

Kimmel, Lauren. 2016. Does the Dakota Access Pipeline Violate Treaty Law? *The Michigan Journal of International Law* 38. Pp?

Kojola, Erik. 2017. (Re)constructing the Pipeline: Workers, Environmentalists and Ideology in Media Coverage of the Keystone XL Pipeline. *Critical Sociology* 43 (6): 893–917.

———. 2019. Bringing Back the Mines and a Way of Life: Populism and the Politics of Extraction. *Annals of the American Association of Geographers* 109 (2): 371–381.

Kulchyski, Peter, and Warren Bernauer. 2014. Modern Treaties, Extraction, and Imperialism in Canada's Indigenous North: Two Case Studies. *Studies in Political Economy* 93 (1): 3–24.

Labor Coalition for Community Action. 2016. *AFL-CIO Constituency Groups Stand with Native Americans to Stop the Dakota Access Pipeline*. September 19, 2016.

Laborers International Union of North America. 2014. *With IG Finding, Keystone XL Opponents are Running Out of Straws to Grasp*.

———. 2015. *LIUNA Lends Expertise at Dakota Access Pipe Line Hearing*.

———. 2016. *Extremists Target Dakota Access Pipeline Construction*. Accessed December 1, 2020. https://www.liuna.org/news/story/extremists-target-dakota-access-pipeline-construction.

LaDuke, Winona. 1999. *All Our Relations: Native Struggles for Land and Life*. Boston: South End Press.

Loomis, Erik. 2015. *Empire of Timber: Labor Unions and the Pacific Northwest Forests*. Cambridge: Cambridge University Press.

Mayer, Brian. 2009. *Blue-Green Coalitions: Fighting for Safe Workplaces and Healthy Communities*. Ithaca: ILR Press/Cornell University Press.

McKenna, Phil. 2017. Standing Rock Tribe Asks Dakota Pipeline Protesters to Go Home. *Inside Climate News*, January 24, 2017. Accessed December 1, 2020. https://insideclimatenews.org/news/23012017/standing-rock-dakota-pipeline-protest-camp.

Mendez, Michael. 2020. *Climate Change from the Streets How Conflict and Collaboration Strengthen the Environmental Justice Movement*. New Haven: Yale University Press.

Milkman, Ruth, and Kim Voss. 2004. *Rebuilding Labor: Organizing and Organizers in the New Union Movement*. Ithaca: Cornell University Press.

Mills, Suzanne E. 2007. Limitations to Inclusive Unions from the Perspectives of White and Aboriginal Women Forest Workers in the Northern Prairies. *Just Labour: A Canadian Journal of Work and Society* 11.

———. 2011. Beyond the Blue and Green: The Need to Consider Aboriginal Peoples' Relationships to Resource Development in Labor-Environment Campaigns. *Labor Studies Journal* 36 (1): 104–121.

Mills, Suzanne E., and Louise Clarke. 2009. 'We Will go Side-by-Side with You.' Labour Union Engagement with Aboriginal Peoples in Canada. *Geoforum* 40 (6): 991–1001.

Mills, Suzanne E., and Tyler McCreary. 2012. Social Unionism, Partnership and Conflict: Union Engagement with Aboriginal Peoples in Canada. In *Rethinking the Politics of Labour in Canada*, ed. Stephanie Ross and Larry Savage, 116–131. Halifax: Fernwood Publications.

Moody, Kim. 2007. *US Labor in Trouble and Transition: The Failure of Reform from Above, the Promise of Revival from Below*. New York: Verso.

Moore, Mik. 1998. Coalition Building between Native American and Environmental Organizations in Opposition to Development The Case of the New Los Padres Dam Project. *Organization & Environment* 11 (3): 287–313.

National Association of State Energy Officials and Energy Futures Initiative. 2019. *The 2019 U.S. Energy and Employment Report*.

National Congress of American Indians. 2015. *Tribal Nations and the United States: An Introduction*. Washington, DC.

Neufeld, Andrew, and Andrew Parnaby. 2000. *The IWA in Canada: The Life and Times of an Industrial Union*. Vancouver: IWA Canada/New Star Books.

New York Nurses Association. 2016. Labor Activists for Standing Rock. *New York Nurse*, November 2016.

Norman, Emma S. 2017. Standing Up for Inherent Rights: The Role of Indigenous-Led Activism in Protecting Sacred Waters and Ways of Life. *Society & Natural Resources* 30 (4): 537–553.

O'Neill, Colleen. 2018. Civil Rights or Sovereignty Rights? Understanding the Historical Conflict between Native Americans and Organized Labor. *Center for Gaming Research, Paper* 43: 1–6.

Obach, Brian K. 2004. *Labor and the Environmental Movement: The Quest for Common Ground*. Cambridge: MIT Press.

Paap, Kris. 2008. How Good Men of the Union Justify Inequality: Dilemmas of Race and Labor in the Building Trades. *Labor Studies Journal* 33 (4): 371–392.

Parnaby, Andy. 2006. 'The Best Men that Ever Worked the Lumber': Aboriginal Longshoremen on Burrard Inlet, BC, 1863–1939. *The Canadian Historical Review* 87 (1): 53–78.

Pasternak, Shiri. 2017. *Grounded Authority: The Algonquins of Barriere Lake against the State*. Minneapolis: University of Minnesota Press.

Pellow, David N. 2017. *What is Critical Environmental Justice?* Cambridge: Polity Press.

Pickerill, Jenny. 2018. Black and Green: The Future of Indigenous–Environmentalist Relations in Australia. *Environmental Politics* 27 (6): 1122–1145.

Prupis, Nadia. 2016. Dakota Access Pipeline Company Attacks Protesters With Dogs and Mace. *Common Dreams*, September 4, 2016. Accessed December 1, 2020. https://www.commondreams.org/news/2016/09/04/dakota-access-pipeline-company-attacks-protesters-dogs-and-mace.

Räthzel, Nora, and David Uzzell. 2011. Trade Unions and Climate Change: The Jobs Versus Environment Dilemma. *Global Environmental Change* 21 (4): 1215–1223.

Restuccia, Andrew. 2012. Labor Union Quits Alliance with Greens over Keystone Pipeline. *The Hill*, January 20, 2012. Accessed December 1, 2020. https://thehill.com/policy/energy-environment/205441-labor-union-leaves-bluegreen-alliance-over-keystone-disagreement.

Roediger, David R. 1991. *The Wages of Whiteness: Race and the Making of the American Working Class*. London; New York: Verso.

Rose, Fred. 2000. *Coalitions Across the Class Divide: Lessons from the Labor, Peace, and Environmental Movements*. Ithaca: Cornell University Press.

Schlosberg, David, and David Carruthers. 2010. Indigenous Struggles, Environmental Justice, and Community Capabilities. *Global Environmental Politics* 10 (4): 12–35.

Service Employees International Union. 2019. *SEIU Statement on Standing Rock Sioux and Dakota Access Pipeline*.

Simpson, Leanne Betasamosake. 2017. *As We Have Always Done: Indigenous Freedom Through Radical Resistance*. Minneapolis: University of Minnesota Press.

Smith, Sherry L., and Brian Frehner. 2010. *Indians and Energy: Exploitation and Opportunity in the American Southwest*. Santa Fe: School for Advanced Research Press.

Steinman, Erich. 2012. Settler Colonial Power and the American Indian Sovereignty Movement: Forms of Domination, Strategies of Transformation. *American Journal of Sociology* 117 (4): 1073–1130.

Stevis, Dimitris. 2012. Green Jobs? Good Jobs? Just Jobs?: US Labour Unions Confront Climate Change. In *Trade Unions in the Green Economy: Working for the Environment*, ed. Nora Räthzel and David Uzzell, 177–195. London: Routledge.

———. 2018. US Labour Unions and Green Transitions: Depth, Breadth, and Worker Agency. *Globalizations* 15 (4): 454–469.

Stevis, Dimitris, and Romain Felli. 2014. Global Labour Unions and Just Transition to a Green Economy. *International Environmental Agreements: Politics, Law and Economics* 15 (1): 29–43.

Stoner, Rebecca. 2019. Why Are Unions Joining Conservative Groups to Protect Pipelines? *Pacific Standard*, May 31, 2019. Accessed December 1, 2020. https://psmag.com/environment/why-are-unions-joining-conservative-groups-to-protect-pipelines.

Sweeney, Sean. 2012. *Resist, Reclaim, Restructure—Unions and the Struggle for Energy Democracy*. Cornell University ILR School, Rosa Luxemburg Stiftung.

———. 2013. US Trade Unions and the Challenge of 'Extreme Energy': The Case of the TransCanada Keystone XL Pipeline. In *Trade Unions in the Green Economy: Working for the Environment*, ed. Nora Räthzel and David Uzzell, 196–213. London: Routledge.

———. 2016. Contested Futures: Labor after Keystone XL. *New Labor Forum* 25 (2): 93–97.

Thompson, Lou. 2018. Tribal-Union Partnerships Address Pipeline Workforce Shortage. *North American Oil & Gas Pipelines*, October 3.

Tuck, Eve, and K. Wayne Yang. 2012. Decolonization is Not a Metaphor. *Decolonization: Indigeneity, Education & Society* 1 (1): 1–40.

Whyte, Kyle Powys. 2017. 'The Dakota Access Pipeline, Environmental Injustice, and U.S. Colonialism.' *Red Ink* Spring: 154–169.

———. 2018. Settler Colonialism, Ecology, and Environmental Injustice. *Environment and Society* 9 (1): 125–144.

Wolfe, Patrick. 2001. Land, Labor, and Difference: Elementary Structures of Race. *The American Historical Review* 106 (3): 866–905.

# 17

# Commoning Labour, Labouring the Commons: Centring the Commons in Environmental Labour Studies

Gustavo A. García-López

## Introduction

Over the last decades, the 'commons' has become a key framing of socioecological transformations and post-capitalist politics. The emerging environmental labour studies field has made significant contributions to understanding the ways labour and nature are similarly threatened by global capital, and identifying the convergence between working-class and environmental struggles (Stevis et al. 2018). But the concept of the commons is nearly entirely absent in this field, which remains centred on the role of labour unions and their environmental attitudes, discourses, alliance, protest actions and policy initiatives. Other forms of working-class organisation, such as indigenous, peasant, fisherfolk, and frontline environmental justice groups—the 'labourers of the earth', to paraphrase Stefania Barca (2014), have been much less studied (see Chaps. 1, 9, 12, 14, 18, 21 in this volume). These organisations are at the forefront of resisting and creating alternatives for environmental justice and a transformative just transition, with the commons at the centre. They are most visible in the Global South and in indigenous and

---

I deeply thank the editors of this volume for their thoughtful, critical engagement and revisions with my text, which significantly improved its final version.

G. A. García-López (✉)
Center for Social Studies, University of Coimbra, Coimbra, Portugal

black-and-brown peripheries of the Global North (the 'South of the North'), regions where environmental-labour studies have not developed substantial research.[1]

In this chapter I discuss what a 'commons frame'[2] can contribute to understanding and advancing our understanding of the relations between labour and environment. Mobilising work at the intersection of eco-feminist, community economy, environmental justice and autonomous Marxist traditions, I argue that the commons are a central element of a particular variety of labour environmentalism that takes place in environmental justice struggles over social reproduction, within 'working-class community ecologies' (Barca and Leonardi 2016, 2018). Through this approach, we can see the inherent relationship between commons' enclosures and environmental injustices, sustained by racist/colonial/patriarchal structures, and the potential of commons-based initiatives that 'aim to radically transform "the economy" based on principles of mutual interdependency between production, reproduction, and ecology' (Barca and Leonardi 2018, 491).

In what follows, I elaborate on the variegated meanings of the commons, and discuss three ways in which a commons frame can contribute to environmental-labour studies: expanding conceptions of 'labour' and the working class through the reproductive labour commons, understanding of working-class environmental subjectivities through commoning and co-becoming, and expanding visions of working-class organisations and 'blue-green' alliances. I then ground these arguments first with a discussion of just transition/GND movements today, and then with a look at these movements in the context of the Caribbean archipelago of Puerto Rico.

## The Commons and Commoning: Key Insights for Environmental-Labour Studies

The original usage of the term 'commons' dates to European feudalism, when it referred to the agricultural and forest lands that were formally owned by the King but to which peasants could use for their livelihoods—for example, growing food, grazing cattle, collecting wood and other forest products. These common lands were sustained by the communities of 'commoners', through

---

[1] A notable exception is South Africa; there is also emerging research in Taiwan and South Korea (Stevis et al. 2018), and Brazil, India, Kenya and Tanzania (JTRC 2018).
[2] I am inspired here by the work of David Bollier and Silke Helfrich and their call to think, learn and act as a commoner.

everyday practices of commons-making, or *commoning* (Linebaugh 2008). Starting in the fifteenth century, these lands were progressively privatised across Europe; peasants, forcibly displaced and then went on to become wage-workers in the new urban factories. These *enclosures* were at the heart of what Marx called original or 'primitive accumulation', foundational to capitalism, but it is a process which has continued to this date (De Angelis 2007; Linebaugh 2008).

From the mid-twentieth century onwards, the 'term commons' was used to refer to natural resources which were characterised by the problem of overuse. Garret Hardin's (1968) famous 'tragedy of the commons' argument used the concept to refer to natural resources—such as a pastureland—which were 'open-access' and destined to be overused and degraded. Ostrom (1990) countered Hardin's 'tragedy' narrative by showing that commons were not always 'open access' and thus overuse was a threat but not an inevitable outcome. Rather, throughout the world, there were plenty of experiences of local communities who were able to engage in 'collective action' and self-organise institutions (rule systems) for sustainable management of the commons—from pastures to forests to irrigation systems to fisheries—on which their livelihoods depended.

The term has progressively expanded over the last decades, focusing on 'new commons' in urban, digital, energy and other spheres, and acquiring new meanings. Here, I emphasise four meanings that are relevant for environmental-labour studies: commons as society's commonwealth ('the common'), including both the material and the immaterial resources produced through cooperative labour (Hardt and Negri 2009; also Hardt 2010);[3] a set of social and natural elements shared by a community, reproduced through relational practices of 'commons-making', that is, *commoning* (Bollier and Helfrich 2015; Bresnihan 2016; Linebaugh 2008); as a political principle rejecting private appropriation, and emphasising cooperative action, solidarity and sustainability; and as a 'paradigm' or framework to reconceptualise how we understand and act in our interdependent world—to 'think, feel and act' as a commoner (Bollier and Helfrich 2015). These meanings are interconnected: commons are 'out there' in the sense that even without us realising, our actions in our environments are interdependent, affecting each other. But commons also 'exist' to the extent they are considered and acted upon as such; otherwise, they are easily enclosed and destroyed. In short, there is no

---
[3] Hardt and Negri (2009) define 'immaterial commons' as 'the languages we create, the social practices we establish, the modes of sociality that define our relationships', and propose that these are the basis of a postcapitalist transformation. This has been challenged by others who emphasise the material-ecological basis of commons (Tola 2015).

commons without a community, and no community without a commons, where community is 'a quality of relations, a principle of cooperation and of responsibility to each other and to the earth, the forests, the seas, the animals', rather than 'a grouping of people joined by exclusive interests separating them from others' (Federici 2012, 228–229).

The praxis of commons is tied to historical collective governance of indigenous and peasant territories, along with their forests, water and agriculture lands, and also to newer initiatives such as urban gardens and food cooperatives; social clinics and collective housing; movements against privatisation and for re-municipalisation of public utilities (water, energy) in cities; open source software, hardware and knowledge platforms (e.g. Wikipedia); digital technologies and creative commons licenses; community renewable energy; urban collaborative laboratories; political parties and governments like *Barcelona en Comú* (Barcelona in common) and the city of Bologna, which have pioneered commons-based initiatives; and regional networks and movements, like the European Assembly for the Commons or the Anahuac Network in Defence of the Commons (Mexico). Indeed, as De Angelis (2017) argues, in an interdependent world, *everything* is a commons (*omnia sunt communia*), waiting to be commonised.

Commons are seen not just as alternative forms of localised collective organisation within capitalism, but as strategic enactments to both resist enclosures and radically transform our societies towards more egalitarian and ecological horizons (Miller 2001). As political principle and transformative paradigm, based on concrete practices and relations, the commons offer a political alternative beyond the dichotomy of 'free-market' and austerity vs. statist neo-Keynesianism, one which favours 'decentralization over central control, democratic over hierarchical management, access over property, transparency over privacy, and sustainability over growth' (Papadimitropoulos 2017, 110). Yet since commoning projects exist 'within' hegemonic social structures and are thus always crossed by racial, gender, class and other categories of difference, they are under threat to become co-opted as 'fixes' for capitalist crises rather than an alternative to capitalism (Caffentzis and Federici 2014). Some speak of 'capitalist' and 'non-capitalist' commons. Commoning can thus be reframed as a dialectic struggle between forces of commoning and forces of enclosures, from the societal to the individual levels: within and against the capitalist state, and embodying transformative alternatives to transcend it (Miller 2001; Cumbers 2015; Routledge et al. 2018).

In what follows, I discuss three ways in which the commons frame contributes to environmental-labour studies and our understanding of 'labour environmentalism'. First, *reproductive commons* and *labour commons* expand our understanding of labour and the working-class, complementing the dominant focus on sectoral wage-labour, with that of rural and non-wage labour, in its *collective* forms. Second, *commoning* points to the processes through which dominant working-class subjectivities can be undone, to foster socio-ecological relations based on care, solidarity and ecology. Finally, *commons movements* push beyond the focus on trade unions and 'blue-green alliances', the workplace as *the* site of contestation, and the state as the interlocutor, towards a focus on grassroots alliances led by frontline communities of reproductive workers, struggling for/from their territories, combining resistance with self-organised (commons) alternatives across scales.

## Expanding Conceptions of 'Labour' and the Working Class: Social Reproduction and the Labour Commons

A key question of environmental-labour studies is how the working class can move society's production towards more socially just and ecologically sustainable forms. While the focus of this field has been mostly on productive labour (wage-labour which generates goods and services for commercial value), a commons frame pushes us to focus on *reproductive* labour and on labour's *collective* and *non-human* dimensions.

Reproductive labour has various definitions. Here I use it to mean labour which is primarily non-waged and is not primarily aimed at capital accumulation (surplus, or 'exchange' value), but rather at producing the subsistence resources (or 'use' value) that the workers need for the reproduction of their own labour-power and that of others, or, more generally, the reproduction of the basic needs of life (see Benjamin and Turner 1992; Federici 2012). While typically associated to household work (childcare, cleaning, etc.), I am particularly interested here in the broader communal *care* activities of 'laboring of the earth' (Barca 2014), including agriculture and environmental management, which involve both physical and emotional dimensions to sustain and restore the commons. For instance: the 'affective labour' of communities protecting and restoring forests in India, which leads to new socio-natural relations and subjectivities (Singh 2013, 2017); the 'convivial labour' of growing, preparing, sharing and consuming of food in urban gardens by working-class immigrants in California, which serve as a space of liberation from the capitalist impositions of 'work' (Valle 2020); or the cooperative work integrated with sharing economies in various aspects of the social reproduction of life—from hunting and fishing to cooking and building houses—essential to black communities' autonomy in rural Belize.

Before the birth of capitalism, reproductive and productive labours were integrated categories of peasant communities, which were done communally and without such marked gendered divisions (Federici 2012; Linebaugh 2008). Labour and the environment were understood as shared commonwealth (i.e. a commons), essential to the reproduction of life, expressing 'relationships in society that are inseparable from relations to nature' (Linebaugh 2008, 44, 289). The birth of capitalism entailed the enclosure of common lands (turned into commercial agriculture or new urban areas), the separation of commoners from their environment and means of reproduction, and the creation of a new division between (male) productive wage-work, and (female) reproductive work. Reproductive work became a way to perform the reproduction of both life and the capitalist labour force, while becoming dependent (by virtue of being separated from the land as means of reproduction) on the male wage-labour (Federici 2012). Capitalism's evolution has increasingly incorporated women into the productive sphere—making women 'doubly exploited' as they now do both productive and reproductive labour. It has also increasingly commodified reproductive labour into the wage relation, through, for example, paid childcare or commercial agriculture. Thus, the enclosures of commons were—and continue to be—central to new and gendered forms of exploitation of labour, as well as of nature.

Understanding labour as a commons can potentially challenge these separations: of commoners from their communality and their environment, of production and reproduction, of women and men, of workers and owners. It points to the ways that labour is 'essential for life', is socially (co)produced through cooperation and interdependency within and between different activities,[4] and requires 'self-governance' through collective rules and participative democracy (Azzellini 2018, 767). This approach also reveals the inherent interdependency between productive/wage-labour (mostly by men) and reproductive/non-wage labour (mostly by women): indeed, both the environment and the labour-power are reproduced through this (reproductive) labour (Benjamin and Turner 1992). Furthermore, labour as a commons implies directly breaking the 'job blackmail' by replacing the capitalist logic of scarcity and growth with a 'logic of abundance': a praxis where enough is produced for all through multiple reciprocal relations, networks of sharing and cooperative governance of basic needs (Helfrich and Bollier 2015). This has the potential to increase social equity and democracy, while reducing environmental impact (ibid.).

---

[4] For instance, the reproductive household labour facilitates the productive wage labour, the labour in electric and water infrastructure or in growing food allows everyone else to carry out their work.

The idea of 'labour commons' envisions an organisational model of 'democratised organisation of productive and reproductive work' which replaces 'the circulation of capital' with 'the circulation of the common', prefiguring alternative post-capitalist futures (De Peuter and Dyer-Witheford 2010, 45; see also Azzellini 2018; Vieta 2016). This model—linked to initiatives such as worker-occupied factories and worker-owned cooperatives—proposes a 'commonly owned and cooperatively based production rooted in solidarity and mutual aid', and resisting hierarchy and coercion, which can also be called *autogestion* or self-management (Vieta 2016). These kinds of initiatives are emerging across the world, building on long legacies of wider popular resistance (and re-existence) movements to neoliberal capitalist exploitation and alienation, often without ties to conventional trade unionism (Vieta 2016).

This approach does not mean that reproductive labour and the labour commons are not exploitative. There are always gender, class and other differences within any communal (commons) relations. Moreover, as noted above, the conditions for reproduction of life are generative of the conditions for productive labour's most important commodities (labour-power and nature), and thus, for capitalist accumulation. The commons framework, however, helps shift our attention towards the 'non-proletarian working class'—the majority of the world's workers—and its care activities in reproducing life (Benjamin and Turner 1992)—the 'labouring of the commons'. Re/productive labour as a commons further highlights the interdependence between workers who co-produce their labour-power, and, more generally, the conditions of life, in interaction with natural or built environments (themselves 'commons'), and suggests the need to shift from the capitalist logic of individualised labour-power to the 'commoning of labour'. It points to struggles which are not mainly about production (e.g. better work conditions), but about '(re)production in common'—which challenge enclosures and seek to re-connect and share labour and wealth collectively through autonomous self-organisation (Federici 2012; also Benjamin and Turner 1992). As such, a commons approach can sow the 'seeds' of different 'forms of production … "from which to reclaim control over the conditions of our reproduction" and "increasingly disentangle our lives from the market and the state" ' (Caffentzis and Federici 2014). Finally, this approach points to the non-human labourers, for example, forests that produce mulch (Singh 2013) or the beavers that build dams (Woelfle-Erskine 2019). This perspective opens up new possibilities of commoning and co-becoming, amongst human workers, and with/in the Earth and all its species (Singh 2017; Tola 2015).

## Deepening Understanding of Working-Class Environmental Subjectivities: Commoning and the Co-Becoming of the 'Worker Commoner'

Environmental-labour studies have documented a long history of working-class struggles to protect nature as a 'common resource vital for human flourishing and survival', as well as to protect workers' environmental health (Stevis et al. 2018, 441). In that sense, there is a connection to commoning as expressing a labour of interconnecting social-environmental spheres. Yet despite this rich history, workers still often oppose environmental protection as a threat. Moreover, trade unions' environmental activism often sees nature through a conservationist lens, as something separate from workers, a 'resource' to be consumed passively through recreation (Stevis et al. 2018) not an integral part of everyday material subsistence. Thus, the question becomes, how do transformative, eco-social worker subjectivities emerge and are nurtured? A commons frame can provide important insights.

This alienation can be seen as a classic commons' dilemma: everyone would be better off if all acted collectively to support a just transition, but are prevented from doing so by their (real or perceived) short-term self-interests. It is a historical product of the enclosure and dispossession of commons and commoning relations, and the concomitant 'proletarianisation' of labour (De Angelis 2007). When mobilised in trade unions' environmental discourses, this subjectivity reproduces the idea of workers as exclusively dependent on the wage relation, which limits their politics to fighting for a better distribution of the costs and benefits of more ecological forms of production (Velicu and Barca 2020). The task is thus to deconstruct this dominant subjectivity and perform others (ibid.).

Commoning processes are a key site for these transformations, a continuous process of 'co-becoming' with each other and nature (Singh 2017). By creating spaces where care, solidarity, sharing, interdependence and non-commodified values are put into practice, commoning can produce both new commons and new counter-hegemonic relations, subjectivities and 'common senses'[5] that replace those of enclosure, extraction and exploitation (Bresnihan 2016; García-López et al. 2017; Tola 2015). Through commoning, there can be a 'production of ourselves as a common subject' (Federici 2012) and a breaking of the division between working-class communities and their ecolo-

---

[5] Common senses refer to dominant ways of perceiving and understanding the world, which emerge through existing relational practices (Gramsci, in Garcia-Lopez et al. 2017). Commoning, by changing everyday practices, can be a key to counter and generating new common senses which are materially and symbolically articulated around commons.

gies. Caring for a degraded or threatened forest (or an abandoned lot turned into urban garden, or a fisheries) can lead to the emergence of communal environmental subjectivities, with visions of the forest not as an external 'nature' to be protected, but as a 'commons' which is part of the community, its identity and its sustenance (Singh 2013)—'forests with people', as put by one organisation (García-López et al. 2017). Community economic projects can 'take back the economy' and 'interrupt' its capitalocentric imaginary through logics of sharing and 'being-in-common' (Gibson-Graham et al. 2016); for instance, social solidarity markets create spaces of encounter and experimentation of cooperative practices and identities between working-class producers, consumers and social movements (Esteves 2020). Commons can also lead to new conceptions and practices of democracy as based on 'popular sovereignty' for the collective good ('the common')—as has happened in struggles over re-municipalising water in Italy and Bolivia (Sauvêtre 2018).

## Broadening Visions of 'Blue-Green' Alliances: The Commons as Common Ground for Environmental-Labour Struggles

Environmental-labour studies has sought to understand what kind of alliances can be forged between unions and environmental organisations to advance transformations towards environmental justice and sustainability, typically centring on trade unions' environmental activism and 'blue-green' alliances. At the same time, scholars and activists alike express the need to continue expanding labour's alliances with environmental, indigenous, women and other movements to confront and transform global capitalism (Stevis et al. 2018). A commons lens proposes to see the struggles over commons and processes of commoning as central articulating element (a 'common ground') to these alliances, and points to other working-class organisations at the centre of marginalised communities' struggles for social and environmental justice, such as community/social unions, environmental justice organisations, producer (e.g. peasant and fisherfolk) associations, community and worker-led cooperatives, land trusts, mutual aid initiatives and community-based natural resource management organisations (see Barca and Leonardi 2018). It shifts the geographic attention of these struggles from the workplace, to the community and territory as the site of these struggles. Finally, the commons underscore the importance of joining resistance and alternatives movements, essential to advancing a counter-hegemonic, emancipatory version of sustainability (cf. Pelenc et al. 2019).

Commons enclosures have always been accompanied by struggles from commoners, linked to broader popular (working-class, peasant, indigenous, subaltern, etc.) struggles (Linebaugh 2008; De Angelis 2014). De Angelis refers to a historically recurrent 'double movement' between the forces of enclosures and those of commons/commoning, which parallel the 'historical rhythms of the class struggle within capitalism' (De Angelis 2014, 300). This is a potential 'virtuous cycle': popular movements make claims to states, seeking public protection of the common good, redistribution and rights to commons; while commons movements promote the creation of new commons (ibid).

Other approaches, such as 'gendered class analysis' (Benjamin and Turner 1992; Turner and Brownhill 2004), 'environmentalism of the poor' and 'environmental justice' (Akbulut et al. 2019), and communalism/autonomism (Escobar 2014; Esteva 2014),[6] have also traced these connections between commons and movements. They suggest that non-proletarian workers are the major mobilised working-class forces confronting capitalism because they are directly dependent on commons for their social reproduction, and these commons are the main source of capitalist accumulation. Furthermore, these movements combine resistance against destruction, with a defence of life and territory and the creation of alternatives that 'extend command over the shared life-ground on which all people and other beings depend' (Giacomini et al. 2018, 5; also Esteva 2014).

This is particularly visible in the Global South: in the Nepalese peasants defending their forests through community forestry as a form of 'accumulation without dispossession' (Paudel 2016); the Mau Mau (Kenya) and Oghaerefe (Nigeria) women who reclaim for 'subsistence commoning' their historically common lands that oil and agribusiness industries had stolen (Benjamin and Turner 1992; Miller 2001; Turner and Brownhill 2004); the Rastafari eco-feminists fighting to protect forests and working to restore the land through communal agroforestry in Trinidad and Tobago (Fox and Smith 2016); the rubber tappers in Brazil who defended the Amazon to protect their subsistence, creating protected 'extractive reserves' for sustainable livelihoods (Barca 2014, this volume); or the Afro-Colombian communities in the Pacific coast which, after achieving collective rights to their traditional territories, are building alternative life-projects (Escobar 2014). We can also find it in the peripheries of the North: in Campania (Italy), where a popular alliance between environmental justice movements resisting contamination of the

---

[6] *Comunalidad* refers to 'a way of being in which the communal condition, the "us," forms the first layer of the meaning of our existence' (Esteva 2014).

territory (which they label *biocide*) and anti-Mafia social cooperatives have reclaimed lands and assets confiscated to Mafia as 'commons', to develop new social and economic production and reproduction strategies; or in Transylvania (Romania), where peasants threatened by a mining project seek to strengthen and diversify their existing livelihoods, rejecting their imposed categorisation as 'workers' and generating an 'alternative imaginary of justice on how to reproduce the socio-ecological conditions of life' (Velicu 2019).

These struggles are clearly not mainly over better health or environmental conditions for wage workers, but about defending or reclaiming reproductive labour from the expansion of wage labour (i.e. proletarianisation). Indeed, they are a response to a context where capitalist policies, ever more violently, continue to attempt to force non-proletarian working people, through enclosures, away from their subsistence livelihoods into wage labour. They seek to defend and reclaim non-state forms of power and territoriality which give primacy to 'the communal over the individual, the connection to the Earth over the separation between humans and non-humans, and *buen vivir* [roughly: the good life] over the economy' (Escobar 2014, author's translation). This represents a particular challenge for environmental-labour studies, predicated on the central role of wage labour, and trade unions in particular, in advancing socio-ecological transformations.

These movements often transcend their particular territorial dimensions to constitute alliances at national and transnational scales. Indeed, in the cusp of the alter-globalisation movements, which saw the popularisation of the idea of 'blue-green' alliances, a 'radical reclaiming of the commons' against neoliberal privatisation was the commonality uniting these disparate movements globally (Klein 2001), and we can observe that these alliances have continued to date. For instance, in Italy, a national network of community water groups (the Italian Forum of Water Movements), created in the aftermath of the G-8 summit in 2006 in Genova, promoted a broad coalition of organisations, including trade unions and environmental and consumer groups, which successfully campaigned for a referendum that ultimately reverted the privatisation of public services, and instituted water as commons (Mattei 2013). The use of the 'commons' frame spread to other struggles including the occupation of theatres in cities across Italy, and the fight to defend the Susa Valley from a high-speed train proposal (ibid.). In Latin America, the commons has also become a framing uniting trade unions with other popular anti-extractivist movements in Latin America. For instance, in Argentina, the State Workers' Association campaigned in 2013 for a referendum on the 'defence of the commons', which—in the context of the government's intentions to increasing oil exploitations—unified self-organised citizen assemblies seeking to 'defend life' against 'the destruction and looting of natural resources' across the country. ([7] Transnational networks of peasant and indigenous organisations have also mobilised a praxis of commons to shape global norms around water and forests).

## Grounding the Commons-Environmental-Labour Nexus in Climate Justice, Just Transition and Green New Deal Debates

The centrality of commons in environmental-labour issues can be grounded by looking at the emerging discussions on 'climate justice' and 'just transition', which reflect on the transition to a fossil-free society to address the climate crisis ([8] Commons and climate justice are mutually reinforcing themes (Perkins 2019). Climate change is simultaneously *the* major commons and environmental justice issue. The planet, as our 'common home', reflects the interdependence of all our actions and how individual (and systemic) self-interest—profit-above-all-else—has generated a social-ecological 'tragedy'. At the same time, the climate crisis evidences the tragedy of enclosures: the climate commons have been appropriated unequally, distributed across racialised, gendered, colonialist divisions that characterise capitalism. Indeed, a central element of the global climate negotiations has been recognising the historical inequities through which the Global North has overexploited both the global climate commons and the Global South labour and nature; and demanding reparations for these historical 'ecological/climate debts'. On the other hand, the climate crisis magnifies such injustices by affecting the same marginalised communities and nations that have been historically exploited. The mainstream solutions proposed by corporations and states—such as carbon markets, carbon storage and corporate-led renewable energy transitions—threaten these same EJ communities and the workers of the fossil-fuel economy. Indeed, agricultural and forest lands are also grabbed now in the name of conservation and other climate mitigation actions. Thus, as Huber (2019) notes, climate change is a class struggle: 'a struggle to build mass social power

---

[7] http://unionsforenergydemocracy.org/unions-in-argentina-call-for-referendum-on-defense-of-the-commons/.
[8] There are different visions of JT/CJ, concerning the nature of the changes needed, the agents of such change, and the centrality of social justice concerns (see Chatterton et al. 2013; Giacomini and Turner 2015; JTRC 2018). Here I focus on the more radical/transformative, working-class and community-centred approach.

to confront some of the wealthiest and most powerful sectors of capital in world history'.

Climate justice movements have sought to confront these structures and advance systemic changes that address the climate crisis and the structural inequalities simultaneously. The 'commons' is embedded in these movements' principles: interconnectedness of all Earth species and more equitable sharing between humans and non-humans; rejecting false 'solutions' based on the commodification of nature (e.g. 'cap-and-trade'); affirming indigenous and peasants rights over their territories, and their participation in government decision-making; promoting solutions based on just transition principles of solidarity and radical democracy (Chatterton et al. 2013). While the mainstream climate movement, led by the 'big' (white and wealthy) environmental NGOs and some trade unions, focuses on institutionalised lobby and electoral politics and endorses neoliberal market 'solutions', these more radical, predominantly working-class and women-led movements emphasise class conflicts, and the links between capitalism, imperialism and the climate crisis. Moreover, they are led by, and centre on, the marginalised 'frontline' communities and workers most impacted by the climate crisis, who are already leading the transition to a regenerative economy with initiatives in zero-waste, sustainable agriculture, energy democracy, land and water stewardship, and community-led renewable energy (Adrar et al. 2019; CJA 2020; Mascarenhas-Swan 2017). Giacomini and Turner (2015, 29–30) see these movements as based in commoning, that is, the 'collective control over the prerequisites of life' through actions that 'stop commodification and elaborate collective, democratic solar alternatives at different links on corporate value chains'. They are not only local but organised polycentrically in interlinked webs of collective actions across multiple scales (from community and municipal to the transnational scales) and interconnected issues—from food and energy, to housing, women and indigenous rights (Tormos-Aponte and García-López 2018).

Based on this praxis, climate justice movements are reclaiming the concepts of Just Transition and Green New Deal, emphasising a societal shift based on a combination of grassroots organising to shut down the extractive economy, and to build the new regenerative economy now in frontline communities. Their theory of change is that transformation occurs from the bottom-up, and that concrete alternatives are a means of addressing the immediate environmental health and climate crises on the ground, preparing for the future, and creating hope and momentum for a movement for just transition. For communities dependent on the toxic fossil fuel economy, support cannot be based purely on future promises; it demands changing the experiences of the basic elements of everyday life, where people can commoning—food, housing, energy, care and so on. In sum, the grassroots JT vision is where people can uphold their collective rights 'to create productive, dignified, and ecologically sustainable livelihoods', as well as rights to nature, which requires governing our shared resources—especially land and money—in common (Mascarenhas-Swan 2017).[9] As Mastini (2020) concludes, the decarbonisation of basic services—transportation, food, energy, housing, care—depends on their decommodification, 'removing them from the market logic and subjecting them to the logic of the commons'.

Global South decolonial/indigenous perspectives are also central to ongoing discussions about JT/GND. Emerging conversations in Latin America propose their own version of the GND—the Eco-social Pact of the South, which brings together indigenous, peasant and urban movements, and highlights cancelling external debt, putting care at the forefront and transitioning towards post-extractivism (Svampa and Viale 2020).[10] Indigenous movements in the United States, for their part, point to how a truly transformative program—within and beyond the GND—needs to counter the colonial logics which systematically treat indigenous territories as disposable 'sacrifice zones' for capitalist accumulation (Red Nation 2020), perpetuating unequal patterns within regions and between the Global North and South. In sum, these perspectives emphasise the need to challenge global and national (neo) colonial-imperial structures, and to put indigenous struggles at the heart of just transition discussions and actions. To further ground these discussions, I now turn to the example of my native home in the Caribbean archipelago of Puerto Rico.

## A View from the Global South: Puerto Rico's Struggle Against the *'colonial* job blackmail' and for a Just Transition

Puerto Rico has been subject to a long history of intensive exploitation of people and environment, along the enclosure and dispossession of commons, first as a colony of Spain (1490s–1898) and then of the United States (1898–present). The local indigenous *Tainos* were exterminated by the Spanish colonisation and its sugar and coffee plantations. Still, their collective, land-based

---

[9] Here we can see a clear connection to scholars such as Vercellone (2015), who centres on 'the democratic reappropriation of the welfare state and the re-socialization of money' as the central aspects of the common.
[10] See also: http://www.pactoecosocialdelsur.com/.

ways of life persisted in enclaves of indigenous and African fugitive and freed slaves at the edges of plantations, in what could be described as 'peasant-proletarian' communities with a 'continuing relation to cropland and other ecologies, autonomous production activities, labour-gang organisation, face-to-face relations among themselves and with foremen, non-cash wage relations, non-union activism and old cultural forms' (Giusti-Cordero 1996, 54).

Starting in the 1930s, and intensifying from the 1950s onwards, the US government shifted from plantations to an intensive militarisation, industrialisation (first with petrochemicals and then pharmaceuticals and transgenic seed experiments), and urbanisation of the archipelago. This was an attempt at the 're-colonisation' of the autonomous spaces of the *jibaros* (Thomas 2019). Yet despite significant changes, peasant-proletarian forms of commoning have persisted in many coastal and mountain communities until today (e.g. García-Quijano and Lloréns 2017; McCune et al. 2018). In this context, environmental justice struggles in Puerto Rico have a long history of interlinking diverse struggles against colonialism (e.g. against military bases), the opposition to industrial contamination and its effects on workers and adjacent communities' health, and the defence and reinvigoration of communal autonomous livelihoods, in a challenge to the colonial-capitalist model of 'development'.

These connections began by organisations such as Misión Industrial (Industrial Mission). Misión emerged in the 1960s at a time of new and rapid industrialisation, and merged socialist, anticolonial and liberation theology ideals with environmentalism (see Chap. 14 in this volume). It went from evangelical activities with workers in the new factories, to advocating for and organising the workers for better health and safety conditions, to mobilising working-class communities living around proposed or existing factories (Meyn 2017). Misión thus became a key actor in the main environmental struggles during the period all across the country, including against proposed extractive projects (an open-pit mine, a nuclear energy facility, a mega-port for petroleum tankers, an agro-chemicals factory) as well as existing industrial pollution (mercury factory; petrochemical industries), and 'wherever the effects of an economic development model that left the trash here and took its fruits elsewhere were felt' (Meyn 2017, author's translation). Other struggles included the fisherfolk movement against the military in the island of Vieques in the late 1970s (McCaffrey 2006). In these struggles, the commons were central: working-class communities facing the colonial industrialisation-militarisation project sought to defend, or reclaim, their historic ties to land and communal subsistence livelihoods, their identity and autonomy, in short their 'reproductive commons'.

Towards the 1990s there was a visible evolution from purely resistance movement, into a combination of resistance and the creation of grassroots commons-based *autogestion* alternatives, as a strategy to achieve social and environmental justice. In a 1998 publication, for instance, Misión Industrial argued that it was 'necessary not only to redefine and restructure our relationship with nature, but to restructure, ... ecologise ... society as such' and that this required communities to take back their 'popular power' by building 'ecological communities' increasingly autonomous from governments and markets, through 'autogestion' of basic needs: energy, water, food and crafts (Meyn 1998). Two cases of working-class community ecology struggles illustrate these dynamics: COPI (Piñones Integrates Corporation), in the eastern coastal town of Loiza, and IDEBAJO (Initiative of Eco-Development of Jobos Bay), in the southern coastal municipalities of Salinas, Guayama and Arroyo.

COPI was created in the 1990s with a focus on promoting the community's cultural, social and economic development, in the context of deep marginalisation. Piñones is a culturally and ecologically rich, predominantly black community, founded by maroon indigenous and African slaves. It has a long history of peasant-proletarian communal relations within its coastal ecosystems, and of struggles against exploitation and displacement, first by sugar plantations, and later, beginning in the 1960s, by a luxury residential-tourist 'developer' (Giusti-Cordero 1996). In the 1980s, the community gained legal recognition to the lands they had occupied for centuries. But in the mid-1990s, the threat of a mega-hotel and its 'coastal gentrification' resurfaced. The hotel threatened not only displacement, but also the destruction of the coastal forest, which protected the community from the sea's surges and other events associated to storms and hurricanes. COPI incorporated the protection of the forest as a central objective, seeing in this protection as both a necessity to protect the community's life, and an opportunity for socioeconomic self-management. With the help of a university architecture group, COPI prepared an alternative proposal, focused on an ecotourism project, a handicraft cooperative and the commercial production of coconut oil. In addition, it began training young people from the community to be employed as tour guides, and later assisted in the creation of two local ecotourism microbusinesses for bike and kayak tours. To coalesce these efforts, COPI reclaimed an abandoned public building to create the Piñones Cultural and Ecotourism Center, from where it has also advanced the revalorisation of the Afro-Caribbean history and culture of Piñones, through a school on the region's *bomba* music.

COPI's struggle attracted a wide range of allies, from environmentalists to leftist politicians, becoming one of the most visible conflicts in the country during the late 1990s to early 2000s. In 2007, COPI finally succeeded in getting the government to protect the threatened coastal lands. Since then, COPI has been developing eco-cultural activities and other environmental care labour, such as clean-ups of the area's mangrove canals, and reforestation. They have also addressed other needs in their community; for instance, they successfully lobbied the government to build new infrastructure for small vendors of the region's typical foods and artisan crafts. In September 2020, COPI started a 'Critical School for Social and Intellectual Justice of Piñones', which seeks 'to develop the different intelligences of young people in our community, to form their critical, reflective and community character' in the areas of visual arts, agriculture, music, and writing, emphasising that 'to defend our spaces, the precious value of the ecological jewel in which we live and our cultural heritage is our mission' (http://copipr.com/author's translation).

IDEBAJO is a coalition of eight organisations from predominantly working-class and predominantly black communities in Jobos Bay, created in the early 2000s. The region has been historically a 'sacrifice zone' within Puerto Rico (Thomas 2019). Today, the region hosts the highly contaminated abandoned petrochemical factories, four electricity-producing plants which generate 50% of the country's electricity, including the largest (oil-gas) plant in the country (Aguirre)—a systematic violator of emissions regulations—and the AES coal plant, which has generated widespread toxic coal ash deposits spread throughout the region. This is the epitome of what De Onis (2018a) calls Puerto Rico's 'energy colonialism' which 'marks certain places and peoples as disposable by importing and exporting logics and materials to dominate various energy forms, ranging from humans to hydrocarbons', impeding the realisation of energy justice and democracy. In addition, there are also a cement kiln, a tire incinerator, a regional landfill, several pharmaceutical companies and thousands of hectares of Monsanto-Bayern-Syngenta transgenic seed experimentation.[11] As a result, the South is one of the most contaminated and poorest regions in the country. IDEBAJO thus emerges from the long history of EJ struggles, first over the colonial plantations and then toxic projects of industrialisation, as well as from a rich experience of communities' organisation for cultural, educational and economic activities in the region. In this context, one of the main goals of IDEBAJO has been to support the construction of grassroots initiatives which integrate community livelihoods, ecological protection and social empowerment. One of their first projects was a

---

[11] Puerto Rico has the second largest per-area concentration of transgenic seed production in US territories, after Hawaii.

community ecotourism enterprise for which they recruited and trained local youth and fishermen to serve as guides.

In 2014, as a direct response to the threat of a proposed offshore methane gas megaport and pipeline (Aguirre Offshore)—which they ultimately defeated—IDEBAJO began developing a community solar project (*Coquí Solar*), not only as a way to provide an alternative source of energy but also to regenerate communal ties and provide employment to local youth. The project moved along slowly for the first few years, but the disaster following Hurricane Maria in 2017 led to a renewed emphasis on it, together with other projects aimed at addressing the community needs which became more pronounced after the disaster. These initiatives included a community kitchen that provided food and supplies after the hurricane, an urban food garden which also serves as a community enterprise, a community open-air market, and a solidarity housing reconstruction initiative (Building Solidarity) which buys materials and trains young people to build the houses of those who lost houses in the hurricane, with the slogan to build 'from mutual aid, love and dedication'. As with Coqui Solar, the house-building training is envisioned as a way to provide dignified work opportunities for the community's youth, most of whom are forced to leave the community (and often the country) in search of employment. These projects have roots in longer histories of mutual aid in the community. Between the 1950s and 1970s, the community organised the Mutual Aid and Self-Help project to address housing infrastructure, based on a process in which community members provided the materials and collective labour.

IDEBAJO explains that the main challenges to achieve broad mobilisation against these toxic industries have been unemployment and emigration, coupled with the job blackmail of residents who depend on 'working in what kills them' (Thomas 2019), the kind of challenge to which working-class environmentalism is poised to respond (Barca and Leonardi 2016, 2018). In this context, the organisation's initiatives are meant to contribute to *autogestion*—to 'build the social fabric and collective body' of the community which can 'manage, develop, structure, and administer' its problems and needs, while developing new sources of livelihoods to break this deadlock and 'liberate' themselves (Thomas 2019). In the colonial context, IDEBAJO's initiatives are also part of a longstanding struggle for liberation—hence why scholars and activists refer to this energy transition movement as an 'energy insurrection' for energy sovereignty (De Onis 2018b; Thomas 2019). Breaking free from the imperialist and toxic fossil cartel through *autogestion* creates the material freedom and demonstrates the possibility of a sovereign people and nation. This is not only a material struggle but also a decolonisation of discourses and imaginaries, a breaking of the 'colonial blackmail': the idea that we are too small, resource-poor and inept to govern and develop ourselves (Berman Santana 1996). Thus, we can see these initiatives as forms of decolonising from below, through actions that transform the socioecological dominant relations of colonial capitalism—shifting from extractive to regenerative economies—particularly on the basic needs of social reproduction: food, energy, water, housing (García-López 2020).

The advances in developing the incipient autogestion efforts, however, are faced with the persistent threat of (trans)national political-economic structures which seek to continue expanding capitalist accumulation through dispossession. In response, IDEBAJO has coupled autogestion with broader efforts for polycentric organising and national mobilisation. In 2018, in the context of government plans to privatise the public electric utility (PREPA) and to significantly expand methane gas energy production in the country, a member of IDEBAJO (Comité Diálogo Ambiental) spearheaded a network called *Queremos Sol* (We Want Sun) together with El Puente-ELAC (another historic EJ organisation), the Anti-Incineration Coalition (a grassroots EJ group that defeated a recently proposed waste incinerator), Sierra Club—Puerto Rico (the local chapter of the largest and oldest US environmental organisation), the PREPA workers' union (UTIER), the PREPA Managers' Association, and Cambio PR (an environmental advocacy think-tank). *Queremos Sol* developed a plan and campaign for a national renewable energy transition, centred on principles of energy as a common good and human right, strong citizens and workers' participation, managing energy demand, public ownership with maximised transparency and accountability, and a comprehensive audit and restructuring of the PREPA debt (Queremos Sol 2020). The participation of UTIER was key since historically, the union had remained silent on questions of environmental justice, and had resisted initiatives to foster renewable energy production, which they saw as part of a slow privatisation of PREPA.

Finally, together with the group Ayuda Legal Puerto Rico (Legal Aid PR), IDEBAJO has launched a national campaign for a 'just recovery', which demands that communities' rights in recovery processes, including the right to meaningful participation in government plans; access to decent housing and the right to remain in place after the hurricane. Through this advocacy, IDEBAJO is positioning its agenda of autogestion at the centre of national political discussions and influencing the use of the significant amount of disaster recovery funds, in a way that advances not just a grassroots-led just recovery but a transformation.

In the context of the deepening of Puerto Rico's colonial crisis, coupled with the global climate crisis, there are indications that the networks of working-class environmental justice are continuing to expand and solidify from the bottom-up. In the aftermath of Hurricane Maria, many 'mutual aid' groups emerged in working-class rural and urban ('peasant-proletarian') communities across the country, to provide immediate food and aid. After the emergency months, many of these groups have developed *autogestion* projects. Agroecology movements with their support networks, including the US Climate Justice Alliance and the Black Dirt Farm Collective, mobilised labour in common through farm-to-farm 'just recovery' brigades, which helped rebuild crops and infrastructure, and organised political trainings and convivial activities (CJA 2019; McCune et al. 2018). Meanwhile, the movement against the AES coal plant in Guayama, which had been gaining ground, became stronger, and was one of the precursive forces to the massive summer 2019 mobilisations that ousted the governor for the first time in history ). In October 2019, a coalition of environmental, agroecology and human rights organisations organised a public forum on 'Just Recovery in the Climate Crisis', and afterwards met in Guayama, to discuss joint strategies for advancing the integration of the struggles for a just recovery with the struggles for a community-led energy transition, food sovereignty and mutual aid, linking to emerging debates of a 'Green New Deal'. The vision advanced by these groups is one where the objective is not recovery of the same system that caused the destruction, but, rather, to have systemic change, led by communities working from their territories, engaged in both resistance and generating grassroots solutions (CJA 2019; García-López 2020). This has resulted in ongoing planning of the first just transition campaign in the country, which will be launched in spring 2021, bringing together these multiple labour-environmental struggles.

## Conclusion: Towards a Decolonial-Commons Perspective on Working-Class Community Ecologies and Just Transitions

In this chapter, I have argued that a commons lens can enrich environmental labour studies and its understanding of working-class environmental (justice) struggles. The commons (as shared resources/wealth, as relations/practices, as ideal/paradigm) help us to rethink the capitalist structures of privatisation, extractivism and accumulation that disproportionally affect people across class, racial, gender and North-South categories of difference. Commons counter these with collective sharing, equality, regeneration and care, within the rights and limits of our common Earth. They invite studying just transitions as commons transitions, based not only on a shift to equitable and sustainable mode of (waged, exchange-value) production, but also on a shift towards a mode of collective *reproduction* (non-waged, use/subsistence value) of shared elements: land, water, energy, housing, food. Grounded in the context of Puerto Rico, I have suggested that the dynamics of this particular variety of labour-environmentalism—led by frontline EJ organisations and other 'reproductive' workers—are focused on the struggles to re-claim commons from capitalist-colonialist enclosures. These grassroots movements do not just resist but enact alternatives through commoning—building a (reproductive) labour commons—to generate an abundance of life through sharing and care for their socio-ecologies/territories, *within-against-and-beyond the state*. Moreover, in the context of (neo)colonial structures, these movements and their initiatives are also a way to build emancipatory decolonial horizons, *within, against and beyond the colony*.

The commons framework suggests several paths for future work in environmental labour studies. Conceptually, environmental labour studies could benefit from engaging more centrally with existing discussions on reproductive labour, reproductive commons, labour commons, nature's labour, *autogestion*, 'co-becoming' and being-in-common, commons movements and commons transition. Geographically it suggests that there is a need to pay more attention to other sites beyond the workplace (i.e. the territory, the 'working-class community ecology'), and to Global South and South-of-the-North regions. It also points to at least three possible empirical research foci related to these new conceptual and geographical sites. First, a focus on the labour-environment dynamics, worker-/community-led cooperatives and other commons organisations in renewable energy, agriculture, forestry, fisheries, housing, eco-development and other issues interrelated with just transitions and 'Frontline Green New Deals'. Second, attention to how new, post-capitalist, eco-social working-class subjectivities emerge and are nurtured, through workers' engagement in these commoning processes. Finally, analysis of the articulating (discursive, relational, material) elements which unite popular 'worker-environmental' movements, beyond the promise of 'green jobs' that holds together the current blue-green (trade union-environmentalist) alliances.

**Acknowledgements** This work was supported by the Stimulus Program for Scientific Endeavors of the Portuguese Foundation for Science and Technology (FCT), under contract CEECIND / 04850/2017 / CP1402 / CT0010.

# References

Adrar, Angela, Olivia Burlingame, Anthony Rogers-White, and Fernando Tormos. 2019. Green New Deal Policies Should Be Fueled by Frontline and Grassroots Power. *Public Administration Review*. http://www.publicadministrationreview.com/2019/07/16/gnd16/.

Akbulut, Bengi, Federico Demaria, Julian-Francois Gerber, and Joan Martínez-Alier. 2019. Who Promotes Sustainability? Five Theses on the Relationships Between the Degrowth and the Environmental Justice Movements. *Ecological Economics* 165: 2–9.

Azzellini, Dario. 2018. Labour as a Commons: The Example of Worker-Recuperated Companies. *Critical Sociology* 44 (4–5): 763–776.

Barca, Stefania. 2014. Laboring the Earth: Transnational Reflections on the Environmental History of Work. *Environmental History* 19 (1): 3–27.

Barca, Stefania, and Emanuele Leonardi. 2016. Working-Class Communities and Ecology. Reframing Environmental Justice Around the Ilva Steel PLANT in Taranto, Apulia (Italy). In *Class, Inequality and Community Development*, ed. Mae Shaw and Marjorie Mayo, 59–76. Chicago, IL: Policy Press.

———. 2018. Working-Class Ecology and Union Politics: A Conceptual Topology. *Globalizations* 15 (4): 487–503.

Benjamin, Craig S., and Terissa E. Turner. 1992. Counterplanning from the Commons: Labour, Capital and the 'New Social Movements'. *Labour, Capital and Society* 25 (2): 218–248.

Berman Santana, Deborah. 1996. *Kicking Off the Bootstraps: Environment, Development and Community Power in Puerto Rico*. Tucson: University of Arizona Press.

Bollier, David, and Silke Helfrich. 2015. *Patterns of Commoning*. Commons Strategy Group and Off the Common Press.

Bresnihan, Patrick. 2016. The More than Human Commons: From Commons to Commoning. In *Space, Power and the Commons: The Struggle for Alternative Futures*, ed. Samuel Kirwan, Leila Dawney, and Julian Brigstockem, 105–124. New York: Routledge.

Caffentzis, George, and Silvia Federici. 2014. Commons Against and Beyond Capitalism. *Community Development Journal* 49 (S1): i92–i105.

Catalina M. de Onís. 2018. Energy Colonialism Powers the Ongoing Unnatural Disaster in Puerto Rico. Frontiers in Communication 3.

Chatterton, Paul, David Featherstone, and Paul Routledge. 2013. Articulating Climate Justice in Copenhagen: Antagonism, the Commons, and Solidarity. *Antipode* 45 (3): 602–620.

CJA – Climate Justice Alliance. 2019. *Our Power PR: Moving Towards a Just Recovery*. https://climatejusticealliance.org/our-power-puerto-rico-report/. Accessed 17 July 2019.

———. 2020. CJA and the Green New Deal: Centering Frontline Communities in the Just Transition. https://climatejusticealliance.org/gnd/. Accessed 15 July 2020.

Cumbers, Andrew. 2015. Constructing a Global Commons In, Against and Beyond the State. *Space and Polity* 19 (1): 62–75.

De Angelis, Massimo. 2007. *The Beginning of History: Value Struggles and Global Capital*. London: Pluto Press.

———. 2014. Social Revolution and the Commons. *South Atlantic Quarterly* 113 (2): 299–311.

———. 2017. *Omnia Sunt Communia: On the Commons and the Transformation to Postcapitalism*. Chicago: ZED Books.

De Onís, Catalina M. 2018a. Fueling and Delinking from Energy Coloniality in Puerto Rico. *Journal of Applied Communication Research* 46 (5): 535–560.

———. 2018b. Energy Colonialism Powers the Ongoing Unnatural Disaster in Puerto Rico. Frontiers in Communication 3.

De Peuter, Greig, and Nick Dyer-Witheford. 2010. Commons and Cooperatives. *Affinities*. https://ojs.library.queensu.ca/index.php/affinities/article/view/6147.

Escobar, Arturo. 2014. *Sentipensar Con La Tierra: Nuevas Lecturas Sobre Desarrollo, Territorio y Diferencia*. Medellin: Ediciones Unaula.

Esteva, Gustavo. 2014. Commoning in the New Society. *Community Development Journal* 49 (suppl_1): i144–i159.

Esteves, Ana M. 2020. Solidarity Economy Markets as 'Mobilizational Commons': Re-Signifying the Market Through the Lens of Cooperation. *Community Development Journal*. https://doi.org/10.1093/cdj/bsaa008.

Federici, Silvia. 2012. *Revolution at Point Zero: Housework, Reproduction, and Feminist Struggle*. PM Press.

Fox, Diana J., and Jillian M. Smith. 2016. Stewards of Their Island: Rastafari Women's Activism for the Forests and Waters in Trinidad and Tobago. *Resilience* 3: 142–168.

García-López, Gustavo A. 2020. Environmental Justice Movements in Puerto Rico: Life-and-Death Struggles and Decolonizing Horizons. *Society & Space Magazine*, February 25. https://www.societyandspace.org/articles/environmental-justice-movements-in-puerto-rico-life-and-death-struggles-and-decolonizing-horizons. Accessed 10 March 2020.

García-López, Gustavo A., Irina Velicu, and Giacomo D'Alisa. 2017. Performing Counter-Hegemonic Common(s) Senses: Rearticulating Democracy, Community and Forests in Puerto Rico. *Capitalism Nature Socialism* 28 (3): 88–107.

García-Quijano, Carlos, and Hilda Lloréns. 2017. What Rural, Coastal Puerto Ricans Can Teach Us About Thriving in Times of Crisis. *The Conversation*, May 31. https://theconversation.com/what-rural-coastal-puerto-ricans-can-teach-us-about-thriving-in-times-of-crisis-76119. Accessed 30 October 2020.

Giacomini, Terran, and Terisa Turner. 2015. The 2014 People's Climate March and Flood Wall Street Civil Disobedience: Making the Transition to a Post-fossil Capitalist, Commoning Civilization. *Capitalism Nature Socialism* 26 (2): 27–45.

Giacomini, Terran, Terisa Turner, Ana Isla, and Leigh Brownhill. 2018. Ecofeminism Against Capitalism and for the Commons. *Capitalism Nature Socialism* 29 (1): 1–6.

Gibson-Graham, Julie-Katharine, Jenny Cameron, and Stephen Healy. 2016. Commoning as a Postcapitalist Politics. In *Releasing the Commons*, ed. Ash Amin and Phillip Powell, 192–212. New York/London: Routledge.

Giusti-Cordero, Juan A. 1996. Labour, Ecology and History in a Puerto Rican Plantation Region: 'Classic' Rural Proletarians Revisited. *International Review of Social History* 41 (S4): 53–82.

Hardin, Garrett. 1968. The Tragedy of the Commons. *Science* 162 (3859): 1243-1248.

Hardt, Michael. 2010. The Common in Communism. *Rethinking Marxism* 22 (3): 346–356.

Hardt, Michael, and Antonio Negri. 2009. *Commonwealth*. Cambridge, MA: Harvard University Press.

Helfrich, Silke, and David Bollier. 2015. Commons. In *Degrowth: A Vocabulary for a New Era*, ed. G. D'Alisa, F. Demaria, and G. Kallis, 75–78. London: Routledge.

Huber, Matt. 2019. Climate Change Is Class Struggle. *Jacobin*, December 19. https://www.jacobinmag.com/2019/12/on-fire-naomi-klein-review-climate-change.

JTRC – Just Transitions Research Collaborative. 2018. *Mapping Just Transition(s) to a Low-Carbon World*. United Nations Research Institute for Social Development (UNRISD).

Klein, Naomi. 2001. Reclaiming the Commons. *New Left Review*, 9, May/June. https://newleftreview.org/issues/ii9/articles/naomi-klein-reclaiming-the-commons.

Linebaugh, Peter. 2008. *The Magna Carta Manifesto: Commons and Liberties for All*. Berkeley: University of California Press.

Mascarenhas-Swan, Michelle. 2017. The Case for a Just Transition. In *Energy Democracy: Advancing Equity in Clean Energy Solutions*, ed. Denise Fairchild and Al Weinrub, 37–56. Washington, DC: Island Press.

Mastini, Ricardo. 2020. A Post-Growth Green New Deal. *Uneven Earth* (blog), February 17. http://unevenearth.org/2020/02/a-post-growth-green-new-deal/. Accessed 15 July 2020.

Mattei, Ugo. 2013. Protecting the Commons: Water, Culture, and Nature: The Commons Movement in the Italian Struggle Against Neoliberal Governance. *South Atlantic Quarterly* 112 (2): 366–376.

McCaffrey, Katherine. 2006. Social Struggle Against the US Navy in Vieques, Puerto Rico: Two Movements in History. *Latin American Perspectives 33* (1): 83–101.

McCune, Nils, Ivette Perfecto, John Vandermeer, and Katia Avilés-Vázquez. 2018. Disaster Colonialism and Agroecological Brigades in Post-Disaster Puerto Rico. *ERPI 2018 International Conference*, The Hague, Netherlands.

Meyn, Marianne. 1998. En busca de una sociedad ecológica [In Search of an Ecological Society]. *El Semillero*. Misión Industrial. https://drive.google.com/file/d/1P4oFAKMnUygFPZFM5JtuPdkNYPIl_mj2/view. Accessed 22 December 2019.

———. 2017. Acerca de Misión Industrial de Puerto Rico [About Mision Industrial]. In *Misión Industrial de Puerto Rico*, ed. Tania d.M. López. https://misionpr.weebly.com/queacute-es-mipr.html. Accessed 22 December 2019.

Miller, Marian. 2001. Tragedy for the Commons: The Enclosure and Commodification of Knowledge. In *The International Political Economy of the Environment: Critical Perspectives*, ed. Dimitris Stevis and Valerie Assetto, 111–134. Lynne Rienner Publishers.

Ostrom, Elinor. 1990. *GOVERNING THE COMMONS The Evolution of Institutions for Collective Action*. Cambridge: Cambridge University Press.

Papadimitropoulos, Vangelis. 2017. From the Crisis of Democracy to the Commons. *Socialism and Democracy* 31 (3): 110–122.

Paudel, Dinesh. 2016. Re-inventing the Commons: Community Forestry as Accumulation without Dispossession in Nepal. *The Journal of Peasant Studies* 43 (5): 989-1009.

Pelenc, J., G. Wallenborn, J. Milanesi, L. Sébastien, J. Vastenaekels, F. Lajarthe, et al. 2019. Alternative and Resistance Movements: The Two Faces of Sustainability Transformations? *Ecological Economics* 159: 373–378.

Perkins, Patricia E. 2019. Commoning and Climate Justice. In *Making the Commons Dynamic*, ed. P. Nayak. London/New York: Routledge.

Queremos Sol. 2020. *We Want Sun: Sustainable. Local. Clean*. https://www.queremossolpr.com/project-4. Accessed 30 November 2019.

Red Nation. 2020. *The Red Deal. Indigenous Action to Save Our Earth*. http://therednation.org/wp-content/uploads/2020/04/Red-Deal_Part-I_End-The-Occupation-1.pdf. Accessed 10 July 2020.

Routledge, Paul, Andrew Cumbers, and Kate D. Derickson. 2018. States of Just Transition: Realising Climate Justice Through and Against the State. *Geoforum* 88: 78–86.

Sauvêtre, Pierre. 2018. Forget Ostrom: From the Development Commons to the Common as Social Sovereignty. In *The Commons and a New Global Governance*, ed. Samuel Cogolati and Jan Wouters. Edward Elgar Publishing.

Singh, Neera. 2013. The Affective Labour of Growing Forests and the Becoming of Environmental Subjects: Rethinking Environmentality in Odisha, India. *Geoforum* 47: 189–198.

———. 2017. Becoming a Commoner: The Commons as Sites for Affective Socio-Nature Encounters and Co-Becomings. *Ephemera: Theory & Politics in Organisation* 17 (4).

Stevis, Dimitri, David Uzzell, and Nora Räthzel. 2018. The Labour–Nature Relationship: Varieties of Labour Environmentalism. *Globalizations* 15 (4): 439–453.

Svampa, Maristella, and Enrique Viale. 2020. A View of the Green New Deal from Argentina. *Jacobin*, June 17. https://jacobinmag.com/2020/06/green-new-deal-argentina-gran-pacto. Accessed 19 June 2020.

Thomas, Roberto. 2019. Ending Colonialism. In *Voices from Puerto Rico: Post-Hurricane Maria*, ed. Iris Morales, 107–112. New York: Red Sugarcane Press.

Tola, Miriam. 2015. Commoning With/in the Earth: Hardt, Negri and Feminist Natures. *Theory& Event* 18 (4) *Project MUSE*. muse.jhu.edu/article/595841.

Tormos-Aponte, Fernando, and Gustavo A. García-López. 2018. Polycentric Struggles: The Experience of the Global Climate Justice Movement. *Environmental Policy and Governance 28* (4): 284–294.

Turner, Terisa, and Leigh Brownhill. 2004. We Want Our Land Back: Gendered Class Analysis, the Second Contradiction of Capitalism and Social Movement Theory. *Capitalism Nature Socialism 15* (4): 21–40.

Valle, Gabriel. 2020. Learning to be Human Again: Being and Becoming in the Home Garden Commons. *Environment and Planning E: Nature and Space*. https://doi.org/10.1177/2514848620961943.

Velicu, Irina. 2019. De-Growing Environmental Justice: Reflections from Anti-Mining Movements in Eastern Europe. *Ecological Economics* 159: 271–278.

Velicu, Irina, and Stefania Barca. 2020. The Just Transition and Its Work of Inequality. *Sustainability: Science, Practice and Policy* 16 (1): 263–273.

Vercellone, Carlo. 2015. From the Crisis to the 'Welfare of the Common' as a New Mode of Production. *Theory, Culture & Society* 32 (7–8): 85–99.

Vieta, Marcelo. 2016. Autogestión: Prefiguring a 'New Cooperativism' and the 'Labour Commons'. In *Moving Beyond Capitalism*, ed. Cliff DuRand, 77–85. London: Routledge.

Woelfle-Erskine, Cleo. 2019. Beavers as Commoners? Invitations to River Restoration Work in a Beavery Mode. *Community Development Journal* 54 (1): 100–118.

# 18

# Agroecological Farmer Movements and Advocacy Coalitions in Sub-Saharan Africa: Between De-Politicization and Re-Politicization

Patrick Bottazzi and Sébastien Boillat

## Introduction: Agroecology and Farmer Movements

Agroecology is increasingly presented as a promising solution to support food security, restore ecosystems and build resilience to climate change (Food and Agriculture Organization of the United Nations (FAO) 2019). Agroecology primarily designates the application of scientific ecology in agriculture; nevertheless, since the 1990s the concept has been appropriated by environmentalist movements, in particular farmer and peasant environmental movements (Wezel et al. 2009). While these movements have started mainly in Latin America (Holt-Giménez 2006; Altieri and Toledo 2011), they have now spread across the globe, particularly to Europe, South Asia and sub-Saharan Africa (SSA). Agroecology is now explicitly promoted by *La* Via *Campesina*, the largest international farmer organization (Rosset and Martínez-Torres 2012). On its website, *La* Via *Campesina* features agroecology as a solution to reach many goals such as ensuring food sovereignty, fighting climate change, empowering small farmers, promoting gender equality and resisting capitalism.

---

P. Bottazzi (✉) • S. Boillat
Institute of Geography, University of Bern, Bern, Switzerland

Centre for Development and Environment, University of Bern, Bern, Switzerland
e-mail: patrick.bottazzi@cde.unibe.ch

Following and interacting with these farmer movements, a growing number of scholars promoting agroecology are advocating for deep socio-economic and political reforms in the agri-food sector (Altieri and Toledo 2011; Gliessman 2013). However, few of their proposals have gone beyond the farm or small community level, and, in academia, a large part of the agroecology community still concentrates on technological rather than socio-political solutions to make agriculture sustainable (de Molina 2013). Furthermore, agroecological farmer movements have been shown to be strongly constrained by their historic constitution process (Meek 2016) and co-optation by governments or powerful private actors (Schiller et al. 2020). In this context, agroecology can be mobilized as a politicizing and emancipatory discourse and programme, but it can also lead to a technicist vision of agriculture that would end up de-politicizing and weakening the movement.

In this chapter, we examine the agroecological farmer movement from a labour environmentalist perspective. We show that this perspective can shed light on the potential emancipatory power of agroecology and its limitations and constraints. As stated in the [*book introduction*], environmental labour studies (ELS) focus on the agency of labour unions to jointly advance social justice and environmental protection agendas. ELS includes 'research on the environmental engagements of workers' organizations like trade unions, but also organizations of small farmers or fishers, like the Via Campesina' (Stevis et al. 2018). Small farmer movements are thus an important domain of ELS. They can be considered a constitutive element of labour environmentalism, as they put an emphasis on workers' intertwined social and environmental claims.

We focus on the context of agroecology and farmer movements in sub-Saharan Africa (SSA), which is emblematic of the blend between technology-oriented development and social emancipation. The region concentrates most of the undernourished world population (Food and Agriculture Organization of the United Nations (FAO) 2015) and faces dramatic food insecurity, unemployment and massive rural to urban migration as well as the consequences of climate change characterized by severe drought and soil degradation. Facing these challenges, agroecology is increasingly presented as a promising solution to reconcile food sovereignty with adaptation to climate change in the region (Food and Agriculture Organization of the United Nations (FAO) 2016). A growing number of farmers across the continent have experimented and adopted agroecological practices with success in fighting hunger, poverty and climate change (Mousseau 2015) and research on the agronomic potential of agroecology in sub-Saharan Africa is flourishing (Giller et al. 2009; Tittonell et al. 2012).

Despite these booming prospects, few academic works have studied these initiatives (Mousseau 2015) and little is known yet on the socio-political aspects of existing agroecological movements in the region. Furthermore, most African governments strongly adhere to a green revolution narrative and favour agricultural development based on industrialization, neo-liberal policies and large-scale agricultural investments (Anseeuw et al. 2012; Edelman et al. 2013; Borras Jr and Franco 2012). Farmer movements have risen to resist these processes and some of them have adopted agroecology as an umbrella concept to defend multiple social and environmental aspects of food production including food sovereignty, adaptation to climate change and farmer autonomy. Despite this, many farmer movements remain strongly dependent on foreign NGO support and face constant weakening due to strong state coercion and co-optation (Hrabanski 2010). The co-existence of dynamics of depoliticization with the emancipatory prospects of agroecology is thus particularly relevant in this context.

This chapter thus focuses on agroecological farmer movements in sub-Saharan Africa, including the agroecological transition movement in Senegal as a case study. The research is based on a set of 35 interviews carried out in Senegal between October 2018 and November 2019, with governmental agencies, NGOs and farmers' organizations, as well as participant observation at advocacy coalition meetings. We enquire about the main actors, their actions and their collaborations with broader transformative coalitions, including the role of international and national organizations who support farmers' actions. In particular, we seek to understand how the appropriation of the concept of agroecology contributes to the joint agenda of social justice and environmental protection.

To do this, we investigate the main repertoires of collective action mobilized by the actors of the agroecological movement in our area of study. Repertoires of collective action build on the concept of 'repertoires of contention' formulated by (Tilly 2010), which he defines as claim-making routines that build on specific performances by claimant actors towards objects of claims. The concept of repertoires of collective action has been widely used to describe recurrent clusters of tactics used by social movements such as general strikes, in different times and places (Johnson 2008). We thus seek to understand how the agroecological narrative has become part of the repertoire of collective action of farmers' organizations and to what extent it structures their capacity to support the resistance of the most vulnerable farmers to environmental pressures and hegemonic threats.

## Agroecology as a Form of Labour Environmentalism

Agroecology initially emerged as the combination of the scientific disciplines of agronomy and ecology aimed at understanding ecological interactions in agricultural systems and the impacts of agriculture on ecosystems (Wezel et al. 2009). The first mobilization of the concept by peasant organizations and engaged scientists started in the 1990s in Latin America (Brazil, Cuba, Central America, Mexico) with the defence of food sovereignty, environmental protection and grassroots empowerment (Altieri and Toledo 2011). Peasants, researchers and civil society organizations promoting agroecology engaged in 'counter-hegemonic' discourse and actions (Figueroa-Helland et al. 2018). On the one hand, these organizations build on modern agronomic science through holistic approaches based on the valuation and development of agroecological specific knowledge, the reskilling of farmers and the development of alternative technologies and market systems (Coolsaet 2016). On the other hand, the movement also seeks to defend and enhance the rights of small-scale farmers and has led to the creation of multiple regional, national and international organizations leading advocacy and peasant resistance actions. These movements eventually contributed to the constitution of *La via Campesina* in 1993, now one of the largest transnational social movements in the world (Rosset and Martinez-Torres 2013).

Building on these developments, agroecology is now understood as encompassing three dimensions: a) a set of ecologically sound agricultural practices (e.g. reducing chemical inputs, preserving soil and water, diversifying crops, trees and livestock production); b) a scientific approach integrating complex ecological processes with agronomy; c) and a social movement combining social and environmental issues (Wezel et al. 2009). Agroecology is increasingly used as an umbrella concept by researchers, policy makers and activists to associate socio-economic equity principles with ecologically sustainable farming practices. It helps positioning the discourse on agricultural development towards a desired situation mainly in favour of most vulnerable producers, consumers and the environment. Concepts such as 'democratic governance', 'market access and autonomy', 'environmental equity' or 'social equity' have now become common characteristics of agroecology (Dumont et al. 2016).

As with other environmental movements, agroecology mobilizes a variety of strategies including discourses and practices to acquire resources, transform behaviour and influence political agendas (Dalton et al. 2003). These discourses envision several overlapping mechanisms of emancipation that build on social justice and environmental protection agendas. First, it has also been suggested that agroecology could contribute to improve the working conditions of farmers (Dupré et al. 2017; Dumont and Baret 2017; Jansen 2000). Timmermann and Felix (Timmermann and Félix 2015) suggest that agroecology could be a 'vehicle of contributive justice' which is understood as the right to meaningful work enhancing farmers' capabilities and leading to their progressive emancipation from global agrarian capitalism. Second, agroecology would also contribute to 'cognitive justice', namely to allow a better recognition of different forms of knowledge including situational and environment-specific knowledge held by farmers or co-produced by farmers and scientists (Coolsaet 2016).

Third, agroecology strives for an economic emancipation from commodified and globalized markets through shorter supply chains, more direct connections between producers and consumers, and the reduction of industrial input use (Holt-Giménez and Altieri 2013). It therefore also contributes to the emancipation of consumers by establishing healthier and more just food systems (Gliessman 2016). This produces some parallels with the degrowth movement, to which agroecology is thought to make a substantial contribution (Boillat et al. 2012; Ruiz López 2018). Fourth, agroecology is also meant to strengthen farmer movements in the first place, making them able to fill claims to ensure access to land, water and a clean environment for farmers, especially smaller producers. Agroecology would thus contribute to 'resilience justice', understood as the ensuring of meaningful social-ecological connexions for agroecological production (Boillat and Bottazzi 2020). Such claims can be linked with territorial claim-based approaches to agroecology, which draw on landscape ecology to emphasize the important connections between ecological zones and thus legitimate territorial land, water and other natural resource access rights raised by groups of farmers at sub-regional levels (McCune 2017; McCune et al. 2017).

All these discourses and strategies taken together underline the importance of agroecology as part of a broader political project linked to a holistic vision of agriculture and farmers' societies (Giraldo and Rosset 2018). What has been considered as a 'way of life', as often argued by agroecological defenders (Rosset and Martínez-Torres 2012), is a particular way to frame the relationship between social and ecological systems in a holistic perspective.

Nevertheless, empirical research has depicted a more complex and controversial picture of agroecological farmers' emancipation processes. In terms of labour, while most agroecological practices are healthier (for workers as well as for consumers) and can offer longer-term production (e.g. through soil conservation), they require a heavier workload; small agroecological farmers may 'self-exploit' themselves and accept painful activities, long working days and wages lower than the regional average (Galt 2013). Furthermore, the discourse of pluralistic knowledge common in agroecology may mask a more technocratic framing in practice, especially in the Global South, where valuing and acknowledging farmers' knowledge and skills remains a challenge (Waldmueller 2015). Another risk is to reduce agroecology to a simple substitution of chemical inputs with certified organic ones (Rosset and Altieri 1997). As for 'agriculture modernization', the input substitution narrative reduces the complexity of agroecosystems and eludes most social and political aspects. Agroecology can be co-opted by the commercialization of organic products and serve the interests of a new marketing discourse (Giraldo and Rosset 2018), thus failing to enable economic emancipation of producers and consumers. Finally, the dependency of farmer organization on external support, especially international funding, may limit the scope of their claims especially regarding counter-hegemonic discourse and actions (Isgren and Ness 2017).

## Agroecology in Sub-Saharan Africa

The challenges discussed above are particularly acute in the SSA region and have been observed through several studies. In SSA the competition between green revolution techniques and traditional agroecological practices places small-scale farmers into a situation of asymmetrical and unequal competition with conventional producers and can lead to higher poverty if no supporting measures are taken (Dawson et al. 2016). In terms of labour, agroecological farmers must usually accept a lower margin of benefits, lower yields and, sometimes, additional pressure to respond to local demand (Dahlin and Rusinamhodzi 2019; Dupré et al. 2017). In the region, there is also a tendency of focusing on strictly agronomic aspects, with international organizations and NGO focusing on proving the technical feasibility of agroecology; this translates into technical staff applying top-down technical protocols with farmers in the field (Bottazzi et al. 2020). A recent comprehensive account of organic certification schemes in SSA has also demonstrated their weak capacity to reach vulnerable workers in terms of income protection, safety norms and labour conditions (Oya et al. 2018).

Furthermore, advocacy for agroecology has had little influence on public policy in the region. Most agricultural policies in SSA strive to achieve the 'green revolution' and implement neo-liberal measures in favour of industry

# 18 Agroecological Farmer Movements and Advocacy Coalitions…

and export (Martín et al. 2019; Koopman 2012). Recent research suggests that 'successful implementation of a reformed and supportive policy context depends on societal debates and social movements that apply pressure to governments and institutions' (Eyhorn et al. 2019). It is only recently that international organizations such as the FAO are showing interest in supporting sustainable change in SSA and have employed civil society consultation forums to promote it in the region (Food and Agriculture Organization of the United Nations (FAO) 2016). Agroecology in the region is also strongly shaped by international funding brought by IGOs and NGOs and raises the question of the autonomy of farmer organizations in this context. A case study in Uganda (Isgren and Ness 2017) shows that the agroecology transition agenda is mainly pushed by NGOs, themselves targets of clear pressures from national governments, facing threats such as deregistration, harassment or arrest. They thus try to remain 'neutral' in what regards issues of social justice and farmers' organization to avoid being perceived too political or 'confrontational'. This makes it difficult to up-scale their actions resulting in NGO-leaded agroecological initiatives remaining limited to 'smallholder-centric' through the promotion of agricultural systems at the plot level, using few capital inputs and avoiding political debates (Isgren and Ness 2017).

## Senegal: Recent Farmer Movements, Productivist Policies and an Agroecology Led by NGOs

In Senegal, the emergence of farmer movements has historically been hindered because of the strong orientation of agriculture around commodity export until the 1970s (Fig. 18.1). Back then, colonial and post-colonial companies directly collected cash crops from family farms reassembled in

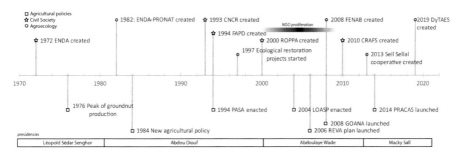

**Fig. 18.1** Timeline of the main milestones in the Senegalese agricultural policies and farmer movements since 1970

cooperatives who were co-opted by traditional and political elites (Ela 1990). These cooperatives were considerably weakened by the decline of groundnut economy and the enactment of the New Agricultural Policy in 1984, issued within the structural adjustment programme, which led to the complete disengagement of the state and the liberalization of the food sector (Duruflé 1995). Farmer unions emerged in this context and were strongly backed up by non-governmental organizations (NGOs) who filled the void left by the disengagement of the State in rural areas (Hrabanski 2010). The current largest and most influential peasant organization, the *Comité national de concertation des ruraux du Sénégal* (CNCR), was created in 1993 during the negotiation process that led to the enactment of the structural adjustment plan for the agricultural sector (PASA) in 1994 (Oya and Ba 2013). The CNCR has then played an important role in negotiations with the State to adopt the 'agro-sylvo-pastoral' law in 2004, which recognizes family-based agriculture and sets development objectives according to the three pillars of sustainable development (Brayer et al. 2008). CNCR is among the founders of the Network of Peasant Organisations of West African (ROPPA) in 2000 and is affiliated to *La Via Campesina* (Pesche 2009). However, the organization has suffered weakening during President Wade's second mandate from 2007 to 2012, due to state coercion and political co-optation (Hrabanski 2010).

The contemporary Senegalese agricultural policy has also been strongly centred on 'special programmes' starting with the Abdoulaye Wade presidency from 2000 onwards. These programmes include the *General offensive pour la nourriture et l'abondance* (GOANA) plan and the *Retour vers l'agriculture* plan (REVA), which encourage industrial agriculture and foreign investments. Current President Macky Sall, elected in 2012, has given continuity to these plans with the 'Programme of agriculture acceleration' (PRACAS). The PRACAS represents the agricultural sector within a larger development plan, the *Plan Sénégal Emérgent* (PSE) (emerging Senegal plan) which aims at achieving 'emerging economy' status by 2035 for the country. These programmes have focused on increasing the national production of commercial crops with little understanding of the social and ecological difficulties that the rural world was encountering (Oya and Ba 2013). They favour conventional agriculture through subsidizing chemical inputs, mechanization and facilitation of land and water access to large-scale investors. Attracting investors has also increased land tenure insecurity for smallholders, since according to the National Domain Law of 1964 most rural land belongs to the state, making expropriation of smallholders and concessions to large investors easier. Since the 1990s, several cases of large-scale land acquisitions, mainly in the Senegal River valley, have given rise to community upsurge and serious conflicts with large companies (Koopman 2012).

In Senegal, the pioneering organization in raising environmental awareness is Environment Development Action in the Third World (ENDA Tiers Monde), an international NGO created in 1972 (Table 18.1). Their environmental branch ENDA-PRONAT created in 1982 started to promote organic agriculture in the Niayes area, the most productive horticultural area near the capital city of Dakar, after they performed a study on the risks related to pesticide use in a context of low education and poverty (German and Thiam 1993). The NGO strongly supported the creation of farmer organizations to implement their projects. Among them, the Federation of Diender Agropastoralists (FAPD), officially founded in 1994, is one of the very first farmer organizations to promote agroecology among its members and advocate for its adoption at local and national levels. It now claims 3000 members operating in the Niayes region and is currently supported by various organizations such as the Swiss NGO HEKS, the Canadian cooperation and the FAO. The FAPD has the objective of establishing a 'sustainable and healthy agriculture' in the Niayes region for 'a better food security' (Ndoye Diop et al. 2019). Similar organizations have emerged in the country with the proliferation of NGOs promoting agroecology after 2000. These organizations have started to create larger structures to advocate for the adoption of agroecology at the national level. This includes the Federation of Organic Producers (FENAB), supported by six NGOs, most with Swiss and German funding. The FENAB includes 44 farmer unions and has the objective of establishing national standards for organic agriculture. In 2019, ENDA led the constitution of an advocacy coalition, the Dynamic for an Agricultural Transition in Senegal (DyTAES), to lobby the government for the adoption of policies favouring agroecology, gathering more than 100 member organizations including NGOs, famer unions and local government representatives from rural areas.

## Farmer Organizations: Repertoires of Collective Action

In the following section, we detail the repertoires of contention used by farmer movements promoting agroecology in sub-Saharan Africa with examples based on our empirical study in Senegal. We identify seven main core repertories of action carried out and promoted by the organizations studied.

Table 18.1 Main organizations cited and their characteristics

| Acronym | Organization (Original name) | Organization (English translation) | Founded | Scope | Membership type |
|---|---|---|---|---|---|
| *Farmer and producer's organizations* | | | | | |
| FAPD | Fédération des Agropasteurs de Diender | Federation of Diender Agropastoralists | 1982 (1994 official) | Local | Individual farmers |
| CNCR | Comité National de Concertation des Ruraux du Sénégal | National Consultation Committee of Rural people of Senegal | 1993 | National | Local farmer organizations |
| ROPPA | Réseau des Organisations Paysannes et des Producteurs Agricoles de l'Afrique de l'Ouest | Network of Peasant Organisations of West Africa | 2000 | Regional | National organizations |
| FENAB | Fédération Nationale pour l'Agriculture Biologique | National Federation for Organic Agriculture | 2008 | National | Local farmer organizations |
| IFOAM | International Federation of Organic Agriculture Movements | | 1972 | International | Local, private and national organizations |
| *Platforms* | | | | | |
| CRAFS | Cadre de Réflexion et d'Action sur le Foncier au Sénégal | Thinking and action environment on land governance in Senegal | 2010 | National | Local and national organizations |
| DyTAES | Dynamique pour la Transition Agroécologique au Sénégal | Dynamic for the agroecological transition in Senegal | 2019 | National | Local and national organizations |
| *Non-governmental organizations (NGOs)* | | | | | |
| ENDA-PRONAT | Environment Development Action—Protection Naturelle des Terroirs | Environment Development Action—Natural Protection of Land | 1972 (ENDA), 1982 (E-PRONAT) | National | |
| HEKS | Hilfswerk der Evangelischen Kirchen Schweiz | Swiss Protestant Churches Aid Agency | 1946 (2006 in Senegal) | International | |

## 'Agroecology-by-Doing'

The concept of 'agroecology-by-doing' mainly consists in transforming local production systems through 'demonstration', with the assumption that technologies that offer advantages in social, ecological and economic terms are likely to be adopted by other farmers and supported by decision-makers. It relies on 'farmer-to-farmer' networks for the diffusion of relevant knowledge, in which farmer unions play a role of event organizers and facilitators. The 'farmer-to-farmer' movement emerged in Mexico to put farmers back in the driver's seat in the production of knowledge, implementation of agroecological practices and collective action (Holt-Giménez 2006) and has spread to most Latin American countries (Rosset and Martínez-Torres 2012). In SSA, farmer-to-farmer extension programmes have flourished as well, some of them not directly tied to agroecology such as seed improvement (Alene and Manyong 2006) and others focusing on diffusing agroforestry practices (Kiptot et al. 2006). In the region, these initiatives are mostly established by research centres or NGOs and have weaker ties with farmer unions.

We also observed this in Senegal, with many agroecological initiatives tied to 'demonstration farms' run with strong support of NGO or private donors (Bottazzi et al. 2020). Furthermore, initiatives that involve farmer unions face strong competition from conventional farming for markets and resources. For example, for more than 30 years the FAPD in Senegal has performed many training and 'farmer-to-farmer' knowledge exchange events focusing on agroecological techniques such as compost making, biopesticides, crop associations, tree nurseries and organic seed production. With support from ENDA-PRONAT and other NGOs, they have also provided small farmers with seeds, irrigation material, revolving funds, preferential credits and the developed organic substitute to chemical pesticide and fertilizers. Though many farmers enrolled in FAPD programmes, only 40 of the currently 1200 active producers are recognized by the FAPD as truly organic, and only 24 have organic certification. Though the FAPD claims that about 600 farmers are in transition, many farmers are discouraged by the domination of conventional farming and the unfavourable socio-economic conditions. In several cases, farmers must sell their labour force to agribusiness in the region. On the other hand, solidarity work among family member and neighbours (commonly called 'santaane' in Wolof) has been progressively altered and farmers need to regularly hire wage workers. Despite these strong processes of labour commodification, agroecological farmer unions, such as the FAPD, have concentrated their activities in the promotion of agroecological practices and more recently on access rights to land and water, with little focus on labour issues (Boillat and Bottazzi 2020).

## Alternative Food Networks

A crucial component of agroecology is to ensure that fair prices are paid for agroecological or organic products, as a key source of compensation for the additional work efforts of the producers. However, in low-income countries most consumers cannot afford more expensive products or are not sufficiently aware of the risks related to conventional products on health. For this reason, the development of organic farming on the African continent has historically focussed on products destined for export markets, and the establishment of local markets for organic products is recent (Bouagnimbeck 2008). These efforts to establish these markets have concentrated on creating niche markets, which target consumers with high purchasing power and consumer awareness. These alternative food networks intend to establish fairer prices through a greater proximity between producers and consumers.

In Senegal, ENDA-PRONAT created the *Sell Sellal* cooperative in 2013 in collaboration with five farmers' federations of the country (among which FAPD). The cooperative purchases products directly from organic farmers and distributes them to three special marketplaces in Dakar. Consumers include mainly expatriates and wealthy national clients. The products are certified by the label *Agriculture Saine et Durable* (ASD) created by ENDA-PRONAT. The cooperative purchases about 30% of the total production of 50 families (in the entire country), with a price premium of 20–40% compared with conventional products. Other initiatives in the country include community-supported agriculture also targeting expatriates and the urban elite of Dakar as consumers (Bottazzi et al. 2020). In this context, niche markets remain too small to allow the up-scaling of agroecological production practices. Until now, the growth of the agroecological and organic sectors has been very limited due to high competition with conventional imports and with products from agribusinesses, who often flood local markets with second-choice products that have been rejected by exporters.

## Supporting Alternative Labelling and Certification

Quality criteria play a key role in the recognition of agroecological farmers and their products. These criteria are usually provided by certification schemes. Internationally, the International Federation of Organic Agriculture Movements (IFOAM) is the most recognized organization providing expertise in matters of organic certification. However, the administrative cost of getting third-party certification is generally too high for small farmers in

Africa. Recent impact evaluations of these instruments in sub-Saharan Africa have shown a limited and mixed effect on improving farmers' incomes and labour conditions (Oya et al. 2018). Since 2004, IFOAM has started to promote participatory guarantee systems (PGS), which are local quality assurance initiatives that involve the participation of producers and consumers and operate outside the third-party certification process (May 2008). PGS are citizen-driven mechanisms of valuation and certification of quality criteria for local markets and have been implemented successfully in Latin American countries such as Brazil (Fonseca et al. 2008) and Mexico (Nelson et al. 2016), and have more recently started in some African countries (Lemeilleur and Allaire 2019). Unlike many organic certifications restricted to input substitution, PGS often include labour conditions and respect of environmental criteria and are increasingly promoted by farmer movements (Niederle et al. 2018).

In Senegal, FENAB has been promoting a formally recognized PGS, the *BioSenegal* label. They have elaborated a detailed *cahier des charges*, a specification document which summarizes the norms for organic certification in Senegal, based on IFOAM recommendations. Though the FAPD has worked in close collaboration with the FENAB to elaborate the *cahier des charges*, we observed that FAPD organic farmers still worked with *ASD* through the cooperative *Sell Sellal*. One barrier to the adoption of the *BioSenegal* standards is the interdiction to allocate part of their productive land to sharecropping (*mbey seddo*), a popular practice in the area. Sharecropping consists in the landowner providing the land and some inputs to an external worker who must then share part of his harvest with him/her. Furthermore, none of these certification schemes has yet received national support.

# Securing Land Rights and Access to Natural Resources

Securing access to land and natural resources is particularly important for the development of agroecological practices as agroecology requires investments in terms of labour and agroecological improvements such as tree plantation, soil enrichment, irrigation infrastructure, terraces and others. In recent years, farmer movements that sought greater autonomy and control over their lands have put a stronger emphasis on agroecology (Rosset and Martínez-Torres 2012). This has led to an increased collaboration between agroecological and food sovereignty farmer movements, in particular after the 2008 financial crisis and the global land demand surge (Martínez-Torres and Rosset 2010).

In SSA, this was reflected in the international forum for agroecology carried out in Nyéléni, Mali, in 2015, which brought agroecology and food sovereignty movements together (IPC 2015).

Questions of land access have been at the core of the debate among Senegalese farmer movements since the enactment of policies favouring large-scale land acquisitions after 2000. In 2010, the CNCR, ENDA-PRONAT and other farmer organizations and NGOs constituted a national forum on land policy, the *Cadre de Réflexion et d'Action sur le Foncier au Sénégal* (CRAFS). In their main position statement (CRAFS 2016), they proposed reforms to the National Domain Law to better protect smallholder farmers and collective land rights. However, their proposal has not been adopted or discussed by the national government until now. There were also actions at local level to secure land rights. In 2015, the FAPD started a programme aiming at securing land rights with the support of a Swiss NGO. The programme had the objective to support farmers in obtaining land use and access right titles, called *délibérations*, that are formally recognized by municipal and traditional authorities. It included participatory mapping, field measuring, training of farmers on land legalization and training of local assessors in GIS. Despite these efforts, the majority of farmers in the area remain without formal land rights and are vulnerable to governmental expropriation and urban expansion. Furthermore, while FAPD has been actively supporting small land holders titling, very little has been done to change the situation for more vulnerable populations such as migrants, youth or women. The latter are forced to sell their labour power working the land of others for low daily wages (about USD 2 per day) or a share of an insecure harvest (in case of sharecropping).

## Resistance and Protest

Incorporating resource access issues into the agroecology framing requires farmer movements to pursue specific actions aiming at filing claims and exert pressure. In this context, the defence of agroecology acquires a dimension of social and environmental (in)justice and therefore requires resistance and advocacy actions (Gliessman 2013). In this context, political freedom acquires a particular significance for the agroecological movement and is likely to be challenging in many SSA countries.

In Senegal, the participation of farmer unions promoting agroecology in resistance actions has been relatively rare until now, but has recently become part of the repertoire of some organizations. For example in 2019, the FAPD played a key role in organizing a demonstration in the municipality of Diender in the Niayes region. They protested against the construction of 11 water drills by the government to supply the city of Dakar with fresh water, which threatens the availability of irrigation water. Following the protest, the government took coercive actions by cutting dialogue with the organization and sending a delegation from the ministry of labour 'to control their activities if they are to conform with government's priorities' (Interview with organizational leader, October 2019). Following this event, the farmer union remains relatively defenceless and needs to continue building strong alliances with other organizations to defend the interests of its members.

## Promoting Territorial Approaches

The sustainable governance of ecosystem resources in a given region is needed to upscale agroecology from the farm level to the landscape or territorial level (Wezel et al. 2009). Territorial approaches seek to protect important and fragile ecosystems such as forested areas or watersheds, providing crucial ecosystem services to rural communities. Such a perspective goes along with the revalorization of the common pool resources such as water sources, pastoral areas or forests that require adapted and robust multi-level governing institutions (Ostrom 1990). For this reason, many agroecological programmes have moved their focus from local ecological restoration to establish closer collaboration with local authorities and natural resource management units (Duru et al. 2015). In this context, territorial governance can become a catalyst for collective action and the strengthening of farmer movements (McCune et al. 2017).

In Senegal, several agroecological programmes have started to improve the governance and the management of ecosystems in their areas of intervention. In the case of the FAPD, the organization has performed ecological restoration activities since 1997, including erosion control and reforestation with the support of a Swiss NGO. More recently, they established an extended collaboration with municipalities to set up territorial governance mechanisms at that level. A few municipal councils are already engaged in this collaboration to set up local territorial planning activities. For example, a watershed management committee has been created with the purpose of defining sustainable rules of water extraction to be adopted by all water users, with the support of a project funded by European cooperation agencies and NGOs. The process is, however, at an early stage and is facing resistance from local industries, which have benefitted from a situation of open access to water for decades.

## National Advocacy Coalition Platforms

In recent years, farmers' organizations, scientific organizations and NGOs have joined efforts through the creation of several networks and platforms aiming at defending the interests of smallholder farmers and advancing the agroecological transition. Advocacy coalitions are groups of actors who collaborate and coordinate actions to enhance the chance that their belief systems get translated into policy outputs and objectives (Sabatier and Weible 2007). These coalitions typically gather around shared values, including fundamental assumptions and worldviews, basic positions on policy, such as what the role of the state should be, and secondary beliefs about specific policies and measures to be implemented (Markard et al. 2016). In this sense, agroecological advocacy coalition platforms are ad hoc platforms aiming at supporting networking and policy transformation and assembling multiple organizations to defend common objectives. Because they specifically aim at reforming policies, such platforms often need to overcome polarization and adapt their discourse to mainstream actors, which in turn influences their own ideas, values and objectives (Bellon and Ollivier 2018).

The Dynamic for the Agricultural Transition in Senegal (DyTAES) created in 2019 is the largest advocacy coalition centred on agroecology in Senegal. The platform has been created and led by ENDA-PRONAT with a group of NGOs and farmer organizations with a long experience in agroecology. It has the overall objective to support the Senegalese national government in implementing an agroecological transition policy. The creation of the platform emerged after President Macky Sall announced the launching of a 'Green' PSE (*PSE Vert*) aiming at increasing investment in environmental protection, particularly reforestation. The DyTAES has carried out a large consultation process in rural areas of the whole country involving farmer organizations, NGOs and local governments, and ended up with a policy proposal that was addressed to the Head of State in January 2020. Though its claims address critical issues to the government, the DyATES proposal remains formal and diplomatic and avoids questioning the main objectives of national development policies. This timidity is due to the fact that most of these platforms are led by foreign and national NGOs that lack legitimacy in Senegalese politics and cannot take the risk of being discarded or 'deregistered'. Furthermore, the participation of government representatives in the popular consultation processes and validation of the documents has also led to the censuring of issues considered too sensitive, such as water rights advocacy and open critique of neoliberal policies.

## Discussion and Conclusion

Agroecology can potentially contribute to the political, social, economic and cultural emancipation of food producers and consumers in many aspects. This includes potentially safer and better working conditions, better recognition of farmers' skills, less dependency on dominant economic actors and a strengthening of farmer and consumer social movements. In sub-Saharan Africa, agroecology also has the potential to increase the resilience of small farmers and vulnerable populations against climate change and external economic shocks.

For these reasons, an agenda for an 'agroecological transition' is being pushed by coalitions including NGOs, farmer unions, local governments and research organizations. The case of Senegal shows a clear dynamism of a current constellation of actors who support agroecology, which is gradually evolving into an advocacy coalition. Nevertheless, these actors remain highly dependent on foreign support, making it difficult to put more emphasis on social justice and the emancipation of farmers as a working class struggle. Examining the repertoires of action of these actors from a labour environmentalist perspective highlights the existence of strong 'depoliticization' mechanisms that challenge the agroecological movement to pursue social justice along environmental sustainability goals.

First, we observe that though farmer unions do not necessarily hold a position of power within the agroecological movement, they are often at its forefront in terms of visibility. In Senegal, farmer organizations depend on international donor support in economic terms but also for their political recognition (Hrabanski 2010). In this context, farmer unions often take a role of intermediaries between international partners, national NGOs and individual farmers. This allows them to clearly enrich their repertoires of collective actions with the development of agroecological practices, the creation of market niches and, more recently, the promotion of territorial approaches. However, the lack of financial independence, the complexity of land, water and natural resource governance issues dominated by political factionalism, and the lack of social embeddedness of agroecological claims explain the limited advocacy capacity of these organizations.

Second, the observed dependency on international funding also limits the scope of national NGOs and other civil society organizations to pursue social justice goals. Since these organizations might be perceived as serving foreign interests as soon as they engage in politically explicit actions, they tend to gloss over crucial issues deemed too controversial. For example, too contentious ideas such as the questioning of neoliberal agrarian policies or water rights issues were removed from the DyATES proposals in the validation meetings. For this reason, NGOs and farmer unions alike tend to frame social justice in terms of poverty reduction and development rather than in redistributive terms. This might lead the agroecological movement to become one movement in a proliferation of development initiatives stemming from different and uncoordinated sources involving NGOs, the national government and international cooperation agencies. In such a case, farmer unions will only play a role of recipients with little confrontational power.

Third, avoiding issues of inequality and social injustice leads to frame agroecology as a trade-off between economic productivity and ecological benefits. This economic-ecological trade-off is often at the core of the debate on agroecology in sub-Saharan Africa, a debate that mainly concerns governments, NGOs and development actors. In response to this framing, agroecological organizations must dedicate substantial resources to prove the economic viability of agroecology at the plot or farm scale. This limits the agroecological movement to a position of technical demonstrators at the micro-level without challenging the broader contextual economic and political drivers that originally hinder its development.

The insistence on proving the economic viability of agroecological techniques at farm level also encourages individualistic versus more holistic approaches of agroecology. This could translate into a reduction of agroecology to input substitution and favour the development of an organic farming industry oriented towards exports and including high-cost organic inputs and certification schemes. Strong lobbies that support the organic industry (Scholten 2014) might push into this direction, leading to the co-optation of agroecology by powerful economic actors and negating its economic emancipatory power. Nevertheless, in the case of Senegal these lobbies have limited their influence on export-oriented sectors.

Fourth, the dominant role of international donors and NGOs also raises questions of 'endogeneity' versus 'coloniality' of agroecology initiatives. Agroecology has been introduced in SSA by NGOs and IOs with the support of foreign research organizations. Senegalese farmers consider agroecology as a 'nassaran' word, meaning created by other-than-Senegalese societies. *Wolof* people used to translate agroecology as *bay bu sell te mucc ayib* (clean agriculture without impurity), a definition also inherited from modernity. Though agroecology has been built from a hybridization of knowledge combining aspects of traditional practices and scientific innovations (Kremen et al. 2012), there is an inherent risk of a top-down knowledge transfer in post-colonial contexts. To ensure that agroecology actually contributes to cognitive justice (Coolsaet 2016), agroecological practices need to find legitimacy at different sociological levels (Schoonhoven and Runhaar 2018). Recent initiatives that involve the *Mouride* Islamic brotherhoods[1] in agroecological projects (Bottazzi et al. 2020) could work in this sense.

Finally, a crucial issue to understand agroecology from a labour environmentalist perspective is the question of the legitimacy of representative structures of advocacy, their modes of selection and to what extent the dominant discourse of the farmer organizations fits the needs and views of its members. At present time, one cannot say that there is a popular movement defending agroecology in Senegal, but rather a coalition of highly educated and dedicated people more or less informed about the everyday realities of farmers. As the example of the FAPD shows, only a minority of farmer union members actually practice agroecology, with most farmers lacking the economic and ecological conditions and the skills to do so.

Furthermore, subaltern categories of workers who do not own land and any other means of production are usually underrepresented, if represented at all, in farmer unions whose leaders are usually small or medium land owners. This configuration highlights the tendency to idealize the 'peasant way of life' (Bernstein 2014) among farmer unions which leads to ignoring internal inequalities and the inherent contradictions of local agrarian structures (Oya 2007). It also explains why labour issues tend to be eluded in farmer organizations. This important aspect requires to constantly keep in mind the most vulnerable farm workers (e.g. youth, migrant and women) who work for agri-businesses but also for small and medium landowners, in order to avoid excluding them from political claims.

Despite these many challenges, opportunities arise with some recent expansions of the agroecological movement in terms of repertoires of action. In particular, framing agroecology beyond the plot and farm level requires the implementation of territorial approaches. In Senegal, we observed a social appropriation of agroecology not only in the technical realm, but also in terms of socio-political issues and access to natural resources, especially in a context where the central government and private companies tend to become the main competitors of rural communities. The reactivation of the 'commons', which have been spaces of collective management in *Wolof, Al-pulaar* or *Sereer* societies, based on lineage modes of production and embedded in customary rules and traditional forms of authority for centuries (Meillassoux 1982), plays a crucial role in this shift from farm to territory.

---

[1] Founded 1883 by Cheikh Amadou Bamba, the Mouride brotherhood is a large Sufi order with headquarters in Touba, Senegal. It has a very large cultural, social, economic and political influence in the country.

These issues have been widely taken up by the Senegalese agroecological movement through its role in land issues via the CRAFS platform and grassroots mobilization aiming at defending access to land and water for smallholders. The more recent attempts to extend agroecology at the territorial level to secure land and water access through an alliance between municipalities, traditional authorities and farmers' organizations appear to be particularly promising and challenging. These territorial reforms can contribute to increase local stakeholders' sovereignty on natural resources and participate to extend the principle of agroecology to a larger share of rural societies. In our sense, a 're-localization' of agriculture will necessarily mean a 're-politicization' of natural resource governance within the deep social structures of SSA societies.

To conclude, we observe that agroecology has a potential to enable emancipation mechanisms for producers and consumers in food systems. From a labour environmentalist perspective, agroecology can build bridges between environmental sustainability and social justice by centring its focus on small farmers in the Global South. In particular, agroecology offers meaningful options to 'bargain for the common goods' (Huber 2019) which include agroecosystems, land and water as well as fair exchange spaces for both producers and consumers. This scope has the potential to find resonance in sub-Saharan Africa where spaces of collective natural resource management still exist and where the commodification of food and labour systems is not complete. Nevertheless, in similar ways observed by Huber (2019) with regard to the Green New Deal in the Global North, agroecology in Africa still lacks a political movement with strong social basis even though it offers a meaningful and emancipatory political vision. The heterogeneous and unequal structure of African peasantries and the dependency of farmer movements on international donors represent key barriers to achieve this. In such context, emancipation can only occur in the longer term through the gradual consolidation and empowerment of farmer movements.

# References

Alene, Arega D., and V.M. Manyong. 2006. Farmer-to-Farmer Technology Diffusion and Yield Variation among Adopters: The Case of Improved Cowpea in Northern Nigeria. *Agricultural Economics* 35 (2): 203–211. https://doi.org/10.1111/j.1574-0862.2006.00153.x.

Altieri, Miguel A., and Victor Manuel Toledo. 2011. The Agroecological Revolution in Latin America: Rescuing Nature, Ensuring Food Sovereignty and Empowering

Peasants. *The Journal of Peasant Studies* 38 (3): 587–612. https://doi.org/10.1080/03066150.2011.582947.

Anseeuw, Ward, Liz Alden Wily, and Lorenzo Cotula. 2012. *Land Rights and the Rush for Land. Findings of the Global Commercial Pressures on Land Research Project*. Rome.

Bellon, Stéphane, and Guillaume Ollivier. 2018. Institutionalizing Agroecology in France: Social Circulation Changes the Meaning of an Idea. *Sustainability* 10 (5): 1380. https://doi.org/10.3390/su10051380.

Bernstein, Henry. 2014. Food Sovereignty via the "Peasant Way": A Sceptical View. *The Journal of Peasant Studies* 41 (6): 1031–1063. https://doi.org/10.1080/03066150.2013.852082.

Boillat, Sébastien, and Patrick Bottazzi. 2020. Agroecology as a Pathway to Resilience Justice: Peasant Movements and Collective Action in the Niayes Coastal Region of Senegal. *International Journal of Sustainable Development & World Ecology* 27 (7): 662–677. https://doi.org/10.1080/13504509.2020.1758972.

Boillat, Sébastien, Julien-François Gerber, and Fernando R. Funes-Monzote. 2012. What Economic Democracy for Degrowth? Some Comments on the Contribution of Socialist Models and Cuban Agroecology. *Futures* 44 (6): 600–607. https://doi.org/10.1016/j.futures.2012.03.021.

Borras, Saturnino M., Jr., and Jennifer C. Franco. 2012. Global Land Grabbing and Trajectories of Agrarian Change: A Preliminary Analysis. *Journal of Agrarian Change* 12 (1): 34–59. https://doi.org/10.1111/j.1471-0366.2011.00339.x.

Bottazzi, Patrick, Sébastien Boillat, Franziska Marfurt, and Sokhna Mbossé Seck. 2020. Channels of Labour Control in Organic Farming: Toward a Just Agroecological Transition for Sub-Saharan Africa. *Land* 9 (6): 205. https://doi.org/10.3390/LAND9060205.

Bouagnimbeck, Hervé. 2008. Organic Farming in Africa. In *The World of Organic Agriculture: Statistics and Emerging Trends*, ed. Helga Willer, Minou Yussefi-Menzler, and Neil Sorensen, 90–96. London: Earthscan.

Brayer, Julie, Jean-René Cuzon, and Bénédicte Hermelin. 2008. Le Contenu et Les Enjeux de La Loi d'orientation Agro-Sylvo-Pastorale Du Sénégal. *GT Politiques Agricoles*.

Coolsaet, Brendan. 2016. Towards an Agroecology of Knowledges: Recognition, Cognitive Justice and Farmers' Autonomy in France. *Journal of Rural Studies* 47: 165–171. https://doi.org/10.1016/j.jrurstud.2016.07.012.

CRAFS. 2016. *Document de Position Du Cadre de Réflexion et d'Action Sur Le Foncier Sur La Réforme Foncière Au Sénégal*. Dakar.

Dahlin, A. Sigrun, and Leonard Rusinamhodzi. 2019. Yield and Labor Relations of Sustainable Intensification Options for Smallholder Farmers in Sub-Saharan Africa. A Meta-analysis. *Agronomy for Sustainable Development* 39 (3): 32. https://doi.org/10.1007/s13593-019-0575-1.

Dalton, Russell J., Steve Recchia, and Robert Rohrschneider. 2003. The Environmental Movement and the Modes of Political Action. *Comparative Political Studies* 36 (7): 743–771. https://doi.org/10.1177/0010414003255108.

Dawson, Neil, Adrian Martin, and Thomas Sikor. 2016. Green Revolution in Sub-Saharan Africa: Implications of Imposed Innovation for the Wellbeing of Rural Smallholders. *World Development* 78: 204–218. https://doi.org/10.1016/j.worlddev.2015.10.008.

Dumont, Antoinette M., and Philippe V. Baret. 2017. Why Working Conditions Are a Key Issue of Sustainability in Agriculture? A Comparison between Agroecological, Organic and Conventional Vegetable Systems. *Journal of Rural Studies* 56: 53–64. https://doi.org/10.1016/j.jrurstud.2017.07.007.

Dumont, Antoinette M., Gaëtan Vanloqueren, Pierre M. Stassart, and Philippe V. Baret. 2016. Clarifying the Socioeconomic Dimensions of Agroecology: Between Principles and Practices. *Agroecology and Sustainable Food Systems* 40 (1): 24–47. https://doi.org/10.1080/21683565.2015.1089967.

Dupré, Lucie, Claire Lamine, and Mireille Navarrete. 2017. Short Food Supply Chains, Long Working Days: Active Work and the Construction of Professional Satisfaction in French Diversified Organic Market Gardening. *Sociologia Ruralis* 57 (3): 396–414. https://doi.org/10.1111/soru.12178.

Duru, Michel, Olivier Therond, and M'hand Fares. 2015. Designing Agroecological Transitions; A Review. *Agronomy for Sustainable Development* 35 (4): 1237–1257. https://doi.org/10.1007/s13593-015-0318-x.

Duruflé, Gilles. 1995. Bilan de La Nouvelle Politique Agricole Au Sénégal. *Review of African Political Economy* 22 (63): 73–84. https://doi.org/10.1080/03056249508704101.

Edelman, Marc, Carlos Oya, and Saturnino M. Borras. 2013. Global Land Grabs: Historical Processes, Theoretical and Methodological Implications and Current Trajectories. *Third World Quarterly* 34 (9): 1517–1531. https://doi.org/10.1080/01436597.2013.850190.

Ela, Jean-Marc. 1990. *Quand l'Etat pénètre en brousse : Les ripostes paysannes à la crise*. Paris: Karthala.

Eyhorn, Frank, Adrian Muller, John P. Reganold, Emile Frison, Hans R. Herren, Louise Luttikholt, Alexander Mueller, et al. 2019. Sustainability in Global Agriculture Driven by Organic Farming. *Nature Sustainability* 2 (4): 253–255. https://doi.org/10.1038/s41893-019-0266-6.

Figueroa-Helland, Leonardo, Cassidy Thomas, and Abigail Pérez Aguilera. 2018. Decolonizing Food Systems: Food Sovereignty, Indigenous Revitalization, and Agroecology as Counter-Hegemonic Movements. *Perspectives on Global Development and Technology* 17 (1–2): 173–201. https://doi.org/10.1163/15691497-12341473.

Fonseca, Maria Fernanda, John Wilkinson, Henrik Egelyng, and Gilberto Mascarenhas. 2008. *The Institutionalization of Participatory Guarantee Systems (PGS) in Brazil: Organic and Fair Trade Initiatives*. 2nd ISOFAR Scientific Conference 'Cultivating the Future Based on Science', Modena.

Food and Agriculture Organization of the United Nations (FAO). 2015. *The State of Food Insecurity in the World 2015*.

———. 2016. *Final Report of the Regional Meeting on Agroecology in Sub-Saharan Africa*.
———. 2019. *Agroecology Knowledge Hub*. http://www.fao.org/agroecology/overview/en/.
Galt, Ryan E. 2013. The Moral Economy Is a Double-Edged Sword: Explaining Farmers' Earnings and Self-Exploitation in Community-Supported Agriculture. *Economic Geography* 89 (4): 341–365. https://doi.org/10.1111/ecge.12015.
German, Paul, and Abou Thiam. 1993. *Les Pesticides Au Senegal: Une Menace ?* Dakar: ENDA.
Giller, Ken E., Ernst Witter, Marc Corbeels, and Pablo Tittonell. 2009. Conservation Agriculture and Smallholder Farming in Africa: The Heretics' View. *Field Crops Research* 114 (1): 23–34. https://doi.org/10.1016/j.fcr.2009.06.017.
Giraldo, Omar Felipe, and Peter M. Rosset. 2018. Agroecology as a Territory in Dispute: Between Institutionality and Social Movements. *The Journal of Peasant Studies* 45 (3): 545–564. https://doi.org/10.1080/03066150.2017.1353496.
Gliessman, Steve. 2013. Agroecology: Growing the Roots of Resistance. *Agroecology and Sustainable Food Systems* 37 (1): 19–31. https://doi.org/10.1080/10440046.2012.736927.
———. 2016. Transforming Food Systems with Agroecology. *Agroecology and Sustainable Food Systems* 40 (3): 187–189. https://doi.org/10.1080/21683565.2015.1130765.
Holt-Giménez, Eric. 2006. *Campesino a Campesino: Voices from Latin America's Farmer to Farmer Movement for Sustainable Agriculture*. Oakland: Food First Books.
Holt-Giménez, Eric, and Miguel A. Altieri. 2013. Agroecology, Food Sovereignty, and the New Green Revolution. *Agroecology and Sustainable Food Systems* 37 (1): 90–102. https://doi.org/10.1080/10440046.2012.716388.
Hrabanski, Marie. 2010. Internal Dynamics, the State, and Recourse to External Aid: Towards a Historical Sociology of the Peasant Movement in Senegal Since the 1960s. *Review of African Political Economy* 37 (125): 281–297. https://doi.org/10.1080/03056244.2010.510627.
Huber, Matt T. 2019. Ecological Politics for the Working Class. *Catalyst* 3 (1): 7–45.
International Planning Committee for Food Sovereignty (IPC). 2015. *Declaration of the International Forum for Agroecology*. https://www.foodsovereignty.org/forum-agroecology-nyeleni-2015-2/.
Isgren, Ellinor, and Barry Ness. 2017. Agroecology to Promote Just Sustainability Transitions: Analysis of a Civil Society Network in the Rwenzori Region, Western Uganda. *Sustainability* 9 (8): 1357. https://doi.org/10.3390/su9081357.
Jansen, Kees. 2000. Labour, Livelihoods and the Quality of Life in Organic Agriculture in Europe. *Biological Agriculture & Horticulture* 17 (3): 247–278. https://doi.org/10.1080/01448765.2000.9754845.
Johnson, Victoria. 2008. Changing Repertoires of Collective Action: American General Strikes 1877–1946. *Politics and Public Policy*, ed. Harland Prechel, vol.

17, 101–34. Research in Political Sociology. Bingley: Emerald Group Publishing Limited. https://doi.org/10.1016/S0895-9935(08)17005-X.

Kiptot, Evelyne, Steven Franzel, Paul Hebinck, and Paul Richards. 2006. Sharing Seed and Knowledge: Farmer to Farmer Dissemination of Agroforestry Technologies in Western Kenya. *Agroforestry Systems* 68 (3): 167–179. https://doi.org/10.1007/s10457-006-9007-8.

Koopman, Jeanne. 2012. Land Grabs, Government, Peasant and Civil Society Activism in the Senegal River Valley. *Review of African Political Economy* 39 (134): 655–664.

Kremen, Claire, Alastair Iles, and Christopher Bacon. 2012. Diversified Farming Systems: An Agroecological, Systems-Based Alternative to Modern Industrial Agriculture. *Ecology and Society* 17 (4). https://doi.org/10.5751/ES-05103-170444.

Lemeilleur, Sylvaine, and Gilles Allaire. 2019. *Participatory Guarantee Systems for Organic Farming: Reclaiming the Commons*. Working Papers MOISA 2019-2, no. 914-2019–3059: 31. https://doi.org/10.22004/ag.econ.292325.

Markard, Jochen, Marco Suter, and Karin Ingold. 2016. Socio-Technical Transitions and Policy Change—Advocacy Coalitions in Swiss Energy Policy. *Environmental Innovation and Societal Transitions* 18: 215–237. https://doi.org/10.1016/j.eist.2015.05.003.

Martín, Víctor O. Martín, Luis M. Jerez Darias, and Carlos S. Martín Fernández. 2019. Agrarian Reforms in Africa 1980–2016: Solution or Evolution of the Agrarian Question? *Africa* 89 (3): 586–607. https://doi.org/10.1017/S0001972019000536.

Martínez-Torres, María Elena, and Peter M. Rosset. 2010. La Vía Campesina: The Birth and Evolution of a Transnational Social Movement. *The Journal of Peasant Studies* 37 (1): 149–175. https://doi.org/10.1080/03066150903498804.

McCune, Nils. 2017. Pedagogical Mediators and the Territorialization of Agroecology. *Revista Praxis Educacional* 13 (26): 252–280.

McCune, Nils, Peter M. Rosset, Tania Cruz Salazar, Antonio Saldívar Moreno, and Helda Morales. 2017. Mediated Territoriality: Rural Workers and the Efforts to Scale out Agroecology in Nicaragua. *The Journal of Peasant Studies* 44 (2): 354–376. https://doi.org/10.1080/03066150.2016.1233868.

Meek, David. 2016. The Cultural Politics of the Agroecological Transition. *Agriculture and Human Values* 33 (2): 275–290. https://doi.org/10.1007/s10460-015-9605-z.

Meillassoux, Claude. 1982. *Femmes, Greniers et Capitaux*. Paris: Maspero.

Molina, Manuel Gonzalez de. 2013. Agroecology and Politics. How To Get Sustainability? About the Necessity for a Political Agroecology. *Agroecology and Sustainable Food Systems* 37 (1): 45–59. https://doi.org/10.1080/10440046.2012.705810.

Mousseau, Frédéric. 2015. The Untold Success Story of Agroecology in Africa. *Development* 58 (2): 341–345. https://doi.org/10.1057/s41301-016-0026-0.

Ndoye Diop, Mor, Matar Ndoye, and Doudou Diop. 2019. *Historique de La Fédération Des Agropasteurs de Diender (FAPD)*. Bayakh: FAPD.

Nelson, Erin, Laura Gómez Tovar, Elodie Gueguen, Sally Humphries, Karen Landman, and Rita Schwentesius Rindermann. 2016. Participatory Guarantee Systems and the Re-Imagining of Mexico's Organic Sector. *Agriculture and Human Values* 33 (2): 373–388.

Niederle, Paulo André, Allison Loconto, Sylvaine Lemeilleur, and Claire Dorville. 2018. *How Do Social Movements Shape Organic Food Markets? Comparing the Construction and Institutionalization of Participatory Guarantee Systems in Brazil and France*. 3rd International Conference Agriculture and Food in an Urbanizing Society. Porto Alegre: UFRGS.

Ostrom, Elinor. 1990. *Governing the Commons. The Evolution of Institutions for Collective Action*. Political Economy of Institutions and Decisions. Cambridge: Cambridge University Press.

Oya, Carlos. 2007. Stories of Rural Accumulation in Africa: Trajectories and Transitions among Rural Capitalists in Senegal. *Journal of Agrarian Change* 7 (4): 453–493. https://doi.org/10.1111/j.1471-0366.2007.00153.x.

Oya, Carlos, and Cheikh Oumar Ba. 2013. Les Politiques Agricoles 2000-2012: Entre Volontarisme et Incohérence. In *Sénégal 2000-2012. Les Institutions et Politiques Publiques à l'épreuve d'une Gouvernance Libérale*, ed. Momar-Coumba Diop, 149–178. Paris: Karthala.

Oya, Carlos, Florian Schaefer, and Dafni Skalidou. 2018. The Effectiveness of Agricultural Certification in Developing Countries: A Systematic Review. *World Development* 112: 282–312. https://doi.org/10.1016/j.worlddev.2018.08.001.

Pesche, Denis. 2009. Construction du mouvement paysan et élaboration des politiques agricoles en Afrique subsaharienne. Le cas du Sénégal. *Politique Africiane* 2009 (2): 139–155.

Rosset, Peter M., and Miguel A. Altieri. 1997. Agroecology versus Input Substitution: A Fundamental Contradiction of Sustainable Agriculture. *Society & Natural Resources* 10 (3): 283–295. https://doi.org/10.1080/08941929709381027.

Rosset, Peter M., and Maria Elena Martínez-Torres. 2012. Rural Social Movements and Agroecology. *Ecology and Society* 17 (3): 17.

Rosset, Peter M., and Elena Martinez-Torres. 2013. *La Via Campesina and Agroecology*. La Via Campesina's Open Book: Celebrating 20 Years of Struggle and Hope. Harare: La Via Campesina.

Ruiz López, Miguel Alfredo. 2018. Agroecology: One of the "Tools" for Degrowth. *Scientia et Technica* 23 (4): 599–605.

Sabatier, Paul, and Christopher M. Weible. 2007. The Advocacy Coalition Framework: Innovations and Clarifications. In *Theories of the Policy Process*, ed. Paul Sabatier, 189–220. Boulder: Westview Press.

Schiller, Katharina, Wendy Godek, Laurens Klerkx, and P. Marijn Poortvliet. 2020. Nicaragua's Agroecological Transition: Transformation or Reconfiguration of the Agri-Food Regime? *Agroecology and Sustainable Food Systems* 44 (5): 611–628. https://doi.org/10.1080/21683565.2019.1667939.

Scholten, Bruce. 2014. *US Organic Dairy Politics: Animals, Pasture, People, and Agribusiness*. New York: Palgrave Macmillan US.

Schoonhoven, Yanniek, and Hens Runhaar. 2018. Conditions for the Adoption of Agro-Ecological Farming Practices: A Holistic Framework Illustrated with the Case of Almond Farming in Andalusia. *International Journal of Agricultural Sustainability* 16 (6): 442–454. https://doi.org/10.1080/14735903.2018.1537664

Stevis, Dimitris, David Uzzell, and Nora Räthzel. 2018. The Labour–Nature Relationship: Varieties of Labour Environmentalism. *Globalizations* 15 (4): 439–453. https://doi.org/10.1080/14747731.2018.1454675.

Tilly, Charles. 2010. *Regimes and Repertoires*. Chicago: University of Chicago Press.

Timmermann, Cristian, and Georges F. Félix. 2015. Agroecology as a Vehicle for Contributive Justice. *Agriculture and Human Values* 32 (3): 523–538. https://doi.org/10.1007/s10460-014-9581-8.

Tittonell, P., E. Scopel, N. Andrieu, H. Posthumus, P. Mapfumo, M. Corbeels, G.E. van Halsema, et al. 2012. Agroecology-Based Aggradation-Conservation Agriculture (ABACO): Targeting Innovations to Combat Soil Degradation and Food Insecurity in Semi-Arid Africa. *Field Crops Research* 132: 168–174. https://doi.org/10.1016/j.fcr.2011.12.011.

Waldmueller, Johannes M. 2015. Agriculture, Knowledge and the "Colonial Matrix of Power": Approaching Sustainabilities from the Global South. *Journal of Global Ethics* 11 (3): 294–302. https://doi.org/10.1080/17449626.2015.1084523.

Wezel, A., S. Bellon, T. Doré, C. Francis, D. Vallod, and C. David. 2009. Agroecology as a Science, a Movement and a Practice. A Review. *Agronomy for Sustainable Development* 29 (4): 503–515. https://doi.org/10.1051/agro/2009004.

# 19

# Working-Class Environmentalism in the UK: Organising for Sustainability Beyond the Workplace

Karen Bell

## Introduction

There is much scope for developing unity across workplace and community on sustainability issues. The production activities of workplaces can have environmental impacts on communities, and community organisations can help to influence environmental policies in the workplace. It will mean trade unions working alongside those not currently engaged with mainstream labour unionism. Though not applying in every case, this could include engaging, for example, unemployed people, disabled people, carers, retired people and those who work in non-unionised and casualised employment sectors.

In order to understand the potential for workplace-community joint organising for sustainability, this chapter draws together relevant existing studies; my own research on working-class environmentalism since 2007; and my personal experience as a woman from a working-class background and a community-development worker for 20 years. The first section gives an overview of 'working-class environmentalism' focussing on struggles that have occurred in the community, rather than the workplace. This is followed by a

K. Bell (✉)
Department Geography and Environmental Management, University of the West of England, Bristol, UK
e-mail: Karen.bell@uwe.ac.uk

reflection on my community-development experience, presented as an autoethnographic case study (Byczkowska-Owczarek 2014). The case study is used to highlight the actual and potential barriers to joint workplace and community organising in the UK. Finally, three ways that unions and communities could join forces to overcome these barriers and achieve a just transition are explored - 'joint health and safety campaigns', 'community unionism' and 'socially useful production'.

However, before going further, it will be useful to offer some definitional clarity around the use of some key terms - 'class', 'community' and 'environmentalism'. Class will be considered here as a synthesis of gradational and cultural definitions, that is, in terms of current and inherited wealth, income, occupation, status, recognition and valuing (as used, e.g., by class analysts, Bev Skeggs 2004 and Lisa McKenzie 2015). It is also important to note 'intersectionality' (Crenshaw 1991), that is, that class cannot be experienced outside of other identities, such as ethnicity, disability, LGBTQIA+, gender and age, and that these will also shape and compound working-class values, perspectives and outcomes. Working-class people do not constitute a homogenous group though there is still a tendency to common experiences in relation to life outcomes, including those related to the environment (Bell 2020).

With regard to the term 'community', it is recognised that communities are constructed through people sharing and interacting with a common purpose (Moseley 2003). A distinction is often drawn between interest communities (e.g. disabled people), where physical proximity is not a requirement, and place-based communities, where neighbourhood residents identify with the place that they live in (Moseley 2003). In this chapter, I focus primarily on place-based communities.

Finally, in relation to the term 'environmentalism', it is acknowledged that this is not a monolithic movement and that it comprises a wide range of perspectives and activities (Wardle et al. 2019; Newell 2020). In particular, the long lineage of indigenous and working-class environmental movements in both the Global North and South is recognised (Montrie 2018; Satheesh 2020).

## Working-Class Environmentalism

Since environmentalism began to develop as a movement, there has been a prevailing view that working-class people are too concerned with meeting their basic everyday needs to be able to think about environmental issues (e.g. Inglehart 1977). For example, Inglehart's (1977) 'post-materialist values' theory argues that citizens of less wealthy societies are more 'materialist', focussed on survival and security, while citizens of more wealthy countries tend to be 'post-materialist', concerned with identity, rights, and quality of life, including the quality of the environment. Similarly, the 'affluence hypothesis' (Franzen 2003; Franzen and Vogl 2013) assumes a direct link between affluence and environmental concern, arguing that environmental quality is a public good for which demand rises with income. Yet, the perceived lack of environmental interest among the working-class clashes with rapidly increasing evidence that low-income communities and countries are often active environmental campaigners and custodians (e.g. see Martinez-Alier 2003; Pellow 2007, 2018; Satheesh 2020) and that economic affluence is not consistently positively correlated with environmental concern (e.g. Dunlap and Mertig 1997; Dunlap and York 2008).

The notion that working-class people are less likely to care about the environment does not take into account different ways of expressing environmental concern. While the generally lower and more insecure incomes of working-class people restrict their ability to carry out green activities which have a direct or indirect financial cost, for example, buying organic food; purchasing longer lasting 'quality' products; eco-tourism holidays; taking time off work to engage in climate protests, there are distinctive forms of working-class environmentalism (see, e.g., Burningham and Thrush 2001, 2003; Brown 2002; Martinez-Alier 2003; Taylor 2016; Bell 2020). For example, research by Burningham and Thrush (2003) on how people living in disadvantaged communities talk about and experience environmental degradation noted a focus on health and safety at home and in the streets around them (Burningham and Thrush 2003). At a global level, it has similarly been noted that working-class people and the global poor have tended to focus on maintaining environments that are adequate for immediate physical survival (Satheesh 2020). Pulido (1998) calls this an 'environmentalism of everyday life' (p. 30).

Martinez-Alier (2003), coining the term 'environmentalism of the poor', has described how the subaltern classes, and the poor in general, can be the most motivated to defend the environment because they are aware that it supports their livelihoods, wellbeing and survival. They may also be 'default environmentalists' because less economic ability to consume means that their overall ecological footprints tend to be lower (see, for example, Pang et al. 2019). Some analysts have suggested that a direct experience of environmental degradation may be a key source of environmental concern (e.g. Dunlap and Mertig 1995; Knight and Messer 2012). Working-class, low-income and other disadvantaged groups often have this direct experience due to their greater likelihood of proximity to environmental harms (Taylor 2014) and so may be much more likely to be troubled when the environment is not healthy.

It is not surprising, then, that working-class people have, for centuries, drawn attention to the environmental contaminants present in their own workplaces and communities. They have campaigned for the adoption and enforcement of environmental policies and regulations to restrict or remove toxic production, through organising and educating. In the remainder of this section, I focus on the working-class environmentalism that has happened in the community, before going on to look at actual and potential environmental organising across workplaces and communities.

There are numerous historical examples over the last century of working-class environmentalism that has occurred within communities, rather than the workplace. For example, in the UK, working-class people in the 1930s had a major impact on our access to the countryside through their actions to gain access to land that had been privatised by the eighteenth-and nineteenth-century Enclosures Acts. In particular, in April 1932, over 400 mostly working-class people participated in a mass trespass onto Kinder Scout in the Peak District. The trespass was for the right to roam on land that was being used exclusively by the wealthy for grouse shooting. The event is widely credited with leading to the later establishment of the UK National Parks, the development of the Pennine Way and many other long-distance footpaths, and securing walkers' rights in the Countryside and Rights of Way Act, 2000 (Hey 2011).

Other historical community struggles occurred in the 1950s in Sydney, Australia. Goodall and Cadzow (2010) describe how working-class activists demanded conservation of native bushland and accessible greenspace. Their stories illustrate how the local river and bushland environments were as important to their working-class identities as their employment.

From the 1980s in the United States, community-based campaigns for 'environmental justice' began when 'People of Color' protested about hazardous and polluting industries being disproportionately located in their neighbourhoods. An emblematic campaign was in opposition to the proposed siting of a toxic PCB-contaminated soil landfill in Warren County, North Carolina (Bullard and Wright 1992). Numerous research studies were then undertaken providing statistical evidence of a significant correlation between the location of hazardous waste sites and the proximity of residential communities of low income or 'People of Color' (e.g. Bullard 1983, 1990; US GAO 1983; UCC CRJ 1987). At its height, the US environmental justice movement was able to prevent the siting of numerous hazardous facilities in working-class communities (see, for example, Pellow 2007; Lerner 2012). Furthermore, the recognition of environmental injustice and related struggles for justice has now been taken up by social movements, and, in some cases, policymakers around the world (Walker 2012).

Alongside the literature on the environmentalism of the poor and environmental justice, there are also are many accounts of the everyday environmental activities of local working-class communities within the academic literature on community development, the voluntary sector and grassroots association (e.g. Harley and Scandrett 2019; Smith 1997a, b). These struggles all exemplify the vibrancy and potential of community-based environmentalism. By way of illustration and for a deeper analysis of the potential input from trade unions, I will now introduce some examples from my own personal experience of environmentalism in working-class communities.

## Autoethnographic Environmental Community Development Experience

I give an overview of this experience here as a form of autoethnography, where the researcher acts as an element of the phenomenon researched (Wall 2008). Autoethnography focuses on understanding oneself as well as others and includes autoreflection and connecting this with the phenomena under study (Gobo 2008).

From 1990 to 2010, I trained and worked as a community development worker in disadvantaged areas in and around Bristol in the South West of the UK. Environmental struggles made up a large part of my work as a community development worker over this period, alongside other activities, such as developing mutual aid services; organising youth and community activities; convening local governance partnerships and organising equalities training and events. The environment is a key issue for communities since it relates to health, wellbeing, jobs and relationships. Some of the environmental issues I worked on alongside the relevant communities were making local derelict land safe and accessible for play and leisure; organising youth camps in the countryside (many of these inner city young people had never been to the countryside) (e.g. Albany Youth and Community Centre, St Pauls); setting up a food coop so that local people could access cheap, healthy food; organising clean ups of the local green spaces (e.g. Hartcliffe Health and Environment Action Group); campaigning for better lighting, safer road crossings and other neighbourhood improvements (Riverside Youth and Community Centre,

Snowhill); organising consultations on regeneration developments; campaigning for accessible public transport (e.g. South Gloucestershire Disability Equality Forum); campaigning against traffic pollution and for improved play facilities (e.g. Hotwells and Cliftonwood Community Association); developing and running a local community market so people could access affordable food and goods locally; campaigning for a car share scheme—mostly only available in well off areas (e.g. Lockleaze Neighbourhood Trust); and campaigning for tree planting and for safety measures in relation to the High Tension pylons that run through the estate (coinciding with high incidence of cancers and depression) (e.g. Lockleaze Environment Group).

Many of these activities were taking place in order to help people cope with their poverty and lack of adequate services. They were not challenging social structures but, even in relation to meeting these basic needs, the community as a whole lacked power. In order to give a more detailed account of how this played out, I will focus here on two of the above examples: the community market (Lockleaze Neighbourhood Trust) and the campaign for tree planting (Lockleaze Environment Group). Both of these took place in Lockleaze where I have lived for the last 25 years and where I also worked, during the period discussed, as a community development worker.

Lockleaze is an outer city estate in Bristol with multiple indicators of deprivation (HCLG 2019). Part of the ward is in the 10% most deprived areas of England (ibid.). There is a high concentration of disadvantaged groups, such as older people, Black, Asian and Minority Ethnic groups, disabled people and people living on low incomes (ibid.). A large proportion of residents live in houses rented from the City Council. Particular problems in the area are unemployment, high levels of road accidents, poor health, low levels of educational attainment and lack of facilities (BCC 2020). Life expectancy is two to three years less than the Bristol average and six years less than the wealthiest wards of the city (ibid.). There are very few shops in the area and little provision of fresh, healthy food. Most people have to go out of the area to shop, yet 29% of people have no car or van (ibid.). In general, there are few amenities in the area, providing little opportunity for people to meet and consequently high levels of isolation. All the local pubs have shut since austerity hit. The area does have more than the average amount of green space for Bristol. However, when the events described below occurred there was very little in the way of trees, landscaping, play facilities, outdoor seating or community activities.

## Lockleaze Community Market (2004–2007)

Members of the Lockleaze Environment Group (LEG), Lockleaze Neighbourhood Trust and other local residents initiated, campaigned for resources and organised this community market (LCM 2004–2007; Bell 2008). The original aims were to meet the local demand for more accessible healthy food and other goods; to stimulate the local economy; to build community capacity; to increase social contacts; to encourage people to use local shops, walk and reduce car use; and to increase self-esteem and confidence (LNT 2006). The City Council agreed to fund the initial markets and a business plan was drawn up to identify how the market would run for the following years. A market management group was formed to steer the project, consisting of myself, as market development worker; volunteer residents and traders; and four paid professional health and regeneration workers from Bristol City Council and North Bristol Primary Care Trust.

Following a series of pilot markets, the monthly markets ran successfully with good attendance for two years (LCM 2004–2007; Bell 2008). Around half the stalls were run by local people and included fresh, organic, locally grown food (ibid.). The market had also begun to be used by local residents to disseminate information and to build other environmental campaigns, such as 'Trees for Lockleaze', discussed below. However, the four paid professionals on the market management group suddenly decided to 'review' the market and took a number of decisions that local residents and traders opposed. We all felt these decisions would be detrimental to the market and would lead to its closure (Bell 2007). The most problematic of these was the decision to move the market indoors and to stop applying for grants. These decisions were made without carrying out any further financial projections or referring to the business plan or the market consultations that had been carried out. None of these changes were necessary to the survival of the market as it was completely within budget and fulfilling all its aims (LCM 2004–2007). The changes immediately led to a decline in the market attendance and number of stalls as local people had predicted, yet the professionals on the management group would not reverse their decisions, despite the protests of all the local residents and traders in the group (LCM 2004–2007; Bell 2008). Eventually, the last market was held on 19 May 2007 and the market was closed without any explanation given to residents for the decisions taken. Subsequent discussions seemed to imply that the professionals felt that there would be more cost-effective ways to encourage local people to eat healthily, revealing a very narrow conception of the impacts of the market.

## Trees for Lockleaze

In 2005, at the same time as the markets were running, residents in Lockleaze began campaigning for more trees and collected several hundred signatures in a petition entitled 'Trees for Lockleaze'. The residents were originally told by the City Council that 'trees are dangerous' even though there are many trees in the wealthier areas of the city. Then, Lockleaze Environment Group (LEG) found out, by chance, that a private company had received a substantial sum from the City Council (£100,000) to carry out environmental improvements in the Lockleaze area (Smith 2006). As the group had not seen any outcomes from this funding, LEG asked if the private company would plant some trees in the communal areas. The private company refused and, when LEG raised this at Council meetings, they were told they could not come to the meetings anymore. LEG never did see results from the £100,000 given to the private company and the City Council refused to investigate the spending (Bell 2008). Again, in subsequent conversations, it seemed that the City Council preferred the green spaces to remain bare so they could be maintained for a lower cost, and also so they could more easily be sold off to developers as brownfield sites. However, some years later, as the result of discussions with a sympathetic council officer, 200 trees were planted across the estate. They are now almost the size of our houses and have pleasantly transformed the entire area.

These examples constitute 'environmental injustices' in terms of the lack of an equitable distribution of environmental 'goods', that is, healthy food and attractive green space (distributive environmental justice); and fair, participatory and inclusive structures and processes of environmental decision-making (procedural environmental justice) (Bell 2014). One interpretation of the environmental injustice shown here, drawing on Duncan and Thomas's work (1999), is that it reflects an asymmetry of power in environmental decision-making between residents and agencies. The residents' best efforts were blocked and those who made decisions on their behalf (councillors, local authority officers, development organisations, etc.) did not work in a participatory way but expected the community to fit in with their own priorities. Pearce and Milne (2010) have also highlighted the lack of recognition for the work of community activists and the sometimes unhelpful attitudes of local agencies, commenting:

> It is important to recognise the hard work of those residents and activists who strive to change their estates. Often this is done quietly, without pay or recognition of financial costs to themselves … The amount of time and effort these resi-

dents invest in their communities is often not recognised by agencies, who sometimes even take the credit themselves for it or disregard it. (Pearce and Milne 2010, 7)

It is also notable in these examples that, through these environmental struggles, the Lockleaze community were not challenging social structures. They were focused on local environmental and social short-term gains, rather than long-term fundamental transitions. Geoghegan and Powell (2009) argued that there are three different kinds of 'community development': *of* neoliberalism, *alongside* neoliberalism and *against* neoliberalism and that few programmes appear to be against. Similarly, in the United States, Holifield (2004) highlights how environmental justice strategies for local communities put the emphasis on data analysis, managed public participation and economic opportunity, deepening the neoliberal project.

The Lockleaze Community Market and Trees for Lockleaze fits with this analysis. It did not challenge neoliberalism. The projects were very much based on improving everyday life in the local area, rather than organising against unjust companies and in support of state policies for wider change. Importantly, the residents were not questioning the underlying social structures of inequality. In 2012, Bunyan, emphasising the need to shift from the micro, to the meso, and eventually the macro level, argued that to galvanise local communities to effect social change, there needs to be 'a hard headed commitment to solidarity and to the building of alternative forms of collective power rooted in but transcending localities and capable of engaging with and unsettling the prevailing institutional order' (Bunyan 2012, 14). The notion of 'rooted in but transcending' in order to build solidarity across social movements seems to be the way forward to ensure transitioning to sustainability works for the benefit of all working-class people, with no one and no place missing out.

Organised trade unions could have supported the Lockleaze Neighbourhood Trust and Lockleaze Environment Group so as to enhance their power. Through their collectivist approach, resources, values and political understanding, trade unions could have aided with funding, negotiating skills, publicity, confidence, feet on the ground, amplification of marginal voices and politicisation (Carriere 2020). For example, they could have supported the development of a 'rooted in but transcending' discourse of solidarity and societal transformation. Both the market and tree initiatives were helpful for the transition to sustainability, in terms of addressing climate change (e.g. encouraging local shopping, tree coverage to absorb $CO_2$). Trade unions could have been an important voice for highlighting these local activities as of global significance (as 'glocal'). They could have politicised the projects by bringing attention to the underlying inequalities and deprivations that made these projects necessary. Furthermore, they could have pushed for respecting the wishes of the working-class community. Saul Alinsky (1989, 100) wrote on community organising that failure is always a failure of respect 'for the dignity of the people' and symptomatic of a disabling 'superior attitude'. This was the failure of these Lockleaze projects.

With all these possibilities for support, it is unfortunate that, while some of the activists for both projects were union and Trades Council members, there was no formal support from their unions as organisations. In order to understand why joint union-community organising around sustainability projects does not occur, the next section reflects upon the possible reasons for the lack of formal labour movement involvement in these projects.

## Joint Workplace and Community Organising in the UK—Potential and Barriers

The Lockleaze examples given here and, indeed, my entire experience as a community development worker for 20 years lead me to believe that, at least in the UK, the vast majority of working-class environmental struggles and activities have tended to take place separately—either in the workplace or in the community—rather than in a unified way. However, it is important to recognise that union members live in communities and that unions can help to create community as well as engage with pre-existing communities. Union activists are also often active within extra-union social movements and networks (Gall 2009). In 2008, a UK Trades Union Congress survey of union reps found that they were involved in a wide array of campaigning external to their own workplaces and were found to be considerably more active in community activities than the members of the general populace (ibid.). The reps were primarily involved in campaigns concerning disability, health and racism, but 25% had been campaigning on environmental issues. As noted in the Lockleaze example, trade unionists were involved in the project, but their unions were not formally involved.

Although, globally, there is a long history of unions working with communities, in the UK these activities diminished dramatically in the late twentieth century. When trade unions were first developing in the UK, they were organised within local communities where the industry was geographically located, such as around the docks, mines, mills and factories (Beynon and Austrin

1994; Dennis et al. 1956; Gilbert 1992; Wills and Simms 2004). People often lived and worked together forming a distinctive working-class culture with trade unions being a critical part of this (Hobsbawm 1987; Savage and Miles 1994; Webb and Webb 1920; Wills and Simms 2004). However, with some exceptions, such as mining communities, this pattern has largely been eroded. With the formation of the Labour Party 'the trade unions helped to re-scale working class politics to the national level' (Wills and Simms 2004, 64). Later deindustrialisation, globalisation, privatisation and Thatcherite assaults on trade union legitimacy weakened the UK trade unions (McBride and Greenwood 2009; Tattersall 2008; Wills and Simms 2004). This heightened the focus on retaining jobs and led to a disconnect between communities and unions in the UK (Simms 2012; Wills and Simms 2004).

Hence, while in the United States and Canada there have been a number of coalitions between trade unions and communities, including Jobs with Justice (JwJ 2020), Sustainable America (SA 2020) and Justice for Janitors (J4J 2020), in the UK, this has been far less developed. In a survey of British trade union reps, while 17% of the survey respondents thought unions already played an active and important role in local communities, 72% felt that unions could develop this role further (Gall 2009). Apart from a few notable examples, such as the TELCO (The East London Community Organization) (Prowse et al. 2017), there has tended to be much less joint working in the twenty-first century (Wills and Simms 2004; Gall and Fiorito 2011; Simms 2012).

Therefore, it fits with this pattern that there was no evident formal union presence in the struggles that occurred in the disadvantaged areas I worked in. In the Lockleaze Community Market and Trees for Lockleaze examples, possibly the main reason for the lack of input from trade unions was that, no one, including myself, thought to ask them. It takes two years' training to qualify as a youth and community development worker and, to the best of my recollection, in the entire two years trade unions were never mentioned at any time. Although I was, myself, a trade union member and come from a strong union-supporting family, I was trained to work with companies, trusts, local authorities and faith groups, rather than trade unions.

It may also be relevant that most of the community activists I worked alongside on these working-class council estates were not in paid employment. They were mainly unemployed, disabled, carers and retired people, not likely to be members of, or thinking about, trade unions. Another possible barrier was that working-class people tend to feel less entitled to ask for support beyond those that they consider are supposed to help, such as the City Council (Piff et al. 2018). They do not believe that anyone else would care about their issues (Bell 2020; McKenzie 2015). In addition, there were no obvious mutual issues to campaign on. Our primary focus was on finding ways to get by. While trade unions are focussed on obtaining good pay and working conditions so that people do not have to live in poverty, community activism is mainly taken up with mutual aid to cope with multi-dimensional poverty, including environmental degradation.

Furthermore, we were not overtly framing our issues in global terms or in terms of social or environmental justice because we had to work towards the goals of our funders. This is a fundamental reason why, as Harley and Scandrett (2019, 9) assert, 'much of contemporary community development is a central instrument in consolidating neoliberalism'. Community activism can be hyper-focused on the local, without consideration for global injustices or attention to building solidarity across borders. Linked to this, we were limited by expectations of political neutrality. If we had wanted to ask for the support of the Labour Party as a way of obtaining trade union support, we would have faced problems as our councillors at that time were Liberal Democrat. Moreover, being in receipt of council and charitable funding meant we could not be seen to be party political. In the UK, groups listed with the Charity Commission must comply with their guidance that political campaigning or political activity 'must be undertaken by a charity only in the context of supporting the delivery of its charitable purposes. Unlike other forms of campaigning, it must not be the continuing and sole activity of the charity' (Charity Commission 2008, 2). This de-politicisation as a result of funding requirements is also the case in other countries, especially those that are dependent on aid (see, for example, Fink 2018 on Bangladesh).

To overcome these barriers the community would have needed outreach from the unions; valuing of our agendas, even if not immediately chiming with their own priorities; and the development of relationships over time, so that we did not have to approach other organisations in a crisis. However, from the union side, it is obvious that unions are limited in the campaigns they can undertake. Low wages, casual contracts, poor working conditions and rampant anti-union policies are now the dominant experience across the world (ITUC 2015a; ILO 2015), limiting the possibility for unions to focus on issues beyond the workplace. The unions would need to have had many more members and activists, as well as more established contacts within the community. The next section discusses three strategies that potentially could overcome these barriers by facilitating outreach, strategic relationship building, opportunities for union recruitment and recognising issues for mutual collaboration.

## Three Strategies for Joint Trade Union-Local Community Environmental Organising

In order to build greater community-workplace unity around environmentalism in the UK, we can look to the examples of where this has occurred, that is, joint campaigns for health and safety, efforts to build community unionism and worker control of production for community benefit, as will now be outlined.

### The Trade Union Health and Safety Movement

There has been some documentation of historical events where trade union health and safety activism have enabled benefits for surrounding communities. For example, Macphee (2014) gives historical examples where workers have developed links with local communities, involving them in their epidemiological studies or strikes over environmental health issues, and forming campaigning coalitions. In one emblematic example, Macphee describes how, in the 1960s, workers and residents in Yellowknife, Canada, suspected that the arsenic released as a by-product of local gold mining was responsible for the increase in local cancer rates. The workers allied with federal and regional First Nations organisations to conduct their own joint study revealing that some of the workers' and residents' samples contained arsenic levels 50 times above the World Health Organization's designated 'safe' standard. As a result of their campaigns to expose these issues, in 1978, the Canadian government created stricter regulations surrounding arsenic emissions from gold mining at a national level. In other examples from North America, in the 1960s and 1970s in the United States, coalitions developed between oil, chemical, atomic, steel and farm workers unions and some environmental organisations, to protect workers and communities against the associated environmental risks (see, e.g., Gottlieb 1993; Rector 2014). In the 1980s in Silicon Valley, California, non-unionised workers in the semiconductor industries facing health issues in relation to chemical hazards linked up with the local communities who were also impacted by these chemicals polluting the water supply. They jointly organised to form a coalition of workers and local people in the Silicon Valley Toxics Coalition.

In Asia, the Asian Network for the Rights of Occupational and Environmental Victims (ANROEV 2020) works for the rights of victims and for overall improvement of health and safety at the workplace across Asia. Formally constituted in 1997, after the industrial disasters of Kader and Zhili, it has brought together victims' groups, trade unions and community groups for overall improvement of health and safety in the workplace. The tactics used have included direct action (demonstrations, street theatre, 'die-ins'); commemoration of 28 April (International Workers Memorial Day); naming and shaming; testimony of victims; and political campaigns (Jenkins and Marsden 2019).

More recently, the UK Hazards Campaign (2020) is an example of environmental organising across communities and unions. Hazards began as a network of radical scientists and workplace health and safety representatives, now including community resource centres, victim support groups, trade union councils, and environmental campaigners, especially those organised in relation to asbestos, pesticides, construction and microelectronics. Among its many activities, it provides training to health and safety reps on pollution and, currently, asbestos in schools. The Hazards campaign led to the formation of an organisation named FACK (Families Against Corporate Killers), a national campaigning network which aims to stop workers and others from being killed in preventable incidents. It was founded in 2006 by members of the Bereaved by Work North West support group and the Greater Manchester Hazards Centre.

## Community Unionism

'Social movement unionism', also sometimes known as 'community unionism', has been strong in the Global South, with the Congress of South African Trade Unions (COSATU) in South Africa being a pioneer of this (Desai 2003). It is now also becoming more prevalent in the Global North (Collins 2012; Mollona 2009; Moody 1997; Waterman 1993). There are many definitions and forms of community unionism (Ellem 2008; Tattersall 2008), though it generally refers to alliances between trade unions and non-labour groups in order to achieve common goals (see Holgate 2009, 2015, 2018; Tapia 2013). As such, they can organise around issues that go beyond a specific workplace. Integral to community unionism is the idea that workers' interests and solidarities reach beyond the workplace (Cranford and Ladd 2003; Wills 2008; Hess 2008; McBride and Greenwood 2009; Weghmann 2019; Kapesea and McNamara 2020). Research by Wills and Simms (2004) indicates that community unionism in the United States and Australia has been particularly successful in allowing trade unions to reach workers in the private, low-paid and high-turnover sectors, particularly women and migrants.

Community unions can now be found in a number of countries, including South Africa, Japan, the United States, Canada, Australia and Zambia (Von Holdt 2002; Barca and Leonardi 2018; Kapesea and McNamara 2020). However, community unionism in the UK, until recently, was considered 'rather ad hoc' (Wills and Simms 2004, 69). There were initiatives, such as the East London Communities Organisation's Living Wage Campaign (Prowse 2016) and the London-based Justice for Cleaners campaign (Wills 2008), but there had been no overarching strategy by the UK unions (ibid.). However, in 2010, the Trades Union Congress (TUC) encouraged trade unions to organise in communities and among vulnerable sections of the working-class (Wright 2010). Since then, a number of UK trade unions have begun to organise among voluntary sector, faith-based groups, students, retired people and those in irregular and precarious employment (Holgate 2015; Weghmann 2019). Some have continued to be project-based or ad hoc (Unison, GMB and PCS), while others (Unite and TSSA) have made strategic decisions to invest resources into their community-based organising (Holgate 2015). As yet, the documented projects have not focussed on sustainability issues, though there is obviously potential for this.

Studies of community unionism indicate that there are a number of issues that need to be overcome for successful organising, some of which are similar to those outlined above with the example of the Lockleaze projects. There are problems resulting from the apolitical nature of much community work; the different organisational structures, with traditional unions tending to be structured more formally than the community organisations; the difference in focus, with community organisations working on local issues or the particular issues that brought their group together, and trade unions focussed on employment issues (Collins 2012); and difficulties in forming equal partnerships between unions and other organisations, and between workers and the unemployed (Suzuki 2008).

## Workers' Control of Production for Socially Useful Production

Worker control of production is briefly introduced here as the final of the three ideas of how to unite communities and workplaces for a just transition to sustainability. Workers' control generally refers to participation in the management of the workplace by the people who work there. Historically, there have been many examples of workers' control around the world, including German Revolutionary Shop Stewards during World War I (Ness and Azzelini

2010); workplace occupations in 1970s UK, beginning with the Upper Clyde Shipyard work-in (Tuckman 1985; Foster and Woolfson 1986); Argentinian recovered factories (Vieta 2020) and workers' councils in Venezuela under Chavez and Maduro (Dobson 2018).

One strand of thinking and motivation that particularly links to workplace-community joint organising is workers' control for socially useful production. An emblematic example of this was the 'Lucas Plan' of 1976 (Wainwright and Elliott 1982). The workers of Lucas Aerospace in the UK developed and published an alternative plan regarding the future of their company. It was mainly a response to the management's intention to cut thousands of jobs in the context of industrial restructuring but the workers saw an opportunity beyond solely the retention of jobs. Around half of Lucas' output supplied military contracts. The workers argued for a shift away from military production and towards socially useful production, including solar panels, wind turbines, electric cars and kidney machines. In promoting their arguments, the Lucas workers sought and attracted support from workers in other sectors, community activists, radical scientists and environmentalists. The plan was rejected by the management, dismissed as being 'in the realm of the brown bread and sandals brigade' (Asquith 2020). It, nevertheless, catalysed ideas for the democratisation of production and innovation for socially useful purposes. Inspired by this, there is now a working group for a New Lucas Plan which is striving for an economy that meets human needs while respecting environmental limits. One of the key strands is Just Transition (Lucas Plan 2020). The response to Covid-19 has demonstrated the speed at which transitions to socially useful production could take place when the will is there. For example, in the UK, companies such as Airbus began to mass-produce respiratory ventilators within weeks of the outbreak of the disease nationally (BBC 2020). Workers' organisations will need to think more about how to make these changes in a context of international solidarity, though there has been some efforts to coordinate this (e.g. see ITUC 2015b; Rosemberg 2014).

## Conclusion

It is evident that working-class people are active on environmentalism and are key to transitioning to sustainability. However, it will be important to continue to build a network of working-class environmentalism that is 'rooted in but transcending' local communities and workplaces in order to build solidarity across all the relevant social movements, including internationally. Historically, there have been many divisions across and between workplaces and communities, locally and globally. However, the examples outlined here, from trade union health and safety activism, community unionism and workers' campaigns for socially useful production, indicate the potential for more unified action to bring about a just transition to sustainability. They may not be appropriate in all cases. Indeed, the only one of these three that would have helped in the Lockleaze case study would be the community unionism model. However, all three are opportunities to develop greater confidence, power and solidarity among working-class people so as to ensure an effective and just transition to sustainability.

Community activities may begin as unpolitical and overly focussed on the short-term and local. By uniting with trade unions, bringing their analysis, skills and collectivist approach, working-class communities can develop much further. In creating a unified struggle to achieve an environmentally healthy and sustainable society, we are effectively considering how to strengthen and combine small resistances with a greater challenge to an environmentally irrational political economy. Local living and working environments are important but mobilising will need to occur with a wider consciousness. There are now many opportunities for increased connection and empathy across the world as a result of globalisation, migration and technology but there is much more to consider regarding how we can build solidarity across the working-class in an age of rampant, competitive individualism. The examples given here point to a way forward but they are often sporadic, vulnerable and disconnected from wider struggles. The first steps to changing this are to learn more about their successes and failures; and to support current initiatives, such as the Hazards group and New Lucas Plan in the UK. While there are still barriers to overcome, there is strong potential to build greater capacity across workplaces and communities to enable an effective and equitable transition to sustainability that benefits all.

# References

Alinsky, Saul. 1989. *Reveille for Radicals*. New York: Vintage Books.
ANROEV. 2020. Asian Network for the Rights of Occupational and Environmental Victims. http://www.anroev.org/
Asquith, Phil. 2020. Former Lucas Aerospace Shop Stewards Combine member speech at 'A New Lucas Plan for Post-pandemic Socially Useful Jobs?' May 13th online event.
Barca, Stefania, and Emanuele Leonardi. 2018. Working-Class Ecology and Union Politics: A Conceptual Topology. *Globalizations* 15 (4): 487–503.

BBC. 2020. Coronavirus: Ventilator Built by Airbus and F1 Approved. https://www.bbc.co.uk/news/business-52309294
BCC. 2020. Lockleaze Statistical Ward Profile, Bristol City Council. https://www.bristol.gov.uk/documents/20182/436737/Lockleaze.pdf/d91c5f24-7a1b-41ec-9c33-b48d1eef946b
Bell, Karen. 2007. 12th March 2007 'Concerns' Email to Lockleaze Market Management Group.
———. 2008. Achieving Environmental Justice in the United Kingdom: A Case Study of Lockleaze. *Environmental Justice* 1 (4): 203–210.
———. 2014. *Achieving Environmental Justice: A Cross National Analysis*. Bristol: Policy Press.
———. 2020. *Working-Class Environmentalism: An Agenda for a Just and Fair Transition to Sustainability*. London: Palgrave Macmillan.
Beynon, H., and T. Austrin. 1994. *Masters and Servants: Class and Patronage in the Making of a Labour Organisation*. London: River Oram Press.
Brown, A.P. 2002. Community Involvement: Findings from Working for Communities. *Development Department Research Programme Research Findings*. 137 Scottish Executive Research Unit.
Bullard, Robert D. 1983. Solid Waste Sites and the Black Houston Community. *Sociological Enquiry* 53 (2–3): 273–288.
———. 1990. *Dumping in Dixie: Race, Class, and Environmental Quality*. Boulder, CO: Westview Press.
Bullard, Robert D., and Beverley H. Wright. 1992. The Quest for Environmental Equity: Mobilizing the African-American Community for Social Change. In *American Environmentalism: The U.S. Environmental Movement, 1970–1990*, ed. R.E. Dunlap and A.G. Mertig, 39–49. New York: Taylor and Francis.
Bunyan, Paul. 2012. Partnership, the Big Society and Community Organizing: Between Romanticizing, Problematizing and Politicizing Community. *Community Development Journal* 48: 119–133.
Burningham, Kate, and Diana Thrush. 2001. *Rainforests are a Long Way from Here: The Environmental Concerns of Disadvantaged Groups*. York: YPS for the Joseph Rowntree Foundation.
———. 2003. Experiencing Environmental Inequality: The Everyday Concerns of Disadvantaged Groups. *Housing Studies* 18 (4): 517–536.
Byczkowska-Owczarek, Dominika. 2014. Researcher's Personal Experiences as a Method of Embodiment Research. *Hexis. Theory, Society and Culture* 1 (1): 11–18.
Carriere, K.R. 2020. Workers' Rights are Human Rights: Organizing the Psychology of Labor Movements. *Current Opinion in Psychology* 35 (Oct.): 60–64.
Charity Commission. 2008. Campaigning and Political Activity Guidance for Charities.
Collins, Jane. 2012. Theorizing Wisconsin's 2011 Protests: Community-based Unionism Confronts Accumulation by Dispossession. *American Ethnologist* 39 (1): 6–20.

Cranford, Cynthia J., and Deena Ladd. 2003. Community Unionism: Organizing for Fair Employment in Canada. *Just Labour* 3: 46–59.

Crenshaw, Kimberlé. 1991. Mapping the Margins Intersectionality, Identity Politics, and Violence against Women of Color. *Stanford Law Review* 43: 1241–1299.

Dennis, N., F. Henriques, and C. Slaughter. 1956. *Coal is Our Life*. London: Eyre & Spottiswoode.

Desai, A. 2003. Neoliberalism and Resistance in South Africa. *Monthly Review* 54 (8): np.

Dobson, Paul. 2018. Venezuelan Constituent Assembly Approves Workers' Councils Law. Venezuelanalysis.com, February 3.

Dunlap, Riley E., and Angela G. Mertig. 1995. Global Concern for the Environment: Is Affluence a Prerequisite? *Journal of Social Issues* 51: 121–137.

———. 1997. Global Environmental Concern: An Anomaly for Postmaterialism. *Social Science Quarterly* 78 (1): 24–29.

Dunlap, Riley E., and Richard York. 2008. The Globalization of Environmental Concern and the Limits of the Postmaterialist Values Explanation: Evidence from Four Multinational Surveys. *Sociological Quarterly* 49: 529–563.

Ellem, Bradon. 2008. Contested Communities: Geo-Histories of Unionism. *Journal of Organizational Change Management* 21 (4): 433.

Fink, Elisabeth. 2018. Transnational Social Movement Unionism as a Vitalization Strategy and Opportunity for Trade Unionists? The Example of the Bangladesh Clothing Sector. *Industrielle Beziehungen*, 2-2018: 188–208.

Foster, John, and Charles Woolfson. 1986. *The Politics of the UCS Work-in: Class Alliances and the Right to Work*. London: Lawrence and Wishart.

Franzen, Axel. 2003. Environmental Attitudes in International Comparison: An Analysis of the ISSP Surveys 1993 and 2000. *Social Science Quarterly* 84 (2): 297–308.

Franzen, Axel, and Dominikus Vogl. 2013. Two Decades of Measuring Environmental Attitudes: A Comparative Analysis of 33 Countries. *Global Environmental Change* 23 (5): 1001–1008.

Gall, Gregor. 2009. *Unions in the Community: A Survey of Union Reps*. London: Trades Union Congress.

Gall, G., and J. Fiorito. 2011. The Backward March of Labour Halted; or What is to be Done with 'Union Organising'? The Cases of Britain and the USA. *Capital and Class* 35 (2): 233–251.

Geoghegan, Martin, and Fred Powell. 2009. Community Development and the Contested Politics of the Late Modern 'Agora': Of, Alongside or Against Neoliberalism? *Community Development Journal* 44 (4): 430–447.

Gilbert, D. 1992. *Class, Community and Collective Action: Social Change in Two British Coalfields 1850–1926*. Cambridge: CUP.

Gobo, G. 2008. *Doing Ethnography*. Los Angeles, CA: Sage.

Goodall, Heather, and Alison Cadzow. 2010. The People's National Park: Working-Class Environmental Campaigns on Sydney's Georges River, 1950–67. *Labour History* 99: 17–35.

Gottlieb, Robert. 1993. *Forcing the Spring. The Transformation of the American Environmental Movement*. Washington: Island Press.

Harley, Anne, and Eurig Scandrett, eds. 2019. *Environmental Justice, Popular Struggle and Community Development*. Bristol: Policy Press.

HCLG. 2019. English Indices of Multiple Deprivation. Ministry of Housing Communities and Local Government. https://www.gov.uk/guidance/english-indices-of-deprivation-2019-mapping-resources

Hess, Michael. 2008. Community as a Factor in Union Organization. *Journal of Organizational Change Management* 21 (4): 497.

Hey, David. 2011. Kinder Scout and the Legend of the Mass Trespass. *Agricultural History Review* 59 (2): 199–216.

Hobsbawm, Eric. 1987. Labour in the Great City. *New Left Review* 166: 37–51.

Holgate, Jane. 2009. Contested Terrain: London's Living Wage Campaign and the Tensions Between Community and Union Organising. In *Community Unionism. A Comparative Analysis of Concepts and Contexts*, ed. I. Greenwood and J. McBride. London: Palgrave Macmillan.

———. 2015. Community Organising in the UK: A 'New' Approach for Trade Unions? *Economic and Industrial Democracy* 36 (3): 431–455.

———. 2018. Trade Unions in the Community: Building Broad Spaces of Solidarity. *Economic and Industrial Democracy*: 1–22. https://doi.org/10.1177/0143831X18763871

Holifield, Ryan. 2004. Neoliberalism and Environmental Justice in the United States Environmental Protection Agency: Translating Policy into Managerial Practice in Hazardous Waste Remediation. *Geoforum* 35 (3): 285–297.

ILO. 2015. *World Employment and Social Outlook 2015: The Changing Nature of Jobs*. International Labour Organisation.

Inglehart, Ronald. 1977. *The Silent Revolution: Changing Values and Political Styles Among Western Publics*. Princeton, NJ: Princeton University Press.

ITUC. 2015a. Scandal: Inside the Global Supply Chains of 50 Top Companies. *ITUC Frontlines Report*.

———. 2015b. Climate Justice: Paris and Beyond. *Frontlines Briefing: October 2015*. Paris: International Trade Union Confederation. http://www.ituc-csi.org/IMG/pdf/ituc-frontlinesbriefing_en.pdf

J4J. 2020. Justice for Janitors, Canada. https://www.justiceforjanitors.ca/

Jenkins, Kathy, and Sara Marsden. 2019. Grassroots Struggles to Protect Occupational and Environmental Health. In *Environmental Justice, Popular Struggle and Community Development*, ed. A. Harley and E. Scandrett, 189–210. Bristol: Policy Press.

JwJ. 2020. Jobs with Justice. https://www.jwj.org/

Kapesea, Robby, and Thomas McNamara. 2020. 'We Are Not Just a Union, We Are a Family' Class, Kinship and Tribe in Zambia's Mining Unions. *Dialectical Anthropology* 44 (2): 153–172.

Knight, Kyle W., and Benjamin L. Messer. 2012. Environmental Concern in Cross-National Perspective: The Effects of Affluence, Environmental Degradation, and World Society. *Social Science Quarterly* 93: 521–537.

Labor4sustainability. 2017. Mission Statement of the Labor Network for Sustainability. https://www.labor4sustainability.org/wp-content/uploads/2017/03/LNS-Mission-and-Principles.pdf

LCM. 2004–2007. Lockleaze Community Market Management Group, Minutes of Monthly Meetings. Bristol: Lockleaze Neighbourhood Trust.

Lerner, Steve. 2012. *Sacrifice Zones: The Front Lines of Toxic Chemical Exposure in the United States*. Cambridge, MA: The MIT Press.

LNT. 2006. 'Lockleaze Community Market Evaluation Report' June 2006. Bristol: Lockleaze Neighbourhood Trust.

Lucas Plan. 2020. http://lucasplan.org.uk/just-transition/

MacPhee, Katrin. 2014. Canadian Working-Class Environmentalism, 1965–1985. *Labour/Le Travail* 74: 123–149.

Martinez-Alier, Joan. 2003. *Environmentalism of the Poor*. Basingstoke: Edward Elgar.

McBride, Jo, and Ian Greenwood, eds. 2009. *Community Unionism: A Comparative Analysis of Concepts and Contexts*. London: Palgrave Macmillan.

Mckenzie, Lisa. 2015. *Getting By: Estates, Class and Culture in Austerity Britain*. Bristol: Policy Press.

Mollona, M. 2009. Community Unionism Versus Business Unionism: The Return of the Moral Economy in Trade Union Studies. *American Ethnologist* 36 (4): 651–666.

Montrie, Chad. 2018. *The Myth of Silent Spring: Rethinking the Origins of American Environmentalism*. University of California Press.

Moody, K. 1997. Towards an International Social-Movement Unionism. *New Left Review*, I, 225 (Sep.–Oct.): 52–72.

Moseley, M. 2003. *Rural Development: Principles and Practices*. London: Sage.

Ness, Immanuel, and Dario Azzelini, eds. 2010. *Ours to Master and to Own: Workers' Control from the Commune to Present*. Chicago, IL: Haymarket Books.

Newell, Peter. 2020. *Global Green Politics*. Cambridge: Cambridge University Press.

Pang, Melissa, João Meirelles, Vincent Moreau, and Claudia Binder. 2019. Urban Carbon Footprints: A Consumption-based Approach for Swiss Households. *Environmental Research Communications*, 2(2020)011003.

Pearce, Jenny, and Elizabeth-Jane Milne. 2010. *Participation and Community on Bradford's Traditionally White Estates*. York: Joseph Rowntree Foundation.

Pellow, David Naguib. 2007. *Resisting Global Toxics: Transnational Movements for Environmental Justice*. Cambridge, MA: The MIT Press.

———. 2018. *What is Critical Environmental Justice?* Cambridge, UK: Polity Press.

Piff, Paul K., M.W. Kraus, and D. Keltner. 2018. Unpacking the Inequality Paradox: The Psychological Roots of Inequality and Social Class. *Advances in Experimental Social Psychology* 57: 53–124.

Prowse, Peter, Ray Fells. 2016. The Living Wage - Policy And Practice. *Industrial Relations Journal* 47 (2): 144–162.

Prowse, Peter, Ray Fells, and Ana Lopes. 2017. Community and Union-led Living Wage Campaigns. *Employee Relations* 39 (6): 825–839.

Pulido, Laura. 1998. Development of the 'People of Color' Identity in the Environmental Justice Movement of the Southwestern United States. *Socialist Review* 26 (3–4): 145–180.

Rector, Josiah. 2014. Environmental Justice at Work: The UAW, the War on Cancer, and the Right to Equal Protection from Toxic Hazards in Postwar America. *Journal of American History* 101 (2): 480–502.

Rosemberg, Anabella. 2014. Climate Change is a Trade Union Issue. Paris: ITUC-TUAC. http://www.ituc-csi.org/IMG/pdf/en_unions4climate.pdf

SA. 2020. Sustainable America. https://sustainableamerica.org/

Satheesh, Silpa. 2020. Moving Beyond Class: A Critical Review of Labor-Environmental Conflicts from the Global South. *Sociology Compass* 14: e12797.

Savage, M., and A. Miles. 1994. *The Remaking of the British Working Class 1840–1940*. London: Routledge.

Simms, Melanie. 2012. Imagined Solidarities: Where is Class in Union Organising? *Capital and Class*. 36 (1): 92–110.

Skeggs, Beverley. 2004. *Class, Culture, Self*. London: Routledge.

Smith, David H. 1997a. The International History of Grassroots Associations. *International Journal of Comparative Sociology* 38 (3–4): 189–216.

———. 1997b. The Rest of the Nonprofit Sector: Grassroots Associations as the Dark Matter Ignored in Prevailing 'Flat-Earth' Maps of the Sector. *Nonprofit and Voluntary Sector Quarterly* 26 (2): 114–131.

———. 2000. *Grassroots Associations*. Thousand Oaks, CA: Sage.

Smith, Lerraine. 2006. Lerraine Smith, Coordinator, Grounds for Change, 8th December 2006. 'Trees for Lockleaze' Email to Karen Bell

Suzuki, Akira. 2008. Community Unions in Japan: Similarities and Differences of Region-based Labour Movements between Japan and Other Industrialized Countries. *Economic and Industrial Democracy* 29 (4): 492.

Tapia, Maite. 2013. Marching to Different Tunes: Commitment and Culture as Mobilizing Mechanisms of Trade Unions and Community Organizations. *British Journal of Industrial Relations* 51 (4): 666–688.

Tattersall, Amanda. 2008. Coalitions and Community Unionism. *Journal of Organizational Change Management* 21 (4): 415.

Taylor, Dorceta E. 2014. *Toxic Communities: Environmental Racism, Industrial Pollution, and Residential Mobility*. New York University Press.

———. 2016. *The Rise of the American Conservation Movement: Power, Privilege, and Environmental Protection*. Duke University Press.

Tuckman, Alan G. 1985. Industrial Action and Hegemony: Workplace Occupation in Britain 1971 to 1981. PhD thesis, University of Hull.
UCC CRJ. 1987. Toxic Wastes and Race in the United States: A National Report on the Racial and Socio-Economic Characteristics of Communities with Hazardous Waste Sites New York. United Church of Christ Commission for Racial Justice.
UK Hazards Campaign. 2020. http://www.hazardscampaign.org.uk/
US GAO (General Accounting Office). 1983. Siting of Hazardous Waste Landfills and Their Correlation with Racial and Economic Status of Surrounding Communities, GAO/RCED-83–168. Washington, DC: Government Printing Office.
Vieta, Marcelo. 2020. *Workers' Self-Management in Argentina: Contesting Neo-Liberalism by Occupying Companies, Creating Cooperatives, and Recuperating Autogestión*. Leiden: Brill.
Von Holdt, Karl. 2002. Social Movement Unionism: The Case of South Africa. *Work, Employment and Society* 16 (2): 283–304.
Wainwright, Hilary, and Dave Elliott. 1982. *The Lucas Plan: A New Trade Unionism in the Making?* London: Allison and Busby.
Walker, Gordon. 2012. *Environmental Justice: Concepts, Evidence and Politics*. London: Routledge.
Wall, S. 2008. Easier Said, than Done: Writing an Autoethnography. *International Journal of Qualitative Methods* 7 (1): 38–53.
Wardle, Paul, Libby Robin, and Sverker Sörlin. 2019. *The Environment: A History of the Idea*. Baltimore, MD: John Hopkins University Press.
Waterman, P. 1993. Social-movement Unionism: A New Union Model for a New World Order? *Review (Fernand Braudel Center)* 16 (3): 245–278.
Webb, S., and B. Webb. (1894) 1920. *The History of Trade Unionism*. London: Longman Green and Co.
Weghmann, Vera. 2019. The Making and Breaking of Solidarity Between Unwaged and Waged Workers in the UK. *Globalizations* 16 (4): 441–456.
Wills, J. 2008. Making Class Politics Possible: Organizing Contract Cleaners in London. *International Journal of Urban and Regional Research* 32 (2): 305–323.
Wills, Jane, and Melanie Simms. 2004. Building Reciprocal Community Unionism in the UK. *Capital and Class* 28 (1): 59–84.
Wright, C. 2010. Swords of Justice and Civic Pillars. The Case for Greater Engagement between British Trade Unions and Community Organisations. London: Trades Union Congress.

# Part IV

Trade Unions and the State

# 20

# A Just Transition Towards Environmental Sustainability for All

Catherine Saget, Trang Luu, and Tahmina Karimova

## Introduction

This chapter will review the concept of just transition towards environmental sustainability for all focusing on two dimensions. First, its scope, that is, whether just transition supports workers in fossil fuel industries to move to green economy, or whether it aims at deeper transformative actions of economies and societies, bringing participation and sustainability to the core of policies. Second, its inclusiveness, that is, the extent to which just transition aims at decreasing inequality created by traditional production processes, in particular vulnerable groups affected by climate change. This will help us to identify the main differences between just transition and other types of restructuring and also different approaches in low-, middle- and high-income economies. The case studies will analyse the main elements of just transition policy packages and provide an overview of what types of just transition are better suited for different socio-economic situations.

---

C. Saget (✉) • T. Karimova
International Labour Office, Geneva, Switzerland
e-mail: saget@ilo.org; karimovat@ilo.org

T. Luu
Geneva School of Economics and Management, University of Geneva, Geneva, Switzerland

## Part 1: Context, Development of Concept and the ILO Approach to Just Transition

### A Just Transition in Dealing with Consequences of Climate Change

Ensuring an environmentally sustainable and fair transition for all means that all groups adversely affected by climate policies are compensated and provided with complementary support to benefit from the green transition. This section sets out the rationale for just transition. The world of work and the natural environment are tightly linked (ILO 2018a, p. 17). Most jobs depend on a healthy environment while human activities affect local environment and the global climate.

Human activities have contributed to accelerated global warming, which has increased the frequency and intensity of natural disasters and extreme weather events (IPCC 2014). Natural disasters destroy jobs and worsen decent work, for example by increasing the unemployment rate in the United States and the number of informal workers in Jamaica (Xiao and Feser 2014; Camilo 2017). Globally, between 2000 and 2015, there were 23 million working-life years lost annually as a result of environmentally related hazards caused or exacerbated by human activity (ILO 2018a, p. 23). Income inequality between countries is widening as the burden is borne heavily by low-income countries.

Rising global temperatures as a consequence of climate change will also make heat stress[1] more common, which would cause a reduction in employment and labour productivity. A recent study highlights that heat stress induced by a scenario of 1.5 °C warming is projected to reduce working hours worldwide by 2.2 per cent in 2030—an equivalent of 80 million full-time jobs and a global GDP loss of US$2400 billion (Kjellstrom et al. 2019). The impact of heat stress is more prevalent among workers carrying out heavy labour outdoors, especially agricultural workers, construction workers and street vendors. In addition, higher frequency and intensity of heat stress point to increasing income inequality. As seen in Fig. 20.1, six out of the ten most affected countries are least developed countries according to the UN definition, including Burkina Faso, Togo, Cambodia, Sudan, Chad and Sierra Leone. Heat stress also worsens existing inequality within and across countries

---

[1] Heat stress refers to heat received in excess of that which the body can tolerate without suffering any physiological impairment.

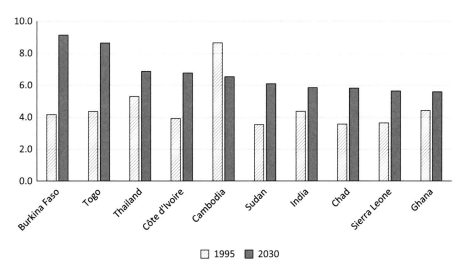

**Fig. 20.1** Percentage of working hours lost due to heat stress, ten most affected countries, 1995 and 2030 (projections). (Source: Kjellstrom et al. 2019)

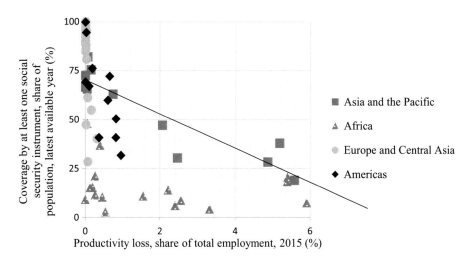

**Fig. 20.2** Correlation between labour productivity loss due to heat stress and social security coverage, selected countries. (Source: Kjellstrom et al. 2019)

in terms of decent work deficits (Kjellstrom et al. 2019). For example, Fig. 20.2 shows the negative correlation between labour productivity loss and social security coverage: countries that are particularly hit hard by heat stress also observe limited social protection, which further increase workers' vulnerability.

Climate change contributes to social instability.[2] Adverse income shocks induced by climate change and extreme weather events could spur violent conflicts. In sub-Saharan Africa, extreme weather events are found to affect both low- and high-intensity conflicts (Burke et al. 2009; Maystadt and Ecker 2014; Almer et al. 2017; Harari and Ferrara 2018). In other parts of the world, for example, rainfall shocks have increased land invasions in rural Brazil (Hidalgo et al. 2010) and civil conflict attacks and fatalities in West Bengal, India (Eynde 2018). Regarding temperature, historical analyses support the positive correlation between extreme cold weather events on political and/or social stability in Europe (Tol and Wagner 2010), Northern Africa and the Near East (Iyigun et al. 2017) and China (Zhang et al. 2006; Zhang et al. 2011). The positive link between cold weather and conflict is channelled through the agricultural production mechanism as agriculture was the dominant livelihood at the time. In urban, modern United States, temperature is found to be positively correlated with crime rates (Ranson 2014).

The interlinkage between the environment and the world of work underscores the urgent need for actions to combat climate change while at the same time increasing decent work and social justice for all. Grievances arising from inequality induced by climate change could spur social unrest and violent conflicts, which could further destabilise fragile societies. Urgent climate actions thus must not exacerbate social-economic inequality. A radical structural transition towards environmental sustainability, while ensuring no one is left behind, is critical for sustainable development.

An environmentally sustainable economy development pathway entails substantial restructuring of the economy in terms of jobs and the skills required. The scope of the transition is not limited to the energy sector but involves structural change within economies, even though the first transition initiatives often focused on energy. A proactive transition towards an energy efficient economy is projected to create 24 million jobs and destroy six millions worldwide, resulting in a net creation of 18 million jobs by 2030 (ILO 2018a, p. 42). While the transition is economically viable, millions of workers will lose their jobs or see shifts in skills demands. Long-term planning, compensatory payments, reskilling and skills updates, public employment programmes and social protection measures, among other measures, would reduce the adverse impacts on affected workers and communities.

Policy responses to climate change will generate distributional impacts across sectors (see Table 20.1). Sectors that see the strongest growth in job demand are construction, manufacture of electrical machinery and copper

---

[2] For a thorough review of the literature, see Burke et al. (2009), Hsiang et al. (2013), Koubi (2019).

## 20 A Just Transition Towards Environmental Sustainability for All

Table 20.1 Sectors most affected by the transition to sustainability in the energy sector

| Industries set to experience the highest job demand growth | | Industries set to experience the strongest job demand decline | |
|---|---|---|---|
| Sector | Jobs (millions) | Sector | Jobs (millions) |
| Construction | 6.5 | Petroleum refinery | −1.6 |
| Manufacture of electrical machinery and apparatus | 2.5 | Extraction of crude petroleum and services related to crude oil extraction, excluding surveying | −1.4 |
| Mining of copper ores and concentrates | 1.2 | Production of electricity by coal | −0.8 |
| Production of electricity by hydropower | 0.8 | Mining of coal and lignite, peat extraction | −0.7 |
| Cultivation of vegetables, fruit, nuts | 0.8 | Private households with employed persons | −0.5 |
| Production of electricity by solar photovoltaics | 0.8 | Manufacture of gas, distribution of gaseous fuels through mains | −0.3 |
| Retail trade, except of motor vehicles and motorcycles; repair of personal and household goods | 0.7 | Extraction of natural gas and services related to natural gas extraction, excluding surveying | −0.2 |

Note: Number of jobs presented is the difference of employment between the sustainable energy and the IEA 6 °C (business-as-usual) scenario by 2030
Source: ILO (2018a, 44)

mining—these are sectors related to the production of renewable energy and manufacturing of electrical vehicles. Sectors that experience the strongest decline in job demand are fossil fuel–based sectors, including petroleum refinery, crude oil and coal extraction. Since the majority of these jobs often concentrate in regions where economic diversity is limited (i.e. coal mining), job losses will also exacerbate local economic development. Other sectors that produce goods based on petroleum (Chisholm 2015) will also see change in employment and skills demand. Although still limited, there have been increasing efforts to evaluate the distributional impact of the green transition and help inform policy actions.[3]

There is hence an imperative for the transition to environmentally sustainable economies. The key question is how to ensure that the transition is just. Such a transition is not achieved by default but requires collaborative efforts from governments, trade unions, employers and workers at local, national and

---

[3] For example, in the Philippines an assessment has been carried out to study the impacts of the Green Jobs Act on winners and losers from the implementation of the law (ILO 2019c).

international levels. Before unpacking the essential elements that make a transition fair and inclusive we offer a brief history of the concept and the ILO's approach to just transition.

## Just Transition in International Instruments and the ILO Approach

### Inclusion of Just Transition in International Instruments

At the international level, the concept of just transition has not been legally defined. However, notions related to just transition have been expressed in various international binding and non-binding instruments. Major climate change instruments and policies such as the United Nations Framework Convention on Climate Change (UNFCCC) of 1992 did not deal with the concept of just transition. However, the UNFCCC and Kyoto Protocol recognised various social dimensions involved in the path towards reaching the objectives of the Convention.[4] Both, and in similar terms, iterate the need to minimise the adverse social, environmental and economic impacts of climate change response strategies particularly in low-income countries.[5]

The significance of these acknowledgements lies in the fact that international climate change frameworks have been constructed acknowledging the social and economic implications of climate change responses. Therefore, they highlight, albeit in general terms, the need for cooperative measures among states. States need to understand the impacts of climate adaptation and mitigation policies on socio-economic processes. This emphasis on the imperative to assemble a coherent picture of the social implications of climate policy is important if adequate policy responses to mitigate adverse social impacts are to be designed. Above all, such an understanding is a first step to addressing issues faced by those affected most by transition, in particular workers.

This recognition of the social implications of climate strategies led to a more pronounced concern for the labour issues within the Paris Agreement[6]

---

[4] Article 4(1)(h) of UNFCCC states that all member states taking into account their common but differentiated responsibilities and their specific national and regional development priorities, objectives and circumstances shall 'Promote and cooperate in the full, open and prompt exchange of relevant scientific, technological, technical, socio-economic and legal information related to the climate system and climate change, and to the economic and social consequences of various response strategies'. UNFCCC, 1992, Article 3.3.

[5] See, for example, Articles 3 and 10, Kyoto Protocol.

[6] The Paris Agreement is an agreement within the United Nations Framework Convention on Climate Change (UNFCCC), dealing with greenhouse-gas-emissions mitigation, adaptation and finance, signed in 2016.

(2016), which embraced the notion of just transition, albeit only in its preamble and without any further elaboration of its content and the specific measures that it should underpin. Despite its rather preambular role in the Paris Agreement, just transition was a major theme at the 24th Conference of the Parties to the UNFCCC held in December 2018, in Katowice, Poland, which has seen the adoption of the 'Silesia Declaration on Solidarity and Just Transition'. The Declaration endeavours to flesh out just transition concerns stipulated by the Paris Agreement. Although non-binding, the Declaration marks a departure from the traditional approach of climate policies in that it provides a broader vision for an equitable and fair response to challenges of transition for communities and countries affected by it.

Another international instrument that refers to just transition is *ILO Recommendation No. 205 on Employment and Decent Work for Peace and Resilience*, 2017 (No. 205), which makes a number of references to just transition as one of the approaches to implement coherent and comprehensive strategies for preventing crises and building resilience (ILO 2017a). The need for a just transition appears as a guiding principle (Paragraph 7(j)) in the Recommendations but also as part of strategic approaches preventing crises, enabling recovery and building resilience (Paragraph 11(e)). The Recommendation, however, does not deal in much detail with the scope and content of just transition but its content would have to be understood in light of the *ILO Guidelines for a Just Transition Towards Environmentally Sustainable Economies and Societies for All* (ILO 2015), which we discuss in the next section.

## ILO Guidelines for a Just Transition Towards Environmentally Sustainable Economies and Societies for All

The concept of just transition,[7] originally born out of the US labour movement discourse in the 1970s (even though not in the explicit just transition language used first in the mid-1990s), essentially expressed the need to find solutions for workers displaced from their work in the process of phasing out polluting industries for the benefit of the environment (Morena et al. 2019). Although this idea and the concept of just transition as such were not yet expressly mentioned in the early stages of the ILO's engagement with the topic of the environment, the need for transitional measures was highlighted already prior to the 1990s when the concept started to permeate international discourse.[8]

---

[7] The explicit use of the 'just transition' language was first used in the mid-1990s.
[8] For an overview, see Olsen and Kemter (2013).

The ILO was actively engaged in the major international discussions on environment.[9] The relationship between the environment and the world of work was articulated in two ILO resolutions: *1972 Resolution concerning Contribution of the ILO to the Protection and Enhancement of the Environment Related to Work (ILO 1972)* and *1990 Resolution concerning Environment, Development, Employment and the Role of the ILO (ILO 1990)*. These resolutions appear to acknowledge trade-offs between employment and the environment and propose to resolve potential conflicts through a balanced economic, social and environmental development.

In 2007, the ILO Director General dedicated a report to the International Labour Conference (ILC) on the promotion of a socially just transition to green jobs as one of the objectives of the Organization (ILO 2007a). Furthermore, the general discussion and conclusions on sustainable enterprises reinforced the insight that 'it is in workplaces that the social, economic and environmental dimensions of sustainable development come together inseparably' and also called for just transitions for workers affected by economic restructuring (ILO 2007b, paras. 3 and 8).

The Green Jobs Initiative, partnered between UNEP (United Nations Environment Programme), ITUC (International Trade Union Confederation), IOE (International Organization of Employers) and the ILO, was launched in 2007 with a mission to promote opportunity, equity and a just transition to sustainable economies and to mobilise governments, employers and workers to engage in dialogue on coherent policies and effective programmes. The initiative published the first global flagship report on green jobs (UNEP et al. 2008). Established in 2008, the ILO Green Jobs Programme has assisted its constituents with knowledge creation, advocacy, capacity building, implementing pilot projects, policy advice and knowledge sharing. The 102$^{nd}$ Session of the ILC in 2013 stated the roles of governments, employers' and workers' organisations and key policy areas concerning sustainable development, decent work and green jobs, which put forward a policy framework for a just transition (ILO 2013).

At the ILO, efforts to concretise the scope and content of just transition including through a dedicated international labour standard on the issue have led to the development of *Guidelines for a Just Transition towards Environmentally Sustainable Economies and Societies for All (Guidelines for a just transition)* (ILO 2015). The Guidelines were to serve as a step towards better equipping

---

[9] These include the UN Conference on the Human Environment of 1972 in Stockholm; the UN Conference on Environment and Development of 1992 in Rio; the World Summit on Sustainable Development of 2002 in Johannesburg and others.

governments, employers and workers to understand the challenges and opportunities of the transition and to take up the active roles that they must play in managing this change.[10] One of the main visions of the Guidelines is to ensure that transition is 'for all' and 'needs to be well managed and contributes to the goals of decent work for all, social inclusion and the eradication of poverty' (ILO 2015, para. 4).

The *ILO Guidelines for a Just Transition*[11] detail a total of nine key policy areas and institutional arrangements where environmental, economic and social sustainability can be addressed: (i) macroeconomic and growth policies; (ii) industrial and sectoral policies; (iii) enterprise policies; (iv) skills development; (v) occupational safety and health; (vi) social protection; (vii) active labour market policies; (viii) rights and (ix) social dialogue and tripartism. The Guidelines particularly address the role of social dialogue in achieving a just transition, including concrete ways to facilitate cooperation between workers' and employers' organisations through collective bargaining and collective agreements regarding specific environmental provisions.

The other overarching objective of the Guidelines is to orient the ILO constituency towards achieving coherence and integration of all dimensions of sustainability: environmental, economic and social, which would provide an enabling environment for enterprises, workers, investors and consumers to embrace and drive the transition towards environmentally sustainable and inclusive economies and societies. These coherent policies also need to provide a just transition framework for all to promote the creation of more decent jobs, including anticipating impacts on employment, adequate and sustainable social protection for job losses and displacement, skills development and social dialogue (including the effective exercise of the right to organise and bargain collectively).

The Guidelines also highlight that just transition measures are not simply an addition, but rather an integral part of broader development policies, if the objectives of the Sustainable Development Goals are to be achieved. The Guidelines specifically stipulate that governments should 'integrate provisions for a just transition into national plans and policies for the achievement of the Sustainable Development Goals and national environmental and climate

---

[10] GB.325/PV, para 494. The Governing Body requested the ILO Director-General to use the *Guidelines* as a basis for organizations activities and outreach.

[11] The Guidelines for a just transition were discussed and agreed upon during the Tripartite Meeting of Experts on Sustainable Development, Decent Work and Green Jobs from 5 to 9 October 2015. Participants included eight experts nominated by the Governments of Brazil, Indonesia, Germany, Kenya, Mauritius, Turkey, South Africa and the United States; eight experts appointed after consultation with employers' organizations and eight others appointed after consultation with the workers' organizations.

change action plans' (ILO 2015, para.15(c)). The Guidelines also take into account the countries' different political, social and economic contexts and recommend that policies and programmes need to be designed in line with the specific conditions of countries, including their stage of development, economic sectors as well as types and sizes of enterprises.

These guidelines are one of the building blocks of policy frameworks related to the environmental sustainability dimension of the world of work. The aforementioned Silesia Declaration endorsed the Guidelines for a Just Transition and identified these Guidelines as the framework to implement just transition reference found in the Paris Agreement, by providing that:

> Taking note of the importance of the International Labour Organization's 'Guidelines for a just transition towards environmentally sustainable economies and societies for all', and its considerations, as appropriate, by Parties while fulfilling their commitments under the Paris Agreement on climate change. (UNFCCC 2018)

## Scope and Inclusiveness of Just Transition from the ILO Approach

Practical approaches distinguish the wide scope of just transition, as well as its focus on inclusiveness and participation of workers' and employers' organisations (Just Transition Research Collaborative 2018). Within that framework, the scope refers to which sectors/workers/other actors are targeted by public policies and enterprises' restructuring packages. Is the support limited to income replacement and retraining programmes and job search services? Or does it also cover people and sectors which are indirectly affected by job displacement, for instance, workers in local services? In this latter case, just transition measures include local development perspectives, such as infrastructure building and incentives for enterprises to move to the region negatively affected by the green transition. Turning to the second dimension—the inclusiveness—just transition measures can be conceived with a view of reducing inequality and supporting the most vulnerable—or leaving the social market forces play their role.

The above framework casts a light on different approaches to just transition among low-, middle- and high-income countries. It is also helpful in identifying some of the institutional factors that will make the transition successful, for example, meeting its policy objectives, in particular social protection, infrastructure development and skills development institutions. However, the framework lacks a fuller perspective of governance of the transition towards

## 20 A Just Transition Towards Environmental Sustainability for All

sustainability and how to analyse collaboration gaps and potential between institutions at the local, regional and national levels. Labour organisations can bring success to the transition. In addition, policies and programmes need to take into account the gender dimension of many environmental challenges and opportunities. Specific gender policies should be considered in order to promote equitable outcomes.[12]

As with any other economic restructuring, international labour standards can promote decent work during and beyond the transition. They provide guidance on direct restructuring (i.e. standards on skills development and social protection policies) and set out a framework to address issues of structural transformation of economies, as well as labour institutions to limit the rise in inequality (i.e. collective bargaining and minimum wages) (ILO 2019a). In addition, international labour standards are centred around social dialogue, freedom of association and the right to collectively bargain. They conceive policy making as based on empirical evidence as well as large consultation of social partners and society. Moreover, labour standards can directly provide for the protection of the environment—such as those regulating occupational safety and health (dealing with dangerous substance and extreme temperatures) and those providing for the imperative of environmental impact studies in development support for indigenous and tribal peoples (ILO 2018a, p. 78).

Turning to national policies, it seems that on the whole, progress has been made in integrating environmental policies with social and employment objectives. This appears to be true of the overarching environmental goals, labour agreements, social protection and skills development (ILO 2018a, p. 132). Regarding social protection, climate change and the increased frequency and intensity of natural disasters have translated into an increased need for effective and tailored measures to protect workers and their families. In particular, unemployment protection schemes can play a key role in supporting a just transition for workers who lose their jobs in the shift to a more environmentally sustainable economy.[13] However at the global level, only

---

[12] For example, in the Philippines, rural women are offered training in organic farming to empower them and improve their income (ILO 2018b); in Guyana, Ruppuni Essence is a cosmetics firm that relies on single mothers to grow lemongrass and promotes business opportunities through cooperatives (ILO 2017c).

[13] For example, the Government of China used unemployment protection measures to provide financial assistance to those who were affected by a ban on logging in natural forests in 1998 but unable to find work elsewhere (ILO and AFD 2016); in Romania, following decision to close two uncompetitive coal mining units by 2018, financial support totalling 54 million euros has been earmarked to provide income support to workers who will become unemployed (see http://europa.eu/rapid/press-release_IP-16-3981_en.htm).

38.6 per cent of the global workforce is covered by unemployment protection (ILO 2017b).

In low-income countries, cash transfer programmes can compensate for the loss of income experienced by households as a consequence of adverse environmental events or structural changes resulting from the implementation of green policies. In Kenya and Ethiopia, for example, cash transfer programmes targeted at poor households are responsive to climatic conditions, in particular droughts and floods (Otulana et al. 2016; Knippenberg and Hoddinott 2017; Ulrichs and Slater 2017). In addition, public employment programmes are important policy tools that combine economic, social and environmental objectives in support of adapting to and mitigating environmental degradation and climate change. In South Africa, one of these programmes is targeted at restoring the natural environment of rivers, while providing income support to the participants (Schwarzer et al. 2016).

These examples prove that traditional tools of social protection aimed at addressing issues in the labour market (i.e. unemployment benefits, cash transfers and public employment programmes) can be effective in ensuring that just transition is effective and inclusive. The reverse is also true: environmental protection measures can also contribute to creating decent work and supporting marginalised population. A classic environmental measure—payment for ecosystem services—can offer cost-effective protection for the environment, while at the same time supporting household incomes (Pagiola et al. 2005).[14] Projections show that policies that extend transfers and strengthen social protection and support green investment are financially viable and conducive to higher economic growth, employment creation and fairer income distribution (ILO 2018a, p. 119).

Turning to skills development policy, some countries have been successful in integrating skills development with environmental policy, particularly in key priority sectors like renewable energy and energy efficiency (ILO 2018a, p. 132, 2019d).[15] Some countries have established bodies to identify skills

---

[14] Payments for ecosystem services (PES) was invented as a market-based mechanism that integrates environmental externalities. More recently, several PES schemes have been designed or expanded to include pro-poor components, such as the Socio Bosque programme in Ecuador and the Bolsa Verde programme in Brazil.

[15] In countries such as Denmark, Estonia, France, Germany, India, the Republic of Korea, the Philippines and South Africa, a number of environmental policies and national development strategies make reference to skills development for the green transition. In some of these countries, skills development strategies have acknowledged the rising demand for skills required for greening the economy. For example, the Philippines' Green Jobs Act of 2016 contains clauses promoting skills for green jobs through initiatives such as identifying skills needs, maintaining a database of green careers, formulating training regulations, skills assessment and certification, as well as curriculum development.

needs[16] while some mainstream environmental sustainability into existing skills development mechanisms. There are also signs of emerging policy coherence in some countries, where environmental policies make explicit reference to skills and/or human resource development or full-fledged skills development policies and legislation for green transition. However, these references are often limited to specific policy areas, target groups, sectors and regions (ILO 2018a, p. 150). The involvement of social partners improves the matching of skills demand and supply, and equity outcomes, including gender equality, although not all countries involved social partners in the discussions to identify skills needs and reflect on new or updated training programmes. It is remarkable that new skills programmes have been adopted in renewables and energy efficiency, often following the enactment of new energy or environmental laws. Progress in other sectors have been slower (ILO 2018a, p. 143).[17]

Regarding policy coherence, most countries have not established sound linkages between their environmental sustainability plans and skills policies (ILO 2019d). Case studies show that processes to facilitate systematic policy coordination across ministries are rare and ministries dealing with education, training and employment are weakly represented in policy-making on climate change issues (ibid.). The lack of knowledge of the environment–skills nexus, the absence of regularly conducted employment projections and of financial mechanisms to promote investments in skills development for the green transition, and the sluggish participation of social partners are still hindering the achievement of an effective transition.

In conclusion, the ILO approach to just transition emphasises the need to mainstream environmental issues in the conception and implementation of the broad policy framework, including macroeconomic and growth policies, industrial- and sector-level policies as well as labour market policies (ranging from occupational safety and health, social protection, skills development to active labour market policies). It also relies on the consultation and participation of social partners in policy making to facilitate the representation of the views of the world of work, and as a vehicle of social justice.

---

[16] In France, the *National Observatory of Jobs and Skills in the Green Economy* conducts regular assessments on the employment trends in green economy

[17] For example trainings in organic agriculture and small-scale farming, harvesting and maintenance of mangroves (Guyana), ecotourism (the Philippines, Colombia) and waste management (Bangladesh).

## Part 2: Just Transition Policies in Practice

This section examines three case studies of different socio-economic contexts and stages of just transition. Spain is a high-income economy that has experienced an aggressive energy transition in which active social dialogue has resulted in a just transition plan in the coal mining sector. South Africa is a unique case study of an upper-middle-income country[18] facing mounting challenges in addressing social-economic issues while aiming to decarbonise the energy sector. The necessity of a just transition has been acknowledged in a variety of its policy documents. Participatory process by different stakeholders has contributed to forming a consensus on how the transition should be managed in the country. Ethiopia represents a low-income country which integrates environmental sustainability and climate resilience in its economic growth strategies. Some socio-economic measures have great potential for integrating just transition into long-term growth and social-economic policies. The cases emphasise the key role played by social dialogue and cooperation among government, trade unions, workers and the private sectors.

# Spain

## Context and Challenges

An EU Directive requires all uncompetitive coal mines in the European Union[19] to cease operations by 1st January 2019, a date when all public funds to support the mines were to come to an end. For Spain, that meant that five regions—Aragón, Asturias, Castilla y León, Castilla-La Mancha and País Vasco—are considered 'coal mining regions' that will experience an energy transition (see Räthzel, Chap. 28). The rationale for the EU to implement the transition was part of the measures to move away from fossil fuel dependence.

On 24 October 2018, the Spanish mining unions concluded a deal, entitled *Plan del Carbon*,[20] for a just transition from coal mining with plans for sustainable development in mining regions (Government of Spain 2018). The plan provides for the closure of all Spanish coal mines that are no longer viable

---

[18] According to the World Bank, upper-middle-income economies are those with a GNI per capita between $3996 and $12,375 (2020 values).

[19] 2010/787/EU: Council Decision of 10 December 2010 on State aid to facilitate the closure of uncompetitive coal mines.

[20] The plan is officially entitled as Framework Agreement for a Fair Transition of the Coal Mining and Sustainable Development of the Mining Countries for the Period 2019–2027.

from an economic point of view. The government and trade unions signed the 250-million-euro deal, with the transition expected to take place between 2019 and 2027.

The agreement was reached after years of trade union struggles as successive governments in Spain have attempted to end subsidies to the mining industry and close mines, without ensuring that this transition will meet the needs of the workers most affected. For example, this led, in 2012, to a 457 km Marcha Negra—Black March—to Madrid by coal miners to protest the loss of jobs that would result from the end of subsidies. Unemployment rate in Spain remains high, averages at 15.3 per cent in 2018 and tops at 34.3 per cent for youth (ILOSTAT 2020).

## Policy Instruments

The *Plan der Carbon* replaced previous subsidies to the coal mining industry with a sustainable development plan. A loss of 1677 jobs was expected by the end of 2019. Around 60 per cent of miners—aged 48 and older, or with 25 years of service—will be able to take early retirement. Younger miners will receive a redundancy payment of €10,000, as well as 35 days' pay for every year of service. The agreement also allocates funds to restore and environmentally regenerate former mining sites. Priority for employment in jobs generated by this will go to former miners. Furthermore, the plan envisages funds to upgrade facilities in the mining communities, including waste management, recycling facilities and water treatment plants, utilities infrastructure and distribution for gas and lighting, forest recovery, atmospheric cleansing and reducing noise pollution (Government of Spain 2018).

The agreement is considered a model among agreements on just transition and provides a package of benefits for miners and their communities. Measures to support just transition are of two types: (i) social protection for workers in coal mining units and (ii) support aimed at covering the closing costs of the production and mitigation of environmental impact. In addition to this landmark Plan del Carbon, Spain is currently considering the adoption of a draft Law on Climate Change and Energy Transition.

Meanwhile, the Parliament of the Balearic Islands approved the Law of Climate Change and Energy Transition of the Balearic Islands in February 2019 (Government of Balearic Islands 2019). The law promoted as a 'green manifesto' paves a way for addressing climate change and implementing a transition to clean energy. The law sets an objective of making the Islands fossil fuel free and opts for a 100 per cent renewable energy future by 2050, and sets a target of reaching 35 per cent of renewable energy, 23 per cent reduction in energy consumption and a 40 per cent reduction of polluting emissions by 2030. This makes the law one of the most innovative in both Spain and Europe.

More generally, the Law's stated objective is to complement international commitments stemming from the Paris Agreement, as well as to pursue a transition to a sustainable energy model, which is socially just, decarbonised, smart, efficient, renewable and democratic. Therefore, the purposes of the legal framework pursued include progress towards the new environmental and energy model following the principles of just transition, taking into account the interests of citizens and the sectors affected by this transition and promotion of employment and training in the new economic sectors that are generated and promoted (Government of Balearic Islands 2019).

# South Africa

## Context and Challenges

Job creation is one of South Africa's priorities. In 2019, the unemployment rate in South Africa was 28.5 per cent while 57 per cent of the youth labour force could not find jobs (ILOSTAT 2020). In addition to high unemployment and inequality, energy poverty remains a challenging issue as 26 per cent of the population do not have access to modern electricity (World Bank 2019). A transition to a low-carbon economy requires structural changes in South Africa's heavily coal-dependent economy. Given increasing production costs, decreasing international market demand in combination with the government's commitment to decarbonise, it is projected that employment in coal mining will decline from 82,000 jobs in 2018 to 21,400 jobs in 2045 (Burton et al. 2018). The burden will be borne by workers employed in coal mines, coal-based utility power plants and communities heavily dependent on coal. More than 80 per cent of coal production is located in the Mpumalanga province where almost half of the households live below the national poverty line (ibid.). Unless carefully managed, the energy transition will have adverse impacts on the region's social-economic development.

The National Development Plan (NDP) acknowledges the need for a just transition in South Africa (South Africa Government 2013). It considers the adverse employment impact of a transition away from coal induced by a stringent international climate policy and seeks to ensure that the transition does not negatively impact jobs and skills. The National Climate Change Response White Paper emphasises the need for policy intervention to address the need of affected workers and communities by acknowledging that 'growth in new sectors alone will be no guarantee of net job creation' and 'government will promote conditions that will increase the mobility of labour and capital out of carbon intensive sectors to greener productive sectors' (Department of Environmental Affairs 2011).

Although the rationale of just transition is well acknowledged by the government, there has been no consensus on what a just transition would mean for South Africa. Social dialogue provides a common platform that allows workers to express concern about the transition. Such a platform provides some direction and helps build consensus on how the transition should be managed (Strambo and Atteridge 2019). Some suggestions gathered from social dialogue include strategies for economic diversification, infrastructure repurposing, training and education programmes, environmental rehabilitation and continued provision of public services.

Although a just transition policy has not been legally formulated, there exist some mechanisms discussed below that could contribute to forming the foundation for just transition integration in future policy design and implementation in South Africa.

## Policy Instruments

The Renewable Energy Independent Power Producer Procurement Programme (REI4P) is a green procurement programme in which private power producers bid for energy projects (ILO 2019b). Bidders have to meet two major deliverables: (i) facilitating local economic activities and (ii) meeting social development needs. In some cases, the programme faces opposition from trade unions blaming the government for giving out contracts to private companies without adequate just transition measures, which would hurt workers in coal-related industries.[21] Ensuring that workers and communities who lose

---

[21] The Congress of South African Trade Unions (COSATU) has criticised the mass roll out of REI4P without just transition measures, including sustainable jobs plan or re-skilling of workers from fossil fuel–based sector to renewable sector (COSATU 2018). The union also contented that local content requirement targets are too lenient; particularly jobs created during the construction phase would be short term, plunging workers deeper into poverty, unemployment and inequality. It urged the government to engage in a process of social dialogue to ensure that jobs are not lost in the process towards renewable energy and embark upon a process of ensuring that the transition towards renewable energy is just with the ownership in the hands of workers and the community. Another union affiliate, the National Union of Metalworkers of South Africa (NUMSA), opposed the programme implementation because buying electricity from the private sector would increase costs and these costs will be passed to users (NUMSA 2012). Moreover, the union criticised the programme as it aims to put energy onto the grid and produce energy security to big corporations instead of provision of energy needs to those who remain off-grid.

out from the energy transition would benefit from REI4P is key for the successful implementation of the programme as REI4P has large potential to enable local employment and skills development (see also Sikwebu/Aroun, Chap. 3).

Another example of inclusiveness is the Expanded Public Works Programme (EPWP), which simultaneously targets green job creation, social protection and skills development. The programme has a cross-cutting social objective and a focus on creating jobs for the most vulnerable, such as women, youth and people with disabilities. EPWP includes sub-programmes that have a strong focus on environmental sustainability, for example, the Working for Water programme, in which participants work on clearing mountain catchments and riparian zones of alien invasive species. The programme grew from just over 6000 jobs in 2012 to more than 50,000 in 2015 (Department of Environmental Affairs 2015). During the same period, more than half of those employed were women and more than 60 per cent of these women were under 35 years old. The programme also targets military veterans and parolees. In addition, skills development targets a largely unskilled and typically vulnerable segment of the population, thus meeting in this particular case both the green and inclusivity objectives (ILO 2019b).

The policy framework employed by the South African government places a strong emphasis on job creation in response to climate change and a transition towards a green economy. The urgent need to address climate change and unemployment through the creation of green jobs has been well recognised by trade unions, environmental activities and social partners (Million Climate Jobs Campaign and AIDC 2017). The policy landscape pays attention to marginalised population and communities vulnerable to climate risks, and several measures such as the EPWP and the REI4P have social development objectives, which refer to the inclusiveness of a just transition. Not much consideration has been paid to workers and communities directly affected by climate policy responses, as in the case of coal mining workers. The delay in forming a consensus on just transition would place more workers at risk as coal mining employment is already declining. Early anticipation of mine closure and long-term planning are necessary to help reduce the adverse impacts of decarbonisation on workers while contributing to South Africa's climate change mitigation goal.

## Ethiopia

Although the concept of just transition has not been explicitly adopted in Ethiopia, the country provides some examples of integrating just transition agenda into its green growth strategy.

### Context and Challenges

Ethiopia is the fastest growing economy in sub-Saharan Africa with an annual growth rate of 10 per cent from 2007 to 2017 (World Bank 2019). Access to modern electricity and energy resources is a challenging issue. Almost 80 per cent of the Ethiopian population live in rural areas and only 44 per cent have access to electricity (ibid.). Climate change and increasing extreme weather events threaten food security and livelihood in Ethiopia.

Ethiopia's policy framework has been designed to meet both growth and objectives of inclusion while addressing the country's environmental and social challenges. The country aims to achieve a middle-income status by 2025 in a carbon-neutral and climate-resilient way. Since 2011, Ethiopia has dedicated a strategic emphasis on green growth, which was reflected in its five-year economic growth plans: the Growth and Transformation Plan 2011–2015 (GTP I) and GTP II (2015–2020). Between the two periods of the GTPs, the Climate Resilient Green Economy (CRGE) strategy was formulated, setting out the vision and action plan for Ethiopia's structural transformation towards a green economy (Federal Democratic Republic of Ethiopia 2011).

The Labour Proclamation provides tools for the implementation of a just transition. It sets out rules for the establishment of trade unions, employer associations and collective agreements as well as workers' and employers' responsibilities with regard to occupational safety and health. The proclamation laid out the foundation for social dialogue in guiding a just transition, allowing unions and employers to participate in all labour matters. More recently, the National Employment Policy and Strategy gives some reference, though minimal, to environmental sustainability (Ministry of Labour and Social Affairs 2017). The strategy consists of five policy priorities, one of which is 'focusing on cross-cutting issues for employment creation'. The priority identifies eight cross-cutting issues, including green jobs to promote employment opportunities through improved environmental protection and natural resource management. The policy, however, remains silent on related implementation measures and does not include any monitoring indicators for green jobs (van der Ree 2017).

## Policy Instruments

Initiated in 2005, the Productive Safety Net Programme (PSNP) provides a practical case of a just transition that can be implemented within a developing country's socio-economic context. The programme provides unconditional cash transfers and cash for works to poor rural households that are chronically food insecure. It combines public employment provision with environmental objectives while providing unconditional cash transfers to vulnerable households unable to provide work. Productive work is conducted in road construction, afforestation, and soil and water conservation activities. Investment in soil and water conservation increases local agricultural and natural-resource productivity, which would help increase farm income in the long term. In addition, the skills gained through the programme (especially soil and water conservation skills) can also be applied on the farmers' own land.

The fossil fuel subsidy had led to increasing budget deficit and the benefits were considered to be captured by higher income groups while a majority of the rural poor does not have access to modern electricity. In 2008, the government decided to replace the fossil fuel subsidy by a food subsidy to offset the adverse impact of increased fuel prices on poor households (see also Houeland, Chap. 21).

# Conclusion

This chapter attempted to canvass the contours of the concept of just transition through the review of the current international discourse, particularly the approach taken by the ILO, to ensure that transition is just in different socio-economic contexts and responds to different challenges across different countries. How the scope and inclusiveness of just transition are applied in practice is reflected in the three selected case studies in the contexts of low-, middle- and high-income economies. Social dialogue and an active role of trade unions, workers, employers and governments are highlighted in the cases of Spain and South Africa, while integrating environmental sustainability policies and social protection in a country's macroeconomic and national development plans is reflected in the cases of Ethiopia and South Africa. It is also important to note that the success of delivering just transition measures in practice also relies on the availability of data on green jobs and green skills as well as the technical capacity of national statistics offices to gather and analyse such data so that winners and losers of the green transition are well identified.

It transpires from our analysis that the concept as originally conceived and its different cases of application do not fully deal with the structural inequalities created by traditional production processes. Nonetheless, the main contribution of the idea is to create an environment that allows for a deeper transformative action. In other words, just transition is about bringing participation and sustainability to the core of policies.

The chapter shows how just transition policies bring to the fore the missing link in the previous attempts to bridge development/growth and environmental sustainability, as well as the intergenerational aspect and the necessity to balance the needs of the present generation without compromising the interests of future generations. Importantly, the thrust of just transition policies is to raise important policy issues, such as how to ensure that the current move towards sustainability leaves no one behind.

The three case studies bring about different contexts, challenges and various development stages of transition policies. The countries vary in terms of economic development: higher, middle and low income. They also differ in terms of institutional capacity to plan, implement and monitor just transition measures. Spain has recently established a just transition plan in the coal mining sector. South Africa acknowledges the role of just transition in a variety of its policy documents and there has been a series of social dialogue among stakeholders, although having been unsuccessful in coming to common understanding, which could contribute to the development of a just transition framework. Ethiopia has not yet provided a legal framework for just transition but several policy measures could be the starting points for integrating just transition strategies in policy design. From the case studies, it can be seen that there is no one-size-fit-all set of measures as it depends largely on a country's political, economic and social context. It is crucial to integrate consultations from different stakeholders to develop just transition plans so that no one is left behind from the transition.

The contribution of just transition to the world of work can be summarised as threefold. In terms of process, it emphasises the participation of social partners and civic society in public debates and the implementation of policies, which can foster social cohesion and reduce inequality. Second, just transition contributes to the success of environmental objectives by ensuring that the right mix of legal frameworks and labour market policies is in place to give people the right sets of skills and social protection measures for the transition. Last but not least, it leads to the creation of green jobs, for example, jobs that are decent in terms of working conditions and also contributes to reducing the negative environmental impact of economic activities.

# References

Almer, Christian, Jérémy Laurent-Lucchetti, and Manuel Oechslin. 2017. Water scarcity and rioting: Disaggregated evidence from Sub-Saharan Africa. *Journal of Environmental Economics and Management* 86: 193–209.

Burke, Marshall B., Edward Miguel, Shanker Satyanath, John A. Dykema, and David B. Lobell. 2009. Warming increases the risk of civil war in Africa. *Proceedings of the National Academy of Sciences* 106: 20670–20674. https://doi.org/10.1073/pnas.0907998106.

Burton, Jesse, Tara Caetano, and Bryce McCal. 2018. *Coal Transition in South Africa—Understanding the Implications of a 2°C-Compatible Coal Phase-out for South Africa*. IDDRI & Climate Strategies.

Camilo, Pecha. 2017. *The Effects of Natural Disasters on the Labour Market: Do Hurricanes Increase Informality?* Working Papers. Inter-American Development Bank. https://doi.org/10.18235/0000944.

Chisholm, Kirk. 2015. 144 Products Made From Petroleum And 4 That May Shock You. Accessed May 30, 2020. https://www.innovativewealth.com/inflation-monitor/what-products-made-from-petroleum-outside-of-gasoline/

COSATU. 2018. COSATU's position on the Renewable Energy Independent Power Producers (REIPP). Accessed June 1, 2020. http://mediadon.co.za/2018/03/14/cosatus-position-on-the-renewable-energy-independent-power-producers-reipp/

Department of Environmental Affairs. 2011. *National Climate Change Response White Paper*. Pretoria.

———. 2015. DEA's Working for Water Programme Celebrates 20 Years of Job Creation & Environment Sustainability.

Eynde, Oliver Vanden. 2018. Targets of Violence: Evidence from India's Naxalite Conflict. *The Economic Journal* 128: 887–916. https://doi.org/10.1111/ecoj.12438.

Federal Democratic Republic of Ethiopia. 2011. *Climate Resilient Green Economy*.

Government of Balearic Islands. 2019. *Law of Climate Change and Energy Transition of the Balearic Islands*.

Government of Spain. 2018. *Framework Agreement for a Fair Transition of Coal Mining and Sustainable Development of the Mining Communities for the Period 2019–2027*.

Harari, Mariaflavia, and Eliana La Ferrara. 2018. Conflict, Climate, and Cells: A Disaggregated Analysis. *The Review of Economics and Statistics* 100: 594–608. https://doi.org/10.1162/rest_a_00730.

Hidalgo, F. Daniel, Suresh Naidu, Simeon Nichter, and Neal T. Richardson. 2010. Economic Determinants of Land Invasions. *The Review of Economics and Statistics* 92: 505–523. https://doi.org/10.1162/REST_a_00007.

Hsiang, Solomon M., Marshall Burke, and Edward Miguel. 2013. Quantifying the Influence of Climate on Human Conflict. *Science* 341: 1235367. https://doi.org/10.1126/science.1235367.

ILO. 1972. *Contribution of the International Labour Organisation to the Protection and Enhancement of the Environment Related to Work.* Geneva: International Labour Office.

———. 1990. *Resolution concerning Environment, Development, Employment and the Role of the ILO.* Geneva: International Labour Office.

———. 2007a. *Decent Work for Sustainable Development.* Report of the Director-General Report 1(A), International Labour Conference, 96th Session. Geneva: International Labour Office.

———. 2007b. *The Promotion of Sustainable Enterprises.* 6. 96th Session. Geneva: International Labour Conference.

———. 2013. *Conclusions of the Committee on Sustainable Development, Green Jobs and Decent Work.* Geneva: International Labour Office.

———. 2015. *Guidelines for a Just Transition Towards Environmentally Sustainable Economies and Societies for all.* Geneva: International Labour Office.

———. 2017a. *Employment and Decent Work for Peace and Resilience Recommendation 205.* Geneva: International Labour Office.

———. 2017b. *World Social Protection Report 2017–19: Universal social protection to achieve the Sustainable Development Goals.* Geneva: International Labour Office.

———. 2017c. *Skills for Green Jobs Study—Guyana.* Report. Port of Spain: Office for the Caribbean—Port of Spain.

———. 2018a. *World Employment and Social Outlook 2018: Greening with jobs.* Geneva: International Labour Office.

———. 2018b. *Skills for green jobs in the Philippines.* Geneva: International Labour Office.

———. 2019a. *Rules of the Game: An Introduction to the Standards-related Work of the International Labour Organization (Centenary edition 2019).* Geneva: International Labour Office.

———. 2019b. *Skills for Green Jobs in South Africa.* Publication. Geneva: International Labour Office.

———. 2019c. *Employment Effects of Green Policies in the Philippines.* Publication. Geneva: International Labour Office.

———. 2019d. *Skills for a Greener Future: A Global View.* Publication. Geneva: International Labour Office.

ILO, and AFD. 2016. *Social Protection and Climate Change: How Can Social Protection Address Regular Climate-Related Risks in the Sahel?* Geneva: International Labour Office.

ILOSTAT. 2020. ILOSTAT Country profiles.

IPCC. 2014. *Climate Change 2014: Impacts Adaptation and Vulnerability.* New York: Cambridge University Press.

Iyigun, Murat, Nathan Nunn, and Nancy Qian. 2017. *Winter is Coming: The Long-Run Effects of Climate Change on Conflict, 1400–1900.* Working Paper 23033. National Bureau of Economic Research. https://doi.org/10.3386/w23033.

Just Transition Research Collaborative. 2018. *Mapping Just Transition(s) to a Low-Carbon World*. United Nations Research Institute for Social Development.

Kjellstrom, Tord, Nicolas Maître, Catherine Saget, Matthias Otto, and Tahmina Karimova. 2019. *Working on a Warmer Planet: The Effect of Heat Stress on Productivity and Decent Work*. Report. Geneva: International Labour Office.

Knippenberg, Erwin, and John F. Hoddinott. 2017. *Shocks, Social Protection, and Resilience: Evidence from Ethiopia*. IFPRI: International Food Policy Research Institute. Washington, D.C.: IFPRI.

Koubi, Vally. 2019. Climate Change and Conflict. *Annual Review of Political Science* 22: 343–360. https://doi.org/10.1146/annurev-polisci-050317-070830.

Maystadt, Jean-François, and Olivier Ecker. 2014. Extreme Weather and Civil War: Does Drought Fuel Conflict in Somalia through Livestock Price Shocks? *American Journal of Agricultural Economics* 96: 1157–1182. https://doi.org/10.1093/ajae/aau010.

Million Climate Jobs Campaign, and AIDC. 2017. *One Million Climate Jobs: Moving South Africa Forward on a Low-carbon, Wage-led, and Sustainable Path*. Cape Town: Alternative Information and Development Centre.

Ministry of Labour and Social Affairs. 2017. *National Employment Policy and Strategy*.

Morena, Edouard, Dunja Krause, and Dimitris Stevis. 2019. *Just Transitions Social Justice in the Shift Towards a Low-Carbon World*. London: Pluto Press.

NUMSA. 2012. Motivations For a Socially-Owned Renewable Energy Sector.

Olsen, L., and D. Kemter. 2013. The International Labour Organization and the Environment. In *Trade Unions in Green Economy: Working for the Environment*, 41–57. Routledge.

Otulana, S., C. Hearle, R. Attah, F. Merttens, and J. Wallin. 2016. *Evaluation of the Kenya Hunger Safety Net Programme Phase 2: Qualitative Research Study—Round 1*. Oxford: Oxford Policy Management.

Pagiola, Stefano, Agustín Arcenas, and Gunars Platais. 2005. Can payments for environmental services help reduce poverty? An exploration of the issues and the evidence to date from Latin America. *World Development* 33: 237–253.

Ranson, Matthew. 2014. Crime, Weather, and Climate Change. *Journal of Environmental Economics and Management* 67: 274–302. https://doi.org/10.1016/j.jeem.2013.11.008.

van der Ree, Kees. 2017. *Mainstreaming Green Job Issues into National Employment Policies and Implementation Plans: A Review*. Working paper. ILO.

Schwarzer, Helmut, Clara van Panhuys, and Katharina Diekmann. 2016. *Protecting People and the Environment: Lessons Learnt from Brazil's Bolsa Verde, China, Costa Rica, Ecuador, Mexico, South Africa and 56 other experiences*. 54. ESS Working Paper.

South Africa Government. 2013. *Our Future—Make it Work: National Development Plan*. The Presidency Republic of South Africa.

Strambo, C, and A Atteridge. 2019. *The End of Coal? Planning a "just transition" in South Africa*. SEI report. Stockholm: Stockholm Environment Institute.

Tol, Richard S.J., and Sebastian Wagner. 2010. Climate Change and Violent Conflict in Europe over the Last Millennium. *Climatic Change* 99: 65–79. https://doi.org/10.1007/s10584-009-9659-2.

Ulrichs, Martina, and Rachel Slater. 2017. *How Can Social Protection Build Resilience? Insights from Ethiopia, Kenya and Uganda*. London: BRACED.

UNEP, ILO, IOE, and ITUC. 2008. *Green Jobs: Towards Decent Work in a Sustainable, Low-Carbon World (Full report)*. Report. Nairobi.

UNFCCC. 2018. *Solidarity and Just Transition Silesia Declaration*. COP24. Katowice.

World Bank. 2019. *World Development Indicators 2019*. Washington, DC: World Bank.

Xiao, Yu, and Edward Feser. 2014. The Unemployment Impact of the 1993 US Midwest Flood: A Quasi-experimental Structural Break Point Analysis. *Environmental Hazards* 13: 93–113. https://doi.org/10.1080/17477891.2013.777892.

Zhang, David D., C.Y. Jim, George C.-S. Lin, Yuan-Qing He, James J. Wang, and Harry F. Lee. 2006. Climatic Change, Wars and Dynastic Cycles in China Over the Last Millennium. *Climatic Change* 76: 459–477. https://doi.org/10.1007/s10584-005-9024-z.

Zhang, David D., Harry F. Lee, Cong Wang, Baosheng Li, Qing Pei, Jane Zhang, and Yulun An. 2011. The Causality Analysis of Climate Change and Large-scale Human Crisis. *Proceedings of the National Academy of Sciences* 108: 17296–17301. https://doi.org/10.1073/pnas.1104268108.

# 21

# Labour Resistance Against Fossil Fuel Subsidies Reform: Neoliberal Discourses and African Realities

Camilla Houeland

## Introduction

Public protests, often led by trade unions, are considered a major hindrance to reforming (i.e. reducing or removing) fossil fuel subsidies (Skovgaard and van Asselt 2018). Fossil fuel subsidies are mainly government actions that lower the costs of production or consumption of fossil products, which skews the energy market in favour of the fossil industry and enforces economic, social and political carbon dependencies. Fuel subsidies contribute to climate change and local air pollution, and they bind a large part of public budgets. 'Despite widespread agreement among experts about the benefits of reforming fossil fuel subsidies, repeated international commitments to eliminate them, and valiant efforts by some countries to reform them, they continue to persist' (Skovgaard and van Asselt 2018).

Popular protests against the removal of subsidies on fuel products are often led by trade unions and mobilise workers and poor people. They are found across continents, regime types and economies, but are particularly common in the Global South. Such protests have been registered in Russia, China, India (Overland 2010), Trinidad and Tobago (Scobie 2018), Nigeria (Houeland 2018a), Brazil (Rentschler and Bazilian 2017), Yemen, Bolivia,

C. Houeland (✉)
Department of Sociology and Human Geography, University of Oslo, Oslo, Norway

Egypt, Nepal, Indonesia (IEA 2014), Sudan, India, Jordan, Iran, Ghana (Lockwood 2015), Cameroon, Venezuela and Bolivia (Rentschler and Bazilian 2017), to mention a few. In light of the dominant discourse on fuel subsidies that emphasises that such subsidies are detrimental to the environment and the climate, as well as to economic and social sustainability (Rentschler and Bazilian 2017; Lockwood 2015), popular protests against fossil fuel subsidy reforms appear not only as threatening climate mitigation, but also as being against workers' own interests.

The apparent contradictions should rather be understood as a reflection of a 'consolidation of neoliberal policy' (Sweeney 2020, 87) on fuel subsidies on the one hand, and an analytical disjuncture between discourse and realities on the other. The 'consolidation of neoliberal policy' refers to how fuel subsidy reforms are framed as climate-mitigation instruments and bring on board environmentalists and others to support neoliberal policies. The dominant discourses on fuel subsidy tend to have a macroeconomic focus based on a 'liberal economic narrative' (Lockwood 2015), with little consideration of political economy, class interests, power relations and specific local contexts and lived realities. The analytical disjuncture between discourse and realities becomes clear when experiences and perspectives from trade unions are considered.

This chapter is organised into three sections. The first maps the dominant discourse on fuel subsidies, climate change and policies globally. The second brings us into realities and experiences in the Global South, exemplified through four country experiences of fuel subsidy removals or removal attempts and labour resistance: Sudan, Zimbabwe, Ghana and Nigeria. These briefly describe the context of the protests against fuel subsidy reforms in terms of political, economic and labour conditions. The third discusses the dominant neoliberal arguments about the socio-economic and work-related aspects of fuel subsidies from the point of view of the trade union informants. Lastly, the final section will draw some conclusions and consider broader insights and reflections.

To do this, I draw on a decade of research on the Nigerian trade union movement, whose history is closely linked to protests against fuel subsidy removals (Houeland 2017). For this chapter, I have carried out an additional literature analysis and interviews with trade unionists from Nigeria, Sudan, Ghana and Zimbabwe.[1] I had working relations with unions in English-speaking Zimbabwe, Ghana and Nigeria as a trade union employee

---

[1] The interviews for this chapter took place in 2019. The chapter has not considered the latest developments in these countries during COVID-19 and following the international fuel price fluctuations since March 2020.

(2006–2011), and it was relatively easy to get access to and communicate in writing or by calling unionists from these countries. Arabic-speaking Sudan, with an ongoing revolution and trade union splinters, is a more unfamiliar and challenging country. The former President of the Sudan Workers' Trade Union Federation (SWTUF), Ibrahim Ghandour, speaks fluent English, and I relied on his help to translate to Arabic when communicating with the current leadership by e-mail.[2]

## The Neoliberal Discourses and Policies

To reach the international 1.5 °C target agreed in the Paris Agreement, it is commonly considered necessary to reform fossil subsidies (Coady et al. 2019). This is also emphasised in the United Nations Intergovernmental Panel on Climate Change (IPCC) report (de Coninck et al. 2018).

A G20 meeting in London in 2009 is considered a turning point in the history of fuel subsidies. Since then, fuel subsidies reform has become a key policy in climate mitigation. Before this, 'the hundreds of billions of dollars channelled to fossil fuels were almost non-existent in the public discourse', according to Skovgaard and van Asselt (2018, 3). Asmelash (2017) also states that the G20 meeting in 2009 'brought the fossil fuel subsidy issue from obscurity to the centre of the global debate over climate change and sustainable development' (10).

In addition to the G20, the main actors in this debate are the World Bank, the International Monetary Fund (IMF), the Organisation for Economic Co-operation and Development (OECD) and the International Energy Agency (IEA) (Lockwood 2015; Skovgaard 2018). There are also specific intergovernmental initiatives such as the Friends of Fossil Fuel Subsidy Reform (FFSR) and a range of civil society organisations and think tanks that lobby for and produce literature and analyses of fuel subsidy reform (Skovgaard and van Asselt 2018; Rentschler and Bazilian 2017). These institutions and organisations tend to be in the Global North, and their reports and research tend to cross-reference each other. It is not uncommon to see the reference that 'all experts agree' about the negative fiscal, distributive and environmental effects of subsidies (Lockwood 2015).

The total cost of fuel subsidies has been estimated to be between $548 billion in 2013 (IEA 2014) and $5.2 trillion in 2017 (Coady et al. 2019).

---

[2] I thank Jacob Høigilt, Professor of Middle East Studies of the University of Oslo, who helped check the translation from Arabic to English.

Cutting fossil fuel subsidies would reduce greenhouse gasses by between 6.4% (Rentschler and Bazilian 2017) and 28% (Coady et al. 2019) depending on how one defines and measures fuel subsidies and their effects. IEA's estimate is based on consumer subsidies, that is, direct financial support to reduce the selling price of fuel and fuel-based consumer goods in relation to market prices. The IMF study by Coady et al. (2019) also includes production subsidies and externalities. Production subsidies are indirect or direct supports to firms in the form of preferential tax treatment, direct government transfers and investments or price regulations. Externalities are health and environmental costs incurred as a consequence of fossil fuel consumption. Coady et al. (2019) argue that a 'right' price of fuel should include taxes to cover such externalities. Most reports do not include such 'post-tax' subsidies or cost-recovery taxation.

Referring to these numbers, which have been widely covered by media across the world, Sean Sweeney (2020, 88) reflects that

> With numbers such as these flying around, it's no surprise that environmental groups have made subsidies a major target and getting rid of them is seen as a potential "D-day" for the climate—a watershed moment in the war between clean and dirty energy.

Also, Lockwood (2015) notes how the climate-mitigation framing of fuel subsidy has brought in actors on the reform agenda who are 'not otherwise inclined to endorse a reform agenda prescribed by the IFIs' (480). This includes civil society, especially environmental organisations, as well as left-oriented politicians. The Norwegian development minister, Heikki Holmås, from the Socialist Left Party (SV), presented a 'white paper' on inequality and redistribution in 2013. While the paper emphasised (radical) re-distribution and a key role for labour and trade unions, it highlighted the need to reform fuel subsidies in a language similar to the neoliberal discourse: 'Government will work to promote reform and the phasing-out of inefficient subsidies that are harmful to the environment and climate, and that counteract fair distribution and undermine fair distribution and undermine sustainable development' (MFA 2013, 51).

At the G20 meeting in 2009, the UK newspaper *The Guardian* reported that US President Barack Obama proposed the 'elimination of the tax breaks, cheap loans and other measures extended to oil, gas, coal and electricity producers. [And that such] elimination of the subsidies would be a "significant down payment" to ending global warming' (Goldenberg 2009). Obama 'emphasised that the US Administration was not opposed to targeted fuel subsidies for the poor, but was seeking to phase out the blanket programmes that also benefit big business and the wealthy' (Goldenberg 2009). *The Guardian* reflected that 'an end to the subsidies would bring world leaders into conflict with powerful fossil fuel lobbies as well as developing nations where the subsidies make fuel affordable' (Goldenberg 2009). In practice, those subsidies that benefit the global energy companies are largely untouched (Bast et al. 2015). Despite being the initiator of reform, the United States continues to be among the world's largest subsidisers (Coady et al. 2019), with heavy production subsidies (Bast et al. 2015).

As the G20 is committed to 'rationalize and phase out [..] inefficient fossil fuel subsidies that encourage wasteful consumption' (Asmelash 2017), both 'inefficient' and 'wasteful' are open to interpretation, and the organisation accepts each country's interpretation. As a consequence, UK tax exemptions and the support of national production by the United States are not considered subsidies (Asmelash 2017). Many countries simply argue that they do not have inefficient subsidies (Skovgaard 2018). However, the cost of production subsidies in the G20 countries combined is estimated at $444 billion (Bast et al. 2015). There is little progress in reducing production subsidies (ISSD 2018), and in practice, G20 countries measure and report on their consumption subsidies (Matsumura and Adam 2018). A similar pattern is clear in the IEA, which is considered the reference authority of fossil fuel subsidy definitions (Asmelash 2017). While the IEA includes producer and consumer subsidies in its definition, thus including a wide range of tax exemptions and state support for research and development (IEA 2014, 315), when measuring and reporting, the IEA focuses singularly on consumption subsidies (see IEA n.d.). 'It is striking that these studies [by the IEA, IMF and more] almost exclusively cover reforms of consumer subsidies' (Rentschler and Bazilian 2017, 900).

The brunt of the subsidy reform is concentrated in the Global South (Skovgaard 2018). Even though it is not happening *in* the north, the subsidy reform agenda is primarily driven *by* northern-dominated actors (i.e. the IMF, IEA, World Bank, G20) (Lockwood 2015). While there is some, albeit limited, normative power and influence that these fuel subsidy debates have over northern countries, the World Bank and the IMF have more power over indebted countries in the Global South through their structural adjustment programmes (Skovgaard 2018). After peaking in 2012, the global total consumption subsidy was halved in 2016, but increased again as international oil prices rose (Matsumura and Adam 2018). The costs of fuel subsidies vary according to the market. It is practically easier to remove subsidies when market prices are low, while in practice, subsidies may be reinstated when market prices increase. Thus, fossil fuel reform may not last.

## African Realities: Contextualising Subsidy Reforms in Zimbabwe, Sudan, Ghana and Nigeria

Contrary to the literature that holds 2009 as the turning point for putting fuel subsidies on the international agenda or into the public discourse, countries in the Global South have decades-long experience of and contestation over fuel subsidy reform, mostly in relation to the structural adjustment programmes of the IMF (Lockwood 2015).

In Africa, there have been three historical waves of protests, where trade unions have been pivotal in most cases. The first wave was of the anti-colonial struggles; the second was of the struggles for democracy and against austerity measures in the 1980s and 1990s; and the current third wave started in 2011, parallel to the Occupy movement and the Arab Spring (Branch and Mampilly 2015; Mueller 2018). The protests against fuel subsidy reforms have been key in the last two waves. During both periods, IMF-backed structural adjustment programmes have included the removal of fuel subsidies.

For workers, the IMF programmes have meant massive job losses (especially in the public sector), reduced welfare benefits and deregulation of the labour market. During the 1980s adjustments, trade unions lost membership and traditional powers. Labour took a 'public turn' and expanded their organisational strategies through social alliances and broadened their agendas to include bread-and-butter and consumer issues, in addition to wages and working conditions (Burawoy 2008). Fuel subsidy is one such consumer issue that has been important for the Nigerian unions to build associational power in social alliances. They remained relevant for workers in formal and informal sectors by bargaining politically about the price of fuel, when collective bargaining power about wages at workplaces was reduced due to job losses and unemployment (Houeland 2018a). Fuel subsidy is considered one of the few welfare benefits and citizen rights in a country where the large oil resources typically profit mainly a small elite (Houeland 2017; Guyer and Denzer 2013; Akanle et al. 2014). The industry creates few jobs but enormous environmental damages that have destroyed alternative livelihoods, such as fishing and farming in the oil region of the Niger Delta, which has fuelled insurgence and violent resistance (Houeland 2015; Obi 2010; Adunbi 2015; Watts 2004).

Support for continued subsidies is particularly strong in countries with little public confidence in the state, in authoritarian countries with weak state capacities and in countries with fossil fuel resources (Skovgaard and van Asselt 2018; Scobie 2018; Lockwood 2015). The four countries in this study fluctuations into these categories in different ways and to varying degrees.

Sudan and Nigeria are petroleum-dependent and vulnerable to the international petroleum market. Both countries are in an economic crisis and have not recovered from the oil price downfall in 2014. Sudan also lost three-quarters of its oil resources when South Sudan seceded in 2011. Ghana has a growing economy. It started petroleum production only in 2008, and petroleum is not dominant in the economy. Zimbabwe has reported some petroleum discoveries but has no production. The country has been in a continuous economic crisis since (at least) the turn of the century. All countries struggle with poverty, inequality, unemployment and socio-political conflicts. In Sudan and Nigeria, some of these conflicts have turned violent.

All four countries are formal democracies, but with varying degrees of democratic practices. The Economist Intelligence Unit's Democracy Index (EIU 2020) ranks Ghana the highest of the four, but it is considered a flawed democracy. Nigeria is a hybrid-democracy, while Zimbabwe and Sudan are at the lowest end and considered as authoritarian states (EIU 2020). The International Trade Union Confederation's (ITUC) Annual Survey of Violations of Trade Union Rights puts the four countries into different categories of violation of rights: Ghana is ranked best of the four, albeit with 'regular violations of rights'; Nigeria comes second with 'systematic violations of rights'. Zimbabwe is categorised as 'no guarantee of rights', and Sudan is at the bottom with 'no guarantee of rights due to breakdown of the rule of law' (ITUC 2019).

The countries here are presented in order of the timing of the most recent fuel protests: Zimbabwe (2019), Sudan (2018), Ghana (2015) and Nigeria (2012).

*Zimbabwe.* In January 2019, popular protests broke out in Zimbabwe after 'massive fuel price hikes, with petrol rising from $1.24 to $3.31 a litre and diesel rising from $1.36 to $3.11—making Zimbabwe the most expensive country in the world to fill a car' (Monks 2019). *The Zimbabwean Congress of Trade Unions* (ZCTU) called for a three-day stay-away, and streets were filled with angry Zimbabweans. Nevertheless, in May, fuel subsidies were removed (AlJazeera 2019).

ZCTU was formed in 1981 by the newly independent Zanu–PF regime under Robert Mugabe. ZCTU became increasingly critical of Zanu–PF after the structural adjustment programmes that started in 1991. In 2005, ZCTU co-founded the opposition party Movement of Democratic Change (MDC) (Saunders 2007). The first reported large fuel price protest took place in 1998 (Kanyenze 2019a), when direct support for fuel prices was removed. While direct support was removed, the state's price controls via fixed foreign exchange systems led to fuel prices that were lower than international market prices. This is also considered a form of consumption subsidy, confirms Founding Director of the Labour and Economic Development Research Institute of (LEDRIZ) Godfrey Kanyenze (2019b). Nevertheless, selling prices increased again, and in 2001 and 2003, ZCTU organised stay-aways from work (i.e. similar to general strikes) to protest price hikes (Sachikonye et al. 2018). Thus, even if the Zimbabwean subsidies were removed in 2019, we may see new protests in Zimbabwe if or when fuel prices increase.

*Sudan.* The ongoing Sudanese 'revolution' started with massive popular protests against rising commodity prices after wheat and fuel subsidies were removed in December 2018. In Sudan, General al-Bashir took power in a military coup in 1989 and soon dissolved not only the government and political parties, but also the trade unions. In 1993, Sudan became a nominally civilian regime, and *The Sudanese Workers' Trade Union Federation (SWTUF)* was re-organised by President Omar al-Bashir with a clear aim to control workers. The SWTUF is widely considered as a trade union 'serving as a government organ' with leadership 'filled by members of the ruling party' (Assal 2016, 8). In 2016, three professional groups, The Central Committee of Sudanese Doctors, The Sudanese Journalists Network and The Democratic Lawyers Association, formed the Sudanese Professionals Association (SPA). They describe themselves as 'a continuation of the long history of Sudanese professionals' persistent attempts to form independent trade unions and bodies to defend their rights and seek to improve their working conditions' (SPA n.d.). Today SPA has 17 mainly white-collar associations as supporters or members (SPA n.d.). It 'may be composed of activists who belong to various political groups, but the appearance and appeal of an unaffiliated, non-ideological body has been instrumental in mobilizing masses of people, reigniting the historically important role of trade unions in Sudanese politics' (Kushkush 2019).

The President of SWTUF, Ali Abdul Kareem Yousif (2019), considers this revolution 'a result of bread and fuel crises':

> SWTUF, while disagreeing with the previous government, did not participate in the protests for several reasons. However, it supported the right of the young people to demonstrate peacefully. It also published statements in support of the

revolution and the agreement that was achieved between the military council and the Movement for Freedom and Change.

Given that SWTUF was considered loyal to al-Bashir, their disagreement is noteworthy. It is also interesting that the former president of SWTUF, Ibrahim Ghandour, is mentioned as both an 'important regime figure' and a 'long-standing opponent' of fossil fuel reform (James 2014).

Despite economic troubles and pressure from outside, the al-Bashir government increased fuel subsidies to reduce the price of fossil products in 1999 (Ghandour 2019). However, when South Sudan seceded in 2011, Sudan lost two-thirds of their oil production that had provided 95% of the state income, and the country experienced multiple economic shocks and high inflation. In relation to IMF adjustment programmes, the government tried to reform fuel subsidies in 2011, 2012 and 2013, all of which met with popular protests. In fact, Ghandour (2019) reports that the only time SWTUF threatened to strike was in 2013, over subsidy reform. The 2013 protest was the biggest popular resistance in two decades (James 2014).

The entrance and role of SPA in taking a lead in the Movement for Freedom and Change has been considered important in explaining why the 2018 protests have led to more pronounced changes than earlier subsidy protests (and war, sanctions or even charges by the International Criminal Court against President al-Bashir) (Elhennawy and Krauss 2019). They led to the downfall of the 30-year-long al-Bashir regime in April 2019 and the establishment of a transitional government in August 2019, of which the SPA is a part. The ongoing Sudanese 'revolution' started with massive popular protests against rising commodity prices after wheat and fuel subsidies were removed in December 2018. In January 2020, the Sudanese finance minister said he was in favour of a gradual lift of the subsidies, while the SPA has threatened to protest against such a reform (Hyde 2020).

*Ghana.* In Ghana, the government announced the removal of fuel subsidies in 2015 (EIU 2015). Although the *Trades Union Congress of Ghana* (Ghana TUC) was against it, there were no major popular protests in 2015, as there had been before.

*Ghana TUC*, with about 500,000 members, was established in 1945 and is the dominant union federation in Ghana (Panford 2011). Ghana TUC has a party-neutral stand, and its constitution does not allow office bearers to hold political positions or agitate for a party (Ghana-TUC 2008). Resistance against austerity and structural adjustment has been important in their history.

The fuel subsidies in Ghana have been declared removed in 2001, 2003, 2004, 2005, 2013, 2014 and 2016 (Reuters 2013; IEA 2014; Kaledzi 2016).

In other words, subsidies have been reinstated several times. Ghana TUC has led protests against fuel price increases several times, such as in 2014 and 2016 (Kaledzi 2016). Most often, these protests have been framed against price hikes and increased living costs. The 2015 removal came at a time when international market prices were low and thus it almost went unnoticed, the Director of the Ghana TUC's Labour and Policy Institute, Kwabena Nyarko Otoo (2019), explains. While Nigeria cut subsidies overnight, Ghana has tried a more gradual reform (Laan and Beaton 2010; Vagliasindi 2012). A report from 2010 has the telling headline reflecting the lessons learnt from Ghana: 'If at first you don't deregulate, try, try again' (Laan and Beaton 2010, 11). When talking to Otoo in 2019, petroleum prices were on the rise again. Asked if the fuel subsidy was permanently removed, he stated, 'We are not happy with the current trend in fuel prices […] People are preparing for a fight—so we have not given up' (Otoo 2019).

Ghana is the only country of our cases that has added taxes on fuel. The union considers these 'unnecessary' (Otoo 2019) and regressive, especially in light of tax reductions for corporations (Ghana-TUC 2016). On 6 September 2019, the General Transport, Petroleum and Chemical Workers' Union of the Ghana TUC issued a press release over concern about rising energy prices, pointing out that workers' salaries could keep up with the price hikes, asking the government to cut energy taxes (GTPCWU 2019).

*Nigeria.* In Nigeria, the massive popular protests and trade union strike against fuel subsidy removal in January 2012 were dubbed 'Occupy Nigeria'. It was possibly the biggest popular mobilisation in the country's history (Branch and Mampilly 2015). Unions, civil society and informal sector workers closed markets, shut down streets and airports, hindered trade flows and occupied public spaces. President Jonathan called the trade unions to negotiations when the oil unions threatened to shut down oil production. After two weeks, the President had to reinstate the fuel subsidy (Houeland 2018a).

The *Nigerian Labour Congress* (NLC), formed in 1978 for blue-collar workers, reports seven million members. The white-collar workers' Trade Union Congress (TUC) claims half a million members. Even if the membership numbers are inflated, with an estimated labour force just below 60 million (CIA 2016) and waged employment estimated at 10% in 2006 (Treichel 2010), it suggests a very high union density in a small formal sector. However, the NLC has historically claimed and been considered to represent beyond their members, including informal sector workers and other social groups, not least through democracy struggles and subsidy protests (Houeland 2017; Beckman and Lukman 2010).

In Nigeria, structural adjustment programmes were introduced in 1986. The following fuel price increase in 1986 did not spur large protests (Ibrahim and Unom 2011). The first large subsidy removal protest was in 1988. Almost all governments since have attempted to remove them and have failed. Democracy protests triggered by fuel subsidies removals even forced President Shonekan to step down in 1993 (Viinikka 2009).

The fuel subsidy remains a contentious issue in Nigerian politics. Knowing the popular power behind fuel subsidies, most politicians typically support them when in opposition, but try to remove them when in power. In the 2019 election, the two main presidential candidates both insisted on keeping fuel subsidies. The trade unions will continue resisting reform, declared the General Secretary of NLC Emmanuel Ogbuaja (2019).

The most glaring disconnection between discourse and reality is that, while framed as a climate-mitigation issue, in practice the design and argument for reform has been and still is fiscal and technical, not for climate mitigation (Lockwood 2015; Rentschler and Bazilian 2017; IEA 2014, 341). Informants in Ghana, Nigeria or Zimbabwe have not encountered climate change arguments for fuel subsidy reform. Owei Lakemfa (2019), who was the acting General Secretary of the NLC during the fuel subsidy protests in 2012, writes, 'As you know, the Nigerian government increases fuel price under IMF-dictates to rake in more money to pay foreign debts and be fritted away. This has nothing to do with climate change'.

The UNFCCC (2006) stresses that 'Africa is not a significant source of greenhouse gas emissions' (1), contributing only 2% to 3% of global greenhouse emissions, but African countries are particularly vulnerable to climate change. Recognising climate change, NLC (2016), Ghana-TUC (2016) and ZCTU (n.d.) have all adopted climate policies, noting the need for reduced $CO_2$ contributions, especially from the extractive industries. Nigeria stands out among the world's highest emissions from gas flaring. Both NLC and the oil unions have long fought this practice, and a small reduction has recently been noted (GlobalData 2020). None of them mention fuel or fuel subsidies.

As much as the unions are concerned about climate change, informants from all countries resonate with Otoo's (2019) reflection:

> The climate discussion is a luxury discussion. There are too many poor people at work, too desperate to care about the climate. The world has it in their power to slow the climate crisis down. When I say the world, I don't mean the poor people. The world, as embodied in UN … please don't put the burden on the poor workers in Africa.

Ghandour (2019)—who was also a Minister of Foreign Affairs from June 2015 to April 2018—recalls that in Sudan the climate issue has in fact been discussed in relation to subsidies. However, he refers to the Paris Agreement (UNFCCC 2015), which states that 'Developed country Parties shall, and other Parties that provide support should, provide information on financial, technology transfer and capacity-building support provided to developing country Parties ...' (17). In the words of Ghandour (2019), 'Without the support of the North, the South cannot get rid of those subsidies and the pollution we see'.

## Fuel Subsidies as 'Socially Inefficient' or a State Benefit for Workers

The IPCC refers to the 'fossil fuel subsidy regimes' as 'socially inefficient' (de Coninck et al. 2018). A much-referenced IMF study shows that the richest 20% consume 43% of fuel, while the poorest 20% consume only 7% (Arze del Granado et al. 2012).

When discussing the effects of fuel subsidy removals on workers, the just transition emphasises the impacts on employment in the fossil industries, although employment in other, both formal and informal sectors, is also affected (Gass and Echeverria 2017). Whereas the support from Nigerian oil workers has been important for the success of the fuel subsidy strikes, they support deregulation of the downstream sector, as they believe it could boost investments and employment in refineries (Esele 2012). However, in the case of Ghana, since oil production started in 2008, the capacity of the Tema Oil Refinery, which used to produce 70% of the Ghanaian petroleum product consumption, has in fact worsened (Vagliasindi 2012), which challenges employment. This is despite deregulation of fuel pricing. The lower refinery output is related to increased corruption, according to Otoo (2019). There are high levels of state-related corruption in all our country cases. In fact, in Nigeria, the subsidy system is one of the key sources of deep corruption.

With the subsidy-related corruption and middle classes as the main beneficiaries of cheap fuel, Collier (2012) believed that the Nigerian fuel subsidy protests 'closely resemble the sad folly of the Tea Party: poor people tricked into lobbying for greedy elites'. By contrast, protesters in Nigeria both dismissed the argument of removing subsidy in the name of anti-corruption and highlighted their vulnerability to fuel price increases with signposts and T-shirts, saying 'Kill corruption, not Nigerians': Cheap fuel can be kept, while corruption is targeted (Houeland 2018a). Otoo (2019) further reflects:

At [the] macro level, such analysis could be right; but at [the] micro level—the argument breaks down. [...] those who run the big cars, the V8s and cross-country vehicles in Accra and Lagos are government officials. The fuel they consume is bought by the state [...] so even if we remove subsidies—they are not affected. [...] If fuel prices go up: it is the poor workers that suffer [...]—whether Nigeria, Sudan or Zimbabwe—that middle class is connected to government. If they increase the fuel to the roof—they don't care.

The recent fuel price hikes have come in the context of decades-long pressure on workers' conditions, with salary levels not keeping up to date with inflation, the minimum wage staying below living wages and inequality rising, all with high external debts in the midst of economic growth (as in Ghana 2015 and Nigeria in 2012), an economic crisis (as in Sudan 2018) and hyperinflation (in Zimbabwe 2019). All informants point to fuel prices as an important workers' issue, mostly due to the effect on costs of living, reduced purchasing power and decreasing real wages, but also as an employment issue.

Fuel price has an immediate and direct impact on transport costs, which is a problem for poor children going to school and their parents going to work, but it also affects the cost of food and other products. The President of NLC, Abdulwahed Omar (2012), stresses that in Nigeria, fuel subsidy removal is most problematic for employment in small enterprises and in the informal sector, as they depend on fuel-driven aggregators and are badly hit by the increased price of transport and goods. The President of ZCTU, Japhet (Moyo 2019), further explains:

In Zimbabwe, the issue of fuel prices is indeed a workers' issue. [Immediately after] the government announced that with effect from May 21, 2019, the price of diesel was pegged at $4.89 [and petrol] at $4.97, there was a knock-on effect on the prices of basic goods and services, worsening the impact of the already-depreciating exchange rate on the general level of prices. Commuter transport fares rose to between $2 and $5; medical aid has more than tripled and funeral insurance premiums tripled, making life unbearable to most Zimbabweans.

From Sudan, Yousif (2019) similarly describes the following:

Fuel prices have a significant impact on workers because fuel is a key driver of the transport sector, industry, agriculture, electricity sector, and by increasing fuel prices, transportation tariffs and electricity and [the prices of] all industrial and agricultural products that workers need in their daily lives are increasing.

In Ghana, Andrews Tagoe (2019), Deputy General Secretary, General Agricultural Workers Union (GAWU) of Ghana TUC, gives details that

> Increase in fuel prices definitely affected all categories of our members. Both self-employed and waged workers. The fraction of fuel prices increment reflects on cost of transport, inputs for production, cost of transportation of goods from farm gate to sale points.

Otoo (2019) indirectly refers to the very definition of the 'inefficiency' of subsidies and the 'wasteful consumption' of fuel when he says that poor people cannot opt out of their dependency on fuel. Sub-Saharan Africa suffers severe energy poverty. Clean energy is not always available (Bridle et al. 2019), as explained by the General Secretary of NLC, Emmanuel Ogbuaja (2019): 'In Nigeria, we don't have electronic cars. We don't have green industries. We don't have electric buses for mass transportation'.

Additionally, the alternative to fossil fuel may be problematic to the environment, as reflected by Moyo (2019): 'most of the developing countries do not have alternative sources of fuel, and rely on the traditional sources that harm the environment'.

Poor people, especially in rural areas, use kerosene for cooking (Rentschler and Bazilian 2017). When fuel subsidy reform includes kerosene, it has a gender dimension. When prices increase, it often leads to women—who mainly do the cooking—reverting to biomass, which has adverse effects on women's time and health (Global Subsidies Initiative–IISD 2019), as well as on the climate.

The production of renewable energy is more labour-intensive than fossil fuel production (Gass and Echeverria 2017), and there are nascent efforts to create green energy alternatives in Africa. For example, in Nigeria, Ogbuaja (2019) states that 'Fuel might not be as important now for the industry as there is a move away from machinery that needs fossil fuel'. However, as much as IMF policies may have helped to reduce energy consumption in the short term, they have not assisted in diversifying and greening the energy mix (Karanfil and Omgba 2019).

Gross domestic products (GDP) and national budgets may be positively affected by fuel subsidy removal, but these removals have a detrimental impact, particularly on poor people's households (Siddig et al. 2014). A study in Ghana holds that the 'negative effect is worst for the poorest group', estimating that almost 400,000 individuals will be pushed into poverty as a result of fuel price increases (Cooke et al. 2016, 105). Although most of the literature that argues for subsidy reform recognises the adverse effects of fuel subsidy reforms on the poor, it still concludes that fuel subsidies are detrimental to the interest of the poor (Rentschler and Bazilian 2017, 896). This is because subsidies are expensive and take up large portions of government budgets, and the literature that supports subsidy reform proposes that this money could be spent on better targeted pro-poor spending. Consumer subsidy costs constitute on average 5% of GDP (IEA 2014). In Sudan, 35% of the budget for 2020 was allocated to fuel subsidies (Hyde 2020), and in Nigeria, it took up a third of the budget in 2011 (Africa Confidential 2012). By releasing these funds, it is possible to mitigate climate change *and* the negative social effects of fuel subsidy removal on the poor. As much as the World Bank, the IMF and others state the need for social mitigation of the adverse effects of fuel subsidy removals on the poor, there are few practical measures to show for (Gass and Echeverria 2017; Asmelash 2017).

Cash transfers to the poor are mentioned as a viable alternative to mitigate the social impact of fuel subsidy removal (IEA 2014; Skovgaard and van Asselt 2018; Rentschler and Bazilian 2017). Cash transfers have been discussed in Zimbabwe, Sudan and Nigeria. In Ghana, the government initiated a conditional cash-transfer programme related to subsidy reform in 2013. This had positive effects on poverty reduction, but would need significant expansion to reduce poverty and compensate for the effects of fuel subsidy removal (Cooke et al. 2016). Cash transfers normally target the poorest, not the 'almost poor', such as the many workers that may be similarly affected by fuel subsidy removal (Lockwood 2015). All informants emphasised the need for real wage increases to compensate for increased living costs as being the most important measure. Yousif (2019) formulates this as follows in an email to the author: 'Our condition for accepting the removal of fuel subsidies is to give the worker a living wage […]. Wages should increase as inflation increases'.

There is a fundamental distrust of governments in all four countries of investigation. Regarding Ghana, Otoo (2019) explains that when the government says they will reallocate funds for social spending, money may be taken away from the subsidy budget while citizens do not see any social improvements. Unionists emphasise the need to see concrete social improvements before accepting any fuel subsidy reforms[3] (Esele 2012). After the January protests in Nigeria, Jonathan's government presented infrastructural and social welfare policies through the *Subsidy Reinvestment and Empowerment Program, SURE-P*, to use savings from reduced subsidy costs to cushion the adverse social effects on poor people. However, the scheme has been ridden by controversies and corruption, and though the programme has had some

---

[3] Reforms are gradual, not necessarily complete removals of fuel subsidies.

achievements, it is insufficient in its design and ineffective in its implementation (Amakom 2013).

Informants emphasise that subsidy reforms have been initiated from outside the country, mainly from the IMF, and without democratic processes. The dominant discourses on fuel subsidies often emphasise the need for information and transparency, mainly to convince people, rather than calling for democratic processes. In Ghana, subsidy reforms have never been negotiated, but were enforced 'through the IMF credit-facilities', reflects Otoo (2019). In 2012, the Nigerian government unilaterally removed the subsidy in the middle of an ongoing stakeholder dialogue with civil society organisations, which only fuelled the popular anger (Houeland 2018a). The tripartite social dialogue in Nigeria, which does not happen regularly, is basically broken (Ogbuaja 2019). However, bargaining over the fuel subsidy, following reforms and protests, is considered by some as part of the social dialogue (Houeland 2018a). In Zimbabwe, the Tripartite Negotiating Forum (TNF) Agreement of 2001 outlines that income, pricing and transport are issues to be discussed and agreed upon in this forum. Thus, when the government unilaterally increases fuel prices, it breaches that agreement (Sachikonye et al. 2018). All the interviewees emphasise the need for democratic bargaining through social dialogue. Also, the labour-driven 'just transition' approach stresses the need to involve the social partners in decision-making processes to ensure social compensation and employment protection regarding the effects of fuel subsidy removals (Gass and Echeverria 2017).

Relatedly, the informants of this study express a lack of trust in governments to implement agreements. 'Even if we did agree to free transport *before* subsidies are removed, we don't trust our government that much', explains Otoo (2019). The SWTUF has been part of regular tripartite dialogue, explains Ghandour (2019), claiming that Sudan has had more salary adjustments than any other African country. He also refers to a lack of implementation of agreements. In 2017, there was a tripartite agreement to gradually reduce subsidies in exchange for an increased minimum salary. When subsidies were removed in December 2018, the minimum salary adjustments were not realised. The Sudan Professional Association (SPA) claimed there had been no minimum salary increase in six years, and in November 2018, they demanded a rise from the statuary monthly minimum wage of $9 to $182, claiming that the cost of living for a family of five people was estimated at $206.30 (Dabanga 2018a, b).

While some senior oil workers in Nigeria have automatic salary increases according to fuel price fluctuations in their collective agreements (Nwafor 2014), the inflation following price increases is rarely reflected in wage adjustments. In Ghana, 'almost every month we see price adjustments, while salaries are fixed annually and there are no measures to adjust those salaries' (Otoo 2019). Tagoe (2019) refers to a reversed example where, in the agricultural sector in Ghana, employers have used fuel price increases 'as a reason to stall workers' salaries with the mere reason that the costs of production have gone high'. In Nigeria, the minimum salary in 1981 was $200 per month. Prior to the attempted fuel subsidy reform in 2011, it was $110. Only in November 2017 was it increased in local currency, but in dollar terms, it was reduced to $67 (see Houeland 2018b). Otoo (2019) suggests that if social issues and economic justice were the aims of the government, then it should rather tax the rich.

## Conclusion

This chapter has engaged with the global discourse on fuel subsidies dominated by neoliberal viewpoints on the one hand, and with African trade union perspectives and workers' lived experiences on the other. We see a neoliberal consolidation in the north, where the framing of fuel subsidies as a climate-mitigation policy has brought leftist and environmental activists into supporting fuel subsidy reforms that trade union representatives in Nigeria, Sudan, Zimbabwe and Ghana describe as detrimental to workers' interests.

There are discrepancies between northern actors' discourse and policy practices in that they call for reforms of fuel subsidies, defined as both consumer and production support to the fossil fuel industry, while in practice the production subsidies are largely untouched. Consumer subsidies in the Global South are the prime targets of reform through IMF-backed austerity packages. Furthermore, when implemented in the south, these reforms are not discussed as climate-mitigation measures. Arguments of social and ecological sustainability appear imposed on an old agenda of financial efficiency, detached from the lived realities of workers and the poor. The argument that subsidies are socially inefficient is based on the view that the upper classes are the prime beneficiaries of subsidies and that the high subsidy costs would be better spent directly on pro-poor policies. The trade unionists point to the poor and workers as disproportionately and negatively affected by increased fuel prices, and removing subsidies will rather deepen existing socio-economic inequalities and injustices. Furthermore, they do not trust that governments will be able and willing to compensate for the negative effects of subsidy removals on workers and the poor. Therefore, they argue that alternative energy and welfare benefits must be in place before reform. Moreover, all measures need to be properly and democratically negotiated.

This chapter identifies core questions of justice in energy transitions for climate change in terms of class and north–south relations. Increased fuel prices are typically the final straw motivating labour and others to protest economic injustices and crises of democracy, which include the lack of welfare, employment, distribution of resources and alternative, green energy. Although there are specific conditions of power relations in the African countries studied, these insights will also be relevant elsewhere in the Global South (as the list of country protests in the introduction and the literature references in this chapter suggest), but also in the Global North. This was highlighted by the French 'Yellow Vests' protests, which was sparked by increased fuel prices after green taxations. Often depicted as protests against climate mitigation, many protesters described them as resistance against austerity politics, exposing social injustices and democratic limitations in neoliberal climate change measures (Kinniburgh 2019).

Fuel subsidy removal is part of an economic liberalisation and austerity agenda that unions see as the very cause of many of the socio-economic and political challenges they face. Austerity policies have for decades weakened trade unions' organisational and bargaining power in workplaces. Fuel subsidy protests, in turn, have been part of the unions' counter-strategy: They have expanded organisational power through social alliances over fuel prices. The price of fuel impacts on job security in the formal and informal sectors, but more than anything, it concerns purchasing power in the face of a downward-spiralling real wage.

This chapter emphasises in new ways that questions of labour and climate go beyond the environment vs. jobs dilemma (Räthzel and Uzzell 2011), when considering trade unions as social movement actors in the public and beyond industrial relations, and workers as consumers not only producers of energy. This has bearing on the studies of the impact on workers of a green transition. Governments' climate efforts focus on changes in consumer behaviour for climate mitigation (Räthzel and Uzzell 2011), as does the neoliberal policy that focuses on macroeconomic efficiency and consumer issues, while they hardly consider the microlevel impact on workers of energy transitions. By contrast, when considering effects or justice for workers in energy transition, studies on labour and the environment and policies based on the labour-initiated 'just transition' approach primarily address workers whose jobs or livelihood depend on fossil fuel production (Evans and Phelan 2016; Gass and Echeverria 2017; Healy and Barry 2017; Piggot et al. 2019; Räthzel and Uzzell 2011), with little consideration of workers as consumers of fossil energy. The findings in this chapter underscore the need for a 'just transition' approach to look at the effects on workers of changes in the whole energy system as well as at the particular injustices in the Global South (Newell and Mulvaney 2013).

# References

Adunbi, Omolade. 2015. *Oil Wealth and Insurgency in Nigeria*. Bloomington: Indiana University Press.
Africa Confidential. 2012. Fuel Fraud Fans Public Anger. Accessed August 11, 2017. http://www.africa-confidential.com/article/id/4474/Fuel_fraud_fans_public_anger.
Akanle, Olayinka, Adebayo Kudus, and Adetayo Olorunlana. 2014. Fuel Subsidy in Nigeria: Contexts of Governance and Social Protest. *The International Journal of Sociology and Social Policy* 34 (1/2): 88–106.
AlJazeera. 2019. Zimbabwe Increases Fuel Prices as Economic Crisis Deepens. May 22. Accessed September 29, 2019. https://www.aljazeera.com/news/2019/05/zimbabwe-increases-fuel-prices-economic-crisis-deepens-190522071225065.html.
Amakom, Uzochukwu. 2013. *Subsidy Reinvestment and Empowerment Programme (SURE-P) Intervention in Nigeria: An Insight and Analysis*. Enugu: African Heritage Institution (AfriHeritage).
Asmelash, Henok Birhanu. 2017. Phasing out Fossil Fuel Subsidies in the G20: Progress, Challenges, and Ways Forward. In *Think Piece*. Geneva: International Centre for Trade and Sustainable Development.
Assal, Munzoul A.M. 2016. Civil Society and Peace Building in Sudan: A Critical Look. In *Sudan Working Paper*. Bergen, Norway: Chr. Michelsen Iinstitute (CMI).
Bast, Elizabeth, Alex Doukas, Sam Pickard, Laurie van der Burg, and Shelagh Whitley. 2015. *Empty Promises: G20 Subsidies to Oil, Gas and Coal Production*. London: Washington Overseas Development Institute/Oil Change International.
Beckman, Björn, and Salihu Lukman. 2010. The Failure of Nigeria's Labour Party. In *Trade Unions and Party Politics: Labour Movements in Africa*, ed. Björn Beckman, Sakhela Buhulungu, and Loyd Sachikonye, 59–84. Cape Town: HSRC Press.
Branch, Adam, and Zachariah Cherian Mampilly. 2015. *Africa Uprising: Popular Protest and Political Change*. London: Zed Books.
Bridle, Richard, Shruti Sharma, Mostafa Mostafa, and Anna Geddes. 2019. Fossil Fuel to Clean Energy Subsidy Swaps: How to Pay for an Energy Revolution. In *GSI Report*. Winnipeg, Canada.
Burawoy, M. 2008. The Public Turn From Labor Process to Labor Movement. *Work and Occupations* 35 (4): 371–387. https://doi.org/10.1177/0730888408325125.
CIA. 2016. *World Fact Book: Nigeria*. Accessed January 23, 2017. https://www.cia.gov/library/publications/the-world-factbook/geos/ni.html.

Coady, David, Ian Parry, Nghia-Piotr Le, and Baoping Shang. 2019. *Global Fossil Fuel Subsidies Remain Large: An Update Based on Country-Level Estimates.* International Monetary Fund.

Collier, Paul. 2012. Should Nigeria be Ruled by the Street? In *Business Day Online*. Accessed January 13, 2012. http://www.businessdayonline.com/NG/index.php/analysis/columnists/31853-should-nigeria-be-ruled-by-the-street.

de Coninck, Helen, Aromar Revi, Mustafa Babiker, Paolo Bertoldi, Marcos Buckeridge, Anton Cartwright, Wenjie Dong, James Ford, Sabine Fuss, and J-C Hourcade. 2018. Strengthening and Implementing the Global Response (Ch. 4). In *Global Warming of 1.5° C. An IPCC Special Report on the Impacts of Global Warming of 1.5° C above Pre-Industrial Levels and Related Global Greenhouse Gas Emission Pathways, in the Context of Strengthening the Global Response to the Threat of Climate Change, Sustainable Development, and Efforts to Eradicate Poverty*, eds. P. Zhai, V. Masson-Delmotte, H.O. Pörtner, D. Roberts, J. Skea, P.R. Shukla, A. Pirani, W. Moufouma-Okia, C. Péan, R. Pidcock, S. Connors, J.B.R. Matthews, Y. Chen, X. Zhou, M.I. Gomis, E. Lonnoy, T. Maycock, M. Tignor, T. Waterfield. IPCC.

Cooke, Edgar F.A., Sarah Hague, Luca Tiberti, John Cockburn, and Abdel-Rahmen El Lahga. 2016. Estimating the Impact on Poverty of Ghana's Fuel Subsidy Reform and a Mitigating Response. *Journal of Development Effectiveness* 8 (1): 105–128. https://doi.org/10.1080/19439342.2015.1064148.

Dabanga. 2018a. Wages to Rise as Fuel and Bread Prices Climb. December 17. https://www.dabangasudan.org/en/all-news/article/wages-to-rise-as-fuel-and-bread-prices-climb.

———. 2018b. Workers' Association Memo for Adjusting Sudan's Minimum Wages. November 23. https://www.dabangasudan.org/en/all-news/article/workers-union-memo-for-adjusting-sudan-s-minimum-wages.

EIU. 2015. Ghana to Remove Remaining Fuel Subsidies. *The Economist*. June 11. Accessed September 29, 2019. http://country.eiu.com/article.aspx?articleid=1223244106&Country=Ghana&topic=Economy&subtopic=Forecast&subsubtopic=Policy+trends&u=1&pid=303218214&oid=303218214&uid=1.

———. 2020. *Democracy Index 2019: A Year of Democratic Setbacks and Popular Protest*. The Economist Intelligence Unit Limited.

Elhennawy, Noha, and Joseph Krauss. 2019. Q&A: Sudanese Face Uphill Climb in Unfinished Revolution. *APNews*, April 26. Accessed March 26, 2020. https://apnews.com/22b15c08f0014624881100e3385accdf.

Esele, Peter. 2012. Personal Communication, August 30. Abuja, Nigeria.

Evans, Geoff, and Liam Phelan. 2016. Transition to a Post-Carbon Society: Linking Environmental Justice and Just Transition Discourses. *Energy Policy* 99: 329–339. https://doi.org/10.1016/j.enpol.2016.05.003.

Gass, Philip, and Daniella Echeverria. 2017. *Fossil Fuel Subsidy Reform and the Just Transition*. Manitoba, Canada: International Institute for Sustainable Development, Global Subsidies Initiative.

Ghana-TUC. 2008. *The Constitution and Internal Regulations of Ghana Trades Union Congress as Amended at the 8th Quadrennial Delegates Congress*. Accra: Ghana-TUC.
———. 2016. *Policies 2016–2020*. Accra: Ghana-TUC.
Ghandour, Ibrahim. 2019. Personal Communication, August 2. WhatsApp.
Global Subsidies Initiative—IISD, BIDS, IRADe and Spaces for Change. 2019. Gender and Fossil Fuel Subsidy Reform: Findings from and Recommendations for Bangladesh, India and Nigeria. Energia.
GlobalData. 2020. Global Gas Flaring Value Approaches US$24bn a Year if Priced at European Prices, Says GlobalData. GlobalData.
Goldenberg, Suzanne. 2009. Obama to Press G20 Leaders to Cut Fossil Fuel Subsidies that Benefit Big Business. *The Guardian*. September 23. Accessed September 9, 2019. https://www.theguardian.com/environment/2009/sep/23/obama-g20-oil-subsidies.
del Granado, Arze, Francisco Javier, David Coady, and Robert Gillingham. 2012. The Unequal Benefits of Fuel Subsidies: A Review of Evidence for Developing Countries. *World Development* 40 (11): 2234–2248. https://doi.org/10.1016/j.worlddev.2012.05.005.
GTPCWU. 2019. *Press Release by GTPCWU of TUC (Ghana) on Recent Energy Hikes Edited by Acting General Secretary Francis Sallah*. Accra: GTPCWU.
Guyer, Jane I., and LaRay Denzer. 2013. Prebendalism and the People: The Price of Petrol at the Pump. In *Democracy and Prebendalism in Nigeria: Critical Interpretations*, ed. W. Adebanwi and E. Obadare, 53–78. Basingstoke: Palgrave Macmillan.
Healy, Noel, and John Barry. 2017. Politicizing Energy Justice and Energy System Transitions: Fossil Fuel Divestment and a 'Just Transition'. *Energy Policy* 108: 451–459.
Houeland, Camilla. 2015. Casualisation and Conflict in the Niger Delta: Nigerian Oil Workers' Unions Between Companies and Communities. *Revue Tiers Monde* 4: 25–46.
———. 2017. *Punching Above their Weight: Nigerian Trade Unions in the Political Economy of Oil*. Department of International Environment and Development Studies (Noragric), Univeristy of Life Sciences (NMBU).
———. 2018a. Between the Street and Aso Rock: The Role of Nigerian Trade Unions in Popular Protests. *Journal of Contemporary African Studies* 36 (1): 103–120.
———. 2018b. The Struggle for a Minimum Wage in Nigeria. *Africasacountry.com*, December 15. Accessed March 27, 2020. https://africasacountry.com/2018/12/the-struggle-for-a-minimum-wage-in-nigeria.
Hyde, Maggie. 2020. Sudan to Tackle Fuel Subsidies as Economy Hangs on the Edge. *APNews*, January 29. Accessed March 26, 2020. https://www.voanews.com/africa/sudan-tackle-fuel-subsidies-economy-hangs-edge.
Ibrahim, Jibrin, and Sam Unom. 2011. *Petroleum Subsidy Issue. Analysis and Stakeholder Mapping*. Abuja. Unpublished report

IEA. 2014. *World Energy Outlook 2014*. Paris, France: International Energy Agency (IEA).

———. n.d. Fossil-Fuel Subsidies. Accessed September 16, 2019. https://www.iea.org/weo/energysubsidies/.

ISSD. 2018. Stories from G20 Countries: Shifting Public Money out of Fossil Fuels. Accessed September 11, 2019. https://iisd.org/library/stories-g20-countries-shifting-public-money-out-fossil-fuels.

ITUC. 2019. Annual Survey of Violations of Trade Union Rights. In *International Trade Union Confederation*. Brussels: ITUC.

James, Laura M. 2014. *Recent Developments in Sudan's Fuel Subsidy Reform Process*. Manitoba, Canada: International Institute for Sustainable Development, Global Subsidies Initiative.

Kaledzi, Isaac. 2016. Ghanaians Protest Rising Fuel and Electricity Tariffs. Accessed September 30, 2019. https://www.dw.com/en/ghanaians-protest-rising-fuel-and-electricity-tariffs/a-18994432-0.

Kanyenze, Godfrey. 2019a. *Giving Social Dialogue a Chance: Review of, and Lessons Learnt from the Tripartite Negitiation Forum (TNF) in Zimbabwe, 1998–2018*. Harare: Ledriz.

———. 2019b. Personal Communication, July 30. E-mail.

Karanfil, Fatih, and Luc Désiré Omgba. 2019. Do the IMF's Structural Adjustment Programs Help Reduce Energy Consumption and Carbon Intensity? Evidence from Developing Countries. *Structural Change and Economic Dynamics* 49: 312–323. https://doi.org/10.1016/j.strueco.2018.11.008.

Kinniburgh, Colin. 2019. Climate Politics after the Yellow Vests. *Dissent* 66 (2): 115–125.

Kushkush, Isma'il. 2019. Protesters in Sudan and Algeria Have Learned from the Arab Spring. *The Atlantic*, April 13. Accessed September 27, 2019. https://www.theatlantic.com/international/archive/2019/04/protesters-sudan-and-algeria-have-learned-arab-spring/587113/.

Laan, Tara, and Christopher Beaton. 2010. *Strategies for Reforming Fossil-Fuel Subsidies: Practical Lessons from Ghana, France and Senegal*. Winnipeg, Canada: IISD and GSI.

Lakemfa, Owei. 2019. Personal Communication. September 25. Messenger.

Lockwood, Matthew. 2015. Fossil Fuel Subsidy Reform, Rent Management and Political. *New Political Economy* 20: 475–494.

Matsumura, W., and Adam, Z. (2018). Commentary: Hard-Earned Reforms to Fossil Fuel Subsidies are Coming Under Threat. October 29. Accessed March 27, 2020 https://www.iea.org/newsroom/news/2018/october/hard-earned-reforms-to-fossil-fuel-subsidies-are-coming-under-threat.html.

MFA. 2013. Sharing for Prosperity: 'Promoting Democracy, Fair Distribution and Growth in Development Policy. Meld. St. 25 (2012–2013) Report to the Storting' (White Paper), edited by the Norwegian Ministry of Foreign Affairs. Oslo.

Monks, Kieron. 2019. How Zimbabwe became World's Most Expensive Place to Fuel a Car. *CNN*, January 16. Accessed September 27, 2019. https://edition.cnn.com/2019/01/16/africa/zimbabwes-impoverished-billionaires-africa/index.html.

Moyo, Japhet. 2019. Personal Communication, August 20. E-mail.

Mueller, Lisa. 2018. *Political Protest in Contemporary Africa*. Cambridge: Cambridge University Press.

Newell, Peter, and Dustin Mulvaney. 2013. The Political Economy of the 'Just Transition'. *The Geographical Journal* 179 (2): 132–140. https://doi.org/10.1111/geoj.12008.

NLC. 2016. *Climate Change Policy 2015*. Abuja: Nigerian Labour Congress.

Nwafor, John. 2014. *Personal Communication, March 10*. Nigeria: Lagos.

Obi, Cyril. 2010. Oil Extraction, Dispossession, Resistance, and Conflict in Nigeria's Oil-Rich Niger Delta. *Canadian Journal of Development Studies / Revue canadienne d'études du développement* 30 (1–2): 219–236.

Ogbuaja, Emmanuel. 2019. Personal Communication, August 29. WhatsApp.

Omar, Abdulwahed. 2012. Personal Communication. Geneva, Switzerland.

Otoo, Kwabena Nyarko. 2019. Personal Communication, August 18. WhatsApp.

Overland, Indra. 2010. Subsidies for Fossil Fuels and Climate Change: A Comparative Perspective. *International Journal of Environmental Studies* 67 (3): 303–317.

Panford, Kwamina. 2011. Trade Unions, Democratic Transition and Organiational Challenges: The Ghana Trade Union Congress, 1989–2009. In *Trade Unions in West Africa: Historical and Contemporary Perspectives*, ed. Craig Phelan, 145–178. Oxford: Peter Lang.

Piggot, Georgia, Michael Boyland, Adrian Down, and Andreea Raluca Torre. 2019. Realizing a Just and Equitable Transition Away from Fossil Fuels. In *SEI Discussion Brief*. Stockholm: Stockholm Environment Institute.

Räthzel, Nora, and David Uzzell. 2011. Trade Unions and Climate Change: The Jobs Versus Environment Dilemma. *Global Environmental Change* 21 (4): 1215–1223. https://doi.org/10.1016/j.gloenvcha.2011.07.010.

Rentschler, Jun, and Morgan Bazilian. 2017. Reforming Fossil Fuel Subsidies: Drivers, Barriers and the State of Progress. *Climate Policy* 17 (7): 891–914.

Reuters. 2013. Ghana Scraps Fuel Subsidy to Reduce Budget Deficit. *Reuters*, May 31. Accessed September 28, 2019. https://www.reuters.com/article/ghana-subsidy/ghana-scraps-fuel-subsidy-to-reduce-budget-deficit-idUSL5N0EC3X920130531.

Sachikonye, Lloyd, Brian Raftopoulos, and Godfrey Kanyenze. 2018. *Building from the Rubble: The Labour Movement in Zimbabwe Since 2000*. Harare: Weaver Press.

Saunders, Richard. 2007. Trade Union Struggles for Autonomy and Democracy in Zimbabwe. In *Trade Unions and the Coming of Democracy in Africa*, ed. Jon Kraus, 157–197. New York: Palgrave Macmillan.

Scobie, Michelle J. 2018. Actors, Frames and Contexts in Fossil Fuel Subsidy Reform. In *The Politics of Fossil Fuel Subsidies and their Reform*, ed. J. Skovgaard and H. van Asselt, 283–302. Cambridge: Cambridge University Press.

Siddig, Khalid, Angel Aguiar, Harald Grethe, Peter Minor, and Terrie Walmsley. 2014. Impacts of Removing Fuel Import Subsidies in Nigeria on Poverty. *Energy Policy* 69: 165–178. https://doi.org/10.1016/j.enpol.2014.02.006.

Skovgaard, Jakob. 2018. International Push, Domestic Reform? In *The Politics of Fossil Fuel Subsidies and their Reform*, ed. Harro van Asselt and Jakob Skovgaard, 100–120. Cambridge: Cambridge University Press.

Skovgaard, Jakob, and Harro van Asselt. 2018. *The Politics of Fossil Fuel Subsidies and their Reform*. Cambridge: Cambridge University Press.

SPA. n.d. About us The Sudanese Professionals Association. Accessed June 12, 2020. https://www.sudaneseprofessionals.org/en/about-us/.

Sweeney, Sean. 2020. Weaponizing the Numbers: The Hidden Agenda Behind Fossil-Fuel Subsidy Reform. *New Labor Forum* 29 (1): 87–92. https://doi.org/10.1177/1095796019893317.

Tagoe, Andrews Addoquaye. 2019. Personal Communication, August 4. Facebook.

Treichel, Volker. 2010. Employment and Growth in Nigeria (Ch. 1). In *Putting Nigeria to Work: A Strategy for Employment and Growth*, ed. Volker Treichel, Francis Teal, and Peter Mousley, 9–60. Washington, DC: World Bank Publications.

UNFCCC. 2006. United Nations Fact Sheet on Climate Change: Africa Is Particularly Vulnerable to the Expected Impacts of Global Warming. Nairobi.

———. 2015. Paris Agreement. UNFCCC.

Vagliasindi, Maria. 2012. *Implementing Energy Subsidy Reforms: Evidence from Developing Countries*. Washington, DC: The World Bank.

Viinikka, Jussi. 2009. 'There Shall be no Property': Trade Unions, Class and Politics in Nigeria. In *Class Struggle and Resistance in Africa*, ed. Leo Zeilig, 122–149. Chicago: Haymarket Books.

Watts, Michael. 2004. Resource Curse? Governmentality, Oil and Power in the Niger Delta, Nigeria. *Geopolitics* 9 (1): 50–80.

Yousif, Yousif Ali Abdel Karim. 2019. Personal Communication, August 22. E-mail.

ZCTU. n.d. Zimbabwe Congress of Trade Unions (ZCTU) Climate Change Policy. Harare.

# 22

# Challenges and Prospects for Trade Union Environmentalism

Adrien Thomas and Valeria Pulignano

## Introduction

This contribution addresses a core issue raised by trade union engagement with climate change mitigation: How are trade unions dealing with their members' differing interests, and how are they articulating these interests and broader societal concerns? Most trade unions throughout the world acknowledge the need to mitigate climate change. They have put forward a range of concepts addressing the challenges raised by the decarbonisation of society and the economy, first and foremost the 'Just Transition' and 'green jobs' frameworks (Silverman 2006; Räthzel and Uzzell 2013; Felli 2014; Hampton 2015). Despite generally acknowledging the need to mitigate climate change, many trade unions face internal dilemmas and tensions over climate politics (Räthzel and Uzzell 2011; Stevis 2018), as fears of possible job losses due to decarbonisation are strong in some sectors of the economy. Looking at trade unions across different socio-geographical contexts, including North–South, this contribution situates and discusses trade union engagement with climate

---

A. Thomas (✉)
Labour Market Department, Luxembourg Institute of Socio-Economic Research, Esch-sur-Alzette, Luxemburg
e-mail: adrien.thomas@liser.lu

V. Pulignano
Centre for Sociological Research, KU Leuven, Leuven, Belgium
e-mail: valeria.pulignano@kuleuven.be

change within the tradition of broader debates of how trade unions articulate short-term and long-term goals, as well as overall objectives and sectoral priorities.

Research illustrates that strategic union action requires the articulation of members' interests and societal concerns, a vibrant union democracy and coordination between different levels of trade union action (Flanders 1970; Gumbrell-McCormick and Hyman 2013) and representation structures (Waddington 2010; Pulignano 2017a). In their seminal contribution on collective bargaining, Walton and McKersie highlighted the role of intra-organisational bargaining in trade unions' definition of members' interests and bargaining objectives (Walton and McKersie 1965). Hyman set out the historical trajectories of European trade unions which continue to influence their interactions with workers, employers and society (Hyman 2001). The resulting union identities (unions as market bargainers, mobilisers of class opposition or partners in social integration) may act as resources or constraints when it comes to trade union engagement with climate change. The current context of de-unionisation and falling collective bargaining coverage represents an additional challenge to the capacity of trade unions to articulate contrasting objectives.

Due to its encompassing nature, climate change extends the range of concerns that trade unions have to take into account. This broadens the debate over the contradictory character of trade union action: Unions negotiate within the existing system of wage labour while striving to transform it, and they defend the interests of their members and simultaneously pursue more wide-ranging social and political aims (Zoll 1976; Hyman 2015). Climate change and environmental sustainability raise the issue of trade unions' capacity to strategically conceive their purpose, questioning the traditional boundaries of union action and the pursuit of corporatist or sectoral interests of specific groups of workers. We argue that this extension revitalises traditional discussions on how to articulate members' interests and societal concerns, an aspect at the core of wider debates on the role of interest groups in democratic societies (Baccaro 2001; Crouch and Streeck 2006; Levi et al. 2009). Drawing upon theories of interest representation and trade union collective bargaining behaviour (Walton and McKersie 1965; Bok and Dunlop 1970; March and Olsen 1989), this contribution analyses the conditions for more strongly internalising societal interests and environmental considerations in trade union decision-making processes on climate change mitigation policies and for countering the much-discussed tendency of organised groups to pursue private gains at the expense of common goods (Olson 1965).

Our analysis is based on three data sources: (a) semi-structured interviews, (b) trade union documents (policy papers, strategy documents) and (c) media sources. Twenty-four semi-structured interviews were conducted between 2017 and 2019 with trade union officials at different levels (international, national and sector/company level) and with employer representatives. A purposive sampling strategy was pursued to ensure that sufficient respondents with relevant information were included (Kumar et al. 1993). Interviewees were selected because they possessed relevant knowledge on debates over trade unions' climate strategies, either because they directly elaborate these climate strategies or because they interact, as employer representatives, with unions over climate policies. The interviews were recorded, transcribed and underwent a qualitative data analysis. Overall, they provide valuable insights into trade union climate strategies.

We start by discussing the impact of climate change mitigation policies on employment. We go on to analyse union responses to decarbonisation policies at international, national, sectoral and company levels. Finally, and in the light of these responses, we examine the core conditions we consider necessary for strategic and cohesive trade union action on climate change.

## Climate Change and Trade Union Dilemmas

The positive and negative impacts of decarbonising the economy are spread unevenly among countries, regions and economic sectors (International Labour Organization and International Institute for Labour Studies 2011). This obviously confronts trade unions with the dilemma of balancing between employment and environmental protection—a dilemma greatly influenced by the configuration of the national economies or economic sectors in which they operate. To stay within the Paris Agreement target of limiting global warming to 1.5 °C, worldwide carbon emissions need to be cut by 45 per cent from 2010 levels by 2030 and carbon neutrality achieved by 2050 according to the 2018 landmark report published by the Intergovernmental Panel on Climate Change (IPCC). The United Nations Environment Programme reported in 2019 that there is no sign of greenhouse gas emissions peaking in the next few years and underlined that every year of postponed peaking means that deeper and faster cuts will be required to stay within the 1.5 degrees target (United Nations Environment Programme 2019).

To achieve the necessary emission reductions, rapid and far-reaching changes are required in energy production, land use, urban infrastructures and industrial systems. Industrial $CO_2$ emissions need to be drastically lowered, and renewable energies rapidly upscaled. While various economic sectors, such as renewable energy generation and green construction, stand to gain from the decarbonisation of the economy, others such as carbon-intensive industries and fossil fuel extraction and processing can be expected to lose out (Organisation for Economic Co-operation and Development 2012; Château et al. 2018; European Commission 2019). By 2030, an estimated 160,000 direct jobs could be lost in the European coal sector, a sector that employed nearly half a million people in direct and indirect activities in 2018 (Alves Dias et al. 2018). Although certain fossil fuel-intensive industries will not completely disappear, tighter environmental regulations may necessitate changes in the production process impacting employment or lead to the offshoring of high-carbon activities. Trade unions from the manufacturing industry in particular emphasise the possible downsides for workers of enforcing stricter carbon emission targets, as for instance seen in their concerns over the offshoring of high-carbon activities, such as blast-furnace steel production, from countries with tighter climate regulations to ones with laxer emission constraints ('carbon leakage') voiced during the 2018 revision of the European Union Emissions Trading System (EU ETS) (Wettestad and Jevnaker 2019).

The main challenge facing trade unions is the fact that many of the jobs with large carbon footprints are in well-unionised sectors covered by comparatively advantageous collective agreements. Unionised workers are indeed disproportionately employed in older and larger workplaces (Scheuer 2011; Schnabel 2013), many in traditional 'brown' industries with a large carbon footprint. In the European Union (EU), the utilities sector responsible for producing and supplying energy and electricity for example features above-average unionisation (Eurofound 2010). By contrast, many newly created 'green' workplaces are not (yet) well unionised. Research into the quality of work in emerging sectors in the EU has shown that jobs in 'green construction' and waste disposal are often characterised by poor working conditions with comparatively low wages and weak interest representation (Holtgrewe et al. 2015). In times of neoliberal market-based policies, dwindling unionisation rates and eroding collective bargaining coverage (Müller et al. 2019; Organisation for Economic Co-operation and Development 2019), this may lead sectoral and national trade unions to defend existing unionised, carbon-intensive jobs. Climate change thus confronts trade unions with a dilemma: If unions support ambitious climate mitigation policies, they risk alienating their core constituency in the unionised manufacturing industry; if, however, unions oppose climate action, they risk a confrontation with environmental movements and losing broader public support.

Concerns over possible job losses have dominated trade union debates over climate change policies, as the examples of the discussions in Polish unions over the phasing out of coal production or debates in German and British unions concerning the switch to electrified cars show (Galgóczi 2019). Worries over potential job losses due to decarbonisation policies thus override concerns over the impact of climate change in terms of deteriorating working conditions, reduced worker productivity (Day et al. 2019) or loss of livelihoods. These risks tend to affect more strongly workers in less industrial countries, whose unions are often weak and thus less able to shape the positions on climate policies adopted by the international trade union movement.

## The Multilevel Engagement of Trade Unions with Climate Change Mitigation

The trade union movement is not homogenous, with different organisational levels (international, national, sectoral and company level) having to come to terms with a variety of interests and identities. As internally contested organisations, trade unions are divided over their primary goals, modes of action and organisational structures (Gumbrell-McCormick and Hyman 2013). The history of the trade union movement is thus rife with quarrels and split-offs over unions' political orientation, the boundaries of unionisation, the internal allocation of resources or the definition of bargaining objectives (Mink 1986; Hyman 2001; Moody 2007). In contrast to analyses of trade unionism focusing solely on the peak union structures and viewing unions as unitary actors, we will look at the three levels at which trade unions engage with climate change mitigation policies: the international, national and sectoral/company levels. We combine the sectoral and company levels because, with specific regard to climate change mitigation, many of the challenges are similar at sectoral and company levels, reflecting the carbon intensity of production and the role of international competition. The international level is particularly relevant as many decisions on climate governance are taken at this level.

### The International Trade Union Movement

Starting in the nineteenth century with the founding of the International Workingmen's Association (a.k.a. the First International) in 1864, trade unions have been intent on initiating common action at international level, organising solidarity against political repression, running political campaigns and promoting programmatic exchanges among trade unionists from different countries (Régin and Wolikow 2002; Stevis 2019). Today, the main international trade union umbrella organisation is the International Trade Union Confederation (ITUC). Representing 207 million members in 163 countries around the world, the ITUC resulted from the 2006 merger of the mainly social-democratic International Confederation of Free Trade Unions (ICFTU) and the World Confederation of Labour composed of Christian trade unions (Gumbrell-McCormick 2013). ITUC affiliates are the various national confederations. In addition, the ITUC closely collaborates with the sectoral global unions (Croucher and Cotton 2009). While the ITUC possesses the expertise and legitimacy to engage with the UN and its agencies, in particular the International Labour Organization, it has no authority over its national affiliates and depends on a limited number of affiliates who provide the bulk of resources (Germans, Nordics, North Americans and Japanese; Cotton and Gumbrell-McCormick 2012).

The ITUC and its predecessor organisation ICFTU took up the issue of climate change mitigation at an early stage through their involvement in the 1992 Rio 'Earth Summit', during which the United Nations Framework Convention on Climate Change (UNFCCC) was negotiated (Silverman 2006). The ICFTU was, however, confronted with reticence among its affiliates over climate policies. As a result, the ICFTU did not endorse the Kyoto Protocol (1997), mainly due to the opposition of the American Federation of Labor and Congress of Industrial Organizations (AFL–CIO). The AFL–CIO opposed the US government's ratification of the agreement on competitiveness grounds as it did not set binding $CO_2$ reduction targets for developing countries. As a result of this reticence, the ICFTU and then the ITUC discourse on climate change initially focused not on environmental issues, but on carbon leakage and concerns over the competitiveness of the industrial sector (Rosemberg 2020).

In a second phase, however, the international trade union movement became more directly involved in mitigating climate change through the Just Transition concept, a concept developed by trade unions in the 1990s and 2000s to underline the importance of taking workers' interests into consideration when implementing the transition to a low-carbon economy. The concept highlights the need to secure the livelihoods of workers and their communities during the low-carbon transition by providing decent work, social protection and training opportunities to workers affected by global warming and climate change mitigation policies (Stevis and Felli 2015).

Initially, the Just Transition discourse within the international trade union movement focused mainly on the challenges of climate change mitigation and on potential job losses caused by decarbonisation policies, and not on adaptation issues and on the hardships resulting from climate change (loss of livelihoods due to changing weather patterns, rising sea levels, droughts, infectious disease patterns). However, the ITUC gradually developed a broader approach, with the participation of trade unions from developing countries, as explained by a former coordinator of the ITUC working group on climate change:

> Some at the ITUC thought our policy was too Europe-centric. This was addressed very early on at the Nairobi COP in 2006, attended for the first time by some fifteen delegates from African trade unions, including a large delegation from Tanzania. They were talking about adaptation issues. We were told the Sahara was moving south, that there were unseasonal rainstorms, floods on the coast, deforestation, all those issues. We quickly understood that we had not worked out how Just Transition was meant to apply to developing countries, how our policies were to be relevant to frontline states. We realised that there was a huge gap in our policy. And it was filled by the thinking brought to the table by African trade unions. (Interview, ITUC official, March 2019)

Developing an international trade union strategy on climate change thus entails coordinating the approaches of trade unions from developed and developing countries, taking account of their differing interests, levels of resource and expertise. While Just Transition provides a general framework for addressing the impact of climate mitigation policies on workers, the definition of its scope and implications varies between national and sectoral trade unions. Trade unions may thus conceptualise the Just Transition approach as entailing their proactive participation in the planning of the decarbonisation of the economy and in the creation of sustainable and decent jobs. However, they may also take a more reactive approach, responding to decarbonisation policies about to be or currently being implemented, and highlighting the need for the state or employers to bear the costs of decarbonisation by providing compensation to workers. An example of such a reactive approach to the transition is the German coal miners' union IG BCE which tried to delay as much as possible the phase out of coal-produced electricity and eventually agreed to it only in the face of strong political pressure and in exchange for retraining and financial compensations for workers (Reitzenstein et al. 2020).

## Trade Unions at National Level

National union confederations, grouping different sectoral and regional affiliates, try to influence public policymaking on subjects of broader importance to their membership and to coordinate the policies of their affiliated entities (Traxler et al. 2001). The broader political economy of nation states and their geopolitical interests, for instance related to energy production, can be expected to have an influence on the policy positions adopted by trade unions. The evolution towards a low-carbon economy will inevitably alter the power and influence of some states and regions relative to others, redrawing the geopolitical map (International Renewable Energy Agency 2019). Likewise, the sectoral membership distribution within national confederations can be expected to affect the positioning of national confederations. National trade union movements with large numbers of members in carbon-intensive sectors are thus more likely to oppose ambitious climate mitigation policies. A trade union official from the European Trade Union Confederation (ETUC) explained the differences in affiliates' approaches to climate change mitigation policies:

> Obviously the ETUC is to a certain extent a reflection of Europe. It is not therefore surprising that we observe within the ETUC the debates we have elsewhere in Europe. Sometimes there are tensions and diverging points of view between national confederations. It is obvious that the Polish trade unions—an often-cited example though they are not the only ones—think differently, given that Poland has an energy mix still strongly dependent on coal. This gives rise to specific positions differing from those voiced in countries such as Sweden where considerable progress has already been made on decarbonisation, or in other countries with an energy mix more reliant on nuclear power or gas, for example. We have the same debate at the sectoral level. The ETUC also represents ten European sectoral federations, including IndustriAll Europe, a federation which represents workers from CO2-intensive sectors. Obviously, they have a particular point of view. (Interview, ETUC official, October 2017)

The national and sectoral diversity of union positions is also related to the ideological identities of trade unions which may be prevalently class-oriented, society-oriented or market-oriented (Hyman 2001). This makes them more or less inclined to take broader societal issues into consideration when taking position on policy issues. While certain national union confederations such as the UK Trades Union Congress (TUC) or the French *Confédération Française Démocratique du Travail* (CFDT) have developed ambitious positions on mitigating climate change, others such as the Polish NSZZ Solidarność oppose decarbonisation policies. The British TUC trained, for instance, environmental representatives in order to promote green workplace initiatives in a number of industries (Hampton 2015). The CFDT signed a 'social and ecological pact' together with social and environmental NGOs in March 2019, and took a leading role in promoting it. This pact, much commented in the context of the Yellow Vest protests in France, advocated the introduction of carbon taxation, an end to the construction of combustion engine cars 'in a time horizon compatible with the Paris Agreement' and social support to mitigate the consequences of the ecological transition on employment. In contrast, Polish trade unions consistently oppose EU climate policies and play a prominent role in advocating a continued strong reliance of Poland on coal-based power generation. With its close ties to the national conservative Law and Justice party PiS, NSZZ Solidarność has developed a discourse in defence of coal-based energy that combines employment concerns and national sovereignty arguments.

In the case of trade unions from the Global South, additional issues come into play. If the situation of unions in developing and transition countries is varied, many of them face considerable difficulties organising the numerous workers operating outside of the formal labour market (Salmon 1999; Bonner and Spooner 2011). In addition, most of these unions have a weak bargaining power and are subordinated to employers, political parties or the state (Clarke 2005; Biyanwila 2011; Gall et al. 2011). A number of these unions are voicing scepticism over the applicability of the Just Transition framework in their respective countries. One ITUC union official in charge of climate policies explained that the core concept developed by the ITUC to address decarbonisation, that is, the Just Transition framework, did not necessarily reflect the reality in a developing country such as Brazil:

> A crucial part of the Just Transition involves having a voice at the social dialogue table. But in countries such as Brazil social dialogue means something completely different than in Denmark. They tell us "we are being shot, we have to move out of the country, the big bosses are mafia, and what do you mean about sitting down at the table with these guys and discussing about the transition?" There is no transition in Brazil. The forests are being cut down, oil exploration goes on everywhere and environmental activists are being killed. There is no transition, even less a just transition. We need a different approach here, a more social activism approach to Just Transition, raising our voices on climate change and asking the bosses what they are doing. (Interview, ITUC official, February 2019)

While this ITUC official, who has a previous professional background in development aid, acknowledges the limits of the Just Transition framework, the ITUC tends in the recent period to stress even more the cooperative dimension of its Just Transition approach. The ITUC's Just Transition Centre, created in 2016, relies indeed strongly on cooperative industrial relations and aims to develop alliances with corporations and initiatives such as the B Team, initiated by the chairmen of Virgin Group and Puma.

The ability of national confederations to develop cohesive positions on mitigating climate change also depends on their ability to coordinate and articulate the policy positions adopted by their various constituents or affiliates at sector and company levels where positions are mainly shaped by employment concerns.

## Trade Unions at Sector and Company Levels

Sector- and company-level unions are engaged in collective bargaining processes and exchanges with employer organisations and large companies. At sectoral and company levels, positions on decarbonisation vary between trade unions operating in carbon-intensive sectors which stand to lose out in terms of jobs (steel-making, fossil fuel extraction and energy production) and sectors which stand to gain, such as regenerative energy and green construction. It has been shown that corporate decarbonisation strategies are shaped by multiple factors such as a company's current carbon intensity, its exposure to international competition and its overall business strategy (Skjærseth and Eikeland 2013). In the case of trade union representatives operating at sectoral or company level, these factors can also be expected to play an important role.

At the sectoral level, the influence of employment-related concerns on trade union policies is manifest. Trade unions in the automotive sector, for instance, opposed the introduction of the stricter emissions standards for cars adopted by the EU in 2019. Without denying the need for decarbonisation, relevant sector-level unions such as IG Metall (Germany), Unite (UK) or *Force Ouvrière*'s autoworkers' federation (France) came out in favour of less strict emissions standards, in concert with employers. While supporting the need for decarbonisation in principle, these unions sought to minimise compliance costs and advocated incremental approaches. Energy sector unions have developed more advanced positions. Contrary to the manufacturing sector, electricity markets are more localised (implying less competition), and the possibility of switching to renewables or of passing on the cost of emissions trading schemes to customers reduces the potential employment impact of decarbonisation (Meckling 2011; Eikeland and Skjærseth 2019). Energy sector trade unions committed to a Just Transition approach in the context of

## 22 Challenges and Prospects for Trade Union Environmentalism

European social dialogue. The European sectoral trade unions EPSU and IndustriAll Europe and the electricity industry employer association Eurelectric thus issued a joint statement, outlining a common understanding of a 'Just Energy Transition' and providing recommendations in terms of policies (mostly focused on retraining) and their financing. The statement defines a 'Just Energy Transition' as a 'combination of plans, policies and investments that enable the sector to decarbonise cost-effectively while ensuring that potential negative impacts on business, employment and living conditions are anticipated and mitigated' (Eurelectric et al. 2017).

Trade unions also deal with climate change at company level. In a way, this is the most challenging level, with the jobs-versus-environment dilemma most pronounced in companies engaged in carbon-intensive production and exposed to international competition. Here, the grassroots interests of members in keeping their jobs visibly clash with trade unions' broader commitment to climate action (Snell and Fairbrother 2011). In addition, company-level union representatives are often engaged in long-term interactions with company management over collective agreements and other matters of common interest, during which they may develop patterns of concession bargaining underpinning disadvantageous arrangements on wages and working conditions in order to maintain employment (Glassner et al. 2011; Pulignano 2017b). The example of steelmaker Arcelor-Mittal[1] is emblematic of this interweaving of company-level social dialogue and climate issues. Arcelor-Mittal actively tried to enlist the support of workers' representatives and trade unions on the issue of the 2018 revision of the EU ETS, a carbon cap-and-trade scheme that sets binding emission reduction targets for industrial facilities and power plants. The management of Arcelor-Mittal discussed the implications of the EU ETS reform in meetings with works councillors and trade unions, mainly framing the issue of emissions trading as a competitiveness issue, according to a member of the corporate board of directors and chairman of the country board of directors of Arcelor-Mittal Luxembourg (Interview, June 2018). A Luxembourg union official in charge of the steel sector, who is himself a former employee from an Arcelor-Mittal production plant, described the company's initiatives at engaging union representatives on the revision of the EU ETS:

> The main argument put forward by the management is that if we have to shoulder these additional costs, we will not be viable, given our current results in

---

[1] Arcelor-Mittal is currently the world's largest steel producer, accounting for 6 per cent of global steel production, and 50 per cent of steel produced in the EU. The company had 209,000 employees in 2020.

Luxembourg. And then the plants will be restructured. That's how things are presented. The big problem is that we here in Europe impose these tight rules on ourselves while the others do what they want, and large multinational companies do not have any problem in offshoring activities. (Interview, sector level union official, April 2018)

In a context of strong interplant competition at Arcelor-Mittal with relocation and headcount reduction threats (Aranea et al. 2018), the interlinking by corporate management of the EU ETS debate to that on competitiveness puts pressure on workers' representatives to support employer demands concerning the EU ETS reform (slower emissions reduction path, relationship between free allocation and auctioning, determination of the best available technologies). The tensions that competitiveness comparisons between production sites generate between workers across a company's production network (Marginson et al. 1995; Greer and Hauptmeier 2016) challenge the capacity for collective action of trade unions. This makes it necessary for unions, which remain primarily local and national in their organisational focus, to develop their capacity for cooperation and coordination across borders.

The articulation between environmental concerns and employment issues is a challenge for the unions at different levels, with approaches to climate change mitigation policies not necessarily homogenous at the different levels. This makes climate change a contested issue within unions, raising the need to identify and discuss the factors affecting strategic union action on climate change mitigation policies.

## Climate Change Mitigation and the Conditions for Strategic Union Action

Contrasting interests shape trade union strategies on climate change mitigation, resulting in inconsistencies between policy discourses, decisions and implementation (Geden 2016). The complexity of climate change as a policy issue for unions stems from the fact that it involves, like other sustainability issues, long time frames, a cross-national spatial dimension, necessary limits to human activities and the irreversibility of changes (Dovers 1996). In order to advance towards a greater policy consistency and integration of ecological considerations, trade unions need to connect more strongly the different levels of trade union action. In this section, we look at three factors affecting

## 22 Challenges and Prospects for Trade Union Environmentalism

trade unions' ability to build up a cohesive strategy when dealing with climate change: the lack of a broader definition of members' interests, bureaucratic decision-making processes and the difficulty to articulate different levels of action in an adverse policy context that gives employers considerably more power than unions.

Strategic action on climate change raises the issue of (re)defining workers' interests. Adopting a long-term perspective, research has stressed the variety of human attitudes towards environmental change (Hulme 2011). As membership organisations, trade unions naturally have to defend member interests. However, these interests are not given, but are social constructs shaped by a wide range of factors, including social interactions and socio-economic structures. For instance, Pierre Bourdieu analysed the 'oracle effect' under which 'the spokesperson gives voice to the group in whose name he speaks, thereby speaking with all the authority of that elusive, absent phenomenon' (Bourdieu 1991, 211). In the same vein, it has been argued that trade unions need to translate and redefine members' demands to establish a consensus (Regalia 1988). Trade union leaders are in particular confronted with the 'two faces' (Flanders 1970) of trade unionism, that is, the short- and the long-term goals of union action. They pursue the sectional or 'corporatist' interests of their constituencies while simultaneously trying to transform the broader socio-economic order (Hyman 2015). As regards decarbonisation, the interests of workers are heterogeneous, depending on their geographical location, sector of work and the timeframe considered. For instance, workers living in a remote geographical area, with a strong community built around a specific sector (e.g. mine workers) and with traditionally high levels of unemployment can be expected to resist decarbonisation policies more strongly than workers from areas with diverse economies and low levels of unemployment (Szpor 2019; Reitzenstein et al. 2020). Thus, workers' attitudes towards decarbonisation policies are shaped by the extent of exposure of their sector of employment to these policies. At the same time, however, workers also have an interest in maintaining a liveable planet, meaning that their longer-term interests also include the proactive planning of the low-carbon transition. Thus, the co-existence of conflicting interests raises the question of which choices unions will make and why.

Entrenched interests within unions and highly unionised member groups may, for instance, exercise a greater influence on specific decisions. This issue is particularly salient, as workers from green economic sectors expected to grow more rapidly under decarbonisation policies are often not yet as strongly represented in trade unions as workers from brown industries. For instance, when positioning itself on the EU ETS reform in 2018, the German metalworkers union IG Metall needed to weigh up the interests of the steelmaking segments of its membership versus those of its members working in renewable energy jobs. The higher organisational weight of the steelworkers helped tip the scales in favour of steel industry interests:

> As is generally the case, our approach to the topic was very industry-specific—in this case, the steel industry. Of course, in theory, our approach could have focused on renewable energies. Then the perspective would probably have been a bit different. But the steel industry was very loud at the time, an industry of particular importance to IG Metall. (Interview, IG Metall official, December 2017)

When it comes to arbitrating between different interests and bargaining objectives, the quality of trade unions' procedural, deliberative and participatory mechanisms is highly relevant, as these can help ensure that all member groups are treated fairly and have the possibility to shape union policies. Larger and conglomerate unions representing a wide variety of member groups are particularly confronted with the need to aggregate and represent the different interests and points of view of their member groups, and to generate consensus among them (Moody 2009; Räthzel and Uzzell 2011). The notions of the administrative and representative rationality of trade unionism help reveal the logics involved in trade union decision-making (Child et al. 1973). While administrative rationality refers to the need for the efficient and centralised functioning of a trade union, democratic rationality refers to the democratic participation of union members, which presupposes taking into account the specific interests of companies or occupations, but also the construction of more extensive syntheses at the level of a branch, a federation or a confederation. Trade union policymaking on an issue such as decarbonisation is very much dependent upon such larger syntheses, given the intertemporal and cross-national implications of climate change (Boston 2017).

The quality of vertical linkages within union structures and the articulation between different levels of trade union action (international, national and sectoral/company) also plays an important role in union policies towards mitigating climate change. The development of cohesive policies requires articulation and coordination between the different levels of trade union organisation. In the absence of such an interlinking, the positions in favour of ambitious decarbonisation policies adopted at the UN climate change conferences by the ITUC and a number of national union confederations are not necessarily taken up by sector- and company-level union representatives. The transformation of many trade unions from oligarchical into 'stratarchical

## 22 Challenges and Prospects for Trade Union Environmentalism

organisations', in which each organisational level pursues distinct self-interests, makes coordination and articulation across levels more difficult (Piotet 2009; Thomas 2017). In the EU's emerging multilevel system of industrial relations (Keune and Marginson 2013), internal fragmentation might be even more pronounced, constituting an additional obstacle to a coherent engagement of trade unions with climate change. In the case of carbon-intensive companies exposed to international competition, the alignment of workplace union representatives with corporate priorities may have a domino effect on other union structures, as highlighted by an ITUC representative in charge of environmental issues:

> A number of unions have made the pretty deliberate choice to stand with employers and basically toe their line. [...] This means that if you have for instance the Arcelor-Mittal unions wanting to follow what Arcelor-Mittal says, that determines the industry federation's course and ultimately the confederation's course. It's only when you have strong leadership in the different segments that you arrive at a different position. (Interview, ITUC official, November 2017)

This raises *inter alia* the issue of the vitality of confederal union structures, be it at international (ITUC, ETUC) or national level (national confederations). Confederal structures can be spaces for coordinating and unifying trade union positions and activities. However, in a context of de-unionisation and falling membership figures, trade union confederations lack the resources and funding to build up capacity on climate change and to invest in the training of union officials and workplace representatives. In addition, the decline of neocorporatist forms of governance in many countries has diminished the relevance of confederations as political interlocutors of political decision-makers (Streeck 1993). In parallel, the decentralisation of collective bargaining has strengthened the role of sectoral- and company-level union representatives (Traxler 1995). In the absence of proper coordination and articulation mechanisms between different trade union levels, it is difficult to counter the tendency of sectoral and workplace union representatives to uphold short-term employment concerns possibly leading them to either outrightly oppose decarbonisation policies or to minimise their extent (Snell and Fairbrother 2011; Sweeney 2013; Galgóczi 2019).

## Conclusion

While the decarbonisation of the economy will obviously impact the labour market, trade union approaches to climate change policies are heterogeneous and dependent on the level of trade union action. Though union organisations at global level tend to support ambitious climate policy goals at UN climate change conferences, national positions are primarily shaped by the interests of national political economies, while attitudes at sectoral and workplace levels are generally shaped by employment concerns and employer preferences. As a result of this lack of cohesion, trade unions may end up either positioning themselves among those calling for ambitious and binding emissions reduction targets or aligning themselves with vested interests and reinforcing political stalemates hindering the effective use of existing technological solutions to reduce carbon emissions (Falkner 2008; Stephan and Lane 2016). The limited empirical evidence currently available suggests that especially sector- and company-level unions in the carbon-intensive manufacturing and fossil fuel-based power generation sectors tend to adopt positions aiming to minimise climate regulations (Räthzel and Uzzell 2013; Galgóczi 2019). Trade unions representing a broader set of workers, and who have a tradition of societal engagement, can be expected to engage more strongly with climate issues. The examples of EPSU (Fischbach-Pyttel 2017) and the International Transport Workers' Union point in that direction (Felli 2014). Sectoral interests thus seem to play a key role, with historically built union identities and conceptions of union democracy acting as meditating factors.

Developing a unified and cohesive approach to decarbonisation raises the issue of strategic trade union action. This requires trade unions to combine a degree of expertise and top-down input with shop-floor involvement. Trade unions need strong internal decision-making processes enabling them to deal with the long timeframe, cross-national spatial scale and irreversibility of changes that need to be taken into account when making decisions on climate change policies. Given the broader implications of climate change, decisions over climate policies should not be left to the sector- and company-level union structures, even if these organisational levels have traditionally the highest legitimacy for dealing with industrial policy issues. Confederal union levels have a broader mandate that should enable them to better take into account the need to balance the interests of presently unionised workers in carbon-intensive industries and the interests of (not yet) unionised workers in the green economy.

The timeframe considered is an essential factor, given that worker interests on mitigating climate change may be contradictory in the short term, but congruent in the long term. Climate change thus highlights the crucial issue of coordinating the different levels of trade union action (international, national and sectoral/company) on the one hand, and articulating trade union demands and broader societal concerns on the other. Interacting and building coalitions with organisations not directly involved in the employment relationship, such as environmental NGOs, may also be a way of furthering the consideration of environmental concerns in trade union decision-making, although the fact that a number of environmental NGOs do not seriously consider employment issues certainly represents an obstacle (Spooner 2004).

In a context in which working lives in many sectors of the economy are characterised by continuous restructuring (Beaud and Pialoux 2012; Raphael 2019), the implications of decarbonising the economy may be perceived by workers primarily as a further attack on their living and working conditions. Trade unions are thus also dependent on support by state institutions willing to develop public policies furthering the low-carbon transition while at the same time enabling the creation of new sustainable jobs and addressing workers' demands for economic security. In addition, the institutional and organisational power resources of trade unions have eroded in many countries as a result of neoliberal market policies. Trade unions in developed countries are confronting the decline of their traditional strongholds in manufacturing industry and the concomitant rise of mostly non-unionised jobs in the service sector. Unions in developing and transition countries are struggling to develop independent policy positions and to overcome their subordination to employers, political parties or the state. These adverse policy contexts increase the uncertainty for trade unions, encourage purely defensive approaches and diminish their power to articulate contrasting interests on climate change policies.

## References

Alves Dias, Patricia, et al. 2018. *EU Coal Regions: Opportunities and Challenges Ahead*. Luxembourg: Publications Office of the European Union.

Aranea, Mona, Sergio González Begega, and Holm-Detlev Köhler. 2018. The EWC as a Management Tool to Divide and Conquer: Corporate Whipsawing in the Steel Sector. *Economic and Industrial Democracy*. Advanced online publication. https://doi.org/10.1177/0143831X18816796

Baccaro, Lucio. 2001. Aggregative and Deliberative Decision-Making Procedures: A Comparison of Two Southern Italian Factories. *Politics and Society* 29 (2): 243–271.

Beaud, Stéphane, and Michel Pialoux. 2012. *Retour sur la condition ouvrière. Enquête aux usines Peugeot de Sochaux-Montbéliard*. Paris, La Découverte.

Biyanwila, S. Janaka. 2011. *The Labour Movement in the Global South: Trade Unions in Sri Lanka*. Oxon: Routledge.

Bok, Dereck C., and John Dunlop. 1970. *Labor and the American Community*. New York: Simon and Schuster.

Bonner, Christine, and Dave Spooner. 2011. Organizing in the Informal Economy: A Challenge for Trade Unions. *Internationale Politik und Gesellschaft* 2: 87–105.

Boston, Jonathan. 2017. *Governing for the Future: Designing Democratic Institutions for a Better Tomorrow*. Bingley, UK: Emerald Publishing.

Bourdieu, Pierre. 1991. Delegation and Political Fetishism. In *Language and Symbolic Power*, ed. Pierre Bourdieu, 203–219. Cambridge: Polity Press.

Château, Jean, Ruben Bibas, and Elisa Lanzi. 2018. *Impacts of Green Growth Policies on Labour Markets and Wage Income Distribution: A General Equilibrium Application to Climate and Energy Policies*. OECD Environment Working Papers No. 137.

Child, John, Ray Loveridge, and Warner Malcolm. 1973. Towards an Organizational Study of Trade Unions. *Sociology* 7 (1): 71–91.

Clarke, Simon. 2005. Post-Socialist Trade Unions: China and Russia. *Industrial Relations Journal* 36 (1): 2–18.

Cotton, Elizabeth, and Rebecca Gumbrell-McCormick. 2012. Global Unions as Imperfect Multilateral Organizations: An International Relations Perspective. *Economic and Industrial Democracy* 33 (4): 707–728.

Crouch, Colin, and Wolfgang Streeck, eds. 2006. *The Diversity of Democracy: Corporatism, Social Order and Political Conflict*. Cheltenham: Edward Elgar.

Croucher, Richard, and Elisabeth Cotton. 2009. *Global Unions, Global Business: Global Union Federations and International Business*. London: Middlesex University Press.

Day, Ed, Sam Fankhauser, Nick Kingsmill, Hélia Costa, and Anna Mavrogianni. 2019. Upholding Labour Productivity under Climate Change: An Assessment of Adaptation Options. *Climate Policy* 3 (19): 367–385.

Dovers, Stephen R. 1996. Sustainability: Demands on Policy. *Journal of Public Policy* 16 (3): 303–318.

Eikeland, Per Ove, and Jon Birger Skjærseth. 2019. Oil and Power Industries' Responses to EU Emissions Trading: Laggards or Low-Carbon Leaders? *Environmental Politics* 28 (1): 104–124.

Eurelectric, IndustriAll Europe, EPSU. 2017. Statement on a Just Energy Transition. https://www.epsu.org/sites/default/files/article/files/JustEnergyTransition%20-%20leaflet.pdf

Eurofound. 2010. *Trade Union Strategies to Recruit New Groups of Workers*. Luxembourg: Publications Office of the European Union.

European Commission. 2019. *Employment and Social Developments in Europe*. Luxembourg: Publications Office of the European Union.
Falkner, Robert. 2008. *Business Power and Conflict in International Environmental Politics*. Basingstoke: Palgrave Macmillan.
Felli, Romain. 2014. An Alternative Socio-Ecological Strategy? International Trade Unions' Engagement with Climate Change. *Review of International Political Economy* 21 (2): 372–398.
Fischbach-Pyttel, Carola. 2017. *Building the European Federation of Public Service Unions: The History of EPSU, 1978–2016*. Brussels: ETUI and European Federation of Public Service Unions.
Flanders, Allan. 1970. *Management and Unions: The Theory and Reform of Industrial Relations*. London: Faber & Faber.
Galgóczi, Bela, ed. 2019. *Towards a Just Transition: Coal, Cars and the World of Work*. Brussels: ETUI.
Gall, Gregor, Adrian Wilkinson, and Richard Hurd, eds. 2011. *The International Handbook of Labour Unions: Responses to Neo-Liberalism*. Cheltenham, UK: Edward Elgar Publishing.
Geden, Oliver. 2016. The Paris Agreement and the Inherent Inconsistency of Climate Policymaking. *WIREs Climate Change* 7: 790–797.
Glassner, Vera, Maarten Keune, and Paul Marginson. 2011. Collective Bargaining in a Time of Crisis: Developments in the Private Sector in Europe. *Transfer: European Review of Labour and Research* 17 (3): 303–322.
Greer, Ian, and Marco Hauptmeier. 2016. Management Whipsawing: The Staging of Labor Competition under Globalization. *ILR Review* 69 (1): 29–52.
Gumbrell-McCormick, Rebecca. 2013. The International Trade Union Confederation: From Two (or More?) Identities to One. *British Journal of Industrial Relations* 51 (2): 240–263.
Gumbrell-McCormick, Rebecca, and Richard Hyman. 2013. *Trade Unions in Western Europe: Hard Times, Hard Choices*. Oxford: Oxford University Press.
Hampton, Paul. 2015. *Workers and Trade Unions for Climate Solidarity: Tackling Climate Change in a Neoliberal World*. London: Routledge.
Holtgrewe, Ursula, Monique Ramioul, and Vassil Kirov. 2015. *Hard Work in New Jobs: The Quality of Work and Life in European Growth Sectors*. London: Palgrave Macmillan.
Hulme, Mike. 2011. Meet the Humanities. *Nature Climate Change* 1 (4): 177–179.
Hyman, Richard. 2001. *Understanding European Trade Unionism: Between Market, Class and Society*. London: Sage.
———. 2015. Three Scenarios for Industrial Relations in Europe. *International Labour Review* 154 (1): 5–14.
Intergovernmental Panel on Climate Change. 2018. *Special Report on Global Warming of 1.5°C*. October.

International Labour Organization and International Institute for Labour Studies. 2011. *Towards a Greener Economy: The Social Dimensions*. https://www.ilo.org/global/publications/ilo-bookstore/order-online/books/WCMS_168163/lang%2D%2Den/index.htm

International Renewable Energy Agency. 2019. *A New World: The Geopolitics of the Energy Transformation*. Abu Dhabi: International Renewable Energy Agency.

Keune, Maarten, and Paul Marginson. 2013. Transnational Industrial Relations as Multi-Level Governance: Interdependencies in European Social Dialogue. *British Journal of Industrial Relations* 51 (3): 473–497.

Kumar, Nirmalya, Louis W. Stern, and James C. Anderson. 1993. Conducting Interorganizational Research using Key Informants. *Academy of Management Journal* 6 (36): 1633–1651.

Levi, Margaret, David Olson, Jon Agnone, and Devin Kelly. 2009. Union Democracy Reexamined. *Politics & Society* 37 (2): 203–228.

March, James G., and Johan P. Olsen. 1989. *Rediscovering Institutions: The Organizational Basis of Politics*. New York: Free Press.

Marginson, Paul, Armstrong Peter, P.K. Edwards, and John Purcell. 1995. Extending Beyond Borders: Multinational Companies and the International Management of Labour. *International Journal of Human Resource Management* 6 (3): 702–719.

Meckling, Jonas. 2011. The Globalization of Carbon Trading: Transnational Business Coalitions in Climate Politics. *Global Environmental Politics* 11 (2): 26–50.

Mink, Gwendolyn. 1986. *Old Labor and New Immigrants in American Political Development: Union, Party, and State, 1875–1920*. Ithaca, NY: Cornell University Press.

Moody, Kim. 2007. *US Labor in Trouble and Transition: The Failure of Reform from Above, the Promise of Revival from Below*. New York: Verso Books.

———. 2009. The Direction of Union Mergers in the United States: The Rise of Conglomerate Unionism. *British Journal of Industrial Relations* 47 (4): 676–700.

Müller, Torsten, Kurt Vandaele, and Jeremy Waddington. 2019. *Collective Bargaining in Europe: Towards an Endgame. Volume I–IV*. ETUI: Brussels.

Olson, Mancur. 1965. *The Logic of Collective Action*. Cambridge, MA: Harvard University Press.

Organisation for Economic Co-operation and Development. 2012. *The Jobs Potential of a Shift Towards a Low-Carbon Economy*. https://www.oecd.org/els/emp/50503551.pdf

———. 2019. *Negotiating Our Way Up: Collective Bargaining in a Changing World of Work*. OECD Publishing.

Piotet, Françoise. 2009. La CGT, une anarchie (plus ou moins) organisée? *Politix* 85: 9–30.

Pulignano, Valeria. 2017a. Articulation and the Role of EWC: Explaining the Social Effects within (and across) Transnational Workplaces. *European Journal of Industrial Relations* 23 (3): 261–276.

———. 2017b. Workplace Inequality, Trade Unions and the Transnational Regulation of the Employment Relationships. *Employee Relations* 39 (3): 351–364.
Raphael, Lutz. 2019. *Jenseits von Kohle und Stahl: Eine Gesellschaftsgeschichte Westeuropas nach dem Boom*. Berlin: Suhrkamp.
Räthzel, Nora, and David Uzzell. 2011. Trade Unions and Climate Change: The Jobs Versus Environment Dilemma. *Global Environmental Change* 21 (4): 1215–1223.
———, eds. 2013. *Trade Unions in the Green Economy. Working for the Environment*. London: Routledge.
Regalia, Ida. 1988. Democracy and Unions: Towards a Critical Appraisal. *Economic and Industrial Democracy* 9 (3): 345–371.
Régin, Tania, and Serge Wolikow, eds. 2002. *Les syndicalismes à l'épreuve de l'international*. Paris: Syllepse.
Reitzenstein, Alexander, Sabrina Schultz, and Felix Heilmann. 2020. The Story of Coal in Germany: A Model for Just Transition in Europe? In *Just Transitions. Social Justice in the Shift Towards a Low-Carbon World*, ed. Eduardo Morena, Dunja Krause, and Dimitris Stevis, 151–171. London: Pluto.
Rosemberg, Anabella. 2020. No Jobs on a Dead Planet: The International Trade Union Movement and Just Transition. In *Just Transitions. Social Justice in the Shift Towards a Low-Carbon World*, ed. Eduardo Morena, Dunja Krause, and Dimitris Stevis, 32–55. London: Pluto.
Salmon, Claire. 1999. Les syndicats dans les pays en développement: leur action sur le marché du travail. *Canadian Journal of Development Studies/Revue canadienne d'études du développement* 20 (4): 661–688.
Scheuer, Steen. 2011. Union Membership Variation in Europe: A Ten-Country Comparative Analysis. *European Journal of Industrial Relations* 17 (1): 57–73.
Schnabel, Claus. 2013. Union Membership and Density: Some (not so) Stylized Facts and Challenges. *European Journal of Industrial Relations* 19 (3): 255–272.
Silverman, Victor. 2006. Green Unions in a Grey World: Labor Environmentalism and International Institutions. *Organization & Environment* 19 (2): 191–213.
Skjærseth, Jon Birger, and Per Ove Eikeland, eds. 2013. *Corporate Responses to EU Emissions Trading: Resistance, Innovation or Responsibility?* Farnham, Surrey: Ashgate.
Snell, Darryn, and Peter Fairbrother. 2011. Toward a Theory of Union Environmental Politics: Unions and Climate Action in Australia. *Labor Studies Journal* 36 (1): 83–103.
Spooner, Dave. 2004. Trade Unions and NGOs: The Need for Cooperation. *Development in Practice* 14 (1): 19–33.
Stephan, Benjamin, and Richard Lane. 2016. *The Politics of Carbon Markets*. London: Routledge.
Stevis, Dimitris. 2018. US Labour Unions and Green Transitions: Depth, Breadth, and Worker Agency. *Globalizations* 15 (4): 454–469.
———. 2019. Global Union Organizations, 1889–2019: The Weight of History and the Challenges of the Present. In *The Internationalisation of the Labour*

Question. Ideological Antagonism, Workers' Movements and the ILO since 1919, ed. Stefano Bellucci and Holger Weiss, 23–49. London: Palgrave Macmillan.

Stevis, Dimitris, and Romain Felli. 2015. Global Labour Unions and Just Transition to a Green Economy. *International Environmental Agreements: Politics, Law and Economics* 15 (1): 29–43.

Streeck, Wolfgang. 1993. The Rise and Decline of Neocorporatism. In *Labor and an Integrated Europe*, ed. Lloyd Ulman, William T. Dickens, and Barry Eichengreen, 80–101. Washington, DC: Brookings Institution Press.

Sweeney, Sean. 2013. US Trade Unions and the Challenge of 'Extreme Energy': The Case of the TransCanada Keystone XL Pipeline. In *Trade Unions in the Green Economy. Working for the Environment*, ed. Nora Räthzel and David Uzzell, 196–213. London: Routledge.

Szpor, Aleksander. 2019. The Changing Role of Coal in the Polish Economy—Restructuring and (Regional) Just Transition. In *Towards a Just Transition: Coal, Cars and the World of Work*, ed. Béla Galgóczi, 33–55. Brussels: ETUI.

Thomas, Adrien. 2017. Conglomerate Unions and Transformations of Union Democracy. *British Journal of Industrial Relations* 55 (3): 648–671.

Traxler, Franz. 1995. Farewell to Labour Market Associations? Organized versus Disorganized Decentralization as a Map for Industrial Relations. In *Organized Industrial Relations in Europe: What Future?* ed. Colin Crouch and Franz Traxler, 3–19. Avebury: Aldershot.

Traxler, Franz, Sabine Blaschke, and Bernhard Kittel. 2001. *National Labour Relations in Internationalized Markets. A Comparative Study of Institutions, Change, and Performance*. Oxford: Oxford University Press.

United Nations Environment Programme. 2019. *Emissions Gap Report 2019*. Nairobi, Kenya: UNEP.

Waddington, Jeremy. 2010. *European Works Councils*. London: Routledge.

Walton, Richard E., and Robert B. McKersie. 1965. *A Behavioral Theory of Labor Negotiations*. Ithaca, NY: Cornell University Press.

Wettestad, Jørgen, and Torbjørg Jevnaker. 2019. Smokescreen Politics? Ratcheting Up EU Emissions Trading in 2017. *Review of Policy Research* 36 (5): 635–659.

Zoll, Rainer. 1976. *Der Doppelcharakter der Gewerkschaften*. Berlin: Suhrkamp.

# 23

# From 'Just Transition' to the 'Eco-Social State'

Béla Galgóczi

## Introduction

Time is running short for us to have a chance to get climate change under control. As a result, the 'climate emergency' is getting more and more recognised both at the political level and in the media. For a realistic chance to keep global warming by 2100 under 1.5°C, the objective of a net-zero-carbon economy must be achieved in the second half of this century at a global level and by 2050 for developed economies. For this, a fundamental revision of our production and consumption model is necessary with far-reaching distributional effects and with a massive impact on the world of work.

To manage such a transformation in a socially balanced way, the concept of 'just transition' has re-emerged. It has been a trade union demand, and it was due to the pressure of trade unions that it became a mainstream concept at the UNFCCC[1] negotiations on climate change. At the same time, with this broadly accepted status, the concept has become complex and multi-faceted often running the risk of becoming meaningless rather than an integral part of the overall environmental-climate and industrial policy framework.

---

[1] United Nations Framework Convention on Climate Change is an international environmental treaty to address climate change.

---

B. Galgóczi (✉)
European Trade Union Institute, Brussels, Belgium
e-mail: bgalgoczi@etui.org

While there is no overall trade-off between environmental and social objectives, transitional policies towards a zero-carbon world do have distributional effects with potential winners and losers, and tensions and conflicts across several dimensions do appear. Just transition strategies that emerge to tackle these tensions should reflect the concrete socio-economic and institutional characteristics of the societies involved. The complexity of these policies also calls for a new integrative welfare concept, often referred to as the 'eco-social state' or 'sustainable welfare' approach (Jakobsson et al. 2017).

This chapter attempts to deconstruct the concept of 'just transition' by discussing the underlying interpretations of inequality and justice and link them to the concept of the 'eco-social state'.

After framing the main challenges in the section "Introduction", section "Scope and Interpretations of a (Green) 'Transition' That Is 'Just'" will provide an overview of the concept of just transition as it has evolved during the last four decades by putting different interpretations of environmental justice, climate justice and energy justice into the centre. Section "Trade Unions and Just Transition" will look at the role of trade unions in just transition at different levels and will argue why in spite of the diversity of their views and attitudes trade unions can play an eminent role in making just transition a success. Section "Cases on the Ground: Coal Phase-Out and the Transformation of the Automobile Industry" will examine the role of actors in the light of case studies from two key sectors, the energy and the automobile sectors in different countries. Section "Conclusions: A New Welfare Concept for Just Transition" will conclude by raising the case for an 'eco-social state'.

## Scope and Interpretations of a (Green) 'Transition' That Is 'Just'

Below I make an attempt to examine the different components and interpretations of just transition separately and then put the pieces of the puzzle together under a framework of environmental labour studies with view to the concept of the 'eco-social state'. First, it matters a lot, how we interpret the 'environmental', 'climate' and 'social' dimensions of justice. All three have developed along different historical pathways and are promoted by organisations based on quite different traditions and core interests (Heffron and McCauley 2018).

## Environmental, Climate and Social Justice

As the transition to a green economy takes place in a socio-economic environment defined by the capital/labour relationship, each aspect of climate (energy) and environmental justice and their link to social justice largely depends on the distributive logic of the capitalist market economy.

When addressing the political economy of just transition, Newell and Mulvaney (2013, 134) described its complexity by pointing out that the goal is: "'to achieve zero-carbon while maintaining equity and justice, in pursuit of 'climate justice' to current and future generations and manage also the potential contradictions that might flow from doing these simultaneously'". More climate ambition does not necessarily create new inequalities, but concrete policies may have regressive distributional effects that need to be addressed. Below I briefly follow how the different dimensions of inequality and their interpretations evolved.

*Environmental justice* has 'distributive' and 'procedural' dimensions. The distributive dimension covers various equity issues related to environmental hazards and toxic substances due to air, water or soil pollution and to the exposure of the local population to these, while the procedural dimension is concerned about participation in related decision-making processes with social dialogue in its centre.

From early on, the concept of environmental justice not only focused on human's harming nature, but also recognised that environmental injustice arises from discrimination (Taylor 2000). Research addressing 'environmental equity' and 'environmental racism', for example, claims that hazardous waste facilities are located disproportionately in minority areas (Anderton et al 1994).

While the predominant agenda of the Environmental Justice movement in the United States has been tackling issues of race, inequality and the environment (Mohai and Saha 2015), in Europe, the focus was on poverty and the environment, tackling also health inequalities and social exclusion (Wilkinson and Pickett 2018). The other dimension of dealing with environmental hazards comes from the trade union movement, that from the beginnings had a core interest in health and safety issues in and around the workplace (Rector 2017).

It needs to be mentioned here that the very idea of 'just transition' emerged from an environmental justice and health and safety context. Mazzocchi (1993) argued that dangerous jobs that produce products which threaten community health and the environment should not be preserved, but workers who lose their jobs should be protected. He also stressed that just transition needs to go beyond traditional welfare policy objectives and pointed to the responsibility of the state that will be a key political economy argument for the emergence of the 'eco-social state' (Jakobsson et al. 2017).

Concepts from environmental, climate and energy justice research increasingly overlap as some authors observe that the environmental justice literature has progressed to include branches of justice studies, including energy and climate justice (Graff et al. 2019).

The interpretations of *'climate justice'* drew on the legacy of 'radical environmentalism' and were initially focusing on the Global South–North perspective (Labour Network for Sustainability 2017). It is a key feature of inequality that while climate change is driven by the expansionist capitalist production and consumption model of the rich societies of the Global North, most of the devastating effects of climate change hit the poorer Global South (Rosemberg 2017).

Inspired by the claims of environmental justice groups (Dorling 2017), the origins of 'climate justice' are rooted in the asymmetrical responsibility between causing climate change on the one hand and in the vulnerability from its effects on the other (Gore 2015). Over time the narrative of 'climate justice' has been extended to also cover the distributional effects of policies that were set up with the goal of mitigating and controlling climate change. These distributional effects have been addressed in theories of climate justice and in models of burden-sharing between rich and poor countries (Koch and Fritz 2014) and within societies. Employment transitions, as a consequence of the implementation of climate policies, have become one of the central issues for trade unions in dealing with climate justice.

Even within societies, different income groups have different responsibilities for causing climate change and are exposed to its effects in an asymmetrical manner. Lower income households tend to have occupations that are more exposed to climate change, for example in sectors such as agriculture, construction, tourism and health care. The housing conditions of the poor make them also more vulnerable (e.g., inner city 'heat islands' vs. green belts). Responsibilities and impacts often work in opposing ways, constituting a 'double injustice' (Walker 2012), since the groups most likely to be affected by climate change are the ones least responsible for causing it. When the costs and burdens attached to necessary climate policies affect lower income groups more, this may even turn into a 'triple injustice'. Feed-in tariffs with higher electricity prices to finance investments into renewables, for example, have a regressive effect, as low-income households are hit hardest (Zachmann et al. 2018). Another example is the unequal access of different income groups to energy efficiency measures, like the retrofitting of buildings.

Within climate justice approaches, *energy justice* has become a separate discipline to focus on the specific effects of the energy transformation on different groups in society (Jenkins et al. 2016). Approaches to energy justice have increasingly emphasised the potential *justice* dimensions of low-carbon energy systems and transitions (McCauley et al. 2019).

There is a complex relationship between climate and environmental justice, with overlapping interpretations going as far as using them as synonyms. The basic pattern of asymmetrical responsibility and affectedness between the rich and the poor is the same, but there are important differences between the two. For local environmental hazards, there is a direct link between the source of pollution and the exposure to its harmful effects with rather clear and identifiable responsibilities. The case is similar with embodied pollution in products traded; once stopped, the effects will diminish.

For human-made climate change, cause and effect are distant both in time and space. Local action in reducing $CO_2$ emissions has a negligible effect on global climate change. The climate system has a long reaction time: Even if humanity had zero emissions tomorrow, global warming would continue for decades. This can only be dealt with through multilateral co-ordinated action at the global level.

Environmental justice concerns and actions often act as a catalyst for climate action, as protests at open-cast lignite mines, in Germany (Bergfeld 2019), or actions against air pollution in cities that played a key role in sparking the 'Diesel scandal' of the car industry showed.

The main principle of *social justice* in the context of climate change and climate policies is fair burden-sharing. It is clear that the colossal transformation from a fossil-fuel-based resource-intensive and linear economic model to a climate-neutral circular economy entails huge costs and burdens for society. Just transition is about how social justice with fair burden-sharing can be applied in the context of controlling climate change, and taking climate, environmental and energy justice with all their dimensions into account. To implement this in real-world societies, an integrative and holistic policy framework is necessary, at best, in the form of an 'eco-social state'. The emerging discipline of 'environmental labour studies' makes an attempt to study the intersections between social and environmental justice, climate change and working conditions and build a bridge between environmental theory and practice (Stevis et al. 2018).

## The Concept of a Just (Green) Transition

The concept of 'just transition' can be seen as an attempt to integrate all of the above claims for justice in the context of the paradigm change towards a zero-carbon world, by applying the principles of social justice and fair burden-sharing. A just transition does not happen in a vacuum, but in the real-life circumstances of current societies and working relations, the quality of which has a decisive effect on what kind of just transition strategies might work. The country and sector examples presented in Section "Cases on the Ground" will illustrate this. There are diverse views regarding what is meant by 'transition', depending on the goals, the level of ambition and the social and economic conditions in a given society. From a functional point of view, just transition has two main dimensions: in terms of 'outcome' and of 'process' (how we get there). The outcome should be an inclusive society in a zero-carbon world with low inequality and quality jobs (ILO 2015), whereby the UN Sustainable Development Goals provide guidance.

From the 'process' perspective, just transition has two main pillars (Galgóczi 2018): one that deals with the distributional effects of climate policies (e.g., how a higher carbon price or electricity price affects different income groups) and one that deals with the management of employment transitions. The latter should be directed not only to the employees whose jobs are being lost or fundamentally transformed during the green transition, but also to those whose livelihoods depend on those activities. This includes active regional restructuring programmes, such as industrial policy and regional development initiatives.

When it comes to the implementation of just transition policies, it also becomes critical under what concrete societal circumstances (e.g., ownership relations, institutions, level of inequality) a just transition is supposed to take place. Present (often entrenched) inequalities in societies need to be addressed. This might even be a precondition for a just green transformation.

Whether the societal and actor-based context of a just transition is conflictual or co-operative, the variety of just transition strategies reflects the underlying variety of societies and real-life situations. There is no silver bullet for a just transition strategy that would fit all societies and societal conditions.

## Interpretations of 'Just Transition'

Just transition is not an abstract concept but refers to actual practices in real workplaces. It is this 'on the ground' perspective that really matters. While the green transition itself is an imperative, only a just green transition will deliver the necessary climate policy targets. While decarbonisation is a common objective, concrete transitions take place in work environments that are determined by the capital–labour relationship. Debates have been intensifying about the meaning and interpretations of the 'just transition' concept. Stevis and Felli (2015) raised the relevant question: 'just transition for whom'? They point to the fact that even among global labour unions, there are different visions, ranging from those that focus on just transition in concrete sectors for their particular members to those that propose fundamental changes in the political economy and make the case for a just transitions for all. Building on terms used by Fraser (2005), they identify views that call for more equity while keeping the existing capitalist political economy framework as '*affirmative*', while those that call for more profound changes and seek a (re)distribution of environmental damage and benefits through a reformed or transformed political economy as '*transformative*'.

According to the authors, all major international organisations (e.g., UNEP, ILO) and trade union confederations (e.g., ITUC, ETUC) follow the '*shared solution*' approach that sees a socially acceptable greening of the economy as mutually beneficial and does not envision transformative changes in the political economy. A transformative approach that also emphasises the *differentiated responsibility* of the actors is more prominent among unions in sectors that are negatively affected by environmental regulations and climate policy, such as energy-intensive industries. This approach demands that the state and capital take responsibility for the workers put at risk during the green transition.

A *transformative approach* of just transition seeks a reorganisation of the relations between state, capital and labour including the democratic control and public ownership of remaining fossil fuel supplies, as represented by the South-African COSATU trade union, the Labor Network for Sustainability (2017) and the Trade Unions for Energy Democracy (TUED) movement (Sweeney and Treat 2018).

When seen from a broader perspective, these three approaches are not necessarily in contradiction with each other but reflect different socio-economic conditions under which the green transformation takes place. There is no inherent conflict between environmental and social concerns, as decarbonisation is a common interest of humanity. But as the transition to a green economy takes place in a socio-economic environment that is defined by the capital/labour relationship, the transformation process towards a shared objective can well be conflictual. As regards employment transitions at workplace level, the context there is defined by the interest conflict between capital and labour.

The public interest is the main objective and the long-term responsibility of the state that distinguishes a green transition from other restructuring processes (as e.g., digitalisation or globalisation). Unlike those, the green transition is a policy-driven process with targets attached to it (zero-carbon). However, in contrast to political declarations and the ILO's Just Transition Guidelines (ILO 2015), concrete just transition policies have not yet become part of an integrated and holistic 'climate-environmental and social' policy framework. The epochal transformation to a new—zero-carbon—economic model requires an overhaul of the traditional welfare and employment policy framework. This is why the claim for an 'eco-social state' is being raised.

## Trade Unions and Just Transition

### Trade Unions and the Historical Capital–Labour Deal

According to a critical environmentalist approach, the post-WWII historical compromise between capital and labour emerged on the basis of the high degree of wealth created through ever-increasing material flows and the depletion of natural resources (Schepelmann 2009, 2014). In established capitalist market economies, trade unions, as interest representation agents of the working people, were thus often seen as an integral part of an—unsustainable—production model. While debates take place within trade unions about whom they actually represent (members versus citizens), they are often seen as 'self-serving' insofar as they are membership organisations with a primary responsibility to protect their members' interests (Spooner 2000).

In this context, an emerging body of literature that identifies itself as 'Environmental Labour Studies' (Räthzel and Uzzell 2013) seeks to overcome the labour–nature divide by focusing on 'labour environmentalism' and 'Social Movement Unionism' where union strategies go beyond the workplace perspective and the membership focus and embrace wider issues. Referring to earlier works (Goldfield 1989; Ross 2007), Stevis and Felli (2015) differentiate between labour unions in terms of political strategy as 'business' or 'social' unionism. 'Business unionists' would limit themselves to obtaining a fair share from a growing economy. 'Social unionists', on the other hand, believe that unions ought to have a say about the organisation of the political economy both because it shapes their material benefits and because they see themselves as engaged citizens (Munck 2002).

## Trade Unions in Renewal

Viewed in historical perspective, trade unions have argued that a consistent defence of their members' interests demands a long-term struggle for a social and political context at national and international levels that is favourable to the wellbeing of people and society as a whole (Spooner 2000).

For capital, labour is one production factor, and natural resources, including the atmosphere, make another. Trade unions fight against capital and, through this, acquire hard-won labour rights. As longstanding opponents of capital, they are precisely in a position to address issues linked to the other production factor exploited by capital, the environment.

As key promoters of a just transition concept (towards a zero-carbon economy), trade unions (in particular the ITUC and the ETUC) have also shown readiness to recognise the limits of material and resource exploitation and argue for a rebalancing of market- and capital-based approaches. This aspect of trade unionism is what was referred to above as 'social (movement) unionism'.

In their traditional role, at the workplace, and framed above as 'business unionism', unions are used to managing change in the production system or work organisations that are driven by the profit motive of capital (Riso and Rodriguez Contreras 2019). Examples include union and works council practices during the economic crisis with concession bargaining practices or shortened working time schemes (Benassi 2016). In these cases, the legitimacy of the change itself was questioned, and at least one way of fending off its effects on employees was also to try to keep the change itself at bay.

In the case of decarbonisation, the situation is different: Getting climate change under control is a commonly shared objective in the interest of humanity. But meeting this objective can well be conflictual, as the transition poses a huge challenge for the world of work and workplace level employment, transitions take place within the traditional capital–labour nexus. These transitions often result in similar patterns of a reorganisation of work as in business restructuring cases (e.g., lay-offs, higher flexibility, higher work pressure, etc.), the core issues against which trade unions used to fight. Moreover, the effects of decarbonisation, technological change and globalisation often add up at the workplace, as for example the case of the automobile industry demonstrates (see below).

In green transformation, there is an apparent contradiction between the main competences of trade unions in managing the consequences of change, while at the same time pushing change forward. As higher climate ambitions raise the pressure on work organisation, it is unions at the plant level that face the consequences, and their task of interest representation becomes harder. This contradiction of roles often appears in tensions between different levels of trade union action.

In the last decades, trade unions at national or supra-national level have promoted the concept of just transition in the context of climate change. But it is trade unions on the ground, at local, regional, sectoral or company level that are confronted with its implementation in the practices of real-life work relationships. The former category of trade union organisations thus take up broader societal issues, such as environmental and climate policy, while the latter are more narrowly focused on their membership. Looking beyond the established production model is more characteristic for the higher-level union organisations; local and plant level unions are more concerned with the immediate challenges posed by the transformation.

## Cases on the Ground: Coal Phase-Out and the Transformation of the Automobile Industry

This section presents results from country- and sector-specific case studies performed in a research project conducted by the European Trade Union Institute with the involvement of external experts. National level practices focus on phasing out coal for energy generation in the two countries that provide two-third of the European coal-based electricity, Poland and Germany. Some insights from France will also be included. To illustrate practices at the company level, the case of the EU's biggest energy multinational, ENEL, will be presented, followed by two case studies from the automobile industry in Germany and France.

The cases demonstrate the main challenges for just transition practices in two key sectors of the European economy: the phasing out coal and managing the change towards a new mobility paradigm. Both transformations demonstrate how just transition politics differ from sector to sector and from country to country. These cases address the 'process' dimension of just transition, with a focus on employment changes and job transitions.

Decarbonisation in both the automobile and the power sector is driven by climate and environmental regulations at European and national levels. There is a major difference between the two sectors concerning both the nature and the magnitude of the challenge: Coal has no future but the automobile does, albeit in quite a different form from the one we know. In the coal-based power sector, the majority of currently existing jobs will disappear in a decade and the regional effects will be harsh (Alves et al. 2018). Even though coal itself does not have a future, workers and their families, as well as the surrounding region, must have one. Employment in the coal sector makes up just 0.15 per cent of European employment but, with its high concentration, the sector is of vital importance for individual regions.

On the other hand, with more than 5 per cent share of total European employment, the automobile sector is a key employer. For the car industry, the demise of the combustion engine and the electrification of the powertrain will require different competences, skills and forms of work organisation. These will have a substantial impact on previously established comparative advantages of nations and manufacturers (Bauer et al. 2018). Both sectors have higher than average wage levels and outstanding trade union organisation.

## Phasing Out Coal

The main focus of both employers and trade unions regarding just transition policies for the coal sector at the national (and regional) policy level is to stretch the phase-out as long as possible, while at plant level defensive strategies such as voluntary redundancy and early retirement are the dominant employment policy tools. The cases of Germany and Poland also highlight important country-specific differences.

## National Level

In *Poland*, there is no genuine social dialogue on coal transition as decarbonisation is not a national policy objective. It is driven by EU policy, as Poland, like any EU member state, is obliged to set up a national energy and climate plan under the EU Energy Union governance system (European Commission 2018). The Polish government sees this as dictatorial by Brussels and complies with the implementation reluctantly.

The coal sector in Poland is highly concentrated and controlled by the state. Despite significant imports of coal, the price of domestic coal is higher than the market price due to long-term guaranteed price contracts by state-owned electricity power plants. The contributions of the coal mining sector to public finances do not cover the state support for the sector. Key domestic stakeholders of energy policy—trade unions and employers—are concentrating their efforts on preserving this status quo against EU climate and energy policy. The CEO of Poland's (and the EU's) largest coal mine recently said: 'Breaking the green trend is practically not possible anymore and we need to adjust. However, we need to do it on our own terms' (Martewicz 2020). Trade union density in the Polish coal sector is almost 100 per cent, and the mobilising power of the unions is legendary (since the 1981 'Solidarity' strike marathon). In February 2020, the unions were on the streets of Warsaw demanding a 12 per cent pay rise (Martewicz 2020).

The result is that strategic documents on the domestic political agenda related to the energy sector do not tackle the problem of reducing the number of coal mines or jobs as this would be instantly opposed by the trade unions and could endanger the stability of any government. The main goal of the Polish energy strategy until 2040 (Szpor 2019) is energy security while ensuring economic competitiveness and energy efficiency alongside the reduction of the sector's impact on the environment. Accordingly, hard coal will still remain the single most important source of electricity production in Poland in 2040, without any further plan for a phase-out.

Given the lack of domestic pressure for more radical change, active transition management is not being seen as necessary. There is no policy framework for managing employment transitions, and the 'contingency measures' addressed to miners are rather limited. They embrace traditional monetary instruments like mineworkers' pensions and on a limited scale, early retirement and redundancy payments. There are virtually no schemes for re-employment in alternative workplaces, and only few projects are addressed to miners' families and local communities. Just transition in the Polish context practically means 'no transition'.

The case of Poland is particularly interesting, as it demonstrates the clash of objectives between policy levels. The main conflict exists between the national and the supra-national (EU) level. In December 2019, the EU 2050 climate neutrality objective was adopted by the European Council without Poland's consent (Morgan 2019). It will come back on the agenda in June 2020. By then the €100bn 'Just Transition Mechanism' (JTM) will be in place (with Poland as the main beneficiary) and could be an incentive for Poland to sign the EU Climate Pact. Pressure from the EU also includes a provision that support from the JTM is conditional on commitments to the EU climate policy targets. The latest EU budget proposal put forward by the President of the

European Council includes a 50 per cent cut for allocations from the JTM for member states not signing up to the EU's 2050 objective (Morgan 2020).

In contrast to Poland, *Germany*, the number one coal burner in Europe, is fully committed to EU climate targets, but employs a cautious, gradual and consensual way of phasing out coal. Building on 60 years of experience in transforming the Ruhr coal and steel region into a modern energy and knowledge-based economy (Galgóczi 2018), Germany's coal phase-out applies three main elements of a just transition approach: slow and gradual transition with a high level of social dialogue; active labour transition management and engagement in industrial and regional development (Gärtner 2019).

The federal government established the Commission on Growth, Structural Change and Employment in 2018 (referred to as 'Coal Commission') to provide recommendations on a gradual reduction of the capacities of existing coal-fired power plants (BMWI 2020). It consisted of policy makers at different levels of governance and all major stakeholders including employers, trade unions, NGOs and experts. The recommendations aim at a gradual and steady reduction of the greenhouse gas emissions in the power sector. The last coal-fired power plant is to be phased out by 2038 (with the option of an earlier exit by 2035), marking the longest farewell to coal in Western Europe. At the same time, the Coal Commission has followed the concept of a just transition along various dimensions (Litz 2019). Firstly, social dialogue has been exemplary, as all stakeholders were consulted; secondly, with a phase-out of two decades, both the regions and the energy companies have been given ample time to transform; and, thirdly, the proposal foresees the provision of comprehensive financial support at a total value of €45bn until 2038 to the stakeholder groups affected.

The Coal Commission's recommendations stipulated that coal regions should be developed into modern energy regions with investment in transport and digital infrastructure, as well as local research and innovation to strengthen the regions' economic potential. The trade union for mining, chemical and energy workers (IG BCE) was the participant most committed to an 'as late as possible coal phase-out' demand and a 'jobs only' focus. The union has also been involved in clashes with environmental activists who came to occupy an open-cast mine in the Hambach Forest (Bergfeld 2019). The IG BCE's general secretary insisted on the need to put jobs first and think about environment, second.

Germany also provides important just transition practices in the coal sector at *regional and plant levels*. The economic diversification of the once coal-dependent Ruhr region that stretched over 50 years had been actively managed by the federal and regional governments, and restructuring processes were embedded in a co-operative industrial relations culture, with a stronger version of co-determination applied in the mining and steel industry. It is noteworthy to mention, however, that strong social dialogue and the active role of the regional government in managing the restructuring process did not automatically lead to success. Hospers (2004) identified a lock-in position of the region into coal and steel up to the mid-1980s where 'institutional lock-in' played a key role. By this, he meant a self-sustaining coalition of local businessmen, politicians and trade unions with a shared interest: the preservation of existing structures.

At the same time, measures taken to facilitate employment transitions of redundant workers were exemplary. In larger companies, this was done via targeted agencies specialised in employment transitions with individualised coaching and training measures. The Ruhr Coal Vocational Training Society was the model case for such institutions and has been widely used during the transformation of the East German economy after German reunification.

To manage manpower restructuring in coal mining, bargaining parties first signed an agreement in 1993 to guarantee a socially responsible downsizing of employment. The agreement about the complete closure of German hard coal mining by 31 December 2018 was negotiated between the German Coal Association and the IG BCE union and came into force in April 2012. It provides a framework for the balance of interests, building on a social compensation plan and various job transition agreements. A key lesson that emerged from the Ruhr transition (Galgóczi 2014) was that a successful transition takes time and requires a strong vision for the future and adequate resources. Worker co-determination has been an important factor in the coal sector, facilitating solutions that embraced social protection and retraining efforts.

By contrast, in *France*, the coal sector is limited to four coal-fired powerplants (but no coal mines) with less than 2000 direct employees. After the declaration of the government in 2018 to close them down by 2022, demonstrations were held by the CGT union, and all four plants started strike actions that continue in repeated waves ever since. CGT and FO unions argue that the government should withdraw the decision, arguing with the low share they have in France's $CO_2$ emissions and with their role in maintaining energy security. Both organisations have also denounced the high social costs of the closure, which could lead to up to 5000 job losses altogether. The third main union, CFDT, while supporting the decarbonisation of the energy sector, denounced the lack of timely consultation and transparency by the government. Social partners had been only involved after the decision had been made with the result of tailored support measures for the affected workers.

## Company Level Practices—The Case of ENEL

ENEL, the number one utility company in Europe by market value with the Italian State as the main shareholder, employs 75,000 people worldwide and 36,000 in Italy. It offers a good-practice case of just transition in phasing out coal in the private sector. Facing up to the challenges of energy transition and the EU's stricter emissions limits from 2021, ENEL announced its 'Future-e Program' to become carbon neutral by 2050. In May 2017, it announced the closure of two large coal power plants by 2018 and a plan to close all of its coal and lignite generation plants by 2030.

ENEL announced the reconversion of 23 power stations (oil, coal, combined cycle and gas) with significant employment implications. The company does not operate coal mines, but owns old lignite mining fields, the reclamation of which is also part of the plan. The company looks back at a good track record in industrial relation practices, including its European Works Council Agreement (2001), the Protocol of National Industrial Relations for the ENEL Group (2012) and its Global Framework Agreement (2013).

ENEL has entered into social dialogue for a just transition framework agreement with its Italian union partners. The framework covers retention, redeployment, reskilling and early retirement for elderly workers. It is an example of a just transition agreement in the power sector that includes a recruitment plan using apprenticeship to ensure the transfer of competences of elderly towards young workers and encouragement of mobility and training for redundant workers for the optimisation of its human resources. The framework also includes dedicated training measures to ensure qualification and employability as well as for the creation of new skills for the development of new business. Trade unions have taken a critical view of the 'Future-e' plan, as they criticised the lack of information and their scarce involvement in the process. In a joint document in February 2017, the unions put forward proposals for a more gradual approach to the process of the decommissioning and adjustment of plants including new investment plants for some (Rugiero 2019). In spite of their criticisms, Italian unions were signing up to ENEL's restructuring plans, in cases where some of their proposals were taken into account. ENEL's plan is also a comprehensive one, as it includes the reclamation of decommissioned power plants for further use in the future in co-operation with the regional and local governments.

The role of trade unions in the coal sector in various countries can be thus regarded as defensive, from defending the coal-based economy status quo (Poland and at plant level in France) to pleading for a lengthy transition process in Germany. Their overall approach is 'business unionism' with focusing on jobs and membership interest. IG BCE in Germany follows a balancing act arguing for 'proper framework conditions' with active industrial policy and job security (IG BCE 2019). These were the Italian unions that moved closest to a 'social movement unionism' approach with forward looking policy proposals.

## The Automobile Industry

The stakes in the automobile industry are much higher than the coal sector, and the transformation is also a more complex one as, besides decarbonisation, the digitalisation of both the production and the product and a reconfiguration of the global supply chains of the industry are proceeding simultaneously. The era of the combustion engine will come to an end with a massive restructuring of the entire industry. There are also several unknown elements in this transformation, as for example how radical changes in mobility patterns will be and what role individual vehicle use will have in the future.

## National Level

At national and European policy levels, the stakeholders in the automobile industry have been playing a controversial role over a long period of time, lobbying for weaker regulations regarding car emission standards, while some manufacturers also took the step of implementing fraudulent practices. Employer associations at national and European levels were lobbying hard for lower standards, and CEOs of the main car manufacturers were regular guests at 'automobile summits' held at the German Chancellor's office, with the silent support of trade unions. However, since 2017, after the diesel scandal, they have at least launched radical restructuring and investment programmes towards electrification. Trade unions at policy level have also been supporting a more gradual transformation. IndustriAll and IG Metall were calling for a balanced emissions reductions policy that considers employment and social aspects. Simultaneously, they introduced a study in the European Parliament that forecasted progressive job losses with higher emission standards. They also argued against penalties for manufacturers who do not meet the emissions standards before zero-emission vehicles acquire a mass market potential.

The case of the post-2021 scenario of EU $CO_2$ emission limits is a good illustrative example. The European Council reached a compromise position of a 35 per cent $CO_2$ reduction by 2030 compared to 2021 levels. Germany first voted for a 30 per cent reduction, with both employers and unions backing the softer target, but agreed finally to the 35 per cent target. Ultimately, the European institutions agreed on a reduction scenario of 31 per cent for vans and 37.5 per cent for cars by 2030 with the addition of an 'impact assessment' in 2023.

While climate policy ambition was not a priority for the actors in the automobile industry, they are now focused on managing the radical transformation in a balanced way, moving from the combustion engine and towards electric cars. Co-operative industrial relations and co-determination practices at main European manufacturers (above all in Germany and France) are great assets to facilitate employment transitions in a forward-looking manner. In the German case, the main task for the trade union IG Metall is to shape location, employment, innovation and investment strategies; conclude agreements to safeguard production locations; and find a development perspective for every plant (Strötzel and Brunkhorst 2019).

## Plant Level Agreements in the Automobile Sector

At the plant level, unions and works councils act as 'business unions' in managing (unavoidable) changes in a co-operative manner in order to secure jobs and manage employment transitions. They are focused on securing the core interests of employees with further agreements about collective bargaining, company pension schemes and profit-sharing. In return, they support the restructuring process.

As a prominent example, the General Works Council of Daimler has reached an agreement about 'Project Future' under which job security for all Daimler employees is extended from 2020 to 2030, including those in logistics and branch offices. The task of the company's innovation committees includes plant management consultations with the works council on future product strategies during which the works council has the right to make its own proposals. Investment commitments of €35bn have been made in German locations over the next seven years dedicated to the areas of e-mobility, mobility services, connectivity and autonomous driving.

Volkswagen is anticipating extensive job cuts as a result of the introduction of new technologies and products. As part of its 'Future Pact', 25,000 jobs will be eliminated, while 9000 will be created. Back in 2016, the works council had already been able to negotiate a job security plan until 2025. This means that the reduction of employment will be achieved in a socially acceptable manner. This includes part-time work for older employees, which is set to be significantly expanded. At the same time, commitments have been made to locate new e-mobility products at German sites. In this way, each department has been given a development perspective over the next few years. Management plans to outsource certain products and logistics or to relocate all new e-components abroad could be fended off. There have also been strategic agreements in the supplier sector linked to the transformation process.

At the German component manufacturer Schaeffler, an agreement about the future has been reached that goes beyond normal employment agreements. Suppliers' products are relatively independent of brand identity, and therefore, the producers can position themselves more flexibly within the new drive and mobility concepts. Therefore, their business models can include diversification strategies that go beyond producing parts for the car.

In the French automobile sector, company trade unions are also in the position of negotiating about issues arising from the company's strategic orientation in the medium-term. Unions may seek to negotiate the types of job categories threatened by economic or technological change, the implementation of employee mobility, sustainable training, the inclusion of young people into the company, the employment of older workers and so on. This negotiation framework, including consultations with the works council concerning strategic orientations, offers the tools to achieve a shared vision of a company's strategy and outlook with the management and to formulate alternative proposals and identify secure development paths (Sonzogni and Schulze-Marmeling 2019).

## Conclusions—A New Welfare Concept for Just Transition?

This chapter attempted to review the concept of 'just transition', examine its environmental, climate (energy) and social justice drivers, and related trade union strategies. In order to make the claim of 'just transition with climate ambition' a reality, a supportive political economy context is necessary: an 'eco-social state' with an overhaul of the concept of the welfare state, as

## 23 From 'Just Transition' to the 'Eco-Social State' 557

proposed by Jakobsson et al. 2017. The task of re-embedding production and consumption patterns into planetary limits requires a 'sustainable welfare' perspective (Koch 2018) that renews the post-WWII welfare policy framework that was built on resource and energy-intensive growth, now to include ecological goals. Social policies also need to address the inequalities and conflicts that emerge during the decarbonisation of production and consumption patterns. A 'sustainable welfare' perspective also argues that the distributive principles underlying existing welfare systems would need to be extended to include those affected in other countries and in the future. Such an alternative eco-welfare governance network would need to redistribute not only carbon emissions, but also work, time, income and wealth (Büchs and Koch 2017). The concept of the eco-social state is thus an ideal type of a welfare state to cope with the challenges of the ecological transformation through a just transition approach.

We also argued that just transition politics do not happen in a vacuum, but in the real-life circumstances of concrete societal and working relations, the quality of which has a decisive effect on what kind of just transition strategies might work. Just transition assumes a policy framework that is matched to the economic and social realities of a given society. Its implementation requires the main actors, including trade unions, to develop new strategies.

The briefly introduced cases of two sectors (energy and automobile producers) pointed to the importance of different policy levels (national and local) and to the role of different actors in shaping just transition policies. Germany, the principal coal burner and car producer in Europe, pursues a cautiously balanced approach to managing the transformation of these two sectors. While the rest of Western Europe will phase out coal by 2030 at the latest, for Germany it will take until 2038. Social dialogue and just transition policies, however, are exemplary at macro level (the Coal Commission), at regional level (Ruhr) and at plant level (the automobile industry), while climate ambitions are modest. Following a 'jobs and competitiveness first' strategy, the German government has been pushing for less ambitious climate and energy policy targets at EU level, while at the domestic level putting high stakes on industrial and regional policy objectives. On the other hand, the Polish government (backed by both employers and unions) is a blocking factor at the European level and resists to implement EU energy policy objectives.

The national cases of energy transition show that none of them comes close to the ideal type of an 'eco-social state'. Poland denies the necessity of a deep ecological transition and resists corresponding EU policies as much as it can. Its welfare state is focused on maintaining the fossil-fuel-based economy. The case of France demonstrates that while the state is committed to an ambitious ecological transformation, the top-down approach to decarbonisation that it implements meets the resistance of workers and unions. Germany comes closest to the ideal with a declared climate policy ambition, genuine social dialogue and the state actively managing the transition. In practice however, it falls short of even its own declarations.

Concerning trade unions, different roles have been observed. Trade unions at supra-national level (ITUC and ETUC) are promoting the concept of just transition and pushing for ambitious climate policy targets based on an understanding of the broader societal implications and thus come closer to the approach of social movement unionism. Trade unions on the ground, at local, regional, sectoral or company level who are confronted with the practices of implementing a just transition within the context of real-life labour relations follow a more cautious approach, along the lines of business unionism.

The cases presented showed that trade unions were mostly acting as 'business unions' in terms of concrete policies also at the national level and in particular in the 'micro-management' of the labour consequences of a green transition at workplace level. Plant level agreements in the car industry have had a forward-looking character by integrating investment, innovation and employment aspects. At ENEL, trade unions elaborated own concepts for employment transitions and developed initiatives to tackle the future after plant closures. These two cases pointed beyond business unionism, but not embracing a full-scale social movement unionism approach. Driving the change forward posed a harder challenge in the energy as well as in the automobile sector, since it was an external force—the EU—that was setting the policy targets and the pace of the change. Thus, unions as well as employers acted defensively at national level as well as in energy-intensive sectors (mining, chemicals).

'Social Movement Unionism' should go beyond the workplace perspective and the membership focus to embrace wider social and ecological issues. Trade unions cannot limit themselves to getting a fair share of a growing economy; they also need to have a say in the organisation of a sustainable political economy. Still, pushing the momentum of change ahead (with more climate policy ambition) poses a huge challenge for the world of work as workplace level employment transitions appear within the traditional capital–labour nexus. Driving the change forward while at the same time fending off its consequences on workers might be contradictory in real-life situations that are often characterised by precarious work and in-work poverty. With such a perspective, the status quo for those in stable and well-paid jobs (e.g., in energy and the automobile sector) remains attractive. A labour market policy framework that supports and facilitates change should also be an essential part of just transition, otherwise policies necessary to achieve decarbonisation targets will meet resistance, as Poland, plant level cases in France and the slow pace of transformation in Germany illustrate. This is why a supportive political economy context in form of an eco-social state is necessary for making just transition with climate ambition a reality.

The contours of such new politics are only visible at EU level to the degree that the European Green Deal seems to embrace this approach. Further research is needed on how and on what concrete policy framework an 'eco-social state' can be established.

# References

Alves, Dias Patricia, et al. 2018. *EU Coal Regions: Opportunities and Challenges Ahead.* Luxembourg: Publications Office of the European Union. https://ec.europa.eu/jrc/en/publication/eur-scientific-and-technical-research-reports/eu-coal-regions-opportunities-and-challenges-ahead. Accessed 18 Jan 2020.

Anderton, D. L., A. B. Anderson, J. M. Oakes, and M. R. Fraser. 1994. Environmental Equity: The Demographics of Dumping. *Demography* 31(2): 229–248.

Bauer, Wilhelm, Oliver Riedel, Florian Herrmann, Daniel Borrmann, and Carolina Sachs. 2018. *ELAB 2.0 Wirkung der Fahrzeugelektrifizierung auf die Beschäftigung am Standort Deutschland.* Stuttgart: Fraunhofer Institut. https://www.iao.fraunhofer.de/lang-de/images/iao-news/elab20.pdf. Accessed 18 Jan 2020.

Benassi, Chiara. 2016. *Extending Solidarity Rather Than Bargaining Concessions.* Policy Brief. Brussels: ETUI. https://www.etui.org/sites/default/files/Policy%20Brief%202016.01%20Benassi.pdf. Accessed 5 Mar 2020.

Bergfeld, Mark. 2019. German Unions Are Waking up to the Climate Disaster. *Jacobin Magazine.* https://www.jacobinmag.com/2019/08/german-unions-climate-environment-fridays-for-future. Accessed 5 Mar 2020.

BMWI. 2020. *Kohleausstieg und Strukturwandel.* https://www.bmwi.de/Redaktion/DE/Artikel/Wirtschaft/kohleausstieg-und-strukturwandel.html. Accessed 25 May 2020.

Büchs, Milena, and Max Koch. 2017. *Postgrowth and Wellbeing: Challenges to Sustainable Welfare.* Basingstoke: Palgrave Macmillan.

Dorling, Danny. 2017. The Rich, Poor and the Earth. *The Internationalist*, 205. https://newint.org/features/2017/07/01-equality-environment. Accessed 5 Mar 2020.

European Commission. 2018. A Clean Planet for All—A European Strategic Long-term Vision for a Prosperous, Modern, Competitive and Climate Neutral Economy, COM (2018) 773 final.

Fraser, Nancy. 2005. Reframing Justice in a Globalizing World. *New Left Review* II (36): 69–88.

Galgóczi, Bela. 2014. The Long and Winding Road from Black to Green: Decades of Structural Change in the Ruhr Region. *International Journal of Labour Research* 6 (2): 217–240.

——— 2018. Just Transition Towards Environmentally Sustainable Economies and Societies for All. ILO ACTRAV Policy Brief. Geneva: International Labour Office.

Gärtner, Stefan. 2019. An Attempt at Preventive Action in the Transformation of Coal-mining Regions in Germany. In *Towards a Just Transition: Coal, Cars and the World of Work*, ed. Béla Galgóczi, 135–155. Brussels: ETUI.

Goldfield, Michael. 1989. *The Decline of Organized Labor in the United States*. Chicago: University of Chicago Press.

Gore, Timothy. 2015. *Extreme Carbon Inequality*. Oxfam. http://oxfam.org/en/research/extreme-carbon-inequality. Accessed 5 Mar 2020.

Graff, Michelle, Sanya Carley, and Maureen Pirog. 2019. A Review of the Environmental Policy Literature from 2014 to 2017. *Policy Studies Journal* 47 (S1): 17–44.

Heffron, Raphael, and Darren McCauley. 2018. What Is the 'Just Transition'? https://papers.ssrn.com/sol3/papers.cfm?abstract_id=3099751.

Hospers, Gert-Jan. 2004. Restructuring Europe's Rustbelt: The Case of the German Ruhrgebiet. *Intereconomics* 39: 147–156. https://doi.org/10.1007/BF02933582.

ILO. 2015. *Guidelines for a Just Transition Towards Environmentally Sustainable Economies and Societies for All*. Geneva: International Labour Office.

Jakobsson, Niklas., Raya Muttarak, and Mi Ah Schoyen. 2017. Dividing the Pie in the Eco-social State: Exploring the Relationship Between Public Support for Environmental and Welfare Policies. Politics and Space Vol (36/2), Sage.

Jenkins, Kirsten, Darren McCauley, Raphael Heffron, Hannes Stephan, and Robert Rehner. 2016. Energy Justice: A Conceptual Review. *Energy Research and Social Science* 11: 174–182.

Koch, Max. 2018. Sustainable Welfare, Degrowth and Eco-social Policies in Europe. In *Social Policy in the European Union—State of Play 2018*, ed. Bart Vanhercke, Dalila Ghailani, and Sebastiano Sabato, 35–50. Brussels: ETUI.

Koch, M., and M. Fritz. 2014. Building the Eco-social State. *Journal of Social Policy* 43/4: 679–703. https://doi.org/10.1017/S004727941400035X.

Labour Network for Sustainability. 2017. Just Transitions -Just what is it? https://www.labor4sustainability.org/files/Just_Transition_Just_What_Is_It.pdf.

Litz, Philipp. 2019. Germany's Long Goodbye from Coal. In *Towards a Just Transition: Coal, Cars and the World of Work*, ed. Béla Galgóczi, 57–81. Brussels: ETUI.

Martewicz, Maciej. 2020. Polish Miners' Strike May Hit The Wall as Coal Is on The Way Out. *Bloomberg*. https://www.bloomberg.com/news/articles/2020-02-17/polish-miners-strike-may-hit-the-wall-as-coal-is-on-the-way-out.

Mazzocchi, Tony. 1993. A Superfund for Workers. *Earth Island Journal* 9 (1): 40–41.

McCauley, Darren, Vasna Ramasar, Raphael Heffron, Benjamin K. Sovacool, Desta Mebratu, and Luis Mundaca. 2019. Energy Justice in the Transition to Low Carbon Energy Systems: Exploring Key Themes in Interdisciplinary Research.

*Applied Energy* 233-234: 916–921. https://doi.org/10.1016/j.apenergy.2018.10.005.

Mohai, Paul, and Robin Saha. 2015. Which Came First, People or Pollution? A Review of Theory and Evidence from Longitudinal Environmental Justice Studies. *Environmental Research Letters* 10 (12): 125011.

Morgan, Sam. 2019. Poland Snubs Climate-neutrality Deal But EU Leaders Claim Victory, Euractiv. https://www.euractiv.com/section/climate-environment/news/poland-snubs-climate-neutrality-deal-but-eu-leaders-claim-victory/.

———. 2020. Poland's Just Transition Bonus Cut 50% Under Latest EU Budget Proposal, Euractiv. https://www.euractiv.com/section/climate-environment/news/polands-just-transition-bonus-cut-50-under-latest-eu-budget-proposal/.

Munck, Roberto. 2002. *Globalisation and Labour: The New 'Great Transformation'*. London: Zed Book.

Newell, Peter, and Dustin Mulvaney. 2013. The Political Economy of the 'Just Transition'. *The Geographical Journal* 179 (2): 132–140. https://rgs-ibg.onlinelibrary.wiley.com/doi/abs/10.1111/geoj.12008.

Räthzel, Nora, and David Uzzell, eds. 2013. *Trade Unions in the Green Economy: Working for the Environment*. London: Routledge.

Rector, Josiah. 2017. Accumulating Risk: Environmental Justice and the History of Capitalism in Detroit, 1880-2015. Wayne State University Dissertations. 1738. https://digitalcommons.wayne.edu/oa_dissertations/1738.

Riso, Sara, and Ricardo Rodriguez Contreras. 2019. *Restructuring: Do Unions Still Matter?* Dublin: Eurofound. https://www.eurofound.europa.eu/sr/publications/blog/restructuring-do-unions-still-matter.

Rosemberg, Annabella. 2017. Strengthening Just Transition Policies in International Climate Governance. https://www.stanleyfoundation.org/publications/pab/RosembergPABStrengtheningJustTransition417.pdf.

Ross, Stephanie. 2007. Varieties of Social Unionism: Towards a Framework for Comparison. *Just Labour: A Canadian Journal of Work and Society*, 11: 16-34. http://www.justlabour.yorku.ca/volume11/pdfs/02_Ross_Press.pdf.

Rugiero, Serena. 2019. Decarbonisation of the Italian Energy Sector—The Case of ENEL. In *Towards a Just Transition: Coal, Cars and the World of Work*, ed. Béla Galgóczi, 109–135. Brussels: ETUI.

Schepelmann, Philipp. 2009. *A Green New Deal for Europe*. Wuppertal Institute for Climate, Environment and Energy. http://www.greens-efa.org/cms/default/dokbin/302/302250.pdf.

Schepelmann, Philippe. 2014. Resource Efficiency, Concepts and Indicators, Paper at the Degrowth Conference Leipzig 2014. https://www.degrowth.info/en/catalogue-entry/resource-efficiency-concepts-and-indicators/u.

Sonzogni, Michel, and Sebastian Schulze-Marmeling. 2019. The French Automobile Industry. In *Towards a Just Transition: Coal, Cars and the World of Work*, ed. Béla Galgóczi, 193–213. Brussels: ETUI.

Spooner, Dave. 2000. *A View of Trade Unions as Part of Civil Society.* Social Development Publications, DFID, UK.
Stevis, Dimitris, and Romain Felli. 2015. Global Labour Unions and Just Transitions to a Green Economy. *International Environmental Agreements: Politics, Law and Economics* 15 (1): 29–43.
Stevis, Dimitris, David Uzzell, and Nora Räthzel. 2018. The Labour–Nature Relationship: Varieties of Labour Environmentalism. *Globalizations* 15 (4): 439–453. https://doi.org/10.1080/14747731.2018.1454675.
Strötzel, Maximilian, and Christian Brunkhorst. 2019. Managing the Transformation of the German Automotive Industry. In *Towards a Just Transition: Coal, Cars and the World of Work*, ed. Béla Galgóczi, 243–273. Brussels: ETUI.
Sweeney, Sean, and John Treat. 2018. *Trade Unions and Just Transition: The Search for a Transformative Politics.* TUED Working Paper, No. 11. New York: Trade Unions for Energy Democracy. http://unionsforenergydemocracy.org/wp-content/uploads/2018/04/TUED-Working-Paper-11.pdf.
Szpor, Aleksandar. 2019. The Changing Role of Coal in the Polish Economy—Restructuring and (Regional) Just Transition. In *Towards a Just Transition: Coal, Cars and the World of Work*, ed. Béla Galgóczi, 33–57. Brussels: ETUI.
Taylor, Dorceta. 2000. The Rise of the Environmental Justice Paradigm: Injustice Framing and the Social Construction of the Environmental Discourses. *American Behavioral Scientist* 43 (4): 508–580.
Walker, Gordon. 2012. *Environmental Justice: Concepts, Evidence and Politics.* London: Routledge. https://doi.org/10.4324/9780203610671.
Wilkinson, Richard, and Kate Pickett. 2018. *The Inner Level: How More Equal Societies Reduce Stress, Restore Sanity and Improve Everyone's Well-being.* London: Allen Lane.
Zachmann, Georg, Gustav Fredriksson, and Gregory Claeys. 2018. *The Distributional Effects of Climate Policies.* Blueprint series 28. Brussels: Bruegel. https://bruegel.org/wp-content/uploads/2018/11/Bruegel_Blueprint_28_final1.pdf.

# 24

# Environment, Labour and Health: The Ecological-Social Debts of China's Economic Development

## Juan Liu

## Introduction

*A coffin sits in a corner of He Quangui's house. He has kept it stubbornly shrouded. But a hacking cough reminds the former gold miner of his mortality. It's been over ten years since he was diagnosed with silicosis, a form of pneumoconiosis in which a worker's lungs are overwhelmed by the dust breathed in years earlier while working in gold, coal or silver mines. It is China's number one occupational disease. He is typical among victims: in his 30s, a sole breadwinner, a migrant worker from a remote, poor village, and facing an uphill trek getting treatment with little money and paltry employment paperwork… out here in hardscrabble rural China, most workers cannot track down mine owners to cover medical bills and don't get treatment until it is too late. Most just give up as costs mount. Some commit suicide.*

This is part of the description for a short documentary film entitled '*Dying to Breathe*' by Sim Chi Yin (2015), a Singaporean photojournalist based in

---

An earlier version of this chapter has been published in Liu, Juan. 2018. "El nexo entre neumoconiosis, salud ambiental y pobreza en la China rural (Understanding the complexity of environmental health and poverty in rural China through the lens of pneumoconiosis)". *Ecología Política,* 56: 30–40.

---

J. Liu (✉)
College of Humanities and Development Studies, China Agricultural University, Beijing, PR China

Institute of Environmental Science and Technology, Universitat Autónoma de Barcelona, Barcelona, Spain

Beijing who witnessed the miner's repeated collapses, one suicide attempt at midnight and what the specific type of pneumoconiosis silicosis does to a person and their family in its final stages. Sadly, He Quangui died from complications related to silicosis on 1 August 2015, two months after the launch of the documentary film. As reported in many other accounts, this single portrait reminds us of the big picture of the most common occupational disease in China, along with its significant achievements in economic growth and poverty reduction.

Caused by the long-term and/or intensive inhalation of dust, pneumoconiosis (meaning dusty lung) is a progressive and irreversible lung disease that affects industrial and agricultural workers with impairment, disability and premature death. Medical treatment is not effective in advanced cases of these diseases, so preventing them, through controlling respirable dust exposure, is essential. Silicosis and coal workers' pneumoconiosis (CWP, commonly called black lung) are two main categories of pneumoconiosis. Other types include asbestosis, byssinosis (also called brown lung), graphite pneumoconiosis and so on. Pneumoconiosis was common in Western industrialised countries during the early twentieth century and has become an urgent environmental-labour-health problem along with the consequences of rapid urbanisation and industrialisation of China.

According to *China Coal Miner Pneumoconiosis Prevention and Treatment Foundation* (CCMPPTF 2015), the reported number of pneumoconiosis cases in China exceeded 720,000 by the end of 2014, of which 62% were concentrated in the coal industry. The number of people affected by this disease is still increasing by around 20,000 every year (China CDC 2019). Pneumoconiosis cases accounted for nearly 90% of the reported cases of occupational diseases in the same year. Also, 90% of the pneumoconiosis sufferers are rural migrant workers (National Health Commission 2015). However, an earlier 'moderate' estimation by an NGO called *Love Save Pneumoconiosis* (2016) stated that there might be 6 million workers from the rural areas[1] who are suffering pneumoconiosis at different stages of the disease. This estimation was based on the fact that many cases were not reported as cases of occupational disease, because most of the affected workers from the rural areas had never signed employment contracts; a contract was still the primary condition of entering the diagnostic process for occupational disease then. In their 2014

---

[1] As different Chinese reports used slightly different terms but represent the same group of people, the English terms of *workers from the rural areas, rural workers, migrant workers* or *rural migrant workers* are also representing the same group of individuals who are rural residents, previously residing in the countryside and engaging in agriculture who subsequently migrate to other areas—but not necessary to the urban areas—to engage in manual labour.

report on living conditions of rural workers, *Love Save Pneumoconiosis* (2014) indicated that, among the rural migrant workers affected by pneumoconiosis, only 6.8% have signed employment contracts, which is much lower than the average rate for migrant workers. In their 2016 report, this rate had increased to 9.5%, which is still quite low (Love Save Pneumoconiosis 2017). By any criteria, this is a large group living in terrible circumstances that demands attention from various stakeholders.

Globally, pneumoconiosis has been studied not only as an occupational health issue that refers to miners' working conditions and hazards as well as medical treatment. It has also been understood as part of the environmental history accompanying the industrialisation process, as a public health issue which demands policy coverage and changes, and sometimes as an environmental justice issue which involves miners' militancy and working-class environmentalism (Chen 2011). In the Chinese context, pneumoconiosis has mostly been studied as an occupational disease which results from unsafe working environments and an unawareness of prevention and protection. Therefore, pneumoconiosis has been framed as a labour issue connected with occupational health that most of the policies or practices try to address (e.g. Liu 2017). However, it is also strongly connected to the extractive and processing industries and associated construction works which also result in natural environmental destruction or degradation.

Based on the empirical evidence from media accounts, academic works, reports from different organisations and the author's preliminary field work, this chapter attempts to understand the complexity of environment-labour-health and its relationship to poverty in rural China through the lens of pneumoconiosis. It will first look back to the drivers which mobilised the peasants to enter workplaces with environmental health risks. The continued section will focus on various cases to show the impacts of pneumoconiosis on the individual body, family life and community development. The third section will examine institutional arrangements and biases and its encounter with environmental labour activism from below. Combining with other environmental labour and health issues in China, the closing section will try to reframe pneumoconiosis as an environment-labour-health complex for analysis and actions. It argues that the extractive processes for natural resources which support the country's industrialisation, urbanisation and modernisation are also extractive processes of rural labour and their physical bodies, while the institutional biases aggravate the reproduction of poverty in those affected households and communities. Besides urging the protection of workers' rights and social justice, this chapter also offers critical reflections on the unsustainable pattern of China's economic growth and indicates the importance of building environment-labour alliances that insert health concerns into the ecological civilisation process.

## Encountering Pneumoconiosis

Although the words 'silicosis' and 'dust prevention' had already appeared in some Chinese reports and policy documents since the 1950s (State Council of PRC 1956), the history of pneumoconiosis in China is closely linked to its economic boom since the 1980s, when reform and the opening up led many peasants to work in industries such as mining, jewellery manufacture, sandblasting, foundries and construction, in which workers are at high risk of developing pneumoconiosis (Fan and Ng 2019). The *Report on Living Conditions of Rural Pneumoconiosis Workers 2014* also revealed that nearly 90% of the pneumoconiosis sufferers had worked in coal mines or other mineral mines, about 7% of them worked in constructions sites as drillers or blasters, while the rest worked in jewellery factories or other industries (ibid., 16).

Since the early 1990s, several hundred villagers from around Leiyang in the central province of Hunan made the 600-kilometre journey south to work in the construction sites of Shenzhen. This was about 10 years after the designation of Shenzhen as China's first Special Economic Zone in 1980. Moreover, the famous 'south China tour' in the spring of 1992 also played a crucial role in guiding and accelerating China's reform and opening, as well as the socialist modernisation process. These migrant workers worked day in, day out, drilling blast holes into the bedrock and establishing the foundations for many of the city's best-known skyscrapers and other building projects, including the subway. They endured this hardship in dangerous, poorly ventilated caverns with little or no protection from the heavy, lung-clogging dust created by their drills, primarily because the pay was good and they were completely unaware of just how debilitating the long-term effects of inhaling rock dust could be. Crucially, neither the building contractors nor the local authorities made any attempt to inform the workers about the hazards inherent in their jobs. In 1999, several workers started to complain of exhaustion, and repeated bouts of influenza. Their strength declined, and many returned home to Leiyang. In 2002, the first of 16 workers from the villages died of pneumoconiosis (China Labour Bulletin 2010).

China has become the world's largest gold producer since 2007, and its production peaked at 463.7 tonnes in 2016 (Gold.org 2016). Its gold mining industry has received increased foreign and domestic investment in recent years, and both projects and gold outputs have increased rapidly especially before 2016, as more discoveries have been made. Like many of his fellow villagers, the protagonist of 'Dying to Breathe', He Quangui's encounter with pneumoconiosis unconsciously started from the late 1990s. As a young migrant worker, he moved out from his village and started working at a gold mine near his hometown.

> At that time, the salary was not high, CNY 800–1200 per month, you earn different wages with different types of work. In order to earn more money, I chose to do some skilled physical work of excavation, and I sucked dust in every day. At that time, I didn't know there would be danger to my body, and I didn't wear a mask. When I started feeling that something was wrong, it turned out that I had contracted pneumoconiosis.

He keeps a notebook to record the names of those who have succumbed to silicosis. Flipping through it, he tells the photojournalist: 'I've watched them die, one by one. I know one day it will happen to me too'. Similar trajectories could be traced for migrant workers from Xihekou and the other townships in Lu'an City of Anhui Province since 1980s, as well as migrant workers from Mashan County of Guangxi since early 1990s. Dreaming of changing their poverty situation through gold mining, thousands of poor peasants emigrated out to Hainan where they ended up contracting the deadly silicosis (Liu 2010; State Administration of Work Safety 2009).

Also, China is the largest producer and consumer of coal in the world and the largest user of coal-derived electricity: 3.45 billion tonnes of raw coal were produced in the country in 2017(Sxcoal.com 2017). Apart from being a risky job due to the sector's poor safety record, coal mining is also disease prone. In August 2017, I met Yuanyuan, a young lady who is 31 years old, in a village in the Southern part of China's Northwest Shaanxi Province and learned the story of her family's experiences with pneumoconiosis. Yuanyuan's father emigrated when she was a child and worked in the coal mines in the neighbouring Shanxi Province in order to earn a living for the whole family. Several years later, he died of lung disease, which was not recognised and not called pneumoconiosis by the villagers at that time. The loss of the breadwinner forced Yuanyuan to emigrate when she was a teenager to support the family until her mother lost her sight in both eyes. As a result, she was obliged to stay at home and marry somebody who would be prepared to marry into and live with her family. The son-in-law then became the breadwinner of the household.

During my field visit, Yuanyuan just received confirmation from the doctor that her husband had contracted silicosis because he had worked in a silica refinery factory for a short period. This reminds Yuanyuan of her deceased father's last years, and she believes that it was also pneumoconiosis. Her husband cannot do heavy physical work, but still takes odd jobs nearby to support the subsistence of the family except during the periods when he doesn't feel very well.

When the land and agriculture proved insufficient to sustain their livelihoods, the rural poor were encouraged to emigrate to seek an alternative source for their livelihoods. Migration, together with agricultural intensification or extensification, was considered as a way of livelihood diversification, which would have been considered as the principal strategies to lift the rural poor out of poverty (Scoones 1998). At the same time, there was a rapidly growing demand for labour to support the booming economy in the country. Many of the rural migrant workers abandoned, partly or fully, agricultural work in the impoverished parts of the country to enter the relatively lucrative industries to which those normally less-educated migrants could sell their manual labour. Following their relatives or fellow villagers, many of them went to the mining sector or construction sites in the urban areas.

At the beginning, when these future pneumoconiosis sufferers were hired, they were enjoying higher incomes and sometimes even quick cash from those sectors compared with their previous involvement in farming. Family members were also satisfied with or even proud of the significant improvements to their daily lives and their increased consumption. After years of non-protective exposure to highly concentrated levels of silica or coal dust, they started to realise that the risks of such inhalation could result in coughing, shortness of breath and influenza-like symptoms in the first stage, followed by losing their strength and eventually all ability to work in stage two, and finally in stage three, death. When medical expenses have wiped out whatever money workers had made through years of working, families that had risen out of poverty found themselves worse off than before. Many of the pneumoconiosis sufferers regretted their decision to enter the industry and taking the job when they found out about their contraction of lung disease, 'I sold my life to Shenzhen. If I had known the danger of pneumatic drilling, I would never have done the work, no matter how poor I am…' a 52-year-old former driller ruefully sighed.

# Painful Breaths: The Impacts of Pneumoconiosis on the Affected Individual, Family and Community

Pneumoconiosis sufferers are bearing the physical pain from pulmonary fibrosis and extreme breathing difficulties; chest distress, coughing and chest pain are the most common symptoms. Many pneumoconiosis patients have to breathe with the help of an oxygen tank; their life is tethered by a plastic pipe just a few meters long. When the disease is at its final stages, many have to kneel down several times a day, even through the night, in order to alleviate the pain in the lungs. Because of the illness, they often have a poor appetite, and it is even difficult for them to have their meals. In the advanced stages, most people have difficulties moving. They will often get out of breath just walking around their homes. They cannot take care of themselves, losing weight quickly, becoming barely skin and bones. This kind of inability creates both economic and emotional pressure leading to the feeling of 'uselessness'.

Pneumoconiosis sufferers are also generally unable to engage in economic activities after developing some apparent symptoms. First, the physical pain and frequent respiratory tract infections prevent them from taking up any full-time paid employment. Most first-stage sufferers continue with their work due to the pressure of making ends meet, but the direct result of working while sick is the rapid intensification of the illness, which will quickly worsen their health. Second, pneumoconiosis sufferers are often dismissed from their workplace when some symptoms have been discovered. It is also difficult for them to pass the pre-employment health checks to get another job and re-enter the labour market. Sometimes the sentiment of 'strangeness' of the sick workers also excludes them from many activities. It's difficult for them to decide whether to expose or conceal their illness as either way they are unable to avoid economic exclusion. Third, workers who had returned to rural areas found that they also could no longer handle farm work that demands heavy physical exertion. After they fell ill, on the one hand, the household lost its labour force, and on the other, living with an incurable disease inevitably creates a serious financial burden for a peasant's household.

Having an unhealthy member in a family means not only an increasing economic burden, but also the redistribution of family obligations and additional caring responsibilities. As most of the victims are male workers (Love Save Pneumoconiosis 2014, 2020), the women, children and elderly in those households have to change or reverse their traditional gender roles or generational roles in a patriarchal society, to share the burden and sustain the family.

Women would take on more physical work that used to be mainly men's jobs or jobs done by women and men together (e.g. the wife of He Quangui, etc.). Children also drop out of secondary or high school to earn a living for the family by migrating themselves (e.g. Yuanyuan and many others). Finally, the elderly continue working on their farmland or re-enter the labour market at a very senior age in order to sustain the life of the patient(s) and the livelihoods in the family (e.g. He Quangui's father is still working on the field at the age of 79). In some extreme cases, the pneumoconiosis sufferer would be disliked or even rejected by family members.

The impacts on family members are particularly distressful for pneumoconiosis sufferers when they feel that they are failing to meet the normative roles and expectations of a man, a husband, a father or a son. They are also frustrated with the situation of being dependent upon their wives for their livelihoods and not being able to uphold filial piety towards their senior parents and to fulfil their roles as husband and father with dignity. They experience a high level of panic and some lament how they have failed to provide a 'better living' or a 'hope for the future' for their wives and families. Their illness creates insurmountable barriers to fulfilling even some basic obligations. All that they can manage is struggling with their illness in order to just stay alive.

The suffering from physical pain, the economic hardship due to the burden of medical costs and debt repayments and the frustration of 'a failed life' lead many advanced stage sufferers to contemplate suicide. A rope, a bottle of pesticide, a pair of scissors or jumping from a high building may end their pain of not being able to breathe, and at the same time ending what ought to have been the prime of their lives (Beijing News 2015).

Some rural towns and villages in the major labour exporting areas across the country have been dubbed 'pneumoconiosis villages/towns' by the media because the disease has weakened or even killed hundreds of villagers who were migrant workers. Another documentary film entitled 'Tears of Shuangxi Village' has shown how a village in Leiyang County was fatally affected by pneumoconiosis being left with poor and broken families and a desolated community.

## Obstacles on the Road Towards Justice

Although in a prominently disadvantaged position, pneumoconiosis sufferers are not passive victims without any agency or silent individuals lacking legal and political efficacy. The media reports have documented various forms of strategies and tactics taken by different groups or individuals to defend their rights and claim compensation. However, the issue had rarely received much attention until Zhang Haichao, a pneumoconiosis sufferer from Henan Province who had acquired the disease at work, took the drastic step of undergoing a thoracotomy at a hospital in Zhengzhou after being rejected by the Xinmi Centre of Disease Control and Treatment. He had been misdiagnosed by the same institution with tuberculosis but not pneumoconiosis in 2009. The operation revealed that the diagnosis was unmistakably pneumoconiosis. Zhang had worked for several years at an abrasive materials factory in Xinmi, near the provincial capital Zhengzhou, breathing in clouds of dust every day. His cough and the tightness in his chest since 2007 took him into several local and national hospitals, all of which confirmed pneumoconiosis. The rejection of legally valid diagnoses from other jurisdictions was one of the obstacles that prevented pneumoconiosis sufferers from beginning a redressment process. Zhang's 'Open-chest surgery' case attracted extensive media coverage, and he eventually received CNY 615,000 (~ USD 95,000) as compensation[2] in a mediated settlement with his former employer, which enabled him to have a double-lung transplant. Few other workers have been as fortunate.

In the reports of the *China Labour Bulletin* (2010) and *Love Save Pneumoconiosis* (2014, 2017), many other obstacles created by the legal and regulatory systems, the employers and the government authorities have been teased out. The *Love Save Pneumoconiosis* 2014 report indicated that only 25.72% of the pneumoconiosis sufferers tried to apply for compensation, and about two thirds of those who applied eventually received some, which accounted for 17.31% of the investigated cases. The vast majority of the pneumoconiosis sufferers did not receive anything. The same report also revealed that it takes an average of 16.9 months to obtain some compensation from the time of applying, while the maximum time for this process is 72 months.

Although there have been some amendments to the *PRC Law on the Prevention and Treatment of Occupational Diseases* in the recent years,[3] for rural migrant workers it's still quite difficult to go through all the steps to claim compensation. For example, the legal provisions dealing with the payment of

---

[2] A later report revealed that the actual compensation he got was almost doubled, CNY 1.2 million, based on a private agreement with his former employer. See China Labour Bulletin 2013.

[3] Amendments were made in 2011, 2016 and 2017, respectively. For instance, the 2011 amendments eliminated the acceptance threshold for occupational disease diagnosis, simplified the labour arbitration process, stipulated the responsibility of the regulatory authority to make judgements on disputed materials under specific circumstances and included workers' self-report as reference for a diagnosis of occupational disease and so on. But until the 2017 amendments, there were still restrictions on the qualification of health care facilities to conduct occupational health checks. After the amendment, health care institutes can conduct occupational health examinations as long as they are designated so.

occupational disease benefits, such as medical expenses and loss-of-earnings compensation, all require proof of a labour relationship with an employer. However, pneumoconiosis has a long latent period. To ask for compensation would result in losing their job at an earlier stage of the disease. Moreover, as reported above, workers often don't have any labour contract to prove prior employment when more serious clinical symptoms appear later. Therefore, they have little or no bargaining power. They either need to bear all the costs themselves or need to accept payments significantly lower than those they would be entitled to if they were still employed or if they could prove the relationship between their disease and their employment. Their expenses are considerable, since they need long-term care and continual treatment just to stay alive. For the vast majority of migrant workers, such resources are beyond their reach. Without proof of the relationship between their disease and their work, it's also difficult for them to go through a judicial review process in the courts when seeking treatment costs or compensation. Such claims are too often ignored or rejected by the courts. However, in contrast, formally employed workers in state-owned enterprises (like state-owned coal mines) who were covered by work-related injury insurance would be entitled to all necessary legal and medical provisions.

The laws stipulate that employers must provide employees with a safe working environment that conforms to national occupational health standards. Employers who cause serious harm to the health of their employees are subject to penalties such as the suspension of operations, fines or closure, and may even bear criminal liability. If the employer has not contributed to the work-related injury insurance fund when an employee contracts an occupational disease, the employer must pay the benefits and related expenses set in the regulations. However, quite often when workers who are in high-dust industries exhibit early-stage symptoms of pneumoconiosis, many employers seek to get rid of them as quickly as possible. Some employers also collude with hospitals to falsify or conceal post-employment medical examinations. Moving to another location, changing the company name, changing investors and managers, and merging with another company are also sophisticated tactics for them to evade their legal responsibilities to the victims of pneumoconiosis. Economic power and political connections are protective umbrellas for the employers. It is easy for them to avoid punitive penalties by delaying proceedings for as long as possible to force the impoverished sufferers to give up the fight for justice.

The doctors, government officials, arbitrators and judges responsible for assessing and ruling on occupational disease cases can be both a help and a hindrance for pneumoconiosis sufferers (China Labour Bulletin 2010). If these actors in positions of authority took a sympathetic and considerate approach, victims could in theory obtain redress relatively quickly. But all too often officials are overly rigid in their interpretation of the law and procedural regulations. Moreover, as noted above, they can be unduly influenced or even bought off by local enterprise bosses. Therefore, a relatively simple process of redress envisioned in the law has become a tangled path of obstacles that many pneumoconiosis sufferers, already debilitated by their disease and crippled with debt, cannot hope to overcome.

Typically, within the temporary and dispatch labour regime, where the 'user of the work' is separated from the 'employer of the person', workers are often protected neither with a formal contract nor with health insurance, pensions, maternity packages, skill training and other benefits (Huang 2017). This kind of informalisation of work and the exclusion of rural pneumoconiosis sufferers from the legal system would lead to the emergence of 'non-legalistic, cellular activism' (Fan and Ng 2019).

Frustrated by the hospital's intransigence and their covering up of the severity of workers' illnesses, about 180 migrant workers from Leiyang who had all been employed as pile-blasting and drilling workers in Shenzhen in the 1990s and contracted pneumoconiosis staged a demonstration outside the Shenzhen Municipal Government on 15 June 2009. They also took the person in charge of the Shenzhen Prevention and Treatment Centre for Occupational Diseases to the building of the Shenzhen Municipal Government (ibid.). The city's top administrators were shocked by this collective action and instructed lower levels to process the victims' demands within the 'legalistic framework with humanistic care'. In July 2009, the Leiyang workers sent a letter to the leadership of the Shenzhen Municipal Government, in which they not only voiced their own plight, but also accused their bosses of illegal employment practices and the authorities of negligence. In practice, street-level bureaucrats urged workers to present evidence of an employer–employee relationship and to apply for a re-diagnosis and certification of their illness. However, because none of the workers had signed a contract with their employers or subcontractors, by the end of July, only 17 of the workers were verified by the government as having actually worked in Shenzhen. For those workers whose employment relationship could not be confirmed, the Shenzhen Municipal Government offered a one-off 'charitable' payment of CNY 30,000 (~USD 4,600) each. Angered by this derisory offer, the workers staged another protest in front of the municipal government building on 30 July. During this demonstration, some of the workers fell seriously ill, one of them died. (EJAtlas 2019a).

On 19 January 2011, several former miners, who had worked for several years in private gold mines close to the Mongolian border, staged a demonstration in Xi'an, capital of Shaanxi province, as representatives of more than 150 pneumoconiosis sufferers from Gulang County in Gansu. They asked to be heard by both the government and society. Before this demonstration, some of them had travelled to Beijing to seek desperately needed medical assistance and compensation from their former employer. The issue had also received attention from national media including CCTV, Global Times, China Economic Times and so on as well as assistance from some civil society organisations and many individuals. Under great pressure from the media and the general public, authorities in Gansu started to mobilise donations as a special fund for the treatments of the pneumoconiosis sufferers in Gulang County.

These protests by workers suffering from pneumoconiosis, as Fan and Ng (2019) commented, could be categorised as 'struggles for compensation', but not 'struggles for progressive reforms' of labour that include signing employment contracts and social insurance coverage as well as workplace environmental health justice. To maintain stability and to be somehow exempted from liability, governments or the employers often choose to mobilise some relief funds and comfort the workers in a 'humanistic' or 'philanthropic' manner instead of paying them a reasonable compensation.

While local authorities and employers often make every effort to divide and rule in order to weaken pneumoconiosis activism, limiting the scale and extent of that mobilised actions or professional advice from NGOs, journalists, scientists, voluntary associations and other intermediate organisations could play in supporting migrant workers to seek justice individually and as a social group, by providing legal and practical advice and training and by helping to advocate for policy changes and so on. For example, on the advice of the Hong Kong-based China Labour Bulletin, the Leiyang migrant workers filed an administrative lawsuit against the Shenzhen Health Bureau, accusing it of having failed in its legal duty to enforce laws and regulations designed to protect workers' health. Therefore, they argued, the Bureau was itself responsible for their occupational illness. This turned out to be the decisive moment in their struggle. In order to avoid getting sued, the Shenzhen Municipal Government arranged for another employment confirmation exercise and proposed a better compensation plan for those who had not been able to confirm their employment relationship. For those workers with a confirmed employment-disease relationship, they were also promised that if their employers had contributed to the work-related injury insurance scheme, they would get compensation from that fund. The field investigation by the Student Investigation Group, which was carried out by some graduate students from universities such as Peking University and Tsinghua University, helped to uncover the problems. The joint statement, written by Professor Shen Yuan of Tsinghua University and five other scholars, was sent to the Ministry of Labour and Social Security (Ministry of Human Resources and Social Security since March 2018) and urged government departments to shoulder the responsibility for the current substandard employment conditions and get responses from the Vice-Prime Minister of that time who instructed the Health Commission of Guangdong Province to 'investigate and verify the conditions of pneumoconiosis-affiliated workers… conduct occupational health examinations, issue occupational disease diagnosis, and provide medical treatment to the suffering workers'. Founded by a famous journalist Keqing Wang, the NGO, *Love Save Pneumoconiosis,* is trying to identify pneumoconiosis sufferers and helping them to breathe. They also tried their best to support them in maintaining their livelihoods and advocate policy changes (EJAtlas 2019a).

A series of petitions and protests were launched at different levels in later years as some workers had not joined those petitioning in 2009 because they did not realise they were sick. In the words of Rob Nixon (2011), the living and working conditions leading to pneumoconiosis are a kind of slow violence that 'occurs gradually and out of sight, a violence of delayed destruction that is dispersed across time and space, an attritional violence that is typically not viewed as violence at all'. In January 2018, around 100 former construction workers and family members affected by pneumoconiosis have been encamped in Shenzhen demanding long overdue compensation for the occupational illness contracted while working in city in the 1990s. The workers and their families are faced with more bureaucratic delays during the petition, yet the long latent period and the separation of working and living spaces make the road towards justice even more difficult. On 16 July 2018, Leiyang pneumoconiosis workers were stopped by Shenzhen police as they began their multi-city campaign for the prevention of the occupational disease: 'we are not here to present grievances, we want to give back to society by raising awareness on disease prevention' (EJAtlas 2019a).

## Pneumoconiosis: An Environment-Labour-Health Complex

The plight of the former gold miner He Quangui and his family can be found across the whole of China. Many of the rural peasants who sought a better living for their families by working in mines, stone quarries, construction sites and gemstone factories returned to their villages sick and dying from pneumoconiosis. The endless spiral of 'migrating out to get rid of poverty and returning to poverty due to illness from migration' is frequently staged in many families and in many villages in poorer areas of the country. To sustain economic and urban growth, which seems to grow without any limits or constraints, not only have enormous natural resources been extracted and the environment both in cities and rural areas has been polluted, but also human beings, as biopolitical subjects, have been extracted through the dangerous, dirty and demanding ('3D') jobs. Poor rural areas, drained by capital as 'open veins' (Galeano 1997) and then left with 'wasted lives' as the 'collateral casualties of progress' (Bauman 2004:15), do not just confront the re-impoverished poverty of the current generation, but also face the similar impasse of potential intergenerational transfer of poverty, although more and more victims have gradually been included in the social security system under China's massive national schemes of poverty alleviation (Love Save Pneumoconiosis 2021).

These kinds of precarious circumstances are made worse because, in their quest for redress, migrant workers face many obstacles created by laws and procedures as well as by the actions of employers and authorities. Moreover, there are institutional biases, and most of the migrant workers are not protected by workers' unions or other organisations representing them.[4] Collective actions or protests might be effective in attracting some attention from the media and the public and help them with their struggle for compensations. However, more progressive reforms are imperative in order to improve workplace environment and safety, ensure the rights of the migrant workers and make certain that occupational disease sufferers are legally entitled to their needed compensations and have easier access to them.

Besides *Dying to Breath* (Sim 2015) and *Tears of Shuangxi Village* (Fan 2010) in which the victims are separated from the working sites, a very recent longer documentary film, *Miners, Groom, and Pneumoconiosis* (Jiang 2020), also

---

[4] The All-China Federation of Trade Unions (ACFTU) is the only official trade union in China. As part of the mass organisation apparatus of the CCP, the ACFTU is hamstrung by its political subordination to the CCP and by its financial dependence on employers, who contribute 2% of their payrolls to the union fund. Migrant workers are normally not covered by this scheme, neither with other legalised social security arrangements like insurances offered by the employers.

focused on the rise and fall of small-scale (illegal) iron ore mining and the sweep of lung diseases that accompanied it. Using a series of metaphors and film techniques, a fourth documentary, *Behemoth* (Zhao 2015), depicts a hellish large-scale landscape as a result of the coal mining industry, ranging from mining to industrial smelting, as well as graveyards with trucks running all the time through the field transporting raw coal to coke-making factories. We also witness workers, immersed in black smoke and struck by wind, continuously taking turns to mine; their skin, due to long-term exposure to ore and sand, is covered with red spots and black dust particles. Wealth accumulates elsewhere, but all living creatures who once inhabited the space have been dispersed and the workers now living there are trapped in poverty. What's left are the continuous explosions, smelting, atmosphere pollution, smog and wastewater—and for the workers coughs, impaired breathing and black and damaged lungs. Ironically, people's desire for the paradise of modern life results in ghost-like cities. This last episode depicts not just the labour and health dimensions of pneumoconiosis, but a broader picture of the disease which involves the environment and the material dimension of the spaces where pneumoconiosis resides.

Documentary films and social studies, reporting on pneumoconiosis and the related activism against it as a labour issue, have increased the awareness of the general public about the miserable stories of the victims. However, its environmental dimension has been overlooked. This makes it difficult to build alliances between workers and environmental activists. Labelled solely as an occupational health and a labour issue, pneumoconiosis is seldom taken into account by mainstream environmental activists in China. When flagged mainly as environmental issues, workers' ordeals and their jobs-health dilemma are generally less visible. There is also evidence showing that those who are working in a polluting plant are less likely to engage in environmental mobilisation, though they might have been affected by the same health impacts or even to a more severe degree (e.g. EJAtlas 2019b). A prominent tension between jobs versus environment, or job blackmail, is one of the obstacles that stops workers from joining forces with environmental activists to create 'blue-green' or 'red-green' coalitions (Bennett 2007; Mayer 2009, 2011; Van Alstyne 2015). Instead, environmental protection measures could result in protests from workers. For instance, in order to achieve the carbon reduction targets of the Paris Agreement, as well as to reduce overcapacity of the steel and coal industries in China, many workers were laid-off under the downsizing plan. Although the government stated that it had allotted CNY 100 billion to help compensate laid-off workers, worker protests against wage arrears and layoffs in the steel and coal mining industries have been on the rise since late 2015, including several high-profile protests involving tens of thousands of workers such as the Heilongjiang Longmay Coal Group protest in early 2016 and the Ansteel strike in Guangzhou (Zhang 2019). Moreover, in some of the environmental conflicts, affected workers enjoy little sympathy from environmentalists or the general public, and they might even be blamed for being eager for money as well as the ones destroying the environment. Some of the pneumoconiosis sufferers are also found entering into a self-blaming trap, 'it's our fate! It's our destiny!'

Running on a large 'treadmill of production (as well as consumption)' (Schnaiberg 1980; Schnaiberg et al. 2002), it is difficult for a society to reject the promises of economic growth, new jobs or steady employment and general prosperity. Pneumoconiosis reminds us that health, both in terms of an individual's health and the health of one's family, can become a common result of environmental pollution. It also indicates that environmental and labour mobilisations should challenge the current unsustainable patterns of economic growth based on the massive extraction, circulation and consumption of resources. Moreover, they should advance social changes and alternative industries to protect workers' health and to improve environmental quality towards sustainability and justice.

**Acknowledgements** The chapter is supported by the Chinese Universities Scientific Fund (2020TC061). The author is part of the ENVJUSTICE project, funded by the European Research Council (ERC) advanced grant (No. 695446).

# References

Bauman, Zygmunt. 2004. *Wasted lives: Modernity and its outcasts*. Cambridge: Polity Press.
Beijing News. 2015. Hunan Leiyang Silicosis Village Investigation: Numerous Painful Suicides. Accessed September 30, 2018. https://mp.weixin.qq.com/s/0tyGeYUzq0fYuK6hC2_6FA
Bennett, Dave. 2007. Labour and the Environment at the Canadian Labour Congress–the Story of the Convergence. *Just Labour: A Canadian Journal of Work and Society*. 10 (3): 1–7.
CCMPPTF. 2015. China Reported more than 720,000 people with Pneumoconiosis, 62% in the Coal Industry. Accessed September 20, 2018. http://politics.people.com.cn/n/2015/0207/c1001-26523549.html
Chen, Lili. 2011. A Review on Pneumoconiosis Studies in the US Historiography since 1980s. *Journal of Historical Science*. 6: 98–107 (in Chinese)

China CDC. 2019. Health Daily: Ten Ministries Join Forces to Fight Pneumoconiosis. Accessed December 26, 2020. http://www.chinacdc.cn/mtbd_8067/201907/t20190722_204195.html

China Labour Bulletin. 2010. The Hard Road: Seeking Justice for Victims of Pneumoconiosis in China. China Labour Bulletin Research Report.

———. 2013. Pneumoconiosis Activist Zhang Haichao Gets Life-saving Double-lung Transplant. Accessed September 30, 2018. https://www.clb.org.hk/en/content/pneumoconiosis-activist-zhang-haichao-gets-life-saving-double-lung-transplant

EJAtlas. 2019a. Urban Construction Sites Provoked Pneumoconiosis Crisis in Shuangxi Village, Hunan, China. Accessed December 26, 2020. https://www.ejatlas.org/conflict/the-shenzhen-pneumoconiosis-crisis-guangdong-china

———. 2019b. Excessive Blood Lead Caused by Haijiu Battery Co., Ltd. in Deqing, Zhejiang, China. Accessed December 26, 2020. https://www.ejatlas.org/conflict/c-332-people-with-excessive-blood-lead-caused-by-a-sino-foreign-joint-venture-in-deqing-county-of-zhejiang

Fan, Jiaju. 2010. *Tears of Shuangxi Village*. Documentary. Accessed May 05, 2017. http://www.iqiyi.com/w_19rrnhyxrh.html

Fan, Lulu, and Kenneth Tsz Fung Ng. 2019. Non-Legalistic Activism from the Social Margin: Informal Workers with Pneumoconiosis in Shenzhen. *China Information*. https://doi.org/10.1177/0920203X18784799.

Galeano, Eduardo. 1997. *Open veins of Latin America: Five Centuries of the Pillage of a Continent*. NYU Press.

Gold.org. 2016. Gold Mining Map and Gold Production in 2016—World Gold Council. Accessed September 20, 2018. https://www.gold.org/about-gold/gold-supply/gold-mining/gold-mining-map

Huang, Philip C.C. 2017. Dispatch Work in China: A Study From Case Records, Part I. *Modern China*. 43 (3): 247–287.

Jiang, Nengjie. 2020. *Miners, Groom, and Pneumoconiosis*. Documentary.

Liu, Cuirong. 2010. Prevalence of Pneumoconiosis in Taiwan and Mainland China and Its Implications. *Taiwan Historical Research*. 17(4): 113–163 (in Chinese)

Liu, Juan. 2018. El nexo entre neumoconiosis, salud ambiental y pobreza en la China rural. *Ecología Política* 56: 30–40. (in Spanish)

Liu, Yuan. 2017. Research on the Plight and Countermeasures of the Rights Protection of Workers with Pneumoconiosis. *Economic Research Guide*. 27: 195–198 (in Chinese)

Love Save Pneumoconiosis. 2014. *Report on Living Conditions of Rural Pneumoconiosis Workers: 2014*. (in Chinese).

———. 2016. The Status of Pneumoconiosis Among Workers from Rural Areas. Accessed October 04, 2017. http://www.daaiqingchen.org/list.php?fid=92

———. 2017. *Report on Living Conditions of Rural Pneumoconiosis Workers 2016* (in Chinese)

———. 2020. *Investigation Report on Pneumoconiosis Migrant Workers in China 2019* (in Chinese)

———. 2021. *Investigation Report on Pneumoconiosis Migrant Workers in China 2020* (in Chinese)

Mayer, Brian. 2009. Cross-Movement Coalition Formation: Bridging the Labor-environment Divide. *Sociological Inquiry.* 79 (2): 219–239.

———. 2011. *Blue-green Coalitions: Fighting for Safe Workplaces and Healthy Communities.* Cornell University Press.

National Health Commission. 2015. National Occupational Disease Report 2014. Accessed September 20, 2017. http://old.chinasafety.gov.cn/zwdt/gwgg/201512/W020171101459125101354.docx

Nixon, Rob. 2011. *Slow Violence and the Environmentalism of the Poor.* Harvard University Press.

Schnaiberg, Allan, David N. Pellow, and Adam Weinberg. 2002. The Treadmill of Production and the Environmental State. *The Environmental State Under Pressure.* 10: 15–32.

Schnaiberg, Allan. 1980. *The Environment: From Surplus to Scarcity.* New York: Oxford University Press.

Scoones, Ian. 1998. Sustainable Rural Livelihoods: A Framework For Analysis. IDS Working Paper.

Sim, Chi Yin. 2015. *Dying to Breathe: The Unseen Cost of Gold Mining.* Accessed April 04, 2017. https://www.nationalgeographic.com/photography/proof/2015/05/15/dying-to-breathe-a-short-film-shows-chinas-true-cost-of-gold/

State Administration of Work Safety. 2009. The Case of Guangxi Mashan Migrant Workers in the Gold Mines of Hainan Getting Silicosis. Accessed September 20, 2018. http://www.esafety.cn/Item/34842.aspx.

State Council of PRC. 1956. Decision on Preventing Dust Hazards in Factories, Mines and Enterprises. Accessed October 10, 2017. http://www.china.com.cn/law/flfg/txt/2006-08/08/content_7060149.htm

Sxcoal.com. 2017. China 2017 Raw Coal Output up 3.2pct, Dec Figure at 24-mth high. Accessed September 20, 2018. http://www.sxcoal.com/news/4567443/info/en

Van Alstyne, Andrew D. 2015. The United Auto Workers and the Emergence of Labor Environmentalism. *WorkingUSA: The Journal of Labor and Society.* 18 (4): 613–627.

Zhang, Lu. 2019. Worker Protests and State Response in Present-day China: Trends, Characteristics, and New Developments, 2011–2016. In *Handbook of Protest and Resistance in China*, ed. Teresa Wright. Edward Elgar Publishing.

Zhao, Liang. 2015. *Behemoth.* Documentary.

# Part V

## Organic Intellectuals

# 25

## Introduction: Trade Union Environmentalists as Organic Intellectuals in the USA, the UK, and Spain

Nora Räthzel, Dimitris Stevis, and David Uzzell

## Introduction

In the following section, we have gathered three chapters which use the same qualitative method. They base their analyses of trade union environmental policies on the recorded life-histories of unionists who played (and in some cases continue to play) a central part in developing such policies and putting them on trade union agendas. This approach makes it possible to understand the relationship between individual trajectories, organisational structures and societal conjunctures not only through documents but also through the personal and interpretive narratives of the actors centrally involved in the processes. Following Gramsci, we see these actors as 'organic intellectuals'. While involved in the practices and politics of their organisations, they are also ahead of them, engaged in transforming their organisations to meet new societal needs, in this case the threat of the global environmental crisis. We brought

---

N. Räthzel (✉)
Department of Sociology, Umeå University, Umeå, Sweden
e-mail: nora.rathzel@umu.se

D. Stevis
Department of Political Science, Colorado State University, Fort Collins, CO, USA
e-mail: dimitris.stevis@colostate.edu

D. Uzzell
School of Psychology, University of Surrey, Guildford, UK
e-mail: d.uzzell@surrey.ac.uk

these chapters together into one section because this permits the comparison and contrasting of the development of environmental policies in the three countries from the point of view of some of their protagonists.

## The Usage of Life-History Interviews

For two of the authors in this section (Uzzell and Räthzel), placing research on labour's responses to the climate crisis in the context of production, rather than consumption (which is the normal focus of climate change mitigation research) revealed the important role of individuals in pushing their unions to face the threats of climate change and global environmental degradation. Such issues had rarely been seen as a priority for trade unions, but it was realised that public, industrial and governmental responses to climate change through policy, legislation or action could have a serious impact on work and jobs, and therefore, it behoved them to be part of the planning and decision-making process to ensure there is a just transition. We met many exceptional people in this early work, and so we decided that a new study should focus on these individuals, in various cultural contexts, to try and understand how they became organic intellectuals. To achieve this, we carried out life-history interviews in six countries: Brazil, India, South Africa, Spain, Sweden and the UK. The chapters on Spain and the UK in this section are derived from this research (for Brazil, see Leandro, Vieira Trópia and Räthzel, Chap. 15).

Encouraged and influenced by the work of Uzzell and Räthzel, Stevis modified his interviews on US, Canadian and global labour-environmentalism and environmental unionism to more fully capture the life stories and work of a number of unionists and environmentalists involved in environmental labour activities. Here, he draws on the material from the USA.

Life-history interviews are insightful for our understanding of change and the subtleties of social, economic and historical influences on social practices over the lifespan. Interviewees were asked to relate their life story beginning with their childhood and early years and including the familial, spatial and societal contexts in which they grew up. They then walked us through their life story. However, while we occasionally sought clarifications and elaborations, we rarely asked questions directly related to the environment, although they knew we were interested in union strategies regarding climate change and individual trajectories into union work and environmental issues. Where interviewees brought up the theme, it was because they regarded it as salient for them. The US data were collected differently. The interviewees were directly asked to relate their life-history in the context of their environmental engagements in the trade unions (or environmental organisations). However, the interviewees were also asked to discuss how other aspects of their life developed over time.

Life-histories provide the opportunity to learn something about the ways in which people experience themselves as actors and act upon what they see as limitations, successes or failures. Ethical principles of research based on life-histories require asking interviewees whether they preferred to remain anonymous or wanted us to use their real name. In the latter case, we could add their position and job description; in the former, we had to omit this, since otherwise they would have been identifiable.

## Comparing and Contrasting Trade Unions in Three Countries

We have chosen Spain, the UK and the USA to explore how the environment and environmental policies are understood and how this relates to the unionists' life stories and the political conjunctures within which they acted. We have chosen these three countries because the history of trade unionism in each is significantly different.

We want to highlight that there is not just one way, and especially not just one successful way, in which unionists can persevere in putting the environment on the trade union agenda. There is a range of possibilities, and they vary according to time, place, individual capabilities and organisational capacity. We follow Macfarlane's (2006) idea that we need both to *contrast and to compare* in order to understand how certain social relations develop. To be able to do this, we need to explore relationships that are similar as well as different. As Marc Bloch emphasised: 'there is no true understanding without a certain range of comparison; provided, of course, that comparison is based upon differing and, at the same time, related realities' (Bloch 2002, 42).

The basic similarities between the UK, US and Spanish unions are obvious: Their role is to negotiate with and employers and the state (whereby the state is sometimes there as an employer) in order to improve the wages and working conditions of their members and the workforce in general (for comparisons of industrial relations, see Frege and Kelly 2020; Bieling and Lux 2014).

Beyond that the three countries differ in terms of the organisation and politics of the union movement. In the USA and the UK, there is only one national confederation, and in Spain, there are three larger and a number of smaller ones. The Trades Union Congress (TUC) in the UK is closely connected to the Labour Party. Trade unions elect 12 of the 32 members of the Labour Party's National Executive Committee and 50% of the delegates to Labour Party Conferences. However, only a minority of trade unions (although the largest ones in terms of membership) are affiliated with the Labour Party and support it financially. In terms of their political perspectives, the federations in the UK cover a broad spectrum, from transformative to business unions. In the USA, the unions are less diverse with most unions being business unionist or weak reformists. The industrial relations in the USA are more liberal and more antagonistic towards unionisation than in the other two countries. In Spain, the Comisiones Obreras (CCOO, Workers' Commissions) can be considered as transformative, while the Union General de Trabajadores (UGT, General Workers' Union) would tend to be more social integrationist in Hyman's (2001) terms, or reformist to use another concept. From another perspective, Köhler and Calleja Jiménez (2013) argue that the Spanish model 'differs from the Anglo-Saxon and the corporatist central and northern European types in terms of its low union density, frequent mobilisation and strike activities' (ibid., 1). There is competitive bargaining in Spain and the state does influence industrial relations through laws and regulations.

The trade unions in the UK emerged out of smaller associations of workers at the beginning of the industrial revolution in the latter half of the eighteenth century, although these were subject to state and employer suppression through the Combination (1799) and Combination(s) of Workingmen Acts (1824; 1825). It was only in 1871 that the Trade Union Act legalised unions for the first time and formed the basis of subsequent UK labour law. The TUC, which is the federation body representing the majority of trade unionists in the UK,[1] was formed in 1868 in the heartland of industrial Britain. It now has approximately 5.5 million members in 48 member unions representing workers across all sectors of the economy in blue- and white-collar jobs, science and the professions (Aldcroft and Oliver 2017; Reid 2005).

In the UK, union density between 1998 and 2018 declined from 30.1% to 23.4%, which was a fall of some 22.3% (OECD 2019). This compares with the two other countries discussed in this section: Spain from 17.9% to 13.6% (a decline of 24%), and the USA from 13.4% to 10.1% (a fall of 24.6%). The rise of neoliberalism as a dominant economic and social model, globalisation and the increasing vulnerability of workers to the effects of these processes have ensured that trade unions in recent decades have been placed at a

---

[1] Not all unions and workers are affiliated to the TUC. Total union membership in the UK is approximately 6.4 million (https://www.tuc.org.uk/national/about-tuc), with a union density of 23%.

disadvantage in the face of increasingly confrontational employers and governments. This has resulted in the relative decline in wages ('a race to the bottom'), growing income inequality, weakened employment rights, a rise in precarious employment conditions and decreasing influence in negotiating wages and conditions—all of which has led to a perceived lack of effectiveness. Paradoxically, this was at a time when the labour force across most OECD countries increased both in absolute terms but also in many countries as a proportion of the population. Union density fell across all industries except 'wholesale, retail, trade and motor repairs' which provides an indication of where the thrust of present-day economic activity lies. While these 'broad-brush' statistics paint a picture of a declining interest and support for trade unionism, underlying these figures is other positive trends. For example, in the UK, trade union density by gender shows that women's trade union membership has remained fairly stable over the last decade and is higher than men's.

US unionism can be traced to the middle nineteenth century (for historical overviews, see Hogler 2004; Katz et al. 2017, Ch. 2). All political tendencies—radical, reformist and liberal—were available from the very beginning. By the end of World War I (WWI), however, the American Federation of Labour (AFL) consisting of craft-based, business unionist unions had emerged supreme. The suppression of the left wing and syndicalist unions by the government, during and after WWI, helped in this. During the 1930s, the AFL's hegemony was challenged by industrial unionists and unions who came together in the Congress of Industrial Organizations (CIO). The CIO brought together reformist and radical elements. After WWII, however, it was purged of its radical elements as part of the deepening Cold War (Luff 2018). The two joined to form the AFL-CIO in the mid-1950s, by which time the AFL had recovered its lost ground. While the CIO unions continued to exert influence during the 1960s and 1970s, they started losing their influence as a result of globalisation—to which the manufacturing sector was more vulnerable. The AFL unions—largely building and construction—emerged dominant but, in recent decades, they have been challenged by service sector and public sector unions, the latter being the most highly unionised, who often represent a more diverse working class. In general, there are no major unions in the USA that are programmatically socialist or communist, and the vast majority of the US labour movement oscillates between liberalism and weak reformism. In addition to its fairly limited political ambition, US unions are also confronted with significant overall membership decline. Density in the private sector is now around 6% of the labour force. Including the public sector, it is about 11%. In either case numbers are well below those in the 1970s when density was around 25%. This combination of limited political ambition and declining membership has translated into diminishing influence in US politics.

US industrial relations are the model of liberal industrial relations and closer to those of Canada and the UK rather than continental European industrial relations which tend to be more corporatist (for US industrial relations, see Katz et al. 2017). To begin with, there are no national level bargaining agreements—involving unions, the state and capital—that establish wages, working hours or other important standards for all workers. Such policies are the result of political pressures and policy-making like every other issue. Also, sectoral negotiations are used only with respect to railroads which are a very small portion of the US working class. During the post-WWII era, and up to the emergence of neoliberalism during the 1980s, there were a number of multiemployer agreements, and these covered all employees in these companies. These agreements have fallen victim to the strong anti-union dynamic in the USA, leading collective bargaining to increasingly devolve to smaller units, such as plants or even particular sections within plants.

In Spain, trade unions were established in the nineteenth century. Today's unions are defined as 'historical' and 'new' unions. Concerning the two largest ones, the UGT belongs to the 'historical workers movement'. During the Franco dictatorship, they were forced into exile and lost contact to labour in Spain. The CCOO is a new union, which formed during the dictatorship, organising strikes for workers' rights but also against the repressive political system. They were forbidden in 1966 and went underground. After the fall of the Franco regime in 1975, they were the only democratic mass movement in Spain (Köhler and Calleja Jiménez 2013). In 1977, the unions, including the UGT who had returned from exile, and smaller unions in the Basque country as well as the Catholic union were recognised as a constitutive part of the democratic process in Spain, something that is enshrined in the Spanish constitution of 1978.

The main three confederations are organised along different political positions. The smallest confederation, traditionally stronger in Catalonia, is the syndicalist Confederación Nacional del Trabajo (CNT). The two larger confederations are the CCOO and the UGT. Both can be considered as being on the left. While the UGT used to be closer to the Socialist Party (PSOE, Partido Socialista de Obreros de España), the CCOO used to be closer to the Communist Party. Both unions were seen and have seen themselves as being in competition to each other. However, today, as their political affiliations have weakened, they work closely together and have for instance jointly organised general strikes (Bieling and Lux 2014).

In 2018 there were approximately 2.2 million union members in Spain with a union density of 13.6%. The CCOO and UGT have around one million members each. The former consists of 11 affiliates from different industrial sectors and territories, and the latter has 10 affiliates (de Amorim et al. 2019). Regarding environmental policies, the CCOO has been and still is considerably more active and committed.

The three countries in focus here range from being oriented more towards mobilisations, to more corporatist to more liberal industrial relations. While we do not present them as ideal types of different varieties of capitalism, combined they provide a balanced portrait of labour politics in the industrial world.

# References

Aldcroft, Derek H., and Michael J. Oliver. 2017. *Trade Unions and the Economy: 1870–2000*. Taylor & Francis.

Amorim, Wilson Aparecido Costa de, Andre Luiz Fischer, and Jordi Trullen. 2019. A Comparative Study of Trade Union Influence over HRM Practices in Spanish and Brazilian Firms: The Role of Industrial Relations Systems and Their Historical Evolution. *International Studies of Management & Organization* 49 (4): 372–388. https://doi.org/10.1080/00208825.2019.1646487.

Bieling, Hans-Juergen, and Julia Lux. 2014. Crisis-Induced Social Conflicts in the European Union—Trade Union Perspectives: The Emergence of "Crisis Corporatism" or the Failure of Corporatist Arrangements? *Global Labour Journal* 5 (2). https://doi.org/10.15173/glj.v5i2.1156.

Bloch, Marc. 2002. *The Historian's Craft*. Repr. Manchester: Manchester University Press.

Frege, Carola M., and John E. Kelly. 2020. *Comparative Employment Relations in the Global Economy*. London; New York: Routledge, Taylor & Francis Group.

Hogler, Raymond. 2004. *Employment Relations in the United States: Law, Policy, and Practice*. Thousand Oaks, CA: Sage.

Hyman, Richard. 2001. *Understanding European Trade Unionism: Between Market, Class and Society*. London; Thousand Oaks, CA: SAGE.

Katz, Harry Charles, Thomas A. Kochan, and Alexander James Colvin. 2017. *An Introduction to U.S. Collective Bargaining and Labor Relations*, 5th ed. Ithaca, New York: ILR Press, an imprint of Cornell University Press.

Köhler, Holm-Detlev, and José Pablo Calleja Jiménez. 2013. *Trade Unions In Spain. Organisation, Environment, Challenges*. International Trade Union Policy. Berlin: Friedrich Ebert Stiftung. http://library.fes.de/pdf-files/id-moe/10187.pdf

Luff, Jennifer. 2018. Labor Anticommunism in the United States of America and the United Kingdom, 1920–49. *Journal of Contemporary History* 53 (1): 109–133. https://doi.org/10.1177/0022009416658701.

Macfarlane, Alan. 2006. To Contrast and Compare. *UC Irvine*, Social Dynamics and Complexity, 94–111.

OECD. 2019. Trade Union Density 1998–2018. OECD.Stat. https://stats.oecd.org/Index.aspx?DataSetCode=TUD

Reid, Alastair J. 2005. *United We Stand: A History of Britain's Trade Unions*. Harmondsworth: Penguin.

# 26

# Embedding Just Transition in the USA: The Long Ambivalence

Dimitris Stevis

## Introduction

In early 2019 Senator Markey and representative Ocasio-Cortez submitted a Green New Deal (GND) Resolution that included just transition, while Senator Sanders had a well-developed just transition policy on his platform (U.S. Congress 2019; Sanders 2020). US national unions, with very few exceptions, have distanced themselves from the GND resolution, while none has adopted a just transition strategy. To the degree that local and state-level unions employ the term, they are the exception rather than the rule.

This rule is underscored by the fact that the BlueGreen Alliance, an organization of environmental and labour organizations that has been in existence since 2006, very hesitantly connects green and just transition. Its recent Solidarity for Climate Action (BlueGreen Alliance 2019) states: 'Working people should not suffer economically due to efforts to tackle climate change. The boldness of any plan requires that the workers and communities impacted are afforded a just and viable transition to safe, high-quality, union jobs....' But that is also the closest that it gets to using the term in a document that recognizes the impacts of social and environmental inequalities.

During this same period, the AFL-CIO, the national confederation of sectoral unions, has produced two major policy documents. The first was the

D. Stevis (✉)
Department of Political Science, Colorado State University, Fort Collins, CO, USA
e-mail: dimitris.stevis@colostate.edu

report of the federation's Commission on the Future of Work and Unions (AFL-CIO 2019). The second is Energy Transitions (AFL-CIO and Energy Futures Initiative 2020) and is the product of collaboration between the federation's Energy Committee, formed in 2013 and led by President Cecil Roberts of the United Mineworkers, and the Energy Futures Initiative, led by Ernest Moniz, Obama's second and last Secretary of Energy. While 'transition(s)' appears frequently in the text, there is no reference to just transition, and the focus is on the preservation and creation of energy jobs as a solution to the challenges of the green transition.

Collectively, these three approaches demonstrate that environmental priorities, of one kind or another, have found their way into US labour (Stevis 2019). But they also demonstrate that the tensions within labour over whether and how to embed just transition within a broader politics remain deep (see Vachon, Chap. 5 in this volume). The first group includes those that seek to embed just transitions within a broader socioecological politics. The second consists of those that tentatively seek to embed justice, if not just transition, within green industrial policy articulated around all forms of energy, manufacturing and physical infrastructure. The third group brings together those that limit themselves to a narrower industrial policy whose goal is to accommodate all forms of energy, including fossil fuels and nuclear power. Hidden in the last group are some unions that actually consider environmentalism to be foreign to labour politics. Over the last 50 years, these tendencies have fought over shaping US labour environmentalism.

In this chapter I will trace the politics of US labour environmentalism and just transition through the experiences and efforts of five people that largely fall in the first two categories. What brings them together are their efforts to steer the US labour movement towards environmentalism. What differentiates them are the ways in which they sought to fuse green and just transitions.

In order to do so, I ask and explore the following questions: What led you to labour environmentalism and just transition? How did you address labour environmentalism and just transition? What were the results? What explains these results? The basic argument of the chapter is that the trajectory discussed here is a combination of organizational politics, historical conjunctures *and* the commitments and efforts of these individuals and their immediate networks. In short, that while individual attributes are not sufficient, they have proven necessary and cumulatively impactful. If nothing else, this research has helped me appreciate the diversity of the views and efforts of these people against quite formidable and, often, dispiriting odds.

## Method: The People and Their Backgrounds

While all of the five labour environmentalists whose lives and work are discussed in this chapter belong to the same age cohort, and have known and even worked with each other, their primary activities related to the environment have largely followed one another chronologically, with one exception, allowing me to trace the politics of labour environmentalism and just transition across time. These people collectively lived, worked for and shaped—in their successes and failures—contemporary US labour environmentalism, including the origination, elaboration and contestation of just transition. The five core interviews lasted from 60 minutes to a number of hours. The shorter interview was complemented by access to that person's archives and two days of both of us going through them and talking about the people and events involved, as well as that person's life story. All of the interviewees were unionists, and four of them are men.

In addition to these core interviews I have interviewed, in great length, another 11 US and Canadian unionists (mostly), environmentalists and environmental justice (EJ) activists that collectively cover the period explored here and fill in a number of the gaps or allow me to better evaluate the core interviews.[1] In the context of two other projects, I also interviewed a significant number of additional unionists and environmentalists, and their views also inform my understanding of contemporary US labour environmentalism.

In many parts of the world it is reasonable to expect that labour activists may have been influenced by some version of socialism or communism. This is not the case for the overwhelming majority of US unionists that have risen up through the ranks. Almost all US unions were purged of communists and radical socialists during the late 1940s with the start of the Cold War, and no leader of a major union ever since has been elected as a communist or a radical socialist. Some social democrats continued to play a role, but social democracy has also remained marginal and divided as an organized movement. Radicalism did not disappear, however, and was reflected in various movements—such as anti-war, environmental, race and gender equality, regulatory reform, and democratic socialism that re-emerged or emerged during the late 1960s and early 1970s. All of the people here have been influenced by one or more of these movements. A second dividing line in the deeply anti-union milieu of the USA is whether someone comes from a union family or not. While this does not always predict one's politics (many US unionists are

---

[1] The Canadians interviewed were in close contact and collaboration with the US unionists discussed here or worked for transnational (US-Canadian) unions.

conservative), it does often influence the way they look at the relations between employers and employees. Moreover, those whose family may have come from the industrial unionism of the Congress of Industrial Organizations (CIO) are likely to have a more class-based approach towards industrial relations than those who come from the American Federation of Labor, which has always been supportive of liberal capitalism that gives workers their 'fair share'. All of the people discussed here have come from the CIO tradition.

## The Rise, Decline and Tentative Resurgence of Just Transition in the USA: 1988–2020

Historians have been pushing back the origins of labour environmentalism (Gordon 2004; Van Alstyne 2015; Montrie 2018; Rector 2018; Dewey 2019). During the 1960s and 1970s, a number of unions, such as the United Automobile Workers (UAW), the United Steel Workers (USW) and the Oil, Chemical and Atomic Workers (OCAW), actively supported clean air and clean water policies. A small number of union environmentalists, such as Tony Mazzocchi of OCAW and Jack Sheehan of the USW, increasingly placed occupational health and safety, a central priority of all unions, within a more comprehensive view of the nature-society relationship (Leopold 2007, 226-227; Bennett 2007; Slatin 2009). During the early 1980s, these labour environmentalists, working with through the Industrial Union Department of the AFL-CIO (IUD), a financially and otherwise autonomous division of the AFL-CIO with its own affiliates, essentially the CIO unions, played a leading role in the creation of the OSHA Environmental Network (Gordon 2004, 311–323; OSHA Environmental Network 1984; interview with Joe Uehlein) whose explicit goal was to expand the environmental horizons of occupational health and safety and amplify the linkage between in-plant pollution and community pollution. The IUD posted a job for field director for the network, and the president of the IUD at the time asked a young Joe Uehlein, one of the five unionists discussed in this chapter, whether he would be interested in it.

> And so I became the field director of the OSHA Environmental Network. And that was one fascinating work because we had active networks in 22 States and we fought for and won the right to know legislation at the state level, meaning that workers would have the right to know what chemicals they're working with in the plant and people in the community would then have the right to know.

### 26  Embedding Just Transition in the USA: The Long Ambivalence    595

And then we won federal right to know. So, it was a very successful operation that ultimately had to be disassembled because of acid rain. I remember that Rich Trumka, at the time president of the Mineworkers union, walked across the street…. He came over and he said to Howard [Samuel, the President of the IUD], and I was in that meeting, you have to take OSHA Environmental Network apart. Cause acid rain is gonna steal all our jobs in the coal industry. So, we killed it [mid-1980s].

This unfortunate development broke a promising route that fused labour environmentalism with occupational safety and health, a connection that most union occupational health and safety directors continue to avoid.

## 1988–2001: The Rise and Decline of Just Transition as an Element of Political Change

Labour environmentalism's contentious road did not come to an end. In 1990 the Steelworkers, one of the largest unions, published their seminal report on climate change that continues to be union policy to this day (USWA Environmental Task Force 1992 [1990]; for brief historical overview, see Moberg 1999). The development most closely associated with just transition, however, was the result of OCAW politics. Tony Mazzocchi had run for the presidency of the union in 1979 and 1981 without success. After the second defeat, he did not pursue another national OCAW office until 1988 when he was invited to support one of the two candidates competing for the presidency of the union. This was facilitated by Robert Wages who was running for Vice President before rising to the union's presidency in 1991. The politics of the OCAW during the critical 1990s, therefore, were a product of convergence of the progressive politics of two leaders, and the networks around them, and a small but active union's resources behind it. This story is being told through the voice of Robert Wages.[2]

Wages had grown up in a working-class family in Kansas. His father worked in the local refinery where he also worked from 1969 through the first part of the 1970s, as a young father and while going to college. During that same time, he was elected president of the local OCAW union, a position he left in 1972 as he embarked on law school. He continued to work in the refinery and remained active in union politics while completing his law degree, and it was

---

[2] Mazzocchi died in 2002. Les Leopold (2007), who worked with him since the 1970s, has written a comprehensive biography of his life. That biography has also served as additional background for this part of the chapter.

during that time that, as a result of the 1973 Shell strike, often referred to as the 'first environmental strike', he

> started paying a lot more attention to the implications of working in a very hazardous industry that also led to the realization that when I looked around my facility in Kansas City and at another sister facility that was in Kansas City, Missouri, I noticed that these plants were always in these industrial areas and they were always surrounded by low-income communities and people of colour.

About a year after his graduation from law school in 1975, he was invited to work for the union at its central headquarters in Denver, Colorado. During the late 1970s and 1980s, he rose up the ranks of OCAW as the union lost many members due to the automation and reorganization of the industry and an amicable separation from its Canadian branch.[3] This led to an internal debate over merging with the United Paperworkers International Union (Paperworkers), a union that was sceptical of environmentalists due to its close ties to the pulp and paper industry. The national profile of OCAW, however, remained high during the 1980s as a result of its long campaign against BASF (Minchin 2002), and OCAW was widely known as a progressive union, which explains the next development.

The national union elections of 1988 promised to be very contentious. In order for the progressive slate to win, Wages persuaded Mazzocchi, whom he had not supported in the latter's efforts to become president of OCAW in 1979 and 1981, to join the ticket in the position of Secretary-Treasurer. Wages's decision was certainly motivated by intraunion politics, but this needs to be placed in context. The OCAW was the product of mergers and had long contained a progressive wing along with a wing that had closely collaborated with the CIA. Wages himself had increasingly moved in a more progressive direction. So, approaching Mazzocchi was a result of both political exigencies and, as the subsequent years demonstrated, political commitment (see Leopold 2007, 428–432). In the words of Wages:

> And so part of getting Tony to run was an agreement that we would put together an agenda that encompassed a Superfund for Workers, support for Labor Party Advocates, and this whole progressive agenda was all in one huge resolution that we spent two days at our convention talking about. We had breakout sessions and educational sessions where we tried to teach folks, you know, what our

---

[3] Many unions had US and Canadian membership. During the 1980s, a period of Canadian nationalism, a number of unions, such as the Autoworkers and OCAW, split up. Others, such as the Steelworkers and the building and construction trade unions, managed to remain binational.

objective was. And the idea was to create, at least within us, a political atmosphere that let you talk about these things. And then hopefully our people would take these things back to the various [AFL-CIO] state federations and local federations and work on it. I mean, it included single-payer health care, it included the Superfund for Workers, it included Labor Party Advocates, it included free college, all of these things that looked like Bernie Sanders's platform because that's who we thought we were and that's how we thought you could protect people and that could respond to environmental crises at a particular facility. We thought those were the measures that you took to be able to respond.

The result was that of committing the resources of a smaller but vibrant union of well-paid workers to broad labour environmentalist agenda (OCAW 1991). This would not have happened without Mazzocchi's involvement and vision (Mazzocchi 1993) but also Wages's personal commitment to progressive politics (see Moberg 1999). Mazzocchi did not run for another Secretary-Treasurer term at the next convention in 1991 but chose a position as an advisor to Wages, who was elected President of the union, so he could work on the Labor Party Advocates (LPA) initiative (see Leopold 2007, Chaps. 16 and 17). Over the subsequent decade, the OCAW spent a lot of energy advocating for such a party. The LPA's 1996 agenda embedded Just Transition[4] within a broader set of priorities that are similar, and occasionally more visionary, to Sanders's platform and the current GND. These priorities included the reduction of bigotry, the design of work and reduction of working time, universal health care and more (Labor Party 1996). The reasons why the LPA initiative failed continue to be debated and will not be discussed here (Dudzic and Isaac 2012; Seidman 2014). One factor, but not the only one, was the merger with the Paperworkers and that union's hostility to environmental and progressive politics and ultimate withdrawal of support.

## The 'BlueGreen Working Group'[5]: Diffusing Just Transition to Environmentalists

Keeping in mind that just transition was part of a broader politics, rather than a stand-alone strategy, the OCAW and its allies pursued two just transition

---

[4] The term 'Superfund for Workers' was replaced by 'just transition' in 1995. For a history and sources, see Stevis et al. (2020, 9–14).
[5] The term 'BlueGreen Working Group' was not used until early 1999. Previously, it was often called the Climate or Climate Change Group. Moreover, there were actually more than one group—one for labour, one for environment and one for both.

initiatives during the second part of the 1990s. One was a more grassroots level of collaboration with environmental justice organizations. The other was their effort to infuse just transition into an alliance of unions and environmentalists in which Wages played a central role. The next part focuses on this effort, but a few words about the first are relevant because of its long-term influence.

During the 1990s, the OCAW paid close attention to the emerging environmental justice movement, particularly since some of these organizations were in frontline communities affected by factories organized by OCAW. The turning point was the formation of the Just Transition Alliance (JTA) in 1997. Over the next several years, the alliance brought together workers and frontline communities in a number of locations, mostly in the Western USA, where the OCAW operated or wanted to unionize. The goal was to fuse these two social forces by demonstrating that they faced the same environmental and social harms. According to an activist that has worked for the JTA since its inception, they employed a training technique developed by the Labour and Public Health Institutes—anchors of Mazzocchi's network. The technique involved small group training events that brought together unionists, community members and environmentalists (Public Health and Labor Institutes 2000). The twin goals were to exercise pressure on employers and, more importantly, to bring workers and their communities together in a manner that fused environmental and social justice. While some of these collaborations did not work out as expected,[6] they helped preserve the concept of just transition during the years from 2001 to the formation of the Climate Justice Alliance in 2013[7]—an alliance which the JTA helped create (Harvey 2018 for an overview).

The BlueGreen Working Group was initiated by John Sweeney, the first President of the AFL-CIO to be elected through a contested election in 1996, and came into existence in the midst of two unfolding dynamics. One was the rising in prominence of climate change, an issue that became central to US union politics as a result of the 1997 Kyoto Protocol which the AFL-CIO eventually opposed. While unions have come to accept the reality of climate change, as discussed in the next part, how to deal with it continues to be a dividing issue amongst unions and between unions and environmentalists (Fellner 1998; also Dougherty 1999). The 1990s were also a period of time

---

[6] And have cast their shadow to the present with communities feeling that the OCAW took an opportunistic approach to some of these collaborations.

[7] But also see the discussion of Bob Baugh's efforts to insert just transition into AFL-CIO policies.

## 26 Embedding Just Transition in the USA: The Long Ambivalence

during which unions and environmentalists collaborated against neoliberal trade agreements, particularly NAFTA and the WTO.

It is during that period Jane Perkins's formative political experiences took place.

> I lived in Harrisburg, Pennsylvania. I bought a house in Midtown. I got active in the neighborhood association (…) I ended up doing stuff where I interacted with the mayor's office (…). I worked for SEIU in Pennsylvania and when Three Mile Island happened, I was kind of known as an activist in town. And so the Three Mile Island people came to me and insistent. Would you just listen to us about nuclear power? And I said, I don't know anything and I'm not making any promises, but I'll listen to you. So they (…) schooled me on nuclear power and I became an anti-nuclear public person. And (…) I got elected to city council and we, I pushed the resolutions to keep the second unit shut. (…) So then I get married and my husband got transferred to DC. And (…) then I got head-hunted to be the President of Friends of the Earth, USA.

So, Perkins was unique in the sense that almost all of the other people that I have encountered in this research had held positions in either the environmental or the labour movements and none had served in an elected political position.

> SEIU [Service Employees International Union] John Sweeney was elected president of the AFL CIO [in 1996]. And he and his staff discovered that they were now involved in all of these environmental lawsuits that various environmental groups had asked the AFL CIO to join on as amicus partners.[8] And they didn't know anything about the issues the groups, the people, whatever and could I come in and help John Sweeney make sense of his new relationship with the environmental movement? So I did that and I was like assistant to the president and I just set out to find out what do these affiliates think about environmental issues and on our side of the table, what are the issues that people would even want to talk to the environmentalists about and Sweeney encouraged me to go ask the questions and to make a report to him and to make recommendations.

Starting in late 1996 there were regular meetings amongst labour, amongst environmentalists and between them that aimed at constructing a common political agenda. Whether intended to or not by Sweeney, Perkins and her closer collaborators, such as Alden Meyer of the Union of Concerned Scientists, steered the BlueGreen Working Group in the direction of building

---

[8] In US law, parties not involved directly in a suit can submit supportive amicus briefs.

a labour-environmentalist politics that included provisions for a just transition and a green industrial policy. A highly visible meeting in April of 1999 reflected this convergence and alarmed those unions that were participating in the Working Group largely to prevent climate policy. The three main dividing lines, on the part of some unions, were the very idea of talking to environmentalists, the implications and content of a just transition policy to deal with climate change[9] and the reality of climate change, itself. The promising synthesis collapsed during the next two years, and relations amongst unions and between some, but not all, unions and environmentalists became increasingly more contentious.[10] A meeting was called for September 11, 2001, to see whether these problems could be addressed. In her words:

> (…) we had a couple of press conferences where all the lefty unions would come and endorse whatever it was around just transition. And finally there was scheduled a meeting of the heads of environmental groups and the heads of affiliated unions for John Sweeney's office (…) And it was like, this is the result of my work. (…) and that meeting happens on 9/11. (…) I was sitting downstairs and I am constantly thinking about what was going on upstairs. And the TV was on—an airplane was flying into the towers of New York, and I'm looking out the window and a black cloud rose up over the Pentagon. No, well that was it. There were no more meetings. I went outside and I watched all of these environmental leaders just file out. (…) because everybody was evacuating downtown DC. And that was it, man. That was it.

Underscoring the deep differences, the meeting was not rescheduled. As discussed in the next section, US unions did come to accept the reality of climate change and a number of them did decide to collaborate with environmentalists. Just transition, however, remains contentious to this day, with most unions avoiding the term. Then, and now, just transition sends a doubly troublesome message, particularly to fossil fuels unions as well as to building and construction unions, the latter largely conservative and business unionist. The first is that transitions do have to take place because some of the products and processes that unions, and their employers, produce are part of the problem. The second was the democratic socialist politics within which the OCAW's and, to a lesser degree, the BlueGreen Working Group's versions of

---

[9] One of the Group's proposals was a comprehensive and far-reaching set of JT principles (Ms in author's possession).
[10] This brief account is based on both my discussions with Jane Perkins and access to the personal archive of a participant.

just transition were embedded. That same tension is evident today in the reaction of most unions towards the GND.

In light of these deep divisions it is unlikely that the version of labour-environmental politics that the BlueGreen Working Group espoused would have survived. However, the nationalism following 9/11 was even more inimical to democratic socialist politics. On top of that, the Paper, Allied-Industrial, Chemical and Energy Workers Union (PACE)—the result of the merger of OCAW and the Paperworkers—stopped supporting everything that OCAW had been trying to accomplish. Wages now recognizes the merger as a major error. We cannot know what would have happened, but we do know that the 2005 merger between PACE and the USW was a difficult marriage between an anti-environmentalist and a more environmentalist union.

## ~2001–2012: The Promise of Green Industrial Policy

The debates in the Working Group were not only between those who supported just transition and those who opposed it. There was a third tendency, or cluster of tendencies, shared by the manufacturing unions, like the USW (a very internally diverse union) and other unions that were affected by globalization. These unions were particularly interested in regulating trade through the inclusion of labour standards with the goal of protecting domestic industry and reshoring manufacturing. In this effort they had found common ground with environmentalists who promoted the inclusion of environmental standards—a collaboration that became very prominent during the 1999 demonstrations in Seattle, against the WTO. Increasingly, this alliance coalesced around a green industrial policy that could well be beneficial for the climate and thus attractive to environmentalists.[11] In addition to manufacturing unions, such a strategy could be attractive to infrastructure unions (such as transportation and communications) as well as some building and construction unions. It is this tendency that emerged dominant, in some form or another, in the decades that followed—variably more focused on energy, manufacturing or infrastructure, depending on the parties involved.

A first iteration of this synthesis was the Apollo Alliance of labour, environmental, political and business leaders who started exploring the formation of a programme that would provide the USA with a vision for the future, similar

---

[11] This emergent synthesis was captured in a report by Barrett et al. (2002) who worked for the EPI and the Center for Sustainable Economy and who had served as professional staff for the BlueGreen Working Group. The report was delayed and finally not published by the AFL-CIO because of opposition by specific unions.

to its space programme. The Apollo Alliance was formally launched in June of 2003 in a major conference which included unions, environmentalists, political leaders and others associated with the centre and centre-left of the Democratic Party. Parallel, but not unrelated to this initiative, there emerged two others that are discussed below through the words of two persons central to each one of them. The AFL-CIO's Energy Task Force involved only unions, while the BlueGreen Alliance brought together unions and environmentalists.

## The Workers Themselves: The AFL-CIO's Energy Task Force

Bob Baugh's political and life choices were shaped by the period and place where he was growing up—Detroit during the late 1960s and early 1970s—when he was studying for his undergraduate degree while also working in steel and auto plants.

> I always loved the outdoors and I loved the lakes in Michigan and was aware of environmental stuff just because of the pollution and stuff there. And I worked in steel mills and I was down at Zug Island, which was a god-awful place. You'd leave after work and you'd be covered in this iron oxide and it'd come out of your nose and you just, that's not good, not good for me. So I was aware of it from that, but ultimately I ended up getting involved in union stuff, becoming aware because of the Detroit race riots of 1967 that were life-changing for me in terms of social justice and engagement. I was 17. I was in the middle of them. I was driving for Interfaith Relief Services and I saw shit that people even today have no idea how big and how violent and how poor people were. I mean, it really changed me. (…) What I really wanted to do was to go and get more knowledge and go to work in the trade union movement—be an asset.

These early experiences led him to both appreciate the environment and realize that without the means to a good life enabled by unions, it was not possible to appreciate it or other important rights. After completing his Master's degree in Oregon in 1976, he moved to St Louis for personal reasons. During his two and half years there, he joined the Board of the Missouri Coalition for the Environment , worked for the Service Employees International Union and helped found the chapter of the socialist New American Movement.[12] His straddling of the labour, environmental and

---

[12] In the early 1980s NAM merged with the Democratic Socialists of America, a party that has grown significantly during the last several years as Bernie Sanders's ideas have received more traction.

progressive movements allowed him to bring environmentalists, unions and community advocates together around some key issues. One was a campaign against the use of lead in housing supported by unions, environmentalists and community activists, particularly African Americans. While in St Louis, he also worked against plant closures, an issue that remained central to his work. By the late 1970s, therefore, he had formulated a worldview and mode of action that was informed by environmental and social concerns.

In 1979 he moved to Oregon to work for the International Woodworkers as their research and education officer (on Woodworkers, see Loomis, Chap. 6 in this volume). At the time the Woodworkers were led by Canadian New Democratic Party leaders were critical of oil and opposed to nuclear energy and negotiated the 1978 Redwood Employee Protection Program, a just transition policy in substance if not in title (DeForest 1999; JTLP 2021). This is all the more important because during the late 1980s and 1990s, after the US and Canadian branches split, the union moved in the anti-environmentalist direction that fuelled the conflicts over logging in the Pacific Northwest of the USA. This was an issue that became central to his work when he was elected to the position of Secretary-Treasurer of the Oregon AFL-CIO and, then, appointed into a public policy position.

> The spotted owl stuff, that came in the middle to the later eighties and I was secretary-treasurer of the [Oregon] AFL-CIO at the time, but then ended up going into government and I ended up being the governor's lead on the spotted owl stuff[13] [see Loomis, ch.6] and the timber task force that worked with the three States out on the West coast of figuring out what's the response that we have to this from an economic development perspective, how do you use every federal agency and every resource possible to assist communities in distress? I mean, you wanna know what just transition was, it was dealing with this…That's exactly it. Then I didn't know it but that's all the plant closure work we did was all around these issues. (…) And environment would often be used as a wedge by corporations when they could use it to peel off unions and, and fight an environmental issue, claiming you have to choose between jobs and the environment kind of stuff.

As noted above, Baugh had been involved in the politics of plant closings during the 1970s and he saw the logging conflicts in the same vein. His

---

[13] During the 1980s and 1990s, the Woodworkers and other unions entered into a major conflict with environmentalists over industrial logging. The pressures for intensified logging came from global competition and demand rather than environmental policies. However, the issue was cast in terms of the protection of an endangered species—the spotted owl—and resulted in deep conflict between unions and environmentalists, something that the industry encouraged.

government policy work, moreover, further reinforced his belief in the need to negotiate economic development policies in collaboration with like-minded politicians, unions and business. In 1993 he was recruited to be a special assistant to the executive director of the national AFL-CIO Human Resource Development Institute, renamed the Working for America Institute in 1996. His portfolio was an extension of his previous work and focused on training, technology and economic development. While he was familiar with the BlueGreen Working Group—and had talked to Perkins about his experience in Oregon—he did not actively participate in its work. In 2003 he took over as executive director of the federation's Industrial Union Council—the weak successor to the Industrial Union Department (IUD).[14] In 2006 he became the chair of the federation's Energy Task Force, and from 2006 until 2013, when he retired, he led the US union delegation to the various COPs.

Baugh was familiar with the development of the Apollo Alliance because one of the key movers behind it came from the Working for America Institute. As I noted earlier, a number of unions, largely in fossil fuels and building and construction trades, were only hesitantly speaking with environmentalists. As a result, the AFL-CIO started discussing the need to have an initiative of its own rather than simply sign onto the initiatives of others. The result was the formation of Energy Task Force which he led as uniquely qualified due to his experiences within the labour movement as well as government. His major accomplishment was to establish consensus that climate change was a problem, one of the three main divisive issues within the BlueGreen Working Group (the other being just transition and collaborating with environmentalists). As he summarized it:

> My argument from the get-go around just transition was that so many people, non-manufacturing people, interpreted just transition, that just means we have to take care of those poor people that lost their jobs. (…) I said, I can't sell that, number one, and it's a bad strategy, number two, you really need to think holistically. Yes, you need to take care of people who've lost their jobs. And the social infrastructure under that that's needed. But if that's where you start, that's the end. You gotta go back to the beginning and say what are we doing about job creation? (…) I mean, you really need an economic strategy, not just one that, kind of takes care of some folks who lost their jobs, which you have to do too.… I was always big on that angle.

---

[14] As noted, the IUD was an autonomous and self-funded department within the AFL-CIO and represented the CIO unions. After its dissolution, it was replaced by the IUC which was neither autonomous nor self-funded.

## 26 Embedding Just Transition in the USA: The Long Ambivalence

The Great Recession, already unfolding during the Fall of 2007,[15] added urgency to the green industrial policy that was at the heart of the Task Force (overviews of its work at Industrial Union Council 2009; AFL-CIO Department of Government Affairs 2009). The 2008 election of Obama and of a Democratic Congress led to the adoption of the American Reconstruction and Reinvestment Act (ARRA) of 2009, which made available significant resources towards a green economy, including green job training, funds for research and development, and loans to green industry. In short, there was an unexpected conjuncture between the AFL-CIO's move towards green industrial policy, articulated around energy, the election of a centrist administration that was somewhat open to unions and environmentalists, and a recession calling for state intervention (Baugh and Rickert 2010).

> Well, I think we took advantage of that. The sad part was we never did a part two. We never got the second bundle of money, which everybody thought, because… here's the stuff we didn't do the first time we really needed to finish or we needed more resources, but it was timely and it was good. The trade union movement was aligned in what we should do and how we could do it. And that there could be a climate bill in the future with resources dedicated towards this stuff, if we did it the right way.

The agreement amongst US unions that climate change was real and that some kind of response was needed also had global implications by helping keep just transition on the global agenda and, less prominently, on the US union agenda. Up to the Bali COP of 2007, climate negotiations were routinely attended by the Mineworkers and the IBEW, who were opposed to climate policy. On the other hand, Joe Uehlein represented the IUD from 1988 through Kyoto and became increasingly discouraged by the AFL-CIO's opposition and the lack of progress (Uehlein 2014). With critical support from particular US activists, Baugh was able to engage more US unions into the COP process leading to a profound improvement in the relations between global unions, unions from other countries, and US unions.[16] Over a period of three years, the goal was to insert just transition language into the climate policy agenda. Based on some progress at the 2009 Copenhagen COP, with

---

[15] The recession 'formally' started in late 2007. Its after-effects were felt through 2011 and, in some cases, beyond.

[16] The ITUC had taken over the cause of just transition and had made it a goal to turn it into the global labour movement's common policy agenda. As early as 1998, just transition language was very prominent in the statement of the ICFTU at the 1998 COP. Subsequent statements tended to highlight the promise of the green economy. The formation of the ITUC and the Trade Union Assembly on the Environment, both during 2006, helped turn just transition into a formal global union policy (Rosemberg 2020, 34–38).

support from the US and Argentinian governments, just transition language was adopted at the 2010 Cancun COP (AFL-CIO 2012). The AFL-CIO and US unions also started discussing just transition domestically, both amongst themselves and with policy makers. The 'thrust of our green technology investments was all about the job creation/ecodevelopment side of just transition'. This was strongly reflected in Resolution 10 adopted at the 2009 Convention of the AFL-CIO (AFL-CIO 2009). Unfortunately, the failure to adopt a climate policy during the 2008–2010 period when Democrats controlled Congress and the Republican victory of 2010 closed that window of opportunity. According to Baugh:

> There were several bills on each side in that period of time. But basically after the 2010 election. (…) I still remember, I went to our big weekly management meeting and the [AFL-CIO] president's assistant turned to me and said, Bob, what's your take on all this. I kind of looked around the table and I said, well, brothers and sisters, let me tell you like this we're…. we are out. I said what's really bad now we're going to get regulation without investment. And what will happen is it will split the trade union movement. It will split labour and the environmental movement where we'd have this big, agreed upon investment agenda.

Baugh recognized the critical role of the state and its resources in moving forward a green transition that included just transition provisions for workers and communities. Alternatively, unions would again line up with employers and environmentalists would return to supporting green capital with limited concern about the social impacts of the green transition. In 2013 the AFL-CIO set up its Energy Committee, chaired by the president of the Mineworkers, a union that had long been opposed to climate policy and just transition, at least publicly. Even though the AFL-CIO did adopt a climate resolution at its 2017 Convention, the 2020 report of the Committee avoids the question of climate policy and just transition. Belatedly the Mineworkers have adopted a just transition platform, while avoiding the term, that inlcudes all energy sources and, incidentally, dismissively associates just transition with environmentalists (United Mine Workers of America 2021).

It is tempting to take a wholly critical approach towards the AFL-CIO's limited ambition towards climate change and its avoidance of the language of just transition domestically. To a large degree, this is due to its weak confederal organization and the negative attitudes of a number of unions in the fossil fuel and building and construction trades. However, Bob Baugh's work has left its imprint in AFL-CIO'S recognition of climate change as a real problem, in helping make just transition a global union narrative, and in showing that just transition, embedded in green industrial policy based on all energy sources, could be allowed, if not accepted, by unions historically opposed to it. The tensions since have not been over whether climate change is a problem but about how to solve the problem when state funding is not available. These disagreements are not secondary and, in fact, may be existential about the future of the US union movement. Its continuing avoidance of the concept of just transition, even as part of a green industrial policy that includes all energy sources, is evidence of that.

## Green Manufacturing to the Rescue: The BlueGreen Alliance

The BlueGreen Alliance of unions and environmentalists was formally launched in 2006 and continues in existence today. Central to the formation of the BGA were the Steelworkers whose 1990 report on the impacts of climate change remains union policy to this day (USWA 1992 [1990]). Not only was climate change real but, combined with globalization, it was a mortal threat to manufacturing unions, unless they turned it into an opportunity. The strategy that the union chose was to look for allies in the fight against trade agreements that eroded its position while crafting a role in the green economy of the future. David Foster played a central role in this.

Foster got involved in union politics during the early 1970s when he was working for the postal service. After a number of years in California and the US West, he moved back to Minneapolis where he worked in various capacities in locations organized by the USW. During the 1980s he was actively involved in a movement within the union to push towards more militant responses to the neoliberalization of the industry which was destroying the union. From that involvement, he ran and was elected Director of District 11 which covered 13 states from Minnesota to the Pacific Coast. Once elected, he was 'very quickly forced to look at why the global economy was putting such pressure on unions in these highly globalized industries like iron, or steel, tire, aluminium…'.

This led him to start looking for allies to reregulate the global economy and

> the only people who really seemed to care about globalisation at the same level that those of us in the manufacturing side of the labour movement did was the environmental movement. You know, they cared about it for many of the same reasons—of course global warming was a global issue—but the mobility of capital created the same problems for environmental regulations as it did for labour

standards. It was a race to the bottom. And in most cases the companies that were the most aggressive in fighting unions were likewise the most aggressive in resisting environmental regulation.

The origin of the BlueGreen Alliance was in the WTO protests of 1999 and the Kaiser Aluminum labour dispute from 1998 to 2000 (NW LaborPress 2000). The labour/environmental collaborations built from the WTO and Kaiser led to the formation of the Alliance for Sustainable Jobs and the Environment (ASJE), co-chaired by David Foster and David Brower, former Executive Director of the Sierra Club, founder of Friends of the Earth and then head of the Earth Island Institute (Paskus 2004). ASJE functioned for several years on the West Coast and was led by former Indiana Congressman Jim Jontz. Foster was familiar with the BlueGreen Working Group initiated by the AFL-CIO, although he was not directly involved until 2002 after the AFL-CIO had dropped out of that effort. In the early years of the 2000s, District 11 managed a pilot project for the rest of the USW, modelling how individuals could become 'associate members' and receive access to services and participate in the union's political campaigns. Environmental issues were key to forming an alliance between the willing amongst unions and environmentalists. During the early years of the millennium, the Steelworkers created state-level BlueGreen alliances to pursue specific political ends, such as a collaborative effort to re-elect Senator Paul Wellstone in Minnesota in 2002 and defuse labour/environmental conflicts over the Boundary Waters and mining regulation. After the Democrats lost the 2004 elections, the USW decided to change direction. According to Foster:

> When the election went against us so profoundly, I asked the Steelworker Executive Board to think about reforming the relationship between the labour and environmental movements.... And the Steelworkers leadership was very supportive of that. So we called a summit meeting of principal environmental leaders in Washington, DC, along with about half the executive board of the Steelworkers union and we spent a day together talking about how we needed to change the relationship between the labour movement and the environmental movement if we were going to have any chance at turning back the conservative anti-regulatory tide in the country.... At the end of that meeting we decided what we were going to form a BlueGreen Alliance, whose purpose was to change the strategic relationship between the labour and environmental movements and I was asked to head up this effort.

After 16 years as Director, he decided not to run for re-election within the Steelworkers union so he could concentrate on the formation of the BGA which started with the USW and the Sierra Club but with the clear understanding that membership would expand in the future (for early years of BGA, see Foster 2010). The great success of the 2008 Green Jobs Conference that took place in Pittsburgh with over a thousand attendees led to the rapid expansion of the BGA. The deepening crisis and the increasing prospects of Obama's election provided a great deal of the motivation behind this enthusiasm because there was certainty that a major stimulus bill would be adopted that would deliver significant numbers of green jobs. In Foster's words

> …the issue then became how green would that stimulus be. We participated globally with the United Nations Environment Programme (UNEP) whose leader coined the original usage of the term "Green New Deal" and with the Center on American Progress (CAP) in the last half of 2008. We worked with CAP with the release of an important report that modelled how many more jobs would be created through investments in renewable energy, energy efficiency, and a variety of other clean energy initiatives than traditional tax cuts and stimulus spending or investment in oil and fossil fuel industry [Pollin and Wicks-Lim 2008]. Achim Steiner, the head of UNEP came to DC to speak at our 2009 Green Jobs Conference and delivered the same message. I spoke at UNEP's biennial conference and delivered the same message in Nairobi. We went from talking about these issues theoretically to asking all governments to invest 25% of their stimulus in clean energy. In the end, the U.S. made the largest green investment of any country as a component part of the Recovery Act.

As with Bob Baugh, David Foster sees climate change as an environmental and economic challenge and an opportunity to rebuild manufacturing in the USA. In his view

> these low carbon technologies can provide a real step forward for manufacturing in the US. So it's an issue of competitiveness to keep energy costs low, and it's an issue of where the manufactured products needed in the 21st century will be made. Whether it's fuel efficient or electric automobiles or a variety of other things, it's all headed in that direction. A successful manufacturing policy in our country in the 21st century is dependent upon embracing these energy efficiency and clean energy technologies and the labor and environmental standards that will ensure domestic production.

The BGA's membership and profile continued to grow with the Green Jobs conferences of 2009 and 2010 especially well attended by unions, environmentalists, corporations, trades associations, administrators from various levels of government and many political leaders. By 2010 the BGA brought together 14 unions and environmental organizations with a combined membership of over 15 million, making it the largest progressive organization in the country. During those years, it was also able to take advantage of the ARRA funds to promote training research and development and support for green industry and the establishment of the Clean Energy Manufacturing Center with visible results (Walsh et al. 2011). The Democratic Party's defeat in the 2010 mid-term elections resulted in the failure to pass climate legislation and, thus, target the additional and massive resources necessary for a green transition. The years that followed proved difficult for the BGA, as they had for the Energy Task Force, highlighting, once again, the need for public policy and resources.

In particular, the BGA was tested by the construction of pipelines, such as the Keystone XL, commissioned in 2010 to transport oil from Alberta's tar sands to the USA. The politics over this, and other pipelines, have been a litmus test for labour and environmentalists (as well as indigenous people and others whose land was affected, positively or negatively) (see Kojola, Chap. 16 and Vachon, Chap. 5, in this volume). In general environmentalists, including the members of the BGA, came out in opposition to the pipeline. A small number of unions, including one member of the BGA, also came out in opposition. On the other hand, a number of unions in the building and construction trades and in manufacturing, including members or collaborators of BGA, were very much in support. As a result, the BGA did not take a position. However, the conflict led to the resignation of one member union and resulted in strict guidelines on how controversial issues were to be handled.

In 2014 Foster was actively recruited to join the US Department of Energy (DOE) in order to build stronger relationships between the Obama Administration, the DOE and the US labour movement and act as Senior Advisor to the Secretary on a range of climate, economic development and energy workforce issues. During his tenure there, the Department formed the DOE Labour Working Group which included most of the unions in the AFL-CIO Energy Committee. During the subsequent years, the BGA faced a number of challenges. According to Foster,

> the inability of the environmental movement to prioritize job preservation and creation as critical to any successful climate strategy led to a necessary shift in BGA's focus. With a Republican Congress, BGA was forced to manage labor/environmental conflicts on the one hand, while identifying and leading successful state initiatives such as the "Buy Clean" Act in California. In 2017, I conducted a series of round table discussions on behalf of BGA in rural, industrial communities in the Midwest with BGA union members who affirmed

climate change as a real problem, but one that had to be solved in concert with preserving and creating jobs.

These discussions were timely and informative because a significant number of Steelworkers and other union members had voted for Trump at the 2016 election. Currently, the organization's Executive Director position is held by Jason Walsh who had helped craft Obama's Partnerships for Opportunity and Workforce and Economic Revitalization (POWER Plus) as close to a just transition plan for the Appalachian Region as there has been. The election of Biden (Obama's vice president) to the Presidency thus places the BGA in a good position to influence national policy (Cohen 2021). As noted at the beginning of this chapter, despite its increasing attention to questions of justice, the BGA has so far (June 2021) avoided the narratives of just transition or the Green New Deal.

## The Difficult Road to a Socioecological Just Transition

Even so, the narrative of just transition has re-emerged in various quarters. One group includes philanthropic foundations and policy think tanks and can be safely considered as minimalist in its ambition. The other group includes environmental justice, local and state-level unions, and left of centre political organizations and networks. While there are differences amongst them, they share a commitment to embedding just transition within a broader political agenda and, in that sense, they are closer to the vision of the OCAW and the BlueGreen Working Group. This group is growing in numbers and prominence and may well push the US union movement towards, at least, the discussion of just transition. Joe Uehlein, a labour environmentalist and musician, and the Labour Network for Sustainability, a small non-governmental organization that he has founded, have long been active in this effort to embed just transition within the broader political economy.

Uehlein was born in Ohio to a working-class family. His father had been an important state leader in the formation of the CIO and Joe followed him in doing industrial work. In the early1970s he got a job building the Three Mile Nuclear Plant. During the late 1970s and much of the 1980s, he largely worked on organizing initiatives, but he became increasingly aware of labour environmentalism. During the late 1970s, for instance, he became close to Tony Mazzocchi, partly because of their share of environmental priorities and partly because Mazzocchi liked to include art in meetings he organized. Uehlein's direct environmental work started unexpectedly in the early 1980s when he became the field director of the OSHA Environmental Network discussed earlier in this chapter (also see Uehlein 2014).

During the rest of the 1980s and early 1990s, his work was largely in organizing, including a strong involvement in the BASF strike as well as most of the other iconic corporate campaigns of the era against the dismantling of post-World War II industrial relations and neoliberal globalization. During the mid-1990s, he was elected to the position of Secretary-Treasurer of the Industrial Union Department and was, thus, familiar with the internal politics of the AFL-CIO.

While the above were an extension of his corporate reregulation work, environmental issues became more and more central. In 1988 he started representing the IUD of the AFL-CIO at the climate meetings, starting with the inaugural meeting of the Intergovernmental Panel on Climate Change, continuing through meetings of the UNFCCC Secretariat and the first Conferences of the Parties—along with the representative from the Mineworkers whose primary goal was to prevent climate policy. During the late 1990s, he also participated in the work of the BlueGreen Working Group. That, and his 1995–1996 involvement in drafting the platform of the Labour Party, had made him familiar with the just transition strategy and the broader ambition within which it was embedded. After the AFL-CIO's highly public rejection of the Kyoto Protocol, he had stopped representing the IUD (which was disbanded in 1998 and eventually replaced by the much weaker Industrial Unions Council) and the AFL-CIO at climate meetings because he felt that the process was not producing any results. He retired from the Federation in 2005 to concentrate on climate policy.

After retirement, he proceeded to talk to over 20 union presidents and scores of other leaders about the subject. Out of that came a short thought piece which, through a complicated process, ended up with a philanthropic foundation which funded one of the proposals in it—the mapping of the power structure within the union movement. At that point (2006), he joined forces with Jeremy Brecher, Brendan Smith and Tim Costello—long-standing activists all—to prepare the power analysis hoping that it helped them figure out

> what we wanted to do and thought we could (…) So what we started to do was try to flip on its head the common understanding in labour that all these answers to climate change are going to steal our jobs as opposed to what does climate change mean to our jobs, which is what they should have been looking at.

Just transition was central from the very beginning, along with a bottom-up organizing strategy, centring around a network rather than a coalition, and strong emphasis on science. It is on these foundations that he started the Labor Network for Sustainability in 2007.

During that same period, he served as a strategic advisor to the BlueGreen Alliance with which he parted ways over the Keystone Pipeline (Uehlein 2011). His opposition to the pipeline may have encouraged some unions to also take a stance against it. For him and the LNS, however, it was a turning point both in terms of tactics and in terms of a resolute shift towards a deeper environmentalist commitment that was willing to challenge the limits of labour environmentalism as it had developed so far in the USA. The Keystone debates propelled LNS to greater recognition and a sharpening of its argument. In this, he was helped by a 2012 practitioner fellowship at the Kalmanovitz Initiative for Labor and the Working Poor which gave him the opportunity to think about strategies, in collaboration with other unionists and environmentalists, leading him to realize the need for an ambitious national programme. A culmination of this process was the initiation of the Labour Convergences on Climate, the first one of which took place in January 2016 and was attended by 60 or 70 unionists and labour activists.

> How did that idea of labour convergence came up? Well, that was my idea… and it came out of the power structure analysis—how labour changes on big social issues. I thought there were analogies to be drawn between the civil rights movement and labour in the fifties and early sixties to climate and labour today…. We looked at civil rights, war, trade, single payer healthcare, immigration, all issues where labour had a conservative position that morphed into more progressive positions. So, the thing that got the convergence in my mind was that the AFL CIO never in history had opposed a war until the Iraq war which would have been around 2003 or something. And that's when [US] Labor Against the War formed. And (…) they called it a labour convergence against the war. And I really liked that because that term implies more activism [than] conference. So for several years I was saying we got to do a labour convergence on climate and we never had the funding or the staff ability to pull it off. We got the first one done without staff or any money at all.

The second Convergence took place in September 2017 and was attended by about 130 people and 17 unions. The 2018 Convergences were regional with the third national Convergence taking place in Chicago in June 2019 and focusing on the Green New Deal. This was attended by over 250 people from 86 different union locals, 9 national and international unions, including the presidents of the Mineworkers and the Association of Flight

Attendants-CWA, 7 labour councils and state federations, 2 worker centres and 33 ally organizations.

Because of its size and the range of entities participating, it can be considered as an important turning point. Another reason was that some important tensions over diversity came to a head, leading to significant internal dialogue and reflection and the diversification of the organization's board and staff.

> So we came out of that third convergence and here we are a year later stronger and better and so on. And that was a turning point for our Green New Deal work for our diversity and that leads us to today, but we're now faced with other challenges around the Green New Deal in that, on the plus side, the concept really caught on, but it caught on in the way that just transition caught on. But what I mean by that is that the Green New Deal now means many different things to many different people. Right. And some people think that's really good. I don't. And so what we find is a weakening of the Green New Deal in the process.... My concern is that if we end up fighting only for what we think we can win, we're selling ourselves short and that's the direction things are going.

At this point, then, LNS is reflecting on the kind of grand vision that is appropriate. Along with discussions on the merits of broader political agendas, it also prioritizes just transition, launching the Just Transition Listening Project (JTLP 2021) in the Spring of 2020. The latter is based on interviews of over 100 unionists and environmentalists involved in transitions, as well as six webinars. An action-oriented report (JTLP 2021) is intended to inform activists and policy makers as well as the discussions leading to the next Convergence that is likely to take place in-person in 2022 or 2023.

During the last several years, LNS has grown and diversified. From a staff of 2–3 people, it has grown to 14, and its visibility has increased. After the second Convergence, it hired a new Executive Director with long and deep involvement in environmental and social justice organizing, including as national director of the Climate Justice Alliance. In addition to the Convergences, the LNS has also led or participated in a number of important on-the-ground initiatives and has launched day-long training programmes for environmental and climate activists on labour union history and functions and for unionists on environmentalism and climate. If the subsequent Convergence follows the growing pattern of the previous ones, then LNS will be at the cusp of another qualitative change in appeal and, quite likely, composition. If so, it will have to address the need for more personnel and resources. How this will happen is important if LNS is to continue its bottom-up strategy in a manner that helps unionists and environmentalists converge around a progressive and actionable political agenda.

## Concluding Thoughts

The AFL-CIO's weak and hesitant approach to climate and environmental politics can well lead one to consider the history presented here as one of failure. That would reflect a misunderstanding of the organizational limitations of this confederation. If one recognizes that and proceeds to have a closer look at the various individual national unions, as well as subnational union organizations, the lesson is different. Labour environmentalism has grown roots within the US labour movement, albeit of various kinds. The Energy Task Force—and Bob Baugh's pragmatic and strategic approach while serving at the upper professional staff of the organization—prevented the closure of the debate over environmental and climate policy that some unions wanted. This he was able to accomplish because of his long experience navigating public policy on economic development on behalf of unions. By placing economic development up front, Bob Baugh was also able to steer the AFL-CIO towards the global union strategy of inserting just transition into climate negotiations while also initiating the discussion of the strategy within the AFL-CIO, a development that is all the more significant in light of the history presented in this chapter. The survival of the BGA is also evidence that labour environmentalism and environmental unionism are here to stay. David Foster, as an elected leader, realized the grave challenge of globalization for US manufacturing unions and was able to find allies around the shared priorities of regulating trade agreements and using green industrial policy to address the climate. Even though the BGA does not use the term just transition, it is engaging environmental and social justice.

After managing to insert an ambitious agenda into the OCAW, turning this small union into a beacon of an alternative labour politics, an unfortunate merger drove Wages, quite prematurely, out of labour politics. Perkins sought to leverage her experience as a unionist and an environmentalist into a political alliance between the two movements. Perhaps because she was on the staff of the AFL-CIO and her main real supporters were small labour unions, like the OCAW, she was not able to bring on board those unions that sat on the fence and thus contain those who were opposed. Joe Uehlein participated in and learned from these experiences. Their individual and collective work have established an important socioecological alternative within which to embed just transition that has survived due to the efforts of the Just Transition Alliance, the LNS, the Climate Justice Alliance, the environmental justice movement, a number of local and state-level unions, and the broader democratic socialist movement. Whether they will be able to converge around a common platform and whether they will be successful at this critical historical juncture remain to be seen. But 10 or 15 years ago, having that discussion would have seemed unrealistic, with the exception of the short window that Baugh was able to open. The Biden Administration's and the BGA's gestures towards a green transition with justice are evidence that justice is politically necessary. Their unwillingness to employ the term just transition and to embed it within a broader socioecological strategy that goes beyond green industrial policy is not evidence that the strategy has been used and found lacking. Rather, it suggests that just transition, like GND and democratic socialism, conveys a more radical politics than the Democratic Party alliance is presently willing to engage.

# References

AFL-CIO. 2012. *The AFL-CIO, ITUC and the COP 17 Negotiations*. With author.
———. 2019, September. *AFL-CIO Commission on the Future of Work and Unions: Report to the AFL-CIO General Board.* https://aflcio.org/sites/default/files/2019-09/Report%20of%20the%20AFL-CIO%20Commission%20on%20the%20Future%20of%20Work%20and%20Unions_FINAL.pdf

AFL-CIO and EFI. 2020 *Energy Transitions*. At https://static1.squarespace.com/static/58ec123cb3db2bd94e057628/t/5f712b93b66cd43eed7d344a/1601252258579/Energy+Transitions+%28LEP%29+FINAL.pdf

AFL-CIO Department of Government Affairs. 2009. *Labor for Green Jobs and the Environment*. With author.

AFL-CIO, Industrial Union Council. 2009. *IUC 2008-2009 Reports*. With author.

Barrett, James, Andrew Hoerner, Steve Bernow and Bill Dougherty. 2002. *Clean Jobs and Energy: A Comprehensive Approach to Climate Change and Energy Policy*. Economic Policy Institute and Center for a Sustainable Economy. ISBN0-944826-97-0

Baugh, Bob, and Jeff Rickert. 2010. Good Green Jobs. *International Union Rights* 17 (1): 3–4, 15.

Bennett, David. 2007. Labour and the Environment at the Canadian Labour Congress—The Story of the Convergence. *Just Labour: A Canadian Journal of Work and Society*, 10: 1–7.

BlueGreen Alliance. 2019. *Solidarity for Climate Action*. At http://www.bluegreenalliance.org/wp-content/uploads/2019/07/Solidarity-for-Climate-Action-vFINAL.pdf. Accessed 14 Nov 2019.

Cohen, Rachel. 2021. Climate Groups Begin Vying for Power in the Biden Era as Pressure for Unity Fades. *The Intercept* (January 21). At https://theintercept.com/2021/01/21/bluegreen-alliance-biden-climate/.

DeForest, Christopher. 1999. *Watershed Restoration, Jobs-in-the-Woods and Community Assistance: Redwood National Park and the Northwest Forest Plan*. Forest Service, United States Department of Agriculture.

Dewey, Scott. 2019. Working-Class Environmentalism in America. *Oxford Research Encyclopedias: American History.* https://doi.org/10.1093/acrefore/9780199 329175.013.690

Dougherty, Laurie. 1999. Between the Devil and the Deep Blue Sea: Workers in the Global Environment. *Dollars & Sense* (July/August). At http://www.dollarsandsense.org/archives/1999/0799dougherty.html

Dudzic, Mark, and Katherine Isaac. 2012. Labor Party Time? Not Yet. *Labor Party.* At http://www.thelaborparty.org/d_lp_time.htm

Fellner, Kim. 1998. Unions and Environmentalists: Will Climate Change Change the Climate? *National Organisers Alliance Newsletter* (June): 5-8, 39-40. In author's files.

Foster, David. 2010. BlueGreen Alliance; Building a Coalition for a Green Future in the United States. *International Journal of Labour Research* 2 (2): 233–244.

Gordon, Robert W. 2004 *Environmental Blues: Working-Class Environmentalism and the Labor-Environmental Alliance, 1968-1985*. Dissertation, Wayne State University.

Harvey, Samantha. 2018. Leave No Worker Behind: Will the Just Transition Movement Survive Mainstream Adoption? *Earth Island Journal*, Summer 2018. http://www.earthisland.org/journal/index.php/magazine/entry/leave_no_worker_behind/. Accessed May 2019.

Just Transition Listening Project. 2021. *Workers and Communities in Transition: Report of the Just Transition Listening Project*. Labour Network for Sustainability. https://www.labor4sustainability.org/jtlp-2021/jtlp-report/

Labor Party. 1996. *A Call for Economic Justice: The Labor Party's Program*. http://www.thelaborparty.org/d_program.htm.

Leopold, Les. 2007. *The Man Who Hated Work but Loved Labor: The Life and Times of Tony Mazzocchi*. White River Junction: Chelsea Green Publishing Company.

Mazzocchi, Tony. 1993, September 8. An Answer to the Work-Environment Conflict? Green Left Weekly 114. https://www.greenleft.org.au/content/answer-jobs-environment-conflict. Accessed Nov 2019.

Minchin, Timothy. 2002. *Forging a Common Bond: Labor and Environmental Activism During the BASF Lockout*. Gainesville, Florida: University of California Press.

Moberg, David. 1999. Greens and Labor: It's a Coalition that Gives Corporate Polluters Fits. *Sierra Magazine* (January/February).

Montrie, Chad. 2018. *The Myth of Silent Spring: Rethinking the Origins of American Environmentalism*. Berkeley, California: University of California Press.

nwLaborPress.org. 2000. Steelworkers Union Announces end to Kaiser Lockout. https://nwlaborpress.org/2000/10-6-00USWA.html. Accessed 19 Jan 2021.

OCAW. 1991. *Understanding the Conflict between Jobs and the Environment. A Preliminary Discussion of the Superfund for Workers Concept*. Denver: OCAW.

OSHA Environmental Network. 1984. *Toxic Earth: The Need to Unite*. Video at https://www.youtube.com/watch?v=qEQiALsDbNU. Accessed 17 Jan 2021.

Paskus, Laura. 2004, May 24. In Search of Solidarity: Will Hard Times Renew Historic Alliances Between Environmentalists and Labour Unions? *High Country News*. https://www.hcn.org/issues/275/14755. Accessed 19 Jan 2021.

Pollin, Robert, and Jeanette Wicks-Lim. 2008. *Job Opportunities for the Green Economy: A State-by-State Picture of Occupations that Gain from Green Investments*. Amherst: Political Economy Institute, University of Massachusetts.

Public Health and Labor Institutes. 2000. *A Just Transition for Jobs and the Environment. Training Manual*. New York: The Public Health and Labor Institutes.

Rector, Josiah. 2018. *The Spirit of Black Lake: Full Employment, Civil Rights, and the Forgotten Early History of Environmental Justice*. Modern American History 1(1): 45–66.

Rosemberg, Anabella. 2020. No Jobs on a Dead Planet': The International Trade Union Movement and Just Transition. In *Just Transitions: Social Justice in the Shift Towards a Low-Carbon World*, ed. Edouard Morena, Dunja Krause, and Dimitris Stevis, 32–55. London: Pluto Press.

Sanders, Bernie. 2020. *The Green New Deal*. https://berniesanders.com/issues/green-new-deal/.

Seidman, Derek. 2014. Looking Back at the Labor Party: An Interview with Mark Dudzic. *New Labor Forum* 23 (1): 60–64. https://doi.org/10.1177/1095796013513238.

Slatin, C. 2009. *Environmental Unions: Labor and the Superfund*. Amityville, New York: Baywood Publishing Company.

Stevis, Dimitris. 2019. *Labour Unions and Green Transitions in the USA: Contestations and Explanations*. Report prepared for Adapting Canadian Work and Workplaces to Climate Change. https://adaptingcanadianwork.ca/wp-content/uploads/2019/02/108_Stevis-Dimitris_Labor-Unions-and-Green-Transitions-in-the-US.pdf.

Stevis, Dimitris, Edouard Morena, and Dunja Krause. 2020. Introduction: The Genealogy and Contemporary Politics of Just Transitions. In *Just Transitions: Social Justice in the Shift to a Low-Carbon World*, ed. Edouard Morena, Dunja Krause, and Dimitris Stevis, 1–31. London: Pluto Press.

U.S. Congress. House Resolution 109. 2019. *Recognizing the duty of the Federal Government to create a Green New Deal*. 116th Congress. https://www.congress.gov/bill/116th-congress/house-resolution/109. Accessed May 2019.

Uehlein, Joe. 2011. Why I'm Marching with Bill McKibben to Protest the Keystone Pipeline. https://www.labor4sustainability.org/post/why-i'm-marching-with-bill-mckibben-to-protest-the-keystone-xl-pipeline/.

———. 2014. Earth Day, Labor and Me. https://www.labor4sustainability.org/post/earth-day-labor-and-me/.

United Mine Workers of America. 2021. Preserving Coal Country. https://umwa.org/wp-content/uploads/2021/04/UMWA-Preserving-Coal-Country-2021.pdf

United Steelworkers Environmental Task Force. 1992 [1990]. Our Children's World: Steelworkers and the Environment. *New Solutions: A Journal of Environmental and Occupational Health Policy*, 2(2): 75-87. Full 1990 report with author.

Van Alstyne, Andrew. 2015. *The United Auto Workers and the Emergence of Labor Environmentalism*. WorkingUSA 18(4): 613–627.

Walsh, Jason, Josh Bivens, and Ethan Pollack. 2011. *Rebuilding Green: The American Recovery and Reinvestment Act and the Green Economy*. BlueGreen Alliance and Economic Policy Institute.

# 27

## Caring for Nature, Justice for Workers: Worldviews on the Relationship Between Labour, Nature and Justice

David Uzzell

## Introduction

The recently published book *Just Transitions* (Morena et al. 2020) has the subtitle *Social Justice in the Shift Towards a Low-Carbon World*. This title affirms the critical link between progressive low-carbon reduction strategies and a transition to new and replacement forms of employment which is not at the expense of workers' health, safety and livelihoods. Moreover, it provides a positive way forward which challenges the 'jobs versus environment' discourse (Räthzel and Uzzell 2011). It is understandable that there has been a focus on energy issues and low-carbon futures, because this is a critical factor in the climate crisis. But the earth systems are collapsing in other ways too, such as through the destruction of ecosystems and biodiversity, and land, air and water pollution. The significance of just transition is that it recognises that while climate change and its origins in carbon emissions are the focus of attention by governments and individuals, the crisis is a reflection of significantly broader causes and consequences. This is most eloquently expressed by Leonardo Boff who talks of our world being subject to two bleeding wounds:

> The wound of poverty breaks the social fabric of millions and millions of poor people around the world. The other wound, systematic assault on the Earth,

D. Uzzell (✉)
School of Psychology, University of Surrey, Guildford, UK
e-mail: d.uzzell@surrey.ac.uk

breaks down the balance of the planet, which is under threat from the plundering of development as practiced by contemporary global societies. (Boff 1997, 104)

Just transition is about creating a social and economic system that heals both wounds, and that is subservient to the interests of a sustainable earth system, not vice versa.

However, while there is much discussion about protecting the environment especially as an integral part of the measures to secure a 'Just Transition', it is rarely thought necessary to specify what exactly is 'the environment' and 'nature'. In the introductory chapter of this handbook (Chap. 1), we discussed some of the different ways in which the words environment and nature are used in the context of work and everyday life and through the lens of different academic disciplines.

In this chapter, union leaders talk about how and when 'nature' and 'environment' emerged as issues in their formative and working lives, how their engagement with the environment has influenced their thinking about the relationship between the environment and labour, and what effect this has had on their work as policymakers and leaders within their respective unions, and as policymakers and leaders.

The study reported here is part of an international project which has sought to study and understand the role of individuals in transforming organisations through the life-histories of environmentally engaged unionists (Räthzel et al. 2015). The research objectives were achieved by investigating how individual life-histories, organisational histories and societal histories intersect to create change. The research was undertaken in six countries across the Global North and South[1] in order to capture both breadth and depth of experiences, histories, influences and contexts that would normally not be possible in a single national study.

## The UK Context

Farnhill's study of the environmental policies of UK trade unions since the 1960s concluded that union interest in the environment 'is long-standing, uninterrupted and overwhelmingly pro-environmental' (Farnhill 2014). This should perhaps be qualified by noting that trade union involvement in

---

[1] 'Moments of danger, moments of opportunity: the role of individuals as change agents in organisations. A qualitative and quantitative study of union officials in national and international unions' (The Swedish Research Council, 2011–2015). Interviews undertaken in Brazil, India, South Africa, Spain, Sweden and United Kingdom.

environmental issues in the latter half of the twentieth century, as we discussed elsewhere (Chap. 1), often came out of a concern for health and safety issues consequent upon production processes affecting not only the workers but their surrounding communities. During the two decades spanning the millennium, the focus on the environment shifted to 'sustainable development' as a consequence of conferences such as the Rio Earth Summit (1992) and Agenda 21 initiatives, and then later to 'climate change' as the work of the UNFCCC entered public consciousness. When the Labour Party was elected to government in 1997, the Department of Environment, Food and Rural Affairs (Defra) worked with the TUC in the following year to establish TUSDAC (the Trade Union Sustainable Development Advisory Committee), the importance of which was signalled by the fact it was jointly chaired by a Government Minister and the Trade Union General Secretary. In 2005, TUSDAC produced its report 'Greening the Workplace' (TUSDAC 2005). It is noteworthy that the authors of the penultimate draft which was submitted to TUSDAC entitled the Report 'A Fair and Just Transition—Research report for Greening the Workplace' (King 2005). The Trades Union Congress (TUC) launched their 'Just Transition' policy in 2008 (TUC 2008) and published a number of documents and reports which made the case for union engagement with the climate crisis. A key figure behind TUSDAC's work was Philip Pearson, then Senior Policy Officer in the TUC's Economic & Social Affairs Department. He played a significant role in supporting national and international efforts, working with the International Trade Union Confederation (ITUC) and chairing its Climate Change Working Group. He helped to organise the UK unions for the first UN-sponsored trade union conference on climate change in Nairobi in 2006 initiated by Sustainlabour (see Räthzel, Chap. 28). He also organised the UK unions' involvement at COP15 in Copenhagen including the global unions World of Work Pavilion (ITUC 2009). This outward-looking internationalist perspective on climate change, and its causes and consequences, stood in sharp contrast to the focus of trade union responses in the 1990s.

Three domestic initiatives should be highlighted. Pearson played a key role in establishing the Green Workplaces projects which are 'workplace-based initiatives that bring together the practical engagement of both workers and management to secure energy savings and reduce the environmental impact of the workplace' (TUC 2010). Moreover, they also provided examples and case studies where specific unions and branches within unions (e.g., PCS, UCU and others) were leading the way in initiating policies and practices that sought to turn around their industries (TUC 2014). Second, TUSDAC recognised early on the need for the establishment of educational programmes in support of trade union climate change responses. This resulted in the launch of an educational work pack (TUSDAC/Carbon Trust n.d.), while the TUC launched its own 'Trade unions and the environment' course in 2005 which has continued to the present (e.g., see UnionLearn 2009, 2020). Third, the One Million Climate Jobs Campaign (Campaign against Climate Change 2010, 2014) was a multi-union initiative which recognised that the climate crisis would lead to many workers losing their jobs. They demanded that the Government had a responsibility to both protect existing and generate new climate jobs. Thus, the language was now moving closer to that which we associate with the just transition movement. It is noteworthy that the subtitle of the later 2014 report changed from 'solving' to 'tackling'.

It was clear from our early research that key organic intellectuals (Gramsci 1999), in various unions and holding different positions of responsibility, have been an important factor in getting the environment onto their unions' agenda. These union leaders recognised that, although there may not necessarily be a groundswell of support on the 'shopfloor' to formulate policies and action, as the climate crisis worsens there would be more pressure on governments to enact regulations and laws to curb carbon emissions and other environmentally damaging actions. This, in turn, would have a significant impact on working conditions and job security. Consequently, workers' relationship to the environment had to be addressed, and through their knowledge, interests, formation and engagement with environmental issues over the course of their lives, these 'organic intellectuals' saw that they could bring something to the formulation of union policies and practices either locally or nationally. This is the focus of this chapter.

## Method

The general research methodology employed in this study has been discussed elsewhere (Chap. 25). In respect of this specific study, life-history interviews were undertaken with 21 unionists in the UK engaging in Just Transition work. Those interviewed came from manual, professional, managerial or administrative-focused unions in the resource extraction, manufacturing and service sectors. Most were senior union officials working in their union's headquarters and had a national responsibility for environmental, sustainability or climate change issues in their union. Others were highly visible activists within their unions, often leaders at the regional level. We feel confident in claiming that we interviewed a majority of the most senior influential unionists working on environmental issues within the UK trade union movement.

It has been argued that although senior union leadership in the UK is male dominated, in contrast the 'average' trade union member is a 40-something, degree-educated, white woman working in the professions and public services (UK Government 2019). Although the current General Secretary of the TUC is Frances O'Grady and women now outnumber men as members of unions, the senior leadership in unions is still male dominated (Ledwith and Hansen 2013; UK Government 2019).

The methodology employed was life-history interviews (Portelli 1997). As the number of unionists involved at a policy and implementation level in environmental affairs is still quite small, interviewees were predominantly male (14) and white. The interviews typically lasted between 1½ and 2½ hours. All interviews were recorded with the interviewees' agreement and transcribed.[2]

In this chapter, I have drawn on just five of the interviews. In order to understand process and change over time, it was decided to focus on a small number of individuals in depth, rather than generalised experiences in breadth. The individuals have been selected because they each had particular formational experiences within their families and communities, as well as contrasting cultural backgrounds. The essence of this analysis was to illustrate how personal histories contribute to (not determine) individuals' understandings and attitudes towards the environment and how these in turn may shape the kind of environment/labour policies and strategies they pursue and promote in their work, including how they may seek to bring about change within their union and community.

## Colin Bedford

The first case is Colin Bedford (a union official in a public service union) who considered that working on environmental and climate issues has been one of the most rewarding and exciting aspects of his trade union work. He was raised in a very 'political' household. His mother was a Catholic at the 'liberal and tolerant' end of the spectrum. His father came out of the Royal Navy at the end of the Second World War as a communist and remained so until the 1950s when he left because of 'Hungary and all that, you know, Khrushchev's speech about Stalin … although he never lost a belief that the Soviet Union

---

[2] The names of all those we interviewed, except Mike Cooley, have been anonymised. This paper was finished shortly after Mike Cooley died (1934–2020). We interviewed Mike for just under five hours which is testimony to his rich life, extraordinary talent, thoughtfulness, inspirational qualities and, of course, generosity.

was worth defending, despite it being a caricature of socialism'. Colin remembers,

> obviously the values of the household were political. You know, there was an awareness of politics. And I remember him actually telling me that he regarded himself as a true Communist and what that meant. So, and it stuck in my mind. And almost from that point I became almost more engaged in political discussions.

Despite the highly political household in which he lived, Colin went to a Catholic Grammar School run by Christian brothers where, 'I just used to think these people are from the Dark Ages, from the Middle Ages'. He joined the Labour Party Young Socialists at 14. On leaving school, he joined the Civil Service and became active in the union. He was elected to be Branch Secretary at 18, and

> first and foremost, I wanted the branch to be active, and there were issues bubbling away, largely around a foretaste of the change in public sector management culture ... I cut my teeth on a number of issues that were of genuine significance to the workforce.

He found that his 'socialist ideas [were] increasingly over the years informed by a sort of ecological perspective ... [which had] ... grown and become more and more an important element of my conception of socialism'. This set him aside from many of his union colleagues because 'neither the Left in the Labour Party nor Militant or many of the Left groups, or even Left Unity, have ever really sort of given the issue a priority'. If the Left were typically not interested in the environment, some environmental group supporters were correspondingly deeply suspicious and antagonistic towards the Left, and thus Colin found himself situated between both.

> The first time I was ever in a meeting or environment where there were more environmentalists [Friends of the Earth] than trade unionists. And we leafleted one of the shifts [about the dangers of nuclear power]. And I've never forgotten the deep hostility of the workers to us.... And they were utterly contemptuous, hostile, regarded us as outsiders threatening their livelihood, that we didn't know what we were talking about, that they'd been convinced that nuclear power was safe, and that we were simply interfering for, you know, extremist reasons and were potentially threatening their livelihood.

What Colin learnt from that experience was that it was not enough for the environmental movement to simply rely on exposing the risks, warn of the dangers, communicate the scientific evidence and highlight the disinformation communicated by industries and their claims to be concerned and caring about the health and safety of the workers and the local environment. Rather, they have to put forward an alternative future. The strategies he was developing not only came from experience but also from a different conception of Marxism and socialism to which some of his colleagues subscribed. His position was

> we're part of the class, and it's the dialectic of the experiences and outlook of the working class with socialist ideas and theory; it's not imparting it down; and it's not like the anarchists, who say it's all about spontaneously arriving at revolutionary conclusions. It's about that interaction.

Thus, he believed that if you are to argue the case against nuclear power which may put 20,000 skilled and well-paid jobs at risk, you have to say to those workers what the alternatives are and provide protection:

> So I always thought unions and socialists need to get involved in environmental issues…. The trade union movement and the organised working class have got an absolutely pivotal role to play. And we need to link the labour movement with the environmental movement at every opportunity.

While he wanted to put these ideas into practice, he 'was never entirely sure when, where or how I could be involved' even though he was beginning to regard himself as a sort of eco-socialist. When he was elected to one of the top positions in the union, he could see there were opportunities to make these links:

> the union had a number of policies at conference that said: 'We should start thinking about climate change' that talked about things like, you know, promoting fair trade, about starting to raise international trade issues, and the relationship of western companies with the developing world.

Few saw these policies as a priority, but he did. By virtue of his new position, he was now able 'to provide a lead and to start directing resources and to start promoting those ideas and to enlist support of activists and officers who shared a similar conviction'. The importance of this was that it was done within the framework of the union's policies and so started to gain traction.

Evidence of this came from the fact that there would be motions and debates on these issues at conferences. This led to the setting up of a task group to discuss how all these issues could be addressed, and from that developed the union's Green Agenda.

It has been suggested that there is a danger that Just Transition initiatives can become framed around energy issues such that justice focuses overly on a low-carbon transition future (White 2020). While this is clearly critical, the case can be readily made that we have to not only reduce carbon emissions, but also question how the economy as a whole operates. As I suggested at the beginning of this chapter, the basis of human life on this planet is being destroyed through causes other than excessive carbon emissions. For Colin, connecting environmental issues to labour issues is about possessing a more holistic worldview and imagining alternative futures where workers' rights globally and a mode of production that takes care of the environment are connected. He not only had a personal interest in the environment that led him to introduce this into the union, but his understanding of environmental issues was informed by his understanding of and approach to socialism; for him, they cannot be separated. Once Colin was in a position of influence in the union, he saw there was an opportunity not simply to promote the environmental cause, but to argue that the environment's interests are the union's interests and by protecting the environment one is simultaneously protecting jobs and the workers' interests. His socialist worldview provided a context and a rationale in support of the kind of issues which he saw intersecting with each other and, as a consequence, what kind of environmental policies should be promoted and which should be rejected. This meant having a 'shopping list' of policies centring on specific environmental issues and developing a broader agenda to create new climate jobs:

> But we're talking about new Climate jobs: its investment in insulation of homes and buildings. So it's investment in clean integrated transport systems. It's investment in renewable technologies, in feed-in tariffs. It's investment in, you know, skills and training and education.

Colin's work and that of his colleagues were effective, he explains: 'it's now quite clearly an important policy issue within [the union]. It's reflected in the conference debates. It's reflected in the expanding network of reps that are interested in green issues'. He went on to talk about how green issues are attracting people to union meetings who have no formal union position. Moreover, 'there was an overwhelming preponderance of younger members and activists—which again just gives you an indication of the possibilities within the trade union movement'. Thus, he sees the environment as an important part of union renewal.

Colin's attitudes and beliefs about the environment and how they are related to his understanding of socialism have influenced his policies and strategies as a union leader and have contributed to making his union one of the leading environmentally active unions in the UK. For him, 'it's not about people superimposing an issue, coming in to members when they're worried about their jobs and conditions', but seeing it as part of a larger longer-term struggle such as setting up the bargaining mechanisms with managements so that they 'consult us about the targets, about what methods are being deployed to achieve the energy efficiency targets and various reductions that they were calling on departments to make', and campaigning in line with conference policies such as feed-in tariffs, calling for Green reps, opposing nuclear power and challenging the idea that science validates investment in so-called clean coal technology.

## Philip Hendon

The second trade unionist, Philip Hendon (a union official in a public service union), illustrates another perspective as to how a set of political beliefs interacted with a concern for the environment. Philip's concerns about nature and the environment (and much else) were and continue to be situated and theorised within a socialist worldview. While it was reinforced knowing his family's history (e.g., his grandfather had been an active member of the Labour Party), it was the study of history and becoming involved informally in Labour politics with friends that was significant:

> the key element in my political development was becoming a socialist, and then developing that, what I understood by being a socialist, in those sort of late-teenage, early-twenties years, and which I still think by and large … I stick to. I think that the views formed at that point are then pivotal to the rest of what I've done. They explain the sorts of other things I've been interested in—that are not related to the environment—over the last two decades! But equally they relate to how I would conceive of environmental issues… So, I think that element of the story's important, because I didn't come to this issue on the environment just on its own terms. I didn't become active in Green politics as such at all. For me it was very much socialist politics, which then led to an engagement with Greens and the arguments with people from, if you like, the Green tradition, as well as mainstream economics.

While what Philip refers to as socialist politics that led to an engagement with the Greens was clearly important, an initial seed was sown by a small but significant encouragement at school, where he won a geography prize and was 'given an atlas which contained several chapters on ecology, the environment and, which for 1985 I guess was fairly early, on climate change'.

A worldview is a perspective that demands 'connectivity and interdependence'. For Hendon, Marxism has such a coherence. However, being a socialist comes first and his environmental position comes after that. This is in contrast to many of his friends who felt that their worldview was informed by environmentalism or ecology.

> And therefore, I would have these discussions with people about, "Well, you know, okay, why can't you just be a socialist who cares about the environment?" and they would say, "Well, why can't you be a Green that just cares about social issues?" and so forth.

The Red–Green connection is regarded as critical as it makes the link between the environment and production, wage labour and capital. And for this reason, 'workers have a direct interest in environmental matters that relates to what they [workers] are centrally about in a capitalist society'.

Philip undertook research for a PhD on the potential role of workers and trade unions in tackling climate change while working for his union. Out of this has come a broader critique of the need and potential for worker engagement in climate issues. He identified three key ways climate change has been approached in the social sciences which are relevant to union positions and policies. First, a neoliberal market-orientated way of looking at climate change which has strongly informed 'elements of the Trade Union movement, certainly trade union leaders, that have adapted too far to that in support for, say, EU ETS [EU Emissions Trading System] support for aviation expansion'. Neoliberalism, he argues, is a problem for unions in terms of privatisation and the retraction of the public sector; and neoliberalism is also at the heart of mainstream climate change politics. Second, 'ecological modernisation' which, while partly influenced by neoliberalism 'certainly the weaker forms, has more emphasis on the state, on regulation, on the involvement of stakeholders, including unions. … Just Transition, for example, I think fits into that mould, certainly in its weaker forms'. The third influence he terms the 'class' pole, where unions extrapolate from core interests and what these mean for the environment. But he sees this as 'not unproblematic: because that doesn't mean just because there's a few jobs in fracking that the unions should throw their weight behind fracking.'

Philip recognises that 'the way you write these things up intellectually is one thing; and then there is how you see the political conclusions for the movement'. He sees himself as part of the labour movement, contributing to its transformational renewal which

> includes taking environmental and climate questions seriously and making them a core part of the mission, and then reeling back from 'the mission' … to 'what does this mean for our strategies, what does it mean for policy, what does it mean for action?

For him, the Vestas dispute[3] (2009) is a good example of where all these issues come together '—industrial action, university and industrial action in many ways, along with Climate solidarity […  and although…] it was a moment where we didn't win, but it was a moment where you could see the potential of workers' environmental action'.

## Jack Cromore

For both Colin Bedford and Philip Hendon, their thinking about socialism framed their approach and understanding of situations, structures and processes. Setting environmental issues in a larger political context and set of beliefs was also important for Jack Cromore (a union official in a general workers union). His starting point was his everyday experience of a world soaked in conflict which led him to think about what sort of world he wanted to see and would be just? His journey begins with his experience of growing up in the conflict-torn society of Northern Ireland (Ulster) during what only the English could understatedly call 'the Troubles' (1968–1998). This experience comprised an awareness of social, economic and environmental injustice. But while beliefs and values are always driven by experiences, these are also framed and interpreted by worldviews and predispositions.

There are two strands to Jack Cromore's story which provide an insight into how a concern about the environment was only partially derived from an experience in which nature is unjustly treated. Another driver of his sense of injustice was the role of discrimination and conflict in his upbringing. He was raised in a country with 300 years of sectarian discrimination, oppression and violence that cut across all aspects of political, social and economic life (Cairns

---

[3] In 2009, Vestas ceased production of wind turbines at its plant on the Isle of Wight (UK) because, the company argued, the blades could be produced cheaper elsewhere. This affected 625 jobs. Approximately 25 workers occupied the administration offices in protest (Macalister 2009).

and Darby 1998). As an Ulster Catholic the Roman Catholic Church represented resistance, because it was a culture under attack. Being a Catholic meant living with a 'siege mentality'. He was brought up by relatives in a rural part of County Down because his parents had died when he was young. He did not belong to any youth organisation such as the boy scouts because they 'were seen as a kind of British establishment thing'; moreover, they were seen to have ties to the established Anglican Church. Many scout groups are associated with their local church and participate in quasi-military parades on Remembrance Sunday (the annual day of the commemoration of those who died in the two world wars). For a Catholic, non-membership was an affirmation of resistance. Jack played Gaelic football, a sport typically associated with the Irish Republic, rather than rugby or soccer which epitomised colonial (i.e., British) culture. In those early years, Jack was not an environmentalist. However, religion shaped his views about 'fairness, equality, stuff like that'. In Northern Ireland, threatening religion threatens identity and one's right to belong, or even exist; a Catholic identity was a statement not only of faith but also of politics.

Although religion ceased to be part of Jack's daily life, its values and precepts became imbedded into his thinking, and continue to inform and guide his actions and views:

> I try to judge them fairly, no matter what their creed, their colour, their sexuality, whatever else it may be... I think that Catholicism did shape my—I suppose my conscience, for want of a better term. It didn't dominate from a Christian ethos; it was more to do with a humanistic thing.

So, while Jack recognises the 'Christian ethos' as decisive, he also seeks to translate it into a different ethos, a humanist one, a worldview which is not necessarily Christian.

It was when he moved to England that his views about 'fairness, equality, stuff like that' started to be translated into environmental issues. Jack was brought up on a farm in rural Ulster and was used to breaking chickens' necks and plucking them, that is, he had a personal experience of the consequences of being a meat-eater, unlike most people whose experience is limited to buying meat in a supermarket. As he said, 'I grew up in a nice green part of the country, you know, and mountains and fresh air and whatever. But it never really occurred to me, to be honest. It became part of the industrialised bit of the job'. But when he moved to Cambridge in England, he became a vegetarian, mixed with anti-hunt activists and animal rights supporters, and developed a concern with 'the ethical aspects of animals and how they were treated, and how they were farmed … and an extension of that is to treat animals fairly, and treat the environment fairly'. And so, this in turn has shaped his views about 'fairness, equality, stuff like that'. While Jack's views on the environment and nature changed as a consequence of going to university, this was more than simply a change in shopping habits or food preferences or even 'coming to this issue on the environment just on its own terms' as Hendon expressed it above. It was contextualised in larger issues about social and environmental justice and growing up in a farming community which brought him face to face with the realities of food production.

Philip Hendon makes an articulated link between ideas and practice and how they emerge and inform not only his day-to-day work but also his own philosophy as the union's policy officer. Jack Cromore does not express such a comprehensive theory within which he can connect workers' environmental and social interests. While explaining that his notion of fairness and equality has its roots in his catholic upbringing, he also 'rejected Catholicism at an early age'. Catholicism did not become a worldview which helped him to connect 'the cry of the poor' with 'the cry of the earth' as in Liberation Theology ((Boff 1997; Gutiérrez 1973) see also Chap. 15 in this volume). It might be that this disconnect leads him to pursue environmental policies not so much as a broader strategy but rather more dependent on what the union in general is putting on the agenda, taking up a number of issues and setting their agenda only cautiously. This is exemplified when asked about his union's position on fracking:

> … officially we don't. Nobody's ever put a motion to our Congress (…) But we've [the environmental group] actually set up a fringe meeting, (…) at Congress this year on fracking. (…) It's going to be a debate, (…) Might get it out in the open, it might get it (…) At least people talking about it, (…) But basically, it's a 'wait and see', to be truthful, because I think currently we wouldn't—we wouldn't necessarily want to go ahead. But (…) in five- or ten-years' time—the situation might change.

This cautious approach to a central environmental issue is also due to the fact that the ideas about fracking are divided within the union, due to the opposing interests of different groups of its members. Cromore is also a pragmatist and recognises that in the day-to-day business of unions, compromises have to be made when one issue or one set of unions' interests (i.e., a faction) may conflict with another.

## Sam Henley

Sam Henley (a union official in a general workers union, but different from Jack Cromore) represents another strain of thinking about the environment and the actions labour can take. Sam's community is in the north of England which has both deep and opencast coalmining and has been an area that has seen massive job losses with the closure of mines in the last 40 years. Having a long family history of living in the area, Sam understood the impact of such closures on the local community and the divisions and hardships caused by the 1974 miners' strike. But he was also highly aware of the relationship between labour and nature such as the environmental devastation caused by mining (e.g., 'blight the area in terms old coal slagheaps[4]') and the community's collective memories and histories of mining disasters such as the nearby 'site of the old Pretoria coalmine, which (…) was the third biggest mining disaster in the country: 344 local people killed at that'.

Sam comes from a coalmining family: His grandparents and relatives were all miners and many were also trade unionists. His grandmother worked as a 'pit brow lass'[5] close to where he now lives. He became interested in the union because his grandfather was President of his local Miners' Federation of Great Britain branch, and his father was Secretary of the local Trades & Labour Council. It's a background of which he is proud: 'So the reason I've got into trade unionism and stuff and socialist politics is because of that background really, being born into it. It was fortunate'. Sam went to University and studied Politics and History. On graduating, he had a variety of jobs and always joined the relevant union. Sam is an organiser of the annual Diggers' Festival which celebrates the life of Gerrard Winstanley (1609–1676), one of the earliest socialist thinkers who was concerned with labour/nature relationships: Winstanley wrote: 'And so the Earth that was made a common Treasury for all to live comfortably upon …' (Winstanley 1649; Winstanley and Benn 2011).

Sam has held senior positions in his union primarily at the regional level (Chair of the local branch, President of the local Trades Council). What comes across from talking to Sam is that he is aware of the practicalities and frustrating difficulties of enabling action at this level—it is one thing to have policies for the union, it is another to convince your fellow workers on the ground that it is in their interests to be concerned about and to engage with these

---

[4] Hills adjacent to mines created from the waste material (*slag*) generated by coal mining.
[5] Pit brow women, until the 1960s, picked stones from the coal after it was hauled to the surface (pit bank or brow).

issues because they will have a crucial effect on jobs and the economic and physical health of the local community.

Sam has long been concerned about climate change and greenhouse gas emissions into the atmosphere, and speaks with passion about anti-nuclear and anti-fracking campaigns and working with environmental groups.

> What I'd thought all along, and colleagues of mine who were already on this climate change agenda, ... human beings are the ones responsible for putting more and more greenhouse gases into the atmosphere—patently obviously so! Cars, you know; you can see Fiddlers Ferry Power Station here! And burning of coal and stuff.

Sam was even opposed to 'open-cast mining all the way around it [the town], which only recently finished, which we opposed—not so much on an environmental basis but on the basis it was undermining deep-mine coal jobs'. Sam talked about the Labour Party always having a clear policy position against nuclear power. The nuclear accident at Chernobyl (1986) brought home, literally, the dangers:

> Because I remember us having a May Day event.... And it actually rained that weekend, but it was only announced on the Sunday by the Government that "there's been a nuclear accident in the Soviet Union today", so we actually got rained on! I don't know whether what consequences it had, but they've certainly killed a lot of sheep off in Wales who got contaminated.

He also cited the radiation leaks from Sellafield Nuclear Power Station (Cumbria) and the higher-than-average incidence of child cancers in the surrounding area (see Davies 2012). Likewise, in recent years, serious 'accidents' surrounding fracking leading to small explosions, earth movements and the pollution of water supplies in north-west England have made the damage inflicted on the environment and local communities personal, immediate, concrete and local. Because these problems are on the doorstep, it becomes easier to get the message across, but it has to be done in the context of jobs: 'I don't care whether it is 100% safe and you can prove it's that; I don't accept it is. But even if it was 100% safe, we don't need it. What we should be doing is putting a million people back to work developing renewables'.

Sam could see that the environment argument may get more traction by appealing to what the local community see as more immediate and concrete: 'But in a way, you could sell a lot of this stuff without tagging the environmental aspect to it. And in some respects, you get better results by doing that because people respond [more] to job creation than they do to environment'. And the same argument applies within the unions. They were cautious at first because of the jobs' implications: 'you need to build up a critical mass of support. And slowly. You know, the Northwest TUC is now opposed to fracking. We've got Unite the union now backing this'.

Perhaps unlike some of the interviewees who expressed their views in more abstract ways, Sam is living in a community which is directly experiencing environmental destruction through production processes, their consequences and possible means of resistance. His is a personal experience that is situated not only in place, but also in history, collective memory, personal memory and a concern for the future.

## Mike Cooley

Mike Cooley (while at Lucas Aerospace chaired the local branch of the Technical, Administrative and Supervisory Section (TASS) and now absorbed into Unite the Union) explicitly references the importance of nature and culture in his early life as a driver for wanting to see change in the world of work and to become part of that change. As in our interview with Jack Cromore, Mike Cooley began his life story by referencing the rigidity, conservatism and the low level of tolerance towards intellectual criticism of traditional Irish catholic education which was dominated by nuns. But unlike the previous interviewees, Cooley spoke at some length about nature and the environment as being foundational for his later understanding and approach to work. His mother 'used to write quite simple but very moving poems about nature'. His father 'loved the countryside as well' and had a small garage and serviced/mended motorcycles and bicycles. In the Second World War, it was difficult to get bikes and he would make them up from recycled metal and so on: 'And that was extraordinary to watch and behold, you know, something coming in a pile of what appeared to be rubbish, and the next thing there was somebody cycling on it down the road!' As a consequence, Mike experienced 'a tension between my artistic interests, if I might put them like that, and my engineering interest'. This was ostensibly a trivial but enduring memory that would later have a profound influence on his view of the relationship between sustainability, production and creativity. In later life, along with his engagement with the countryside, this became a creative tension leading to a melding of these interests in order to provide intellectual, conceptual and hard design solutions to both production processes and products.

But his vision extended more widely to other cultural views about the relationship between people and the land. Mike was able to cite, some 50 years later ('it's burnt in my memory'), the 1854 speech of Chief Seattle (1786–1866). He even wrote to the Duwamish people to check that what Chief Seattle is alleged to have said was correctly attributed. This speech, for him, encapsulated 'the way I feel about certain things':

> Every part of this country is sacred to my people. Every valley, every hillside and every glade is hallowed in the memory and experience of my tribe. Even the soil on which you stand responds more lovingly to our footsteps than to yours. But the soil is rich with the life of my people. Our religion exists in the hearts and the minds of our people; your religion was written by the iron finger of an angry god.[6]

Cooley went on to tell the story of Chief Seattle's response to the then US Government's claim that they will educate the Suquamish and Duwamish people and give them a written language: 'And how do you educate your people? Do you tell your young, as we do, that man did not weave the web of life; he has just put one strand in it. Who destroys the web destroys himself'. Without a symbiotic relationship to the earth, humans will destroy it, and in so doing destroy themselves. Although there is some debate as to the authenticity of Chief Seattle's speech (Abruzzi 2000), it became foundational in the early stages of the environmental movement in the 1960s as an appeal for respect of Native American rights and environmental values. Subsequently, it became a manifesto for human rights and environmental activists.

The Suquamish and Duwamish worldviews, along with William Morris' ideas of socially useful work and socially useful products (Morris 1883, 1908), were a source of inspiration and informed Cooley's thinking in the development of the Lucas Plan (Wainwright and Elliot 1982; Räthzel et al. 2010). This was a collective project overseen by the Lucas Shop Stewards Combine Committee. Mike Cooley, a leading figure on the committee, was the archetypal organic intellectual (Gramsci 1999) who went on to develop a philosophy of production that sought to transform existing production processes, using technology that capitalised on the skills of workers rather than replacing them. Cooley worked for a transformation that relied on workers' capabilities to create socially useful products, thus respecting the world's resources and the world's labour. Cooley believed that we need challenging visions to transcend

---

[6] There is no single version of this speech. The original was in the Suquamish dialect of central Puget, and there are variations of the translation.

given roles and taken-for-granted paths (Cooley 1987). The Lucas Plan was one attempt to achieve this:

> The way of production and the way we were being defined by things, that definition was much too narrow, and we intended breaking out of it. And indeed, we did. But also, the sort of environment in which the company regarded it as legitimate to operate.

The innovation of the Lucas Plan was also expressed in terms of how they conceived the production process:

> [We] realised that there was a lot of knowledge on the shop floor and elsewhere which is not articulated verbally, or even in the written form—it's people describing and giving of their intelligence through what they do and how they do it, rather than the way they talk and so on about it.

When you had craftsmen talking to and working with designers,

> they were able to do profound work but one at the craft level of building prototypes, the others at the design and development stage and produce such things as a road-rail bus which is the sort of transport solution needed now'. Also, prototypes were built '… in order to develop the ideas more widely outside the factory, so that management couldn't say it was not feasible and we would be left in isolation to be picked off.

Many of the 'socially useful products' were the result of a concern for the environmental impacts of products and production (see Chap. 36). Cooley and the Lucas workers were ahead of their time in recognising the need for sustainable production methods, even if such a concept was not widespread in 1976. In today's terms, their argument was for a 'Just Transition'. They saw that it was important in adapting production for different needs and markets to ensure that workers work alongside designers and engineers to construct not only an alternative production plan but also an alternative future for themselves. The Plan identified 150 product ideas including heat pumps which were efficient in saving waste heat; solar cells and fuel cells; a road-rail public transportation vehicle which would be lightweight using pneumatic tyres on rails; electric vehicles; a combined internal combustion engine/battery-powered car which could give up to 50% fuel savings while reducing toxic emissions; braking systems; undersea exploration technology and remote control (telecheric) devices for people with disabilities. Although this was over 50 years ago, many of the ideas look advanced today in terms of production for sustainable ways of life. The Lucas Plan was rejected by the Lucas management. A New Lucas Plan Group has now been established to promote these ideas (The New Lucas Plan Working Group 2020), for example, where workers in an Airbus factory have used their skills to produce ventilator components in response to the Covid-19 crisis (Wainwright 2020).

Mike Cooley took his ideas for socially useful production around the world through honorary academic appointments and publications like *Architect or Bee: the Human Price of Technology* (Cooley 1987), which has been translated into six languages. He and the Combine Committee were awarded international prizes including the Alternative Nobel Prize.

## History, Society and Personal Biography in Labour/Environmental Transformation

In this chapter, I have focussed on those unionists for whom recognising the importance of and caring for the environment are an integral part of their worldview which informs not only their personal lives but also professional work. The purpose of the life-history interviews was to understand what it was in their formation as personalities and unionists that led to the importance given to the 'environment'. Following this, how does the integration of environmental concerns in their essentially socialist worldview, in which justice is a strong feature, inform their work as trade union leaders? One could undertake a policy analysis and attempt to draw more precise links between the personal and the policy. But as we know, policies are not formed in isolation. This was well illustrated when we tried to interview a trade unionist who was charged with taking his membership forward environmentally in what is considered a major carbon-emitting industry. Despite a number of requests, he refused because he argued that policies and decisions in unions are made collectively; the individual's view in terms of shaping policy is irrelevant. We fully respected his position. But what this chapter has tried to show is that while collective agreements and actions are crucial, these have to come from somewhere, and that is from the energy, commitment, knowledge and enthusiasm of individuals. This is not to make an individualistic thesis, but it is to recognise that unions like other societal organisations need 'organic intellectuals'. As Leandro and Tropia (Chap. 15) argue, organic intellectuals help to create a new hegemony challenging existing power relations, 'The mode of being of the new intellectual can no longer consist in eloquence, (…) but in active participation in practical life, as constructor, organiser, "permanent persuader'" (Gramsci 1999). How they become organic intellectuals has to be looked for, in part, in their personal biographies. Few illustrate this better than Mike Cooley. All individuals live in a social world, and we can only understand that world when we understand the intersection of history, society and personal biography (Mills 2000). In turn, this can provide us with an insight into the implications of their worldviews for their work as union leaders trying to get environmental issues onto their unions' agendas.

All interviewees were trade unionists first and environmentalists second. Nevertheless, in the case of Cooley and Cromore, who were both brought up in rural Ireland and Ulster, the environment was an important part of their upbringing. Hendon was brought up in a politically 'vibrant' household, but became involved in labour and environmental issues through Marxist theory which he saw as having 'connectivity and interdependence' which enabled him to incorporate the environment. He also traces his interest in the environment back to being given an atlas. This was not a direct encounter with the environment, but a mediated encounter through reading about environmental threats to the world. One felt that Sam Henley's engagement with the environment was a visceral experience embedded in what he witnesses every day, and what he brings to that from within his collective and familial memory.

Both Cooley and Cromore, coming from a divided Ireland, were highly aware of issues of conflict, justice and inequality. Cromore, Bedford and Cooley were all educated in religious institutions which were reactionary and whose representation of the political order stood in contrast to what was assumed to be progressive in their own homes. The importance of this was that it created a tension—it forced them to confront opinions and see that there were alternative ways of looking at the world.[7]

Seeing injustice in one domain (e.g., discrimination against minorities, those without power) often leads to a realisation of and a transference of the principles/concerns to another (e.g., the environment, animal welfare). As one unionist expressed it, his 'understanding of environmental issues was informed by his understanding of and approach to socialism'. And the corollary of this for him was that there is a need to link the labour movement and environment movement. In Jack Cromore's case, his experience of daily conflict and injustice on the streets and in the institutions of Northern Ireland made him sensitised to justice issues across other domains, so that when he saw how meat was farmed, his beliefs about justice were applied to animals and he became a vegetarian. This was not simply a change of habits. His

---

[7] We also interviewed trade unionists whose family background was religious and progressive, including being sympathetic to social and environmental issues (Uzzell and Räthzel 2019).

environmental views were contextualised in a broader set of beliefs about social, economic and environmental justice. But while having a worldview which sees the interlinkages between issues of equity and justice can be the foundation for more transformational change, Cromore reminds us that the pragmatics of union politics can be a significant impediment to progressive change.

## Bringing About Change in an Organisation

What these accounts draw attention to is that the unionists took an ecological perspective on the world. Ecological, not necessarily in the environmental sense but in the sense that underlying their accounts is an argument that we can only begin to have an understanding of the world if we see how all its parts interact and create mutual interdependencies. As a consequence, this had an impact on the way they approached the issue of change in their organisation and across the labour movement more generally. These unionists had clear views about not only how you need to bring about change in organisations, but also how one should think about the environment, and this guided their strategies. They were also aware of the problem that while their union leadership may be signed up to the adoption of climate change policies, this does not mean the membership is aware or supportive of those policies. Colin Bedford, Mike Cooley and Sam Henley raised the issue as to how do you take people with you?

Should leaders work with a top-down model in which instructions are communicated and those below are meant to follow? Or should they adopt a more ground-up, participatory approach in which the needs, the enthusiasm and the skills to take issues forward are discussed, collectively formulated and owned? Colin Bedford realised that although he was now in a leadership position, this did not mean he could or should drive change from the top-down. For him, it was about creating conditions which would encourage members to see that change was necessary and beneficial. This could be realised by finding resources and motivating within the framework of union policies. One way to achieve this was encouraging discussion about the issues within the union. His approach had an immediate focus, focussing on the union rather than being potentially transformative for society at large. For example, he saw incorporating environmental issues into union policy as part of union renewal since they are now attracting younger people to union meetings.

Mike Cooley too argued that you cannot be dogmatic and drive change from above, especially where solutions are likely to put people out of work.

Mike Cooley and the Lucas Shop Stewards Combine Committee demonstrated in 1976 one important way of ensuring that workers are involved in planning, designing and implementing their future, and the future of production. While the Lucas Plan was rejected by the company and received little support from either the Labour Government or the TUC, the process did have a transformative impact on the workers themselves and provided an inspiration for all those who see the necessity of transforming production in order to confront the environmental crisis. What came across from talking to Mike was his limitless faith in the ingenuity of human beings to solve the social and environmental problems we face. This, he argued, will not be achieved 'top-down' but by a participatory process from the ground-up drawing on the creative imagination, skills and craftsmanship of workers. He talked about how this participatory approach not only generated more ideas, but enhanced workers' sense of self-efficacy and comradeship: 'And people came in and they were pouring out ideas! I mean, they were full of ideas. Because it was amenable, the context in which it was being discussed. It was their own mates who'd made this road-rail vehicle, so it kind of belongs to us'. He thought these ideas would 'endure' not necessarily in an organisational sense (i.e., creating a 'Lucas Aerospace identity') 'but rather that it would be a way of viewing the world and viewing change and viewing our relationship with technology'.

Finally, Sam Henley's involvement in environmental issues taught him that by experiencing the interaction and consequences of industrial production on the lives of workers and the communities in which they live, it is possible to develop comprehensive environmental policies with workers and their communities. Sam captures well what the union officer who may be charged with promoting environmental action has to take on board, and how to take people with you:

> So it seems quite ironic that, as late as the Nineties we're fighting to save pits stay open, and now I'm more or less fighting to stop them reopening! The thing is, (…), I supported the miners in Spain recently, and somebody might say, "Well, how does that fit in?" And for me it's like they're being forced into unemployment by a government that isn't saying, "Oh, let's replace what you're doing with something more sustainable"—they just want to put these people out of work. So my argument would be: well, I'm sure if you went to a coalminer in Spain and said, "How about working in renewable industries …" they would probably vote for that. So generally, in terms of the campaigning work we've been doing, instead of just opposing things we generally try to put a positive campaigning alternative towards it.

## A Participatory Approach to Transformational Change

It is clear from these accounts by senior trade unionists working on environmental strategies, policies and actions at the national and regional levels that their worldview and its formation influenced how they think about their work and how the union should engage with environmental and climate issues. Their kind of socialist perspective proposed that unions should not just have environmental policies which support legal, regulatory, financial or technological responses, but transformational kinds of policies which address larger societal problems. Cooley, for example, by insisting that workers must be actors and create alternative products is questioning the role of management and the private ownership of the means of production. It was not because the Plan was not feasible or even profitable that it failed, but because the company refused to recognise the legitimacy of the Combine Committee and saw this as absolving them from a responsibility to even discuss the Plan, or as one MP expressed it in Parliament 'it did not need the Combine Committee to tell them [i.e. Lucas Aerospace]' (Wainwright and Elliot 1982, 232).

The final chapter of Wainwright and Elliot's book *The Lucas Plan: A New Trade Unionism in the Making* (1982) begins with what has to be seen as an ironic quote from the Combine's Corporate Plan:

> It is certainly not the assumption of this Corporate Plan that Lucas Aerospace can be transformed into a trail blazer to transform large scale industry in isolation. Our intentions are much more modest namely, to make a start to question existing economic assumptions and to make a small contribution to demonstrating that workers are prepared to press for the right to work on products which actually help to solve human problems rather than create them.

The work of Cooley and the members of the Combine Committee illustrate that this questioning came out of a worldview that saw the interconnection of things and the need to avoid reducing social problems to technical issues. To this, one might add that climate policies are not just a question of reducing carbon emissions. Those charged with responsibility for reducing emissions often default to policy and intervention responses such as technological-fix solutions or governmental regulations, or simply abrogate responsibility to the individual. But too often these lead to inequality and injustice. As we have seen in these interviews, some of the most committed and imaginative union leaders recognise that the carbon-emission and nature-destroying problems generated by their industries are problems situated in societal and economic structures. These union leaders' analysis is informed by a worldview that has moved 'upstream' from the immediate manifestation of cause and effect to acknowledging the relationship between economic assumptions, workers' rights and the impact of production including the way it transforms nature.

**Acknowledgements** The author is grateful to the Swedish Research Council (Vetenskapsrådet) for funding this project.

# References

Abruzzi, William S. 2000. The Myth of Chief Seattle. *Human Ecology Review* 7 (1): 72–75.
Boff, Leonardo. 1997. *Cry of the Earth, Cry of the Poor*. Ecology and Justice. Maryknoll, NY: Orbis Books.
Cairns, Ed, and John Darby. 1998. The Conflict in Northern Ireland: Causes, Consequences, and Controls. *American Psychologist* 53 (7): 754.
Campaign against Climate Change. 2010. *One Million Climate Jobs: Solving the Environmental and Economic Crises*. London: Campaign against Climate Change. https://campaigncc.org/sites/data/files/Docs/one_million_climate_jobs_2014.pdf.
———. 2014. *One Million Climate Jobs: Tackling the Economic and Environmental Crises*. London: Campaign against Climate Change. https://campaigncc.org/sites/data/files/Docs/one_million_climate_jobs_2014.pdf.
Cooley, Mike. 1987. *Architect or Bee? The Human Price of Technology*. London: Hogarth.
Davies, Hunter, ed. 2012. *Sellafield Stories*. London: Constable.
Farnhill, Thomas. 2014. Environmental Policy-Making at the British Trades Union Congress 1967–2011. *Capitalism Nature Socialism* 25 (1): 72–95. https://doi.org/10.1080/10455752.2013.879196.
Gramsci, Antonio. 1999. *Selections from the Prison Notebooks of Antonio Gramsci*. Eds. Q. Hoare and G. Nowell Smith. London: ElecBooks.
Gutiérrez, Gustavo. 1973. *A Theology of Liberation: History, Politics, and Salvation*. Maryknoll, NY: Orbis Books.
ITUC. 2009. *Trade Unions at the UN Framework Convention on Climate Change UNFCCC – COP15*. Brussels: ITUC. https://www.ituc-csi.org/IMG/pdf/COP15_ITUC_report_final-2.pdf.
King, Charles. 2005. *A Fair and Just Transition – Research Report for Greening the Workplace*. London: TUSDAC. http://www.cts.fra.utn.edu.ar/xframework/files/entities/contenidos/96/AFairandJustTransition.pdf.
Ledwith, Sue, and Lise Lotte Hansen, eds. 2013. *Gendering and Diversifying Trade Union Leadership*. 1st ed. Routledge Research in Employment Relations 29. New York: Routledge.

Macalister, Terry. 2009. Vestas Protest: What's It All About? *The Guardian*, July 24. https://www.theguardian.com/vestas/vestas-closure-protest-workers.

Mills, C. Wright. 2000. *The Sociological Imagination*. Oxford, England; New York: Oxford University Press.

Morena, Edouard, Dunja Krause, and Dimitris Stevis. 2020. *Just Transitions: Social Justice in the Shift Towards a Low-Carbon World*. London: Pluto Press.

Morris, William. 1883. *Useful Work versus Useless Toil*. The William Morris Internet Archive. https://www.marxists.org/archive/morris/works/1884/useful.htm.

———. 1908. *News from Nowhere*. London: Longmans. https://www.marxists.org/archive/morris/works/1890/nowhere/nowhere.htm.

Portelli, Alessandro. 1997. *The Battle of Valle Giulia: Oral History and the Art of Dialogue*. Madison, WI: University of Wisconsin Press.

Räthzel, Nora, and David Uzzell. 2011. Trade Unions and Climate Change: The Jobs versus Environment Dilemma. *Global Environmental Change Part A* 21 (4): 1215–1223. https://doi.org/10.1016/j.gloenvcha.2011.07.010.

Räthzel, Nora, David Uzzell, and Dave Elliott. 2010. Can Trade Unions Become Environmental Innovators? *Soundings* 46 (46): 76–87.

Räthzel, Nora, David Uzzell, Ragnar Lundström, and Beatriz Leandro. 2015. The Space of Civil Society and the Practices of Resistance and Subordination. *Journal of Civil Society* 11 (2): 154–169. https://doi.org/10.1080/17448689.2015.1045699.

The New Lucas Plan Working Group. 2020. *An Idea Whose Time Has Come*. A New Lucas Plan. http://lucasplan.org.uk/.

TUC. 2008. *A Green and Fair Future for a Just Transition to a Low-Carbon Economy, Touchstone Pamphlet 3*. London: Trades Union Congress. https://www.tuc.org.uk/sites/default/files/documents/greenfuture.pdf.

———. 2010. *GreenWorks: TUC GreenWorkplaces Project Report. 2008–2010*. London: Trade Union Congress.

———. 2014. *The Union Effect: Greening the Workplace*. Economic Report Series. London: Trades Union Congress. https://www.tuc.org.uk/sites/default/files/The_Union_Effect_Greening_The_Workplace_Covers_2014_All.pdf.

TUSDAC. 2005. *Greening the Workplace: A Report by the TUSDAC Unions*. London: TUC/DEFRA.

TUSDAC/Carbon Trust. n.d. *Environmental Education for Trade Unionists: A Guide for Students*. Didcot: Envirowise/Carbon Trust. http://www.wrap.org.uk/sites/files/wrap/GG488B_final.pdf.

UK Government. 2019. *Trade Union Membership: Statistical Bulletin*. London: Department for Business, Energy and Industrial Strategy, UK Government. https://www.gov.uk/government/collections/trade-union-statistics.

UnionLearn. 2009. *Targeting Climate Change*. London: Trade Union Congress.

———. 2020. *TUC Education Offers Green Training*. https://www.unionlearn.org.uk/news/tuc-education-offers-green-training.

Uzzell, David, and Nora Räthzel. 2019. Labour's Hidden Soul: Religion at the Intersection of Labour and the Environment. *Environmental Values* 28 (6): 693–713.

Wainwright, Hilary. 2020. *From Airplane Wings to Ventilator Parts*. A New Lucas Plan. April 27. http://lucasplan.org.uk/2020/05/06/from-airplane-wings-to-ventilator-parts/.

Wainwright, Hilary, and Dave Elliot. 1982. *The Lucas Plan: A New Trade Unionism in the Making*. London: Allison and Busby.

White, Damian. 2020. Labor Centered Design for Just and Sustainable Transition. In *Just Transition Research Collaborative Webinar Series Justice in Low-Carbon Transitions*. United Nations Research Institute for Social Development. https://www.unrisd.org/jtrc-webinar-intro.

Winstanley, Gerrard. 1649. *The True Levellers Standard Advanced: Or, The State of Community Opened, and Presented to the Sons of Men*. https://www.marxists.org/reference/archive/winstanley/1649/levellers-standard.htm.

Winstanley, Gerrard, and Tony Benn. 2011. *Gerrard Winstanley: A Common Treasury*. Revolutions. London; New York: Verso.

# 28

# Individuals Transforming Organisations: Spanish Environmental Policies in Comisiones Obreras

Nora Räthzel

## Introduction

How do trade unions develop environmental policies that go beyond their traditional domain of health and safety issues at the workplace, including global threats like the climate crisis? Trade union action develops through a number of processes feeding off each other. There are exogenous processes like the destruction of the environment itself, political pressures from governments, parties and corporations as well as endogenous processes like pressures from their members, the ways in which decisions are taken, their organisational structures, political fractures and the history of their creation. A number of such processes and conditions that have led to trade union environmental policies—or the lack of them—have been analysed. They include the economic sector in which unions operate, the societal, political and economic pressures they experience, and the histories, which define their political ambitions (see chapters in Parts 1, 2 and 4 in this volume, and Felli 2014; Farnhill 2016; Snell and Fairbrother 2010; Vachon and Brecher 2016; Morena et al. 2020).

What has not been discussed is the role that individual unionists might play in creating an environmental agenda in their union. To understand the role of individuals in transforming organisations is the purpose of this

N. Räthzel (✉)
Department of Sociology, Umeå University, Umeå, Sweden
e-mail: nora.rathzel@umu.se

chapter. To investigate this turned out to be a delicate matter. Some unionists declined to be interviewed, arguing that individuals are not essential for the way in which unions act. What counts is collective action. Others thought we had 'nailed it', describing how an agenda could collapse if the individuals engaged in creating it left.

The strength of trade unions lies in their character as collective organisations, which can harness the power of workers to fight for their rights. Concepts like 'collective bargaining' signify this perspective. Where individual unionists are portrayed, they tend to be presented as lone heroes achieving their goals through their exceptional character. Feminist researchers have critically discussed this perception where union leaders are still predominantly 'stale, male, pale' (Ledwith and Hansen 2013, xiii), although the membership has become more female and diverse. Discussing the role of individuals in organisations does not mean to disregard collective strength. By placing life-histories of unionists within their organisational and historical context, I try to avoid structuralist determinism and individualistic reductionism. Understanding leading individuals as 'organic intellectuals' will help to achieve this.

The philosopher and co-founder of the Italian Communist Party, Antonio Gramsci, created the concept of 'organic intellectuals' in difference to 'traditional intellectuals'. The concepts are embedded in his definition of new social movements. Their aim, he argued, was to gain cultural and political hegemony for a transformation of society, which would replace the ruling block and democratise society, including the mode of production. To achieve such a hegemony, he regarded 'organic intellectuals' as essential, situating them within their societal context as representatives of their class. He emphasised the interrelationship between an individual's capacities, the social relations within which they were developed and the societal context within which they become influential:

> Can one find a unitary criterion to characterise equally all the diverse and disparate activities of intellectuals and to distinguish these at the same time and in an essential way from the activities of other social groupings? The most widespread error of the method seems to me that of having looked for this criterion of distinction in the intrinsic nature of intellectual activities, rather than in the ensemble of the system of relations in which these activities (and therefore the intellectual groups who personify them) have their place within the general complex of social relations. (Gramsci 1999, 139)

The 'intrinsic nature of intellectual activities' do not define an organic intellectual, he argued, since, 'All men are intellectuals' (…) in any physical work, even the most degraded and mechanical, there exists (…) a minimum of creative intellectual activity' (ibid.). What is decisive for the definition of an intellectual is therefore not their activity as such, but their activity within a context where they have the 'function' of an intellectual:

> The mode of being of the new intellectual can no longer consist in eloquence, which is an exterior and momentary mover of feelings and passions, but in active participation in practical life, as constructor, organiser, "permanent persuader" and not just a simple orator. (ibid., 142)

The difference between an eloquent orator and a 'permanent persuader' marks the difference between the traditional and the organic intellectual. Translating Gramsci's idea of an organic intellectual into our context needs the addition that the function of environmentally engaged unionists consists not only in persuading those *outside* the union but also those *within* the union of the need to broaden their agenda. It means creating a new hegemony, where environmental issues, like the climate crisis, that go beyond occupational health and safety concerns will be regarded as a self-evident trade union issue. In order to succeed, such unionists would need to 'actively participate' in the union's 'practical life', by connecting the struggle for workers' rights with the struggles for 'nature's rights'. They would have to help create a structure within the union that entered its different sectors and involved members and union officials alike. In our research project,[1] we found many examples of unionists who were introducing environmental policies into their unions with varying success (Räthzel and Uzzell 2019; Räthzel et al. 2018; Lundström et al. 2015; Uzzell and Räthzel 2013). The most comprehensive practices were developed by unionists in the *Comisiones Obreras*, one of the two largest trade union confederations in Spain. This chapter presents three moments of environmental policies in CCOO: its beginnings in 1991, its crisis after 2008 and its revival since 2017. It is based on 13 life-history and informative interviews with its main protagonists conducted between 2009 and 2020 as well as documentation obtained from interviewees and at the webpage of the department for the environmental of CCOO. Three of the interviewees (all men) were born towards the end of the 1940s and beginning of the 1950s, and the other

---

[1] Moments of Danger, Moments of Opportunity. The Role of Individuals as Change Agents in Organisations: Trade unions in Sweden, UK, Spain, Brazil, South Africa and India. Project members: Nora Räthzel, David Uzzell, Ragnar Lundström, Beatriz Leandro, Payoschni Mitra, Nilanjan Pande, Piya Chakraborty, Tamara Walker. Funding: Swedish Research Council (VR), DNR: 2010-1990.

six (two women and four men) at the beginning of the 1960s. Presenting their life-histories shows how they developed their perceptions of the labour-nature relationship and how these informed their practices. Since they elaborated systematic strategies to put environmental issues at the centre of their union's policies, I speak of their work as an 'environmental project'. The next section discusses how the first generation of environmentalists developed their ideas about the labour-nature relationship. The following two sections describe the progress of their environmental project within CCOO and subsequently at the European and the global level. Section 'How Did the Project Succeed?' steps back to explore the individual, organisational and societal conditions that allowed the project to flourish, while section 'How Did the Project Implode?' discusses the reasons for its crumbling. In section 'The End or Dormant Seeds Waiting to Re-flourish?' I describe the recent revival of the project, ending with a conclusion that reflects on what can be learned from this history about the role of individuals for the transformation of organisations.

## Life-Histories of Spanish Trade Union Environmentalists

### Growing Up Under a Dictatorship

The first generation of union environmentalists grew up under the Franco dictatorship, and all but one attended religious schools or convents, since the public schools were considered bad quality by their parents. The experience was, nevertheless, not the same for all. In one case, the convent was reactionary and abusive. In other cases, the unionists became acquainted with priests who subscribed to liberation theology or were at least 'liberal' and 'respectful' towards students who were not religious. All of the men were active in organisations while at school, either members of the scouts, of youth groups engaged in political resistance or both. The women were told by their parents to stay out of organisations since they were organised by the regime or they did not have the time, since, as one of the women said, 'I was a very hard-working student'.

Between 1991 and 2014, when we conducted the life-history interviews, they were or had been responsible for environment and occupational health in different positions: Joaquín Nieto had been on the executive board of CCOO and the first director of the department for the environment in 1991 and remained there until 2008. Manuel Gari Ramos had been director of the Institute for Research on the Environment and Occupational Health (ISTAS) since its creation in 1996 until 2011. Carlos Martínez Camarero remains advisor and member of the CCOO Department of environment in Madrid since 1992. Llorenç Serrano was responsible for Environmental Policies in Catalonia between 2002 and 2015, and Manel Ferri Tomás worked in Catalonia together with Serrano, creating the department of 'Sustainable Mobility' between 2004 and 2015. Mariano Sanz Lubeiro is the present confederal secretary of environment and mobility at CCOO headquarters since 2017. Three interviewees wished to remain anonymous which is why we do not describe their positions, since that would disclose their identities. We call them Adrián, Maria and Amalia.[2]

Nieto, Gari Ramos, and Martínez belong to the initiators of environmental policies. To understand their politics and their worldviews, one needs to know that CCOO started at the end of the 1950s as a resistance movement against the Franco regime, organising workers against appalling working conditions and the regime in general. It was founded by the communist party and became a clandestine organisation in 1967 when it was forbidden. While being organisers of strikes and demonstrations, the unionists had to watch out not to be caught and imprisoned. They had to meet in different places and change houses regularly. Two of them were imprisoned. One of them told us that when, after the end of the Franco regime, his charges were dropped, he told his wife that she did not know him under his real name. 'Well', she answered, 'you do not know me under my real name either'.

## Perceptions of the Labour-Nature Relationship

When CCOO was established legally in the Spanish democracy, Marcelino Camacho, the famous leader simultaneously of the Communist Party and the CCOO, wanted to open up the union further to other political groups. The Communist Party had played an important role in the union while it was clandestine, but other left-wing groups had also been active in it. It was in the interest of the union to broaden its political space if it was to represent workers from different walks of life. Thus, Gari Ramos, who belonged to a small left-wing group, which he had joined because it was more democratic, was asked by Camacho to work for him in the union and agreed. He became one

---

[2] The quotes used in this text were sent to the interviewees for their approval.

of the first unionists who was freed from his employer (a bank) to work for the union.

The integration of different political groups went further under its next general secretary Antonio Gutierrez. Joaquín Nieto was the leader of a left-wing faction in CCOO, promoting a more radical and open unionism. Gutierrez told him that they wanted to integrate the political tendencies by electing their leaders to the executive board and giving them a responsibility. Asked, which responsibility he wanted to assume, Nieto answered he wanted to create a department of the environment:

> At that time my thinking was already formed but remained theoretical. My influences were the economic Marxism of the workers movement, and the pacifist ecologism and feminism of the social movements at the time. While they valued the economic and social vision of Marxism, they also saw gaps, for example, concerning the natural resources. Petra Kelly and the German Green Movement were central. The pacifist movement of which I was a part, was responsible for peace and disarmament in the Catalan branch of CCOO and it had an influential green intellectual element. In this context I started to think about the relations between society and nature and how what we did to the environment affected society. (…) And then there was my concern about occupational health. I had always been close to the world of work, concerned with health, I had been working many years in construction and seen many accidents and the dangerous working conditions. (…) And when you go into occupational health you immediately begin to see the whole picture: how the contaminators that destroy the environment are the same as the ones affecting the workers. That the species that needs to be protected is the worker and the human species. You then acquire a more profound insight about how the environmental dimension inserts itself into production and into its surrounding, the workers, the community. (…) With the help of environmental literature, I realised that the world that is going to come, the way it is configured, cannot continue. Not only because of inequality, and social injustice but because if the relationship between society, the economy and nature does not change, there will be no viable future. The natural resources are limited, and we use them at a pace that is much faster than the time they need to regenerate, and we use resources that are not renewable. (…) With these basic ideas I went to Gutierrez and said, "the world is changing, (…) and one of the most important changes will come from ecology. As unions we are outside of that, but the workers are not. Everything that is happening is happening to them. (…) It is important to find out what we need to do (…) and that is why I want to take on the responsibility to explore that question."

I have quoted Nieto's statement at length because it exemplifies how the unionists of his generation came to merge the ecological with the workers' issue as a theoretical framework that enabled them to make sense of their experiences as workers and as political activists. It is this connection between theory and practices, their being emerged in different but connected struggles that formed their capacities as organic intellectual. Gari Ramos gives a similar example:

> Being active in the pacifist movement I came in contact with ecologism and I was fascinated by the fact that the ecologists said very interesting things, they questioned many things about productivism, they made us re-think Marxism (…) insisting that we did not only have to change the world but the way we live as humans, address the problem of our relationship to nature. So, I took up a thread within Marxism and began to rethink all those issues that I had left in standby, namely that Marxism itself related that the enrichment of the Bourgeoisie rests on two feet: the exploitation of the worker and the exploitation of the earth.

Being Marxists but also engaged in the new political movements, these unionists came to perceive nature as one of the two elements that makes society possible: work and nature. Their perception resembled the ideas of the German philosopher Ernst Bloch who developed the concept of an 'alliance technique' (*Allianztechnik*) within and with nature. Only by developing a nature-labour alliance, he argued, can humans stop living in nature as if in enemy territory, thereby facing the constant threat of self-extermination (Bloch 1975, 251). Whether they argue for the need to adjust the pace of production to the pace of nature's recreation or retrieve Marx' analysis that Capital productivity is based on the destruction of the worker and the earth,[3] the nature-labour relationship becomes the centre of their environmental trade union project and simultaneously a regeneration of a Marxist understanding that includes this relationship, as Martinez formulates it:

> I believe that the solution of the problems needs to result from a change of the present model of production (…) we even have to address the issue of private property, to transform it into public property, to share work, share employment and many more things. Less usage of materials, less usage of fossil fuels. That is

---

[3] …, all progress in capitalistic agriculture is a progress in the art, not only of robbing the labourer, but of robbing the soil; all progress in increasing the fertility of the soil for a given time, is a progress towards ruining the lasting sources of that fertility. The more a country starts its development on the foundation of modern industry, like the United States, for example, the more rapid is this process of destruction (Marx 1998, XV–726).

obvious and that can include jobs in many sectors, we need to make a transition, and we also have to address the issue of sharing employment. (…) For instance, parts of CCOO talk about growth, not about development. We environmentalists have never succeeded in defeating that idea that growth is necessary in order to solve the employment problem.

Unlike many Marxists at the time and today, they rejected the productivist model, falsely attributed to Marx, while questioning the growth model. Employment can be decoupled from material growth and instead embedded in a vision of reduced but shared employment. As the previous two chapters have shown, there are also unionists in the UK and the USA subscribing to a Marxist or Socialist ecology. But they cannot be found in the leadership of any union. It was from a position of union power that the environmental policies in CCOO took off as shown in the next section.

## Strategies for Environmental Trade Union Policies

When elected Director of the department for the environment in 1991, Nieto set out to develop environmental policies by learning from other unionists in his position in other countries:

> I had the illusion that I would look around in Europe and in the world at countries like Germany, Brazil, (…) to find many things about work and the environment. I was surprised that I didn't find anything. I thought, surely there was something, but it was hidden, because there were partial, emerging experiences but I did not find anything of relevance. So, I said to myself: "Joaquín, you are on your own. You have to start from scratch".

By studying ecological literature and contacting environmental groups, Nieto defined the areas in which CCOO would have to develop its environmental practices: 'climate, energy, and chemical risks, and (…) biodiversity and water'. This was a broad delineation of environmental politics, a genuine ecological programme, that included but went beyond health and safety risks at the workplace. It was in tune with the political and scientific developments of the time.

The historian Spencer Weart (2008) notes that in the 1960s the conviction among scientists grew that $CO_2$ was producing global warming. The issue was settled at the beginning of the 1990s when solar activity plunged but the Earth's temperature increased. This showed that the influence of solar activity on the Earth's climate was weak. The Intergovernmental Panel on Climate Change (IPCC) was founded in 1988, and in 2001, it declared the consensus that the earth was warming. In 1992, world leaders met in Rio de Janeiro and in a Framework Convention signed by more than 150 nations promised to work on preventing dangerous anthropogenic interference with the climate system. Thus, Nieto took up his position when the climate crisis had begun to be accepted as a global threat. While some unions across the world will have discussed the issue and started to develop broader environmental policies including climate change (Waddington and Hoffman 2000), there was no European union that had appointed a member of its executive board responsible for environmental union policies.

Nieto's and his colleagues' work included setting up a network of union representatives for health *and* the environment across Spain and a trade union research institute for environment and occupational health (ISTAS) in 1996. All shop stewards responsible for health and safety were educated in environmental topics, enabling them to detect environmentally dangerous elements in production processes. Roughly 60,000 shop stewards were trained this way, and 350 union officials across Spain became trained environmental specialists. ISTAS employed over 100 technicians, social scientists and engineers conducting research on environmental and health risks as well as investigating the relationship between environmental transformation and employment. It became the first trade union organisation in Europe to investigate the effects of a fossil-free industry on jobs:

> In 2000 we were the first union showing that with a programme to mitigate climate change, with renewable energies, you could actually create jobs. People were fascinated. It was something new. Until then the unions had always thought about climate change measures as job killers, this was the first time they learned that such measures could actually create jobs. (Amelia)

Since then, campaigns for 'Climate Jobs' have developed in different parts of the world building on this argument (see Chap. 10).

## Beyond the Jobs Versus Environment Dilemma

The main reason why trade unions hesitate to embrace environmental policies remains their and their members fear that they will cost jobs. To overcome this perceived (Räthzel and Uzzell 2011) and *in the short run* real dilemma, protecting jobs and the environment is one of the aims of the campaigns for climate jobs. They show that an environmentally sound economy creates more jobs than the present fossil fuel–based ones. The Spanish unionists radicalised such arguments, like for instance Llorenç Serrano, one of the younger environmentalists. Until 2015, he was responsible for the environment at the branch of CCOO in Catalonia. Shortly after the demise of the Franco regime, at the age of 16, he co-organised a youth movement against mining on one of the region's most emblematic mountains. Democratic movements were seen as positive, especially led by young people who were regarded as apolitical. With the support of environmental movements, they succeeded in convincing the Catalan government to withdraw the mining rights and pass a law for the creation of natural reserves. From this time, Serrano explained, came his sensibility for environmental issues. His argument for another model of production turned the tables on the job vs. environment dilemma:

> We discussed (…) what are the necessary conditions of companies in order to have employment? (…) We argued that what was necessary now was to produce sustainably. If you do that there will be jobs in 20 years' time, if not you will not have jobs. Either you produce sustainably (…) or you will have to close down. We are not discussing whether we can stop the change. The change will come. (…). We need to be the motor of change, not those trying to block it.

This argumentation departs from a defensive position in which workers are reassured that the transition, while costing jobs, will also create more new ones. Instead, it argues that it is the existing mode of production which is the actual job killer while and that an environmental transformation is the progress that will secure jobs, not a misfortune from which workers need to be protected. Connecting environmental transformation with economic progress, Serrano creates a link between caring for workers and caring for the environment and implicitly discards the idea that environmentalists want to send us back to the stone age. It generates a discursive constellation where those supporting fossil fuel industries need to defend why they are putting nature *and* employment in jeopardy. Consequently, environmental unionists, developed alternatives at all levels of the production process, releasing the ingenuity of workers, who, assisted by technical staff, learned to detect dangerous working conditions and how they could be improved. Worker involvement became an important part of environmental union strategies:

> We do not only do research within ISTAS, but also together with workers in the factories. For instance, we did a study about the dangers of detergents and developed the usage of ecological alternatives for cleaning buildings and machines.

> We won a prize for this study from the department of health and safety of the European Union. And those who went to collect the prize were the women workers who had participated in the research. (Gari Ramos)

Such initiatives were only possible with the consent of management and dealt with health and safety and environmental issues (contamination of communities by factories nearby) locally. Broader environmental policies, like the reduction of $CO_2$ emissions, were part of the general CCOO demands discussed and approved at their annual congresses. In 1998, CCOO founded the journal *Daphnia* for Environment and Labour that still exists. Our research has not assessed the effects of CCOO's environmentalism at workplaces, but we can argue that their activities, the number of unionists involved and the organisational structures created are not matched by any trade union federation we know of.

## Sustainable Mobility: Broadening the Notion of the Environment

One example of the breadth of the CCOO's environmental engagement is their model of 'sustainable mobility'. This was the work of a unionist who belonged to the younger generation of environmentalists in CCOO. Manel Ferri was born in 1960. When he began his engagement with energy and mobility issues in 1976, the Franco regime had just ended. Asked about the beginning of his environmental interest, he answered:

> As a boy I worked and when I got up at five, I already saw how the city was contaminated, the black fumes, the contaminated ground. I was also a member of the hikers. That is where my environmental conscience comes from. (…) I was 16 and I read a book about urbanism, the noise, the cars. This defines you and you start to think about the majority, how to live, the theme of urbanism. Conceptually, how to improve the public spaces, less contamination and noise. And that environmental sensibility then also led to politicization, to resistance against nuclear energy.

Ferri loves the city he lives in, Barcelona. Thus, his engagement with urban issues is also a way of improving his living conditions. His definition of 'environment' reflects his interest in the lived environment.

> People did not share our vision. If you take a survey and ask people about the environment what they think about are flowers, little birds. But when you say

the environment is the noise, petrol, garbage, all that, people do not associate this with the environment.

While for the people around him the environment was an escape from daily life for recreation, Ferri wants to make the place of daily life a place of recreation. In 2004, he created the department of sustainable mobility within the department of environment in Catalonia. That he is able to create such a department within the regional union leadership is due to the federal organisational structure of the CCOO, constituted by economic sectors and territories. As a territorial union, CCOO cares for the issues pertaining to the workers' lives outside the factory. Nevertheless, a centre of Ferri's work was the journey to the workplace and included organising workers to fight for public transport to their workplaces. He relates how CEOs and administrators with whom he negotiated were surprised to find that somebody who had just been blocking the streets would know more about urban issues than they did.

Becoming an international authority in his area, Ferri was asked in 2008 to work in Madrid. First, he created a 'sustainable mobility' department at ISTAS.[4] In 2009, Serrano became the new director for the environment in Madrid and Ferri joined him. However, due to financial constraints after the financial crisis, CCOO had to abolish the department in Madrid and in Catalonia in 2015, and he moved back to his job at the Barcelona administration continuing to work on sustainable mobility ever since. In 2019, his plans were published by the Spanish ministry of ecological transition and the ministry of the interior.[5]

## Transcending Borders

The union environmentalists developed links with the European Trade Union Confederation (ETUC) and numerous interested unionists in French, German, Belgian and other unions to create a European network of environmentally engaged unionists. In 2002, during the Spanish presidency of the European Union, they organised a conference on unions and the environment in Sevilla. Its results were accepted by the ETUC as a common European strategy.

In preparation of the World Summit of Sustainable Development in Johannesburg, 2002, relations with unions in different parts of the world were

---

[4] http://www.gencat.cat/transit/2014_VI_Congres/documents/ManelFerriENG.pdf.
[5] www.idae.es/sites/default/files/la_movilidad_al_trabajo_un_reto_pendiente_dgt_idae_junio_2019.pdf.

developed as well as with Lucien Royer, who was responsible for environmental issues at the International Confederation of Free Trade Unions (ICFTU). In 2003, with the support of The United Nations Commission for Sustainable Development (CSD) and ETUC, ICFTU and CCOO, Nieto co-founded *Sustainlabour*, the 'International Labour Foundation for Sustainable Development', based in Madrid. Its purpose was to reach out to unionists across the world, especially in the Global South, to offer training about climate change, workplace contamination and strategies of mitigation and adaptation. With the support of the ILO and United Nations Environmental Programme (UNEP), *Sustainlabour* organised the first international trade union conference on labour and the environment in Nairobi 2006, which was also the year in which the International Trade Union Confederation (ITUC) was formed out of a merger of the ICFTU and the World Confederation of Labour. One of the organisers recounts:

> Jean and I (…) knew our union but we had to (…) find out about the unions all over the world, get on the phone and call people, solve conflicts between unions in different countries. And Jean was only 24 years old and we are still laughing that we had the audacity. We knew that this work was important and therefore it was important to do it well. We needed to identify the unions and their leaders and tell them, for instance, you need to come because you are progressive.

They had resources from UNEP but only one year to organise the conference. Creating international working groups with different themes, their aim was to develop analyses and alternatives. When we interviewed unionists in the UK, Brazil and South Africa in 2009, many of them mentioned the Nairobi conference as a turning point, where they realised the importance of climate change and the need to develop a mitigation and adaptation programme in their union.

Three years after Nairobi, *Sustainlabour* supported the ITUC in organising the first 'Pavilion of the world of work' (WOW) at the COP15 in Copenhagen. The ITUC declared:

> the trade union body representing trade union confederations globally, is organising a space where trade unions and trade union-related organisations from different regions, sectors and interests will be able to present their experiences and share, debate and plan with other actors (governments, social and environmental movements, enterprises, UN and regional agencies) how a new society can be built, a society where climate is protected, where workers and trade

unions are integral parts of decision making and where an environmentally friendly and fair sustainable future is no longer an illusion but is a reality.

The holistic view of this declaration, where planning to protect the climate is embedded in planning for a new society, is noteworthy. In later years, especially since 2019, ITUC declarations have narrowed their ambitions. The COP15 was the beginning of a broader environmental engagement of unionists across the world. This was predominantly the work of Anabella Rosemberg. She worked with Royer for four years, and after he retired in 2009, she became responsible for environmental issues at the ITUC. She communicated with unionists across sectors and political affiliations in a way that motivated many of them to get involved.

# How Did the Project Succeed?

In what follows I take a closer look at the conditions for the success of the environmental project at CCOO, which lasted roughly until 2010, when it started to come under pressure due to the 2008 financial crisis.

## The Individual Conditions

I begin with an answer to the introductory question: How do the unionists perceive the labour-nature relationship, and how does that influence their environmental strategies? The three unionists we interviewed developed their concept of the labour-nature relationship as Marxists fighting against the Franco regime. Through participation in new social movements, they came to critically rethink their Marxist worldview as one lacking the insight that not only labour but labour and nature sustained societies. Without explicitly using Bloch's concept of the nature-labour alliance, they saw the human-nature relationship as a mutually supportive one. While nature cannot be destroyed, nature as a condition for human life can. In this perception, there is no place for the idea that one can decide between protecting the environment and protecting jobs. Thus, unions had to embrace the question of the environment as a genuine trade union issue.

While their vision of the labour-nature relationship as inseparable motivated their goal to transform their union, it was also decisive that this vision was part of how they saw themselves and their role in society:

In fact, my professional life, or my life as a social activist is marked by (…) the relation of my work to the environmental world. (…) For 20 years it has been the most important commitment of my life to take the environment to the world of work and to the environmental world the social dimension. That is my life. (Nieto)

That the labour-nature alliance became part of their worldview and simultaneously part of what they saw as their role in society enabled them to often 'win the argument' and withstand internal conflicts. Their identification with their work also explains why, when they left the union, they continued connecting the world of work with the world of the environment.

It was also this holistic worldview that enabled them to avoid a reductionist approach to climate change policies linking issues like contamination, water, air, mobility, energy (climate change) and biodiversity. This included the creation of worldwide networks, since it was clear that none of these issues could be solved in one country alone. While international networks strengthened their position at home, being an environmental avantgarde also strengthened their influence internationally. Their policies were recognised not only by European unions but also by some in the USA like the Oil, Chemical and Atomic Workers union (Leopold 2007) and supported by the ILO and UNEP.

The holistic theoretical-political framework within which they understood the labour-nature relationship enabled them to step out of a defensive position and connect environmental measures with progress and the lack of them with job losses.

## The Organisational Conditions

In spite of resistance from more traditionally oriented unionists, the history and the structure of the CCOO facilitated the creation of a confederal secretary of environment at the level of the executive board as well as the creation of ISTAS and Sustainlabour as organisational resources. Having evolved as a social movement that organised workers for their rights at the workplace and in society at large, the idea that unions needed to engage in improving the whole way of life of workers was part of the union's identity reflected in their organisational structure of sectoral and territorial units. The aim to reconcile different political factions in CCOO by including their representatives in the executive body became another facilitator of environmental policies but only because there were individuals wanting to take on that responsibility. While having to convince his fellow members of the executive board and the general secretary and getting approval from the national congress, the secretary of environment did not have anybody above him. He was authorised to issue directives as opposed to receiving them. This institutional position of power explains the breadth and depth of the environmental project at CCOO.

While during the first four years Nieto was occupied with developing a strategy for environmental union policies, he then accepted the additional responsibility for occupational health which allowed him and his colleagues to connect broader environmental issues with the everyday concerns of workers. Developing arguments and strategies for an alternative model of production, they also created the necessary staff, institutions and research to support workers in their struggle for decent, non-contaminating working conditions locally. This did not mean that workers necessarily also bought into environmental policies, but it meant that they supported the people, who helped them solve their daily troubles. As Nieto put it: 'I had a strange love—hate relation with the miners: they loved me as secretary of occupational health, but they hated me as secretary of environment because I was "going to take away their jobs"'.

## The Societal Conditions

Although the heydays of social movements after the demise of the Franco Regime were over, Comisiones Obreras and other social movements were still strong and anchored in society. The socialist government of Felipe Gonzalez was in power (1982–1996), and the climate for left-wing policies was favourable. Internationally, the United Nations were preparing the conference for sustainable development in Rio (1992). While Nieto did not find his homologue in any European Union, he found environmentally engaged unionists in Europe and other parts of the world.

In sum, the project of environmental policies went through many conflicts but also from success to success roughly between 1991 and 2010 because of what Althusser would have called an overdetermination (Althusser 1971), an accumulation of conditions at different societal levels strengthening each other.

When I went back to Sustainlabour and interviewed environmentalists at different CCOO branches in 2014, things were looking bleak. In what follows I will discuss how and why the environmental practices of the CCOO 'imploded' as Ferri put it.

## How Did the Project Implode?

In 2008, the financial crisis hit the world. Spain was one of the most severely hit countries because its economy was centred on a 'construction bubble'. When people lost jobs and could not pay their mortgages, when companies in the construction sector lost their money and the cities filled with 'urban corpses' of unfinished buildings, the unions experienced a disastrous loss of members and thereby resources. As a result, the environmental project crumbled on all levels.

On the individual level, because some of the first generation of environmentalists left or were dismissed. After 15 years, Nieto left his position in 2008 arguing that it was time to let the next generation continue the work. In 2009, Llorenç Serrano became the new confederal secretary of environment and occupational health. However, the mood in CCOO had changed, and Serrano did not stay long. The people I interviewed were the architects of the great expansion of environmental practices in CCOO. Their disappointment with the crisis of the project needs to be seen in that context. For them, the project, the work they had invested, crumbled. At the same time, the department for environmental and health issues remained in spite of financial constraints, and Pedro Linares served as its secretary through difficult times. Although with less staff, ISTAS also remained an essential organisation and a reference point for unionists concerned with environmental and health issues.

A number of interviewees emphasised the importance of individuals for the success and in turn the loss of momentum, Amalia explained:

> When you are a leader and you change an organisation, you also create enemies. For instance, Nieto was in the public eye all the time, (…) everybody wanted to talk to him. This created envy and some hatred with other unionists. Therefore, when he left, they began to sabotage the issue of climate change as a kind of revenge. Not only because of that but also because they were not convinced the issue was a trade union issue and now saw their time coming, when they could get rid of the question, or at least diminish its importance.

At the same time, younger environmentalists were not able to enter into important positions, as older ones were leaving:

> We thought we were strong, but part of our strength derived from the fact that we had some power at the top. In big organisations you have some complicated dialectics. There were hundreds of people out of which 15 or 20 could have taken over. But what happens if these people do not access the higher levels of

the organisation? Then you lose the opportunity and they lose terrain. (…) Many people are not there anymore, either because they were fired (…) and that was very stupid because they fired a person of high capacities, and they now have, I believe, a mediocre team. And the result was that many people quit, since (…) they had worked there with less money because there was a project and there was a good working climate. (Gari Ramos)

There are some good people, but a generational shift did not take place because the project that we were constructing was not realised. (…) CCOO had a project for the country: to expand social protection, dignified work, to protect the environment. But with 27% unemployment this project has fallen apart. (Maria)

These stories do not only highlight the vulnerability of environmental policies in times of resource crises, but also highlight the vulnerability of organic intellectuals when the conditions for their success disintegrate. They are made to pay individually for the crises, while their success is defined as a collective effort.

In Ferri's view, the backlash also happened because the union environmentalists had not succeeded to convince their comrades that another model of production was necessary, the transformation had not gone far enough:

CCOO has been the best union, and now it is imploding, everything is imploding that is the right word. (…) They haven't been able to promote the idea that the energy companies need to be nationalised. (…) There is no failure, we have achieved things. Only at the moment the situation is very hard.

The project crumbled on the organisational level due to the economic crisis and the departure of environmentalists. Both events strengthened the position of those who thought the environment was not a trade union issue and the union should concentrate on its original duties, protecting workers. This was understandable since the situation was serious as Maria explained: 'every week you have a company that wants to throw half of its workers on the street. The whole union has to concentrate on this'. However, like all interviewees, she emphasised that in times of crisis it is an error to stick to the old model of production. The lack of resources hit Sustainlabour hard. It was closed down in 2016, after neither the ITUC, nor any of the unions worldwide, to which it had supplied knowledge and resources to attend the COPs during more than 10 years, stepped in to allow its survival. The ITUC did not create an archive for the important research reports that Sustainlabour had created for unions worldwide.

Interviewees were surprised how quickly the results of their work seemed to disappear:

> Leaving such a significant work, we did not believe that it would be that easy to destroy it through manoeuvrings in the union and that people would take over the positions who did not know anything about the environment, about occupational health, and didn't give a damn about it.

However, people like Martínez did not consider the project lost. Their comprehensive worldview provided them with resolve:

> It is a constant game of advances and backlashes. I will never have the situation that I have lost. It is a fight that will never be over. We will never lose, and we will never succeed. There will always be a situation of steps forward and backward.

On the societal level, it did not help that the socialist government lost the elections in 2011. Its minister of the environment had financially supported research projects of ISTAS and set up a round table in which the government, employers, trade unions and environmental NGOs discussed environmental policies. A law was adopted making these meetings compulsory at least twice a year. The Partido Popular (PP) government discontinued the meetings. It was hostile not only towards unions but also towards environmental issues.

Not only the economic base of society had changed but also its political and cultural texture as Amalia reflects:

> There are three interrelated crises that are leading to a demise of the environmental issue: the economic crisis, the fact that Nieto and other people left as well as the crisis of trade unions and political institutions in general. The union is not seen as an institution that can solve problems. They are part of the traditional institutions. And they are not trying to change their image. Although union members take part in other social movements, they do not want to say they come from the union because people would see them negatively.

The social movements, young people who fought against austerity politics and evictions, did not see unions as their allies. On the other hand, the younger generation of unionists had not made the experience of being part of a broader social movement like those fighting against the Franco regime. One interviewee complained that CCOO was becoming more like other unions, leaving its social movement history and identity behind and with it the conviction that social issues, including the environment, were trade union issues.

## The End or Dormant Seeds Waiting to Re-flourish?

In 2014, almost all our interviewees conveyed with different degrees of certainty that the demise of the project would be a passing period and that because of the gravity of the climate crisis times would change and the union would pick up where they had left since the structures and resources, while weakened, could become the seeds of a new movement of environmentally engaged unionists:

> We endured a few more years. With less resources, reducing activities. (…) At the moment the idea that CCOO could be the motor of change doesn't exist. (…) But the reality will force itself upon it. (…) I had more success in convincing the ecologists that there needs to be a just transition than to convince my union that there needs to be a transition. But although we have not convinced the whole organization, we left some resources which will regain weight.

Nieto, answering to the question how he sees the future, differentiated between the future of the labour-nature relationship and the future of environmental policies in the union:

> The future? Badly. I think the difference between environmental challenges and societal challenges is that the former are irreversible. Societal situations can be changed, you can improve the conditions of work. (…) Neither societal backlash nor progress are irreversible. But in the case of the environment, no. Many of the environmental damages are irreversible. In this sense all the indicators you see point to a disaster, and I think we are going towards disaster. (…) Will climate change destroy the human species? No, but society? Yes. In this sense I am very pessimistic. But I also think of what Gramsci said, pessimism of the intellect and optimism of the will. Reason makes you a pessimist, but you need the optimism of the will to hope you can avoid this undesired future. (…) From the point of view of the environmental agenda, it will soon be very high up there because the manifestations of what we have been talking about will be so evident, so hard, they will require a reaction, a late reaction, yes. Far too late. Governments cannot continue looking to the other side because of their interests and that of their friends in the oil industry, or similar sectors. They will have to face the problem and take action and probably with enormous social costs. Because those decisions will have to be taken very quickly, very late, creating difficulties for processes of a social just transition. The environmental agenda will surge and (…) whoever is concerned about the environment will have a future. But not for good reasons, but because the issues are there and cannot be

hidden. (…) But when decisions are finally taken (…) one has to be there, each of us has to continue from where they are.

In the next section, I will explore whether the hopes of our interviewees have come true by presenting the state of the environmental policies of the CCOO at the time of writing, 2020.

## Elements of the Revival

As our interviewees foresaw in 2014, the environmental question came back on the union agenda not least because the effects of the environmental crisis are becoming ever more evident. The CCOO website shows that Environment and Occupational Health are now two separate departments. The former does not need to piggyback on the latter anymore. One driver of a renewed environmental engagement was a demand from the government of the PP to send a representative to the commission for energy transition, which they created. Sanz, who became confederal secretary of environment and mobility in 2017, brought representatives from all the union's sectors and territories together, and for the first time, they agreed on a plan for a 'just energy transition' (Confederación Sindical 2018). This work and its result gave the environmental work in the union a boost and increased its basis among unionists. Sanz emphasises the inclusion of unionists at the workplace into the process of environmental change because they are the ones, who know the work processes best. In 2020, CCOO conducted a survey among 4530 unionists of which 98% agreed that there is a climate crisis, and 94% thought it was manmade. Eighty per cent agreed that something needs to be done by the union. This is a good foundation for environmental work as well as the support of the general secretary of CCOO for the issue.

In 2020, the COVID 19 pandemic is occupying the minds of civil societies and governments worldwide. While all the unions we investigated in Sweden, the UK, Brazil and India feature COVID 19 on their homepages, only CCOO describes the pandemic as an opportunity to rebuild the economy as an environmentally sustainable one.

A remarkable document was published in May on the CCOO homepage, demanding that the post COVID recovery has to be developed within the parameters of environmental sustainability. The document describes a range of demands that are familiar from other trade union publications for a 'just transition' but also goes beyond them:

The de-escalation of the health crisis towards a recuperation of economic activities needs not only to include measures to protect social needs and employment, including a guaranteed minimum income, the redistribution of wealth through fiscal measures, the sharing of employment through the reduction of working days and an increased public control of the economy, but also a change of the production model to confront the climate emergency. (Secretaría Confederal de Medio Ambiente y Movilidad de CCOO 2020, 6. Translation: author)

These perspectives, specifically the demand to 'reduce the consumption of energy, water and superfluous goods in the contexts of scarceness of raw materials and of global chemical contamination' (ibid., 8), set a different agenda than the calls for a 'green economy' which assume that 'green growth' is possible, thus disregarding the planet's physical limits (Räthzel and Uzzell 2019). The environmental and economic impacts of COVID 19 lend themselves to a vision of an economic recovery that transcends business as usual. The document incorporates perspectives from other social movements like the reduction of paid employment, discussed in feminist literature (Soper 2020; Haug 2011). It is also exemplary in connecting the issues of energy, biodiversity, mobility (a CCOO trademark since the work of Ferri and his colleagues), a circular economy and waste management, advocating an approach reminiscent of the earlier union environmentalists. Does this mean that the structures and resources built up in the 1990s are indeed flourishing again?

An interview with Mariano Sanz taught us that a simple yes or no is not possible. He emphasised the commitment of CCOO and his department to a profound societal ecological transformation. His department develops strategies for the reduction of $CO_2$ emissions, the further development of renewable energy, against the loss of biodiversity and for a just transition for all workers, especially those now affected by the closure of coal mines and nuclear plants:

> As trade unions we cannot remain at the margins of environmental problems because the transformation of an economic and productive model based on fossil fuels has societal and economic repercussions and consequences for workers. We need to develop suggestions that respect the rights of workers and enable a renewal of the affected territories and economic sectors. (…) In alliance with civil society we are mobilising for ambitious measures against climate change nationally and internationally. We belong to the *Alliance for the Climate (Alianza por el Clima: https://alianza-clima.blogspot.com)*, an umbrella association containing more than 400 organisations, of which we are co-founders, and in which we remain active.

The Alliance for the Climate was built in 2015 in preparation for the COP 21 in Paris. The membership of CCOO in the alliance shows that the union has overcome its previous reservations against other social movements immersing itself into civil society. The mentioning of territories and sectors underlines the double structure that CCOO owes to its history and also indicates a source of conflict: Traditionally, the territorial organisations are more open to environmental issues, while especially the economic sectors that need to either disappear or thoroughly transform are more averse to them. Thus, internal conflicts persist, Sanz says, but his department is ready to take them on.

This would be the 'yes' part of the answer, the seeds laid by the first environmentalists are germinating and the environmental agenda has taken off to a new start. However, the material conditions under which this is happening are far less favourable than those created in the 1990s. The department of the environment consists of four people working at the headquarters of CCOO in Madrid and four working at ISTAS. One person responsible for environmental issues exists in each territorial and sectoral office. During Nieto's times, there were at times 40 unionists working at headquarters and hundreds across the country. ISTAS needs to acquire 80% of its resources through external funding. Nevertheless, Sanz and his staff are determined to push the environmental agenda forward.

> I see a future full of complications, but an exciting one. We will face a lot of difficulties and contradictions concerning the social dialogue that we demand of the government and companies. Despite its weaknesses, the present government is a breath of fresh air with its dedication to an ecological transition, of which we have to take advantage. (…) I am convinced we will make progress. How much and how is another matter. It will depend on many variables that we cannot fully control. But I think we are at a moment where we are sowing the seeds for another model of development, that will respect people and the environment leaving the planet in a better state for the coming generations. I think the moment is now or it will not be. We are joining all our resources to make it happen.

The urgency is obvious. The moment is now because given the COVID pandemic, there is a need to re-construct the way to work and live. Not only in Spain and not only within trade union movements, there is a demand to 'build back better'. Some surveys show that the COVID crisis is motivating a majority of people to rethink our ways of working and living (Ferguson 2020). In that sense, CCOO's present environmental work towards renewal is reflecting a prevalent mood in society. However, the problem is that the changes discussed in society at large are not based on the fact that the corona virus and climate crisis have the same causes (Quammen 2020). Therefore, what change means is only superficially discussed in terms of an environmentally sound recovery. The argument goes that COVID is a possibility for 'greening' the economy, not that it is itself proof that a production model based on profitability does not only destroy the worker and the soil, but the future of humans and non-human species.

With a Spanish government that is—in spite of being a minority government and facing internal conflicts—convinced that the economy needs to become socially and environmentally sound, there is nevertheless a window of opportunity, however small, to create a society that is an ally, not an enemy of the earth's natural support system.

## Concluding Reflections

The surge as well as the depth and breadth of environmental policies in the CCOO, but also its crisis and its revival, are difficult to understand without taking the role of organic intellectuals as persuaders and activists seriously. While the role CCOO played in organising the resistance to Franco's regime provided a fertile ground for its left perspectives to be respected in society, people were needed who were able to take advantage of the societal affordances. Taking part in the workers' movement as well as in anti-nuclear and peace movements, they embodied the link between struggles that were otherwise fought separately, even at times adversely. This enabled them to merge insights from their practices with diverse theories, thus (re)connecting Marxism with the notion of a necessary labour-nature alliance. They integrated Gramsci's notion of organic intellectuals as persuaders and activists. While coming from the workers' movement and representing it, they also went beyond it, aiming to integrate other movements and ideas into the movement's identity. This was their strength but also their vulnerability. The importance of their presence became evident when they left. Traditional intellectuals clinging to a narrower definition of trade union identity got the upper hand, their arguments gaining authority due to the financial crisis of 2008 that devastated the Spanish economy and affected predominantly the working-class and thereby unions. The COVID crisis is different in that it shows—to those that want to see—the threats of the present mode of production and thereby opens a window for change. Moreover, the present Spanish government is sympathetic to unions and aims to reduce the material hardships of workers and businesses.

Without disputing the importance of material needs, what the implosion of environmental policies in the CCOO in times of crisis also demonstrates is a fear of change, of leaping into unknown territory. In spite of winning the argument, environmentalists did not convince a majority of their fellow unionists that an environmental transformation of the mode of production is a guarantee of jobs, not a threat.

One can only speculate whether the environmentalists could have stemmed the tide against their project had they still been present or had they created a generational transition. It is unlikely. Societies change, organisations change and with them what people see as worthwhile. The stronger immersion of CCOO in civil society organisations today is an indicator that while the environmental project might re-emerge, it also needs to be different, to adjust itself to the present. Organic individuals are indispensable, but their success is dependent on societal and organisational conditions they cannot control.

The main insight we gain from the Spanish case is that organisations change when there are organic intellectuals who take advantage of structural affordances. To do this, they need to be capable of creating comprehensive concepts that connect practices with theories and link experiences of their own movement with those of other social movements. Inclusion, not exclusiveness, is the path to hegemony and thereby to societal transformation. That individual capacities and structural affordances need to go hand in hand sounds trivial. However, environmental labour studies (and labour studies and environmental studies for that matter) are yet to create research that enables us to understand this relationship better.

# References

Althusser, Louis. 1971. Ideology and Ideological State Apparatus. In *Lenin and Philosophy and Other Essays*, 127–187. New York and London: Monthly Review Press. http://www.marx2mao.com/Other/LPOE70.html#s5.

Bloch, Ernst. 1975. *Experimentum Mundi: Frage, Kategorien d. Herausbringens, Praxis.* His Gesamtausgabe; Bd. 15. Frankfurt (am Main): Suhrkamp.

Confederación Sincical de CCOO. 2018. *Propuestas de CCOO Para La Transición Energética Justa.* CCOO. https://www.ccoo.es/ebc1375a411344d-ed377311728ebb201000001.pdf.

Farnhill, Thomas. 2016. The Characteristics of UK Unions' Environmental Activism and the Agenda's Utility as a Vehicle for Union Renewal. *Global Labour Journal* 7 (3). https://doi.org/10.15173/glj.v7i3.2536.

Felli, Romain. 2014. An Alternative Socio-Ecological Strategy? International Trade Unions' Engagement with Climate Change. *Review of International Political Economy* 21 (2): 372–398. https://doi.org/10.1080/09692290.2012.761642.

Ferguson, Donna. 2020. Only 12% Want a Return to the Old "Normal" Britain after Covid-19. *The Guardian*, July 12. https://www.theguardian.com/world/2020/jul/12/only-12-want-a-return-to-the-old-normal-britain-after-covid-19.

Gramsci, Antonio. 1999. *Selections from the Prison Notebooks*. Essential Classics in Politics: Antonio Gramsci. London: The Electric Company Ltd.

Haug, Frigga. 2011. *Die Vier-in-einem-Perspektive: Politik von Frauen für eine neue Linke*. 3. Aufl., dt. Orig.-Ausg. Hamburg: Argument-Verl.

Ledwith, Sue, and Lise Lotte Hansen. 2013. *Gendering and Diversifying Trade Union Leadership*. New York: Routledge.

Leopold, Les. 2007. *The Man Who Hated Work and Loved Labor: The Life and Times of Tony Mazzocchi*. White River Junction. VT: Chelsea Green Publishing.

Lundström, Ragnar, Nora Räthzel, and David Uzzell. 2015. Disconnected Spaces: Introducing Environmental Perspectives into the Trade Union Agenda Top-down and Bottom-Up. *Environmental Sociology* 1 (3): 166–176. https://doi.org/10.1080/23251042.2015.1041212.

Marx, Karl. 1998. *Capital. A Critique of Political Economy*. 1887th ed., Vol. 1., 3 vols. London: ElecBook.

Morena, Edouard, Dunja Krause, and Dimitris Stevis. 2020. *Just Transitions: Social Justice in the Shift towards a Low-Carbon World*. https://search.ebscohost.com/login.aspx?direct=true&scope=site&db=nlebk&db=nlabk&AN=2294678.

Quammen, David. 2020. *We Made the Coronavirus Epidemic*. January 28. https://www.nytimes.com/2020/01/28/opinion/coronavirus-china.html.

Räthzel, Nora, and David Uzzell. 2011. Trade Unions and Climate Change: The Jobs versus Environment Dilemma. *Global Environmental Change* 21 (4): 1215–1223. https://doi.org/10.1016/j.gloenvcha.2011.07.010.

———. 2019. The Future of Work Defines the Future of Humanity and All Living Species. *International Journal of Labour Research ILO* 9 (1–2): 145–171.

Räthzel, Nora, Jacklyn Cock, and David Uzzell. 2018. Beyond the Nature–Labour Divide: Trade Union Responses to Climate Change in South Africa. *Globalizations* 15 (4): 504–519. https://doi.org/10.1080/14747731.2018.1454678.

Secretaría Confederal de Medio Ambiente y Movilidad de CCOO. 2020. *Aspectos Medioambientales En Los Planes de Recuperación Post-Covid19*. Comisiones Obreras. https://www.ccoo.es/b755cd0cf4703aa21ce7e827a78f256d000001.pdf.

Snell, Darryn, and Peter Fairbrother. 2010. Toward a Theory of Union Environmental Politics: Unions and Climate Action in Australia. *Labor Studies Journal* 36 (1): 83–103.

Soper, Kate. 2020. *Post-Growth Living for an Alternative Hedonism*. London: Verso. https://rbdigital.rbdigital.com.

Uzzell, David, and Nora Räthzel. 2013. Local Place and Global Space: Solidarity across Borders and the Question of the Environment. In *Trade Unions in the Green*

*Economy. Working for the Environment*, 241–256. London; New York: Earthscan Routledge.

Vachon, Todd E., and Jeremy Brecher. 2016. Are Union Members More or Less Likely to Be Environmentalists? Some Evidence from Two National Surveys. *Labor Studies Journal* 41 (2): 185–203.

Waddington, Jeremy, and Reiner Hoffman. 2000. *Trade Unions in Europe. Facing Challenges and Searching for Solutions*. Brussels: ETUI. https://www.etui.org/sites/default/files/Challenges.pdf.

Weart, Spencer R. 2008. *The Discovery of Global Warming*. Rev. and Expanded ed. New Histories of Science, Technology, and Medicine. Cambridge, MA: Harvard University Press.

# Part VI

## Rethinking and Broadening Concepts

# 29

# The Commodification of Human Life: Labour, Energy and Money in a Deteriorating Biosphere

Alf Hornborg

## Introduction

My aim is to provide an overview and synthesis of some essential theoretical contributions on the relation between labour and energy over the course of human history. I shall review and critically discuss some central attempts to approach the labour/energy nexus in ways that defamiliarise our conventional understandings of these concepts. Such deconstruction of established notions of 'labour' and 'energy,' I will argue, is essential to grasping the conditions of what we have come to perceive as technological progress. It prompts us to rethink not only the global history of labour and energy technologies since the Industrial Revolution, but also the current ecological impasse of the so-called Anthropocene.

To accomplish such a rethinking, we must bring together two very different approaches to the labour/energy relation, in order to suggest the outlines of a third. The first approach adopts a physicalist perspective by discussing the use of labour as a source of energy among other energy sources with which it can be compared in terms of a historical progression of technologies (Debeir et al. 1991; Crosby 2006; Smil 2017). From this perspective, energy is approached in a straightforward way as a physical feature of the universe which does not require much epistemological reflection. The second, however, is a

A. Hornborg (✉)
Department of Human Geography, Lund University, Lund, Sweden
e-mail: alf.hornborg@hek.lu.se

constructivist approach to the very concepts of 'labour' and 'energy,' focusing on how these categories emerged in tandem in nineteenth-century Europe (Illich 2013b [1983]; Mirowski 1988, 1989; Rabinbach 1990; Daggett 2019). In juxtaposing historical accounts of material practices and constructivist accounts of the terms in which they were understood, we may surmise how economic history and the history of ideas have been recursively intertwined. This third approach allows us to grasp the significance of the seemingly neutral concepts of 'labour' and 'energy' as categories that naturalise societal processes of asymmetric exchange and accumulation. They serve as a scheme of classification filtering our instrumentalist understanding of the world which, although not scientifically flawed in the sense of not being efficacious, directs our attention and agency towards maximum exploitation of humans and non-human nature alike.

The harnessing of inorganic energy sources in nineteenth-century Britain was not just promoted by global processes (Inikori 2002; Allen 2009; Beckert 2014) but was itself an intrinsically global phenomenon, in the sense that the expansion and viability of the technologies for doing so were inextricably contingent on the relative prices of labour and other commodities on the world market. These circumstances prompt us to question conventional understandings of technologies propelled by inorganic energy as exhaustively explained in terms of human inventiveness applied to natural forces (Landes 1969; Marsden and Smith 2005). The alternative account offered here is that such technologies are paradigmatically *socionatural* phenomena, bridging the social/natural binary that has organised the history of science since the Enlightenment and sequestered the discourses of economics and ecology from each other. Both fields—economics and ecology—are committed to the common pursuit of universal currencies, yet their ontological assumptions are worlds apart. The disciplines of conventional economics and ecology provide two complementary halves of an integrated account of the Industrial Revolution, the essence of which was to replace and transcend muscular labour by means of machines. In juxtaposing these two perspectives, it is possible for us to see that ostensibly neutral facts of engineering properly belong within the inevitably ideological realm of political economy.

In hindsight, the cataclysmic consequences of the Industrial Revolution need to be fundamentally reconceptualised (cf. Barca 2011). The harnessing of fossil energy for mechanical work created a new kind of connection between social and ecological systems. From now on, the accumulated solar energy that had been deposited in geological sediments over hundreds of millions of years became accessible to some segments of world society as a substitute for muscular labour. The products of ancient photosynthesis entered into the metabolism of human society. The solar energy that for billions of years had flowed through organisms and ecosystems now flowed through—and reorganised—human societies at a global scale. This new form of socioecological coupling has been narrated as 'development' and 'progress' and understood as the result of scientific discoveries of the inherent potential of nature. Such conventional narratives do not acknowledge the decisive role of global societal relations in prompting the shift to fossil energy. This shift was made feasible and can only be comprehensible in relation to the relative world market prices of labour and other resources in different parts of the world-system. This means that discourses on energy technologies, although seemingly detached from questions of global political economy, must be reconceptualised as only *ostensibly* neutral in relation to such political considerations. No less than the field of modern economics, the neutral-sounding form that such engineering discourses tend to assume conveys an aura of incontrovertibility that generally precludes critical contestation. Nevertheless, in substituting for human labour and completely transforming the output and market pricing of commodities—including the labour time that is embodied in their production—the formation and application of new technologies is always imbued with sociopolitical implications.

At a very general level, technologies can be understood as 'socionatural' strategies for redistributing embodied human labour time and natural space. Although discourses on energy conceal such distributive dimensions of world society, this is not to suggest that they have been intentionally designed for the purpose of mystification. Rather than a grand and insidious conspiracy, these modes of deliberating about the social organisation of resource use represent the inherently limited perspectives of social actors positioned at specific points in the global economy. It is incumbent on historical social science to demonstrate why engineering discourses need to be reassessed in the light of global political economy. Such demonstrations cannot be complete without addressing the societal implications of capital. Capitalism is the aggregate logic of the capacity of general-purpose money to organise the incentives and agency of humans. We must thus also consider the pivotal role of this peculiar artefact in prompting the historical shift to fossil energy as a strategy for replacing, channelling and aggrandising the agency of labour. But the imperative of rethinking the categories of engineering from the perspective of political economy requires an interdisciplinary approach that integrates aspects of social and natural science. The requisite human ecology might be called historical political ecology.

# Energy Technologies Viewed as Historical Progress

From a physicalist perspective, the history of human civilisation is often narrated in terms of a technological progression from various modes of organising muscular energy to increasingly efficacious ways of harnessing inorganic energy sources. The role of energy in human societies has been a central topic of discussion since the emergence of thermodynamics in the second quarter of the nineteenth century. Prompted by the so-called energy crises of the 1970s, Jean-Claude Debeir and his co-authors (1991 [1986]) sketch the transformations of energy systems from pre-agricultural hunter-gatherers over 35,000 years ago to nuclear power in the 1980s. They define 'energy systems' as composed of 'ecological and technological' features, on the one hand, and 'social structures' for their appropriation and management, on the other (p. 5). This distinction between technology and society reflects the ontological boundary that modern people tend to draw between physical objects and social relations. It is fundamental, for example, to the division of labour between the fields of engineering and economics, and it is equally evident in the Marxist distinction between 'productive forces' and 'relations of production.' Although the technical/social binary continues to guide most of our thinking on the history of technology, it tends to obscure the fact that technological infrastructures are in themselves socially constituted. I mean this not only in the sense that technologies reflect their social contexts, but in the sense that they are *literally* constituted through societal exchange relations. To achieve this perspective, we must acknowledge that the global transfers of biophysical resources—the material net inputs into technology—are socially organised. It is of course possible to describe the shift to fossil energy in early industrial Britain simply as a technical breakthrough, but the nineteenth-century investments in steam technology were also contingent on asymmetric global flows of embodied labour, land, materials and energy. Steam technology, in other words, was a manifestation not only of British ingenuity but also of the physical metabolism of the British Empire.[1]

The material efficacy of the steam engine and its various successors ultimately powered by fossil energy has made such technology immune to social analysis. Machines are classified as belonging to an extra-social realm in which natural forces are harnessed through politically neutral feats of engineering. In

---

[1] To be sure, a steam engine can operate without slavery, but not without the asymmetric resource flows of which slavery was an eighteenth-century expression. The market is as effective an institution for organizing such flows as whips and chains.

## 29 The Commodification of Human Life: Labour, Energy and Money...

the current historical moment, at least, we tend to be unable to discern the continuities between the metabolism of the Roman Empire, the British Empire and the modern world-system of neoliberal capitalism. Debeir and his co-authors write that the splendour of imperial Rome 'rested on the exploitation of immense rural territories supplying it with energy, in the form of slaves and food, mainly corn' (Debeir et al. 1991, 34). A 'driving force of the Roman energy system,' they continue, was 'the exploitation, on a scale never equalled before or since, of hundreds of thousand [sic] of slaves' (ibid., 36).[2] Importantly, they suggest that 'it was the crisis of the labour supply in the fourth century that speeded up the replacement of animal and slave treadmills by hydraulic wheels in Rome' (ibid., 39).[3] To some extent, in other words, the first machines appear to have served as replacements for slaves. In eleventh-century Britain and France, there was an average ratio of one mill to about 250 people (Debeir et al. 1991 [1986], 75). The water-mill raised the productivity of labour but 'was also indisputably a tool of the lords' exploitation of the peasants' (ibid., 77). Water-power initiated the Industrial Revolution in the eighteenth century and was eclipsed by steam in the nineteenth (Malm 2016). The production of cotton textiles in steam-powered factories in early industrial England was largely geared to markets generated in Africa and America by the trans-Atlantic slave trade (Inikori 1989, 2002). The slaves were not only bought and clothed with cotton textiles, they also harvested the raw material for the cotton mills. The British foreign trade in cotton in 1850 entailed highly asymmetric flows of embodied labour and land from the periphery to the core of the world economy (Hornborg 2006, 2013). But although Debeir and his co-authors occasionally apply the concept of 'unequal exchange' to characterise British trade (pp. 94, 96), they do not define it. Nor, although they are centrally concerned with the relation between energy and power, do they reflect on our propensity to let energy sources—such as coal—exhaustively account for the operation of specific technologies, obscuring the extent to which the existence of those technologies is contingent on social relations of power and inequality. In Fig. 29.1, vertical arrows represent energy sources, while horizontal arrows indicate the flows of commodities, money, labour and resources that underlie the accumulation of technological capital. Modern technologies are no less societal strategies for displacing workloads and environmental burdens than was ancient Roman slavery. The energy sources that

---

[2] Actually, the authors refer to an estimate of about three million slaves in Italy in the year 28.

[3] In the first century, the sources of the slave trade 'began to dry up, leading to a rapid price increase; by the end of the first century AD, they were already ten times more expensive than they had been centuries earlier' (Debeir et al. 1991 [1986], 39).

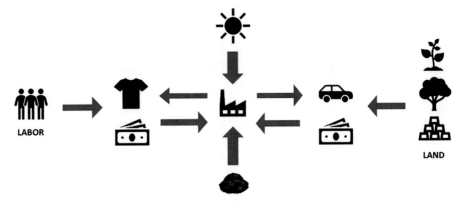

**Fig. 29.1** Industrial technology as capital accumulation: how energy sources (vertical arrows) obscure social exchange relations (horizontal arrows)

are harnessed to propel them are 'available' only to the people who can afford the requisite infrastructure.

This argument is certainly not meant to denigrate the millennia of human ingenuity that have been invested in harnessing various sources of energy. Vaclav Smil's (2017) richly illustrated history of human energy use traces the emergence of new techniques and inventions in great detail, excelling also in providing comparative, quantitative data on inputs and outputs of various energy systems. My point is that such concerns with the physical design and metabolism of energy technologies—particularly after the Industrial Revolution—need to be complemented with accounts of the requisite global contexts that made them possible. Blueprints of steam engines do not include the cotton plantations with which they were inextricably entwined. The 'system boundaries' of steam technology, in other words, should include the vast geographical spaces required to generate the capital to finance it.[4] Without the world markets and processes of capital accumulation of the eighteenth and nineteenth centuries, the expansion of steam technology would have been unthinkable. This *societal* component of industrial technology should be part of its definition, but remains very difficult to incorporate into our conventional worldview distinguishing the social from the technical.[5] In the next

---

[4] This means that estimates of the 'power density' (Smil 2015) of a technology would arrive at much lower figures than calculations of the energy output per square metre of technical infrastructure.

[5] Significantly, technological systems are not included among Vaclav Smil's (2017, 429) examples of 'macroparasitism,' defined as 'a variety of social controls of energy flows relying both on coercion—ranging from slavery and corvée labor to military conquest—and on complex (and partly voluntary) relationships among unequal groups of people.' Yet, many modern technologies should be recognised as vortices in—and inseparable from—precisely such parasitic flows.

section, I shall review some epistemological reflections on the concept of 'energy' that might be helpful in this regard.

## The Social Construction of Labour, Energy and Technology

In 1983, in the wake of the 'energy crises' that had spawned so much debate over the previous decade, Ivan Illich's (2013b [1983]) sprawling indictment of the modern work/energy nexus contains some incisive insights on how the categories that organise our perception of reality may mystify social relations of power and inequality. He recognised that the mid-nineteenth-century European obsession with energy as 'nature's ability to perform work' (ibid., 110) could be viewed as a search for 'something like a gold standard in nature' (ibid., 109). 'Through the imputation of energy,' he observes, 'nature was recast in the image of the newly constituted human as worker' (ibid., 110). Illich indicates that '[t]he simultaneous invention of…two distinct potentials for work, energy and labour-power, deserves to be explored' (ibid., 111). He illuminates the link between the social science of economics and the natural science of physics by presenting 'nineteenth-century energetism' as an attempt to 'reduce value to energy' (ibid., 114). Building on Adam Smith's labour theory of value, Illich reminds us, David Ricardo distinguished between 'live work' and 'past labour bound up as capital that could be put to work' and recognised that 'capital, in the form of machinery, could replace live labor' (ibid., 116). The steam engine, Illich notes, 'lurked behind all reality' (ibid., 110). Reiterating Marx's theory of surplus value, he recognises the 'parallel between the potential to work possessed by nature and by the proletariat' (ibid., 116). While the 'right to work' has become an incontrovertible demand inadvertently subscribing to the constructs of capitalist ideology, the persuasiveness of the word 'energy' is based on 'the myth that what it expresses is natural' (ibid., 117). Labour is simultaneously an expenditure of energy and a means of acquiring access—through wages and consumption—to the energy embodied in commodities.[6] An implication of Illich's argument is that the

---

[6] Given the huge global differences in the exchange-value of labour, I would add, the conversion of expended labour in the North into increasingly greater quantities of embodied labour from the South is an obvious index of widening inequalities. But the mainstream understanding of development is that such inequalities represent differences in historical time rather than societal space. Thus, for instance, the prominent historian of energy use Vaclav Smil (2017, 4016) remarks that 'only about one-fifth of the world's 200 countries have accomplished the transition to mature, affluent industrial societies supported by a high per capita consumption of energy.'

very concepts of 'work' and 'energy' serve to naturalise—that is, to neutralise—this parasitism of some humans on others.

Somewhat later, Philip Mirowski (1988, 1989) published systematic and meticulously referenced discussions of the historical relation between the concepts of 'work' and 'energy.' He demonstrates that the Marginalist Revolution[7] in economics was as inspired by physics as physics had been inspired by economics in identifying energy as an analogue to the economic concept of 'work.' Given the widespread and deeply rooted consensus among classical economists on work as the source of economic value, the physicists' urge to establish the concept of 'energy' inevitably also implicated the notion of value. 'It does seem inexorable,' Mirowski (1988, 816) observes, 'that the two concepts of energy and economic value are perdurably twinned.' As Georgescu-Roegen (1975, 352) concluded, 'thermodynamics is at bottom a physics of economic value.' Against this background of a common preoccupation with universal currencies—monetary value in economics, energy in physics—it is paradoxical that a long line of critics of neoclassical economics have objected that 'the only 'true' economic value is energy' (Mirowski 1988, 811).[8] In other words, although deriving its basic theoretical framework from physics,[9] economics is criticised for not being sufficiently concerned with physical phenomena.

Particularly paradoxical in these deliberations is Sergei Podolinsky's effort in 1880 to convince Marx and Engels that energy is a more fundamental source of surplus value than embodied labour (Martinez-Alier and Naredo 1982; Martinez-Alier 1987). While Marx had offered a more material measure of value (embodied labour) than that of the neoclassical economists (utility as measured in market price), Engels' negative response to Podolinsky[10]

---

[7] The Marginalist Revolution refers to the establishment in the 1870s of neoclassical economic theory as the mainstream approach within economics.

[8] Mirowski's (1988) list of such critics includes Wilhelm Ostwald, Patrick Geddes, Alfred Lotka, Frederick Soddy, Frederick Taylor (the father of 'Taylorism'), the Technocracy movement of the 1930s, ecological anthropologists following Leslie White and ecologists such as Howard Odum. The historical genealogy of what is now known as 'ecological economics' has been traced in great detail by Martinez-Alier (1987).

[9] Mirowski (1989, 3–4) writes that the neoclassical economic theorists 'did not imitate physics in a desultory or superficial manner; no, they copied their models mostly term for term and symbol for symbol, and said so. ... Although it was ultimately called "energy" in physics and "utility" in economics, it was fundamentally the same metaphor, performing many of the same explanatory functions in the respective contexts, evoking many of the same images and emotional responses, not to mention many of the same mathematical formalisms.' However, as highlighted by Georgescu-Roegen (1971), the theoretical framework of neoclassical economics remains 'helplessly locked into the physics of circa 1860' and is incompatible with the concept of entropy and the implications of the second law of thermodynamics (Mirowski 1989, 389–394).

[10] Engels famously rejected Podolinsky's suggestion that the Marxian theory of surplus value could be understood in terms of energy flows (cf. Martinez-Alier and Naredo 1982).

demonstrates that the Marxist theory of surplus value could not be reduced to thermodynamics. Although Marx (1967 [1867], 215) had conceded that 'labour-power itself is energy,' his understanding of surplus value thus occupies an ambiguous position between economics and physics (Hornborg 2019b). As Mirowski (1988, 817) notes, '[m]any of the arguments tendered for an energy theory of value are strangely reminiscent of those tendered in Marxian economic theory, especially with regard to an *a priori* common denominator of all commodities.' This observation is confirmed by Lonergan (1988), who shows that the analytical structure of labour and energy theories of value are virtually identical. By implication, Nicholas Georgescu-Roegen's (1979) decisive rejection of an energy theory of value[11] could be extended to a labour theory of value. Indeed, there is no logically consistent argument for deriving the exchange-value of a commodity—measurable in money—from the quantities of some biophysical input (whether land, energy or labour) embodied in its production.[12]

A year after Mirowski's (1989) *magnum opus* on the work/energy nexus was published, Anson Rabinbach pursued a similar argument in his book *The Human Motor: Energy, Fatigue, and the Origins of Modernity* (1990). Rabinbach shows how human labour in nineteenth-century Europe was perceived as an example of the same force of nature that propelled industrial machines. From this perspective, he writes, '[t]he human body and the industrial machine were both motors that converted energy into mechanical work' (ibid., 2). The concept of 'labour power' expressed 'the fundamental imperative that links society and nature in nineteenth-century thought' (ibid., 3). This meant that 'the word *work* was universalised to include the expenditures of energy in all motors, animate as well as inanimate' (ibid., 4). Rabinbach argues that 'the scientific language of labour power and the hegemony of productivism were not merely "corruptions" of Marxism or of the labour movement, but integral aspects of the intellectual framework of nineteenth-century materialism' (ibid., 16). His focus on representations of bodily fatigue illuminates 'a widely shared belief that conserving the energy of the working body held the key to both productivity and social justice' (ibid., 18).[13]

---

[11] Georgescu-Roegen (1979, 1048) concluded that it is 'perfectly clear that in *absolutely* no situation is it possible for the energy equivalents to represent economic valuations.'

[12] After centuries of tenacious but futile efforts to identify the substance of 'value' (cf. Mirowski 1989), the edifice of economic discourse on the topic assumes a vain and esoteric appearance cognate to that of Medieval theology.

[13] The extent to which it was held that the alleviation of fatigue could be accomplished through mechanisation raises crucial questions about the societal *distribution* of heavy work—and about whether industrial technology is to be understood primarily as a *re*placement or *dis*placement of such work. Nineteenth-century British machinery had implications not only for British factory workers but also for African labourers harvesting cotton on American plantations (Hornborg 2006).

Rabinbach observes that work, '"cleansed" of all of its social and cultural dimensions and considered exclusively as a form of energy conversion … could be applied to nature, technology, and human labor without distinction' (ibid., 46). He notes that there was a 'semantic shift in the meaning of "work," [as] all labor was reduced to its physical properties, devoid of context and inherent purpose' (p. 47).[14] Hermann von Helmholtz's use of the concept of labour power (*Arbeitskraft*), '[u]niversalized as the demiurge present in all nature … redefined the principle of motion in the universe in terms of its power to "perform work"' (p. 55). The invention of steam power gave rise to a new image of 'the unity of society and nature in the process of mechanical—or industrial—production, and ultimately of the universe as a vast productive machine' (ibid., 59). In the early 1860s, Helmholtz acknowledged that '[t]he concept of work for machines or natural processes is taken from the comparison with the work performance of human beings, and is … comprehensible through a comparison with human labor' (quoted in Rabinbach 1990, 59). Such insights were fundamental to the theories of Karl Marx, 'whose later work was influenced and perhaps even decisively shaped by the new image of work as "labour power"' (ibid., 70). By the early 1860s, writes Rabinbach, Marx had 'shifted his focus from the emancipation of mankind *through* labor to emancipation *from* productive labor by an even greater productivity' (ibid., 73). His political vision was to 'release human beings from the burden of labor by the productive forces of technology' (p. 74). The foundation for this shift of focus, according to Rabinbach (1990, 80), was Marx's discovery of energy.

This understanding of technological progress in Marxist thought reflects the optimism of many Europeans in the mid-nineteenth century. Steam technology was perceived as an ingenious revelation and harnessing of natural forces that *in itself* had no political or distributive implications. In Marx's view, technological development under capitalism was a means for commodified, abstract labour to transcend itself. Rabinbach (1990, 79) summarises Marx's vision: 'Labour power is the motor of history; its goal, however, is the replacement of men with machines.' This Promethean view of technology continues to be fundamental to some strands of Marxism to this day (cf. Bastani 2019).[15] As in mainstream discourses in economics and engineering,

---

[14] This decontextualisation of technologically mediated work is a crucial condition for modern confusion regarding the purported agency of technological artefacts. I have elsewhere argued that agency should be reserved for entities that have *purposes* (Hornborg 2017). Posthumanist attributions of agency to artefacts (Latour 2005) illustrate how the mechanisation of work tends to obscure the *human* purposes (and agency) that have been delegated to and embedded in the technologies.

[15] Other Marxists vehemently reject Prometheanism, for instance John Bellamy Foster (2017).

## 29 The Commodification of Human Life: Labour, Energy and Money...

it effectively insulates the phenomenon and concept of technology from global social analysis. But even technologies that are locally perceived as emancipatory may be founded on exploitation elsewhere. The immunity of technology to deconstruction no doubt derives from its *materiality*—its capacity to serve as an index of physical nature, a capacity which becomes paradoxical once we grasp the extent to which our notion of an autonomous, physical nature is itself a social construction. Of course, this is not to say that the material efficacy of steam-powered factories in early nineteenth-century Britain was illusory, but to acknowledge that the large-scale operation of those technologies for harnessing steam power was *dependent* on the relative world market prices of African labour, American land, raw cotton, cotton textiles and coal. The technological infrastructure for harnessing the chemical energy of coal was thus a *socionatural* phenomenon: It was as much a product of societal processes of unequal exchange and capital accumulation as it was a means of freely harvesting nature.[16]

In the final chapter, Rabinbach (1990) reflects on the nineteenth-century emergence and establishment of a 'laborcentric' perspective on the universe and recent visions of how this worldview might be transcended. Much as Illich deplored the modern claim to a 'right to work,' Rabinbach's book illuminates how the political landscape of modern society is ultimately an accommodation to the societal conditions established by nineteenth-century industrialism. Given no alternative, labour must reproduce the exploitative inertia of industrial capitalism, that is, the exploitation of their own labour power.[17] Citing a number of sociologists—from Ralf Dahrendorf to André Gorz—contemplating 'the end of the work-centered society,' Rabinbach observes that the technologically induced decline of work as a source of meaning and identity has had major consequences for the traditional working class, the labour movement and trade unions. Yet, given economic globalisation, the conclusion that '[t]echnology was making corporal work obsolete' (ibid.,

---

[16] Both Mirowski (1989) and Rabinbach (1990) observe that Marx's central concept of labour power is ambiguously defined. 'Both a social and a physiological magnitude,' says Rabinbach (1990, 74–75), 'it is a measure of value and a measure of energy.' Although obviously intent on anchoring his theory of capital in natural science, Marx could not circumambulate its pivotal reliance on relative exchange-values. With Engels, Rabinbach (1990, 82) suggests, political economy 'took its place among the other economies of energy: physiology, physics, and chemistry.' The fundamentally ambiguous position of Marxist theory—torn between natural and social science—is summed up by Rabinbach (ibid.): 'Energy is the universal equivalent of the natural world, as money is the universal equivalent of the world of exchange.' This formulation succinctly captures the point that our modern ontology of nature since the Industrial Revolution is to some extent a reflection of capitalist society.

[17] With globalisation, of course, this phenomenon has become global in scope. Although workers in the Global South may have average wages that are a fraction of European or American ones, they will naturally demand that their factories keep exporting their embodied labour to the Global North.

297) appears myopic. The experience that work-centred society is disappearing is clearly *positioned* in the traditional cores of the capitalist world-system. To discuss the circumstances of that experience without reflecting on the simultaneous displacement of work to low-wage areas of the world economy is to avoid problematising the distributive dimensions of technological development. For workers in sweatshops and extractive sectors in the Global South, it is not as clear that technology is making work obsolete.

In the footsteps of Mirowski and Rabinbach, Cara New Daggett (2019) has recently conducted a similar investigation into the genealogy of the concept of 'energy' as an expression of the 'mania to put the world to work' (ibid., 4). She endorses 'the nascent field of energy humanities,' which 'analyzes energy … as more than a set of fuels and their associated machines, but also as *a socio-material apparatus that flows through political and cultural life*' (p. 3; emphasis added). This rather diffuse phrase no doubt intends to express the central point made decades ago by Illich, Mirowski and Rabinbach that the *concept* of energy is a cultural construction with political implications—a means of approaching nature viewed through the lens of political economy. In this sense, as they convincingly demonstrate, concepts and political economy are recursively intertwined. Some passages in Daggett's book indeed echo the insights of her predecessors: 'Just as energy became tightly bound by the governing logic of work, so too work increasingly came to be governed through the metaphors and physics of energy' (ibid., 4). Like Illich, she notes that 'work' has become 'necessary to becoming a worthy citizen' (ibid., 83) and that we have become addicted to 'the ideology of work' (ibid., 100).[18] Like him, she problematises the fact that work 'earns humans the right to consume what they will' (p. 206). Many of Daggett's conclusions duplicate those of Rabinbach—including advocacy for a 'post-work' society—but she is more attentive to the *global* contexts of the European obsession with energy, with regards both to the historical emergence of industrialism and to the contemporary politics of energy. Importantly, she highlights how the imperative 'to produce energy [is] a political rationality that justifies extractivism and imperial capitalism' (ibid., 5). At the end of a paragraph attempting to chart a posthumanist course between physicalist and constructivist approaches to nature, Daggett (2019, 6) reaches the pertinent conclusion that '[t]he dominant figuration of energy cannot be detached from the sociomaterial context in which it emerged, which was the convergence of bodies, fossil fuels, and steam engines in imperial Europe and its factories.' This succinct statement is

---

[18] Daggett (2019, 145) notes that the existence of a wage 'became an indicator in the British public's imagination that demarcated the line between slavery and free labor.'

an advance on Mirowski and Rabinbach in that it positions the history of science in relation not only to industrial capitalism, but also to the global context of European industrialism. Such instances of analytical clarity render much of Daggett's posthumanist jargon superfluous.[19] The only thing that must be added is that the sociomaterial context of energy science included very many bodies *outside* of Europe and its factories. The steam engine and the accompanying science of energy had just as momentous implications for labourers on the cotton plantations as for those in the cotton mills.

Nevertheless, it is obvious that Daggett's approach to the work/energy nexus is influenced by the frequently amorphous efforts of posthumanists[20] to articulate the disorienting insight that the lens through which we view the world will always be refractive in some sense. The fact that this lens can be understood as a social construction does not mean that it cannot serve as a materially efficacious guide to successful action.[21] But '[r]ather than accept the master narrative of energy's discovery and diffusion as objective knowledge,' Daggett (2019, 7) is 'interested in parochializing energy, troubling its claims to universality.' The challenge posed by this ambition is to combine Daggett's (2019, 8) insights on 'thermodynamics as an imperial science' with an acknowledgement of the 'thermodynamics of imperialism' (Hornborg 1992, 2001). We need to understand that the new science of thermodynamics not only served the interests of imperialism but simultaneously provided us with the conceptual tools to reveal the mystified logic of its operation. This narrow path between constructivism and objectivism is difficult to follow, but the one we must pursue. The productivist worldview that continues to prompt us to incessantly search for new sources of energy can also help us expose the logic of ecologically unequal exchange (Hornborg 2019a). The sense in which 'empires functioned as living organisms … [and] energy fueled their metabolism' (Daggett 2019, 8) was not just a biased, parochial assumption, but a crucial (albeit insufficient) observation with potentially revolutionary, political implications. In jettisoning the master narratives of natural science, however, posthumanism cannot provide us with the tools to accomplish such a

---

[19] Daggett's struggle to straddle the 'distinction between energy-as-knowledge and energy-as-fuel' leads her to imagine historical time 'in loops and spirals' and to evoke 'the "widening gyre" of falcons' flight and things falling apart in William Butler Yeats's 'Second Coming' (ibid., 25).

[20] Daggett acknowledges the influence of posthumanists such as Donna Haraway, Bruno Latour, Jane Bennett and Timothy Morton.

[21] What is to be considered 'successful,' however, can be debated. As Andreas Malm (2016) and others have highlighted, the exorbitant economic and military 'success' of the nineteenth-century turn to fossil energy simultaneously initiated the global warming that is currently threatening to make much of the planet uninhabitable. In hindsight, then, the science that promoted that turn must be understood as fundamentally insufficient and thus ultimately flawed.

critical unmasking of global social metabolism. It is an exaggeration to say that 'energy and entropy are epistemological constructs' and that 'things called energy and entropy do not actually exist' (ibid., 69). Such ontological relativism is fashionable in the social sciences but difficult to reconcile with the claim that energy fuelled the metabolism of empires.

When Daggett (2019, 8) writes that it was 'energy knowledge' that 'had made possible the *specific activities* by which Europeans had advanced,' she omits the dependence of technological superiority on the asymmetric global resource flows (embodied labour, land, materials and energy) that were intrinsic to the operation of the world market established under imperialism. In other words, she conceives of the expansion of steam technology in imperial Britain as exhaustively accounted for by advances in energy science, apparently ignoring that it was made *physically* possible by the market prices that orchestrated global transfers of embodied biophysical resources such as labour and land (Hornborg 2006, 2013). Daggett (2019, 9–10) observes that disciplines such as economics and thermodynamics served 'the interests of planetary industrialization, having helped to justify European imperialism by externalizing its ecological and social injustices.' However, she fails to see that the very phenomenon of industrial technology *is itself* an imperialist strategy for externalising injustices. Although she incisively notes that '[t]he goal for scientists was to naturalize engines' (ibid., 47), she does not pursue the implication that engines are anything but natural. Similarly, her observation that '[t]he geo-theology of energy was simply another method for tackling the much older problem of labor governance' (ibid., 82) is never elaborated into a critical exposure of the distributive and political dimensions that are intrinsic in the very concept of 'technology.' If posthumanist approaches challenging us to rethink 'the material' are to deliver any revolutionary insights, they would do well to begin with the machine (Hornborg 2001). What Daggett (2019, 196) calls 'techno-fixes' are mystified societal strategies for burden-shifting—global displacements of work and environmental loads that are naturalised by disciplines such as economics and engineering.

The machine would not be possible without money.[22] The alarming inequalities and unsustainability of world society are not simply products of human politics but of the inexorable logic of the seemingly neutral artefact of all-purpose money, which veils the asymmetric resource flows that make

---

[22] The extent to which money obscures the relation between technology and labour is illustrated by the juxtaposition of two passages from Alfred Crosby's (2006) book *Children of the Sun*. He notes that power in the cotton plantations 'was still supplied by the primitive prime mover of muscle, in the form of slavery, an institution whose slumping status was revived by the profits to be made from the new textile mills' (p. 78), and three pages later remarks: 'Power had been about muscle for all of human history, and the most effective way to marshal it had been by assembling serfs and slaves. Now, by golly, the best way was to get yourself a steam engine.'

modern technologies possible. All-purpose money and industrialisation combined in the early nineteenth century to obscure asymmetric global flows of embodied labour and land from the peripheries to the core of the British Empire. To this day, the fictive reciprocity of money prices conceals such asymmetric flows of labour and resources. Although preindustrial, mercantilist trade had been mostly in what Wallerstein called preciosities, since the nineteenth-century volumes of trade in bulk goods such as food and materials have increased enormously. The feasibility of such global flows was the *sine qua non* of industrialisation.

Daggett (2019, 90) concedes that heat engines 'were functionally analogous to any other organ or body that transformed energy from heat into motion, and they could be governed as such,' but she does not mention that the viability of engines hinged on particular ratios of exchange on the world market. While it is crucial to recognise, with Daggett (2019, 102), the 'singular ruling logic of energy' in 'the feverish frenzy by which every person, every mountain, every river, and every clod of soil was to be put to work in the service of fossil capital,' the 'drive to put the world's materials to use for human profit' ultimately does not derive from the concepts of 'work' and 'energy' but from the logic of money. It is thus an incomplete analysis to conclude, as a 'key argument,' that 'our commitment to growth and productivity has been reinforced by a geo-theology of energy that combines the prestige of physics with the appeal of Protestantism in order to support the interests of an industrial, imperial West' (ibid., 190).

The continuing preoccupation with 'work' and 'energy' derives from the fact that human time and natural space are commodified—that they can be bought with money. Although implicit in passing references to 'capital' and 'capitalism,' the artefact that continues to generate this logic—the ubiquitous incentive to make a good deal—remains the elephant in the room. It is the cultural construct of all-purpose money that opens opportunities for capitalist commodification and accumulation. A 'post-work' and 'post-energy' society will only be possible if the design of that artefact is challenged. The notion of a universal basic income (cf. Daggett 2019, 203)—if paid in a currency that does not imply universal commensurability[23]—is a step in that direction.

---

[23] That is, a currency that cannot make claims on embodied low-wage labour and resources from anywhere in the world.

## The Relation Between Labour and Energy Technologies as Political Economy

To compare various forms of human labour in terms of the quantities of energy mobilised by different technologies is a particular way of *describing* production processes. Such comparisons are *explanatory* only in the sense that they help account for the relative success of populations applying new inventions enabling them to harness more energy. They may account for the *consequences* of technological advantages for historical power relations between different nations or groups, but they do not explain how those advantages emerged out of specific historical circumstances. Technologies for harnessing energy are not exhaustively accounted for as fortuitous breakthroughs in engineering that provide access to intrinsic features of nature. A technology has metabolic requirements of its own: Its operation is contingent on inputs of resources such as metals, fuels and lubricants, which are in turn contingent on ratios of societal exchange. The appearance of new technologies should thus not be approached simply as causal factors that help explain social and historical processes, but always also as conditioned by such processes to begin with. In other words, the relation between technology and the structure of world society is recursive. Technological progress is conventionally approached as propelled by individualised factors such as ingenuity and the drive for profits, but the conjunctures of the global market are rarely understood as macro-scale determinants of technological transitions. The shift to a new technology in a certain location can be conceptualised as a form of global arbitrage.[24]

In 1973, Ivan Illich (2013a [1973]) observed that the emergent concept of an 'energy crisis' was inextricably entangled with questions of equity. At that time, he estimated, 250 million Americans, 'for the sole purpose of transporting people,' burn 'more fuel than is used by 1.3 billion Chinese and Indians for all purposes' (ibid., 78). He famously calculated that '[t]he model American puts in 1600 hours to get 7500 miles: less than five miles per hour' (p. 82). He shows that each technological increase in transport velocity dramatically increases the demands on space represented by that technology, and that this relation pivots on '[t]he exchange-value of time': 'As societies put price tags on time, equity and vehicular speed correlate inversely' (ibid., 84). 'Beyond a certain velocity,' he concludes, 'passengers become consumers of other people's time, and accelerating vehicles become the means for effecting a net transfer of life-time' (ibid., 85). Illich's criterion for an equitable

---

[24] Arbitrage is a term used in economics for transactions taking advantage of varying prices in different markets.

technology—such as that of the bicycle—is that it does not put 'undue claims on the schedules, energy, or space of others' (ibid., 98). Although frequently more indignant than systematic, Illich's perspective offers analytical tools for understanding technologies as social mechanisms for redistributing time and space.

Modern technologies are not unique in redistributing time and space—they are merely the most recent social strategy for doing so. Daggett (2019, 20) suggests that '[i]t is through thermodynamics that energy became ... a unit of work that was amenable to technical governance,' but Lewis Mumford (1967) traces what he called the 'megamachine'—the large-scale social regimentation of work—to ancient Egyptian slavery. The grand architectural projects of ancient Egypt or Rome also required their specific sciences of engineering, bodies of knowledge that are as analytically distinguishable from their energy source—slave labour—as thermodynamics is from fossil energy (Hornborg 2016, 9–10). In neither case can the science of engineering be excised from political economy, but the organisation of social inequality—whether by means of whips or market prices—is founded on distinct principles. Drawing on Malm (2016), Daggett (2019, 30) correctly notes that 'the advantages of coal had more to do with human power than with mechanical power,' but fails to emphasise that the exercise of human power by means of steam technology was not just a matter of controlling factory workers in Britain—it was ultimately made possible by the relative world market prices of embodied labour time and natural space (Hornborg 2006, 2013) and was thus ultimately also a matter of controlling labourers on plantations on other continents.[25]

The identification of asymmetric global resource flows associated with imperialism necessarily presents a conundrum to posthumanists, as it does to most economists, because their indignation over such material asymmetries will be difficult for them to substantiate. Daggett (2019, 132) asserts that '[f]ossil fuel-driven capitalism required an unjust circulation of materials and bodies; the concentration of wealth in some sites occurred at the expense of other people and things,' but even if such injustices have been 'exhaustively catalogued by postcolonial theorists,' it is not clear by which criteria they would define 'unjust circulation' or wealth accumulation occurring 'at the

---

[25] I have estimated that a British textile manufacturer in 1850, by selling £1.000 worth of cotton textiles on the world market and buying raw cotton for the same sum, was able to exchange the product of 14,233 hours of British factory labour for that of 20,874 hours of plantation labour and that of less than one hectare of British land for that of 58.6 hectares of plantation land (Hornborg 2013, 91). Such calculations exemplify how historical processes of industrialisation have been contingent on the unequal exchange of embodied labour and land.

expense of other people.' Is there an essential difference between the asymmetries and injustices of imperialism, on the one hand, and the current operation of globalised free trade, on the other? How can we theorise an 'unjust circulation of materials' without according some validity to the quantitative measures of material flows developed by natural sciences, and using such measures to challenge the fictive reciprocity of the market? If 'energy fuelled the metabolism of the imperial organism' (Daggett 2019, 137), must we not grant energy an existence beyond that of an 'epistemological construct'? In encouraging relativism, has not posthumanism jettisoned all possible means of challenging neoclassical economics? Conversely, if we say that European industrialisation depended on 'underpaying colonized peoples' (ibid., 138), does this intuitively legitimate assertion imply that we can determine the correct market price of their labour? Or are our notions of 'price' and 'value' reified constructions that mystify not only asymmetric resource flows but even benevolent attempts to expose them?

## Conclusions

A general conclusion from these reflections on the work/energy nexus is that both concepts have been generated by the logic of general-purpose money—that is, capitalism. However, this insight should not imply a retreat into constructivism, as embodied labour and embodied energy are very real, biophysical phenomena that we need to highlight as essential to the metabolism of technological infrastructure, whether in imperial Britain or the modern world. Such metrics are finally the only tools we have to reveal the material asymmetries obscured by the fictive reciprocity of money prices.

As demonstrated by Debeir and his co-authors (1991), Crosby (2006) and Smil (2017), it is perfectly feasible to narrate human history in terms of a progressive succession of energy technologies. On the other hand, Illich (2013b [1983]), Mirowski (1988, 1989), Rabinbach (1990) and Daggett (2019) have shown that the concept of 'energy' itself has a history that is closely intertwined with that of the concept of 'labour.' In other words, the lenses through which we view history are themselves historical products. This does not mean that they are incorrect, only that they reflect the social and political systems in which they have emerged. The study of material asymmetries in such systems must be combined with the study of how such asymmetries are ideologically obscured and defused. In the modern world order that was consolidated in the nineteenth century, a central means of obscuring sociopolitical asymmetries has been to represent them as 'natural.' In this way, we have been taught to conceive of energy technologies as revelations of nature. My main point has been to argue that we are mistaken in understanding 'energy' and 'technology' as entirely neutral, physical phenomena purified of political economy. Modern energy technologies are imbued with global asymmetries. Ever since the Industrial Revolution, they have relied on continuous net transfers of embodied labour and land from the peripheries to the cores of the world-system.

Energy is perceived as a natural force that can be harnessed by humans. Modern people tend to see it as a free gift of nature, waiting to be harvested. But harnessing energy requires technology, and technology is a manifestation of capital accumulation. The conceptual distinction between energy technologies—what Marx called 'productive forces'—and capital appears to be misguided. Energy is conceived as a substitute for human labour, but the components of energy technologies are embodiments of labour power expended elsewhere. No less than slavery or peonage, the harnessing of 'natural' (i.e., non-human) energy—whether steam, hydroelectric, nuclear, solar or wind—is thus a *social* arrangement. It is a strategy for displacing work and environmental loads—that is, appropriating human time and natural space—between stratified segments of world society. Importantly, the feasibility of such 'technological progress' is contingent on the relative market prices of two kinds of labour—the labour embodied in the machine and the labour that it displaces.

# References

Allen, Robert C. 2009. *The British Industrial Revolution in Global Perspective*. Cambridge: Cambridge University Press.
Barca, Stefania. 2011. Energy, Property, and the Industrial Revolution Narrative. *Ecological Economics* 70: 1309–1315.
Bastani, Aaron. 2019. *Fully Automated Luxury Communism: A Manifesto*. London: Verso.
Beckert, Sven. 2014. *Empire of Cotton: A Global History*. New York: Vintage Books.
Crosby, Alfred W. 2006. *Children of the Sun: A History of Humanity's Unappeasable Appetite for Energy*. New York: W.W. Norton & Co.
Daggett, Cara N. 2019. *The Birth of Energy: Fossil Fuels, Thermodynamics, and the Politics of Work*. Durham: Duke University Press.
Debeir, Jean-Claude, Jean-Paul Deléage, and Daniel Hémery. 1991. *In the Servitude of Power: Energy and Civilization Through the Ages*. London: Zed Books.
Foster, John B. 2017. The Long Ecological Revolution. *Monthly Review* 69 (6) https://monthlyreview.org/2017/11/01/the-long-ecological-revolution/.

Georgescu-Roegen, Nicholas. 1971. *The Entropy Law and the Economic Process*. Cambridge, MA: Harvard University Press.
———. 1975. Energy and Economic Myths. *Southern Economic Journal* 41 (3): 347–381.
———. 1979. Energy Analysis and Economic Valuation. *Southern Economic Journal* 45: 1023–1058.
Hornborg, Alf. 1992. Machine Fetishism, Value, and the Image of Unlimited Good: Toward a Thermodynamics of Imperialism. *Man (New Series)* 27: 1–18.
———. 2001. *The Power of the Machine: Global Inequalities of Economy, Technology, and Environment*. Walnut Creek, CA: AltaMira.
———. 2006. Footprints in the Cotton Fields: The Industrial Revolution as Time-Space Appropriation and Environmental Load Displacement. *Ecological Economics* 59 (1): 74–81.
———. 2013. *Global Ecology and Unequal Exchange: Fetishism in a Zero-Sum World*. Revised Paperback Version. London: Routledge.
———. 2016. *Global Magic: Technologies of Appropriation from Ancient Rome to Wall Street*. Houndmills: Palgrave Macmillan.
———. 2017. Artifacts Have Consequences, Not Agency: Toward a Critical Theory of Global Environmental History. *European Journal of Social Theory* 20 (1): 95–110.
———. 2019a. *Nature, Society, and Justice in the Anthropocene: Unraveling the Money-Energy-Technology Complex*. Cambridge: Cambridge University Press.
———. 2019b. The Money-Energy-Technology Complex and Ecological Marxism: Rethinking the Concept of 'Use-Value' to Extend Our Understanding of Unequal Exchange, Part II. *Capitalism Nature Socialism* 30 (4): 71–86.
Illich, Ivan. 2013a [1973]. The Energy Crisis. In *Beyond Economics and Ecology: The Radical Thought of Ivan Illich*, ed. Sajay Samuel, 73–104. London: Marion Boyars.
———. 2013b [1983]. The Social Construction of Energy. In *Beyond Economics and Ecology: The Radical Thought of Ivan Illich*, ed. Sajay Samuel, 105–123. London: Marion Boyars.
Inikori, Joseph E. 1989. Slavery and the Revolution in Cotton Textile Production in England. *Social Science History* 13 (4): 343–379.
———. 2002. *Africans and the Industrial Revolution in England: A Study of International Trade and Economic Development*. Cambridge: Cambridge University Press.
Landes, David S. 1969. *The Unbound Prometheus: Technological Change and Industrial Development in Western Europe from 1750 to the Present*. Cambridge: Cambridge University Press.
Latour, Bruno. 2005. *Reassembling the Social: An Introduction to Actor-Network-Theory*. Oxford: Oxford University Press.
Lonergan, Stephen C. 1988. Theory and Measurement of Unequal Exchange: A Comparison Between a Marxist Approach and an Energy Theory of Value. *Ecological Modeling* 41: 127–145.

Malm, Andreas. 2016. *Fossil Capital: The Rise of Steam Power and the Roots of Global Warming*. London: Verso.

Marsden, Ben, and Crosbie Smith. 2005. *Engineering Empires: A Cultural History of Technology in Nineteenth-Century Britain*. Houndmills, Basingstoke: Palgrave Macmillan.

Martinez-Alier, Joan. 1987. *Ecological Economics: Energy, Environment and Society*. Oxford: Blackwell.

Martinez-Alier, Joan, and José M. Naredo. 1982. A Marxist Precursor of Energy Economics: Podolinsky. *The Journal of Peasant Studies* 9 (2): 207–224.

Marx, Karl. 1967 [1867]. *Capital*. Vol. 1. New York: International Publishers.

Mirowski, Philip. 1988. Energy and Energetics in Economic Theory: A Review Essay. *Journal of Economic Issues* 22 (3): 811–830.

———. 1989. *More Heat than Light: Economics as Social Physics, Physics as Nature's Economics*. New York: Cambridge University Press.

Mumford, Lewis. 1967. *The Myth of the Machine*. Vol. 1. New York: Harcourt Brace Jovanovich.

Rabinbach, Anson. 1990. *The Human Motor: Energy, Fatigue and the Origins of Modernity*. New York: Basic Books.

Smil, Vaclav. 2015. *Power Density: A Key to Understanding Energy Sources and Uses*. Cambridge, MA: The MIT Press.

———. 2017. *Energy and Civilization: A History*. Cambridge, MA: The MIT Press.

# 30

# Workers, Trade Unions, and the Imperial Mode of Living: Labour Environmentalism from the Perspective of Hegemony Theory

Markus Wissen and Ulrich Brand

## The Labour-Environment Divide

There is growing recognition of the need to reduce the levels of resource consumption and emissions in areas such as industrial production and transport of goods. On the one hand, however, it is unclear what exactly is at stake: is it sufficient to increase eco-efficiency in order to decouple economic growth from the consumption of natural resources (for a discussion see UNEP 2011a; Parrique et al. 2019)? Should efforts focus on replacing non-renewable with renewable resources and on changing the material basis of the generation of wealth (cf. Fücks 2013)? Or is it a matter of challenging the paradigm of economic growth as well as resource- and emission-intensive increases in prosperity itself, of fundamentally transforming the underlying patterns of

---

The text is based on our paper '*Working-class environmentalism* und sozial-ökologische Transformation. Widersprüche der imperialen Lebensweise', published in *WSI Mitteilungen* 72 (1), 2019, 39–47, https://doi.org/10.5771/0342-300X-2019-1-39, which was substantially revised and further developed for the purpose of this handbook. We wish to thank the handbook editors for their patience and helpful suggestions. Furthermore, we are grateful to Bert Preiss for the English translation, to Barbara Jungwirth for the language editing and to Christopher Beil for editorial assistance.

---

M. Wissen (✉)
Department of Business and Economics, Berlin School of Economics and Law, Berlin, Germany
e-mail: markus.wissen@hwr-berlin.de

U. Brand
Department of Political Science, University of Vienna, Vienna, Austria

production and consumption and thus ultimately transcending the capitalist growth imperative (Brand and Wissen 2018a, 2021)?

On the other hand, the question is what role workers and trade unions could play in the socio-ecological transformation. The much-discussed, 420-page transformation report by the German Advisory Council on Global Change (WBGU 2011) makes frequent mention of 'pioneers of change', 'global governance', and 'the shaping state'. A search for the term 'trade unions', however, yields zero results. At best, they appear as 'employee organizations' in chap. 5 under the subheading 'Opposing Forces and Resistance: Lobby and Interest Groups'. There, the criticism is that in the crisis after 2008 unions, together with the employer organisations, had successfully campaigned for a bonus granted to everybody who scrapped her or his old car in order to buy a new one (ibid., 201–202).[1] This and other criticisms, such as the one addressing the resistance by the Union for the Mining, Chemical and Energy Industries (IG BCE) to the phase-out of lignite, point indeed to a crucial problem that Räthzel and Uzzell (2011) have described as the 'jobs versus environment dilemma' (see also Barca 2012; Brand and Niedermoser 2019). From a Marxist point of view, the roots of the latter lie in the separation of workers from the means of subsistence, a process that Marx has discussed in the chapter on the so-called primitive accumulation in *Capital* (Marx 1976 [1867], chap. 24). Standing at the beginning of capitalist industrialisation and being repeated in countless acts of valorisation, this process 'has torn the relationship between labour and nature apart' (Stevis et al. 2018, 440; see also Burkett 2014, 79). Given the aggravation of the ecological crisis, the extreme inequality and the impoverishment of workers due to the Corona pandemic as well as the crisis of social reproduction under the conditions of a neoliberal capitalism, it will be of utmost importance to overcome this divide. For social scientists, a major task consists in seeking conceptual and empirical links between labour and the environment and identifying the conditions under which these links can be practically established (Hürtgen 2020, 185–186).

In the following, we want to contribute to this task by asking to what extent experiences of crisis encourage defending a resource- and emission-intensive mode of production and living and under which conditions could they be politicised in favour of a socio-ecological transformation

---

[1] As part of its 2009 'Recovery Package', the German government promoted new auto purchases with an 'environment bonus' (arguing that new cars produced less emissions) of 2500 Euro (totalling 5 billion Euro). About 1.75 million new cars were bought through this programme between January and September 2009. In Austria, 30,000 new vehicles were purchased with the government subsidy of 1500 Euro per car in an effort to prop up the automotive industry.

that actively involves workers and trade unions. We do not attempt an exhaustive answer here. Instead, we are primarily concerned with conceptual considerations. Some empirical evidence has emerged from research on the labour-environment relationship and initial conclusions can be drawn with regard to union policy.

We start by examining the tension between labour and the environment in developed capitalist societies. This is done in the light of our concept of the imperial mode of living (Brand and Wissen 2018a, 2021), which we believe explains both the cross-class normalcy of ecologically destructive patterns of production and consumption and the class character of the respective practices (Section 'The Imperial Mode of Living'). Subsequently, we analyse the societal deepening of the imperial mode of living, its international spread, and the more recent manifestation of its contradictory nature (Section 'Contradictions of the Imperial Mode of Living'). The latter has given rise to various (new) attempts at linking labour and the environment. These will be systematised and discussed in Section 'Overcoming the Labour-Environment Divide'. Our objective is to identify strategies for overcoming the imperial mode of living through reconceptualising the economy from the perspective of social reproduction. As we will argue, this requires a democratic organisation of economic key sectors under an active participation of workers and unions.

## The Imperial Mode of Living

Our basic assumption is that the deeply rooted patterns of production and consumption, which predominate above all in the early industrialised capitalist societies, presuppose the disproportionate access to nature and labour power on a global scale. Developed capitalism is characterised by the fact that it requires a less developed or non-capitalist geographical and social 'outside', from which it obtains raw materials and intermediate products, to which it shifts social and ecological costs, and in which it appropriates both paid labour and unpaid care services (Biesecker and Hofmeister 2010; Salleh 2017). From a socio-ecological point of view, the imperial mode of living cannot be generalised. It is exclusionary and exclusive and it presupposes an imperialist world order, which at the same time is normalised in countless acts of production and consumption so that its violent character is rendered invisible for those who benefit from it.

We understand the *imperial mode of living* as a concept of hegemony theory in the tradition of Antonio Gramsci, which connects the everyday life of people with the social and international structures and reveals those prerequisites of capitalist patterns of production and consumption that cannot be generalised in socio-ecological terms. As such, it also refers to the way of working and producing in capitalist societies. Exploitation of nature and labour power is not only a structural feature of the relationship between the Global North and the Global South. Instead, it takes place in the class societies of the Global North itself, where significant social and spatial inequalities exist and—we will turn to this below—have increased in recent decades. What we want to emphasise, however, is that the exploitation of labour power in advanced capitalist countries is inherently linked to, and mediated by, exploitative structures elsewhere.

Alf Hornborg has pointed to this in his work on 'unequal ecological exchange' (Hornborg 2017, 2019). Accordingly, the high degree of labour productivity in advanced capitalist countries is due not only to scientific discoveries and domestic social conflicts resulting in a sophisticated technology of production, but also to asymmetric material transfers on a global scale that have allowed the development of the productive forces and the increase of relative surplus value. Seen from this perspective, it becomes understandable that and how the 'technologically advanced sectors of the world-system have increased the rate of exploitation of its own workers […] by simultaneously increasing the rate of net imports of resources from elsewhere' (Hornborg 2019, 80). The intermediate products and raw materials from other regions make (re)production cheaper; a large part of the added value in transnational corporations is created in the centres; together with the structural, organisational, and institutional power of the labour force,[2] this enables relatively high wages, a well-developed public infrastructure, and services of general interest.

The imperial mode of living implies a hierarchy on a global scale: Since the onset of colonialism, the working and living conditions in the economies of the Global South, with their predominant forms of resource extraction, industrial or service production, have been largely geared to the economic needs of the capitalist centres. Domestic class, gender, and racialised relations are not exclusively, but essentially, oriented towards these needs. This is the core of the concept of the 'coloniality of power' developed by the Peruvian sociologist Aníbal Quijano (2000). The fact that Europe became the supposed centre of modernity is therefore due to a long historical process imbued with power that constituted certain forms of division and control of labour in the respective societies and on an international scale. In the course of colonisation, *race* became 'the fundamental criterion for the distribution of the world

---

[2] For a typology of power resources, see Schmalz et al. (2018).

population into ranks, places and roles in the new society's structure of power' (ibid., 535) of the colonised countries. Hierarchical identities related to skin colour were created, and a 'systematic racial division of labour' was imposed (ibid., 536).[3]

The concept of the imperial mode of living sheds light on these dominant interdependencies both between the Global South and the Global North and within the societies concerned. Above all, it aims to show and explain how domination, power, and violence are normalised in neo-colonial North-South relations, in class and gender relations, and by racialised relations in the practices of consumption and production, so that they are no longer perceived as such. The term is not intended to make the social contradictions *within* the Global North and the Global South disappear in favour of a seemingly superimposed imperialist North-South divide. Instead, the upper (and middle) classes of the Global South have to be understood as important forces of the imperial mode of living. Not only do they tend to adopt, and benefit from, Northern patterns of consumption (see Myers and Kent 2004; OECD 2019), but as the dominant forces of their societies, they also organise the extraction of resources or foster resource-intensive patterns of industrial development.

In the Global North, the infrastructures of everyday life in areas such as food, transport, electricity, heat, or telecommunication to a large extent rely on material flows from elsewhere, on the workers who extract the respective resources, and on the ecological sinks on a global scale that absorb emissions produced through the operation of infrastructural systems. Workers in the Global North draw on the latter not just because they consider them as components of a good life, but because they *depend* on them (cf. Lessenich 2019, 34). Mostly, it is not an individual choice that makes workers purchase cheap 'food from nowhere' (McMichael 2009), drive a car, or light their homes with electricity that is generated by burning fossil fuels. Rather, they have to do so in order to nourish their families, to get to work, or because the utility does not offer renewable alternatives. Thus, they are forced into the imperial mode of living simply because the latter is materialised and institutionalised in many of the life-sustaining systems of the Global North.

Of course, capitalists are also forced by competition to socially and ecologically destructive practices—at least there is a strong incentive to do so which is due to the structural tendency of the capitalist mode of production to generate 'negative externalities' (Wright 2010, 59–60). Yet they assume a

---

[3] Georg Jochum (2016) has emphasized the connection between coloniality and work in an even more systematic manner than Quijano. Of course, this is not to deny the struggles in the colonized countries and the social forces fighting racism and capitalist exploitation. As Beverly Silver (2003) and Zhang Lu (2015) have pointed out, social conflicts follow the relocation of capital.

*dominant* position in this process. Workers who process raw materials extracted elsewhere in the production process, who use fossilist infrastructures (energy supply, automobility) or who produce mass consumer goods at high energy and material costs mostly do so because they lack alternatives, that is, because they have nothing else to sell but their own labour power. The buyers of this labour power equally benefit from its exploitation as well as from the exploitation of nature and labour power elsewhere in the world. In other words, workers participate in the imperial mode of living and reproduce it as *subalterns*.[4] In addition, as consumers, they benefit materially from this mode of living to a much lesser extent. Due to the quantity and the way of their consumption, they also produce and externalise lower socio-ecological costs than the middle and upper classes (Chancel and Piketty 2015).

## Contradictions of the Imperial Mode of Living

The workforce in the industrialising countries of the nineteenth and early twentieth centuries experienced the health consequences of industrialisation at the workplace and in their residential areas much more than members of the middle and upper classes. On the consumption side, they were only marginally involved in the imperial mode of living. And, as shown by the example of sugar imported from the colonies, which acted as an energy supplier for the over-exploited workers (Osterhammel 2011, 338), their involvement was hardly for their own benefit. The workers' individual consumption, even of products based on colonial exploitation, was 'productive for the capitalist and the State, since it produces the power that creates their wealth' (Marx 1976 [1867], 719).

This situation changed with the emergence of a fossil energy regime, which made it possible to extend some of the benefits of the imperial mode of living to larger parts of the working class. As long as this regime was based on the use of coal, its production facilities and infrastructures were highly centralised and thus susceptible to disruptions and strikes. The workers were able to use targeted actions to withhold from society the products necessary for its reproduction, that is, to interrupt the energy supply. Their *production power* thus increased, while at the same time their *organisational power* was enhanced by their spatial concentration in vertically integrated large-scale factories and the extractive industry. As Timothy Mitchell (2011) and Michelle Williams

---

[4] We use the term 'subalterns' in a broad, that is, Gramscian sense as opposed to the dominant classes of society. It thus also includes workers in capitalism.

(2018) have argued, the working class used its increased power as a lever in the struggle for political and social rights. Thus, after World War II, these rights were expanded throughout the countries of the Global North. This link between production power and energy source weakened with the increasing importance of oil as the central energetic resource of Fordism; extraction and transport became more capital-intensive, and the spatially extensive network of storage facilities, pipelines, and tanker fleets was far less vulnerable to labour struggles than the infrastructures of the coal-based energy regime. However, petroleum revolutionised the *mode of living* of the working class (Wissen 2016, 50–51): A petroleum-based norm of consumption emerged, which enabled unprecedented increases of prosperity and tied workers more closely than ever before to capitalism and its growth logic through the individual possession or consumption of goods from mass production (cf. Aglietta 1979, 152–169). The availability of cheap oil became a crucial moment in the reproduction of the working class. 'Petroleum not only was the material basis for countless products themselves (e.g., plastics, clothing, and medicine), but also its centrality as transportation fuel ensured that even if products were not made with petroleum, they were distributed and consumed via petroleum-based modes of mobility' (Huber 2013, 180–181).

The societal generalisation of the imperial mode of living as a structural condition, which in the early stages of industrial capitalism was limited to productive consumption and the (luxury) consumption of the upper classes, was accompanied by an intensification of ecological problems, due to the significant increase regarding the extraction of natural resources and combustion of fossil fuels. Furthermore, the societal generalisation of the imperial mode of living went hand in hand with neo-colonial domination over the countries which had been decolonised in the nineteenth and twentieth centuries. Finally, patriarchal gender relations were perpetuated for example by the car-centred spatial structure of suburbanisation, which made it almost impossible to combine wage labour and reproductive work.

The famous dictum by Marx (1976 [1867], 638) that capitalist production is 'undermining the original sources of all wealth—the soil and the worker', thus needs to be modified to the effect that the undermining of the living conditions of workers in the Global North, that is, the socio-ecological contradiction of capitalism, was superimposed by the access to cheap nature and labour power in the Global South as well as by the unpaid reproductive labour of women in the Global North itself. The Fordist class compromise therefore also rested on a patriarchal and colonial global and social order, which was essential for processing the class contradiction in the Global North. Whereas the imperial mode of living has always been both an imperative *and* a promise, a constraint *and* an expansion of opportunities, with Fordism the enabling elements gained importance for workers in the Global North (Brand and Wissen 2021, chap. 3; Brand and Wissen 2018a, b).

This, however, should not tempt us to deny the class character of the imperial mode of living and to perceive ecological destruction and health burdens primarily as 'risks' across classes (Beck 1992). A glance at the living and working conditions in the coal and steel regions and industrial centres of Fordist capitalism shows that the increases in prosperity were only the other side of environmental pollution, which was particularly detrimental to the health of workers and their families (Barca and Leonardi 2018). Today there are clear signs that what Alain Lipietz (2000) diagnosed as the 'family similarity' of social and ecological contradictions and movements is once again coming to the fore. Although the promise of an imperial mode of living for material prosperity still carries a strong global and cross-class appeal, to many its redeemability is becoming increasingly questionable. Thus, it is crucial to understand the destruction of nature and that of the worker as interrelated parts of the same social relationships of power and domination (Hürtgen 2020).

On the one hand, this is due to the rising social inequality in the Global North, which, in contrast to Fordism, makes it impossible for more and more poor people to access the material benefits of the imperial mode of living to an extent that is even roughly comparable to that of the middle and upper classes.[5] On the other hand, the North-South conflict over participation in the imperial mode of living has intensified, as can be seen not least in the recent refugee and migration movements.[6] The latter demonstrate that an ever-increasing number of people from the Global South are no longer prepared to let their lives be destroyed by the externalisation effects of the imperial mode of living of the Global North and instead aspire to participate in the conveniences of this mode of living themselves. Another symptom is the economic rise of countries like China and India and the resulting spatial expansion of the patterns of production and consumption that constitute the

---

[5] See Dörre (2018, 13, our translation): It can be 'justifiably doubted that the 45 richest households in the Federal Republic of Germany, which have a share of assets roughly equivalent to that of the poorer half of the population, are connected by a common mode of living, a hegemonic promise of happiness, with the one million people who never managed to get out of the benefit system since the existence of the Hartz IV basic security scheme'. See also Nachtwey (2018) and, with regard to energy, Dahm and Bannas (2011). The fact that more and more people are locked in the framework of the imperial mode of living and appealed by its ongoing attractiveness and at the same time are deprived of the capacities to benefit from it in the same way as they were used to can be seen as one of the root-causes of the rise of the social and political right (see below and Dörre et al. 2018; Sauer et al. 2018).

[6] Migration is a multifaceted phenomenon, because it refers to, for example, people who temporarily migrate within Europe in order to achieve a higher income in other countries, but also to people who have to flee under life-threatening circumstances (cf. Lang 2017).

imperial mode of living. The industrialising countries themselves have now become dependent on an outside source from which they obtain raw materials, the working capacity which they can access, and to which they can shift their socio-ecological costs.[7] They are thus becoming competitors to the early industrialised countries of the Global North, not only economically but also ecologically. As a result, the eco-imperial tensions intensify, leading to fierce competition for raw materials and sinks.

The guiding principles, policies, and everyday practices dominating in the Global North, their diffusion into the Global South and the demands for participation, which for many can only be realised through flight or migration, show that the imperial mode of living is still an attractive possibility and promise. However, the aggravation of crisis phenomena such as climate change and the increasing conflicts around the world over $CO_2$-sinks and fossil, metallic, and agricultural raw materials leave little room for doubt that the promise can only be fulfilled in an ever more exclusionary and exclusive manner. This applies not only in geographical terms, that is, in the North-South relationship, but also in social terms, that is, within the Global North itself. The more the eco-imperial tensions that result from the generalisation of the non-generalisable intensify, the more the ecological upheavals in the centres will also become apparent in economic and social terms (Dörre and Becker 2018).

The Corona pandemic can also be seen in the context of the contradictions that become manifest with the generalisation of the imperial mode of living. This applies both to its causation and to its consequences. Although its outbreak zones might be locally identifiable, a comprehensive understanding of the disease requires taking into account the unequal exchange within the 'relational geographies' which have been created by global capital circuits. Agricultural business is particularly important here. Its extractivist activities contribute to destroying natural habitats through deforestation, thus fostering the zoonotic emergence of diseases that are then distributed through global value chains. Driven by, and driving, socio-ecologically destructive consumption patterns, 'commodity agriculture serves as both propulsion for and nexus through which pathogens of diverse origins migrate from the most remote reservoirs to the most international of population centers' (Wallace et al.

---

[7] China, for instance, is shifting ecological costs in form of greenhouse gas emissions to other countries as part of the 'New Silk Road' project. See the instructive essay by Federico Demaria and Joan Martinez-Alier (2017). The authors point out that in the course of this project Chinese companies have already invested in 240 coal-fired power plants and thus contributed to spatially shifting environmental damage and conflicts. See also Hoering (2018).

2020). Here they often encounter strong social inequalities which, in turn, are strengthened by them (Davis 2020).

Countering the increasingly manifest contradictions of the imperial mode of living mainly by an ecological modernisation[8] falls short (although it is without doubt preferable to the authoritarian solutions which have become so popular precisely because they deny the environmental crisis and promise an exclusive and exclusionary stabilisation of the imperial mode of living). Ecological modernisation strategies address the energetic and material basis of capitalism, but not its political economy and its inherent growth logic (Altvater 1996; Rilling 2011), that is, they leave the 'structural unsustainability' (Sommer and Welzer 2015, 37 and chapter 4) of the capitalist mode of production intact. Therefore, they will hardly be able to decouple economic growth entirely from resource consumption and environmental impact (Wiedmann et al. 2013). In a 'green' capitalist economy, the externalisation logic of the imperial mode of living is unlikely to be overcome but rather will be shifted to other areas, for example, to metals, which are becoming increasingly important in comparison to fossil fuels (see, e.g., Exner et al. 2015; Groneweg et al. 2017). Moreover, the *green jobs* envisaged in the relevant modernisation strategies are often of a poor quality, particularly those held by women. Their proponents hardly raise issues like the reduction in working hours and the need to overcome the gendered division of labour (Littig 2018, 567–569).

Therefore, it remains a crucial challenge to develop strategies that question the prevailing patterns of production and consumption much more fundamentally and put an end to the imperatives of growth associated with the capitalist mode of dynamic stabilisation (cf. Rosa 2019). This is an imperative of environmental justice and the *just transition* demanded by the international trade union movement (cf. Stevis and Felli 2015; Sweeney and Treat 2018), and it addresses the contradictions that currently manifest themselves, that is, the fact that even in the Gobal North the ecological question is increasingly becoming a social issue and the ecological implications of class and gender differences as well as racist discrimination are becoming more visible again. For *political ecology* as a critical scientific perspective and for many movements that work for environmental justice in the Global South as well as in parts of the Global North there has always been a constitutive relation between the social and the ecological (Perreault et al. 2015; Görg et al. 2017). Their interest and protest were sparked by the 'unequal distribution of environmental

---

[8] See, for example, the 'green economy' initiative of UNEP (2011b) and the concept of a 'European Green Deal' (European Commission 2019).

costs' (Barca and Leonardi 2018, 489–490). Drawing on these experiences and concepts, we finally explore the possibilities of a socio-ecological transformation that takes into account issues of labour and the environment similarly and acknowledges their constitutive relationship.

## Overcoming the Labour-Environment Divide

The point of departure of our considerations is that labour is the medium of metabolism between society and nature: 'as a primary agent of energy and matter transformation through the labor process, workers—broadly defined as those performing physical labor, including non-paid housekeeping and life-supporting work—are the primary interface between society and nature' (Barca 2012, 75; cf. Räthzel and Uzzell 2019, 146). This is not trivial. Rather, as Stefania Barca points out, it constitutes the particularity of a *working-class environmentalism* as a specific manifestation of the *environmentalism of the poor*, a term coined by Joan Martinez-Alier (2002). This differs from an *upper-* and middle-class *environmentalism* in that it is less concerned with mere nature conservation—that is, the protection of nature from destructive influences of 'humankind'—but with the inextricable link between social reproduction, human health, and an intact environment. In this respect, the concern for nature is less the result of abstract, often global and future threats than of the awareness and experience that (capitalist) control of nature endangers the lives and health of subalterns in a very tangible, local, and immediate way. The subalterns bear much less responsibility for this threat, yet at the same time they are much more exposed to it than the middle and upper classes. In this context, domination of nature is often inextricably connected with societal domination (Görg 2011; Bryant and Bailey 1997, 38–47). This can be seen, for instance, in the experience of peasants, whose existence is threatened by the monopolistic control of seeds by industrial agriculture. Similarly, it is reflected in the experience of workers, who are exposed to toxic substances and emissions from industrial plants not only in production but also in their residential areas. From their perspective, the struggle against the destruction of nature could become the decisive starting point for social emancipation and vice versa (although the objective exposure to environmental risks, of course, does not necessarily result in their progressive politicisation).[9]

---

[9] See the cartography and analysis of socio-ecological conflicts conducted by the EU FP7 project 'Environmental Justice Organisations, Liabilities and Trade' (http://www.ejolt.org/).

As seen, the relationship between societal domination and the domination of nature varies spatially and temporally. It is contingent on the struggles for improved living and working conditions, on the prevailing energy regime, and on a country's position in the international division of labour. As pointed out above with reference to Timothy Mitchell's work on (Fordist) *carbon democracy*, an increasing domination of nature can sometimes be accompanied by an expansion of social and political rights. This requires, among other things, that the societies concerned assume a privileged position in the international division of labour which enables them to shift the socio-ecological costs of destroying nature to other, weaker actors or marginalised areas. Wage earners in the Global North can thus be in a position to partially externalise certain negative preconditions and consequences. However, as we have also discussed, such a constellation never completely eliminates the link between social domination and the domination of nature, even in the externalising societies themselves. Moreover, experience to date suggests that it is not a permanent state. The Fordist class compromise provoked counter-movements long before it was destroyed by capital. They attacked the unequal gender, nature, and North-South relations on which the class compromise rested. Likewise, they politicised the relations of power and dominance inherent in the Taylorist factory system and the corresponding housing and living conditions of the wage earners (cf. Gehrke and Horn 2018).

In the emancipatory movements since 1968, there have been repeated indications that even in the Global North the social and ecological questions can no longer be grasped and tackled as separate issues. As seen, this is more topical now than ever. Nevertheless, the link between the two is not automatically established. Rather, it presupposes the productive handling of dilemmas by trade unions and other progressive forces politicising, for example, environmental and gender issues (cf. Brand and Niedermoser 2019). For this purpose, it is indispensable to reveal the global premises as well as the unequal distribution of the consequences of the imperial mode of living. This applies in particular to the materially prosperous countries, where the imperial mode of living tends to be widely accepted. It would imply to address and confront the racism that can be found in parts of the Northern working class itself, to critically reflect on the unions' role in fostering strategies of local, regional, or national competitiveness, to challenge the *competitive corporatism* and to constitutively incorporate international dimensions into trade union strategies. The project of a just transition, that is, a socio-ecological transformation of economy and society that is not carried out on the back of workers but actively involves and strengthens their demands for social security and decent work, must be conceived and implemented beyond national economies and corporatist structures.

There are several respective approaches pursued by workers and unions in many parts of the world. They can be distinguished according to the way in which labour and the environment are linked to each other.[10] Often, there is not one single way in the same union, but a union can be internally divided regarding the understanding and the significance of the labour-environment relationship: The union's heads may be more or less progressive than its rank and file; workers' views in different industries may diverge although they are represented by the same union; and there may be locally varying assessments and policies which are due to the different socio-economic profiles of the respective localities (Stevis 2018). Therefore, the three approaches of understanding and politicising the labour-environment relationship that we distinguish in the following, cannot necessarily be assigned to individual unions. Often, more than one approach can be observed in a union, pointing to the (increasingly) contested character of the ecological question in union politics. The three approaches thus concern ideal-typical abstractions, illustrated by concrete cases, which serve to identify and assess strategies of how to deal with the labour-environment dilemma.

The first approach can be called *instrumental*. Environmental issues are considered a necessity that has to be addressed in order to safeguard as much as possible of an otherwise environmentally destructive business-as-usual that has increasingly got under pressure. This can be observed in sectors like coal-mining and car manufacturing which are economically and/or politically confronted with the necessity to reorganise themselves, to shrink or even to phase out. Here, the climate crisis has become 'an issue that directly affects the "core business" of trade unions when environmental regulations (or the lack thereof) affects wages, conditions and the level of employment' (Felli 2014, 374). The immediate threat of job losses may bring unions in these sectors to see the problem in climate *policy* rather than in the climate *crisis*. Apart from denying the urgent character of the latter, they may resort to (often reactive) strategies of an ecological modernisation that help to mediate between two seemingly contradictory goals, namely safeguarding jobs and protecting the

---

[10] Stevis (2018) and Stevis et al. (2018) systematize union approaches to labour and the environment according to the criteria *depth* (what significance does the labour-environment relationship have for a union, what is the character of the proposed socio-economic changes and how far do they reach?), *breadth* (the scale and scope of the measures considered necessary) and *agency* (is the union strategy proactive or reactive?). For our purpose, the depth criterion is the most important one, although we also include aspects of the other two.

environment (Stevis 2018, 460–463; Cock 2018, 222–223; Sweeney and Treat 2018, 18–30).[11]

Secondly, labour and the environment can be linked in a *strategic* manner. This is the case where unions enter into alliances with environmental organisations in order to enhance their social power, or, in the words of Hans-Jürgen Urban, member of the board of the German metalworkers' union IG Metall, 'to complement individual power resources through cooperation with other movements, initiatives and organisations' (Urban 2018, 347, our translation). The aim may be to foster green job creation and green infrastructure provision or, more ambitious, to develop 'a *new culture of mosaic-leftist alliances* […] ultimately with the intention to transform the system' (ibid., 347). Examples include the BlueGreen Alliance in the United States (Stevis 2018, 458–460), the rapprochement between Austrian unions and environmental organisations which, on the side of the unions, has been triggered both by the growing acknowledgement of an energy transition and 'the need to build strong alliances with other progressive forces in society to counteract the neoliberal shift in general, particularly in energy policy' (Soder et al. 2018, 526), and the cooperation between IG Metall and the environmental NGO German Nature Conservation Ring in the early 1990s, which resulted in a programmatic document on the socio-ecological transformation of the transport system (IG Metall and Deutscher Naturschutzring 1992).[12]

A strategic connection between labour and the environment may increase the social power of progressive forces in their respective fields through alliances that allow for a common politicisation of issues related to wage labour and ecology. This is important because it can result in concrete improvements that should not be underestimated (see ILO 2018 and Galgóczi 2019 for several examples). Yet, there is a danger that these improvements will not extend beyond an ecological modernisation of the imperial mode of living. This is where a third way of linking labour and the environment comes into play. It is about an *organic* connection which re-establishes the constitutive relationship between labour and nature torn apart by the capitalist mode of production. In contrast to the instrumental view on labour and the

---

[11] In Germany, the miners' union (IG BCE) has to be mentioned in this respect, whereas, for example, the metalworkers' union (IG Metall) is more ambivalent. Being confronted with a massive structural change in the automotive sector, the central domain of IG Metall's organisational power, some union representatives call for an ecological modernisation in the form electro-automobility. Others however strive for a more fundamental transformation that aims to overcome the car-centred mobility system. For a recent overview of the German debate on labour and ecology, see Schröder and Urban (2018).

[12] Although this initiative was rather short-lived—actually it was marginalised in the course of German reunification and by the socio-ecologically destructive effects of neoliberal globalisation—it has become a point of reference in recent union debates again.

environment, the organic concept considers the climate *crisis* rather than climate *policy* as the principal threat for workers and their communities. Like the strategic approach, it draws on strong alliances between unions and environmental organisations, but places more emphasis on developing a common agenda for a fundamental socio-ecological transformation than on mutually supporting particular organisational interests. Without doubt, the latter, and with them the workplace-related 'economic' functions of unions, remain crucial. However, they are formulated in a way that they form part of a comprehensive strategy of systemic change.

The organic link between labour and the environment stands in the tradition of social movement unionism (Moody 1997). Starting from the experience of unions in the Global South that have successfully transcended the separation between narrower 'economic' and wider 'political' struggles, social movement unionism has confronted Northern unions with the challenge 'to reinvent themselves as social movements, not only responsible for the working conditions of their members, but for their general living conditions as well' (Räthzel and Uzzell 2011, 1221). It has, however, been rather silent on ecological issues—a void that has been filled by more recent contributions. Thus, Ashley (2018) describes the South African Million Climate Jobs Campaign, which was formed in 2011 by activists from trade unions, social movements, non-governmental organisations, and academics, as an attempt to combine the creation of decent work with the protection of the environment and the provision of communities with essential infrastructure (see also Satgar 2015 on the climate justice approach of the South African metalworkers' union NUMSA, see Sikwebu and Aroun, Chap. 3, Pillay, Chap. 4, Cock, Chap. 8).

Cha and Skinner, together with a working group composed by unionists, have developed an ambitious climate jobs agenda for New York state that aims 'to simultaneously reduce greenhouse gas emissions and reverse inequality by protecting workers and creating good, family-sustaining jobs' in the building, energy, and transportation sector (Cha and Skinner 2017, 3; see also McAlevey 2019). In addition, Felli (2014, 389–391) analyses the climate change-related activities of the International Transport Workers' Federation (ITF) as a paradigmatic example for a socialist strategy of addressing the ecological crisis, centred around an ecologically transformed and democratically controlled mobility system.

A common goal of the authors arguing for an organic connection between labour and environment is to rethink labour and the whole economy from the point of view of social reproduction, that is, to raise the questions of '*what, how* and *how much* to produce' and to demand 'that these questions be answered through mechanisms other than that of price—that is, by non-market institutions' (Felli 2014, 391). This is an issue, first, of rethinking and reorganising the economy from the perspective of what is socially useful and necessary; second, of reshaping the society-nature relationship in a way that takes the reproductive necessities of human and non-human nature into account; third, of redefining work in a comprehensive way that does include not only wage labour but also care, political work, and individual development; and, fourth, of equally redistributing the four dimensions of work between women and men (Haug 2009; Littig 2018). It is against this background that, as Jacklyn Cock (2018, 216) has argued, 'a social reproduction approach can potentially validate and link separate struggles'.

For unions this means that they, rather than striving mainly for a redistribution of the wealth which is produced within a socio-ecologically destructive economic system, attack the very *form* of wealth production and place emphasis on the satisfaction of human needs in a socio-ecological sustainable manner, in other words, on the 'foundational economy' of care and life-supporting infrastructure systems (cf. Foundational Economy Collective 2018). This is not an easy undertaking, since it requires learning processes and organisational efforts that go far beyond those already organised in unions. And the more industries are concerned whose future is questionable on ecological grounds, the more difficult it becomes. Thus, the Corona pandemic, through reinforcing the structural crisis in key sectors like car manufacturing, has given rise to traditional positions within unions which conceive the labour-environment relationship in a merely instrumental way. Like in the crisis of 2008–2009, officials of the German metalworkers' union, in concert with business representatives, for example, call for the state to stimulate the demand for ecologically modernised cars. The idea of a comprehensive transformation of the car-centred transport system here is sacrificed to a certain extent for a supposed job security of the union's core membership.

Without denying the dilemma trade unionists are confronted with, this is a narrow and short-sighted strategy which, while aiming at protecting workers against crisis policy, deepens the crisis itself and enhances workers' vulnerability in the long run (cf. Ashley 2018, 290). It is, however, contested in the union itself: months before the outbreak of the Covid-19, Katharina Grabietz and Kerstin Klein, two secretaries of the IG Metall, emphasised that workers in car manufacturing are 'not solely employees of a climate-damaging industry, they also live in cities without clean air and have children who take to the streets on Friday fighting for their right to a liveable planet. [...] The discussion about transformation and decarbonisation is linked to fundamental questions of how we want to live, produce and do the economy' (Grabietz and

Klein 2019, 37, our translation).[13] The two authors demand a democratic discourse about the transformation of production on the level of the firm, the branch, and society at large. 'This discourse must be explicitly critical of capitalism, if it shall not merely fight symptoms but strive for abolishing the systemic causes of job losses and environmental destruction' (ibid., 37).

The Covid-19 crisis will probably serve as a catalyst for reshaping the labour-environment relationship. Without doubt, there are strong forces fostering a socio-ecological roll-back, both within and outside the unions. However, there is also much evidence that unions have a lot to win from organically linking labour and environment from the perspective of social reproduction. Public ownership and democratic control of the means of production and subsistence are important preconditions. Reclaiming them and searching for the respective social and international alliances will be crucial for overcoming the imperial mode of living in favour of a good life for all.

# References

Aglietta, Michel. 1979. *A Theory of Capitalist Regulation. The US Experience*. London: Verso.

Altvater, Elmar. 1996. Der Traum vom Umweltraum. Zur Studie des Wuppertal Instituts über ein 'zukunftsfähiges Deutschland'. *Blätter für deutsche und internationale Politik* 1: 82–91.

Ashley, Brian. 2018. Climate Jobs at Two Minutes to Midnight. In *The Climate Crisis. South African and Global Democratic Eco-Socialist Alternatives*, ed. Vishwas Satgar, 272–292. Johannesburg: Wits University Press.

Barca, Stefania. 2012. On Working-Class Environmentalism: A Historical and Transnational Overview. *Interface* 4 (2): 61–80.

Barca, Stefania, and Emanuele Leonardi. 2018. Working-Class Ecology and Union Politics: A Conceptual Topology. *Globalizations* 15 (4): 487–503. https://doi.org/10.1080/14747731.2018.1454672.

Beck, Ulrich. 1992. *Risk Society. Towards a New Modernity*. London: Sage.

Biesecker, Adelheid, and Sabine Hofmeister. 2010. (Re)Productivity: Sustainable Relations Both Between Society and Nature and Between the Genders. *Ecological Economics* 69 (8): 1703–1711. https://doi.org/10.1016/j.ecolecon.2010.03.025.

Brand, Ulrich, and Kathrin Niedermoser. 2019. Overcoming the Impasse of the Current Growth Model and the Imperial Mode of Living. The Role of Trade

---

[13] See the similar position of the Korean Power Plant Industry Union (KPTU) which supported the phase-out of older coal-fired power stations with the following words: 'Although our hearts are heavy, we welcome the shutdown of worn out coal power plants because we are clear about what kind of country we want to leave for our descendants' (statement of the KPTU, quoted in Sweeney and Treat 2018, 39).

Unions in Social-Ecological Transformation. *Journal of Cleaner Production* 225: 173–180. https://doi.org/10.1016/j.jclepro.2019.03.284.

Brand, Ulrich, and Markus Wissen. 2018a. *The Limits to Capitalist Nature. Theorizing and Overcoming the Imperial Mode of Living*. London: Rowman & Littlefield.

———. 2018b. What Kind of Great Transformation? The Imperial Mode of Living as a Major Obstacle to Sustainability Politics. *GAIA* 27 (3): 287–292. https://doi.org/10.14512/gaia.27.3.8.

———. 2021. *The Imperial Mode of Living. Everyday Life and the Ecological Crisis of Capitalism*. London: Verso.

Bryant, Raymond L., and Sinead Bailey. 1997. *Third World Political Ecology*. London: Taylor & Francis Ltd.

Burkett, Paul. 2014. *Marx and Nature. A Red and Green Perspective*. Chicago: Haymarket Books.

Cha, J. Mijin, and Lara Skinner. 2017. *Reversing Inequality, Combatting Climate Change. A Climate Jobs Program for New York State*. Buffalo, Ithaca, New York and Rochester: The Worker Institute at Cornell. https://archive.ilr.cornell.edu/sites/default/files/InequalityClimateChangeReport.pdf. Accessed 13 June 2020.

Chancel, Lucas, and Thomas Piketty. 2015. *Carbon and Inequality: From Kyoto to Paris. Trends in the Global Inequality of Carbon Emissions (1998–2013) & Prospects for an Equitable Adaptation Fund*. Paris: Paris School of Economics. http://piketty.pse.ens.fr/files/ChancelPiketty2015.pdf. Accessed 8 May 2020.

Cock, Jacklyn. 2018. The Climate Crisis and a 'Just Transition' in South Africa: An Eco-Feminist-Socialist Perspective. In *The Climate Crisis. South African and Global Democratic Eco-Socialist Alternatives*, ed. Vishwas Satgar, 210–230. Johannesburg: Wits University Press.

Dahm, Daniel J., and Stephan Bannas. 2011. *The Decline of the Fossil Age Is the Rise of Distributive Justice. International Development Policy*, Policy Briefs. Working Papers. http://journals.openedition.org/poldev/835. Accessed 14 March 2020.

Davis, Mike. 2020. The Monster Enters. *New Left Review* 122: 7–14.

Demaria, Federico, and Joan Martinez-Alier. 2017. China Has a Plan – Peak Coal and the New Silk Road. *The Ecologist*, July 25. https://theecologist.org/2017/jul/25/special-report-china-has-plan-peak-coal-and-new-silk-road. Accessed 13 June 2020.

Dörre, Klaus. 2018. Imperiale Lebensweise - eine hoffentlich konstruktive Kritik. Teil 1. Thesen und Gegenthese. *Sozialismus* 6: 10–13.

Dörre, Klaus, and Karina Becker. 2018. Nach dem raschen Wachstum: Doppelkrise und große Transformation. In *Gute Arbeit. Ökologie der Arbeit – Impulse für einen nachhaltigen Umbau*, ed. Lothar Schröder and Hans-Jürgen Urban, 35–58. Frankfurt a. M.: Bund-Verlag.

Dörre, Klaus, Sophie Bose, John Lütten, and Jakob Köster. 2018. Arbeiterbewegung von rechts? Motive und Grenzen einer imaginären Revolte. *Berliner Journal für Soziologie* 28: 55–89.

European Commission. 2019. *Communication from the Commission to the European Parliament, the Council, the European Economic and Social Committee and the*

Committee of the Regions. *The European Green Deal*. COM(2019) 640 Final. Brussels: European Commission.
Exner, Andreas, Christian Lauk, and Werner Zittel. 2015. Sold Futures? The Global Availability of Metals and Economic Growth at the Peripheries: Distribution and Regulation in a Degrowth Perspective. *Antipode* 47 (2): 342–359. https://doi.org/10.1111/anti.12107.
Felli, Romain. 2014. An Alternative Socio-Ecological Strategy? International Trade Unions' Engagement with Climate Change. *Review of International Political Economy* 21 (2): 372–398. https://doi.org/10.1080/09692290.2012.761642.
Foundational Economy Collective. 2018. *The Infrastructure of Everyday Life*. Manchester: Manchester University Press.
Fücks, Ralf. 2013. *Intelligent wachsen: Die grüne Revolution*. München: Hanser.
Galgóczi, Béla, ed. 2019. *Towards a Just Transition: Coal, Cars and the World of Work*. Brussels: ETUI.
Gehrke, Bernd, and Gerd-Rainer Horn, eds. 2018. *1968 und die Arbeiter. Studien zum "proletarischen Mai" in Europa*. Hamburg: VSA.
Görg, Christoph. 2011. Societal Relationships with Nature: A Dialectical Approach to Environmental Politics. In *Critical Ecologies. The Frankfurt School and Contemporary Environmental Crises*, ed. Andrew Biro, 43–72. Toronto: University of Toronto Press.
Görg, Christoph, Ulrich Brand, Helmut Haberl, Diana Hummel, Thomas Jahn, and Stefan Liehr. 2017. Challenges for Social-Ecological Transformations: Contributions from Social and Political Ecology. *Sustainability* 9 (7): 1–21. https://doi.org/10.3390/su9071045.
Grabietz, Katharina, and Kerstin Klein. 2019. #FairWandel. Für eine Industriegesellschaft, die weder Mensch noch Klima auf der Strecke lässt. *Sozialismus* (6): 36–38.
Groneweg, Merle, Hannah Pilgrim, and Michael Reckordt. 2017. Diesseits der Dematerialisierung. Der Ressourcenbedarf der Industrie 4.0. *PROKLA* 47 (4): 623–633. https://doi.org/10.32387/prokla.v47i189.60.
Haug, Frigga. 2009. The "Four-in-One Perspective": A Manifesto for a More Just Life. *Socialism and Democracy* 23 (1): 119–123.
Hoering, Uwe. 2018. *China's Long March 2.0. The Belt and Road Initiative as Development Model*. Hamburg: VSA.
Hornborg, Alf. 2017. Political Ecology and Unequal Exchange. In *Routledge Handbook on Ecological Economics. Nature and Society*, ed. Clive L. Spash, 39–47. London and New York: Routledge.
———. 2019. The Money-Energy-Technology Complex and Ecological Marxism. Rethinking the Concept of 'Use-Value' to Extend Our Understanding of Unequal Exchange, Part 2. *Capitalism Nature Socialism* 30 (4): 71–86. https://doi.org/10.1080/10455752.2018.1464212.
Huber, Matt. 2013. Fueling Capitalism: Oil, the Regulation Approach, and the Ecology of Capital. *Economic Geography* 89 (2): 171–194. https://doi.org/10.1111/ecge.12006.

Hürtgen, Stefanie. 2020. Arbeit, Klasse und eigensinniges Alltagshandeln. Kritisches zur imperialen Lebensweise – Teil 1. *PROKLA* 50 (1): 171–188. https://doi.org/10.32387/prokla.v50i198.1832.

IG Metall, and Deutscher Naturschutzring, eds. 1992. *Auto, Umwelt, Verkehr. Umsteuern, bevor es zu spät ist. Verkehrspolitische Konferenz der IG Metall und des Deutschen Naturschutzrings*. Köln: Bund-Verlag.

ILO. 2018. *Just Transition Towards Environmentally Sustainable Economies and Societies for All*. ILO ACTRAV Policy Brief. Geneva: International Labour Office. www.ilo.org/actrav. Accessed 13 June 2020.

Jochum, Georg. 2016. Kolonialität der Arbeit. Zum historischen Wandel der durch Arbeit vermittelten Naturverhältnisse. In *Nachhaltige Arbeit. Soziologische Beiträge zur Neubestimmung der gesellschaftlichen Naturverhältnisse*, ed. Thomas Barth, Georg Jochum, and Beate Littig, 125–149. Frankfurt a. M./New York: Campus.

Lang, Miriam. 2017. Den globalen Süden mitdenken. Was Migration mit imperialer Lebensweise, Degrowth und neuem Internationalismus zu tun hat. *Movements. Journal for Critical Migration and Border Regime Studies* 3 (1): 179–190.

Lessenich, Stephan. 2019. *Living Well at Others' Expense. The Hidden Costs of Western Prosperity*. Cambridge: Polity Press.

Lipietz, Alain. 2000. Political Ecology and the Future of Marxism. *Capitalism Nature Socialism* 11 (1): 69–85. https://doi.org/10.1080/10455750009358898.

Littig, Beate. 2018. Good Work? Sustainable Work and Sustainable Development: A Critical Gender Perspective from the Global North. *Globalizations* 15 (4): 565–579. https://doi.org/10.1080/14747731.2018.1454676.

Martinez-Alier, Joan. 2002. *The Environmentalism of the Poor: Study of Ecological Conflicts and Valuation*. Cheltenham: Edward Elgar.

Marx, Karl. 1976 [1867]. *Capital*. Vol. 1. London: Penguin Books Ltd.

McAlevey, Jane. 2019. Organizing to Win a Green New Deal. *Jacobin*, March 26. https://www.jacobinmag.com/2019/03/green-new-deal-union-organizing-jobs. Accessed 8 May 2020.

McMichael, Philip. 2009. The World Food Crisis in Historical Perspective. *Monthly Review* 61 (3). https://monthlyreview.org/2009/07/01/the-world-food-crisis-in-historical-perspective/. Accessed 8 May 2020.

Mitchell, Timothy. 2011. *Carbon Democracy. Political Power in the Age of Oil*. London and New York: Verso.

Moody, Kim. 1997. *Workers in a Lean World. Unions in the International Economy*. London: Verso.

Myers, Norman, and Jennifer Kent. 2004. *The New Consumers. The Influence of Affluence on the Environment*. Washington: Island Press.

Nachtwey, Oliver. 2018. *Germany's Hidden Crisis. Social Decline in the Heart of Europe*. London and New York: Verso.

OECD. 2019. *Global Material Resources Outlook to 2060: Economic Drivers and Environmental Consequences*. Paris: OECD Publishing.

Osterhammel, Jürgen. 2011. *Die Verwandlung der Welt. Eine Geschichte des 19. Jahrhunderts*. München: C.H. Beck.

Parrique, Timothée, Jonathan Barth, François Briens, Christian Kerschner, Alejo Kraus-Polk, Anna Kuokkanen, and Joachim H. Spangenberg. 2019. *Decoupling Debunked. Evidence and Arguments Against Green Growth as a Sole Strategy for Sustainability*. European Environmental Bureau. https://eeb.org/library/decoupling-debunked/. Accessed 12 May 2020.

Perreault, Tom, James McCarthy, and Gavin Bridge, eds. 2015. *The Routledge Handbook of Political Ecology*. London: Routledge.

Quijano, Anibal. 2000. Coloniality of Power, Eurocentrism, and Latin America. *Nepentia. Views from the South* 1 (3): 533–580.

Räthzel, Nora, and David Uzzell. 2011. Trade Unions and Climate Change: The Jobs Versus Environment Dilemma. *Global Environmental Change* 21 (4): 1215–1223. https://doi.org/10.1016/j.gloenvcha.2011.07.010.

———. 2019. The Future of Work Defines the Future of Humanity and All Living Species. *International Journal of Labour Research* 9 (1–2): 145–171.

Rilling, Rainer. 2011. Wenn die Hütte brennt… "Energiewende", green new deal und grüner Sozialismus. *Forum Wissenschaft* 28 (4): 14–18.

Rosa, Hartmut. 2019. *Resonance – A Sociology of the Relationship to the World*. Cambridge: Polity Press.

Salleh, Ariel. 2017. *Ecofeminism as Politics. Nature, Marx and the Postmodern*. London: Zed Books.

Satgar, Vishwas. 2015. A Trade Union Approach to Climate Justice: The Campaign Strategy of the National Union of Metalworkers of South Africa. *Global Labour Journal* 6 (3): 267–282. https://doi.org/10.15173/glj.v6i3.2325.

Sauer, Dieter, Ursula Stöger, Joachim Bischoff, Richard Detje, and Bernhard Müller. 2018. *Rechtspopulismus und Gewerkschaften. Eine arbeitsweltliche Spurensuche*. Hamburg: VSA.

Schmalz, Stefan, Carmen Ludwig, and Edward Webster. 2018. The Power Resources Approach: Developments and Challenges. *Global Labour Journal* 9 (2): 113–134. https://doi.org/10.15173/glj.v9i2.3569.

Schröder, Lothar, and Hans-Jürgen Urban, eds. 2018. *Gute Arbeit. Ökologie der Arbeit – Impulse für einen nachhaltigen Umbau*. Frankfurt am Main: Bund-Verlag.

Silver, Beverly J. 2003. *Forces of Labour. Workers' Movements and Globalization Since 1870*. Cambridge/New York: Cambridge University Press.

Soder, Michael, Kathrin Niedermoser, and Hendrik Theine. 2018. Beyond Growth: New Alliances for Socio-Ecological Transformation in Austria. *Globalizations* 15 (4): 520–535. https://doi.org/10.1080/14747731.2018.1454680.

Sommer, Bernd, and Harald Welzer. 2015. Transformation Design: A Social-Ecological Perspective. In *Transformation Design. Perspectives on a New Design Attitude*, ed. Wolfgang Jonas, Sarah Zerwas, and Kristof von Anshelm, 188–201. Basel: Birkhäuser.

Stevis, Dimitris. 2018. US Labour Unions and Green Transitions: Depth, Breadth, and Worker Agency. *Globalizations* 15 (4): 454–469. https://doi.org/10.1080/14747731.2018.1454681.

Stevis, Dimitris, and Romain Felli. 2015. Global Labour Unions and Just Transition to a Green Economy. *International Environmental Agreements: Politics, Law and Economics* 15 (1): 29–43. https://doi.org/10.1007/s10784-014-9266-1.

Stevis, Dimitris, David Uzzell, and Nora Räthzel. 2018. The Labour-Nature Relationship: Varieties of Labour Environmentalism. *Globalizations* 15 (4): 439–453. https://doi.org/10.1080/14747731.2018.1454675.

Sweeney, Sean, and John Treat. 2018. *Trade Unions and Just Transition. The Search for a Transformative Politics*. Trade Unions for Energy Democracy, Working Paper No. 11. http://unionsforenergydemocracy.org/resources/tued-publications/tued-working-paper-11-trade-unions-and-just-transition/. Accessed 14 March 2020.

UNEP (United Nations Environment Programme). 2011a. *Decoupling Natural Resource Use and Environmental Impacts from Economic Growth*. https://www.unenvironment.org/resources/report/decoupling-natural-resource-use-and-environmental-impacts-economic-growth. Accessed 12 August 2019.

———. 2011b. *Towards a Green Economy. Pathways to Sustainable Development and Poverty Eradication*. https://sustainabledevelopment.un.org/content/documents/126GER_synthesis_en.pdf. Accessed 12 August 2019.

Urban, Hans-Jürgen. 2018. Ökologie der Arbeit. Ein offenes Feld gewerkschaftlicher Politik? In *Gute Arbeit. Ökologie der Arbeit – Impulse für einen nachhaltigen Umbau*, ed. Lothar Schröder and Hans-Jürgen Urban, 329–349. Frankfurt a. M.: Bund-Verlag.

Wallace, Rob, Alex Liebman, Luis Fernando Chaves, and Rodrick Wallace. 2020. Covid-19 and Circuits of Capital. *Monthly Review*. https://monthlyreview.org/2020/04/01/covid-19-and-circuits-of-capital/. Accessed 8 May 2020.

WBGU – German Advisory Council on Global Change. 2011. *World in Transition. A Social Contract for Sustainability*. Berlin: WBGU.

Wiedmann, Thomas O., Heinz Schandl, Manfred Lenzen, Daniel Moran, Sangwon Suh, James West, and Keiichiro Kanemoto. 2013. The Material Footprint of Nations. *Proceedings of the National Academy of Sciences of the USA* 112 (20): 6271–6276. https://doi.org/10.1073/pnas.1220362110.

Williams, Michelle. 2018. Energy, Labour and Democracy in South Africa. In *The Climate Crisis. South African and Global Democratic Eco-Socialist Alternatives*, ed. Vishwas Satgar, 231–251. Johannesburg: Wits University Press.

Wissen, Markus. 2016. Jenseits der carbon democracy. Zur Demokratisierung der gesellschaftlichen Naturverhältnisse. In *Transformation der Demokratie – demokratische Transformation*, ed. Alex Demirović, 48–66. Münster: Westfälisches Dampfboot.

Wright, Erik Olin. 2010. *Envisioning Real Utopias*. London/New York: Verso.

Zhang, Lu. 2015. *Inside China's Automobile Factories. The Politics of Labour and Worker Resistance*. New York: Cambridge University Press.

# 31

# André Gorz's Labour-Based Political Ecology and Its Legacy for the Twenty-First Century

Emanuele Leonardi and Maura Benegiamo

## Introduction

In setting the stage for a new field of inquiry called Environmental Labour Studies (ELS), Nora Räthzel and David Uzzell pointed to a specific theoretical rationale, that of investigating 'the way in which nature and labour are intrinsically linked and equally threatened by globalising capital' (Räthzel and Uzzell 2013, 10). From a different but convergent perspective, the same authors and Dimitris Stevis suggested that ELS are particularly interested in 'the investigation of both internal and external dynamics of labour environmentalism' (Stevis et al. 2018, 443), where 'internal' refers to ecological politics within unions and 'external' denotes its impacts on the broader economy.

Our aim in this chapter is to discuss the relevance of André Gorz (1923–2007) for such an endeavour by focusing on his labour-based political ecology and its implications for assessing how the link between capital and nature has changed over the last half a century, and how profoundly such mutation affected union politics. It is our conviction that Gorz's work can significantly contribute to foster debates about ELS and, more broadly, to

E. Leonardi (✉)
Department of Humanities, Social Sciences and Cultural Industries, University of Parma, Parma, Italy
e-mail: emanuele.leonardi@unipr.it

M. Benegiamo
Department of Political and Social Sciences, University of Trieste, Trieste, Italy

connect (and transform) labour militancy, political ecology and environmentalist advocacy. We believe so despite the fact that, with the partial exception of *Letter to D.* (2006/2010)—a remarkable editorial success which has allowed a new generation of activists and academics to discover his thought—the name of Gorz does not often pop up in ELS analyses at a transnational level. This is a rather curious occurrence given his diverse international connections, amongst which of particular importance were those with German and Swedish Metalworkers' Unions and with the Italian New Left—especially its *workerist* segment (Gianinazzi 2016). Actually, we believe his interest in the linkage between labour and ecology and his efforts to deepen the environmental politics of unions make him an ELS scholar *ante litteram*. Thus, the goals of this chapter are to present some features of his original account of socio-environmental crises, to highlight their reliance on a labour-based approach, and to explore some elements of his legacy in the twenty-first century.

Sections "André Gorz as a Precursor of Critical ELS Scholarship" and "The Oil Shock as a Crisis of Reproduction and the Green Capitalism Debate" articulate an overview of Gorz's 1970s writings on environmental issues, with particular regard for what we define as his *labour-based political ecology*. We focus on Gorz's interest in workers' struggles as the cornerstone of his ecological theorizing. Section "André Gorz on Cognitive Capitalism" discusses Gorz's 2000s writings on cognitive capitalism and especially the implications for a labour-based political ecology that addresses the dynamics of the twenty-first century. The concluding section elaborates on the political implications of Gorz's legacy and proposes possible lines of further research. In particular, we argue that Gorz provides valuable insights but also shows problematic shortcomings. In order to overcome the latter, and revive the former, we refer to a second author, Melinda Cooper, whose research highlights the capitalist attempt to turn physical limits to growth from obstacle into drivers of accumulation. Such an attempt, we argue following her insights, rests on an unprecedented relationship between the sphere of production and that of reproduction. This leads to a general rethinking of labour-based political ecology and union practices: a rethinking that Gorz had long hoped for.

## André Gorz as a Precursor of Critical ELS Scholarship

The compass of Gorz's theoretical trajectory has been Jean-Paul Sartre's Existentialism, with its peculiar way of linking individual-centred, humanist freedom and structure-focused Marxism (Gollain 2018). As explicitly stated in a 2005 interview with the *EcoRev'* journal, Gorz always believed that the autonomy of each person justified opposition to domination by social megamachines (2008/2010). In this context, the purpose of this section is to frame Gorz's contributions as anticipations of ELS' threefold focus on ecological deterioration, workers' agency (or lack thereof) and capital's metabolic intrusiveness. In our view this is justified because, from the 1970s onwards, Gorz's thought revolved around the relationships between *environmental crises*, the processes of *labour transformations* and the violence of *capitalist development*. Yet this set of themes needs to be carefully examined against the background of his changing interests and analytical foci—especially expressed in the controversial *Farewell to the Working Class: An Essay on Post-industrial Socialism* (1980/1987). Despite the intellectual continuity which Gorz himself highlighted (2008/2010), his trajectory and legacy have been shaped, as much, by the aspects of his thought that interpreters have chosen to privilege—whether in emphasizing one element over the remaining two or in the ways in which they have framed the relations between the various elements.

With the above in mind, Gorz's early ELS insights can be first detected during his politico-ecological phase (1971–1980)—best expressed in the seminal essay *Ecology as Freedom* (1978/1980). His focus on political ecology can be understood both as a thematic interest during this specific period and—more importantly—as the general framework from which different and apparently unrelated issues were approached later on in his intellectual journey. A first key element is the influence of the wave of May 1968 over Gorz's interest for political ecology: in fact, the years immediately following the uprising should be understood as a gestational period. The role played by 1968 can be appreciated in two different ways. First, Gorz was one of the very few influential thinkers on the Left who actually understood the epoch-changing character of the ecological crisis and immediately grasped the challenge it posed for labour-based strategies of emancipation.[1] For Gorz, it is the new politics unleashed

---

[1] There were, of course, notable exceptions, amongst which a few names ought to be mentioned: Rudolf Bahro, Murray Bookchin, Barry Commoner and Laura Conti. For a thorough investigation on these issues, see Barca (2014, 2017).

by May 1968 that provide the background against which the workers' movement, as well as its unions, can renew itself *ecologically*.

Second, this deep appreciation of ecology played a crucial role in his critique of some traditional pillars of Marxism (in particular productivism). After all, the *civilization of liberated time*—the Gorzian eco-socialist utopia *par excellence*—combines a recognition of biophysical limits with worker control over the productive process, in collaboration with self-managed local communities. Tellingly, Gorz defines such desirable future as 'society of re-appropriated work' (1998/2017, 8—translation modified). At least two remarks, however, seem necessary at this point. Firstly, such a vision works only insofar as Marxism is denuded of any prophetic element. Since the 1950s, in fact, Gorz had been sceptical towards the philosophy of history, and especially critical of the historicist tendencies of dialectical materialism pointing to the idea of inexorable stages from Feudalism to Communism, through Capitalism. Relatedly, it is in this linear, automatic sequence of the modes of production that one can locate the historical affinity between the 'official' labour movement (unions and communist parties) and productivism.

This is why Gorz's political ecology can be labelled as *labour-based* (and socialist): its ultimate goal is to foster workers' power during the unprecedented political conjuncture inaugurated by May 1968 and marked by the increasing centrality of environmental issues during that period. To properly understand such an ambitious objective, it is important to focus on Gorz's politico-ecological phase of 1971–1980. This phase, already in gestation since 1968, begins in 1971 with a series of articles for *Le Nouvel Observateur* and finds a formal systematization in 1973 with the publication of two books—one more theoretical (*Critique de la division du travail* [*The Division of Labor*], 1973a/1976) and the other more journalistic (*Critique du capitalisme quotidien* [*Capitalism in Crisis and Everyday Life*], 1973b/1977).[2] Both books assume that the emerging ecological crisis is a crucial challenge for rethinking a technically feasible and politically desirable socialist strategy. Although it is fairly simple to detect within them an echo of the Club of Rome's *Limits to Growth* report, which had a significant impact on European public opinion, it should be stressed that Gorz had come to recognize the centrality of environmental deterioration on his own. Such independence is shown by a text

---

[2] The distinction between theoretical inquiry and journalistic investigation is relevant. The author's name (born in Austria, in 1923, under the name Gerhart Hirsch) was modified to Gérard Horst in 1930, following his father's conversion to Catholicism, most probably due to profound antisemitism + in the country. Later in life, the author used the pseudonym André Gorz for theoretical works and Michel Bosquet for journalistic works. It is not by chance that *Ecology as Freedom* was the first book to be 'authored' by both. The text (1971/1978)—which was read during the conference—is included in the French and Italian editions of *Ecology as Politics* but not translated into English.

written in 1971 and addressed to the West German Metalworkers' Union (IG Metall)[3] on occasion of its April 1972 conference 'Aufgabe Zukunft: Qualität des Lebens' (Task Future: Quality of Life)—to which Gorz had been invited.[4] This close relationship with industrial unions enables us to further clarify an important dimension of Gorz's theoretical elaboration: it is through constant engagement with the working class and its conflictual agency that Gorz develops his particular form of political ecology.[5] Moreover, up until the mid-1970s, Gorz's voice is widely discussed in union circles, both in France and in Northern Europe (Gianinazzi 2016). The main argument of Gorz's text in 1971 is that since economic crises are increasingly due to 'rising costs of reproduction' (1971/1978, 70)—for example, environmental safeguard (*ex ante*) or clean up (*ex post*)—the IG Metall should firmly engage in a civilizational project grounded not only on wage-centred collective bargaining, but also (even more importantly) on 'extra-economic, "qualitative" claims' (1971/1978, 72—our translation).

In *The Division of Labor* (1973a), Gorz re-interpreted such 'qualitative' claims in an ecological fashion and argued that they militated against the 'heteronomous character of labour-employment [*travail-emploi*][6]' (Méda 2017, 202). Gorz's political ecology can be considered *labour-based* insofar as the possibility of mending the breach between society and nature—a breach which is produced by the capitalist organization of productive processes—rests on the simultaneous, yet tensive, liberation *from* wage labour and liberation *of* work, in general.

Such tension between *refusal of labour* (aimed at eradicating exploitation) and *liberation of work* (aimed at eliminating alienation) is important in our analysis and needs to be clarified. In Gorz's view, in fact, the cycle of struggle of the 1970s expressed *qualitative* claims against the capitalist organization of production that went well beyond the mostly *quantitative* tradition of

---

[3] The text (1971/1978)—which was read during the conference—is included in the French and Italian editions of *Ecology as Politics* but not translated into English.

[4] The conference was extremely well-attended—1250 attendees—and featured renowned political figures such as West German Minister Erhard Eppler and Swedish Prime Minister Olof Palme.

[5] This aspect is compellingly discussed—although with no reference to Gorz—by Flemming and Reuter (2020).

[6] As Dominique Méda (2017) aptly points out, Gorz advances a key distinction between *work* as an anthropological category concerning the nature-society metabolism and *labour-employment* as a specifically modern form of heteronomous activity. Heteronomy here is due to three factors: the capitalist framework of society (centrality of private property); the wage-form (subalternity of workers); and the fragmentation of the productive process operated by the division of labour (standardisation of tasks and replaceability of workers). According to Méda, Gorz's emphasis is decidedly on the third element. We agree with her when it comes to his thinking and publications during the 1980s and 1990s, whereas we believe a strong connection amongst all the elements characterise the politico-ecological phase and the later part of his work (especially 2005–2007).

post-World War II collective bargaining. They raised political questions regarding: (a) the meaninglessness of alienated labour; and (b) an existence shaped by capitalist temporalities—conducive to productivism and impermeable to autonomy and collective self-management. Thus, it comes as no surprise that Gorz developed his thought in close connection with the likes of Herbert Marcuse and Ivan Illich, with whom he maintained constant intellectual exchanges (Gianinazzi 2016).

Nonetheless, the above-mentioned tension keeps manifesting itself: if one struggles to eliminate alienation, her objective is to make the wage-form 'bearable', so to speak, by exerting workers' control. To the contrary, if one struggles to eradicate exploitation, her objective is to flee from the wage-form. Such seeming contradiction is mirrored in the notion of *self-management*. Firstly, by self-management Gorz means the re-appropriation, by citizens and workers, of those creative capacities that have been subsumed by capital (through commodification) and atrophied by the State (through bureaucratization). The reclamation of those capacities is expressed by 'subordinating industrial technologies to the continuing extension of individual and collective autonomy, instead of subordinating this autonomy to the continuing extension of industrial technologies' (Gorz 1980/1987: 40). These words are consistent with Gorz's critique of the European Left's productivist attitude, in particular of their exclusive focus on wage-related demands—hence on the *quantitative distribution* of value—at the expense of what should be an equally important struggle, namely, the *qualitative definition* of the composition of production—hence of what should be produced, and why, where, when, by whom (Leonardi 2019a).

In order to illuminate the idea of self-management, while amplifying its political potential, Gorz employs the heuristic of utopia—despite a widespread scepticism within the Marxist tradition.[7] In fact, the last chapter of *Ecology as Freedom* is entitled 'Possible Utopia'. In the Gorzian utopia, the President of the French Republic proposes three policies to be immediately implemented: 'We shall work less'; 'We must consume better'; 'We must reintegrate culture into the everyday life of all' (1978/1980: 44–45). Moreover:

> To translate these principles into practice, the Prime Minister said it was necessary to rely on the workers themselves. They would be free to hold general assemblies and set up specialized groups, following the system devised by the workers of Lip, where planning is done in specialized committees, but decisions are taken by the general assembly. The workers should allow themselves a month,

---

[7] Again, there are notable exceptions, especially Bloch (1954/1986).

the Prime Minister estimated, to define, with the assistance of outside advisers and consumer groups, a reduced range of product models and new sets of quality standards and production targets […] During this first month, said the Prime Minister, production work should be done only in the afternoons, the mornings being reserved for collective discussion. The workers should set as their goal the organizing of the productive process to meet the demands for essential goods, while at the same time reducing their average worktime to twenty-four hours a week. The number of workers would evidently have to be increased. There would, he promised, be no shortage of women and men ready to take these jobs. (46–47)

This excerpt mentions the Lip watch factory and, thus, allows us to introduce a historical example of self-management. In 1973, threatened to be fired *en masse*, workers at the Besançon-based factory occupied all facilities and seized machinery and products to demand that no one lose their job. Led by the CFDT union, male and female workers alike formed committees to bring every aspect of the production process under their supervision. Workplace democracy finally became an actual, if temporary, reality. Student activists and engaged citizens supported labourers involved in the struggle, which quite surprisingly lasted throughout the decade. The Lip strike, which historian Donald Reid rightly defines as 'the last national expression of workers as the creators of a new world' (2018, 11), has been mostly analysed through the lenses of working-class feminism and of blue-collar autonomy. Gorz's originality, in this case, lies in his attempt to grasp Lip self-management experimentation from the perspective of political ecology. Again: a *labour-based* political ecology, since a veritable solution to environmental degradation requires an oppositional stance towards capitalism.

In other words: it is because the Lip occupation could be envisaged as a 'vacation from alienated life' (129) that its interpretation could acquire an ecological dimension—in the Gorzian sense. Yet, such an ecological dimension is immediately torn between the liberation *of* work and liberation *from* labour. As Arno Münster (2008) notices, Gorz was correct in highlighting how the strikers constantly oscillated between the pride of practising workers' control over the productive process and the desire to escape the constraints of wage labour and enact the Situationist slogan 'Never Work!'+. Working through this dilemma—workers' control vis-à-vis alienation *vs* workers' exodus vis-à-vis exploitation—from a politico-ecological perspective has been Gorz's main endeavour throughout the 1970s—and, arguably, in his last years. To properly assess the outcome of such endeavour, however, it is important to detail his original interpretation of the 1973 Oil Shock as a *crisis of reproduction*.

## The Oil Shock as a Crisis of Reproduction and the Green Capitalism Debate

Gorz's labour-based political ecology, we argue, is of great significance for ELS. Another important element is evident in chap. 4 of *Ecology as Freedom* (1978/1980) entitled 'Ecology and the Crisis of Capitalism'. Here, the ecological crisis is placed at the intersection between social and environmental contradictions and is viewed as a metabolic breakdown induced by the profit-imperative which drives the accumulation of capital. In short: the ecological crisis is the environmental expression of a socially dysfunctional dynamic. That does not mean, however, that physical limits to growth are unimportant. Rather, Gorz argued that the best way to face these limits is not to worship Nature as a divine entity or conform to its putatively 'eternal' order, but to develop a theory of the relationship between the capitalist mode of production (grounded on *labour exploitation*) and its surrounding environment (enacted by *resource appropriation*):

In his words (1978/1980, 21) 'Nature is not untouchable. The 'promethean' project of 'mastering' or 'domesticating' nature is not necessarily incompatible with a concern for the environment. All culture encroaches upon nature and modifies the biosphere. The fundamental issue raised by ecology is simply that of knowing:

- whether the exchanges, which human activity imposes upon or extorts from nature, preserve or carefully manage the stock of non-renewable resources; and
- whether the destructive effects of production do not exceed the productive ones by depleting renewable resources more quickly than they can regenerate themselves' (Gorz 1978/1980, 21).

Thus, according to Gorz, ecology and Marxism are perfectly compatible: while the latter focuses on and criticizes the *internal limits* to productive activities, the former deals with its *external limits* and denounces their crossing when that is the case.

On this basis Gorz elaborates a twofold theory of the 1973 Oil Shock—a major crisis of the Fordist *growth paradigm* that had characterized much of the twentieth century (Schmelzer 2016). In order to unpack the notion of the Fordist growth paradigm, we need to underscore Gorz's interpretation of the history of proletarian struggles in which wage labour 'is not just a way for capital to grow' but also 'a means of dominating the workers' (Gorz 2005/2010, 151). In the 'heroic age of trade-unionism', during mid-nineteenth-century mercantilist capitalism, most proletarians actively *refused* the institutional implications of wages, preferring a common 'norm of *sufficiency*' (*Ivi*, 153)—a salary decent enough to satisfy their shared needs and those of their families. In this period, capital faces an already formed productive network, centred around the workshop and workers' handicraft, and internalizes it as its own base. From the perspective of class composition, the key subjective figure is the *professional worker*. These craftworkers 'stopped work when they'd earned *enough* to live in a manner to which they were accustomed' (ibid.). However, according to Gorz, after the New Deal in the US (1930s) and the Marshall Plan in Western Europe (post-WWII), dynamics that had been set in motion during the late nineteenth century became dominant. The capitalist goal of expanding markets to absorb an ever-increasing volume of commodities required a new type of consumer whose purchases 'were motivated less and less by *shared needs* and more and more by *differentiated individual desires*' (*Ivi*, 154). Although a bit simplistic at times, Gorz's reconstruction has the advantage of showing that the social inclusion of waged workers (and their families) through access to the mass-consumption and protection provided by the welfare state is one of the defining features of Fordism.

Here the model of production is the large-scale factory, whose output of standardized goods for mass-consumption implies a polarization of workers' knowledge and skills This, in Gorz's view, involves a strict division between standardized and replicable working tasks (sphere of execution) and planning skills (sphere of conception). From the perspective of class composition, the prevalent subjective figure is the *mass worker*. What counts the most, for Gorz, is the intertwining of social and ecological aspects within Fordism as a regime of regulation. On the one hand, it 'succeeded in combining raising wages, greater social benefits and public expenditures and, above all, increased production and employment [...] With the exception of a minority Left-wing section of the trade unions, the labour movement didn't criticize the nature and orientation of this expansion but called, rather, for it to be speeded up' (ibid., 155). On the other hand, 'from the ecological standpoint, speeding up the circulation of capital leads to excluding everything that reduces profit in the immediate term. The continual expansion of industrial production thus entails an accelerated pillaging of natural resources' (*Ivi*, 156). As Gerard Strange compellingly remarks, within Fordist conditions there is a trade-off between the 'democratisation of consumption' (which is compatible with capitalism) and 'environmentalism as politics' (which assumes an anti-capitalist posture) (1996, 84–85).

It is against such a background that it is possible to define Fordism—especially when coupled with the Cold War—as an *entropic device* (Leonardi 2019b), namely as an institutional arrangement grounded on a social pact—working-class heteronomy in exchange for social protection and inclusion. As Lucio Magri put it: 'the Yalta agreement had enshrined certain constraints as well as guarantees' (2009/2011, 43)—which premised its stability on continuous (and environmentally destructive) growth. In this sense, at least until the mid-1970s, growth allowed for the provision of full employment and higher wages, serving as the means to buy social peace. Sociologist Claus Offe (1992) named the twin societal goal of full employment and perpetual growth the *productivist nexus* which served to displace social antagonism from the qualitative composition of production (what, where, when is to be produced, and how, by whom, for whom?) to the terrain of quantity. As Matthias Schmelzer argued, the growth paradigm 'promised to turn difficult political conflicts over distribution into technical, non-political management questions of how to collectively increase GDP […] It helped integrate labour and the political Left, rendered rearmament feasible without a decline in living standards, it helped stabilize the Bretton Woods system, and in the context of global inequalities it offered the (post)colonial countries in the Global South a possible route out of poverty towards what came to be defined as "progress"' (2016, 266).

This Fordist productivist nexus was tested by the 1973 Oil Shock which, according to Gorz, was the first example of a *double crisis*.[8] On the one hand, starting from the Marxist views regarding the tendency of the rate of profit to fall—as living labour, which can be exploited, is increasingly replaced by machinery—there is an outcome of *overproduction*. Capital reacts to this through a number of strategies, amongst which are the planned obsolescence of commodities and the creation of artificial needs—both intended to stimulate consumption. Additionally, Gorz sees a *crisis of reproduction* due to the ever-increasing costs that capital has to bear to regenerate the environment (up to that point used as a free landfill) so that it can be polluted again—a circuit whose consequence is the higher price of final products. In a fundamental passage of *Ecology as Freedom*, Gorz writes:

---

[8] For a historical contextualization of the 1973 Oil Shock and its aftermath, see Bini et al. (2016) and Basosi et al. (2018).

This forward flight [planned obsolescence and artificial needs] which was in any event bound to culminate in economic crisis, came to a stop with the so-called oil crisis. The latter did not cause the economic recession; it merely revealed and aggravated the recessionary tendencies which had been brewing for several years. Above all, the oil crisis revealed the fact that capitalist development had created absolute scarcities: in trying to overcome the economic obstacles to growth, capitalist development had given rise to physical obstacles. (*Ivi*, 24)

The emergence of such absolute scarcities is interpreted by Gorz as the inauguration of a new phase for labour-based political ecology. If, at the beginning of the 1970s, the social force acquired by the 1968-inspired movements and newly aggressive unions alike opened up a window of opportunity for liberation *of* work and liberation *from* labour, during the second half of the decade the double crisis of overproduction and reproduction produced two distinct but interrelated outcomes. On the one hand, social movements began to fold onto themselves, eventually collapsing into identity politics; on the other, unions were forced to shift to defending existing jobs rather than envisaging new forms of working—non-alienated and ecologically sound.

In short, for socialism to be truly emancipatory it must break capital's hegemony on economic policies and productive tools. Policies that promote growth as an all-encompassing panacea serve to obscure social inequality (since class polarization has never ceased to get deeper and deeper), but also trap political imagination within the misleadingly neutral terrain of *quantity*. By the time Gorz ends writing his *Farewell to the Working Class* (1980/1987), his collaboration with industrial unions was far less intense than it used to be. Moreover, his discussion of a *non-class of non-industrial workers* appears at best fragile, from a sociological perspective. It must be acknowledged, however, that regardless of such shortcomings, Gorz's research never endorsed neoliberal buzzwords such as 'human capital' or 'self-entrepreneurship'. His effort remained that of mapping the changing profile of class composition to better equip the Left in its struggle for emancipation.

## André Gorz on Cognitive Capitalism

The goal of this section is to discuss the new political ecology of labour as delineated by the transition from a Fordist to a post-Fordist regime of accumulation, which Gorz considers to be a major turning point (Gorz 2005/2010). The politicization of environmental issues in the 1970s took the form of the physical limits to growth, as argued by the Club of Rome. Yet, to turn such limits into an emancipatory political strategy, an additional element was needed: that of the *socially desirable* character of the ecological transition. This is a central issue, especially when posed against the background of the economic landscape of the 1990s: techno-scientific changes driven by information technology with direct impacts on labour—with a specific focus on the digitization of work (Gorz 1997/1999, 2003/2010). Gorz paid particular attention to notions such as knowledge-based society, or learning- and data-economy, because '*the most important form of fixed capital is now the knowledge stored in, and instantly available from, information technologies, and the most important form of labour power is brainpower*' (Gorz 2003/2010, 6).

This new relationship between value, capital and knowledge is interpreted by Gorz as capital's response to the widespread social unrest and to struggles of workers during the immediate post-1968 period. It takes the form of a capitalism that, having declared (and won, by the 1980s) the war against the Fordist working class, seemed prepared to commodify the social fabric *as a whole* (Gorz 2005/2010). According to Gorz, due to the possibilities opened up by technological development, capitalism managed to turn into commodities the products of social cooperation (as shown by the data economy) or to assume them as central components of labour exploitation (as embodied in the notion of human capital). Moreover, the transition to this knowledge-based economy also drove the relocation of the manufacturing of commodities into the Global South. In this manner, this post-Fordist class project eventually undermined the working class in the Global North and, with this, partially its political struggles. Gorz, therefore, acknowledges that socio-economic transformations, induced by information technologies, allowed capitalism to overcome the crisis of legitimacy experienced by the Fordist model while containing the criticisms advanced by working class and civil society alike. Nonetheless, he was still interested in working through the legacy of such struggles to envision a post-capitalist scenario. The cognitive character of labour in Western economies, fostered by mass education and the diffusion of information technology, was thus assumed by Gorz as a particular terrain of struggle, characteristic of the historical conjuncture of the late 1990s and early 2000s.

In order to make sense of how Gorz conceived of the interaction between this new socio-economic reality and political ecology, it is important to connect his struggle-based method with his understanding of cognitive capitalism as both a particular form of capitalist crisis and a concrete possibility to make the exit from capitalism *desirable*. Gorz's perspective rests on the following assumption: technological development and 'immaterial' labour embody the crisis of the historical categories upon which capitalism had functioned until that point, such as that of labour-value (measured in discrete units of time) (Gorz 2003/2010). Such crisis is due to the contraction of socially necessary labour time, implying the reduction of abstract labour's value-potential. This in turn translated into increased unemployment and declining purchasing power. Thus, Gorz—way too optimistically—perceived capitalism as *virtually depleted*. In parallel, automation processes were supposed to (automatically) undermine commodification both in the sphere of production and in the sphere of reproduction.

The question concerning the 'immaterial' needs further unpacking. It is important to stress that Gorz moves from a somewhat orthodox Marxist crisis-theory to the hypothesis of cognitive capitalism as developed in the Italian and French workerist debate (Corsani et al. 2001). That hypothesis was based on the then new theories on human capital and the driving role of knowledge for corporate growth by referring to the Marxian category of *general intellect* (Negri 1979/1991; Vercellone 2007). This is outlined in the 'Fragment on Machines' (Notebook VII of the Grundrisse): 'It is neither the direct human labour he [the worker] himself performs, nor the time during which he works, but rather the appropriation of his own general productive power, his understanding of nature and his mastery over it by virtue of his presence as a social body—it is, in a word, the development of the social individual which appears as the great foundation-stone of production and of wealth' (Marx 1973/1939, 704). Accordingly, as knowledge production is not limited to the factory's space and the learning process (of both algorithms and human beings) is turned into a life-long endeavour, two main considerations come to the forefront. First, the general intellect opens up the opportunity to overcome labour time as the measuring unit of value. Second, as this knowledge is mostly produced outside of the workplace and independently from wages and employment contracts, different and unprecedented mechanisms of valorization are required (i.e. monopoly, life patents and new forms of enclosing rights, financialization).[9]

In this sense, the idea of a knowledge-based economy stresses the importance of the intellectual qualities of labour power as a key resource in the valorization process. Quite differently, the notion of cognitive capitalism adopts a

---

[9] Contemporary patenting is not a novelty in the history of capitalism. However, its role evolved, becoming a central commodity under the economy of techno-science. Thus, it represents a veritable paradigm that took hold in the second half of the twentieth century, particularly in the 1960–1970s period (Echeverría 2005). Many authors read this process according to the framework of bio-cognitive capitalism (Birch 2017). In this context, and compared to the nineteenth and early twentieth centuries, both legislation and biotechnologies developed in line with the potentialities offered by new gene-sequencing techniques. An important turning point, in the 1980s, was the shift from patenting involving the inanimate world to patenting targeted at living products, subsequently portrayed as human-made inventions.

critical stance to shed light on the growing privatization and commodification of such knowledge. In Gorz's view, this tension between knowledge as the cornerstone of the productive process in general and knowledge as a privatized commodity will eventually be solved for the better: immaterial enclosures will prove unreasonable while the disalienating potential of new technologies will spread through society, eventually moving it beyond capitalism. In other words, the capitalist attempts to appropriate human knowledge and the products of cooperatives and self-managed practices allowed by new ITCs (Mason 2015), for the purpose of producing surplus value is bound to fail as these subjectivities will somehow come to realize that they do not need capitalism either to interact or to increase their productivity.

This line of argument focuses on capital's interest in the value of knowledge and especially of that knowledge whose productivity rests on the socialized cooperation of human brainpower. For Gorz, however, knowledge is a peculiar object: not only does it not get exhausted when consumed (unless it is commodified), but also gets enriched when broadly exchanged. Hence, the only way to make a profit from it is through coercive privatization (i.e. intellectual property rights). Yet, such subalternity to profit constrains rather than drives further development. Capitalism can survive, but it is forced to modify its *raison d'être*. It no longer aims to reduce natural scarcity through productivity gains, but now seeks to impose scarcity on a naturally abundant product—collective knowledge. This is why Gorz argued that the question of the exit from capitalism had never been timelier; actually, such an exit could well be already in place (Gorz 2005/2010). The attention Gorz devoted to hacker communities and free software movements drew precisely from his conviction that an autonomous use of technology is not only possible, but also able to reconcile production and reproduction on the basis of alternative and socially and ecologically sound desires (Gorz 2003/2010). However, more than ten years after Gorz' death, it should be admitted that, rather than providing the basis for *Reclaiming work, beyond the wage-based society*—as stated by the title of the English translation of *Misères du present: richesse du possible* (Gorz 1997/1999)—the above-mentioned mutations have instead contributed to the rise of precarious work, the diffusion of delusional self-entrepreneurship and the dismantling of the welfare state.

## Conclusions: Gorz's Legacy and the Critique of Contemporary Biocapitalism

Our analysis in this paper was aimed at making the case that Gorz's thought—especially his labour-based political ecology—belongs to Environmental Labour Studies as a cutting-edge field of inquiry. Particularly, we believe that Gorz's understanding of the relations between capitalism, nature and the history of working-class struggles is very useful to assess how labour-based political ecology has changed over the last decades. It also allows us to identify in the ecological crisis, understood as an exacerbation of social metabolism, both a limit of capitalist development and a potential new terrain for value-creation. It is against this background that, according to Gorz, the productivist nexus that strongly supported capital's expansion throughout the Fordist period entered into crisis. Such crisis impacted both sides of class struggle: that of the protests against exploitation and that of the mechanisms for appropriating surplus value.

Though these are pretty powerful insights, we also believe that, before concluding, it is worth discussing how Gorz's legacy for the twenty-first century should take into account both acute insights and inevitable shortcomings. These are, in our opinion, mainly related to an excessively optimistic and mechanistic view of the rise of a post-capitalist society, the only one capable of facing the urgency of the ecological crisis. From this perspective, a key feature of contemporary accumulation processes must be recalled: alongside the increasing importance of knowledge, emotions and human capital, biological and genetic processes are also brought to the foreground. We consider this a direct outcome of the crisis in the sphere of reproduction leading to the capitalist attempt to turn such sphere from a *condition* to a *source* of valorization. This is pursued not only through the commodification of intimacy and care but also by means of a new focus on the materiality of bodies, genes and the environment. In order to properly grasp this shift, Gorz's contribution cannot be underestimated: he had well anticipated the increasingly tight alliance between techno-science and capitalism within a post-Fordist regime. Nonetheless, his intuitions are not sufficient as he could not have predicted to what extent the peculiar relationship between knowledge and technology has been oriented towards what can be considered as a new, capital-based political ecology. In this context, the work of Melinda Cooper (2011) is particularly helpful in shedding light on the close correlation between scientific developments in the domain of life sciences (from biotechnology to genetic engineering) and transformations in the forms of economic growth carried out by the neoliberal counter-revolution. In this context, the primary valorization of materials derived from human bodies and other living organism has been addressed under the concept of *biocapitalism* that expands the idea of cognitive capitalism (see also Fumagalli et al. 2019) to grasp the transformations of accumulation strategies and related working practices of exploitation, induced by these new forms of commodifying nature. In particular, by shifting the focus from social to biological reproduction, Cooper concentrates on how the connections between knowledge, technology and the environment demand that we revisit the same notion of surplus value under Neoliberalism.

In the wake of Foucault's research, she highlights how progress in the life sciences has gone hand in hand with the establishment of a new field of knowledge that takes life itself, in its generative and reproductive capacities, as a paradigm for economic action. It is within this paradigm that a neoliberal ambition of manipulating life to redraw capital's ecological limits can eventually become thinkable. In Cooper's words: 'What is at *stake* in the *accumulation* of *capital* today is the *regeneration of the biosphere*—that is, the *limits of the earth itself*' (Cooper 2011, 30). The unprecedented centrality of biological processes for value-production has strongly influenced the political economy of nature-society relations. This can be clearly appreciated by taking into account how economic thought has reframed the issue of ecological limits to growth. Consider, for example, the role played by the biological dimension with regard to the emergence of new regulatory paradigms, such as 'bioeconomy', or more recently, 'climate-smart agriculture'. These are attempts to reconcile environmental restoration with the imperative of growth or to directly turn ecological crises into a source of profit, according to the rhetoric of the green economy (Smith 2007; Pellizzoni 2015).

What Cooper adds to this line of argument is an analysis of how these processes are associated with the development of a global economy based on debt, which in turn has a profound influence on the casualization and feminization of work. She demonstrated that through the expansion of debt-relations, biology and economy mutually reinforced each other in the elaboration of an alternative paradigm that does not shy away from complexity, but rather speculatively foreshadows its chaotic deployment. Here, the idea of biological life as intrinsically expansive and structured in complex open-systems that, as a result, functions without predetermined biochemical limits, reflects the speculative and future-oriented temporal order of debt-based financial capitalism.[10] From this perspective, the new capital-based political ecology also

---

[10] By the same token, the peculiar relationship between debt and speculation that underlies financial capitalism is also mirrored in the speculative character that innovation takes on in the context of knowledge economies and the biotech sector with the proliferation of high-risk start-up companies.

influences the specific way in which the economy addresses social risks. Indeed, the outcomes of such processes were the erosion of the welfare state, the transformation of labour subjectivities and the emergence of new fields of social conflict. Such reasoning is carried out by Cooper through three books. In her first book (Cooper 2011), she questions the new form of capitalism that has emerged since the Oil Shock and its crisis-ridden aftermath, when life and debt became the preferential bases of value-extraction. In the second book (Cooper and Waldby 2014), she interrogates the new forms of labour that emerge from this shift, namely feminized labour and biolabour. Finally, in her last book (Cooper 2017), she focuses on the persistency of the family form, at the very heart of capitalist accumulation, and in the context of the post-Fordist political economy.

In particular, two key intuitions for a labour-based political ecology stems from Cooper's work. Especially relevant here is the thesis that neoliberal biopolitics deeply reworked 'the value of life as established in the welfare state and New Deal model of social reproduction' (Cooper 2011, 9). This is mirrored in the casualization of labour following the economic measures undertaken by highly indebted governments, as shown by Saskia Sassen (2003a, b) and Mike Davis (2006a, b). In a nutshell, neoliberalism exacerbates the violent hierarchies between expendable lives and those whose reproduction should ensured. The neoliberal disinterest in the social impacts of growth is reinforced by the idea that collapse is always partial and never a final phase. The casualization of labour also engendered the growing feminization of work, with women involved in what Saskia Sassen calls the counter geographies of globalization: 'they include the illegal trafficking in people for the sex industry and for various types of formal and informal labor markets as well as other cross-border migrations, both documented and not, which have become an important source of hard currency for governments in home countries' (Sassen 2003b, 59).

Thus, we believe that a new, capital-based political ecology is also emerging and deeply influencing the neoliberal economic landscape. Should we then conclude that Gorz's theory of the 1970s capitalist crisis was wrong? To address this issue we can consider how, in the course of his reflections, Gorz has answered, in different ways, a key question concerning the relationship between capital and nature, namely: *can capitalism incorporate the ecological crisis not as an obstacle, but rather as a driver of accumulation*? A first option can be labelled as a 'visionary yes'. In 1972, during a debate with Sicco Mansholt at the *Nouvel Observateur*, Gorz said: 'In a not so distant future […] there will be a new cycle of accumulation based on the capitalization of Nature itself, on the capitalist subsumption of the totality of factors and conditions which allow life on Earth [...] Even air will become a commodity' (1973/1977, 185). Shortly after, however, Gorz aligned himself with the conventional wisdom of the 1970s—according to which environmental nuisances were to be conceived, in capitalistic terms, as pure costs (O'Connor 1973)—and turned to a sort of 'apocalyptic no', a scenario in which the alternative is between eco-socialist conviviality or techno-fascist dictatorship: 'Either we agree to impose limits on technology and industrial production so as to preserve natural resources [...] or else the limits necessary to the preservation of life will be centrally determined and planned by ecological engineers' (1978/1980, 16–17). Very interestingly, and quite surprisingly, Gorz changes perspective in the later part of his reflection and proposes a paradoxical answer to the question at hand: 'yes but no'. In a 2007 article, in fact, he argues that knowledge-based societies can indeed manage the ecological crisis, but can do so precisely because they are grounded on non-capitalist principles:

> 'It is capitalism itself that, unwillingly, works towards its own extinction by developing the tools for a sort of high-tech craftmanship, which allows the fabrication of basically everything with a productivity which is higher than that of industry and a requirement of natural resources which is lower'. New technologies (e.g. rapid prototyping) and new subjectivities (e.g. digital fabricators) disclose 'an economy beyond wage labour [*travail-emploi*], beyond money and beyond the commodity-form. What is possible is an economy based on the generalized availability of the outcomes of an activity immediately perceived as a commons: an economy of gratuity.' (2006/2010, 61–62—translation modified)

Although Gorz was too optimistic and could not foresee the progressive subsumption of the sphere of reproduction in the value chains, the transformation he desired is becoming a shared goal in the anti-capitalist camp. However, his vision of a non-capitalist society needs to be directly linked to the emergence of a 'new' link between capital and nature. Such an idea requires a change of perspective; in Cooper's words: 'In a real sense we are living on the cusp between the petrochemical and biospheric modes of accumulation, the foregone conclusions of oil depletion and the promise of bioregeneration. An effective ecological counterpolitics therefore needs to operate on both levels' (2011, 49). This also demands a general rethinking of the status of the sphere of reproduction. At stake, for labour-based political ecology and union practices alike, is an enlarged concept of the working class. This extension aims at pushing class politics to previously unexplored areas—for example, decolonization, anti-extractivism and feminist struggles (Barca and Leonardi 2018). This is the case, we argue, in platforms or campaigns—for example, Just Transition (Stevis and Felli 2015) or the Green New Deal (Klein 2019)—that recognize a common root (the accumulation of capital) for different forms of violence (colonial, extractivist, gender-related, environmental, exploitation-based, etc.). Large alliances like these have the potential to go beyond the productivist nexus, towards a more radical labour-based political ecology.

## References

Barca, Stefania. 2014. Laboring the Earth: Transnational Reflections on the Environmental History of Work. *Environmental History* 19 (1): 3–27. https://doi.org/10.1093/envhis/emt099.

———. 2017. Labour and the Ecological Crisis: The Eco-modernist Dilemma in Western Marxism(s) (1980s–2010s). *Geoforum* 83: 91–100. https://doi.org/10.1016/j.geoforum.2017.07.011.

Barca, Stefania, and Emanuele Leonardi. 2018. Working-Class Ecology and Union Politics: A Conceptual Topology. *Globalizations* 15 (4): 487–503. https://doi.org/10.1080/14747731.2018.1454672.

Basosi, Duccio, Giuliano Garavini, and Massimiliano Trentin. 2018. *Counter-Shock: the Oil Counter-Revolution of the 1980s*. London: Tauris.

Bini, Elisabetta, Giuliano Garavini, and Federico Romero. 2016. *Oil Shock: The 1973 Crisis and Its Economic Legacy*. London: Tauris.

Birch, Kean. 2017. Rethinking Value in the Bio-economy: Finance, Assetization, and the Management of Value. *Science, Technology, & Human Values* 42 (3): 460–490. https://doi.org/10.1177/0162243916661633.

Bloch, Ernst. 1954/1986. *The Principle of Hope—Vol. 1*. Boston: MIT Press.

Bosquet, Michel. [Gorz, André.]. 1973/1977. *Capitalism in Crisis and Everyday Life*. Brighton: Harvester.

Cooper, Melinda. 2011. *Life as Surplus: Biotechnology and Capitalism in the Neoliberal Era*. University of Washington Press.

———. 2017. *Family Values: Between Neoliberalism and the New Social Conservatism*. New York: Zone Books.

Cooper, Melinda, and Catherine Waldby. 2014. *Clinical Labor: Tissue Donors and Research Subjects in the Global Bioeconomy*. Durham: Duke University Press.

Corsani, Antonella, Patrick Dieuaide, Maurizio Lazzarato, Jean-Marie Monnier, Yann Moulier-Boutang, Bernard Paulré and Carlo Vercellone. 2001. *Le capitalisme cognitif comme sortie de la crise du capitalisme industriel: un programme de recherche*. Conference proceedings, Forum de la Régulation. Paris, Ecole Normale

Supérieure, 9–10 March. Accessed September 12, 2019. https://rechercheregulation.files.wordpress.com/2012/12/01_corsani_et_alii.pdf.

Davis, Mike. 2006a. *Planet of Slums*. London: Verso.

———. 2006b. *City of Quartz: Excavating the Future in Los Angeles*. New ed. London: Verso

Echeverría, Javier. 2005. La revolución tecnocientífica. *Conference proceedings, CONfines de relaciones internacionales y ciencia política* 1 (2): 09–15.

Flemming, Jana, and Norbert Reuter. 2020. Who can Afford to Degrow? In *Degrowth in Movement(s): Exploring Pathways for Transformation*, ed. Corinna Bukhart, Matthias Schmelzer, and Nina Treu. New York: Zero Books.

Fumagalli, Andrea, Alfonso Giuliani, Stefano Lucarelli, and Carlo Vercellone. 2019. *Cognitive Capitalism, Welfare and Labour: the Commonfare Hypothesis*. London: Routledge.

Gianinazzi, Willy. 2016. *André Gorz. Une vie*. Paris: La Découverte.

Gollain, Françoise. 2018. *André Gorz, une philosophie de l'émancipation*. Paris: L'Harmattan.

Gorz, André. 1971/1978. Movimento operaio e progetto di civiltà. In *Ecologia e politica*. Bologna: Cappelli.

———, ed. 1973a. *The Division of Labor*. Brighton: Harvester.

———. 1973b. *Capitalism in Crisis and Everyday Life*. Brighton: Harvester.

———. 1978/1980. *Ecology as Politics*. Boston: South End Press.

———. 1980/1987. *Farewell to the Working Class: An Essay on Post-industrial Socialism*. London: Pluto Press.

———. 1997/1999. *Reclaiming Work: Beyond the Wage-based Society*. London: Polity Press.

———. 1998/2017. Quel type de travail prend fin? *EcoRev'* 45: 5–16.

———. 2003/2010. *The Immaterial*. Chicago: University of Chicago Press.

———. 2005/2010. The Exit from Capitalism Has Already Begun. *Cultural Politics* 6 (1): 5–14.

———. 2006/2010. *Letter to D. A Love Story*. London: Polity Press.

———. 2008/2010. *Ecologica*. Chicago: University of Chicago Press.

Klein, Namoi. 2019. *On Fire: The (Burning) Case for a Green New Deal*. New York: Simon & Schuster.

Leonardi, Emanuele. 2019a. The Topicality of André Gorz's Political Ecology. In *Towards a Political Economy of Degrowth*, ed. Stefania Barca, Ekaterina Chertkovskaya, and Alex Paulsson, 55–68. London: Rowman and Littlefield.

———. 2019b. Bringing Class Analysis Back In: Assessing the Transformation of the Value-Nature Nexus to Strengthen the Connection between Degrowth and Environmental Justice. *Ecological Economics* 156: 83–90. https://doi.org/10.1016/j.ecolecon.2018.09.012.

Magri, Lucio. 2009/2011. *The Tailor of Ulm: Communism in Twentieth Century*. London: Verso.

Marx, Karl. 1973/1939. *Grundrisse. Foundations of the Critique of Political Economy*. London: Penguin Books.
Mason, Paul. 2015. *Postcapitalism: A Guide to Our Future*. London: Allen Lane.
Méda, Dominique. 2017. La réduction de l'emprise du travail et son partage constituent-elles encore un objectif réalisable ? In *Le moment Gorz*, ed. Christian Fourel and Alain Caillé, 200–219. Lormont: Le bord de l'eau.
Münster, Arno. 2008. *André Gorz ou le socialisme difficile*. Paris: Nouvelles Editions Lignes.
Negri, Antonio. 1979/1991. *Marx Beyond Marx. Lessons on the Grundrisse*. London: Pluto Press.
O'Connor, James. 1973. *The Fiscal Crisis of the State*. London: Routledge.
Offe, C. 1992. A non-productivist design for social policies. In: Van Parijs, P. (Ed.), *Arguing for Basic Income*. Verso, London, pp. 61–78.
Pellizzoni, Luigi. 2015. *Ontological Politics in a Disposable World: The New Mastery of Nature*. Farnham: Ashgate.
Räthzel, Nora, and David Uzzell. 2013. Mending the Breach between Labour and Nature: A Case for Environmental Labour Studies. In *Trade Unions in the Green Economy*, ed. Nora Räthzel and David Uzzell, 1–12. London: Routledge.
Reid, Donald. 2018. *Opening the Gates: The Lip Affair, 1968–1981*. London: Verso.
Sassen, Saskia. 2003a. Global Cities and Survival Circuits. In *Global Woman: Nannies, Maids and Sex Workers in the New Economy*, ed. Barbara Ehrenreich and Arlie Russell Hochschild, 254–274. London: Granta Books.
———. 2003b. The feminization of survival: alternative global circuits. In *Crossing Borders and Shifting Boundaries*, ed. Umut Erel, Mirjana Morokvasic, and Kyoko Shinozaki, 59–77. Wiesbaden: VS Verlag für Sozialwissenschaften.
Schmelzer, Matthias. 2016. *The Hegemony of Growth*. Cambridge University Press.
Smith, Neil. 2007. Nature as Accumulation Strategy. In *Coming to Terms with Nature*, Socialist Register, 43, ed. Leo Panitch and Colin Leys, 16–36. Monmouth: The Merlin Press.
Stevis, Dimitris, and Romani Felli. 2015. Global Labour Unions and the Transition to a Green Economy. *International Environmental Agreements: Politics, Law and Economics* 15 (1): 29–43. https://doi.org/10.1007/s10784-014-9266-1.
Stevis, Dimitris, David Uzzell, and Nora Räthzel. 2018. The Labour-Nature Relationship: Varieties of Labour Environmentalism. *Globalizations*, 15(4): 439-453. https://doi.org/10.1080/14747731.2018.1454675
Strange, Gerard. 1996. Which Path to Paradise? André Gorz, Political Ecology and the Green Movement. *Capital & Class* 20: 81–102. https://doi.org/10.1177/030981689605900107.
Vercellone, Carlo. 2007. From Formal Subsumption to General Intellect: Elements for a Marxist Reading of the Thesis of Cognitive Capitalism. *Historical Materialism* 15 (1): 13–36. https://doi.org/10.1163/156920607X171681.

# 32

# Rethinking Labour/Work in a Degrowth Society

Anna Saave and Barbara Muraca

## Introduction

From an ecological perspective, economic growth can be linked directly to environmental destruction. All visions of a green growth notwithstanding, so far, the only year that successfully achieved a reduction in $CO_2$ emissions was 2009 after the global financial crisis and in 2020 as a short-term effect of the corona crisis. Efficiency innovations are eaten up by the increasing overall production and consumption that are necessary to keep growth going (*rebound effect*: Santarius et al. 2016). The ecological impact of digitalization is much higher than commonly thought and is not going to save the day (Lange and Santarius 2020). Environmental interventions embedded into the hegemonic paradigm of growth are bound to fail because, as the degrowth movement has been articulating for decades, capitalist growth itself is part of the problem and not of the solution.

---

A. Saave (✉)
Department of Sociology, Friedrich Schiller University Jena, Jena, Germany
e-mail: anna.saave@uni-jena.de

B. Muraca
Department of Philosophy, University of Oregon, Eugene, OR, USA

With respect to work/labour,[1] taking a degrowth perspective implies a significant challenge to the way in which work is framed because the types of work available in capitalist economies are intimately linked to economic growth, typically measured in terms of increasing rate of the Gross Domestic Product. People don't just have jobs that more or less support their existence, but that are designed to sustain a growing economy. Thus, to overcome economic growth dependency, rethinking work and labour must be at the heart of the degrowth discourse.

Degrowth has established itself over the last 20 years as an increasingly influential discourse, both as an academic research field inspired by Ecological Economics and Political Ecology (Martinez-Alier et al. 2010; Muraca 2013; D'Alisa et al. 2015) and as (a slogan for) social movements (Petridis et al. 2015; Burkhart et al. 2020). Within its heterogeneity, degrowth articulates a sophisticated critique of capitalist growth not only from an environmental perspective, but also with respect to the root causes of the environmental crisis, that is, the structural function of growth for the stabilization of modern, capitalist societies (Rosa et al. 2017; Muraca 2015; Kallis et al. 2018).

While very different perspectives on work and labour have been articulated so far within the degrowth discourse, a systematic analysis of their assumptions and implications is still lacking. In this chapter, we address this challenge from the specific perspective of materialist ecofeminism.

## Degrowth: An Overview

The Degrowth discourse has a multifaceted history that cuts across different scholarly traditions, political discussions, and activism (Muraca and Schmelzer 2017). Rooted in the critique of economic growth from different perspectives, it developed recently into a heterogeneous and fluid social movement. Schmelzer and Vetter (2019) identify seven (plus one), partially overlapping streams of growth critique: ecological, socio-economic, cultural, critique of capitalism, feminist critique, critique of industrialism, critique of developmentalism, plus other forms of critique of growth that are (so far) external to the degrowth discourse. A recent book introduced the felicitous slogan 'degrowth in movement(s)' to address the dynamic heterogeneity that characterizes degrowth, which 'questions the hegemony of the growth paradigm,

---

[1] Because the two terms are used differently in the literature, it is difficult to maintain a coherent use in the chapter. Generally, however, work is used as a general term referring to socially necessary activities, while labour indicates the specific form of work under capitalism aimed at generating income and profit, both as wage labour and as independent labour (thus including precarious forms of self-employment).

bringing together quite diverse and sometimes contradictory currents and positions. What they all have in common, however, is that they criticize the technological optimism of the discourse around sustainable development that has prevailed since the 1990s, with its promise of decoupling growth and environmental consumption' (Burkhart et al. 2020).[2]

While lacking the homogeneity of a political programme and a clearly identifiable organizational structure, degrowth operates as a provocative slogan (Petridis et al. 2015). As such, it hits the core of the neoliberal restructuring of societies after the social-ecological crisis of the 1970s (Muraca 2020). The critique of capitalist growth combines the analysis of the environmentalist crisis with a profound critique of its most prominent root causes: (a) the structural function that growth has played for the dynamic stabilization of Welfare democracies in Atlantic Fordism and its crises (Rosa et al. 2017; Muraca 2015; Kallis et al. 2018) and (b) the functioning of growth as a 'mental infrastructure' (Welzer 2012), a mode of subjectivation (Muraca 2020; Eversberg 2014), or a culturally hegemonic mode of living (Brand and Wissen 2018). Against the neoliberal re-articulation of sustainable development as green growth under the sign of a win-win-win solution for employment, environment, and development, degrowth represents a renewed and more radical politicization of the sustainability discourse (Muraca and Döring 2018; Asara et al. 2015) and an alternative path for a radical social-ecological transformation of society (Muraca 2013).

The historical roots of the term degrowth can be traced back to the turmoil that the publication of the Report to the Club of Rome, 'Limits to Growth' (Meadows et al. 1972) provoked in the political climate of the early 1970s, especially in France (for conceptual roots see Demaria et al. 2013). The French term *décroissance* was introduced by André Gorz to address the incompatibility between the capitalist system and the earth balance and later established by Georgescu-Roegen as an alternative to zero-growth or steady-state (D'Alisa et al. 2015; Muraca 2013).

In Germany, a vivid debate around the 'future of work' in societies, in which growth was no longer considered a feasible or desirable option, sparked alternative imaginaries of liberation *of* and *from* wage labour, including contributions from the materialist-feminist perspectives of the Bielefeld School (Bennholdt-Thomsen et al. 1988). In 1972, the German Steel Workers Union IG Metall discussed, for example, during the Oberhausen Congress not only

---

[2] Degrowth is close to the tradition of strong sustainability (Muraca and Döring 2018), but strongly critical of the neoliberal direction that sustainable development has taken (Gómez-Baggethun and Naredo 2015). Sustainability and Sustainable Development descend from two different traditions of thought. The latter is the more growth-oriented approach (Caradonna 2014).

the environmental crisis but also perspectives for a 'humanization of the working life beyond growth' (Muraca and Schmelzer 2017, 186).

During the neoliberal restructuring of the global economy in the 1980s and 1990s that responded to the announced doom of *Limits to Growth* with a massive programme of expansion and deregulation (Muraca 2020), the critique of growth almost disappeared behind the new hegemonic paradigm of a *sustainable development*, which more adequately mirrored the new managerial approach of neoliberalism (D'Alisa et al. 2015; Muraca and Döring 2018). Under the new win-win promise linking ecology and economy, the core obstacle on the way to sustainability was no longer growth, now euphemistically replaced by development, but poverty: eradicating poverty through neoliberal modernization became the chief goal for a new expansive economy (Gómez-Baggethun and Naredo 2015).

It took almost 20 years for a come-back of growth critique under the new umbrella of the contemporary degrowth discourse. This happened through the re-emerging of a radical critique of globalization in the Global North linked to indigenous and peasants' movements in the Global South, which identified in the growth ideology the devastating face of the Western model of development and called for resistance and alternatives to its pervasive hegemony (Muraca 2013).

Faced with the diagnosis of limits to growth and the plausible scenario of long-term shrinking economies characterized by stagnation, social inequality, and pauperization, a conservative discourse emerged after the global financial crisis of 2008 that tried—so far unsuccessfully—to enter the degrowth discourse by stressing the impending end of economic growth as unavoidable destiny under business-as-usual conditions (Miegel 2011). Against this reductive and passive reading of degrowth in terms of economic shrinking, degrowth activists forged the slogan 'your recession is not our degrowth' to claim that degrowth does not mean simply turning GDP-growth upside down, all other things being equal. Rather, it embodies a project for a radical transformation to liberate societies from their structural dependency on the logic of growth that does not only overexploit natural resources and sinks, but also human creative energy, and that destroys social relations. While also calling for a reduction of the material size of the economy, degrowth goes well beyond advocating a society with a smaller metabolism under unchanged conditions and explores alternative ways of living together besides and beyond the capitalist growth diktat. What is at stake is not so much doing less of the same, but envisioning 'a society with a metabolism which has a different structure and serves new functions' (D'Alisa et al. 2015, 4).

Work and labour have to be re-conceived in the larger context of this radical transformation. In this respect, degrowth supporters see in the crisis of economic growth a unique opportunity for a liberation from alienation and oppression and a path towards a better life for all beyond the diktat of productivism that structures the wage labour system and precludes alternative imaginaries. Following Gorz (1999), wage labour would be drastically reduced in favour of what he called *true* or *real* work—the creative activity of humans not necessarily oriented towards production. While we observe an overall lack of consideration of workers as agents in the degrowth discourse, at least for some recent perspectives within the degrowth mosaic, the alliance between struggles in the sphere of production and in the sphere of reproduction converge (Leonardi 2017). While it is thanks to feminist movement and scholarship (i.e. regarding the 'wages for housework' campaign) that the sphere of social reproduction has been considered a field of social transformation, degrowth scholars only recently, and still insufficiently, have started acknowledging explicitly their indebtedness to feminism.

## Degrowth and Labour/Work: Reconstruction of a Discourse

In the following sections, we propose a classification of approaches to labour/work that partially, albeit not entirely, overlap with Schmelzer and Vetter's (2019) typology of different streams within the degrowth discourse: (1) the first approach questions the (global) division of labour and suggests degrowth as a path of re-localization of production chains; (2) the second approach envisions a significant reduction of wage-based labour as a *result of* or as a *condition for* degrowth that would lead to an increase in individual happiness; (3) the third approach claims a necessary increase of working time for a degrowth scenario; (4) the reformist approach stresses the institutional continuity within the division of labour, but a qualitative shift towards types of labour with a lower environmental impact; (5) more radical approaches are grouped together under the headline of a transformation of society that requires a reconceptualization of work/labour and its collective meaning. Here we include perspectives that question the separation between production, consumption/use, and reproduction. We dedicate a detailed analysis in the last section to radical feminist alternatives.

## Beyond the (Global) Division of Labour: Post-growth-Economy and Re-localization

A prominent thinker within the degrowth spectrum is the German economist Niko Paech, who developed what he calls a 'post-growth-economy' (Paech 2012). Because for Paech growth is the chief driver of the current global environmental crisis, a post-growth-economy is the only effective way to reduce carbon emissions worldwide, while guaranteeing a decent quality of life. In his analysis, the global division of labour, technological innovation, and the mechanisms of interests and debts have contributed *structurally* through unleashed competition and *culturally* through effective mental infrastructures (Welzer 2012) to the pervasive logic of increase and acceleration that characterizes modern societies (Rosa et al. 2017) and brought about the environmental crisis. Accordingly, this is due to a threefold process aimed at removing limits to material exploitation: (1) *physically*, by replacing human labour with resource-intensive machines guaranteeing productivity increase; (2) *spatially*, by extending global value-chains and relocating polluting, heavy-duty, and undesirable activities to remote locations; (3) *temporally*, through the pillage of future generation by shifting the burden of financial and ecological debt onto them.

From this analysis, Paech derives the two central foundations for an alternative post-growth-economy: (1) *structurally*, a massive de-globalization and re-localization of production through what he calls 'a proximity economy' that would render regions largely independent from global supply chains; and (2) *culturally*, a general reduction in the use of material flows through frugal, individual life-styles in order to tackle the exponential spiral of constructed needs and desires that support consumerism and production growth.

The proximity economy is described as a mixed model, in which small-scale market production and locally rooted business coexist with a wide range of subsistence-oriented activities and cooperatives. This would require re-appropriating manual proficiency and creative skills combined with shared use of tools and products in order to extend their life-cycle and reduce their total amount. Locally based production should be rooted in non-monetary exchange or through local currencies free of interest. Long-distance supply chains should only deliver necessary products that cannot be built locally.

To address inequalities, Paech proposes a cap on income and wealth combined with a massive reduction of working hours dedicated to wage labour, albeit without wage adjustment (which would foster consumption and nullify the envisioned cultural transformation towards voluntary simplicity) and a redistribution of working time across the whole population. Work is generally re-appropriated as socially useful and creative activity mostly aimed at subsistence production and coordinated by local communities in combination with more traditional wage-labour relations under the cultural framework of frugality. It remains unclear who is in charge of implementing the structural adjustments and how this would work, as Paech does not directly address the role of institutions, but seems to assume current societal configurations. In his early work he does not explicitly address the gender-based division of labour nor the question about how (re)productive activities would be organized and maintained.

## Working Less While Being Happier?

Within the degrowth discourse, a scenario of reduced working hours is fairly widespread, although the arguments articulated to support this assumption vary, mostly because of diverging understandings of work/labour. Moreover, while some scholars see in the reduction of the total amount of work/labour a necessary *condition* to shrink production and implement degrowth as a normative goal to tackle the environmental crisis, others consider it an unavoidable *consequence* of the impending end of economic growth due to ecological constraints and policies addressing climate change.

Kallis (2013) identifies three main positions supporting the reduction of working hours in a degrowth scenario:

1. If the total scale of the economy will be restricted by effective political interventions to tackle climate change, further increases in productivity will inevitably bring about large-scale unemployment (Victor 2019).[3] Only a corresponding reduction of working hours and a redistribution of paid work will stabilize society and keep it on a sustainable path by reconciling employment and climate policies.
2. The shift of socially relevant activities from the unpaid to the paid sector has led to an increasing commodification and marketization of daily life with fatal consequences for social cohesion and quality of life (Latouche 2009; Sekulova et al. 2013). From this point of view, degrowth as a radical alternative can decommodify labour and liberate spaces of self-determination

---

[3] This position is rooted in a lack of trust in a green growth path and implies an overall smaller economic scale. While a supposedly growing green economy would be less capital intensive than a fossil-based economy and possibly create more jobs, a mere shrinking of the economy under current conditions implies a loss of jobs.

and solidarity within communities and society. If (a post-growth) society will be running out of work, this will occur in favour of an increase in happiness and quality of life.
3. The distribution of and access to goods and services does not have to depend on wage or salary. Decoupling income and labour, for example, through the introduction of an unconditional basic income or similar policies is proposed as a solution to unemployment and a path of decommodification (in synergy with the former approach) (Liegey et al. 2013). In a similar vein, Alcott (2013) suggests decoupling work and income by considering work as a political right and introducing a Job Guarantee programme.

Against the assumption that society will soon be running out of work, Sorman and Giampietro (2013) stress that a degrowth society will necessarily require more work. As addressed in the next section, resource and energy scarcity would have to be replaced by an increase in human or animal work in order to generate the same output.

Part of the diverging positions around whether degrowth will imply more or less work results from unclear assumptions about what is meant by 'work'. As Kallis (2013) rightly claims against Sorman and Giampietro, the reduction he and other scholars advocate for a degrowth society refers to paid labour and not to work as a socially necessary activity in general. Even if work in general might increase due to a metabolic shift in a degrowth society, Kallis claims that it will not be organized as commodified labour, but through alternative forms rooted in self-determination, solidarity, and orientation towards the common good. As Kallis suggests, even with a reduced use of resources, productivity will likely continue to increase—at least in the short term: 'We can afford to work less, at least for the time being' (ibidem, 96). Moreover, he stresses, degrowth advocates do not argue that the production outcome should remain the same in order to guarantee the same level of affluence. Rather, 'working less in the paid sector now will lead to a shift of values and perceptions that will make the downscaling of desired functions easier' (ibidem).

Following up on these assumptions, Kallis et al. (2013) conducted a study based mostly on neoclassical economic modelling to analyse whether a reduction of working hours can 'provide for employment, earning capacity and a healthier economy without leading to a growth of material production and consumption' (ibidem, 1563). An important conclusion of the study is that the positive correlation between reduction of working hours and environmental benefits is not obvious. A reduction of working hours would actually fail to lead to environmental benefits if not embedded in a framework of specific

## 32 Rethinking Labour/Work in a Degrowth Society

policies targeting consumers' behaviour and the way in which workers might ultimately use their free time. Therefore, Kallis et al. propose the reduction of working hours as an interim measure that would temporarily relieve unemployment and to reconsider this strategy in the long term in combination with further structural changes in taxation or access to infrastructures.

In similar perspectives within the degrowth spectrum, it is assumed that reducing working hours in the paid sector would more or less automatically lead to an increase in quality of life because of the value of free time for more self-determined and meaningful activities. Empirical studies demonstrating a negative correlation between long working hours and happiness are invoked as evidence (Sekulova and van Den Bergh 2013). Accordingly, people report to be happier when having more free time for social interactions with friends even if this implies a lower income. Besides a general critique of surveys based on self-reported happiness, which is vulnerable to cultural expectations, easily influenceable, and dependent on adaptive preferences (Nussbaum 2001), the empirical surveys invoked as evidence lack a methodical analysis of social milieus and risk reproducing the unquestioned social background of the majority of degrowth activists and scholars, that is, academic middle-class radicalism (Eversberg and Muraca 2019), while neglecting the perspective of the (precarized) working class.

The degrowth's idea of society 'running out of work' goes back to Gorz's analysis of the crisis of social and ecological reproduction of capitalism in the 1970s (1982). Gorz saw in this a unique opportunity not only for a liberation *of* work as struggle against alienation, but also *from* wage labour as a path for autonomy and self-management of production that would improve quality of life and liberate human creativity. However, degrowth scholars do not acknowledge Gorz's later discussion of the neoliberal response to the social-ecological crisis of Fordism. Gorz brilliantly identified how work was thereby re-instituted in a new, destandardized, and demassified fashion: besides or beyond wage labour as 'the basis of social belonging and rights, and the obligatory path to self-esteem of others' (1999, 5). According to Gorz, '[t]he order to be "active subjects", but to be so in the service of an Other whose rights you will never contest, is in fact the accepted lot of all those creative individuals *with a real, but limited, subjugated sovereignty*, the jobbing producers of ideas, fantasies and messages'. By engaging with their whole person in gratifying activities, people 'become the venal and eager instruments of an alien will: in which they *sell themselves*' (ibid., 42) (see also Leonardi and Benegiamo, Chap. 31).

The promise of a good life beyond wage labour in a degrowth society is not automatically immunized against the neoliberal commodification of life's creativity in all its forms (Lazzarato 2004) and the pressure to invest in one's own self-enhancement or future employability (Leonardi and Chertkovskaya 2017). Exploitation extends from wage labour to the production of subjectivity in the neoliberal framework, which, according to Leonardi and Chertkovskaya, 'conceives of productive subjects as individual enterprises who compete to make the best investment out of their human capital, and do so in an increasingly financialised and debt-driven way' (ibidem, 113). Little attention is paid within the degrowth discourse to this neoliberal version of a liberation *from* traditional wage labour.

Finally, the generally accepted assumption that degrowth would imply less work and therefore more happiness fully disregards the feminist critique, by assuming that, in a non-productivist society, all work-related activities could simply be reduced, including time dedicated to care activities. The gender-related burden embedded in the idyllic idea of a happy life beyond work was for a long time a blind spot in the mainstream degrowth discourse.

## Degrowth as Shrinking Scenario Will Require a Higher Workload

By considering the flows of energy demand and supply of societies, Sorman and Giampietro discuss a scenario of reduced input of energy flows and resource use—as depicted by degrowth—and correlate it, under current conditions, to the amount of required working hours. They employ *Social Metabolism Analysis* to study society's energy flows and to analyse the effect that changes in the flow might have on the structure of society (2013). It seems obvious that, by looking at 'past trend of changes in the metabolic patterns of developed countries, we can clearly see that a reduction in the number of hours of paid work in the Primary and Secondary sector has been possible only because [of] an increase of the energy consumption in these sectors' (ibidem, 90). This is evident with respect to agriculture, in which mechanization freed 60–70% of the workforce for the other sectors. The reduction of working hours and its shift to the service sector in *developed*[4] countries were rendered possible by the massive energy intensification of the productive sectors as well as the relocation of productive activities to *developing* countries and the dependence on imported goods.

For Sorman and Giampietro, reducing working hours in order to redistribute employment and enable a decent standard of living, while shifting the

---

[4] While we adopt the language used by the authors, we italicize the terms because we are critical of developmentalist terminology.

benefits of greater productivity from GDP to more leisure time, does not take adequately into account biophysical conditions. Accordingly, it seems highly improbable that a society that uses less capital, less energy, and fewer resources per hour of labour 'could provide an income capable of maintaining a decent salary to its workers, based on fewer working hours. The ability to pay high wages to labour depends on the surplus made available to society from the activities carried out during work' (ibidem, 91).

Whether reducing working hours with corresponding reduction of income would indeed be replaced by alternative satisfiers such as leisure time or social relations is a complex question for the authors. They understand degrowth more in terms of a future shrinking scenario than as a project of social transformation and reject Victor's idea of degrowth by design to avoid disaster (Victor 2019). Instead, they suggest that downscaling will occur as any other social change through 'an unplanned and a self-organized process', that we can study in order to be prepared to 'develop flexible management strategies whilst investing the remaining high-quality energy in wise alternative energy options to make a smoother transition' (Sorman and Giampietro 2013, 92). In their analysis, the authors remain confined to the alternative between top-down design and flexible adaptation to a shrinking economy and do not consider other possibilities. A third option, however, might lie in the transformative power of the radical alternatives (Muraca 2014) embodied by *degrowth in movement(s)*. This path operates as a fractal, indeed unplanned and self-organized process seizing counter-tendencies in the meanders of radical alternatives.

## Reforming Labour and Welfare Systems

According to reformist approaches, the structural interdependence between growth, productivity, and labour, which is framed as the root cause of the environmental crisis, can only be challenged through incremental reforms of basic institutions of liberal Welfare States towards a green and sustainable post-growth society (Seidl and Zahrnt 2019). From an ecological-liberal perspective (Schmelzer and Vetter 2019), market-instruments flanked by tax reforms and institutional frameworks are the key instruments of change. Wage labour is considered not only as a key driver of growth (fuelled by productivity increase) but also as the funding source for welfare systems. This is why, according to Seidl and Zahrnt (2019), reforming the system of wage labour is necessary to lift the structural dependence of (liberal welfare) societies from growth. To achieve this goal, the financing of the welfare system should shift from taxes on labour to the taxation of pollution and emissions, as well as the taxation of property, capital, or wealth (Köppl and Schratzenstaller 2019). Furthermore, such a reform requires the reorganization of social protection through redistribution based on, for example, $CO_2$ tax dividends or profit sharing in conventional enterprises or cooperatives, incentives for a redistribution of time dedicated to reciprocal care (time or life-time accounts), and social security (Kubon-Gilke 2019).

In a similar vein, Jackson and Victor claim that escaping the productivity trap requires work sharing and a structural shift of labour to low productivity jobs in the personal and social service sectors, including jobs in the areas of repairing, recycling, care, creative activities, and so on—what Jackson calls *Cinderella economy* (Jackson 2009; Jackson and Victor 2011).

These reformist positions do not challenge wage labour as such, but only insofar as it is structurally entangled with increasing productivity as driver of economic growth—and therefore of the ecological crisis—and as it financially sustains social security systems. Neoliberal transformations of the labour market are considered favourably as far as they increase flexibility and autonomy, thus creating spaces for community and care-oriented activities that complement welfare interventions. Labour productivity is seen as the key driver of economic growth, while working-class struggles are considered as a component in the entrepreneurial drive to increase productivity (Seidl and Zahrnt 2019). Liberal States in their regulating function and the business sector are the leading agents of transformation towards a post-growth-society. A critical analysis of capitalist accumulation under neoliberalism and its relation to State power is absent.

## Rethinking Work as a Project of Radical Transformation of Society

Of the different streams within the degrowth discourse, Schmelzer and Vetter (2019) identify three transformative currents: (1) oriented towards commons and alternative economies, (2) feminist, and (3) rooted in a more radical critique of capitalism and globalization. Transformative approaches address the very idea of work and its symbolic meaning for modern societies and question the underlying assumption of separation between production, consumption/use, and (re)production. Despite specific differences, the goal of transformative approaches is not only a reduction 'in size' of the economic performance, but a qualitative change of its role in society. In their heterogeneity, these approaches include the analysis of feasible alternatives as well as a variety of niche experiments connected by horizontal networks of cooperation and solidarity (Burkhart et al. 2020). In the following section, we limit our analysis to radical feminist perspectives.

## Limits to Work: Objections from an Ecofeminist Political Economy Perspective

We conclude our analysis by dedicating attention to a perspective that has only recently been considered by degrowth scholars: the feminist critique of work in relation to the social and ecological reproduction of society. In the degrowth discourse, the analysis of the social and ecological crisis brought about by economic growth had been largely conducted without *explicit* reference to and acknowledgement of the feminist materialist tradition. When feminist critique has been incorporated into degrowth, it was often without thorough consideration of its complexity. Nonetheless, the feminist critique of growth, productivism, and capitalism is one of the most significant and more sophisticated inspirational sources of the degrowth discourse. Only recently, the obvious cross-fertilization between material feminism, feminist (ecological) economics, and degrowth has been more explicitly articulated in some scholarly works (Barca 2019) and in activist practices, such as an initiative to connect feminisms and degrowth via the 'Feminisms and Degrowth Alliance' (Saave-Harnack et al. 2019; FaDA 2020).[5] However, the conceptualization of work and labour within the degrowth discourse is overall still lacking the thorough analysis that materialist ecofeminism has to offer regarding the structural link between capitalist growth imperatives, the ecological crisis, and the devaluation and subalternization of (re)productive work.[6] We aim here at counteracting its longstanding peripheral presence. Moreover, a feminist perspective reminds us that social and economic transformation have gendered and racialized consequences that are often neglected. The ecofeminist political-economic critique of the economy is essential for the project of a just, sustainable, and equitable degrowth society.

---

[5] See for example: www.degrowth.info/en/feminisms-and-degrowth-alliance-fada/.
[6] Biesecker and Hofmeister (2010) have coined the term (re)production to point out the (material) inseparability of human so-called productive activities as well as human and non-human reproductive activities and processes.

## The Ecofeminist Political-Economic Critique of the Economy

In ecofeminist political-economic theory, two fields, feminist materialism and ecofeminism, come together. Ecofeminism (e.g. Mies 1986; Salleh 1997) is rooted in women's joint political struggles for autonomy, health, and environmental protection. Against the allegation of being essentialist and thus reinforcing social inequality by advocating that women are closer to nature, many counter-arguments and historical explanations have been articulated in favour of a non-essentialist ecofeminism (e.g. Salleh 1997; Gaard 2011). From an *empirical* perspective, women are rendered 'closer' to nature because of the material relations to nature that they embody due to the socio-economic position they (have been forced to) occupy. This takes a very concrete and material meaning in some communities, where in many cases women are 'typically' responsible for fetching water or firewood and taking care of 'small-scale subsistence farming ensuring the survival of their families' (Oksala 2018, 218). Because of such activities, women are likely the first to be affected by environmental destruction. This empirical and material connection alone suggests the strategic combination of political struggles for women's rights and environmental protection, as 'protecting the environment also directly improves the lives of poor women' (ibid., 219) as well as their livelihood, sovereignty, and autonomy.

Materialist ecofeminists working in the tradition of Marx's critique of the political economy claim furthermore that the connection between women and nature is not simply empirically given but enforced and perpetuated by capitalist relations. Accordingly, social and ecological reproduction activities, that is, so-called women's work[7] as well as what is constructed as 'nature,' have similar functions for the accumulation of capital: both are productive in the sense of producing use values, but not exchange values. They do not become visible in the value or commodity form and are formally excluded from commodity markets. Accordingly, capitalist accumulation does not only rely on the direct *exploitation* of surplus value extracted from wage labour, but also on the *appropriation* of (women's) reproductive labour and natural resources or, in other words, 'on the ongoing and violent expropriation of women, indigenous peoples, nonhuman animals, and the biosphere' (ibidem, 222). Women's bodies and reproductive capacities are appropriated and put to the

---

[7] By women we do not mean a biological or essentialist category, but a social role that is functional to the reproduction of capitalist relations in the bourgeois family and for subjectivities in general (Federici 2012). This includes care, emotional, and sexual work.

service of accumulation. 'This process is masked and legitimized by an essentially ideological process that "naturalizes" women—understands them as being less civilized, less rational, and closer to nature' (ibidem). This applies to nature in the reversed form as well: it has been put to the service of accumulation and is conceived of as *feminine* and therefore as an exploitable resource (Merchant 1982).

## Ecofeminist Relation to Degrowth

This materialist ecofeminist perspective draws attention to the relations between the economy and what is actively excluded or externalized from it (Bennholdt-Thomsen et al. 1988; Biesecker and von Winterfeld 2014). In doing so, materialist ecofeminism points to the varieties of activities and processes, in addition to resources, that capitalism depends on beyond the wage relation. It calls for an extended account of the socio-ecological metabolism which encompasses various processes and activities (e.g. reproductive labour and nature) that are not typically coded as labour within a capitalist economy. Being essential for social reproduction and thus any sort of economy, they have to be accounted for when conceptualizing (human and non-human) ways of being active in a degrowth society, if the economy is to be transformed. From this perspective, degrowth scholars and activists are challenged to systematically ask how to frame externalized activities directed at maintaining reproduction within a degrowth economy and how to distribute and organize them in society.

Scholars with a political-economic, ecofeminist background have discussed the question of economic growth long before degrowth became a well-established term. Maria Mies (1986) analyses the dependence of economic growth on society-nature-relations and gender-relations to delineate a future economy. According to Mies, only by including all relations of production created by capitalist patriarchy and by adopting a global and holistic approach, scholars and activists can develop a vision of a future society, in which women, nature, and colonized others are no longer exploited in the name of development and growth (ibidem). Mies does not only unveil the growth diktat as a major driving force for unsustainable and unjust societal relations. She also shows its patriarchal roots, a perspective that has been only marginally considered in the degrowth discourse.

The externalization of (re)productive (Biesecker and Hofmeister 2006) activities and their denial in capitalist economies is, accordingly, the actual root cause of the environmental crisis. Rethinking work and labour cannot be reduced to just a side aspect of any serious degrowth project. If the environmental crisis is a crisis of (re)production, then reconceptualizing work and labour and their metabolic structure is a necessary condition for a radical degrowth transformation.

Despite the rather obvious contributions to degrowth by Mies, von Werlhof, and other scholars from the Bielefeld School, their work was largely ignored in the early all-male history of degrowth. Feminist accounts have long been a blind spot that has only recently been slowly recuperated (Gregoratti and Raphael 2019; Saave-Harnack et al. 2019). Because of this neglect, the degrowth discourse oftentimes lacks attention to how everyday practices perpetuating growth are gendered and racialized. Any vision for a degrowth society that does not account for gender roles as well as gendered and racialized institutions is part of the problem rather than of the solution.

## Focus on Paid Labour

Most approaches to labour/work in the degrowth discourse, with few exceptions, tend to focus more or less implicitly on paid labour only and, while acknowledging the importance of a wider understanding of work, lack a thorough analysis of what this would imply for the transformation they envision. In her pioneering work, Mies describes the economy as an iceberg, of which only a small top is considered as labour and becomes visible as paid work, while the bulk of activities lies 'under water' (Mies 2009). Paid labour is only a fraction of all activities and processes that facilitate human and non-human life on this planet, including the work of peasants, housewives, and the wider field of subsistence production (von Werlhof 1991). All these processes and activities are materially and energetically linked (Biesecker and Hofmeister 2010). However, a so-called monetary boundary (Dengler and Strunk 2018) separates this totality and is a signifier of the metabolic rift that breaks apart the (necessary) material links across these processes and activities. By doing so, it veils their interconnection through the selective application of monetary valuation or the commodity form.

Even in their sole focus on paid labour, degrowth scholars have not yet taken fully into account the conditions of the global working class, which still depends chiefly on income generated through fossil-based activities. As Barca (2019) argues, dismissing the struggles of the working class and its organizations does not help the degrowth project. According to her, alliances are possible between parts of the labour movement and environmental justice movements, as 'there exist, at this historical conjuncture, concrete possibilities for articulating degrowth and labour politics in new ways, via grassroots mobilizations in community unionism and social movement unionism, pushing labour organizations toward a radical critique of the growth paradigm' (ibidem, 212). The road ahead requires struggling for a de-alienation of work and actions, including the collective re-appropriation of its products (ibidem).

## Intermediation of Social-Ecological Processes

While expanding the perspective from paid work to labour/work in general, including (re)productive activities, a materialist (eco)feminist perspective also highlights the social inequalities embedded in it and considers it as an integral part of the socio-ecological metabolism. According to Barca (2019), work is a gendered and racialized mediator of social metabolisms. By understanding labour/work as already practically embedded in social-ecological relations, it is possible to make visible power relations not only between workers and capitalists, but also within households, between humans and 'nature', and between production and social reproduction. It is also possible to articulate the constant and complex mediation of all social-ecological processes, including, for example, ecosystem services. Far from being simply flows coming from 'nature', ecosystem services are the product of mediated relations of maintenance, care, and co-production—relations that are often invisibilized because they are performed by peripheral social actors (Farrell 2014; Comberti et al. 2015; Biesecker and Hofmeister 2010). They are, in other words, co-generated, both performatively and materially within and by societal relations to nature. While the degrowth discourse generally shares the idea that social and economic processes are embedded in ecological ones, it does not sufficiently address how society-nature relations are gendered and racialized.

Barca (2019) proposes a radical reorganization of social metabolism, in which surplus is collectively managed. An increasing 'emancipatory, ecological class consciousness' (ibidem, 213), fuelled by the awareness that climate change and environmental violence are 'the newest form of class war—as always, articulated with gender and racial domination' (ibidem)—can form an alliance within what Salleh (2010) calls *meta-industrial labour*.[8] Salleh's position is different with respect to the potential revolutionary subject. In her account, peasants, indigenous people, (re)productive workers, and all those who forcibly inhabit the externalized dimension of capitalist reproduction

---

[8] Salleh calls a certain group of people *meta*-industrial, because they work not within industrial production, but before and beyond it, as its very condition. Through their work, they mediate the society-nature metabolism and contribute largely to the production of use values from/with nature.

could join a common struggle against the destruction of livelihoods. For Barca, instead, the political subject of a degrowth revolution remains the global working-class fighting for de-alienation and communing rather than 'an ecologically minded global middle class willing to reduce consumerism and work-addiction, and/or to engage in direct action to express its disappointment with economic/environmental policies' (Barca 2019, 214).

## Rethinking Labour/Work in a Degrowth Society: A Feminist Proposal

In the recent attention to materialist ecofeminism by degrowth, some proposals of how to conceptualize labour/work in a degrowth society have been rediscovered. They range from assuming the totality of work (*das Ganze der Arbeit*—Biesecker and Hofmeister 2006) to a proportionate distribution of time for leisure, wage work, care work, and political participation, as exemplified in the 4-in-1 perspective (Haug 2011). To conclude, we present here one of the few concrete proposals that have been developed explicitly as a feminist degrowth perspective.

In her book *Subversion Feminista de la Economia*, Amaia Pérez Orozco proposes degrowth as a path for a radical post-capitalist, commons-based and self-governed restructuring of care. She does so in a detailed analysis of the modus operandi of heteropatriarchal capitalism, in which the separation between paid and unpaid work, production and (re)production is a necessary condition for accumulation and follows not only heteropatriarchal but also anthropocentric and neo-colonialist principles. She understands care not in the narrow sense of 'care services', but in terms of a radical rethinking of life tout-court, not as an exploitable resource (in both ecological and social terms) but as the foundation and creative engine of communities and society.

According to Pérez Orozco, the capitalist organization of work and labour (both within direct exploitation and via the appropriation of reproductive activities) is aimed at appropriating life by destroying it in all its forms: not only the living activity of production and regeneration, that is, the life of workers and of those accomplishing (re)productive activities, but also the living processes beyond human life. This leads to a fundamental, structural contradiction between accumulation processes and processes that sustain life, or, in other words, to a radical opposition between capital and life (Pérez Orozco 2014). Not all life, obviously, is exposed to destruction, as power and resources are to be accumulated around one single life, that is, the life of the white, bourgeois, heterosexual male, separated from the community of the living. In

## 32 Rethinking Labour/Work in a Degrowth Society

opposition to the capitalist destruction of life, degrowth can embody a radical alternative, both *materially* by subtracting power and resources from the circuit of capital accumulation and by questioning private property and financialization and *symbolically* by challenging the very logic of capitalist accumulation (ibidem). Socialization of care, democratization, and depatriarchalization of the households are important, but not sufficient steps for a radical subversion of the economy.

The goal of an ecofeminist degrowth has to be de-privatizing and de-feminizing the responsibility of sustaining life. This implies recuperating the sense of horizontal interdependence and vulnerability of all life collectively embodied in a *buen convivir* (good living together) rooted in autonomy and self-determination. If paying for (re)productive work might in the short term obtain visibility and social relevance, in the long run the redistribution of all necessary activities and forms of work (paid and unpaid, pleasant and unpleasant ones) has to take place across the social and geographical spectrum (rural versus urban areas, unequal and unjust distribution across classes, genders, racialized groups).

The very concept of socially necessary work has a different meaning from the point of view of ecofeminist degrowth: socially necessary are those activities that serve the interdependence and care for the vulnerability of life—of all life. As it is already the case in the myriad of self-managed projects and social experiments worldwide, care should be organized neither through the mediation of the market nor by the State, but in self-managed and self-governed structures linked together through horizontal and autonomous networks of solidarity. This also requires a reconfiguration of elective families, reorganized on the ground of communes or other forms of living together (ibidem). The goal is to change the (re)productive matrix so that it is composed of different entities and sectors, in which the responsibility of sustaining life is collectively shared. Pérez Orozco calls her vision not a *U-topia*, a non-place, but a *Hyper-topia*, constituted by a plurality of decentralized and networked, self-governed and collectively organized structures aimed at guaranteeing the necessary conditions for living well together. Crucial for this radical proposal is degrowing the sphere of accumulation while at the same time democratizing the households and eliminating the sexual (and racial) division of labour. In a concentric model, democratized households, rooted in 'intimacy networks' and elective families, would support and be supported by 'proximity networks' organized in a commons-based way and embedded in collective institutions with a larger reach that constitute the public sphere.

Both radical proposals of a commoning of social provisioning and of a self-governed, collective, and solidarity-based provision of care can be understood as conceptions of work that truly envision *being* and *doing* beyond the wage relation and capitalist accumulation. Reciprocal and collectively shared responsibilities of care for the vulnerability of life in all its forms embody what Pérez Orozco calls an ecofeminist subversion of the economy towards a depatriarchalized degrowth society.

While Orozco's analysis offers an original example of how the degrowth critique could look like after fully adopting a feminist materialist perspective and taking seriously the gendered and racialized dimensions of growth-fuelled capitalism, her proposal deserves some critical considerations. A thorough analysis of the role of the State or any other functional equivalent in charge of the coordination across self-organized communities is marginal, as well as the question about how to rethink the role of the Welfare State. It is unclear who would be the transformation subjects and whether, how, or why the global working class or the meta-industrial-labour alliance would struggle for the implementation of these projects. Her *hyper-topia* risks reproducing the romanticization of small communities and fails to discuss the role that technological development and innovations play in the reframing of future work/labour. Although this critique is relevant, it should have become clear, at least after the Covid-19 pandemic, that care activities are still the very foundation of any type of production, no matter how innovative, and that commons-based, open-source innovations instead of the corporate privatization of ideas are essential for a just and sustainable future for all.

## Conclusion

As we have shown in this chapter, a promising path to rethink labour/work in a degrowth society follows the steps that have already been taken in materialist ecofeminism. This implies addressing not only how the transition to degrowth affects paid labour, but also how it affects work in general. Materialist ecofeminism linked to degrowth unveils gendered notions of nature and the assumption that reproductive work is a *natural* responsibility of women. Moreover, it shows how these assumptions are rooted in the accumulation of capital. Rethinking work beyond the wage relation, the (re)production of social-ecological processes, and the economy in terms of an economy of provisioning lays the foundations for a radical degrowth alternative that truly tackles the root causes of the environmental crisis.

From a materialist ecofeminist perspective, the felicitous structural critique of capitalism brought forward by Rosa et al. (2017) lacks a *decisive sting* (Neusüß 1985, 233) if it continues ignoring gendered and racialized relations of power inscribed in the drivers of economic growth. In this chapter, we argue that degrowth itself can be read as a reformulation of the feminist critique of capitalism, with a specific focus on the environmental and social impact of economic growth and on how social and ecological reproduction are intimately linked. The path forward requires a radical social-ecological transformation of society. As a component of this transformation, we discussed Pérez Orozco's proposal of commoning social provisioning of care through self-governed, collective, and solidarity-based structures. Degrowth activists and scholars can lead the way by collectively asking and answering Pérez Orozco's key question: What are the conditions needed to support the sustainability of life? Ensuring that the sustainability of life and caring relations take centre stage in economic thinking and action is a necessary condition for a radical transformation towards a democratic and equitable degrowth society.

# References

Alcott, Blake. 2013. Should Degrowth Embrace the Job Guarantee? *Journal of Cleaner Production* 38. Degrowth: From Theory to Practice: 56–60.

Asara, Viviana, Iago Otero, Federico Demaria, and Esteve Corbera. 2015. Socially Sustainable Degrowth as a Social–Ecological Transformation: Repoliticizing Sustainability. *Sustainability Science* 10: 375–384.

Barca, Stefania. 2019. The Labour(s) of Degrowth. *Capitalism Nature Socialism* 30 (2): 207–216.

Bennholdt-Thomsen, Veronika, Maria Mies, and Claudia von Werlhof. 1988. *Frauen, die letzte Kolonie. Zur Hausfrauisierung der Arbeit*. Reinbek bei Hamburg: Rowohlt-Taschenbuch-Verlag.

Biesecker, Adelheid, and Sabine Hofmeister. 2006. *Die Neuerfindung des Ökonomischen: ein (re)produktionstheoretischer Beitrag zur sozial-ökologischen Forschung*. München: Oekom Verlag.

———. 2010. Focus: (Re)Productivity. Sustainable Relations Both between Society and Nature and between the Genders. *Ecological Economics* 69 (8): 1703–1711.

Biesecker, Adelheid, and Uta von Winterfeld. 2014. Extern? Weshalb und inwiefern moderne Gesellschaften Externalisierung brauchen und erzeugen. *Working Paper der DFG-KollegforscherInnengruppe Postwachstumsgesellschaften* 2: 1–16.

Brand, Ulrich, and Marcus Wissen. 2018. *The Limits to Capitalist Nature. Theorizing and Overcoming the Imperial Mode of Living*. London and New York: Rowman & Littlefield.

Burkhart, Corinna, Matthias Schmelzer, and Nina Treu. 2020. *Degrowth in Movement(s). Exploring Pathways for Transformation.* Winchester and Washington: Zero Books.

Caradonna, Jeremy. 2014. *Sustainability. A History.* Oxford: Oxford UP.

Comberti, Claudia, Thomas Thornton, Victoria de Echeverria, and Trista M. Patterson. 2015. Ecosystem Services or Services to Ecosystems? Valuing Cultivation and Reciprocal Relationships between Humans and Ecosystems. *Global Environmental Change* 34: 247–262.

D'Alisa, Giacomo, Federico Demaria, and Giorgos Kallis. 2015. *Degrowth: A Vocabulary for a New Era.* London and New York: Routledge.

Demaria, Federico, François Schneider, and Filka Sekulova. 2013. What is Degrowth? From an Activist Slogan to a Social Movement. *Environmental Values* 22: 191–215.

Dengler, Corinna, and Birte Strunk. 2018. The Monetized Economy Versus Care and the Environment: Degrowth Perspectives on Reconciling an Antagonism. *Feminist Economics* 24 (3): 160–183.

Eversberg, Dennis. 2014. Die Erzeugung kapitalistischer Realitätsprobleme: Wachstumsregimes und ihre subjektiven Grenzen. *WSI-Mitteilungen* 67: 528–535.

Eversberg, Dennis, and Barbara Muraca. 2019. Degrowth-Bewegungen: Welche Rolle können sie in einer sozial-ökologischen Transformation spielen? In *Große Transformation? Zur Zukunft moderner Gesellschaften*, ed. Klaus Dörre, Hartmut Rosa, Karina Becker, Sophie Bose, and Benjamin Seyd, 487–501. Berlin: Berliner Journal Für Soziologie.

FaDA. 2020. *Collaborative Feminist Degrowth: Pandemic as an Opening for a Care-Full Radical Transformation.* On: degrowth.info.

Farrell, Katharine N. 2014. Intellectual Mercantilism and Franchise Equity: A Critical Study of the Ecological Political Economy of International Payments for Ecosystem Services. *Ecological Economics* 102: 137–146.

Federici, Silvia. 2012. *Revolution at Point Zero: Housework, Reproduction, and Feminist Struggle.* Oakland: PM Press.

Gaard, Greta. 2011. Ecofeminism Revisited: Rejecting Essentialism and Re-placing Species in a Material Feminist Environmentalism. *Feminist Formations* 23 (2): 26–53.

Gómez-Baggethun, Erik, and José Manuel Naredo. 2015. In Search of Lost Time: The Rise and Fall of Limits to Growth in International Sustainability Policy. *Sustainability Science* 10: 385–395.

Gorz, André. 1982. *Farewell to the Working Class.* London: Pluto Press.

———. 1999. *Reclaiming Work: Beyond the Wage-based Society.* Cambridge: Polity Press.

Gregoratti, Catia, and Riya Raphael. 2019. The Historical Roots of a Feminist 'Degrowth': Maria Mies's and Marilyn Waring's Critiques of Growth. In *Towards a Political Economy of Degrowth*, ed. Ekaterina Chertkovskaya, Alexander Paulsson, and Stefania Barca. Lanham: Rowman & Littlefield.

Haug, Frigga. 2011. Die Vier-in-einem-Perspektive als Leitfaden für Politik. *Das Argument* 53: 241–250.

Jackson, Tim. 2009. *Prosperity without Growth: Economics for a Finite Planet*. London: Earthscan.

Jackson, Tim, and Peter Victor. 2011. Productivity and Work in the "Green Economy": Some Theoretical Reflections and Empirical Tests. *Environmental Innovation and Societal Transitions* 1: 101–108.

Kallis, Giorgos. 2013. Societal Metabolism, Working Hours and Degrowth: A Comment on Sorman and Giampietro. *Journal of Cleaner Production* 38: 94–98.

Kallis, Giorgos, Michael Kalush, Hugh O'Flynn, Jack Rossiter, and Nicholas Ashford. 2013. "Friday off": Reducing Working Hours in Europe. *Sustainability* 5: 1545–1567.

Kallis, Giorgos, Vasilis Kostakis, Steffen Lange, Barbara Muraca, Susan Paulson, and Matthias Schmelzer. 2018. Research on Degrowth. *Annual Review of Environment and Resources* 43: 291–316.

Köppl, Angela, and Margit Schratzenstaller. 2019. Ein Abgabensystem, das (Erwerbs-)Arbeit fördert. In *Tätigsein in der Postwachstumsgesellschaft*, ed. Irmis Seidl and Angelika Zahrnt, 207–226. Marburg: Metropolis.

Kubon-Gilke, Gisela. 2019. Soziale Sicherung in der Postwachstumsgesellschaft. In *Tätigsein in der Postwachstumsgesellschaft*, ed. Irmis Seidl and Angelika Zahrnt, 193–206. Marburg: Metropolis.

Lange, Steffen, and Tilman Santarius. 2020. *Smart Green World? Making Digitalization Work for Sustainability*. New York and London: Routledge.

Latouche, Serge. 2009. *Farewell to Growth*. Cambridge: Polity Press.

Lazzarato, Maurizio. 2004. From Capital-Labour to Capital-Life. *Ephemera* 4: 187–208.

Leonardi, Emanuele. 2017. *Lavoro Natura Valore. André Gorz tra marxismo e decrescita*. Nocera Inferiore: Orthotes.

Leonardi, Emanuele, and Ekaterina Chertkovskaya. 2017. Work as Promise for the Subject of Employability. Unpaid Work as New Form of Exploitation. *Sociologia del Lavoro* 145: 112–130.

Liegey, Vincent, Stéphane Madelaine, Christophe Ondet, and Anne-Isabelle Veillot. 2013. *Un projet de Décroissance. Manifeste pour une Dotation Inconditionnelle d'Autonomie*. Paris: Les Editions Utopia.

Martinez-Alier, Joan, Unai Pascual, Franck-Dominique Vivien, and Edwin Zaccai. 2010. Sustainable De-growth: Mapping the Context, Criticisms and Future Prospects of an Emergent Paradigm. *Ecological Economics* 69: 1741–1747.

Meadows, Donella H., Dennis L. Meadows, Jørgen Randers, and William W. Behrens III. 1972. *The Limits to Growth*. New York: Universe Books.

Merchant, Carolyn. 1982. *The Death of Nature: Women, Ecology and the Scientific Revolution*. London: Wildwood House.

Miegel, Meinhard. 2011. *Exit. Wohlstand ohne Wachstum*. Berlin: List.

Mies, Maria. 1986. *Patriarchy and Accumulation on a World Scale: Women in the International Division of Labour*. London: Zed Books.

———. 2009. Hausfrauisierung, Globalisierung, Subsistenzperspektive. In *Über Marx hinaus*, ed. Marcel van der Linden and Karl Heinz Roth, 255–290. Hamburg: Assoziation A.

Muraca, Barbara. 2013. Décroissance: A Project for a Radical Transformation of Society. *Environmental Values* 22: 147–169.

———. 2014. *Gut Leben: Eine Gesellschaft jenseits des Wachstums*. Berlin: Wagenbach.

———. 2015. Wider den Wachstumswahn: Degrowth als konkrete Utopie. *Blätter für Deutsche und Internationale Politik* 2: 101–109.

———. 2020. Possibilities for Degrowth: A Radical Alternative to the Neoliberal Restructuring of Growth-Societies. In *The Cambridge Handbook of Environmental Sociology*, ed. Katharine Legun, Julie Keller, Michael Bell, and Michael Carolan. Cambridge: Cambridge UP.

Muraca, Barbara, and Ralf Döring. 2018. From (Strong) Sustainability to Degrowth. A Philosophical and Historical Reconstruction. In *Routledge Handbook of the History of Sustainability*, ed. Jeremy Caradonna, 339–362. New York and London: Routledge.

Muraca, Barbara, and Matthias Schmelzer. 2017. Sustainable Degrowth: Historical Roots of the Search for Alternatives to Growth in Three Regions. In *History of the Future of Economic Growth: Historical Roots of Current Debates on Sustainable Degrowth*, ed. Matthias Schmelzer and Iris Borowy. London and New York: Routledge.

Neusüß, Christel. 1985. *Die Kopfgeburten der Arbeiterbewegung oder Die Genossin Luxemburg bringt alles durcheinander*. Hamburg: Rasch und Röhring Verlag.

Nussbaum, Martha. 2001. Adaptive Preferences and Women's Options. *Economics and Philosophy* 17: 67.

Oksala, J. 2018. Feminism, Capitalism, and Ecology. *Hypatia* 33 (2): 216–234.

Paech, Niko. 2012. *Befreiung vom Überfluss*. München: Oekom.

Pérez Orozco, Amaia. 2014. *Subversión Feminista de la Economía*. Madrid: Traficantes de Suenos.

Petridis, Panos, Barbara Muraca, and Giorgos Kallis. 2015. Degrowth: Between a Scientific Concept and a Slogan for a Social Movement. In *Handbook of Ecological Economics*, ed. Joan Martinez-Alier and Roldán Muradian, 176–200. Cheltenham: Edward Elgar.

Rosa, Hartmut, Klaus Dörre, and Stephan Lessenich. 2017. Appropriation, Activation and Acceleration: The Escalatory Logics of Capitalist Modernity and the Crises of Dynamic Stabilization. *Theory, Culture & Society* 34: 53–57.

Saave-Harnack, Anna, Corinna Dengler, and Barbara Muraca. 2019. Feminisms and Degrowth—Alliance or Foundational Relation? *Global Dialogue* 3 (1): 29–30.

Salleh, Ariel. 1997. *Ecofeminism as Politics: Nature, Marx and the Postmodern*. London: Zed Books.

———. 2010. From Metabolic Rift to "Metabolic Value": Reflections on Environmental Sociology and the Alternative Globalization Movement. *Organization and Environment* 23 (2): 205–219.

Santarius, Tilman, Hans Jakob Walnum, and Carlo Aall. 2016. Rebound Research in a Warming World. In *Rethinking Climate and Energy Policies. New Perspectives on the Rebound Phenomenon*, ed. Ibid, 1–16. New York: Springer.

Schmelzer, Matthias, and Andrea Vetter. 2019. *Degrowth/Postwachstum zur Einführung*. Hamburg: Junius.

Seidl, Irmis, and Angelika Zahrnt. 2019. *Tätigsein in der Postwachstumsgesellschaft*. Marburg: Metropolis.

Sekulova, Filka, Giorgos Kallis, Beatriz Rodríguez-Labajos, and Francois Schneider. 2013. Degrowth: From Theory to Practice. *Journal of Cleaner Production* 38. Degrowth: From Theory to Practice: 1–6.

Sekulova, Filka, and Jeroen C.J.M. van Den Bergh. 2013. Climate Change, Income and Happiness: An Empirical Study for Barcelona. *Global Environmental Change* 23: 1467–1475.

Sorman, Alevgul H., and Mario Giampietro. 2013. The Energetic Metabolism of Societies and the Degrowth Paradigm: Analyzing Biophysical Constraints and Realities. *Journal of Cleaner Production* 38: 80–93.

Victor, Peter A. 2019. *Managing without Growth: Slower by Design, Not Disaster*. 2nd ed. Cheltenham: Edward Elgar.

von Werlhof, Claudia. 1991. *Was haben die Hühner mit dem Dollar zu tun?* München: Verlag Frauenoffensive.

Welzer, Harald. 2012. *Mental Infrastructures: How Growth Entered the World and Our Souls*. Berlin: Heinrich Böll Stiftung.

# 33

# Labour and Societal Relationships with Nature. Conceptual Implications for Trade Unions

Thomas Barth and Beate Littig

## The Role of Labour in Times of Sustained Unsustainability

Social scientists in some areas of sustainability research have long reached the conclusion that a real turn towards more sustainability is still not foreseeable, despite at least three decades of proclaimed sustainability policy (Barry 2012; Blühdorn 2016). The relationships between society and nature, that is, society's demands for natural resources and its related interactions with the environment, are in deep crisis. The threats to all life, human and non-human, are becoming increasingly evident. If a high standard of quality of life—so far only afforded to the richer members of the world's population—is to be maintained and, above all, if the basic needs of the majority of human beings are even to be satisfied then, given the inevitable effects on the environment,

---

The work on this text was funded by the Vienna Science and Technology Fund (WWTF) through project ESR17-067.

---

T. Barth (✉)
Department of Sociology, LMU Munich, Munich, Germany
e-mail: Thomas.barth@lmu.de

B. Littig
Socio-Ecological Transformation, Institute for Advanced Studies, Vienna, Austria

society must fundamentally change the way it dominates and appropriates nature. In this chapter, we follow the approach adopted in environmental labour studies, that is, that labour should be brought into the focus of the analysis in order to identify the causes of the prevailing unsustainability and create practical paths to more sustainable socio-ecological relations.

In our view, the obstacles to sustainability and the social tensions encountered in the transition to a low-carbon economy result above all from the way societies organise, distribute, and value labour. This does not only highlight just how radical a transformation towards sustainability would have to be, that is, far more than is frequently assumed, but it also creates new starting points for such change.

When critical issues in the relationships between society and nature are viewed from the point of view of the organisation of labour, it immediately becomes evident that environmental and climate protection are in essence social issues. The way societies regulate their relationships with nature, the purpose of environmental protection and/or preventive measures, the distribution of the costs and positive effects of such measures, and the allocation of the corresponding responsibilities cannot be separated from the way in which societies arrange their exchange with nature via the organisation of labour. Accordingly, bringing the topic of labour into the sustainability debate turns the focus of attention to fundamental questions regarding power structures in society. How are labour and its outcomes distributed and valued in society? Who actually makes such decisions? In our opinion, such questions belong at the centre of sustainability research and policy because a socio-ecological transformation of society towards sustainability goes hand-in-hand with a fundamental transformation of the—(post)industrial capitalist—'society of work'.[1]

In the next part of the chapter (section "Societal Relationships with Nature. A Social and Political Ecology of Labour"), we present the theoretical concept of Societal Relationships with Nature (SRN).[2] The SRN approach builds in particular on concepts of social and political ecology and emphasises the co-constitution of society and nature, the material and symbolic dimensions of these relationships, and the arrangement of society and nature through labour.

---

[1] By 'borrowing' this term, its German equivalent being *Arbeitsgesellschaft*, we seek to draw attention to three things. *First*, to satisfy human needs, each society is dependent upon the appropriation of nature through work (*Arbeit*). *Second*, modern capitalist societies can be fittingly described as societies based on paid work (*Erwerbsarbeitsgesellschaften*), which means that only this kind of work is defined as work. Nevertheless, these societies remain, thirdly, always structurally dependent on access to unpaid reproductive work.

[2] We use this term in the way suggested by Görg (2011). For more on the difficulty of translating the German term *gesellschaftliche Naturverhältnisse*, see Görg et al. (2017, 2).

We argue that by placing a stronger focus on the topic of labour, the close link between the societal organisation of labour and the present unsustainability of the SRNs will become clear.

In section "Sustainable and Unsustainable Work", we focus more specifically on the modern-day debate on sustainability, outlining the principles behind the United Nations' recently adopted goal of sustainable work (UNDP 2015) and its reception in the social sciences. The UN's shift of perspective towards addressing the prevailing unsustainability pushes the role of work promisingly into the centre of the sustainability debate. Nevertheless, a solid grounding in socio-ecological theory, required to analyse the obstacles to genuinely sustainable work effectively, is still missing. Thus, we discuss the UN's concept of sustainable work from the SRN perspective.

In the concluding section, we follow the line taken in environmental labour studies research (Stevis et al. 2018) and examine the consequences of this theoretical perspective for trade unions as central players in the design of sustainable work (section "Conclusions"). A discussion of labour and SRNs will help us to understand some of the predicaments that trade unions face in their environmental activities. It will also help to identify some starting points for further activities in this field. To illustrate this point, we take a look at the situation for selected industrial trade unions in Germany (IG Metall) and Austria (PRO-GE) based on recently published academic research and material obtained from or published by these trade unions themselves.

## Societal Relationships with Nature. A Social and Political Ecology of Labour

As our theoretical base, we draw on approaches to *social and political ecology*[3] which accord a central role to the concept of societal relationships with nature. In our view, this extended family of concepts is characterised by *two assumptions*, which we will first briefly outline before a more detailed examination of the conceptualisation of the current relations between labour, nature, and society.

The *first* of these assumptions is that the prevailing socio-ecological crisis can only be analysed in an interdisciplinary and transdisciplinary research setting. Doing so requires competencies, experience, and know-how that are

---

[3] These include above all the prominent approaches developed in the German-speaking world at the *Institute for Social-Ecological Research* (ISOE, Frankfurt, Germany) and the *Vienna School of Social Ecology* (Austria) (for an overview, see Kramm et al. 2017b).

spread across various academic and non-academic players. The *second* is that while the goal of sustainability forms a more or less explicit normative point of reference, it should also be understood and reflected upon in a way that permits the analysis of any relations of power and inequality—the central element in political ecology approaches—that have been legitimised by the reference to (purported) sustainability.

Accordingly, we will first discuss the material and symbolic dimensions of SRNs. Since work as a concrete form of the appropriation of nature is central to SRNs, we will examine the specifics of work in capitalist societies and their implications for the corresponding relations to nature. This will include addressing the global distribution of labour, gender relations, and the significance of everyday practices.

## The Concept of Societal Relationships with Nature

A key point to remember when it comes to sociological debates on how to adequately explain the 'environmental crisis' is that there are a number of diverse theoretical approaches. Some of these, for example, take a more scientific realist and others a more constructivist perspective. Each can contribute to different aspects commensurate to their own particular possibilities and limitations.[4] We concur in essence with the notion that nature and society should not be conceptualised as separate entities and each considered as scientifically appropriate objects of study in their own right. On the contrary, it is to be assumed that human and non-human existences, society and nature, are co-produced and co-evolve (Smith 1984).

From the SRN perspective, society and nature are basically assumed to be mutually constitutive (e.g. Görg 2011), that is, that which appears to us to be 'nature' cannot be grasped independently of society, and 'societal' processes cannot be sufficiently understood without their ecological basis. Societies enter the relationships with nature that they require in order to reproduce, that is, to at least meet basic human needs, and these relationships have both a material and a symbolic dimension (ibid., 26). This is exemplified in the aforementioned 'production of nature', which occurs in a material, technical, and practical sense in the transformation of nature as well as, in a cultural and scientific sense, in the construction of 'nature' (ibid., 49).

There continues to be much debate on how precisely to define the relationship between nature and society when both the construction/production of

---

[4] For an overview of various such theories, see, for example, Robbins et al. (2014).

'nature' and the materiality of environmental crises that is experienced firsthand (for instance, in the devastating consequences of anthropogenic climate change) should be co-considered in the analysis.[5] As far as concepts of SRN are concerned, we adhere in this debate to critical theory based on the *non-identity* (Adorno) of nature and society (Brand and Wissen 2018, 26). Nature cannot be subjugated infinitely to human purposes, it cannot be changed at will, and it does not follow human rules—it has its own materiality (Görg 2011, 52f.). The crises shaping society's relationship with nature, be it global warming or the SARS virus, clearly expose this materiality as well as the limits of the 'mastery of nature' (Görg 2011, 49). At the same time, 'nature' is both the object and outcome of societal agency and also the starting point for powerful attempts of naturalisation, that is, to discredit and control 'others' (e.g. women, indigenous communities) as part of nature.

Given the reciprocal relationship between nature and society in its material and symbolic dimension, the necessary, ongoing appropriation of biophysical natural resources takes different forms in different times and places. The inherent and transformed gestalt of the natural environment thus influences the form in which it is appropriated, in the same way that certain cultural images of nature shape its material processing and vice versa. This transformation—that is, production and destruction—of nature is mediated through labour. Labour as an appropriation of materials and the shaping of nature thus forms the basis of any SRN. This material dimension is captured aptly in Karl Marx's definition of labour as an 'activity that assimilated particular natural materials' (Marx 1976, 133). By comparing historical and international patterns of resource use, this definition is applied by socio-ecological approaches to systematically analyse the material and energetic input and output of specific societies across history. The results can then be used to identify vast differences in the 'socio-metabolic regimes' of societal formations when the changeover was marked by radical 'socio-ecological transitions' (Fischer-Kowalski and Haas 2016). In other words, they can be used above all to quantify the appropriation and transformation of natural resources and the corresponding effects on the environment, by hunter and gatherer, agricultural and industrial societies (ibid.).

If we combine those concepts of social ecology that focus primarily on material, socio-metabolic relationships with a social theory perspective, we can also adequately grasp both the significance of the change in the shape of

---

[5] A comprehensive discussion of this long-standing discourse would exceed the scope of this chapter since it is a separate theoretical debate in its own right, the value of which is ultimately determined by its contribution to practical research (e.g. see Braun and Castree (1998), Haraway (2008), or Malm (2018), who offers a harsh critique of the elimination of the difference between nature and society).

labour in the transition to industrial, wage-based, fossil fueled, socio-metabolic regimes as well as its enormous, expansive momentum. To do so, we first need to take a closer look at how labour is organised in capitalist economies and how the inherent crisis-ridden relationships between capitalism and nature are stabilised to some extent through regulation.

## Labour and the Regulation of Capitalist SRNs

The first thing we must bear clearly in mind is that the switch to the use of paid labour as the dominant form of labour is part of the capitalist expansion and accumulation process: the goal of capitalist production is to generate and acquire surplus value in order to reinvest it in light of the omnipresent competition structure into a new and expanded cycle of valorisation.

Karl Marx refers to three related aspects of the unsustainability of the capitalist mode of production: (1) In the capitalist economy, each commodity has a 'dual character' (Marx 1976, 131), namely an *exchange value* and a *use value*. Since capitalist production is profit-oriented, that is, focused on the exchange value, it systematically abstracts from the material requirements of the production process. (2) The use values of commodities, that is, their natural forms, remain however the necessary 'material bearers [*Träger*] of … exchange value' (Marx 1976, 126) albeit in subordinate form. Labour likewise only figures in the value calculation as abstract labour and not as what it invariably also is, namely a concrete use-value generating praxis of transforming materials. The *perpetual cycle of valorisation*, driven permanently by the imperative to reinvest profits, thus stands at odds with the limited resources and absorption capacities of the biophysical system. (3) The underlying compulsion in a *competitive environment* to produce more profitably than the competition—'on pain of going under' (Marx 1992, 353)—results in the need to constantly raise productivity, whereby efficiency gains are generally overcompensated by growth effects (so-called rebound effects, see Hickel and Kallis 2019).

The class-based organisation of labour thus constitutes the basis of the relationships to nature in capitalist societies outlined above: while the owners of capital are 'forced' by the structures that are in place to exploit labour and nature, the workers, who generally have no property of their own, remain subject to a 'particular vulnerability to what has been called the job blackmail' (Barca and Leonardi 2018, 3). They are generally faced with the 'choice' of having no income or earning their keep from work that destroys nature—internally or externally.

While society's relationships with nature are always precarious, they are inherently crisis-ridden in capitalist economies. Hence, approaches to SRN assume that societal reproduction requires a successful *regulation* of these relationships with nature. 'Successful' is, of course, difficult to categorise. In line with Brand and Wissen (2018, 27), we would like to note that even crisis-ridden, unsustainable relations to nature, like those mentioned briefly above, can be institutionally stabilised for a certain length of time. In other words, ecological or planetary limits do not have an immediate effect, but capitalism and the adaptability of individuals are able to a certain extent to shift borders through 'capitalist land seizure' (*Landnahme*) processes (Dörre 2018).

The term regulation here refers to regulation theory and signals a course that has already been taken convincingly by Görg et al. (2017), who outlined a link between social and political ecology[6] approaches under the conceptual umbrella of SRN. Inequality, power, and domination in the relations between capitalist societies positioned in the centre and periphery of the global economy as well as between society and nature emerge thereby as the central factors of influence in the appropriation of nature. Giving stronger consideration to the political and economic dimensions of SRNs and their embeddedness in power and exploitation contexts on a global scale and in gender relations, modes of living, and everyday practices extends the analysis to more differentiated patterns of appropriation and destruction of nature. Accordingly, we will now take a closer look at the following three aspects of societal relations with nature and the respective role played therein by labour: (a) the global distribution of labour, (b) gender relations, and (c) the significance of everyday practices.

1. The systemic imperative to produce more profitably than the competition encourages the use of machines and thus leads to an accelerated throughput of energy and materials and a constant expansion of the markets. It is precisely these imperatives that lead companies to continually search for new sources of raw materials and labour to reduce their production costs as well as for new markets in which to sell their goods. Accordingly, and in line with the definition used by Jason Moore (2015), capitalism can be seen as a 'world-ecology system' that relies from the outset on the dominat-

---

[6] While we concentrate in the following on highly industrialised northern nations like Germany and Austria, 'classic' political ecology focuses primarily on conflicts in the appropriation of nature in the Global South (e.g. Peet and Watts 2004). Barca and Bridge (2015), however, rightly demonstrate that the political ecology perspective should also be applied to the industrialised nations. This would allow the specific forms of regulating relations to nature encountered in the rich capitalist nations of the North to be analysed with regard to the inevitable links between both regions of the world in the global division of labour context.

ing, and frequently also racist and violent, appropriation of 'cheap work' and 'cheap nature' (Patel and Moore 2018).[7] The huge expansion in global socio-metabolic processes in the past 150 years identified by social ecologists (Krausmann et al. 2016) is thus a consequence of the capital-driven growth of the global division of labour. The early industrial centres of the world economy have profited through this ecologically unequal exchange both in economic and ecological terms. By relocating some of their extractive and particularly environmentally intensive activities to semi-peripheral and peripheral countries, they have profited economically because the labour costs in these countries are lower and ecologically because the decisive environmental costs of the goods produced for export are not exported with them (Jorgenson and Givens 2014; Lessenich 2019). The costs of unsustainable work are thereby externalised in a large part to distant lands, jobs, and natures. Conversely, numerous 'good' jobs in industry that have been secured by strong trade unions are reliant on previous work that is both socially and environmentally unsustainable.

2. But precisely these jobs and the workers represented by the industrial trade unions continue to be also characterised by a high degree of gender inequality (Ledwith 2012; Barca and Leonardi 2018). Feminism and eco-feminism experts have already demonstrated that this is not a coincidence but rather an underlying characteristic of the capitalist mode of production.[8] The crucial argument for our perspective on the sustainability of work lies in the fact that the unsustainable capitalist relationships between labour and nature discussed above are also based on the unpaid exploitation of care work, which is predominantly a female domain (Fraser 2016). The long hierarchical tradition of attaching less value to 'women's' reproductive work than to 'men's' paid productive work facilitates the economic exploitation of women and lowers the value of labour as a commodity, that is, the reproduction costs of workers. The eco-feminist Ariel Salleh sees this situation as a capitalist reproduction crisis that results from a nexus of 'nature-woman-labour', that is, the gender-specific exploitation of work and labour with all its ecologically destructive consequences (Salleh 1995). Only an expanded concept of labour that goes beyond the classic productivism perspective recognises such links and can also illuminate the con-

---

[7] Patel and Moore (2018) distinguish a total of seven related 'cheap things' that facilitate the global, capitalist exploitation: nature, money, work, care, food, energy, and lives.
[8] While a discussion of the long-standing and differentiated feminist critique and development of Marxist concepts would exceed the scope of this chapter, further details on this topic can be found, for example, in Salleh (1997), Saave/Muraca, Chap. 32, Räthzel, Chap. 34.

nections between formal paid labour practices and their informal counterparts in the private sphere, that is, everyday practices.

3. A practice theory extension to the perspective laid out so far allows us to see two things: *first*, unsustainable economic structures are only reproduced through the everyday actions of the members of a society, and *second*, the labour force must be viewed thereby not only in their capacity as wage-earners but also in their reproductive practices of everyday life. Everyday practices are attributed high relevance from an SRN perspective. It is vital hereby to connect the crucial recognition of the 'rootedness of unsustainability in everyday practices' (Görg et al. 2017, 5) with the function and context of production and labour, that is, also with the global exploitation of labour and nature. As Brand and Wissen (2018, 28) note 'the regulation of societal nature relations takes place via institutions, norms, values, processes of subjectivation and normalized practices that often bring to the fore new strategies of capital valorization'.

Unsustainable SRNs are thus the result of the complex interaction of institutions and political and regulatory structures, which create and reproduce powerful interests in unsustainability, and long-established everyday practices that correspond to actual (e.g. infrastructural) reality (Kramm et al. 2017a; Görg et al. 2017; Fuchs et al. 2016). This is consistent with practice theory concepts, which see, for example, a collective political process of systematically depriviliging the automobility system as a condition for establishing a sustainable mobility practice (see, for instance, Watson 2013). In contrast, the often-stated appeal to people to act more sustainable as individuals seems to be a powerful individualisation and depoliticisation of a political question that should in fact be answered collectively, namely by a collective decision about what should be produced, why, and how. Otherwise, the nexus of everyday practices-labour-production—or the connection between the mode of production and the mode of living (as Brand and Wissen (2018) refer to it) recedes from view.

We can thus demonstrate two central connections between everyday practices, including work and unsustainable SRNs: *first*, paid labour is a decisive determinant for the everyday practice of the monetarised exchange of commodities, which 'is internalised into people's way of life and, above all, stabilised by hegemonic bundles of practices of working, life and consumption' (Jonas 2017, 123f.). *Second*, the example of automobility makes apparent the close bond between dominant practices (of automotive transport) and the complex web of infrastructures, space/time relationships, laws, planning, and so on that actually facilitate it (ibid., 120f.) and to which powerful valorisation interests and jobs are likewise bound. In this case, the connection between unsustainable everyday practices and labour implies that concrete jobs are often tailored to unsustainable practices. On the one hand, these jobs secure the reproduction of wage-earners while, on the other hand, produce the necessary infrastructures and means for unsustainable practices.

In this section, we have presented an approach to analysing the relationships between society and nature, emphasising the particular relevance of labour. In the next section, we will focus on how the topic of labour is addressed in the current discourse on sustainability and what contribution the SRN concept can make to this debate.

## Sustainable and Unsustainable Work

Both discourse in the social sciences and sustainability policies have hitherto focused primarily on sustainable production and consumption; the topic of labour has only played a marginal role.[9] Consequently, the main target groups for sustainability policies have been companies and consumers, whereas governments should in any case establish the necessary framework. Yet the approaches which have dominated to date have proved less than successful. The agenda for ostensibly sustainable consumption has lacked, for instance, the 'actual need for unmitigated reduction' (Fuchs et al. 2016) and has thus led more to the growth of a pseudo-sustainable consumer goods sector and the fuelling of increased individualisation and distinction (Brunner et al. 2020). The associated depoliticisation of production, which also applies to companies that produce sustainably on the basis of voluntary agreements, tends to enshrine existing power structures. It is only very recently that the latter have been called into question. The actual labour force, in contrast, seems to have few opportunities to exert any influence. Workers are often still seen as objects to be protected in a transformation towards sustainability rather than as active drivers of socio-ecological change (Barth et al. 2019).

As shown in the previous section, this neglects the fact that the sphere of labour is central to the sustainability debate: the way companies produce goods and services determines also the way nature is appropriated through work processes. Furthermore, people don't just work, be it on a paid, unpaid, formal, or informal basis, purely to satisfy their needs and make a living. Employment, and above all paid employment and especially the undesired

---

[9] This is revealed, for instance, in the topics and indices found in relevant environmental sociology handbooks (see, for instance, Redclift and Woodgate 2010; Dunlap and Brulle 2015). Aside from a few limited exceptions (see footnote 6), only eco-Marxist and eco-feminist works take a different path.

lack thereof (unemployment), is still a central factor in determining a person's status in the social hierarchy and their subjective experience of the life world.

While there have been periodic efforts by researchers and politicians in recent decades to study the links between sustainability and labour in Germany, these have proved to be the exception, not the rule (Warsewa 2016; Krüger 2002). However, as it has gradually become clearer that the changes needed to bring about a 'sustainability transformation' will have a significant impact on social inequality, the topic of labour as a socio-ecological issue has returned to the agenda. A good example in Germany, and to some extent also in Austria, are the structural and associated changes caused by decarbonisation in both the energy production sector as well as the automobile industry and its supply chain. (See also Galgoczi in this volume.) The primary concerns here are the threat of job losses, the lack of compensation, and the uneven distribution of the costs, that is, the social sustainability of the changes as a whole. These key strategic issues, which reflect the principles of the *Just Transition* framework (Morena et al. 2020), have also been acknowledged by trade unions like IG Metall (2019a) in Germany. Indeed, international trade unions already began linking the topic of jobs to that of climate change in the early twenty-first century, reinstating it on their agenda under the keywords *green jobs, green economy*, and the aforementioned *just transition* (Räthzel and Uzzell 2013; Felli 2014).

We, however, would like to consider a different approach to the links between labour and sustainability—one that stems from the current supranational sustainability discourse. Our starting point is the notion of *sustainable work* used in the 2015 United Nations Development Programme report 'Work for Human Development' (UNDP 2015). While green growth, green economy, and green jobs ultimately also form the background in this concept, the report uses a more comprehensive definition of sustainable work as 'work that promotes human development while reducing or eliminating negative externalities that can be experienced over different geographic and time scales. It is not only critical for sustaining the planet, but also for ensuring work for future generations' (ibid., 37). It makes work the focus of sustainability policies and thus goes further than the UN's Sustainable Development Goals (SDGs), although these do also refer specifically to work in Goal 8 ('Decent Work and Economic Growth'). However, sustainable work systematically links the threat to the natural environment to the sustainability of the society of work, ties environmental targets with their development policy counterparts, and does not restrict the concept of work from the outset to paid work in a growth-dependent economy, as SDG 8 suggests.[10]

The concept of sustainable work shares two core ideas with the SRN approach discussed in this chapter. *First*, it makes labour and workers a central topic and starting point for sustainability policy. *Second*, by considering human development and ecological limits together, it emphasises the combination of social and ecological sustainability: sustainable work is not just decent work *or* environmentally sustainable work, it is a combination of both (see also Ytterstad in this volume).

That being said, the UNDP report does not go beyond the frequently criticised and weak sustainability concept of a *green economy*, that is, a growth-based, full-employment society that cares for the climate and the environment. While it does adopt the expanded perspective in which 'the notion of work is broader and deeper than that of jobs or employment' (ibid., 29) and also includes 'care work, voluntary work and creative expression' (ibid.), it does not relate these definitions either to the structures of capitalist production and the division of labour or to gender relations (Littig 2018). The implicit call in the UNDP report for a concept of work that encompasses 'labour as a "whole"' (Biesecker and Hofmeister 2010, 1707) will thus remain unanswered without an analysis of the causes of the dominance of paid labour. After all, a redefinition of the relationship between paid labour and other spheres of work and activity presupposes a critique of the growth-oriented economy and the private appropriation of profits.

A debate on sustainable work that does not consider the organisation of labour in society thus seems truncated and reflects an understanding of sustainability that Longo et al. (2016) characterise as 'pre-analytic' insofar as it naturalises capitalist societal relations and thus ignores key drivers of unsustainability.

A critical sociological perspective on sustainability and labour clearly reveals that the entire structure of the industrial, capitalist society of work needs to be revised within a transformation process towards sustainability. This applies not least to the relation between paid and unpaid work and the corresponding gender differences, as they are institutionalised in political and legislative regimes, and the global division of labour.

---

[10] The German government's 2018 updated sustainability strategy also uses GDP growth and the employment rate as indicators for Goal 8 and thus remains within the limited parameters of a classic industrial society labour regime (Deutsche Bundesregierung 2018, 55). The ILO (2019) likewise refers explicitly to SDG 8 in its outline of 'decent *and sustainable work*'. It thus also seeks to reach this goal primarily through investments in a green economy, increased competitiveness and growth (ibid., 49): '[h]uman centred growth, [...] decent jobs, gender equality and sustainable development' (ibid., 46) should primarily be achieved through the formalisation of work and therefore through the expansion of paid work.

The hegemonic understanding of sustainable development, which has spread precisely because it does not question aspects of structural unsustainability like the orientation on economic growth or capitalist exploitation of labour and nature, has rightly been the subject of much criticism in the past (Longo et al. 2016). But as Fuchs (2017) argues, this does not rule out the possibility of a 'critical theory concept of sustainability' (ibid., 449). Such a critical understanding is basically oriented towards the goals of social *and* ecological sustainability. It seeks to analyse the socio-structural obstacles to realising these goals and the ideologically legitimised reduction of prevailing sustainability practices to, for instance, actions that support growth and relate to the consumption of goods. In such a theory, the concept of sustainable labour, provided it adheres to a 'strong' definition of sustainability, that is, one that grasps the 'existence of ecological limits and planetary boundaries' as 'cornerstones to analyses of sustainability' (Longo et al. 2016, 5), has the advantage of linking the diagnosis of capitalist social and ecological unsustainability to the emerging discourse on labour and sustainability. However, an extended interdisciplinary research agenda on sustainable work (Jochum et al. 2019) requires a positioning of the concept of sustainable work in social theory as we have outlined above. Only on this basis can we analyse the societal relations which keep unsustainable labour alive.

## Industrial Trade Unions in Germany and Austria: Promoting Sustainable Labour or Sustained Unsustainability?

In this final section, we examine the current approach to socio-ecological transformation adopted by German and Austrian industrial and manufacturing trade unions from the SRN and sustainable labour perspectives. Thereby we orient ourselves on three previously identified aspects of current unsustainable SRNs: (a) the dominant role of economic growth, (b) the class-based organisation of labour and the dominance of paid work, and (c) everyday practices and modes of living as a potential area of trade union policy. We provide corresponding examples relating to trade unions in the fields of energy/climate change and automobility, which focus on a general policy level rather than specific activities.[11] We will, however, begin by outlining the parameters of trade union engagement in the area of discussion.

---

[11] Obviously 'the' trade unions as such do not exist, not least because of the different business sectors, specific national trade union cultures and different logics applied at different levels within the trade union

The *German* metalworkers' union *IG Metall* first began taking an interest in environmental protection back in the early 1970s. Over the years, it has regularly entered into corresponding alliances—not all of which were conflict-free—with various environmental NGOs (Flemming 2018; Krüger 2002). This largest sectoral trade union in the world in terms of membership has a history of relatively progressive environmental policies. Some of its officials have far-reaching ideas. One of its most recent debates centred on the concept of a 'labour ecology', which grasps socio-ecological transformation as an explicit trade union task and links it to ambitious democratic demands for shaping the economy (Urban 2018).

However, as Prinz and Pegels (2018) show in their study of the role of trade unions in the energy transition, trade unions act both as drivers of and as barriers to the renunciation of the fossil fuel regime. Since IG Metall represents workers in both fossil fuel and renewable energy sectors, it is internally heterogeneous, that is, represented in both 'camps'. It therefore lacks a clear position on climate policy and its practical implementation and thus, despite its size, also its influence compared in particular to the smaller IG BCE mining, chemicals, and energy trade union in Germany, which acts more as an impediment to the energy transition (ibid., 216f.).

Since the *Austrian* trade union landscape and its parameters are described at length in other publications (e.g. see Brand and Pawloff 2014; Soder et al. 2018; Brandl et al. 2019), we will restrict ourselves here to a few key relevant points. In contrast to IG Metall, Soder et al. (2018, 530) report that 'Austrian trade unions rarely succeed in formulating independent environmental and climate policies'. Relationships between Austrian trade unions and environmental NGOs have also tended to be far more conflict-ridden than those of their German counterparts. Indeed, only recently new alliances have been formed in Austria between the trade unions, NGOs, and the Federal Chamber of Labour to promote, for instance, new welfare models that go beyond economic growth (ibid., Niedermoser 2017).

The causes of these frequently conflict-ridden relations between labour and environment lie in a 'corporatism at the expense of the environment' that is characteristic of both Austria and Germany (Brand and Pawloff 2014). In other words, a social partnership between the state, capital, and labour that frequently puts the mutual interests of the actors participating in economic growth and their competitiveness on the international stage before environmental concerns. Allan Schnaiberg's (1980) 'treadmill of production' theory traces the dilemmas faced by trade unions like IG Metall back to their

---

apparatus, including this level of detail for each trade union would exceed the scope of this chapter.

structural embeddedness in the modern, growth-oriented economic model. And there can also be no denying that this structural embeddedness ultimately plays a role in maintaining the unequal relationship between countries in the Global North and South—as can be seen within international trade unions (Uzzell and Räthzel 2013).

1. We can thus state with regard to *the first of the three topics mentioned above*, namely whether trade unions can overcome the dogma of economic growth, that some trade unions have partly long since departed from a policy that is oriented solely towards economic growth. IG Metall, for instance, already began focussing on '*qualitative* growth' and quality of life in the early 1970s. However, it is also true that, in the case of Germany and Austria, ecological approaches which see the combination of green growth and green jobs as the solution to the 'jobs vs. environment' dilemma dominate (Brand and Niedermoser 2019; Flemming 2018; Soder et al. 2018). On the whole, it has frequently been stated—and as Niedermoser (2017, 137) concludes for trade unions in Austria—that, given the corporate embeddedness of trade unions, commitments to the 'party line' on climate protection can quickly dwindle when concrete industry developments or local production sites are at stake. This observation brings to light a structural dilemma in trade unions that was pointed out by Offe (1981, 76ff.). Trade unions face this dilemma as a result of the class relations in society. No matter how radical their demands, for example regarding wages, workers always also have to consider the economic well-being of their employers, since it is the success of the latter that ultimately ensures they can pay their workers.

2. This brings us to the *second topic*, that is, the extent to which trade unions question the class-based organisation of labour and the dominance of paid work identified in the second section of this chapter as key driving forces for unsustainable SRNs. The trade unions considered in this chapter are clearly committed to the Paris Climate Protection goals, and IG Metall (2019a, b) calls explicitly for a social-ecological turn that is advocated by the state, government, and manufacturers alike, for example, in the field of electric mobility. In 2016, it also called upon the automobile industry to actively tackle the changes induced by climate policy and to see 'climate protection as an opportunity' (IG Metall 2016). Now, at least before the COVID-19 pandemic, it sees *itself* more as a driving force in this transformation, for example, by advocating that the workforce should actively play a co-determining role in this process. This should, of course, guarantee above all that the jobs threatened by such changes are at least cushioned by

investments in environmentally friendly sectors with promising futures (IG Metall 2019a). But it also indicates that the union is embracing a more active role in shaping sustainable production—one that is more oriented on benefitting society as a whole.

Yet despite their demands for extended co-determination regarding democratic decisions over production and production sites, when it comes to the discussed requirements for sustainable work, trade unions challenge neither the focus on international competitiveness nor the capitalist ownership and power structures per se. Moreover, they barely extend their focus beyond paid work. Efforts by trade unions to politicise the economic parameters, and thus also the relationships between labour and nature, are still marginal, as Hans-Jürgen Urban (2018, 329), member of the IG Metall steering-committee, states.

3. This, in turn, takes us to the *third aspect*, that is, the extent to which trade unions view and act on behalf of their members as subjects outside the sphere of paid work. This is a vital aspect because it relates to the actual interests of workers and how these should be represented. If we apply the concept of 'working class ecology' (Barca and Leonardi 2018) and our considerations of the role of work as an activity that mediates between society and nature, then we can clearly see that a separation of work from its everyday living environment is an artificial one. 'Working-class people are intrinsically ecological subjects' (ibid., 3), whose communities are as dependent on an intact environment as their workers are on income from work that frequently destroys nature. Therefore, trade unions might be able to extend their support base by acknowledging that work, nature, and everyday life are mutually constitutive of each other.

Automobility is an excellent case in practice here, since it relates to a sector of industry that is of central relevance in both Austria and Germany. Not only does the complex 'car system' shape the everyday practice of mobility to a large extent (Urry 2004), the automobile industry is also a sector in which the industrial trade unions are traditionally strong and which provides a relatively high income to many people in both these countries. Flemming (2018, 185) notes that the core union business of protecting jobs is clearly the priority for IG Metall, and little thought is extended to sustainable mobility that does not involve automobiles. This applies all the more the closer the union representatives are to concrete production sites. Although the relevance of public transport has recently been raised significantly as part of a necessary ecological and economic turnaround in mobility (IG Metall 2019a, 747), the automobile industry is focussing above all on electric vehicles. The goal of mass-producing electric vehicles thus dominates over ideas for sustainable mobility beyond personal motor vehicles and is being pursued largely without regard for the socio-ecological disadvantages that this creates in other parts of the world (e.g. lithium extraction). There is still an unresolved tension between the preservation of automobile production structures and jobs and an as yet vague notion of a 'mobility turnaround' for which it remains unclear how far it will have to reach.

According to Brand and Niedermoser (2019, 177), who refer in this regard to Segert (2017), such a 'contested shift' from the trade union demand for 'cars for all' to 'public transport for all' has already occurred in Austria. Segert describes this shift, which is also reflected in the PRO-GE union's call for large-scale promotion of public transport (PRO-GE 2018, 78), as one of the historic dimensions from a trade union perspective (Segert 2017, 64). Yet according to Brand and Niedermoser (2019, 177), the trade unions still see themselves as agents in the sphere of production and affirm the separation of the world of work and 'leisure time'. This leads to them also viewing environmental behaviour as a private matter (i.e. not work-related). While the unions do traditionally have a duty to shape social as well as labour policy, they have been cautious (especially in Austria) about taking a position when it comes to matters relating to the mode of life. They consider such issues at best in debates on the sustainable consumer behaviour of their members (Niedermoser 2017) and do not link them to a specific mode of production.

This applies even though mobility beyond the car would, in fact, be entirely in the interests of wage-earners in their capacities as citizens, fathers, mothers, cyclists, and so on. In short, trade union support for sustainable work would thus also embrace alternative production and everyday practices that would be based on less resource-intensive infrastructures and—as part of the 'foundational economy' (Foundational Economy Collective 2018)—would be accessible to people in all income brackets.

The converging transformation processes in the automotive sector (the climate crisis, the digitalisation of production and consumption, and the growing relevance of the emerging markets) might also be indications of far-reaching change. In Germany and Austria, for instance, new or revived alliances between environmental and climate activists and the labour movement are (re-)emerging (IG Metall et al. 2019; Soder et al. 2018). Despite the still dominating focus on cars, the IG Metall has also partly accepted the socio-ecological transformation as inevitable and is getting on board with demands for greater cooperative decision making in shaping its form (IG Metall 2019b). Moreover, the synchronicity of the transformation processes highlights common interests, for instance, concerning the reduction of working hours, redistribution of labour and gender equity.

## Conclusions

The dialectic understanding of nature and society and the identification of socio-metabolic regimes presented in this chapter, that is, the linking of symbolic and material forms of the appropriation of nature to the social structures in specific societal formations, deliver a key insight: the forms of appropriation of nature encountered in the current 'ecological crisis' are clearly systemic (Kramm et al. 2017a, 4). We argue that the societal form and organisation of labour constitute a crucial point of departure for the analysis of the prevailing unsustainability and for developing a fundamental shift towards socio-ecologically sustainable relationships with nature.

The fact that labour has been hitherto neglected in sustainability policy and research is now being increasingly recognised: the transformation to sustainability can only succeed if the roles of work and labour are included as a central aspect. The sustainable work model adopted by the UN in 2015 (UNDP 2015) is further confirmation of this trend. If we apply the SRN approach, the following picture emerges: the transformation towards sustainable relations between labour and nature requires a radical change in the way modern societies appropriate nature. From a labour perspective, this means, above all, the way in which labour is organised, distributed, and valued in society.

We argue that such renegotiations must extend to the capitalist organisation of labour and the appropriation of surplus value, the global division of labour and gender relations, as well as the dependency of central practices of living, nutrition, mobility, and care on wage earnings. These are all genuine political topics that are shaped and influenced by power structures and conflicts. As our consideration of selected trade unions in Germany and Austria reveals, such organisations in these wealthy nations of the Global North currently rarely support such far-reaching positions. The reason for this lies to a large extent in their interdependency on resource-intensive, growth-oriented, and globally uneven modes of production. Nonetheless, there is much to suggest that the socio-ecological transformation is already evident at least in some sectors of industry and that some trade unions are seeking to actively embrace this challenge. 'Active' trade union engagement means not questioning the fundamental need for a shift towards sustainability but rather emphasising the inherent link between social and ecological issues and controlling democratically what is produced and how it is produced from the point of view of their members (i.e. the workers).

To adapt such a perspective, trade unions would have to reinvent themselves to some extent. Therefore, this can be expected to be an open and conflict-ridden process. Conflicts are to be expected not only when it comes to the owners' interests in maintaining the unsustainable capitalist relationships with nature but also within the trade union organisations themselves. After all, they are dealing with fundamental issues here, namely: which of their members' interests do they need/want to represent, how do they want to do this, and should the trade union mandate extend far beyond the sphere of paid employment.

# References

Barca, Stefania, and Gavin Bridge. 2015. Industrialization and Environmental Change. In *Routledge Handbook of Political Ecology*, Routledge International Handbooks, ed. Thomas Albert Perreault, Gavin Bridge, and James McCarthy, 366–377. Abingdon, Oxon and New York, NY: Routledge.

Barca, Stefania, and Emanuele Leonardi. 2018. 'Working-Class Ecology and Union Politics: A Conceptual Topology.' *Globalizations* 15 (4): 487–503.

Barry, John. 2012. *The Politics of Actually Existing Unsustainability: Human Flourishing in a Climate-Changed, Carbon-Constrained World*. New York: Oxford University Press.

Barth, Thomas, Georg Jochum, and Beate Littig. 2019. Machtanalytische Perspektiven auf (nicht-) nachhaltige Arbeit. *WSI-Mitteilungen* 72 (1): 3–12.

Biesecker, Adelheid, and Sabine Hofmeister. 2010. Focus: (Re)productivity. *Ecological Economics* 69 (8): 1703–1711. https://doi.org/10.1016/j.ecolecon.2010.03.025.

Blühdorn, Ingolfur. 2016. Sustainability—Post-Sustainability—Unsustainability. In *Oxford Handbook of Environmental Political Theory*, ed. Teena Gabrielson, Cheryl Hall, John M. Meyer, and David Schlosberg, 259–273. Oxford, NY: Oxford University Press.

Brand, Ulrich, and Kathrin Niedermoser. 2019. The Role of Trade Unions in Social-Ecological Transformation: Overcoming the Impasse of the Current Growth Model and the Imperial Mode of Living. *Journal of Cleaner Production* 225: 173–180. https://doi.org/10.1016/j.jclepro.2019.03.284.

Brand, Ulrich, and Adam Pawloff. 2014. Selectivities at Work. Climate Concerns in the Midst of Corporatist Interests. The Case of Austria. *Journal of Environmental Protection* 5: 780–795.

Brand, Ulrich, and Markus Wissen. 2018. *The Limits to Capitalist Nature: Theorizing and Overcoming the Imperial Mode of Living*. London and New York: Rowman & Littlefield International.

Brandl, Jana, Beate Littig, and Irina Zielinska. 2019. Urbaner Klimaschutz und Arbeit. Zu den qualitativen und quantitativen Beschäftigungsauswirkungen der Emissionsreduktionziele am Beispiel der Stadt Wien. In *Gute Arbeit und ökologische Innovationen—Perspektiven nachhaltiger Arbeit in Unternehmen und Wertschöpfungsketten*, ed. Guido Becke, 279–296. München: Oekom.

Braun, Bruce, and Noel Castree. 1998. *Remaking Reality: Nature at the Millennium*. New York: Routledge and Chapman & Hall.

Brunner, Karl-Michael, Michael Jonas, and Beate Littig. 2020. Capitalism, Consumerism and Democracy in Contemporary Societies—Towards a Sustainable Future? In *Routledge Handbook of Democracy and Sustainability*, ed. Basil Bornemann, Henrike Knappe, and Patrizia Nanz. Abingdon: Routledge. Forthcoming.

Deutsche Bundesregierung. 2018. *Deutsche Nachhaltigkeitsstrategie*. Aktualisierung 2018. Berlin.

Dörre, Klaus. 2018. Europe, Capitalist Landnahme and the Economic-Ecological Double Crisis: Prospects for a Non-Capitalist Post-Growth Society. In *The Good Life Beyond Growth. New Perspectives*, ed. Hartmut Rosa and Christoph Henning, 241–249. London: Routledge.

Dunlap, Riley E., and Robert J. Brulle, eds. 2015. *Climate Change and Society: Sociological Perspectives*. New York, NY: Oxford University Press.

Felli, Romain. 2014. An Alternative Socio-ecological Strategy? International Trade Unions' Engagement with Climate Change. *Review of International Political Economy* 21 (2): 372–398.

Fischer-Kowalski, Marina, and Willi Haas. 2016. Toward a Socioecological Concept of Human Labor. In *Social Ecology*, ed. Helmut Haberl, Marina Fischer-Kowalski, Fridolin Krausmann, and Verena Winiwarter, 169–196. Cham: Springer International Publishing.

Flemming, Jana. 2018. Jobs kontra Umwelt? Gewerkschaften als Brückenbauer für eine sozial-ökologische Transformation. In *Ökologie der Arbeit: Impulse für einen nachhaltigen Umbau*, ed. Lothar Schröder, Hans-Jürgen Urban, Nadine Müller, Klaus Pickshaus, and Jürgen Reusch, 176–191. Frankfurt: Bund Verlag.

Foundational Economy Collective. 2018. *Foundational Economy: The Infrastructure of Everyday Life*. Manchester: Manchester University Press.

Fraser, Nancy. 2016. Contradictions of Capital and Care. *New Left Review* 100 (July/Aug.): 99–117.

Fuchs, Christian. 2017. Critical Social Theory and Sustainable Development: The Role of Class, Capitalism and Domination in a Dialectical Analysis of Un/Sustainability: Critical Social Theory and Sustainable Development. *Sustainable Development* 25: 443–458.

Fuchs, Doris, Antonietta Di Giulio, Katharina Glaab, Sylvia Lorek, Michael Maniates, Thomas Princen, and Inge Røpke. 2016. Power: The Missing Element in Sustainable Consumption and Absolute Reductions Research and Action.

*Journal of Cleaner Production* 132: 298–307. https://doi.org/10.1016/j.jclepro.2015.02.006.

Görg, Christoph. 2011. Societal Relationships with Nature: A Dialectical Approach to Environmental Politics. In *Critical Ecologies. The Frankfurt School and Contemporary Environmental Crises*, ed. Andrew Biro, 43–72. Toronto: University of Toronto Press.

Görg, Christoph, Ulrich Brand, Helmut Haberl, Diana Hummel, Thomas Jahn, and Stefan Liehr. 2017. Challenges for Social-Ecological Transformations: Contributions from Social and Political Ecology. *Sustainability* 9 (7): 1045. https://doi.org/10.3390/su9071045.

Haraway, Donna. 2008. *When Species Meet, Posthumanities*. Minneapolis: University of Minnesota Press.

Hickel, Jason, and Giorgos Kallis. 2019. Is Green Growth Possible? *New Political Economy* 25 (4): 1–18. https://doi.org/10.1080/13563467.2019.1598964.

IG Metall. 2016. Strengere Abgasnormen können eine Chance sein. November 23. Last modified April 7, 2020. https://www.igmetall.de/autoindustrie-und-klimaschutz-24170.htm

———. 2019a. Beschluss des 24. Gewerkschaftstags der IG Metall. Leitantrag 1 'Aktionsprogramm zur Mobilitäts- und Energiewende'. Frankfurt, pp. 742–751.

———. 2019b. *MANIFEST. Die IG Metall in einer neuen Zeit. Miteinander für morgen—solidarisch und gerecht*. IG Metall Vorstand, Frankfurt.

IG Metall, BUND and Nabu. 2019. *Die Klima- und Mobilitätswende gestalten. Gemeinsame Eckpunkte von IG Metall, NABU und BUND*. July 2019. Berlin and Frankfurt.

ILO, and Global Commission on the Future of Work. 2019. *Work for a Brighter Future*. Genf: International Labour Office.

Jochum, Georg, Thomas Barth, Sebastian Brandl, Ana Cardenas Tomazic, Sabine Hofmeister, Beate Littig, Ingo Matuschek, Stephan, Ulrich, and Günther Warsewa. 2019. *Sustainable Work—The Social-Ecological Transformation of the Working Society*. Position paper of the Working Group 'Sustainable Work'. German Committee Future Earth.

Jonas, Michael. 2017. Transition or Transformation? A Plea for the Praxeological Approach of Radical Socio-ecological Change. In *Praxeological Political Analysis*, ed. Michael Jonas and Beate Littig, 116–133. Abingdon: Routledge.

Jorgenson, Andrew K., and Jennifer E. Givens. 2014. The Emergence of New World-Systems Perspectives on Global Environmental Change. In *Routledge International Handbook of Social and Environmental Change*, ed. Stewart Lockie, David A. Sonnenfeld, and Dana R. Fisher, 31–44. London and New York: Routledge.

Kramm, Johanna, Melanie Pichler, Anke Schaffartzik, and Martin Zimmermann. 2017a. Societal Relations to Nature in Times of Crisis—Social Ecology's Contributions to Interdisciplinary Sustainability Studies. *Sustainability* 9 (7): 1042. https://doi.org/10.3390/su9071042.

———, eds. 2017b. *Social Ecology: State of the Art and Future Prospects*. Sustainability 9 (7). Special Issue. Basel u.a.: MDPI.

Krausmann, Fridolin, Anke Schaffartzik, Andreas Mayer, Nina Eisenmenger, Simone Gingrich, Helmut Haberl, and Marina Fischer-Kowalski. 2016. Long-Term Trends in Global Material and Energy Use. In *Social Ecology*, ed. Helmut Haberl, Marina Fischer-Kowalski, Fridolin Krausmann, and Verena Winiwarter, 199–216. Cham: Springer International Publishing. https://doi.org/10.1007/978-3-319-33326-7_8.

Krüger, Sabine. 2002. *Nachhaltigkeit als Kooperationsimpuls: Sozial-ökologische Bündnisse zwischen NGOs und Gewerkschaften*. Münster: Westfälisches Dampfboot.

Ledwith, Sue. 2012. Gender Politics in Trade Unions. The Representation of Women between Exclusion and Inclusion. *Transfer: European Review of Labour and Research* 18 (2): 185–199.

Lessenich, Stephan. 2019. *Living Well at Other's Expense: The Hidden Costs of Western Prosperity*. Cambridge: Polity Press.

Littig, Beate. 2018. Good Work? Sustainable Work and Sustainable Development: A Critical Gender Perspective from the Global North. *Globalizations* 15 (4): 565–579.

Longo, Stefano, Brett Clark, Thomas Shriver, and Rebecca Clausen. 2016. Sustainability and Environmental Sociology: Putting the Economy in its Place and Moving Toward an Integrative Socio-Ecology. *Sustainability* 8 (5): 437. https://doi.org/10.3390/su8050437.

Malm, Andreas. 2018. *The Progress of This Storm: Nature and Society in a Warming World*. London and New York: Verso.

Marx, Karl (Ben Fowkes, Translator). 1976. *Capital: Volume I*. Penguin Books.

——— (David Fernbach, Translator). 1992. *Capital: Volume III*. 3rd ed. Penguin Classics.

Moore, Jason W. 2015. *Capitalism in the Web of Life: Ecology and the Accumulation of Capital*. London: Verso.

Morena, Eduardo, Dunja Krause, and Dimitris Stevis, eds. 2020. *Just Transitions: Social Justice in the Shift Towards a Low-carbon World*. London: Pluto Press.

Niedermoser, Kathrin. 2017. 'Wenn wir nicht mehr wachsen, wie verteilen wir dann um?': Die Rolle von Gewerkschaften bei der Gestaltung eines sozial-ökologischen Wandels. *Österreichische Zeitschrift für Soziologie* 42: 129–145. https://doi.org/10.1007/s11614-017-0261-y.

Offe, Claus. 1981. Die Institutionalisierung des Verbandseinflusses—eine ordnungspolitische Zwickmühle. In *Verbände und Staat: Vom Pluralismus zum Korporatismus*, ed. Ulrich von Alemann and Rolf G. Heinze, 2nd ed., 72–91. Opladen: Westdt. Verl.

Patel, Raj, and Jason W. Moore. 2018. *A History of the World in Seven Cheap Things: A Guide to Capitalism, Nature, and the Future of the Planet*. Carlton: Black Inc.

Peet, Richard, and Michael Watts, eds. 2004. *Liberation Ecologies: Environment, Development, Social Movements*. 2nd ed. London and New York: Routledge.

Prinz, Lukas, and Anna Pegels. 2018. The Role of Labour Power in Sustainability Transitions: Insights from Comparative Political Economy on Germany's Electricity Transition. *Energy Research & Social Science* 41 (July): 210–219.
PRO-GE. 2018. *Arbeitsprogramm. 3. Gewerkschaftstag der Gewerkschaft PRO-GE.* 6–8 June 2018. Austria Center, Vienna.
Räthzel, Nora, and David L. Uzzell, eds. 2013. *Trade Unions in the Green Economy: Working for the Environment.* New York: Routledge.
Redclift, Michael R., and Graham Woodgate, eds. 2010. *The International Handbook of Environmental Sociology.* Cheltenham, UK and Northampton, MA: Edward Elgar.
Robbins, Paul, John Hintz, and Sarah A. Moore. 2014. *Environment and Society: A Critical Introduction.* 2nd ed. Hoboken, NJ: Wiley Blackwell.
Salleh, Ariel. 1995. Nature, Woman, Labor, Capital: Living the Deepest Contradiction. *Capitalism Nature Socialism* 6: 21–39.
———. 1997. *Ecofeminism as Politics: Nature, Marx and the Postmodern.* London: Zed Books.
Schnaiberg, Allan. 1980. *The Environment. From Surplus to Scarcity.* New York: Oxford University Press.
Segert, Astrid. 2017. Gewerkschaftliche Strategien für nachhaltige Mobilität. In *Gewerkschaften und die Gestaltung einer sozial-ökologischen Gesellschaft*, ed. Ulrich Brand and Kathrin Niedermoser, 59–91. Wien: ÖGB-Verlag.
Smith, Neil. 1984. *Uneven Development: Nature, Capital, and the Production of Space.* New York, NY: Blackwell.
Soder, Michael, Kathrin Niedermoser, and Hendrik Theine. 2018. Beyond Growth: New Alliances for Socio-Ecological Transformation in Austria. *Globalizations* 15 (4): 520–535. https://doi.org/10.1080/14747731.2018.1454680.
Stevis, Dimitris, David L. Uzzell, and Nora Räthzel. 2018. The Labour–Nature Relationship: Varieties of Labour Environmentalism. *Globalizations* 15 (4): 439–453.
UNDP. 2015. *Work for Human Development. Human Development Report 2015.* New York: United Nations Development Programme.
Urban, Hans-Jürgen. 2018. Ökologie der Arbeit. Ein offenes Feld gewerkschaftlicher Politik? In *Ökologie der Arbeit: Impulse für einen nachhaltigen Umbau*, ed. Lothar Schröder, Hans-Jürgen Urban, Nadine Müller, Klaus Pickshaus, and Jürgen Reusch, 329–349. Frankfurt: Bund Verlag.
Urry, John. 2004. The 'System' of Automobility. *Theory, Culture & Society* 21 (4–5): 25–39.
Uzzell, David, and Nora Räthzel. 2013. Local Place and Global Space. Solidarity Across Borders and the Question of the Environment. In *Trade Unions in the Green Economy: Working for the Environment*, ed. Nora Räthzel and David Uzzell, 241–256. New York: Routledge.
Warsewa, Günter. 2016. Vom 'Ende der Arbeitsgesellschaft' zum 'Peak Capitalism'— Ein kurzer Rückblick auf die deutsche Forschungsliteratur zu Arbeit und Umwelt.

In *Nachhaltige Arbeit. Soziologische Beiträge zur Neubestimmung der gesellschaftlichen Naturverhältnisse*, ed. Thomas Barth, Georg Jochum, and Beate Littig, 33–54. Frankfurt and New York: Campus Verlag.

Watson, Matt. 2013. Building Future Systems of Velomobility. In *Sustainable Practices: Social Theory and Climate Change*, ed. Elizabeth Shove and Nicola Spurling, 117–131. Abingdon, Oxon and New York, NY: Routledge.

# 34

# Society–Labour–Nature: How to Think the Relationships?

Nora Räthzel

## Introduction

There is an endless amount of literature devoted to thinking the relationship between society and nature or, more generally, between humans and nature, including almost all disciplines in the social sciences and humanities. Given the centrality of this relationship in the age of the capitalocene (Moore 2016), where capitalist relations of production are on a trajectory to destroy their own foundations, the worker and the soil, a history of these debates would help us to understand how and to what degree the thinking of this relationship has accompanied, supported, or confronted the destructive power of a production system based on enrichment at all costs.

I cannot provide such a history in this chapter, nor do I present a literature review of the most important conceptualisations in the area. Instead, what follows is a reflection on contrasting ways of understanding the society–nature relationships, which I have selected because they serve me as a springboard to suggest a way of connecting these perspectives with notions of the labour–nature relationship and gender relations.

In the second section, I put several ways of thinking the society–nature relationship in dialogue with each other: nature as social, humans and nature as mutually producing each other, capital as producing nature and being

N. Räthzel (✉)
Department of Sociology, Umeå University, Umeå, Sweden
e-mail: nora.rathzel@umu.se

produced by nature, capital as alienating nature and humans. In the third section, I propose that we think of work as a necessary mediator between humans and non-human nature. In the last section, I suggest possibilities of bringing Moore's concept of capital as producing and being produced by nature in a process of exploitation and appropriation together with Haug's concept of gender relations as relations of production.

## The Society–Nature Relationship

### Nature as 'Inescapably Social'

As Noel Castree maintains in his introduction to the book he edited together with Bruce Braun, *Social Nature*, geography has a long tradition of investigating the society–nature interface (Castree and Braun 2001). In their book, they assembled authors who saw nature as 'inescapably social', arguing that at the time of publication this was a new and surging perspective.

There are three dimensions along which the societal character of nature was perceived. A first dimension concerns the way in which nature is 'socially defined and determined', 'often in order to serve specific, and usually dominant, social interests'. This meant, Castree argues, that 'the social and the natural were intertwined in ways that made their separation—in either thought or practice—impossible' (ibid., 3). A second dimension concerns knowledge of nature. 'Knowledge of nature', Castree writes, 'is invariably inflected with the biases of the knower/s' and therefore there can be 'only particular, socially constituted knowledges, in the plural' (ibid., 10). It is important to stress that this is not a relativist statement in the sense that 'we can know nothing', which is often held to be a social constructivist position, but that our knowledge is never complete, it can never reflect the world around us (nor who we are for that matter) in a one-to-one relationship.

A third dimension on which nature is seen as social is the way in which seemingly natural events like storms, droughts, and floods have different effects on different societies and social groups. Those who have the least resources to protect themselves are hit hardest. Or to put it differently, a drought is only a drought for those whose lives depend on working the soil in a way that requires water. An example is the climate crisis, which is created by richer but affects poorer countries much stronger. Thus, technological fixes do not solve the problems because they do not address the deeper issues of societal inequality. In the same vein, relations of power determine the ways in which societies intervene in nature as when hazardous facilities are sited where people have the least resources to contest them. Climate Justice is the concept and practice aiming to address such inequalities.

These ways of understanding the society–nature relationship as necessarily social and determined by social power relations which have material effects, Castree argues, make it necessary to deconstruct given knowledges of nature. That is, 'showing them to be social products arising in particular contexts and serving specific social or ecological ends that ought to be questioned' (ibid., 12). He criticises the ways in which scholars who see nature as not being social speak about society 'impacting on' or 'destroying' environments. Against such wording, critical geographers argue that the social and the natural cannot be physically disentangled. All there is, Erik Swyngedouw, for instance, maintains is 'socionature' (Swyngedouw 1999, 443).

The question then arises: if 'impacting on' nature is not possible because nature and society are one, how can it be said that human actions are a decisive cause for the climate crisis? Castree argues that 'physical opportunities and constraints of nature' (ibid., 12) are relative, that, for instance, the Amazon rainforest will have different physical attributes depending on who wants to use it. This seems almost trivial. Certainly, the rainforest presents itself differently to communities that live within and through it than to those who see the forest as nothing else than a source of timber—to give just two examples. But does that mean that the physical characteristics of nature are 'not fixed' (ibid., 13) or rather that out of a range of physical characteristics, different ones have different meanings for different actors, depending on how they want to use those characteristics and on the social power they have to pursue their interests?

On the dimension of knowledge, Castree maintains, 'We must live with this inability to know nature "as it really is," while still remaining committed to the idea that some knowledges of, and practices on, nature are better or worse than others' (Castree and Braun 2001, 16). What we learn from this brief examination of geographical notions of the society–nature relationship is the difficulty to find a language that accurately represents the complicated dialectics of a relationship between society and nature. Castree's text is an example of this difficulty when he assures us that we cannot talk about humans 'acting on' because nature and society are one, while writing a few pages later that 'some (…) practices *on nature* are better or worse than others' (emphasis added). There is a dilemma between conceptually absorbing nature into society while still maintaining that society (humans) acts towards nature. The following scholars handle this dilemma differently, suggesting new language to overcome it.

## Finding a New Language—A Feminist Approach

From the perspective of a feminist historian of science and technology, Donna Haraway does not shy away from inventing a new language to help us understand that

> what used to be called nature has erupted into ordinary human affairs, and vice versa, in such a way and with such permanence as to change fundamentally means and prospects for going on, including going on at all. (Haraway 2016, 40)

This could be a description not only of the way in which the crisis of the society–nature relationship is changing the lives of all human and non-human species on the globe but also a description of how nature has erupted into Haraway's thinking. In the 1980s, she became famous with her 'Cyborg Manifesto' arguing that we need not fear but celebrate the fusion of human bodies and machines. Much of her following work dealt with the way in which nature was 'invented' by scientists, for example, primatologists, reproducing narratives of gendered and racialised power relations by 'discovering' them in the behaviours of animals. In her more recent work, Haraway has re-directed her gaze to investigate the relationship between humans and non-humans paying attention not only to how humans 'invent' nature but on nature as an active agent as well. Numerous authors have thought of nature (and things in general) as active,[1] many of which Haraway discusses in her book. I discuss only Haraway here as an example of finding a language, which creates a bridge to the concept of work/labour, in which I am interested.

Reflecting on the work of philosophers like Isabelle Stengers and the latest insights into microbiology, she suggests words that describe the process in which humans and non-humans act together to create a common world. One word she suggests to understand this process is 'sympoiesis':

> Sympoiesis is a simple word; it means "making-with." Nothing makes itself; nothing is really autopoietic or self-organizing. (…) That is the radical implication of sympoiesis. Sympoiesis is a word proper to complex, dynamic, responsive, situated, historical systems. It is a word for worlding-with, in company. Sympoiesis enfolds autopoiesis and generatively unfurls and extends it. (ibid., 58)

Although Haraway has used similar words like Swyngedouw's socionature when speaking about natureculture (Haraway 2008), she transcends the idea

---

[1] Authors thinking about nature as active adhere to quite different epistemological schools of thought. To name just a few: Latour (2000), Bloch (1973, Ch. 37), Bennett (2010), Barad (2007), and Stengers (2010).

of nature as being always already social. Sympoiesis, meaning 'working with', thinks of non-human nature as truly active. This overcomes the problem of simply absorbing nature into society which, as I have argued, makes it conceptually impossible of speaking about humans acting on nature, while wanting to differentiate between better and worse practices on nature.

With humans and nature as actors working together, it becomes possible—and necessary—to understand their relationship more concretely by examining the different kinds of agencies that humans (as part of nature) and non-human nature have. Haraway's reference to Isabelle Stengers' description of nature's agency can be useful at this point. Stengers, Haraway quotes from an email she received from her, describes the earth using the term Gaia as

> an intrusive event that undoes thinking as usual. "She is what specifically questions the tales and refrains of modern history. There is only one real mystery at stake here: it is the answer we, meaning those who belong to this history, may be able to create as we face the consequences of what we have provoked." (ibid., 44)

I see the importance of this narrative in the way it describes Gaia as simultaneously provoking an answer from humans while being itself an answer provoked by human activities. While her phrasing emphasises the violence within this unequal relationship, it can also be understood as describing a process in which humans and non-humans are 'necessary to each other's becoming' as Haraway puts it (ibid., 64).

Thinking of the human–non-human relationship as a process, where both are necessary to each other and simultaneously separate from each other would allow us to conceptualise this relation not only as a destructive one, but also as a (potentially) mutually constructive one. We become able to envision a perspective in which transforming nature (which is and will always be necessary for human survival) can become a fruitful interaction, as opposed to an exploitative extraction.

There is, nevertheless, a certain ambiguity between the notion of sympoiesis and the notion of Gaia as an event provoking human answers while its actions are also provoked by humans. Saying that humans and non-humans are necessary to each other's becoming suggests an equality in their interdependence. However, while humans can create an alliance with non-human nature deciding to work responsibly within it, non-human nature has no responsibility and as Haraway explains:

> Gaia does not and could not care about human or other biological beings' intentions or desires or needs, but Gaia puts into question our very existence, we who have provoked its brutal mutation that threatens both human and nonhuman liveable presents and futures. (Haraway 2016, 44)

There is a clear difference between Gaia's actions and the actions of 'us', who have provoked the mutations against which Gaia, (the Earth) reacts. This difference implies that while humans need nature to survive, nature does not need humans. One can argue that once humans people the earth and need to transform it, natures need humans who perform this transformation in alliance with her, tending to nature's needs when they tend to their own. While in this process different natures and humans consistently change each other, I do not think the formulation 'needing each other' is accurate. While humans can change and destroy elements of nature, non-human nature cannot be destroyed by humans and does not need humans to exist.

Haraway's concepts, with which she denotes the symbiotic relationship between human and non-human nature as being necessary for each other's becoming, avoid the problem of completely absorbing nature into society. However, another problem emerges when the activity of humans as work/labour, that is, of consciously transforming nature for the need of humans, is not differentiated from the activity of non-human nature. This is the whole point of post-humanist theorisation (from which Haraway takes a certain distance in this book). However, trying to overcome the notion of humans as the dominant, more valuable species by erasing the difference between humans and other natures flies directly into the face of that honourable endeavour. Erasing the specificity of humans as consciously and deliberately acting species theoretically absolves humans from their responsibility for the way in which they co-construct themselves through nature. When Gaia does not care about intentions or desires, it is because she is neither a caring nor a revengeful system. However, as humans we need to care, we need to practice response-ability (as Haraway writes it, thus emphasising the ability) for our role in the process of sympoiesis, of 'working together' not to become destructive. After all, humans are the only species with the ability of destroying their own natural support system on a global scale.[2]

---

[2] In this short space, it is not possible to do justice to Haraway's book since I am selecting those aspects that suit my argument. So, I call on the reader to turn to the book itself. It has considerably more to say and says it in a much more complex, multidimensional way than I have been able to convey here.

## Finding a New Language—A Marxist Approach

The environmental historian and geographer Jason Moore (2015) emphasises the need for a language able to overcome the century-old division between 'Nature' and 'Society', which, as he maintains, is older than, but central to, capitalism. Descartes' separation of mind and body and his idea of a rational universe, Moore argues, 'can be viewed as both symptomatic of, and contributing to, the seventeenth century's massive reorganization of power, capital, and nature' (ibid., Introduction).

Like the geographers presented above, Moore criticises conceptions seeing Society (or Capitalism) 'acting on' nature because they fail to understand that 'the species-specificity of humans is already co-produced within the web of life' (ibid.). Here he is close to Haraway, while his language is different. Moore talks of the 'web of life', the 'flow of flows in which the rest of nature is always moving through us', 'life making (…) that views the boundaries of the organic and inorganic as ever-shifting', about 'world within worlds' and 'webs within webs' (ibid., introduction) and it becomes clear to the reader that finding a way to articulate how human and non-human lives co-produce each other is not an easy undertaking. Reducing his flow of words, Moore suggests one concept with which he wants to express that capitalism needs to be seen as a producer as well as a product of nature. The term he offers is the oikeios. It originates from the Greek, meaning household, and includes humans, non-humans, as well as the way in which they are organised:

> … the oikeios is a relation that includes humans, and one through which human organization evolves, adapts, and transforms. Human organization is at once product and producer of the oikeios: it is the shifting configuration of this relation that merits our attention. (ibid.)

This concept serves Moore to define the difference between his understanding of the society–nature relationship and those of many others who have already claimed that humans are part of nature. Oikeios provides a bridge between such philosophical claims and his historical method, that is, an investigation of the specific ways in which the nature–human relationship is constantly changing historically. What changes he maintains is the ways in which specific aspects of humanity, such as civilisations, '"fit" within nature' (introduction). It is indeed his historical method that differentiates Moore's work from other attempts to make sense of the society–nature relationship and makes it, in my view, especially insightful. It allows him to pose new questions:

Instead of asking what capitalism does to nature, we may begin to ask how nature works for capitalism? If the former question implies separation, the latter implicates unification: capitalism-in-nature/nature-in-capitalism. (introduction)

These inverting processes constitute what Moore calls the 'double internality', that is, nature and capitalism (the specification of the transhistorical notion of society) as always already inside each other.

In spite of a language which at times seems to reiterate the equation of 'human and extra-human work' (chap. 2) problematised above, Moore does mark the difference between capitalism-in-nature and nature-in-capitalism when he explains the double internality:

> ... the first is capitalism's internalization of planetary life and processes, through which new life activity is continually brought into the orbit of capital and capitalist power. The second is the biosphere's internalization of capitalism, through which *human-initiated projects* and processes influence and shape the web of life. (ibid., chap. 2, emphasis author)

This is a crucial recognition of the 'uneven process' of the double internality: while capitalism internalises planetary life by changing it, the biosphere's internalisation of capitalism is an effect of *human-initiated projects*. Similar differences are made in other parts of the book. In the co-constitution of capitalism and the biosphere, it is capitalism that is responsible and needs to be taken to account for the ways in which it has transformed nature. The challenge for understanding the nature–society co-construction is to stick with the simultaneity *and* the inequality of the processes. While capitalism is the deciding, intentional part of the processes, at the end of the day, nature is the more durable part: Capitalism may transform and thereby destroy parts of nature as we know it. However, capitalism may destroy the natural basis of the survival of humans (as Moore demonstrates in his historical account) but non-human nature will live on.

## The Question of Alienation

While there are and have been many authors connecting Marx and ecology,[3] John Bellamy Foster has become the most prominent of them, especially his concept of the 'metabolic rift' (Foster 2000), which he has deduced from Marx's analysis of the dialectical conflict created by capitalist industrialisation:

---

[3] See, for example, O'Connor (1994), Harvey (1996), Burkett (1999), Salleh (2017), and Saitō (2017).

it concentrates populations in big centres and while this enables the capability of societies to develop, it 'disturbs' the metabolism between humans and earth. The elements that have been used in producing food and clothes cannot be returned to the soil and this diminishes soil fertility (Marx 1998, XV–725).

What I want to discuss here is an element of the debate between Moore and Bellamy Foster, because it allows me to discuss the relationship between Marx's concept of alienation and Moore's concept of the oikeios.

In chap. 3 of his book, Moore (2015) delivers a critique of the 'metabolic rift' stating that Foster and those who picked up the concept are guilty of a dualistic separation of nature and society. In turn, Foster has claimed that Moore has now left the realm of Marxism and has moved 'to the other side', to capital. The conflict centres on whether one needs to overcome a dualistic conception of society and nature or rather, as Foster argues, capitalism needs to be understood as a mode of production, which violently separates humans from nature (Bellamy Foster 2016). He explains that Marx did not only use the concept of alienation to analyse how capitalism separates workers from the means of production and their product but also humans from nature (Foster 1999, 2000). While Moore wants to replace the concept of metabolic rift with metabolic shift to emphasise the way in which nature, class, and capital are united, that does not mean that he denies the alienation of humans from nature and of workers from production. In a review of Paul Burkett's book, Foster defined the alienation of workers from nature and production as 'two sides of a single contradiction' (Foster 2000). Moore interprets Foster's formulations as a possibility to see the history of capitalism as a process in which nature is not only a 'consequence but constitutive'. He claims: 'Foster's enduring contribution, then, was to suggest how we might read Marx to join capital, class, and metabolism as an organic whole. From this perspective, all social relations are spatial relations and relations within the web of life' (chap. 3).

How can we understand the alienation of workers from nature and production as a process in which 'capital, class, and metabolism' are joined as an 'organic whole'? To understand this, it is crucial to reflect a moment on the words separation and alienation. While their meanings overlap, they also differ. According to an etymological dictionary, alienation derives from 'transfer of ownership, the action of estranging' (Etymological Dictionary Online). This meaning reflects what Marx analyses in Capital as the condition of the production of surplus value—in present economic wording, the production of profit. Marx writes in the English version:

> The separation of labour from its product, of subjective labour-power from the objective conditions of labour, was therefore the real foundation in fact, and the starting-point of capitalist production. (Marx 1998, XXIII–818)

But what does that mean? Obviously not that labour and the conditions of labour (which according to Marx include the earth as the condition and source of all raw materials, tools, and machines) are spatially separated or cut off from each other. It means that workers do not own those conditions, therefore can control neither the working process in which they are engaged nor the trajectory of the products they produced. In the same paragraph, Marx clarifies:

> Since, before entering on the process, his own labour has already been alienated from himself by the sale of his labour-power, has been appropriated by the capitalist and incorporated with capital, it must, during the process, be realised in a product that does not belong to him. (Marx 1998, XXIII–819)

The meaning of alienation is here the loss of ownership of the means of production and thus of control on the side of the worker. This is what allows capital to appropriate the workers' product. This is also why workers come out of the production process as they went in: without anything else to sell but their labour-power. Is it therefore not fruitful to understand alienation (i.e. the creation of a working class owning nothing than its labour-power) as a historical process that unifies capital, class, and metabolism, into an 'organic whole' as Moore argues? While the notion 'organic whole' might connote harmony, in this case it must be understood as describing a process in which all elements are actively involved, albeit including violent alienation and unsolvable contradictions

The harshness of the conflict between Foster and Moore is not comprehensible for an outsider. It would also be possible to understand both positions as complementary, resulting from different points of departure. Moore is interested in analysing the historical process of the capitalism–nature relationship, insisting on the simultaneous but unequal agency of capitalism and nature and the ways in which they co-construct each other. Foster puts the emphasis on analysing how capitalism destroys humans and nature alike. If one reads the book co-authored by Moore and Patel (Patel and Moore 2018), it becomes obvious that this destructive character of capitalism is at the centre of their story. Understanding the way in which capitalism is a product *and* a producer as much as the biosphere allows them to show (a) how capitalism produces the natures it needs to work for it and (b) how these produced natures shape and limit capitalist accumulation and production. In the next section, I interpret this further. Moore's historical method is useful for environmental labour studies because it provides a theoretical lens to connect, in an overarching theoretical framework, two areas of work that are not only separated spatially and socially today but also studied separately in labour and feminist studies. It is therefore to the subject of work to which I turn in the next section.

## What Is Work?

One common feature of the three approaches to understand the society–nature relationship presented above is that work[4] does not feature centrally in their conceptualisations. Surely it appears in their texts, but it does not contribute meaningfully to their discussion of the society–nature relationship. I begin with Marx's well-known definition of work as

> … in the first place, a process in which both, human beings[5] and nature participate, and in which human beings of their own accord start, regulate, and control the metabolism (Stoffwechsel) between themselves and nature. Humans face natural matter as a force of nature. They set in motion the natural forces of their own bodies, arms and legs, head and hands, in order to appropriate natural matter in a form suitable to their own lives. By thus acting on the external world through their movements and changing it, humans simultaneously change their own nature. They develop their own slumbering capacities and subject their forces to their sway. (Marx 1998, chap. 5, own translation)

This process is what Marx calls an eternal necessity for humans to survive. No matter the societal mode of production within which work takes place at different times, in different places, and within different relations of power and domination, work will always be a process in which humans transform nature

---

[4] Work and labour have different meanings varying historically, according to cultures, and to the disciplines and scholars using them (see our comments in the introduction). I use the words mostly interchangeably here. I prefer work when I want to signify specifically the process through which workers transform nature into products (into which I include services, transport, and even human beings as the section will show). The English translation of Marx uses the term labour, but since there is only one word in German, one might as well translate *Arbeit* with work.

[5] The English translation distorts the German original, using the word 'man' where Marx speaks of 'Mensch'. In German 'Mann' (man) and 'Mensch' (human being) are two different words and the latter is male only grammatically. I have therefore changed the English translation also to remain closer to the German original in terms of the relationship between human work and nature. The English might sound clumsy.

to meet their needs of survival.[6] His formulations have motivated authors to interpret this passage as a description of humanity's domination of nature. One can see it that way. But since nature is at the same time defined as the opposite of *and part of* 'human's own forces' and since both, internal and external nature, transform each other in the process, this is a definition of work in which both, humans and nature play an active part. However, an important difference between Marx's and Haraway's notions of human and non-human nature 'working together' remains: while human and non-human nature both change their own nature in the process, for Marx it is human beings who 'start, regulate, and control' the process. He then goes on to introduce the means of the work process, earth and water being there first without the work of humans. They turn into 'natural resources' once they have been transformed (e.g. excavated) to serve in the work process. And finally, tools are produced which humans put between themselves and the nature they want to transform to improve the process.

If we understand the work process in this general way, it is possible to understand the work of, say, metal workers as having common features with, say, the work of subsistence farmers. The difference between their work is not that they are transforming their external and internal nature but that one is working under conditions of alienation as explained above, while the other can control their work process and their product.

Work as the mediator of nature and human needs and, consequently, work and working conditions as the mediators of nature and society could serve as a conceptual start for a dialogue between research on environmental justice, which investigates environmental conflicts predominantly in the Global South and 'working-class environmentalism' or 'labour environmentalism', investigating environmental policies and protests of predominantly trade unions in the Global North. I think that such a dialogue is necessary since workers in these areas are already closely connected through the system of globalised capitalist production.

Broad definitions are useful because they are inclusive and allow us to analyse connections and create dialogues between different areas of investigation and political practices. On the other hand, broad definitions also risk erasing processes that might be included theoretically but are excluded as a result of the relations of domination-structuring societies. The erasure of vast areas of work is an issue that has been and is investigated, criticised, and denounced by feminist scholars at least since the 1970s (Fox 1980; Federici 1975; Dalla

---

[6] This does not mean that one cannot think of work processes which are superfluous, only destructive and not serving any human need. But such a judgement would be of a historically concrete process.

Costa and James 1975). One of the first critiques was launched against Marx because, while one could argue that his general definition of work includes unpaid work performed predominantly by women at home, his empirical work does not account for it, nor does his central theory, the theory of exchange value, since this only includes paid work.

Since then, feminist scholars have investigated and proved the significance and value of unpaid work performed predominantly by women (76.2% of it worldwide, according to the ILO International Labour Office 2018)[7] in what has been conceptualised as the private sphere (Waring 1990, Hirway 2017). Ecofeminists have broadened the notion of unpaid work by including the work of colonised peoples (Mies 2001, Oksala 2018). To connect the different kinds of unpaid work not integrated into the capitalist production system, Ariel Salleh (2017) has used the notion of meta-industrial workers, into which she includes subsistence workers, who work the earth with care. More recently, the feminist investigation of work has broadened into a research about care work more generally, whether paid or unpaid. The crisis of care, feminist scholars have argued, demonstrates the impossibility of capitalist profit-oriented production to provide for the basic needs of people to secure 'social reproduction' (Luxton and Bezanson 2006, Fraser 2016).

In the next section, I want to suggest that paid and unpaid care work, the subject of feminist research, and waged work, the subject of labour studies and environmental labour studies can be understood not only as connected but also as being different realisations of the general process of work. Like any other work, care work transforms nature (human and non-human nature) to meet the needs of human life.

## Bringing Schools of Thought Together

Marx's transhistorical concept of work can in principle be used to understand the basis of any process by which humans transform non-human nature to adapt it to their needs. This includes processes, which we might not define as work, like writing a piece of music or a poem, taking care of oneself or of other human beings. All these processes include a transformative relation

---

[7] 'Estimates based on time-use data (…) for 64 countries with time-use data on both paid work and unpaid care work, representing 66.9 per cent of world's working age population, show that time spent in unpaid care work (own-use provision of services) accounted for 16.4 billion hours per day, with women contributing more than three fourths (76.2 per cent) of the total (…). This is equivalent to 2.0 billion people working on a full-time basis (i.e. 40 hours per week) without pay' (International Labour Office 2018, 43).

between humans and non-human nature. Firstly, the musician, in order to exist and have the force to compose must have eaten, be clothed, be housed, and so on. Secondly, the means used to write down a piece of music will not have been produced by the musician. However, without them a musician might be able to sing or whistle their composition, but not to write it down, much less have it performed. The same can be said of the process of care: it is undertaken by people who themselves require nature-transforming work to exist and the process itself requires materials, tools, machines, all the result of transformed nature, while process itself transforms nature, including human nature. This is why, when we develop a perspective of environmentally and socially sustainable production, every single production process (which is work including materials, tools, and to date almost always machines) needs to be scrutinised to examine how it transforms nature and that includes its value chain.

However, a lot more theory and analyses are needed to understand work processes within specific societal relations, what Marx called the *relations of production*. In what follows I want to suggest that connecting the thoughts of Moore and the Marxist Feminist scholar Frigga Haug could create a framework that serves to better pay attention not only to the differences and contradictions between different kinds of work, paid or unpaid, but also to their similarities and interdependencies within a global system of various kinds of capitalist relations of production.[8] To link these different strands of thought that are either oblivious of or even adverse to each other makes sense to me as an effort to understand the interdependencies of different kinds of work, which are usually investigated independently of each other: the production of life (including care in the broadest sense), and the production of the means of life (including food, clothing, housing, transport, services, and the tools and machines needed to produce these). Within the latter we find yet another division, namely subsistence forms of production and industrialised forms of production. With the latter I mean mass production on a high level of mechanisation and automation and based on waged labour. These two forms of production occur in all sectors of the production of the means of life and increasingly in the area of the production of life.

Moore seeks to synthesise 'the core insights of Marxism and environmental historiography' (2015, chap. 1). The result is a theory of the accumulation of capital that interweaves Marx's theory of the exploitation of waged labour

---

[8] To remain within the limits of this chapter and keep it simple, I cannot include the broad literature of feminist and other scholars on Marx's concept of value and capital accumulation or the broad literature on gender relations and work.

# 34 Society–Labour–Nature: How to Think the Relationships?

with the appropriation of unpaid labour and nature outside of the commodification process for which he draws on feminist and ecofeminist work. Frigga Haug synthesises Marx's concept of the relations of production with feminist theories of gendered power relations, suggesting that we should think of gender relations as relations of production.

Moore argues that

> ...value as abstract labor cannot be produced except through unpaid work/energy. (…) The "commodification of everything" can only be sustained through incessant revolutionizing—yes, of the forces of production, but also of the relations of reproduction. The relations of reproduction cut across the paid/unpaid work and human/extra-human boundaries. In this, the historical condition for socially necessary labor-time is socially necessary unpaid work. (ibid., chap. 2)

In other words, the value form, the representation of the socially necessary time needed to produce a commodity (= abstract labour) does not include *all* the work needed for the production of this commodity, namely the work that does not form part of the process of commodification and is therefore not included in the exchange value—like the unpaid work of humans (women and racialised workers) and non-human nature. This means that the capitalist relations of production (alienating workers from the conditions of their work) subsume unpaid human work and the unpaid activity of non-human nature.

Feminist scholars have shown that the cost of this work is 'externalised' by capital. Moore conceptualises this process differently, arguing that this externalisation is part and parcel of the *internal* process of capital accumulation: exploitation of waged labour and appropriation of non-waged labour are two parts of the same process. Within this understanding, the construction of the category of women, indigenous peoples, and people of colour as 'nature' (critically analysed by feminist scholars; for more detail, see Saave and Muraca, Ch.29) can be understood as part of the process of *appropriation*: since nature can be appropriated for free, everything that is appropriated for free needs to be constructed as nature. Thus, it is possible to understand Moore's claim that capital produces nature and nature produces capital, both being producers and products in the following way: capital as a system (making use of the economic and ideological systems of patriarchy, racism, and colonialism) produces women, colonised and racialised peoples *as part of nature*. Naturalising these categories of humans and lumping them together with extra-human nature as objects of appropriation for free, these so produced natures produce capital and its profit.[9] If we follow this argumentation it would mean that appropriation of unpaid 'work' (of humans and non-human nature) is an essential part of the creation of profit.[10] As I have shown above, the sheer value of unpaid work would support this argument as well as what has been discussed by feminists as the 'crisis of care'.

This resonates with Marxist feminist scholar Frigga Haug's concept of gender relations as relations of production (Haug 2015; Haug 2005).[11] After critically analysing the lack of gender relations within Marx's theory of the exploitation of labour, she does find some 'flashes of inspiration' (Haug 2005, 284) in Marx, which are close to Moore's idea of the double process of exploitation and appropriation:

> However, it still remains true that to replace them [the commodities functioning as elements of capitalist industrial production] they must be reproduced, and to this extent the capitalist mode of production is conditional on modes of production lying outside of its own stage of development. But it is the tendency of the capitalist mode of production to transform all production as much as possible into commodity production. (MECW 36, 108, quoted in Haug 2005, 284)

The examples Marx gives for work existing outside of the capitalist mode of production are the work of slaves, serfs, or communities. The unpaid work of women does not feature. However, in another 'flash of inspiration' he differentiated between:

> On the one hand, the production of the means of life, of the objects of food, clothing and shelter and the tools necessary for that production; on the other hand, the production of human beings themselves, the propagation of the species. (MEW 21, 27, quoted in Haug 2005, 287)

These points are taken up by Haug to develop her theory of gender relations as relations of production. Firstly, it needs to be stressed that both the production of the means of life and the production of life itself are two dimensions of a process of production that is the basis for human survival. Every production process includes the reproduction of resources, tools, machines,

---

[9] There is no space here to discuss what Moore sees as the consequences of this process, namely the 'frontiers' of capital accumulation, when cheap labour and cheap natures are no longer available.
[10] Moore's concepts differ from Harvey's notion of accumulation through dispossession, which the latter sees as a new form of accumulation and imperialism as a result of over-accumulation (Harvey 2003).
[11] With a similar intention, authors write about reproduction (Biesecker and Hofmeister 2010).

and workers. Every process defined as social reproduction, caring for other humans in the broadest sense, are forms of production, the production of healthier, happier, revitalised, better-educated human beings. The process of care will reproduce their capabilities, produce new ones, and also produce humans, who are more or less different than the ones they were before. As mentioned before, both production processes include the transformation of nature. The majority of (not only) feminist research using the term reproduction for what Marx calls (paraphrasing Morgan), the production of life itself runs the danger of splitting off care work from other kinds of work. By emphasising that care work, the production of life and the production of the means of life are both different but intertwined dimensions of production *and* reproduction, it is easier to see that both these dimensions are dependent on each other as well as taking place in similar relations of production.

Neither is external to the other, using Moore's concept, there is a double internality at work here. Both dimensions are internal to the capitalist relations of production. In addition, both produce each other. The production of life produces the workers necessary for the production of the means of life, while being dependent on these means for its own process. Whatever happens in one dimension has repercussions in the other. One could say they constitute two communicating vessels. That care work, whether paid or unpaid is not external to capitalist reproduction and exploitation can be captured with Moore's concept of capital accumulation as exploitation and appropriation. If we merge this with the idea that gender relations are relations of production, that the work of producing human life is appropriated by capital for free like non-human nature and necessary for its own reproduction and the creation of profit, we can also understand that it is not a realm of peace and loving care. Yes, it is this as well, but the logic of caring is subsumed under the logic of exploitation.

This creates specific contradictions, since the time needed for care and the emotional energy needed to care for others (not only children, the sick, and the old, as it is often described, but other people in general) subsumed under the logic of efficiency, the need for success on the labour market, constantly distorts the societal relations of the production of life. In fact, these conflicts are quite similar to the conflicts in the dimension of the production of the means of life. There as well, workers who develop a producer's pride (Sennett 2008; Räthzel et al. 2014), wanting to do their job well for its own sake, work with care and run up against the logics of profit, which forces them to use as little time as possible, since 'time is money'.[12] Similarly, parents, who desperately want their children to succeed, aim, if they can afford it, for the best schools, in the best urban areas with no 'immigrants' in them. Not to speak of commodified care-work, in which, for instance, carers in homes for the elderly are subordinated to time frames by the minute to do their demanding jobs (e.g. 15 minutes for taking a person out of bed, washing, and clothing them).

Thinking of gender relations as relations of production and inserting them, with Moore's help, into the process of capital accumulation as exploiting labour and appropriating unpaid labour and the workings of nature as 'unvalued value' ('In other words: value does not work unless most work is not valued', Moore, chap. 2) helps us to see the interdependencies between the environmentalism of the poor, resisting appropriation of their labour and lands, often led by women (Martinez-Alier et al. 2016), and labour environmentalism of workers and unions in industries, resisting the exploitation of their labour and threats to their health.

To be sure, seeing the interdependencies between working people's subsumption under exploitation and appropriation does not mean becoming blind to their different situations and their different situated knowledges, which lead to different forms of resistance against environmental degradation—or the lack of it. For instance, the perception of the labour–nature relationship may differ markedly between them (Räthzel and Uzzell 2019). Where workers in agriculture and fishing can develop a holistic concept of the labour–nature relationship, especially if they are part of cultures that defend the rights of 'mother earth', workers in industrial processes need theoretical knowledge to connect their experiences of drought, floods, weather changes to forms of global environmental destruction like the climate crisis, the loss of biodiversity, and so on. What they immediately experience are threats to their health resulting from dangerous production processes they are part of. These different positioning within the nature–labour matrix is one (not the only) explanation for the reluctance of industrialised workers in the Global North and South to join the environmental struggles of food-producing farmers and peasants.

Connecting the conceptual frames of Moore and Haug allows us to resolve lacunae in each of them. Moore's conceptualisation of the oikeios omits the

---

[12] In the more than 40 years that I have been talking to workers in industrial sites and elsewhere, I have been surprised again and again by the love and pride with which they talk about their work. I will never forget the worker in the bus factory of a transnational corporation, who explained to me in detail what he did and how important that was for the safe functioning of the bus. Having next to no technical understanding, I must have looked nonplussed because he then asked me for my notebook and carefully drew up the element that he makes for the bus and where it is placed in it.

## 34 Society–Labour–Nature: How to Think the Relationships?

concrete differences of appropriated unpaid work of producing life and the exploited work of producing the means of life. As Haug (and others, e.g. Adam and Groves 2011) explains, the production of life has another rhythm of time, it is slower, long-term, and when subordinated to the rhythms of profit-oriented efficiency, cannot achieve its goals. Moore's theorisation does not account for those differences. Haug's notion of gender relations as relations of production, in turn, does not account for the way in which capital produces and is the product of human and extra-human nature. Her consideration of the destruction of the natural basis of human life comes as an add-on, not as an integral part of the differentiation between the production of life and the production of the means of life. Her strength is that her focus on gender relations avoids a concept of women as an ontological fixed position.

By merging the two conceptual frameworks we achieve a multidimensional system. The production of the means of life includes the exploitation of waged work and the appropriation of unpaid work of humans and non-humans. Similarly, the production of life consists of exploited waged work and appropriated unpaid work of humans and activities of non-humans. Analysing how the two forms of production constitute each other we can connect research into the production of the means of life with research into the production of life and investigate how the former dominates the latter but also how both these forms of production are subsumed (in different ways) under the capitalist logic of profit to the detriment of workers, non-human nature, and the products themselves. This would allow to investigate how care processes in the private sphere function as a subordinated area of the production of the means of life. It would mean to investigate it as a realm of contradictions, where processes of care are interwoven with destructive processes. Like the production of the means of life, the production of life itself is implicated into the logic of profit to a degree that it risks the survival of humans and non-human species. However, we also need to be careful not to reduce all processes to one logic. Violence against women, for instance, cannot be reduced to the logic of profit—but the way in which the two spheres of production are separated from each other is one of its causes.

If we broaden our perspective towards global processes of the production of the means of life, the ways in which workers and natures are interwoven are much more obvious.

To give just one example as illustration: on 30 September 2020, the German magazine, *Der Spiegel* (Klawitter 2020), reported how cattle owners in Paraguay provide the leather for the car seats of German car producers. The cattle are held on land gained through legal and illegal forest clearing, threatening an area which is home to an indigenous community. The forests in this region disappear more quickly than almost anywhere else. Every two minutes an area of the size of a football field is cleared, mostly for cattle to satisfy the international demand for meat and leather.

Nothing here is new; bookshelves can be filled with such examples. One way to understand this is to analyse how workers in industrialised production (in this case in the Global North) owe their work and their comparatively comfortable way of life to the destruction of natures and other peoples (see Wissen and Brand, chap. 27). This can be rightly seen as a contradiction between workers in the cattle business, car production, and indigenous communities. But it is also possible, using the idea of interconnectivity suggested in merging Moore's and Haug's frameworks, to pose a different set of questions. How could workers, geographically separated and yet existentially connected, transform this connection into a common strength? The trade union movement has experience in actions of solidarity across borders, striking against the treatment of workers by their company in other areas of the world. Indigenous communities have experience in connecting across borders to fight against the dispossession of their lands. What if these experiences could be combined, as opposed to creating contradictions between these different categories of workers, since it is ultimately the same power that exploits and dispossesses them? In the aforementioned example, we do not learn anything about the ways in which the production of life is implicated in burning the forest to create land for cattle, producing the German cars with the leather from Paraguay. What would change in our understanding of such processes if gender relations were always part of the analysis of global production processes and their contradictions? That is, if they would be part of the bigger picture as opposed to be added-on or analysed separately? What kind of new forms of resistance would be possible if workers in all geographical areas and all areas of production could see themselves as allies among themselves and with nature?

# References

Adam, Barbara, and Chris Groves. 2011. Futures Tended: Care and Future-Oriented Responsibility. *Bulletin of Science, Technology & Society* 31 (1): 17–27. https://doi.org/10.1177/0270467610391237.

Barad, Karen Michelle. 2007. *Meeting the Universe Halfway: Quantum Physics and the Entanglement of Matter and Meaning*. Durham: Duke University Press.

Bellamy Foster, John. 2016. In Defense of Ecological Marxism: John Bellamy Foster Responds to a Critic. *Climate and Capitalism* (blog).

https://climateandcapitalism.com/2016/06/06/in-defense-of-ecological-marxism-john-bellamy-foster-responds-to-a-critic/.
Bennett, Jane. 2010. *Vibrant Matter: A Political Ecology of Things*. Durham: Duke University Press.
Biesecker, Adelheid, and Sabine Hofmeister. 2010. Focus: (Re)Productivity. *Ecological Economics* 69 (8): 1703–1711. https://doi.org/10.1016/j.ecolecon.2010.03.025.
Bloch, Ernst. 1973. *Das Prinzip Hoffnung 3*. Frankfurt am Main: Suhrkamp.
Burkett, Paul. 1999. *Marx and Nature: A Red and Green Perspective*. 1st ed. New York: St. Martin's Press.
Castree, N., and B. Braun. 2001. *Social Nature. Theory, Practice and Politics*. London: Wiley-Blackwell.
Dalla Costa, Mariarosa, and Selma James, eds. 1975. *The Power of Women and the Subversion of the Community*. 3rd ed. Bristol: Falling Wall Press Ltd.
Federici, Silvia. 1975. *Wages Against Housework*. Bristol: Falling Wall Pr.
Foster, John Bellamy. 1999. Marx's Theory of Metabolic Rift: Classical Foundations for Environmental Sociology. *American Journal of Sociology* 105 (2): 366–405. https://doi.org/10.1086/210315.
———. 2000. *Marx's Ecology: Materialism and Nature*. New York: Monthly Review Press.
Fox, Bonnie, ed. 1980. *Hidden in the Household: Women's Domestic Labour Under Capitalism*. Toronto: Women's Educational Press.
Fraser, Nancy. 2016. Contradictions of Capital and Care. *New Left Review* 100: 99–117.
Haraway, Donna Jeanne. 2008. *When Species Meet*, Posthumanities 3. Minneapolis: University of Minnesota Press.
———. 2016. *Staying with the Trouble: Making Kin in the Chthulucene*, Experimental Futures: Technological Lives, Scientific Arts, Anthropological Voices. Durham: Duke University Press.
Harvey, David. 1996. *Justice, Nature, and the Geography of Difference*. Cambridge, MA: Blackwell Publishers.
———. 2003. *The New Imperialism*. Oxford and New York: Oxford University Press.
Haug, Frigga. 2005. Gender Relations. *Historical Materialism* 13 (2): 279–302.
———. 2015. Marxistische Refundierung Des Marxismus, Feministische Des Marxismus. In *Wege Des Marxismus-Feminismus*, ed. Das Argument, vol. 314, 517–526. Hamburg: Argument Verlag.
Hirway, Indira. 2017. *Mainstreaming Unpaid Work: Time-Use Data in Developing Policies*. Oxford: Oxford University Press.
International Labour Office. 2018. *Care Work and Care Jobs for the Future of Decent Work*. Geneva: ILO.
Klawitter, Nils. 2020. Abholzung für europäische Autositze. In: *Der Spiegel*, 30.9.2020.

Latour, Bruno. 2000. When Things Strike Back: A Possible Contribution of "Science Studies" to the Social Sciences. *The British Journal of Sociology* 51 (1): 107–123. https://doi.org/10.1111/j.1468-4446.2000.00107.x.

Luxton, Meg, and Kate Bezanson, eds. 2006. *Social Reproduction: Feminist Political Economy Challenges Neo-Liberalism*. Montreal: McGill-Queen's University Press.

Martinez-Alier, Joan, Leah Temper, Daniela Del Bene, and Arnim Scheidel. 2016. Is There a Global Environmental Justice Movement? *The Journal of Peasant Studies* 43 (3): 731–755. https://doi.org/10.1080/03066150.2016.1141198.

Marx, Karl. 1998. *Capital. A Critique of Political Economy*. Vol. 1. 1887th ed. London: ElecBook.

Mies, Maria. 2001. *Patriarchy and Accumulation on a World Scale: Women in the International Division of Labour*. London: Zed Books.

Moore, Jason W. 2015. *Capitalism in the Web of Life: Ecology and the Accumulation of Capital*. 1st ed. New York: Verso.

———. 2016. *Anthropocene or Capitalocene? Nature, History, and the Crisis of Capitalism*. Oakland, CA: PM Press.

O'Connor, Martin, ed. 1994. *Is Capitalism Sustainable?: Political Economy and the Politics of Ecology*, Democracy and Ecology. New York: Guilford Press.

Oksala, Johanna. 2018. Feminism, Capitalism, and Ecology. *Hypatia* 33 (2): 216–234. https://doi.org/10.1111/hypa.12395.

Patel, Raj, and Jason W. Moore. 2018. *A History of the World in Seven Cheap Things: A Guide to Capitalism, Nature, and the Future of the Planet*. London and New York, NY: Verso.

Räthzel, Nora, and David Uzzell. 2019. The Future of Work Defines the Future of Humanity and All Living Species. *International Journal of Labour Research, ILO* 9 (1–2): 145–171.

Räthzel, Nora, Diana Mulinari, and Aina Tollefsen Altamirano. 2014. *Transnational Corporations from the Standpoint of Workers*. Basingstoke: Palgrave Macmillan.

Saitō, Kōhei. 2017. *Karl Marx's Ecosocialism: Capitalism, Nature, and the Unfinished Critique of Political Economy*. New York: Monthly Review Press.

Salleh, Ariel. 2017. *Ecofeminism as Politics: Nature, Marx and the Post Modern*. 2nd ed. London: Zed Books.

Sennett, Richard. 2008. *The Craftsman*. New Haven: Yale University Press.

Stengers, Isabelle. 2010. *Cosmopolitics*, Posthumanities 9–10. Minneapolis: University of Minnesota Press.

Swyngedouw, Erik. 1999. Modernity and Hybridity: Nature, *Regeneracionismo*, and the Production of the Spanish Waterscape, 1890–1930. *Annals of the Association of American Geographers* 89 (3): 443–465. https://doi.org/10.1111/0004-5608.00157.

Waring, Marilyn. 1990. *If Women Counted: A New Feminist Economics*. San Francisco: Harper.

# 35

# Labour-Centred Design for Sustainable and Just Transitions

Damian White

## Introduction

The relationship between design, labour and the many waves of socio-ecological critique that have washed over the affluent world since the late 1960s has often been surprisingly uneven. The Viennese designer and environmentalist Victor Papanek famously opened *Design for the Real World* (1971) by denouncing the whole design industry for what he viewed as a near criminal addiction to producing trivial and toxic goods for an irrational consumer economy that was slowly destroying people and planet. The choice was simple, for Papanek, design must either stop 'defiling the earth itself with poorly designed objects and structures and become "…an innovative, highly creative, cross disciplinary tool responsive to the needs of men", or it should "cease to exist"' (Pananek 1971: x). Some three years later, Harry Braverman's Marxist classic *Labour and Monopoly Capital* (1974) highlighted how technological transformations and designs driven forward by management are invariably used as tools to increase efficiency and control of workers. Papanek was a key figure in the evolution of humanitarian design and ecological architecture, Braverman taught a generation of radical designers (see Cooley 1980; Ehn 1990) to recognize that forms of design that become enrolled in

D. White (✉)
Division of Liberal Arts, Rhode Island School of Design, Providence, RI, USA
e-mail: dwhite01@risd.edu

supporting 'scientific management' have been used to relentlessly expand the powers of capital over labour. These two interventions would seem to suggest fruitful political possibilities for environmental labour studies to think about how design could be systematically critiqued but also possibly reconstructed to serve other purposes. Establishing these connections between labour, design and ecology though has not always been easy to make or sustain.

If we just take a passing glance around our material culture, it's really not too hard to make a comprehensive environmental and labour-focused *critique* of our design economies (Julier 2017) and the design industry more generally (see Boehnert 2018). All forms of mainstream design—whether we consider fashion or the food industry, consumer electronics or digital platforms have vast resource and waste impacts and contribute enormously to pollution, land use change, carbon emissions and biodiversity loss. These impacts and the human and non-human modes of suffering and exploitation at every point across the supply chain are then very carefully and intentionally hidden—*by design and through design* (Ceschin and Gaziulusoy 2020). Human-Computer interaction design (HCI) and user experience design (UX) are not only dependent on energy and carbon intensive digital infrastructures but continually furthering modes of algorithmic governance that ensure labour can be watched, monitored, controlled from morning to night (see Irani and Silberman 2016; Scholz 2016; Benjamin 2019). Despite the potential for thinking about the ways in which design, ecology and racialized, classed and gendered forms of labour intersect, many important currents of design that are focused on questions of sustainability have tended to sidestep these issues. Projects seeking to expand the capacity for design to contribute to economic democracy have often side-lined environmental hazards in the workplace or left unexplored the broad environmental impacts of capitalist industry. Why is this?

This chapter will attempt to construct such a conversation between design and environmental labour studies through a number of moves. Designs for labour and designs for sustainability have not always moved in lockstep and we will explore certain tensions between these currents. I will also argue there are traces and threads of possibility that anticipate the contours of what a labour-friendly eco-design might look like. The proposition that we could redesign the workplace, the home, our material economies and our urban and rural relations to bring about different socio-ecological relations have been at the forefront of some startling schemes and dreams woven by various designers, social theorists and activists across the last 150 years to think the world differently. We will also find in this chapter that if we shift our horizons beyond telling stories of professional designers and the objects they fashion to focus more attention on the vast amount of gendered, racialized, classed and other modes of *invisible labour* (Crain et al. 2016; Allen 2019) involved in the design and maintenance, care and repair of our material culture (see Barca 2012; Dharia 2015; Mattern 2018; Akama and Yee 2019), possibilities may exist for thinking a design politics in more encompassing terms. Finally, we consider how some of these discussions might impact attempts to think about designs for the Green New Deal.

## Theorizing the *Labours of Design*

'Are you an environmentalist or do you work for a living?' is the name of a seminal intervention in environmental labour studies made by the historian Richard White (1995). The essay title drew from a bumper sticker widely circulated in the Pacific Northwest in the late 1980s and early 1990s, amongst loggers and extractive industry workers. The sticker marked the rise of a particular kind of divisive 'hippy versus hard-hat' sloganeering that captured the beginning of a new wave of corporate funded anti-environmental backlash politics. White went on to press an uncomfortable point though. Whilst unfair and unwelcome, he argued, the slogan did touch on some deep tensions in the cultural politics of certain kinds of romantic white environmentalist discourse circulating in the United States circa 1990. Specifically, White suggested that the proposition that 'Nature seems safest when shielded from human labour', prevented many environmentalist currents from fully investigating the relations between work and labour, class and race, history and ecology. Certain kinds of environmental discourse, he argued, had tendencies to downgrade or even erase the knowledge and pleasure gained from productive work through the transformation of nature. Notably, not only did this frame force an unhelpful dualist 'humans versus nature' binary on environmental politics but it re-enforced the dominance of upper class white environmental narratives at the expense of interventions which explored the environmental histories of the multi-racial working-class and indigenous people. This in turn delimited the possibilities of such groups seeing environmental struggles *as their own struggles*.

Since White wrote his essay, there has been a remarkable wave of scholarship in environmental justice studies, environmental history, indigenous studies and political ecology recovering the labour histories of environmental transformations. We have come to see all kinds of landscapes and urbanscapes, from the national parks to the Amazon rain forest, as in some part, social ecologies, that have been co-produced through non-human natures metabolizing with racialized, gendered, classes forms of labour (see Taylor 2009; Barca 2012; Finney 2014; White et al. 2016; Escobar 2017; Moore 2015). Yet acknowledgement of this labour is still largely missing from the core of environmental design studies and design education (see Deamer 2015). The very concept of labour-centred ecological or sustainable design could be interesting here for it suggests a terrain of engagement for disrupting these dualisms and blind spots. But it is also a frame that contains some critical ambiguities which need further attention.

Let us start here with the matter of *design*. Are we speaking here of design as that which is done by people who *professionally identify as designers* and would this include architects, planners and engineers? Or rather, are we trying to capture the activities of these people but also the sweat, toil and craft of a much broader set of designers, makers and creators—from coders to contractors, model makers to building managers—whose *invisible labour* across diverse geographies sustains all kinds of design projects (see Allen 2019).

The question of *labour* invites further reflection on the *kinds of labour* that a labour-centred design should focus on. One approach would be to demarcate our inquiry to the plight of *industrial labour* as has often been the case in some of the most remarkable demands for *worker-orientated design* that have run through all manner of revolutionary, socialist and social democratic designerly projects across the twentieth century; from those advocated by Alexander Bogdanov in revolutionary Russia (Wark 2015) to Pelle Ehn's call for work-orientated designs in Scandinavia (Ehn 1990). A more expansive approach though might include such approaches but also seek to map the diverse forms of material and immaterial, direct and affective, physical, cognitivte and emotional labour that the diverse multi-racial and multi-gendered contemporary working class continually contribute to designing, maintaining and repairing our worlds (see Dharia 2015; Mattern 2018). The latter approach might force us to think about the labours of design in relation to questions of gender, race, ability, age, caste and class, sex, ecology and coloniality?

When we speak of 'environmentally friendly' design or 'eco-design'—are we speaking of labour-friendly designs that are seeking to derive *lessons from a vision of external 'Nature'* conceptualized as a series of organic relations or interlocking systems that stand apart from history, culture and power? Alternatively, would we be better off following recent eco-Marxist, feminist, post-human and decolonial scholars who have suggested that the categories of natural history and social history need to be understood as intertwined and invariably meshed with a broader web of life (Moore 2015)? Would it be advantageous to acknowledge that this broader web of life has been shaped by racial, class, gendered and colonial relations (Finney 2014) and all matter of non-human agencies, apparatus and technologies that continually pushes back, frustrates and surprises us (White et al. 2016; Wark 2015)?

Finally, it should be asked, does any attempt to ally labour with ecological or sustainable design involve attempts to construct alliances between labour and the kinds of 'small is beautiful' (Schumacher 1973) and localist design imaginaries that shaped 1970s environmentalism in the affluent world? Are we still seeking to build 'convivial technologies' (Illich 1973) and acupunctural strategies that follow 'soft energy paths' (Lovins 1977) bringing us into balance with nature—as much of the degrowth movement suggests today? Alternatively, and as many climate hawks argue today—does an acknowledgement that we must decarbonize as quickly as possible require a rethinking of red-green design imaginaries and a spatial and scalar step change in thinking of the potential terrain of action (see Cohen 2019; Fleming 2019)? Could we envisage a design politics that is multi-scalar, dynamic and combines both productivist and degrowth imaginaries into something new (Aronoff et al. 2019; Pevzner 2019; White 2020)?

Clearly definitional starting points carry political outcomes. They can produce rather different understandings of what might be entailed by bringing ecology, labour and design together to fashion a politics. Let us begin here though in a fairly conventional (and admittedly Eurocentric) fashion with the iconic figure of William Morris (1834–1896), whose life and politics were continuously struggling to reconcile the relation between design, craft and labour, socialism and nature, meaningful work and useless toil.

## 'A Factory as It Might Be'—Building the Googleplex for All!

An invitation to think of the relationship between socialism, ecology and design immediately invites images of the great dreary blight of universal modernism poured concrete and bureaucratic inertia. There have been currents of socialist and anarchist thinking from the late nineteenth century that tried to think about the ascetic forms of life and their potential for opening up new, more fecund relations with the natural world in very different ways to the Promethean traditions of socialism that dominated design in the first half of the twentieth century. William Morris's essay 'A Factory as It Might Be' written in 1884 stands as a precursor to this moment.

In this remarkable essay, Morris proposes designs for a workplace for 'the days when we shall work for livelihood *and pleasure* and not for "profit"'. The central claim that is explored in the piece is that there is no inherent reason why manufacturing processes or facilities should alienate human labour from each other or from the natural world. Rather, Morris presents us with a vision of a workplace in a socialist future that is 'built with pleasure by the builders and designers' and designed with a range of further facilities such as a 'dining hall, library, school, places for study of various kinds, and other such structures' to further expand the cultural lives of labour. The work occurring in the factory will be 'useful and therefore honourable and honoured' because it will not turn out trash. The 'manufacture of useless goods, whether harmful luxuries for the rich or disgraceful makeshifts for the poor' would end. The workplace of the future would strike a balance between automation 'machines of the most ingenious and best approved kinds will be used when necessary, but will be used simply to save human labour' and craft. Work would be 'neither burdensome in itself nor of long duration for each worker; but furthermore, the organization of such a factory, that is to say of a group of people working in harmonious cooperation towards a useful end, would of itself afford opportunities for increasing the pleasure of life'. He argues that industrial facilities should be situated in the broader contexts of living landscapes and beautiful gardens that are co-operatively maintained by workers on a voluntary basis. This reconfigured workplace would 'make no sordid litter, befoul no water, nor poison the air with smoke'.

'A Factory as It Might Be' provides us with a utopian proposition which never entirely resolves the tension running across Morris's thinking between craft and efficiency, between his celebration of the freedom that comes from non-alienated work through craft/handmade production and the freedom from toil and the possibilities of plenty that is offered by advanced manufacturing. His assertion that 'For a socialist, a house, a knife, a cup, a steam engine must be either a work of art or a denial of art' is a stunningly demanding vision of what a liberatory eco-aesthetic might look like. How to put this into practice and avoid a design economy that merely produces 'nice things for rich people' is another matter. How we might identify ways of socially organizing design, engineering and manufacturing to combine the gains of technological innovation and craft production without fostering alienation, deskilling and ecological pollution is a question that troubles all matter of design schools from this point on: from the Bauhaus to the Garden City movement, the Russian Constructivists to contemporary advocates of peer production and the commons. Nevertheless, and for all its limits of time and place, Morris's essay can hold up a mirror to current times.

## 35 Labour-Centred Design for Sustainable and Just Transitions

'A Factory as It Might Be' comes close to anticipating what we would now recognize as the basic project of industrial ecology to build closed loop production systems. Yet, this essay asks that we think about modes of manufacturing that are devoted to an emphasis on the quality and not simply the quantity of production, *better* rather than simply *more* in ways that are entirely missing from contemporary environmental engineering. His request that we attend to *the kinds of products* that are made in workplaces is at the centre of numerous important contemporary eco-design interventions from McDonough and Braungart's *Cradle to Cradle* (2002) to Jonathan Chapman's innovative and important call for *emotionally durable design* (Chapman 2005). However, unlike these pragmatic literatures which largely see a new dematerialized ecological culture as driven by enlightened green CEOs and leading-edge private companies, Morris seeks to place *labour* in the driving seat of these changes and asks us to make far-reaching political and aesthetic judgments about the kinds of material culture that contribute to a good human life.

'A Factory as It Might Be' provides broader guidelines for the design of the humane workplace that bears some resemblance to the kinds of best practice that are still offered in certain elite spaces of higher education or certain privileged tech-worker spaces like the Googleplex with their generous food and leisure facilities, time off for personal projects and so on. Yet, Morris asks that we see the right to the humane workplace as a right for all not simply, as we have come to see it today, as a benefit offered to certain special kinds of elite workers in the affluent core. His proposition that the design of workplaces of the future should aspire to bring the garden into the workplace has many contemporary resonances with the kinds of ecological architecture interventions that have been proposed and built by progressive architects like Michael Singer (https://www.michaelsinger.com/) to introduce living systems into the architecture and design of the sustainable workplace. Morris's call to bring nature into the workplace moves this discussion further forward by forcing us to also ask questions about the political organization of work, the political geographies of factories and their extended estates that they are intimately and reciprocally connected to. He can lead us to the observation that calls for the ecological redesign of the workplace at home are going to be of little value they are built off the backs of labour exploitation and environmental displacement at the plantation, colony, mine and sweat shop elsewhere.

If contemporary research programs on industrial ecology or green material production has helped us think well beyond Morris at the level of *technology*, it is still striking how few currents of sustainable design pose the political question that Morris asks so simply and elegantly. 'A Factory as It Might Be' squarely asks us to think about a mode of design that reconfigures the relations between capital, labour, ecology and life at multiple scales. As Kristin Ross (2015) has argued, much like the Paris Communards of 1871, Morris is asking us not simply to think about a workplace of the future marked by changing patterns of ownership and control. Rather, he is inviting us to think about the institutions of a revolutionary social ecology of the workplace that could possibly erase the distinctions between art and work, work and pleasure.

## Designs for Sustainability: From Kropotkin to Cradle to Cradle™

The rise of attempts to make design 'ecological', sustainable or 'regenerative' has many potential genealogies (see Ceschin and Gaziulusoy 2020). Yoko Akama and Joyce Yee (2019) and Julia Watson (2020) have most recently argued that there are a multitude of indigenous design practices cultivated across human history that have constructed robust and enduring material cultures that we can still learn from today. This is certainly true and a persistent and valuable theme of most iterations of design for sustainability. Working within the more conventional vein of design history we might acknowledge a range of further influences and currents. Early twentieth-century points of inspiration would include Kropotkin's *Fields Factories and Workshops*, Gandian discussions of 'village technology', Ebenezer Howard's visions of *Garden Cities for Tomorrow* and the broader visions of regional planning articulated by Patrick Geddes and Lewis Mumford. In the post-war era, the convergence of the radical design movements of the 1960s and 1970s and the environmental movement have also been critical (see Boyle and Harper 1976). Touchstones of this latter moment would include the more technocentric interventions of Buckminster Fuller, the DIY sensibilities of the *Whole Earth Catalogue* and Amory Lovin's (1977) advocacy of decentralized and distributed soft energy paths. The more socially and politically inflected writings of Pananek (as previously mentioned), E.F. Schumacher's influential *Small Is Beautiful* (1973), Bookchin's (1971) call for 'liberatory technologies' and Ivan Illich's (1973) hugely influential call to distinguish 'tools of domination' from 'tools for conviviality' are also key (see White 2020). Rybcynski has additionally and usefully reminded us that many currents of the appropriate technology movement in the West were also heavily influenced by a range of developments in the Global South: from Maoist experiments with small industry and rural technology in China to the Indian community development movement (Rybcynski 1991: 67–82).

These foundational literatures offer all manner of insights and provocations which are still central to thinking about the design forms that just and egalitarian post-carbon futures need to take. We would not have mainstream research programmes in green materials, industrial ecology, green chemistry, sustainable product, systems and service design or the current renewables energy industry more generally without the radical technology movements of the 1970s. Reading this literature through the lens of environmental labour studies is instructive because it does also reveal the unevenness of how these traditions have engaged with questions of labour.

If we trace the precursors of design for sustainability through Morris, Kropotkin, Howard, Mumford and Bookchin, we can certainly identify important utopian socialist and anarchist engagement with design running across much of the early to mid-twentieth century. As we have seen, Morris explicitly tried to think design with labour and ecology. The father of landscape architecture, Frederick Law Olmstead, was deeply influenced by the utopian socialist Charles Fourier (Hayden 1981). All these thinkers had some concerns with the ways in which design and planning might augment the capacities for the self-organization of labour (even if race and gender was often dealt with in uneven ways in their analysis). Some of these commitments remain amongst those traditions of radical technology sympathetic to workers co-operatives (see Schumacher 1973; Boyle and Harper 1976) and currents of social ecology committed to community or municipal control of factories and workplaces (see Bookchin 1971). Perhaps more influential though in the post-war era has been currents of design for sustainability that have been largely *technocentric* (focused on environmental/energy performance/product, service or systems innovation), *managerial* (foregrounding green business, logistical re-organization as primary avenues for change) and marked by high degrees of *naturalistic reductionism* (the assumption that 'nature' often understood as a system separate from history and politics, offers some kind of direct normative basis for organizing social life or directing design).

For example, Steward Brand's hugely influential *Whole Earth Catalogue* and *Co-Evolution Quarterly* may have had a progressive utopian-hippie veneer the political content of these design-political interventions have often been highly ambiguous. As Fred Turner (2006) has observed, emphasis on developing 'tools to construct new worlds' suggests Left-libertarian sympathies but, in Brand's hands, the focus is individualist rather than radical. *Whole Earth* was at the end of the day happy to champion a somewhat politically incoherent mix of the insights of Ivan Illich, Milton Friedman and Ayn Rand (see Turner 2006). Brand's subsequent interventions beyond this—whilst not without insights—still generally affirm a fairly technocentric design worldview that presents a sustainable future as owned and produced by a melange of Silicon Valley digital capitalist solutionist interventions with moments of ecomodernism and market-friendly eco-entrepreneurial approaches.

If we trace further the pragmatic and market-friendly traditions of sustainable, regenerative and green design that run from Amory Lovins (1977) call for soft energy paths to *Natural Capitalism* (Hawken et al. 1999); from the managerialist vision of the circular economy promoted by the Ellen MacArthur Foundation https://www.ellenmacarthurfoundation.org/ and *Project Drawdown* (Hawken 2017) https://drawdown.org/ to the now vast literature on design for sustainability (see Ceschin and Gaziulusoy 2020 for an excellent review), it is striking how many of these influential currents combine a technical focus on environmental performance, a social and ecological imaginary largely drawn from systems theory, engineering and management studies with a business consultancy model of politics. Such observations should not undercut the importance of this work. A focus on low environmental impact or non-hazardous materials, material life extension and lifespan optimization, design for disassemblage and lifecycle analysis are all vitally important ongoing research programs. Attention to organizational dynamics, logistics and management structures has its role in facilitating designs for decarbonization. Designers have to make a living in a capitalist economy. Yet, the impact of this work is significantly blunted by the lack of engagement with racialized or colonial histories and historical political ecologies, a marked tendency to naturalize market relations, largely uncritical readings of organizational theory and the lack of interest in excavating the power relations and pathologies that can delimit design processes, imaginaries and praxis within a broader capitalist political economy. As such, the political frame of this work tends to linger in a space of technocratic innocence, somewhere between philanthrocapitalism, green market thinking, EU/UN technocracy and Clinton-Blair third way positions. It is green business owners and shareholders, designers and 'green consumers' that are lionized and presented as true agents of change. The sustainable workplace is presented as a happy, healthy place but still a site of unquestioned private power. The gendered, racialized and classed labour though that sustains the closed loop eco-workplaces of the future, that are going to build and maintain the low-carbon energy structures, that labour in the organic fields, the lithium mines and the sustainable call centres of the future, are nowhere to be seen.

It could also be observed that whilst the last decade and more has seen an effervescence of attempts to reground eco-design on a radical, anti-systemic and ecological basis, much of this work has had surprising little to say about the politics of labour either. Discussion of redirective practice (Fry 2009), transition design (Kossoff et al. 2015), adversarial design (DiSalvo 2017) and design 'where everyone designs' (Manzini 2014) have extended many of the vital insights of 1970s radical design. Many of these contemporary currents follow the lead of 1970s anti-capitalist libertarian traditions focusing on developing design strategies and tactics that address culture, lifestyles and citizen needs *outside the workplace*. Much interesting communitarian design interventions and propositions have flowed from these moves. However, the question of whose racialized, gendered and classed labour sustains such design interventions and at what cost is rarely discussed. The question of how green, sustainable or eco-design might contribute to labour and power struggles in workplaces or help facilitate labour solidarities across supply chains is rarely attended to. How eco-design and planning might help to strategically identify pressure points that could not only help decarbonization but further contribute to the de-commodification and democratization of production and consumption (see Aronoff et al. 2019) has been displaced from view. More generally, how eco-design might relate to public agencies and public institutions *that could operate at scales beyond the local* or how we might develop eco-design strategies that augment the voice, knowledge, material security and power of diverse working people *across the supply chain* is largely unclear.

So, what are the alternative design histories that might help us bring the labours of design into view?

# Worker-Centred Design and Scandinavian Designs for Industrial Democracy

One contribution that environmental labour studies can bring to contemporary design discussions is to remind us of how often the long history of struggles for industrial democracy across the twentieth century has given rise to debates not just about the social relations of production but the material forms that an altered mode of production would produce. As Immanuel Ness and Dario Azzellini and their colleagues remind us, 'In the past hundred years workers have occupied factories, and other workplaces and formed workers' councils and self-managed enterprises in almost all regions of the world' (Ness and Azzellini 2011: 1): from the German work councils of 1914–1918 to Turin in 1919–1920, the Spanish Revolution/Civil War of 1936–1937 to post-colonial eruptions in Java, Indonesia 1945–1946, Allende's Chile (1970–1973) and to more contemporary struggles that have occurred in

Bengal and Argentina. As these movements have erupted, disputes and questions have continually emerged time and again about the ways in which a redesign of the workplace, a redesign of the goods made in the workplace and a redesign of the knowledges brought to bear in the production process might open and expand the capacities for new forms of democracy.

The participatory design movement is important here with its origins in Scandinavian dialogues and practices that occurred within the trade union movement during the 1970s. Participatory design—in its initial radical formation—was specifically focused on attempts to implement labour-friendly modes of design and innovation at scale. The aspiration was that modes of design could be developed and propagated that would not only be capable of resisting Taylorism and deskilling but actively expand industrial democracy in the workplace. The movement converged at the highpoint of Scandinavian Social Democracy as a wave of 'democracy at work' legislation was passed in Norway and Sweden. This legislation allowed for: (i) worker representation on company boards; (ii) the right by trade unions to stop production dangerous to worker health and (iii) the Swedish Joint Regulation act of 1977 which (in theory) outlined workers and trade unions the right 'to co-determination in production issues such as design and use of new technology' (Ehn 1990: 256). The 1970s and 1980s in Norway and Sweden thus opened up a fairly unique moment for publicly funded designers to work with organized labour to think not only about 'freedom from management prerogatives' (Ehn 1990: 254) but also modes of work-orientated designs that could aid worker-friendly redesigns of industry. As Pelle Ehn documents in his Magnum Opus *Work-Oriented Design of Computer Artifacts* (1990), two projects here are significant.

DEMOS (Democratic Planning and Control in Working Life—On Computers, Industrial Democracy, and Trade Unions) was established in 1975 (Ehn 1990). For four years it drew together a set of inter-disciplinary design teams with workers and trade union representations at four different Swedish enterprises. Influenced by but significantly adjusting Enid Mumford and the Tavistock Institute's 'socio-technical' approach to humanizing work, Braverman's labour process theory (1974) focused on alienation and deskilling with Paulo Freire *Pedagogy of the Oppressed* (1970), a central aim of this research project explored the ways in which unions and workers could influence the design and use of technology in the workplace. Rather than construct the research agenda in terms of humanizing work via 'gifts from management' the focus of the Scandinavian approach was to explore 'the importance of the employees themselves having the right to determine the content of humanization by real and meaningful participation' (Ehn and

Goranzon cited in Ehn 1990: 269). A follow-up, the UTOPIA Project, explored whether the design process could be democratized through workers and designers developing modes of collaboration via 'designing tools and environments for skilled work and high-quality products and services' (Ehn 1990: 279).

Both projects had to address multiple complexities, management resistance and disappointments. Nevertheless, they produced openings (Ehn 1990: 279). They demonstrated that under certain circumstances workers through their trade unions could be brought into the design of a product through the establishment of 'mutual learning processes' between workers, social scientists and engineers (Ehn 1990: 332). DEMOS and UTOPIA also led to further discussions in Sweden and Norway about the need for a national technology policy for democratization and the need for designers and workers to receive broader forms of education and training to make industrial democracy a reality. The projects helped consolidate the proposition that employers had to negotiate with local unions before making 'major changes in production'.

DEMOS and UTOPIA are, of course, examples of *labour* rather than *ecology*-centred design projects. They can, in some senses, be seen as produced by the unique political conditions of the time. The call for worker-orientated design in Scandinavia emerged from a broader political culture underpinned by very high levels of unionization, a large dominant social democratic party and corporatist trade union relations that allowed for centralized negotiations with the government and employers (Ehn 1990: 254). Nevertheless, the issues that emerge from these attempts to think about the relations between design and labour reverberated widely further afield.

For example, the far-sighted attempts by workers at Lucas Aerospace in the late 1970s also offers insights for our current discussions. The Lucas aerospace workers in the late 1970s faced redundancy. Rather than simply accept this fate though, shop stewards opened up a comprehensive discussion with workers concerning how the factory might transition to making new goods beyond defence production. The consultation process that followed generated a rich abundance of proposals for what we would now call 'transition designs'. Proposal ranged from healthcare and design for disability to ecological technologies (see Cooley 1980; Räthzel et al. 2010). A lack of support from both management and the broader labour movement ultimately unravelled the capacity of the Lucas Plan to be implemented. Edward Cohen-Rosenthal's advocacy of red-green worker-friendly industrial ecologies and worker-friendly eco-industrial districts in the 1990s makes for one further important moment here (see Cohen-Rosenthal and Musnikow 2017). The rise of recent proposals to transform the extractive sharing economy through the development of worker-owned and managed platform co-operatives marks something of a revival of this moment and it could have considerable promise (see Scholz 2016).

It is important to acknowledge that 1970s socialist visions of a worker-centred design—whether situated in Scandinavia or Latin America—confront very significant challenges for addressing the contemporary moment. Attempts by worker-centred designers to make consequential impacts on the organization of the workplace in the 1970s were in large part premised on site-specific struggles that occurred within the broader context of (relatively) nationally bounded Fordist economies. The neo-liberal globalization of supply chains, the logistics, software and information technology revolutions, automation, the geographical dispersal of manufacturing and service industries and shifts in the global division of labour has altered the terrain of struggle (see Chua 2018). The racialized and gendered histories of exclusion within labour organizing are deep seated and require constant interrogation. They can reassert themselves when calls for 'green jobs' are only focused on achieving just transitions for fossil fuel worker jobs but ignore the vast amount of gendered and racialized labour involved in social reproduction, care and service economy work and the potential of this sector to be repurposed to advance low carbon futures (see Aronoff et al. 2019).

A viable labour-centred eco-design for these different times clearly needs to understand these shifts and work with new spatial, scalar and political imaginaries. The ongoing struggle for democratic and ecological modes of design that can contribute to the empowerment of workers in local grassroots struggles is vital. But we also need design platforms, tools and imaginaries that can help to build solidarities and recognition across struggles, empower workers across supply chains and open up platforms for organizing labour and environmental currents within and across different sectors of the economy. Recent discussions of the green new deal have underlined the importance of thinking about the critical role that regional, state and federal interventions could also play in building robust modes of designs and planning for decarbonization. Aronoff et al. (2019) have compellingly argued a strategic focus on proposing *national industrial policies* for low carbon restructuring and innovation, *federal performance standards* targeted at key high carbon points of production and high carbon sectors, federal legislation that guarantees workers' rights to organize and attempts to fundamentally shift the balance between work and leisure are all critical to democratize, decarbonize and de-commodify production.

## Socialist-Feminist Labour-Orientated Designs

If struggles around the democratic design of the formal workplace constitutes a significant moment of radical design history, it is also worth acknowledging that many other radical voices turning their attention to the gendered labours of design have argued that we need to open up spaces for redesign at very different sites and scales. Dolores Hayden's (1981) socialist-feminist history of design is critically important here and serves as a reminder that many late nineteenth- and early twentieth-century radical thinkers and movements did not simply delimit their explorations of the labours of design to the factory and class relations but were fully attentive to the labour of social reproduction more generally. They often directly connected such concerns with what we understand today as environmental issues.

*The Grand Domestic Revolution—A History of Feminist Designs for American Homes, Neighborhoods and Cities* documents how between the 1860s and the 1930s, numerous 'materialist feminist' movements in the United States raised critically important environmental and labour questions though exploring the design politics of cooking and cleaning, food preparation and childcare and through asking broader questions about how the social, political and financial subordination of women to men could be overcome. Central to the rise of these material feminists was not simply the call for economic remuneration but many proposals for 'a complete transformation of the spatial design and material culture of American homes, neighbourhoods and cities' (Hayden 1981: 1).

Hayden's material feminists adopted the earlier communitarian socialist focus on spatial analysis 'to accompany economic analysis' (Hayden 1981: 8) but they moved beyond the village to embark on probing and imaginative investigations of the future of urban form more generally. The 'overarching' theme of much of this materialist feminist advocacy was focused on overcoming the 'split between domestic life and private life created by industrial capitalism as it affected women'. Hayden argues two key aspects of industrial capitalism that came under critical scrutiny here: (1) 'the physical separation of household space from public space'; and (2) 'the economic separation of the domestic economy from the political economy' (Hayden 1981: 1).

The design politics proposed by material feminists in the United States had its limitations. As Hayden fully acknowledges, the racial politics of many late nineteenth century, predominantly white material feminists, was not always enlightened. Like the communitarian socialists before them, the material feminist traditions had an insufficient analysis of power, class relations or race relations. Nevertheless, the most developed discussions did outline how changes in social relations, design forms and ownership structures could facilitate social emancipation. Feminist activists across this period argued for wages for housewives, abolition of the private kitchen in favour of communal kitchens and community restaurants. They championed socialized childcare and proposed designs for communal workplaces. As urban and municipal infrastructures of water, fuel and electricity began to develop across US cities, many material feminists further argued that women should own these services to give them financial independence from men; 'and use them as their base of economic power' (Hayden 1981: 12).

Many feminist campaigners (much like Morris) had a nuanced view of automation. They were deeply concerned with the eradication of toil and the need to address socio-technical changes that would address the loneliness and isolation of domestic world. As Hayden notes: 'Material feminists believed the solitary housewife doing her ironing or mixing dough could never compete with the group of workers employed in well-equipped laundries or hotel kitchens…Neither could the isolated home compete with the technological and architectural advantages offered by larger housing complexes' (Hayden 1981: 16). A very different view of the configuration of more social and ecological domestic technologies could also have beneficial impacts to increase residential densities generating social and ecological gains but also further facilitating power relations. As she notes:

> Devices such as elevators, improved gas stoves, gas refrigerators, electric suction vacuum cleaners, mechanical dishwashers, and stream washing machines which were designed for use in large enterprises such as hotels, restaurants and commercial laundries, could also be used in large apartments. Because this technology was first developed at the scale suitable for fifty to five hundred people, any group interested in mechanizing domestic work simply had first to socialize it and plan for collective domestic consumption by organizing households into large groups inhabiting apartment hotels, apartment houses, model tenements, adjoining row houses, model suburbs or new towns. What was unique about the material feminists was not their interest in these technological and architectural questions, which also attracted inventors, architects, planners, speculators, and efficiency experts but their insistence that these economic and spatial changes should take place under women's control. (Hayden 1981: 17)

In this respect, Hayden argues that the particular forms that suburbanization took on in the United States in the mid-twentieth century can be viewed now as not simply re-enforcing racial and gender divides but providing the carbon and energy inventive infrastructures that we have to deal with today.

## Invisible Labour and the Hidden Histories of Black, Brown and Decolonial Labour

Hayden's work is important for the ways she disrupts conventional understanding of the labours of design. Her socialist-feminist interventions can be productively read alongside recent moves by Carolyn Finney (2014), Ijlal Muzaffar (2013) to highlight the racialized histories of the fields of architecture, planning and design. Such discussions of the racialization of labour can open up many new promising avenues for environmental labour studies.

For example, if we are to write the environmental labour histories of design in the twentieth-century United States, it is a reasonable question to ask why so much conventional design history has focused intently on exploring and celebrating the genius of (mostly white and male) individual architects and designers. Much less attention has not only been paid to Black and Brown architects and designers that seemingly exist 'outside history' but the multiplicity of many other creative workers that play a vital role in contributing to the project of design. For example, it could be noted that most forms of design education tend to render invisible *not only* the draughtsmen and draughtswomen, the model makers and construction workers who play such an important role in the 'formally designated' design industries but many other kinds of designer workers that are involved in labours of maintenance and repair (see Dharia 2015; Allen 2019; Mattern 2018). After the architect and designer has moved on, our material culture *is still being designed* by all matter of *design workers*—mechanics and milliners, tailors and janitors, custodians and cleaners that are central to social and ecological production. As Carolyn Finney has observed in *Black Faces, White Spaces*, stories of the rise of landscape design and architecture miss much if they do not attend to the vast creative design labour of African American, Latinx and working-class women and men that tended the land as groundskeepers and gardeners and then often found themselves disposed and displaced by the owners when they were no longer useful (Finney 2014).

The shift in design studies from a sole focus on *object making* to attending to *the design of systems and services* could potentially have many complementary convergences with a concurrent shift occurring in design studies that we have sketched above, that decentres professional designers and sociologically re-grounds the many forms of *design labour* that is involved in maintaining our material culture. Notably, it allows us to think more broadly about how political processes of design have often been enacted across twentieth century by a wide variety of *non-credentialed designers* yet these have largely been erased from design history and design education.

Scholarship by Dorceta Taylor (2009), Jessica Gordon Nembhard (2014) and Ted Jojola and his colleagues (see Walker et al. 2013) can be seen as close allies to this work in demonstrating how a recovery of the design histories of indigenous, African American and Latinx driven experiments in design and planning can open up a much more encompassing understanding of the labours of design and their potential contribution to building sustainable futures. For example, a good deal of attention has been given recently to how the sociologist, novelist, public intellectual and philosopher W.E.B. DuBois made vital contributions to the social sciences and helped steer the Harlem Renaissance but additionally made important contributions to constructing politicized forms of graphic design and data visualization (Battle-Baptiste and Rusert 2018). To take a further example, on the West Coast of California, between 1966 and 1972, the Black Panther Party combined mass organizing of African American communities, revolutionary education for self-defence with the cultivation of community art, design and media under the leadership of the artist and designer Emory Douglas. Their 1972 Ten-Point Program is interesting for the ways it demanded 'land, bread, housing, education, clothing, justice, peace and people's community control of modern technology' and further sought to develop a range of community mechanisms to achieve these ends (Newton 1999 cited in Gaiter 2018). Many of the community interventions that the Panthers developed between the late 1960s and the early 1970s bear some considerable resemblance to the kinds of communitarian modes of social design innovation that are advocated by service designers today. Arguably, the Panthers model of social movement-orientated community design was rather more successful at scaling than a good deal of contemporary design innovation. It is notable that the Panther's experiment in community food provisioning at its high point was running programs that fed 20,000 children across 19 cities (Gaiter 2018).

## Conclusion: The Design-Labours of the Green New Deal and the Just Transition

The concept of just transitions has provided an important point of engagement between movements pushing for labour, indigenous rights and environmental/climate justice over the last decade (White 2020). In the United States, this has emerged in the form of calls for a Green New Deal which presently stands as an influential frame for moving this discussion forward (see Aronoff et al. 2019). As Billy Fleming (2019), Daniel Cohen (2017) and Nicholas Pevzner (2019) have argued, the Green New Deal is an imaginary which invites, perhaps for the first time in many generations, not simply the promise of a labour-focused environmental politics but the necessity of thinking about this project, at least in some part, in terms of design. Such thinkers are fully aware of the dangers of designers repositioning themselves as 'climate saviours' (Fleming 2019). Nevertheless, they have compellingly argued that design will be central for enacting a Green New Deal. Whether we consider the need to facilitate post-carbon energy infrastructures and low carbon transportation to the need to build sustainable housing, the need to build climate resilient urbanisms and thriving rural economies that sequester carbon, a healthcare system that provides protection from a warming planet to an economy that works for the multi-racial and multi-gender working class, the search for post-carbon futures seems to invite a design politics. Will the design aspects of a green new deal though seek to ignite a *design politics for labour*, a *design politics by labour* or *a design politics with labour* that opens, expands and reconfigures the whole field of design?

As this chapter has demonstrated, there are many radical traditions of *design for labour* that have emerged sporadically out of professional design circles over the last 100 years and more which involve calls for professional designers to ally themselves with workers and environmentalists to build new futures. Such calls have largely followed the logic that designers can and should produce beautiful spaces for working people to live in and that it is the job of professional designer to creatively design artefacts, contexts, conditions and circumstances so that the lay public can go about their lives and think about other matters. From William Morris to the Garden City movements, from Red Vienna to the building of the Scandinavian social democratic welfare state, there have been some extraordinary gains delivered by this particular understanding of design. The first New Deal used design, architecture and planning at many scales to build parks and schools, municipal facilities, large public infrastructures, rural electrification and so on. The New Deal was of course an imperfect moment and came with errors, mistakes and was unquestionably marred by segregation. Yet, the WPA put graphic designer and artists, craftsman and construction workers, photographers and theatre directors to work across the nation. The Civilian Conservation Corps employed nearly three million people in various forms of environmental conservation (see Pevzner 2019). Exploring what can be learned from moments in history when design and planning have been creatively enrolled with public agencies and public institutions is critically important.

We have also seen in this chapter though that there are radical traditions of design which have also warned of the ways in which design allied to state power can and has given rise to modes of administrative rationalism which can re-enforce technocratic and paternalist politics. Research by Irani and Silberman (2016) has demonstrated that even well-intentioned radical designers can generate 'worker-friendly' design project which re-centres designers at the expense of workers. This has led Ruha Benjamin to wonder whether an excessive or overly expansive understanding of design is itself dangerous because it has its own *colonizing logic*. As she argues:

> Whether or not design-speak sets out to colonize human activity, it is enacting a monopoly over creative thought and praxis. Maybe what we must demand is not liberatory *designs* but plain old liberation. (Benjamin 2019: 179)

Whilst we must acknowledge, as Benjamin notes that design can erase 'the insights and agency of those who are discounted because they are not designers' and 'capitalize on the demands for novelty across fields' (Benjamin 2019: 179), we have encountered civic republican, populist/anarchist and decolonial traditions of radical design in this chapter that have accented the importance of approaches that acknowledge the design abilities and knowledges of different kinds of working people. Many of these traditions anticipate the work of contemporary transition designers that seek to cultivate a design politics as a collective dialogue between designers and publics and maximally argue that the multi-racial and multi-gendered working class have the capacities to be designers in their own right and potentially self-authors of their own material worlds.

The gains of these latter traditions are that they acknowledge much broader groups of people that have been involved in projects of design, repair, maintenance and new forms of care than is often the case in radical interventions that are still primarily addressed to professional designers (see Mattern 2018). De-centring professional designers from design history allows us to see the labours of designs in much more encompassing ways. Such accounts of a design politics have some resonance with those aspects of degrowth research that is concerned with constructing alternative economies, the focus of the decolonial design group on repair and the more communitarian calls for 'design where everyone designs' (Manzini 2014). Participatory and populist approaches to design though come with their own challenges. Posing design solutions purely at the communitarian level in some contexts may not be helpful or lead to empowering interventions that are meaningful.

We have also seen that from William Morris to the contemporary project of decolonial designers, the question of where stuff comes from and where it is going to? and who and what is disaggregated along the way? has to be foregrounded in any meaningful labour-centred design. A Green New Deal that makes claims to just transitions in the affluent core cannot possibly be built off the backs of workers in factories, fields and lithium mines of the so-called periphery. At the same time, the hard-won gains achieved by labour and environmental movements in the core cannot be glibly dismissed by Malthusian calls for collective sacrifice and eco-austerity for all (see Aronoff et al. 2019: 139–169). The stakes of the conversation around designs for post-carbon futures then could not be higher. Real dilemmas abound. Mistakes will be made. But still we must press on and environmental labour studies may well be a critical current to help us realign more labour-friendly and ecology-friendly modes of design that are orientated to achieving just transitions.

# References

Akama, Y an Yee, J 2019 Special Issue: Embracing Plurality in Designing Social Innovation Practices, Design and Culture, 11:1, 1–11, https://doi.org/10.1080/17547075.2019.1571303.
Allen, Matthew. 2019. The Death of the Genius: An Alternative History of Computation Lays Bare the Problem of Invisible Labor in Architecture. *Harvard Design Magazine* 34: 1–6.
Aronoff, Kate, Alyssa Battistoni, Daniel Cohen, and Thea Riofrancos. 2019. *A Planet to Win: Why We Need a Green New Deal.* London: Verso.
Barca, Stefania. 2012. On Working-Class Environmentalism. A Historical and Transnational Overview. *Interface. A Journal for and About Social Movements* 4 (2): 61–80.
Battle-Baptiste, Whitney, and Britt Rusert, eds. 2018. *W. E. B. Du Bois's Data Portraits: Visualizing Black America*. Princeton: Princeton Architectural Press.
Benjamin, Ruha. 2019. *Race after Technology*. Cambridge: Polity Press.
Boehnert, Joanna. 2018. *Design, Ecology, Politics: Towards the Ecocene*. London: Bloomsbury.

Bookchin, Murray. 1971. Towards a Liberatory Technology. In *Post Scarcity Anarchism*, 89–103. New York: Ramparts Press.
Boyle, Godfrey, and Peter Harper, eds. 1976. *Radical Technology*. London: Penguin.
Braverman, Harry. 1974. *Labor and Monopoly Capital: The Degradation of Work in the Twentieth Century*. New York: Monthly Review Press.
Ceschin, Fabrizio, and İdil Gaziulusoy. 2020. *Designs for Sustainability*. London: Routledge.
Chapman, Jonathan. 2005. *Emotionally Durable Design*. London: Earthscan.
Chua, Charmaine. 2018. Logistical Violence, Logistical Vulnerabilities. *Historical Materialism* 24 (4): 167–182.
Cohen, D.A 2017. "The Last Stimulus" Jacobin 26, 83–95.
Cohen, D.A. 2019. A Green New Deal for Housing. *Jacobin*. https://jacobinmag.com/2019/02/green-new-deal-housing-ocasio-cortez-climate.
Cohen-Rosenthal, Edward, and Judy Musnikow. 2017. What is Eco-Industrial Development? In *Eco-Industrial Strategies*, 14–29. London: Routledge.
Cooley, Mike. 1980. *Architect or Bee?* Slough, UK: Langley Technical Services.
Crain, Marion, W. Poster Winifred, and Miriam Cherry. 2016. *Invisible Labor: Hidden Work in the Contemporary World*. Oakland: University of California Press.
Deamer, Peggy. 2015. *The Architect as Worker: Immaterial Labour, the Creative Class, and the Politics of Design*. London: Bloomsbury Academic.
Dharia, Namita Vijay. 2015. *Scaffolding Sentiment: Money, Labor, and Love in India's Real Estate and Construction Industry*. Doctoral diss., Harvard University, Graduate School of Arts & Sciences.
DiSalvo, C. 2017. Adversarial Design. Cambridge, MA: MIT Press.
Ehn, Pelle. 1990. *Work Orientated Design of Computer Artifacts*. Portland: Lawrence Erlbaum.
Escobar, Arturo. 2017. *Design for the Pluriverse*. Durham: Duke University Press.
Finney, Caroline. 2014. *Black Faces, White Spaces Reimagining the Relationship of African Americans to the Great Outdoors*. Durham: University of North Carolina Press.
Fleming, Billy. 2019. Design and the Green New Deal. *Places Journal*. https://places-journal.org/article/design-and-the-green-new-deal/.
Fry, T. 2009. Design Futuring. London. Bloomsbury.
Gaiter, Colette. 2018. Visualizing a Black Future: Emory Douglas and the Black Panther Party. *Journal of Visual Culture* 17 (3): 299–311.
Gordon Nembhard, Jessica. 2014. *Collective Courage: A History of African American Co-operative Thought and Practice*. University Park, PA: Penn State Press.
Hawken, Paul. 2017. *Drawdown: The Most Comprehensive Plan Ever Proposed to Reverse Global Warming*. New York: Penguin Books.
Hawken, Paul, Amory Lovins, and Hunter Lovins. 1999. *Natural Capitalism: The Next Industrial Revolution*. London: Earthscan.
Hayden, Dolores. 1981. *The Grand Domestic Revolution*. Cambridge, MA: MIT Press.
Illich, Ivan. 1973. *Tools for Conviviality*. Harper & Row.

Irani, Lilly, and Silberman, Max. 2016. Stories We Tell About Labor: Turkopticon and the Problem with 'Design'. *Proceedings of the 34th Annual ACM Conference on Human Factors in Computing Systems*, ACM.

Julier, Guy. 2017. *Economies of Design*. London: Sage.

Kossoff, Gideon, Cameron Tonkinwise, and Terry Irwin. 2015. Transition Design Provocation. *Design Philosophy Papers* 13 (1): 3–11.

Lovins, Amory. 1977. *Soft Energy Paths: Toward a Durable Peace*. Cambridge, MA: Ballinger Publishing Company and Friends of the Earth. International.

Manzini, E. 2014. Design When Everyone Designs. Oxford: Oxford University Press.

Mattern, Shannon. 2018. Maintenance and Care. *Places Journal November*. https://placesjournal.org/article/maintenance-and-care/.

McDonough, William, and Michael Braungart. 2002. *Cradle to Cradle: Remaking the Way We Make Things*. New York: North Point Press.

Moore, Jason. 2015. *Capitalism in the Web of Life*. London: Verso.

Morris, William. 1884. A Factory as it Might Be. https://www.marxists.org/archive/morris/works/1884/justice/10fact1.htm.

Muzaffar, M.I. 2013. "Fuzzy Images: The Problem of Third World Development and the New Ethics of Open-ended Planning at the Joint Center of Urban Studies at Harvard and MIT," in A Second Modernism: Architecture and MIT in the PostWar, A. Dutta ed. Cambridge MA: The MIT Press.

Ness, Immanuel, and Dario Azzellini. 2011. *Ours to Master and to Own. Workers' Control from the Commune to the Present*. Chicago, IL: Haymarket Books.

Papenek, Victor. 1971. *Design for the Real World*. New York: Pantheon Books.

Pevzner, Nicholas. 2019. The Green New Deal, Landscape, and Public Imagination. *Landscape Architecture Magazine*. https://landscapearchitecturemagazine.org/2019/07/23/the-green-new-deal-landscape-and-public-imagination/.

Räthzel, Nora, David Uzzell, and David Elliott. 2010. Can Trade Unions Become Environmental Innovators? *Soundings* 46: 76–87.

Ross, Kristen. 2015. *Communal Luxury: The Political Imaginary of the Paris Commune*. New York: Verso.

Rybcynski, W. 1991. Paper Heros: Appropriate Technology: Panacea or Pipe Dream? London: Penguin.

Scholz, Trebor. 2016. *Platform Cooperativism: Challenging the Corporate Sharing Economy*. New York: Rosa Luxemburg Stiftung.

Schumacher, E.F. 1973. *Small is Beautiful: Economics as if People Mattered*. London: Harper Collins.

Taylor, Dorceta. 2009. *The Environment and the People in American Cities 1600s–1900s: Disorder, Inequality and Social Change*. Durham: Duke University.

Turner, F. 2006. From Counterculture to Cyberculture: Stewart Brand, the Whole Earth Network, and the Rise of Digital Utopianism Chicago: University of Chicago Press.

Walker, Ryan, Ted Jojola, and David Natcher, eds. 2013. *Reclaiming Indigenous Planning*. Montreal-Kingston, Canada: McGill-Queens University Press.

Wark, Mackenzie. 2015. *Molecular Red*. London: Verso.
Watson, J. 2020 Lo—TEK, Design by Radical Indigenism. Cologne, Germany. Taschen.
White, Richard. 1995. Are You an Environmentalist or Do You Work for a Living? In *Uncommon Ground: Rethinking the Human Place in Nature*, ed. William Cronon, 171–185. New York: W.W. Norton & Company.
White, D.F. 2020. Just Transitions/Design for Transition: Preliminary notes on a Design Politics for a Green New Deal. *Capitalism Nature Socialism* 31 (2): 20–39.
White, Damian, Alan Rudy, and Brian Gareau. 2016. *Environments, Natures, and Social Theory: Towards a Critical Hybridity*. London: Palgrave.
Yoko, Akama, and Yee Joyce. 2019. Embracing Plurality in Designing Social Innovation Practices. *Design and Culture* 11 (2): 1–11.

# 36

# Technology and the Future of Work: The Why, How and What of Production

David Elliott

## Technology and the Future of Work

Technological change and industrial activities can have negative social and environmental impacts. To reduce or avoid them, new consumer technologies, products and service systems are being developed. In the energy supply field, some progress has been made with renewable energies. But in many other fields, progress has been more uncertain. One of them concerns the development of new *production* technologies. Some manufacturing processes have been cleaned up, but mostly things in factories are made in the same old way. However, this is already changing, based on the spread and further development of automatic systems and cybernetic controls—automation.

Some see automation as a more efficient way to make things, and so as inherently 'green'. It can be, for example, by cutting energy use and material wastage, the associated cost savings often being a major motivation behind the drive to automate production. However, there is a larger societal picture on which I will focus in this chapter. A major attraction of automation is that it can remove the need for people, replacing them with technically efficient machines.

---

Due to space restrictions, this chapter focuses on manufacturing processes. Other areas of work are discussed in Parts III and VI of this volume.

---

D. Elliott (✉)
School of Engineering and Innovation, The Open University, Milton Keynes, UK
e-mail: d.a.elliott@open.ac.uk

## Automation

Some people have seen automation in 'liberatory' terms, freeing mankind from the curse of backbreaking, mindless work, a view that has underpinned many utopian visions from the mid-nineteenth century onwards. Modern variants, from a radical post-capitalist perspective, have ranged from Bookchin's 'liberatory technology', enabling a shift to a 'post-scarcity' future (1965), to Bastani's vision of 'fully automated luxury communism' (2019).

Others have seen automation as destroying jobs, technical skills, economic remuneration, and social satisfactions associated with productive work. Dire warnings about widespread job losses and technological unemployment emerged regularly from the mid-twentieth century onwards, although it was also argued that, given its high cost, industrial automation was only likely to spread slowly (Elliott 1975; Elliott and Elliott 1976).

Although short of full automation, some new technologies did spread, and there were warnings about the likely impacts of, for example, microchips, computers and Information and Communication Technologies (Hines and Searle 1978). Their advent was increasingly seen as part of an overall trend to introduce automation in an expanding range of sectors, not just industry. That has happened, for example in retail, office work and much of the service sector, leading to a wide-ranging debate about 'the future of work' (Rifkin 1995) and the social and ethical problems raised by technological change (Lawrence et al. 2017).

Advanced robotics and the prospects for systems with Artificial Intelligence capabilities have moved the debate on further—the whole role of humankind is now in question. Some see the likely changes as threatening, others, while recognising that there will be challenges, tend to look to the opportunities and benefits (WEF 2018).

The substantial literature on this debate was recently reviewed by Benanav (2019). Looking at the history of automation and reactions to it so far, he identifies two main themes. Firstly, there have been frequent predictions of a coming age of 'catastrophic unemployment and social breakdown', which could be prevented only if society were reorganised. Secondly, there is the view that, while the new technologies promised an enormous bounty, 'there is no economic law that says that all workers, or even a majority of workers, will benefit from these advances' (Brynjolfsson and McAfee 2014). Moreover, as one study put it, while the coming of advanced robotics was inevitable, 'there is no necessary progression into a post-work world' (Srnicek and Williams 2015).

The anti-automation camp points to the negative impacts of new production technologies in the past, such as work rate speed-up and increased output demands as the effect of Fordism (Pizzolato 2013). Automation, it is argued, could continue and extend this process, eliminating the need for skilled workers in some areas, while putting more pressure on those remaining in the production system. This prognosis is sometimes contrasted with an (arguably mythical) golden past, when life was fuller and work was less alienated. It is true that we have lost much that was valuable. However, there was little that was romantic about the hard-graft drudgery that most people faced. Technology has lessened that, or at least changed it into new forms of drudgery. The issue now is whether automation will create something new or more of the same?

Both sides in the automation debate agree that, for good or ill, automation will happen, with the debate now being over the pace and the driver of changes. The main driver is seen as competitive market pressures, requiring an increase in productivity to maintain profit levels and market shares. Based on this analysis, it is a matter of capitalist economics. There are limits to how much workers can be exploited to increase labour productivity, not least since, when effectively organised, they are able to fight back seeking better working conditions and higher pay (Silver 2003). Those who own and control industrial production systems have therefore begun to look at other approaches like increased *capital* productivity, higher returns on their investment.

In this context, the introduction of new technology is one option. It can be used to increase labour *and* capital productivity, investing in new skills *and* machines. That is done in some high technology sectors, although the trend is usually to replace low skilled with a few high-skilled workers. In some roles, some skills are hard, or too expensive, to replace and low-skilled workers may be cheaper than machines, for example, in some partly automated assembly lines, so that, for the moment, those jobs are left unchanged. However, technology is improving and automation may eventually spread to most sectors, offering those who own and control the production system a way to radically cut back on the troublesome labour component.

The primary reason for its introduction would then be to improve the economic and technical efficiency of production, not a concern for working conditions, much less to ensure the availability of work. So *socially beneficial* automation outcomes for workers may not emerge.

Given that context, rather than following through the twists and turns of the contemporary academic debate on automation and what might drive it, I want instead to focus on what might be done to ensure that the direction that technology takes is a positive one for workers and the wider society. My reference point is the idea of 'human-centred' production, as promoted in the 1980s and subsequently by a leading UK trade unionist, academic and activist, Mike Cooley.

## Human-Centred Technology

Cooley argued that we can make sure the new technology is *human-centred*, so that it *enhances* skills rather than replaces them, enabling people to engage in meaningful and creative work (Cooley 1982). This might be seen as a defensive approach, harking back to the artisan culture that William Morris and the Arts and Crafts movement tried to resuscitate in the UK in the late nineteenth century. Although Morris did hope for wider changes, his approach only involved a small group of artisans producing a limited range of cultural products for a mostly affluent minority of consumers, with the role of new technology being limited. Cooley argues that human-centred technology could achieve much more than that, so that *all* can participate in a broader form of 'socially appropriate' production on a wide scale, should they wish to.

His approach is partly based on the work he and others did over the years on *machining*. The industrial revolution owed much to the developing skills of those who learnt how to use machines to work with metals and other materials. As industry has developed, some of these machining skills have been threatened. For example, modern semi-automated lathe tools are run using pre-programed instructions, based on designs created using computers. One of Cooley's early studies was of Computer Aided Design (CAD). CAD may give designers some freedom to create new ideas, but the whole subsequent production system, including the work then done by machinists, tends to squeeze out the tacit knowledge of materials and how to work with them that typifies craft work (Cooley 1972).

That approach has often shaped the use of new technology in industry generally, with implications both for the production process and for the products. What the competitive market-driven production system wants is speed and reliability, so as to reduce costs and increase productivity. So it has often gone for standardised products made in standardised ways, with little opportunity for originality or diversity, inventiveness or innovation.

Cooley, wanted production processes to re-connect with imagination, intuition and creativity, as well as skill and tacit knowledge, with more flexible worker-controlled production leading to new innovative products. His 'human-centred' machines would allow the machinists to take back control, enabling them to use their skills to the full to make a wider range of better, more useful products (Cooley 1983). This may result in lower scores in narrow economic 'productivity' terms, but Cooley argued, as well as leading to a more satisfying and productive experience for the producer, the more creative approach could lead to products of higher social use value, designed to meet social needs.

He was not talking about making high *economic* value 'luxury' products or unique cultural artefacts, of the sort beloved by Morris. In Cooley's view, with workers being able to identify social needs and make products to meet them, these could have high *social* rather than commercial exchange value. Some products might have both social and exchange value, and there will probably always be a market for traditional craft-based products. However, his approach was more concerned with end-use value and worker involvement. His emphasis was on 'socially useful production' (Cooley 1982).

Cooley sees this approach as being *generalisable* to sectors beyond factories, trades and professions, where deskilling is a threat. Rather than trying to extract expertise, skills and tacit knowledge from expert practitioners and codify them in so-called expert systems, with central computer systems then being used by non-experts, he suggests to spread expertise and offer opportunities for all to participate, inputting their own ideas for products and processes.

This may seem Utopian. But Cooley pointed to the burst of creative ideas that emerged from 'ordinary people' when given the chance to contribute to the development of the 1970s Lucas workers' alternative plan for socially useful production, with which he was closely involved.

The Lucas workers campaign has become a much-celebrated part of labour movement history, but it is still relevant. Workers at that time were facing similar problems of new technology, deskilling, automation and wider production and environmental issues. Although the job insecurity issues they faced were mostly related to higher-level corporate *product* policy, there were also issues related to the work *process*, including deskilling and the two levels interacted. The genius of the Lucas workers' response was they made a link between *process* and *output*, asking what is to be made *and* how should it be made.

## The Lucas Plan and Beyond

The Alternative Corporate Plan produced by shop stewards from the 13 Lucas Aerospace plants in 1975/1976 outlined a series of 150 or so 'socially useful products' that they felt could meet social needs and secure employment on a sustainable basis (Wainwright and Elliott 1982). The company was reliant on defence orders and the Labour government of the time was cutting back on defence spending. There were threats of large-scale redundancies. The cross-plant, cross-union, shop stewards 'Combine' committee thought that, to avoid job losses, the company's resources and the workers' skills could be used on alternative products. They asked their members for ideas via a cross-plant questionnaire. The diversification plan they produced as a result, with inputs from the shop floor, included the development of advanced medical aids, new transport systems and alternative energy technologies, including solar and wind technologies, heat pumps and fuel cells, novel at that time. One strategic idea was that the public funding saved from the defence cuts could be retargeted to socially useful production along these lines, with local council-linked projects aimed at meeting local social needs.

The emphasis in the plan was on products and their social use, including also a concern for the environmental impacts of products and production, as well as the impacts of the production processes on the workforce. Much of the work that was going on in Lucas involved batch production, not large-scale continuous mass production, so, in principle, a conversion to new types of human-centred production could have been easier (Cooley 1987; Rosenbrock 1989). Cooley later proposed a formal set of criteria for socially useful production stressing all these aspects (Cooley 1982), which were implicit in the Lucas Plan itself.

The political point was that this group of workers could identify what was needed in the communities to which they belonged and they had the skills to make the technology. But they did not have the power or money to make it happen. Though they tried, by using grassroots trade union power, sadly they failed. The Labour government offered warm words, but little help. The official trade union bureaucracy was also less than helpful. The cross-union Combine Committee was an unofficial grassroots organisation which they did not recognise. In 1979 Margaret Thatcher was elected as Prime Minister and launched a major attack on the trade union movement, culminating in the defeat of the coal miners' union. In the new anti-union climate, many of the Lucas activists were sacked or moved on, and the battle for radical product diversification was lost.

Lucas was not alone. There had been some other similar 'workers plans', following on from the Lucas plan, for example, in power engineering (Clarke Chapman and Parsons, in Newcastle, backing Combined Heat and Power) and in defence (submarine maker, Vickers, in Barrow, with wave power being one idea), but they too were sidetracked (Wainwright and Elliott 1982).

Some of the product ideas from the Lucas plan were taken up by radical local authorities, notably by the Greater London Council (GLC) via its local Technology Networks. They sought to provide technical and financial support for new enterprises and job creation. Sadly, these initiatives were also seen off as politics swung to the right in the 1980s (Elliott and Mole 1987; Smith 2015).

The 'workers plan' and allied campaigns may have failed, but the idea lives on. What the Lucas and other campaigns showed was that it is possible for the workforce themselves, rather than external experts or technocrats, to develop better plans for the future, more attuned to needs rather than profits. Indeed, it is often the case that the best ideas come from outside the system, from the fringe. That has happened in the case of the early days of renewable energy development (Elliott 2019).

However, moving from the 'ideas' stage to actual wide-scale production is often difficult, especially with novel products. The capitalist system is meant to rely on risk-taking entrepreneurship, but often investors have become risk adverse. Nevertheless, many of the product ideas that emerged from the Workers' Plan movement have become part of the product repertoire of companies, especially in relation to new energy systems. That change may have happened anyway, but the Workers' plans helped to put the need for progressive technological change on the agenda and have continued to inspire similar initiatives.

For example, there has been a UK trade union-linked Campaign for 1 million Climate Jobs, focused on green energy and transport options (CACC 2014; see Chaps. 6 and 10 in this volume). Another UK focus has been on the development of alternatives to employment on projects like the proposed Trident nuclear submarine system renewal (CND 2016). There have also been moves to develop alternative plans for specific companies faced with closure threats, in relation to a GM car plant in Canada (Perry 2019) and shipbuilders Harland and Wolff in Northern Ireland (Wainwright 2019).

## What Now— for Products and Processes?

The Lucas workers also called for new forms of production, to avoid deskilling, and to shift to a human-centred production process. The issue is still with us. Although some companies may be producing green products, most of them are made in standard ways, with all the problems that can entail: work subdivision, speed up pressures and the drive for higher worker productivity. From the point of view of the workforce employed in conventional enterprises, not much has changed. It may be more satisfying, and more ethical, to work on green products, but as one trade unionist put it, 'a green boss is still a boss'. Much of the boom in low-cost solar PV energy has been due to the mass production of PV modules often in non-union plants, where wage levels and working conditions were often poor. In the 2010s, as market competition intensified, PV production mostly moved to China, where conditions and pay could be even worse, at least initially (Elliott 2015). Instances of 'poor factory labor standards', along with other social and environmental problems, have also been noted in the US solar industry (Mulvaney 2019). A shift to green products on its own is not enough. Attention also has to be paid to how they are made, including workers' control of production.

The argument for more 'workers control' has underlain grassroots campaigns and conflicts over the centuries, from the Luddites onwards (Elliott and Elliott 1976; Luddites 200 2013). When new technologies are imposed in industry, there can be resistance, which is often defensive, seeking to protect existing skills and pay levels. During periods when grassroots militancy and confidence is high, there can also be more *positive* initiatives, as in the case of the Lucas workers plan, seeking to expand the level of worker control to include what it produced and how it is produced.

While academic interests in human-centred (May et al. 2015) and socially useful production (Smith 2014) continued, there are as yet no examples of fully fledged *alternative production* approaches. Some grassroots initiatives may begin to blur the distinction between products and production, and who engages in designing and making things, as in 'Fablabs' and 'make spaces'. These are community-based centres set up to help people to explore 'new tech' ideas, often based on digital technology (Fablab 2019). One option is for initiatives like this to use 3D printing, which offers an accessible new tool for inventive 'bespoke production' of a wide range of products.

While there are some solar co-ops and community groups actively engaged with local not-for-profit solar and wind energy power generation projects, they mostly use equipment made elsewhere. Moreover, industrial PV solar module production is based on standard production methods, with the familiar occupational health and safety and environmental pollution issues, standard management control, and work organisation patterns. Due to competitive market rules operating on a global basis, job security can be unstable in this area, as in others. With cheap Chinese solar PV cells being available, there has been a trade war, with the USA imposing import tariffs, which some see as having contributed to the loss of 20,000 solar jobs in the USA (Ellsmoor 2019).

Although the social effects of market competition will remain an issue, with global demand for green energy systems growing rapidly and new enterprises opening up in the renewable energy field around the world, there may arise an opportunity to try to organise production in a different way. If so, what could be done? Staying with the solar PV cell example, some factory-based solar system design and production could benefit from the sort of human-centred approach outlined above. PV systems need to be configured for wide variety of end uses, and new applications are emerging, so there is opportunity for creative design and development work. However, for most module assembly work, if we want to get away from tiresome, mostly low-skilled mass production, there is only the option of shifting to automation. That is also the likely route for cell production work, with limited human involvement. Cell fabrication is already a capital-intensive, high technology process involving few technical specialists.

The situation will be different in each sector, with more or less human involvement. In some, full automation may make sense, in others limited automation could leave humans doing more interesting work, while in some cases they could be central to the process, with technology *adding* to their capacities, engendering a new form of improved social productivity and changed work patterns. The emphasis is on *change*, rather than on defending existing skills and jobs. Such technology would *upgrade* the skills of users.

That was the line taken by Mike Cooley. He looked to a new socially productive, skill-enhancing path ahead, which may well be viable in some sectors. At the same time, he also talked of 'minimising that part of work that is backbreaking or soul destroying', and most people would be in favour of that (Cooley 2018). Most standard mass-production can and arguably should be 100% automated.

While new jobs and skills might emerge, it could be that the total number of jobs, or at least hours of work required, might decrease. That opens up the wider issue of the future of work, and whether reducing the need for work is a good thing or not. That depends on how it is done.

## The Future of Work

Given the way our society and economy is organised, will more creative and inventive jobs be available, perhaps in non-manufacture sectors for those made redundant? A small elite apart, that is not what happens now: for most people, all that is on offer is part-time casual work on zero-hour contracts, or retail, warehouse and low-grade service jobs, with technology adding more pressure, until those jobs also disappear. Breaking out of this and working in local co-ops or community organisations is an option for some. There is much that can be done at the community level. For example, community drop-in centres can offer a range of support, advice and self-help and local trading/start-up opportunities (Monbiot 2019).

Workers co-operatives and retail co-ops offer new areas of employment. They have social benefits and, according to the United Nations Secretariat (2014), co-operatives employ over 12.5 million persons or roughly 0.2% of the world's population. In the UK, the UK Labour Party has talked of expanding the co-operative sector, that is, backing the creation of local community energy projects within the context of a wider publicly owned energy system (Labour Party 2018). However, there may be limits to how much employment might be supported via local co-ops.

The battle is no longer just one involving paid workers. Recent commercial developments have led to increased 'self servicing' since many goods now require final assembly by consumers (Glucksmann 2013). As new digital information and communication technology (ICT) has spread, the distinction between workers and consumers, producers and users, has changed even further. In social media systems, like Facebook and Twitter, the product is created by the user: the supplier just provides and manages the medium and the consumers create value without being paid. Instead, they have to pay to use the system buying and managing the operating systems.

Some see the production of physical consumer goods as becoming almost an ancillary activity in cash/profit terms, with the new high-value products increasingly being information and services. Although this new industry is different, there are many workers in this sector, that is, 'creative content' producers and Amazon-style retail/distribution staff, having to cope with poor working conditions (Semuels 2019; Guendelsberger 2019).

Nevertheless, with much consumption now being of information, broadly conceived, including education as well as entertainment, some falsely see us moving to a 'dematerialised' world, signalling a potential end not only to work, but to capitalism as we know it. Digital products and the information-based economy change the markets' rules, and, it is claimed, open possibilities for a post-capitalist future (Mason 2015). Some assume that an economy based on trade in information would reduce the use of energy and materials, as also could wider automation of production (Bastani 2019).

There are limits to Utopian visions, even leaving aside ownership and control issues and what the redundant workers might do, and how they might pay for goods and services. The information economy has limits, including environmental and resource limits. It takes energy to shift and process information, and to make the equipment to do that. For example, it has been claimed that, globally, about as much energy is now being used by Information and Communication Technology (ICT), including mobile devices and servers, as for air transport (BBC 2018).

It is not clear that a fully automated high-technology future would in fact deal with our wider environmental and resource problems. We will still need material and energy resources, even with the most technically efficient system, and automation may feed the growth of consumerism, making it easier to produce and to buy more things. So environmental impacts will still rise. There is a strong environmental case against endless material growth (Trainer 2019; Hickel and Kallis 2019). In theory, some form of 'stable state' work-free society could emerge in which consumerist growth was absent, with automated production then just meeting the more limited demand efficiently, but, in the current competitive market context, the drive to market expansion, and more sales and more products, is strong, and, unless radically redirected, automation would feed it.

What most people still spend most of their money on is basic necessities, such as food, housing, transport and energy, all with significant environmental and resource costs. Nevertheless, some of that is changing. Although the costs savings may not always be passed to consumers, the generation and environmental cost of energy is falling, as new technologies like zero-carbon PV solar spread: PV costs have fallen by 76% since 2009. It has even been claimed that, at some point, green power will be so ubiquitous and cheap it will not be charged for directly (Clark 2018). That is some way off: energy and most other things still have costs, and most people still need to work to be able to afford goods and services. Moreover, most of their work seems likely to involve producing goods and running services, including public transport and health care. It is hard to see that changing rapidly, with many people still working in low- or semi-skilled jobs, and in hard-to-automate jobs, including in social care, an area requiring special skills.

Nevertheless, increasingly, *some* of these employment opportunities may disappear, in production especially. If some form of economic exchange system is to survive, and support the remaining producers, ways have to be found to enable people no longer needed in paid production jobs to continue to participate in society and the economy.

## A Way Forward to a Secure Future?

Due to the oncoming pressures of increased automation, new ways have to be found for people to have a good living in an equitable and sustainable way. In the short term, reduced working hours may help provide a respite, if the remaining work can be fairly shared, but in the longer term, one option is a basic income for all, regardless of circumstances and capacities.

Proposals like this have figured in many recent responses to automation (Stollery 2018; Ford 2019). Within that model, some will continue to work in interesting, improved, jobs, and others will find creative ways to go beyond normative behaviours and roles. Such a future has its attractions, but there would be many implementation issues (Martinelli 2017) and we would need to rethink the basic concepts of work and employment.

One possible starting point, in this regard, is to look back to earlier definitions of what constitutes work and how it should be rewarded. Clause 4 of the UK Labour Party constitution talked of aiming

> to secure for the workers by hand or by brain the full fruits of their industry and the most equitable distribution thereof that may be possible upon the basis of the common ownership of the means of production, distribution, and exchange, and the best obtainable system of popular administration and control of each industry or service.

Although abandoned in the 1990s by *New Labour* under Tony Blair, that aim is still a good aim! Leaving the prospects for worker-owned companies aside (it is a big separate issue), a more general point is the distinction made between workers by hand and by brain, implying there is a fixed and permanent separation. Cooley's argument, taken to its logical conclusion, is that the artificial distinction between 'brain' and 'manual' work ignores the valuable tacit knowledge and skills associated with what is called manual work. Recognising that, and also the social value of non-commercially orientated work, could go some way to creating a society in which the social benefits of automation (less dull work, more time for living, caring and engaging in society) would outweigh its social costs in terms of job losses.

Historically, there has been a split between skilled trade and craft unions, growing out of the craft guild tradition, and general unions, covering unskilled mostly manual workers, but increasingly also 'white collar' clerical and administrative office workers. The skilled trade and craft unions, like the scientific and technical workers' unions that emerged later, tended to be better paid and protective of their jobs and skills, but the general and clerical unions had larger numbers, and sometimes that gave them more political clout. Levels of militancy and political commitment on each side have varied, but one of the successes of the Lucas Combine Committee was to bring both sides, blue and white collar, skilled and unskilled, together. That is vital to bridge the gap between 'hand and brain', and to widen the agenda concerning the nature of work, what is produced and how.

An area in which change is possible is in terms of commitments to the greening of industry. Trade unions in the UK and elsewhere have welcomed the growth of the 'green jobs'. However, they have sometimes been passive or reactive, worrying mainly about pay and conditions, in defensive mode. An early influential TUC report 'A Green and Fair Future' said that union support for environmental policies was 'conditional on a fair distribution of the costs and benefits of those policies across the economy, and on the creation of opportunities for active engagement by those affected in determining the future wellbeing of themselves and their families' (TUC 2008).

As the climate crisis has deepened, the Trade Union movement has promoted the idea of a 'Just transition', and Frances O'Grady, the current TUC General Secretary, backed the 'socially useful' technology approach in an introduction to a 2016 edition of Cooley's book *Architect or Bee*. However, the emphasis has mainly been reactive, focused on protecting workers rights and conditions in response to national and international policies on climate change (ETUC 2018).

That stance is understandable: why should workers bear more than their fair share of any costs incurred from making the necessary changes? Resistance to aspects of the proposed wider transition to a sustainable future is also understandable. For example, the transition to renewables will mean the loss of jobs in conventional energy industries, where unions are often well established. That issue can lead to conflicts between environmentalists and workers, for example in relation to coal mining and resistance to proposals for energy source changes, as recently in South Africa (Beetz and Bellini 2018; see Sikwebu/Aroun, Chap. 3, Pillay, Chap. 4, Cook, Chap. 8) and the USA (Marinucci and Kahn 2019).

While it may be true that in the longer term there will be new, replacement, jobs in new energy technologies, in the short term there could be painful disruptions. The change has to be carefully handled, with the necessary support being provided (IPPR 2019). The UK Labour party, in its Green New Deal plan, said it aimed to create guaranteed work in the new zero carbon economy '*for those whose current roles are set to change*', and it stressed the need for a 'just transition' and for '*a sustainable future based on good, secure, unionised jobs*' (Labour GND 2019). In the USA, Bernie Sanders stressed that under his Green New Deal plan, displaced workers in the fossil fuel and other carbon intensive industries would receive, strong benefits, a living wage, training, and job placement', adding:

> We will guarantee five years of a worker's current salary, housing assistance, job training, health care, pension support, and priority job placement for any displaced worker, as well as early retirement support for those who choose it or can no longer work. (Sanders 2019)

While these commitments are welcome, they seem defensive, seeking to avoid trade union opposition to the proposed energy transition. However, while it is necessary to provide guarantees, more has to be done. From the workers' perspective, the new jobs need to be better than the old jobs. The powerful extra message from the Lucas campaign was not just jobs at all cost, but better jobs making better things, in better ways. However, positive outcomes like that have to be fought for.

As technology changes, the nature of jobs will change, and this is resisted or resented when imposed from above. Even without the benefit of Lucas-style workers' plans for positive change, changing patterns of employment and work roles can, however, open up opportunities, as well as creating problems. It may often be hard to deal with change, but there are examples of creative efforts to rethink the approach to changed circumstances. For example, postal workers in France, faced by falling demand for mail services due to the spread of the internet, have taken on a limited social welfare role, given that they visit outlying areas. For many old age pensioners in rural France 'the postie' is one of the few regular visitors. For a small fee, the postal workers keep an eye on how they are faring, as an extension of their primary job (Chrisafis 2018). However, if this approach was extended, paid social/care workers may not be happy about what they may see as professional encroachment, and there may also be legal implications. Special training might be needed. Issues like that bedevil many attempts to redefine 'social' and 'paid' work and to open up new routes to voluntary social participation.

That is not to say that conventional employment rights do not need defending strongly, though the nature of what is being defended may have to change.

Many trade unions have been campaigning for a 'just transition' and for what are sometimes called 'Just Jobs'—green jobs which are sustainable and safe as well as properly paid. The International Trade Union Confederation has emphasised the production process as well as products, as part of a transition to a 'fairer, environmentally responsible society that respects human and labour rights' (ITUC 2010) and it has continued to campaign on that (ITUC 2012, 2019).

There are huge obstacles to making any of that viable. Moreover, there are even larger issues ahead in relation to growth. It is sometimes argued that, in order to deal with climate change and other environmental constraints, there will have to be a reduction in the level of economic growth: we have to move away from 'quantity' to 'quality' and seek prosperity without growth (Jackson 2017). That is easy to say, but there are many who are struggling just to live even at a subsistence level, so redistribution for them is vital, as also is economic growth. At the same time, there is a strong environmental and climate case for limiting or even halting growth (Trainer 2019). However, for a while, in some parts of the world, some growth is vital. That does not mean it has to be growth of the type we currently have: it can hopefully be a different kind of growth, prefiguring some of the new approaches to work, economy and life, while reducing environmental impacts (Pollin 2018).

There are some large issues to resolve, not least whether, on a planet with finite resources, we can move to an environmentally sustainable future, while still consuming more energy and other resources (Elliott 2020). While some 'eco-modernists' argue that technology will allow us to expand continually, their critics see this as not just utopian, but as dangerous, failing to recognise the global social, economic and environmental limits to growth that will not be overcome by technical fixes (Boehnert and Mair 2019).

While technology may be available to limit some impacts on the ecosystem, it is hard to see how the growth of material consumption can continue indefinitely. One implication is that we will have to move away from an economy in which growth is fundamental. That would require profound social and economic changes, including a move away from consumerism. This would have major impacts not just on lifestyles, but also on work. In a putative low or zero growth economy, there would presumably be less work to do, much as in the high-tech, high growth, fully automated economy proffered by some eco-modernists.

## Conclusions

A common starting point in the analysis of the relationship between technology and employment is the distinction between capital-intensive and labour-intensive projects. In the former, the hardware dominates, and in the latter, workers dominate. In this respect, as I have argued elsewhere (Elliott 2015), we are in the process of moving from 'manual' to 'high-tech' automated systems, with capital constantly replacing labour.

However, while labour is replaced in some sectors, need for it expands in others, at least until some of those areas are automated. The debate over the impacts of automation from the 1970s onwards has assumed that this process would continue (Elliott 1975; Kaplinsky 1984; Rifkin 1995; Benanav 2019). Optimists looked to a future of leisure with reduced work hours and a cornucopia of automated production, pessimists to a grim future of mass unemployment and deskilling, driven by a triumphant capitalism, benefiting mainly the elite and the affluent.

In the event, what has emerged has not been quite like that. Capitalism has triumphed, but so far has managed to spread affluence to some degree in the Global North by accelerating growth in both production and consumption, using advanced technology. We have seen the creation of mass consumerism and global markets, often based on new advanced products, with new groups of workers in newly industrialising countries taking over from the earlier unskilled workforce, and the rise of a technically skilled workforce alongside a vast new service and retail sector, but leaving many poor unskilled people on the margins, disproportionately in the Global South.

Apart from the likely elimination of many employment opportunities, and the consequent loss of consumer buying power, given its potentially massive environmental impacts, whether this system can continue to expand indefinitely has long been the subject of debate. The exploitation of natural resources and the planet has deepened, and it has been argued that there may be fundamental economic contradictions in the overall growth-based process. Karl Marx saw this in terms of the falling rate of profit, as rival chunks of capital tried to expand, forcing capitalists to reduce wages. The vast increase in productivity through technology has mostly avoided that outcome for the moment: some of the benefits have spread in the Global North. However, there may be limits to how much further this process can go: the cost of each high tech-based increase in capital productivity has been rising, often with diminishing returns. It may be that some new technologies can deliver economic outputs at lower direct economic cost, but it remains to be seen if the indirect social and environmental costs can also be reduced.

Some green optimists look to renewable energy and other sustainable technologies to allow for needs to be met with lower environmental impacts, while also supporting jobs. Some even hope that these technologies will allow for continued economic growth, including better employment opportunities. That could be attractive, given that there is a vast underclass of sweated labour, who barely enjoys any benefits, especially in the Global South.

While some have benefitted and become more affluent, the growth we have is also leading to more inequality, more environmental impact and ever-more stressful and insecure employment. However, there are some ideas for alternative approaches to the use of technology in production and to the use of the skills and knowledge of workers. From a trade union perspective, the availability of what the International Labour Organisation calls 'decent work', based on congenial, safe, properly remunerated and sustainable employment, could be seen as an ethical requirement, a basic right, as well as a driver for change in how technology is used more generally. In terms of growth, the main growth areas will hopefully be in sustainable energy technology development and use, a sector that already employs over 10 million people globally (Elliott 2020). In terms of responding to automation, one way forward might be to insist on a new approach to the divisions between 'hand and brain', and between 'social' care and 'paid' work. There is a need to recognise the value of all types of work, whether conventionally paid or not, including so-called domestic work (see Barth/Littig, Chap. 33 and Saave/Muraca, Chap. 32).

That opens up the issue of gender divisions, which have many implications for who does what type of work (paid or otherwise) and how that might or should change in future (Urban and Pürckhauer 2016). Within the paid work area, new technology has already had a massive impact on the sort of jobs women have been employed to do, including in factories and offices, and that will continue as automation spreads (Madgavkar et al. 2019; Roberts et al. 2019).

However, within the residual paid employment world, there could be new patterns of work with a more human-centred approach, engaging a wider range of skills, tacit knowledge and social orientations. Technology is thus used to enhance work, not just eliminate it. A new concept of 'social productivity' might emerge, with the *social use value* of products and production being fully recognised. Given that some jobs will and should disappear, for those less directly involved with production or service roles, or enjoying more time off work as working hours are cut, there may be opportunities for wider involvement in society, including in social care, an area of increasing importance as we now tend to live longer. At the very least, we need to move to a fairer balance in pay levels between those employed in these areas and those involved with industrial and commercial activities. We all need to engage in and value social care as part of a wider social responsibility and community commitment.

The reordering of social and economic values implied above would probably require a radical political transformation and a change in world view. That would include a new vision of the role of human-orientated technology, as Cooley (2018) has argued in his analysis of what he describes as 'an odd relationship with technology and scientific reductionism', stressing the potential for human liberation and creative freedom (Cooley 2020).

That level of change may be hard to make. There are practical issues in relation to the idea of widening social participation in shaping the future, including in relation to technology and work. Can we expect 'ordinary people' to solve problems that have beaten their so-called betters? I well remember at one of the Lucas workforce meetings, someone standing up from the floor and saying it's 'all very well the shop stewards talking about these grand ideas, but to us ordinary folk it's all new'. Some of it is down to confidence. We are mostly educated and trained to fit into our allotted roles, rather than being enabled to explore wider more creative ways of living and working. That can and should change. Social mobility can also help, but it's rare these days, and potentially divisive and marginal—a few escapees do not change much except themselves. Of more value is the creation of a widespread culture of 'higher expectations' in terms of worthwhile and sustainable life opportunities (Clark 2019).

Will those relieved of work just end up 'down the pub', watching TV, or playing computer games, once automation replaces the daily grind of conventional employment? There is nothing wrong with enjoying, and indeed expanding, opportunities for entertainment and similar leisure pursuits, although an endless diet of passive consumption may have its limits. However, the extent of hobbyist interests, artistic endeavours, participation in sports and other active recreation, educational enthusiasm and voluntary work, as well as the enthusiasm which most people show for family and community life, suggests that they also have more active things they want to do and ideas aplenty.

Will that be enough to sustain a creative, engaging society when paid work has diminished? That may depend on the sort of society we create. At present, gainful employment, within an increasingly competitive market context, is the only option for most. That determines how we work and the technologies we work with, use and produce. If we redefine what is meant by 'gainful' and 'productive' work, and open up new ways of living, based on a wider range of social engagement and a culture which values wider ranges of social contribution, the role of gainful employment could change. There will still be production, and some people will still work supporting that, but technology can be developed that has low social and environmental impacts and creates opportunities for new forms of social engagement and new skills used to meet new social ends.

**Acknowledgement** Tam Dougan provided useful insights and advice on some key aspects of this chapter.

# References

Bastani, Aron. 2019. *Fully Automated Luxury Communism; A Manifesto*. New York: Verso.
BBC. 2018. Climate Change: Is Your Netflix Habit Bad for the Environment? *BBC Realty Check Team*, October 12. https://www.bbc.co.uk/news/technology-45798523
Beetz, Becky, and Emiliano Bellini. 2018. Unbelievable: Coal Puts Halt to South Africa's Renewables Industry. *PV-magazine*, March 13. https://www.pv-magazine.com/2018/03/13/unbelievable-coal-puts-halt-to-south-africas-renewables-industry/
Benanav, Arron. 2019. Automation and the Future of Work-1. *New Left Review*, 119. https://newleftreview.org/issues/II119/articles/aaron-benanav-automation-and-the-future-of-work-1
Boehnert, Joanna, and Simon Mair. 2019. Techno-fix Futures Will Only Accelerate Climate Chaos—Don't Believe the Hype. *The Conversation*, October 31. https://theconversation.com/techno-fix-futures-will-only-accelerate-climate-chaos-dont-believe-the-hype-125678
Bookchin, Murray. 1965. Towards a Liberatory Technology. *Anarchy*, 78. https://theanarchistlibrary.org/library/lewis-herber-murray-bookchin-towards-a-liberatory-technology
Brynjolfsson, Erik, and Andrew McAfee. 2014. *The Second Machine Age: Work, Progress, and Prosperity in a Time of Brilliant Technologies*. London: Norton.
CACC. 2014. One Million Climate Jobs. *Campaign Against Climate Change Trade Union Group*. https://www.cacctu.org.uk/sites/data/files/Docs/one_million_climate_jobs_2014.pdf
Chrisafis, Angelique. 2018. Care Package: The French Postal Workers Helping Lonely Older People. *The Guardian*, November 23. http://www.theguardian.com/world/2018/nov/23/care-package-french-postal-workers-helping-lonely-older-people

Clark, Bryan. 2018. Renewable Energy Will be 'Effectively Free' by 2030. *The Next Web*, August 14. https://thenextweb.com/insider/2018/08/14/analyst-renewable-will-be-effectively-free-by-2030/

Clark, Tom. 2019. The Social Mobility Trap. *Prospect*, 284, pp. 49–53. https://www.prospectmagazine.co.uk/magazine/the-social-mobility-trap-education-schools-equality-jobs-work

CND. 2016. Trident and Jobs. *Campaign for Nuclear Disarmament*. London: CND. https://cnduk.org/wp/wp-content/uploads/2018/02/Trident-and-Jobs.pdf

Cooley, Mike. 1972. Computer Aided Design—Its Nature and Implications. *AUEW (TASS) Publications*, reproduced in Cooley, M. (2020).

———. 1982. *Architect or Bee?* London: The Hogarth Press (new edition Spokesman 2016). First self-published by Langley Technical Services in 1980.

———. 1983. The New Technology: Social Impacts and Human-Centred Alternatives. *Technology Policy Group*, Occasional Paper 4. Milton Keynes: The Open University.

———. 1987. Human Centred Systems: An Urgent Problem for Systems Designers. *AI & Society Journal*, reproduced in Cooley, M. (2020).

———. 2018. *Delinquent Genius*. Nottingham: Spokesman Books.

———. 2020. *The Search for Alternatives: Liberating Human Imagination*. Nottingham: Spokesman Books.

Elliott, David. 1975. *The Future of Work*. Man Made Futures course file, T262/7. Milton Keynes: The Open University.

———. 2015. Green Jobs and the Ethics of Energy. In *Ethical Engineering for International Development and Sustainability*, ed. Marion Hersh, 141–164. London: Springer.

———. 2019. *Renewable Energy in the UK: Past, Present and Future*. Basingstoke: Palgrave Macmillan.

———. 2020. *Renewable Energy: Can it Deliver?* Cambridge: Polity.

Elliott, David, and Ruth Elliott. 1976. *The Control of Technology*. London: Taylor & Francis.

Elliott, David, and Veronica Mole. 1987. *Enterprising Innovation: Technology for People*. London: Belhaven.

Ellsmoor, James. 2019. Under Trump's Tariffs, The US Lost 20,000 Solar Energy Jobs. *Forbes*, February 24. http://www.forbes.com/sites/jamesellsmoor/2019/02/24/under-trumps-tariffs-the-us-lost-20000-solar-energy-jobs/

ETUC. 2018. A Guide for Trade Unions: Involving Trade Unions in Climate Action to Build a Just Transition. *European Trade Union Confederation*. Brussels: ETUC. https://www.etuc.org/en/pressrelease/involving-trade-unions-climate-action-build-just-transition

Fablab. 2019. Fablab Foundation web site: https://www.fablabs.io and a Fablab guide: https://www.fablabni.com/what-fablab.html

Ford, Martin. 2019. Why a Universal Basic Income is the Answer to Job Automation. *Medium.com*, with links to TED talk. https://medium.com/@MFordFuture/why-a-universal-basic-income-is-the-answer-to-job-automation-d111d7eac430

Glucksmann, Miriam. 2013. Working to Consume: Consumers as the Missing Link in the Division of Labour. Centre for Research in Economic Sociology and Innovation (CRESI) Working Paper 2013-03, *University of Essex*. http://repository.essex.ac.uk/id/eprint/7538

Guendelsberger, Emily. 2019. I Worked at an Amazon Fulfillment Center; They Treat Workers Like Robots. *Time*, July 18. https://time.com/5629233/amazon-warehouse-employee-treatment-robots/

Hickel, Jason, and Giorgos Kallis. 2019. Is Green Growth Possible? *New Political Economy*, April, 1–18. https://doi.org/10.1080/13563467.2019.1598964.

Hines, Colin, and Graham Searle. 1978. *The Chips are Down*. London: Earth Resource Research.

IPPR. 2019. A Just Transition: Realising the Opportunities of Decarbonisation in the North of England. *Institute for Public Policy Research*. London: IPPR. http://www.ippr.org/research/publications/a-just-transition

ITUC. 2010. Resolution on Combating Climate Change through Sustainable Development and Just Transition. *International Trade Union Confederation*. Brussels: ITUC. http://www.ituc-csi.org/resolution-on-combating-climate

———. 2012. Growing Green and Decent Jobs. *International Trade Union Confederation*. Brussels: ITUC. https://www.ituc-csi.org/IMG/pdf/ituc_green_jobs_summary_en_final.pdf

———. 2019. ITUC Frontline Campaigns and Four Pillars for Action 2019. *International Trade Union Confederation*. Brussels: ITUC. https://www.ituc-csi.org/ituc-frontline-campaigns-and-pillars

Jackson, Tim. 2017. Prosperity without growth. Sustainable Development Commission, 2nd edn. London: Routledge.

Kaplinsky, Raphael. 1984. *Automation: The Technology and Society*. Harlow: Longman.

Labour GND. 2019. UK Labour Green New Deal Outline. https://www.labourgnd.uk/gnd-explained

Labour Party. 2018. Bringing Energy Home. UK Labour Party. http://www.labour.org.uk/wp-content/uploads/2019/03/Bringing-Energy-Home-2019.pdf

Lawrence, Mathew, Carys Roberts, and Loren King 2017 Managing Automation Employment, Inequality and Ethics in the Digital Age. *Institute for Public Policy Research*. London: IPPR. https://www.ippr.org/files/2017-12/cej-managing-automation-december2017-1-.pdf

Luddites 200. 2013. Brief History of the Luddites. Luddites 200 web site. https://www.luddites200.org.uk/theLuddites.html

Madgavkar, Anu, et al. 2019. The Future of Women at Work: Transitions in the Age of Automation. McKinsey Global Institute (with 9 authors). https://www.mckinsey.com/~/media/McKinsey/Featured%20Insights/Gender%20Equality/The%20future%20of%20women%20at%20work%20Transitions%20in%20the%20age%20of%20automation/MGI-The-future-of-women-at-work-Exec-summary-July-2019.ashx

Marinucci, Carla, and Debra Kahn. 2019. Labor Anger over Green New Deal Greets 2020 Contenders in California. *Politico*, June 6. https://www.politico.com/states/california/story/2019/06/01/labor-anger-over-green-new-deal-greets-2020-contenders-in-california-1027570

Martinelli, Luke. 2017. Assessing the Case for a Universal Basic Income in the UK. Institute for Policy Research, University of Bath. https://www.bath.ac.uk/publications/assessing-the-case-for-a-universal-basic-income-in-the-uk/attachments/basic-income-policy-brief.pdf

Mason, Paul. 2015. The End of Capitalism has Begun. *The Guardian*, July 17. https://www.theguardian.com/books/2015/jul/17/postcapitalism-end-of-capitalism-begun

May, Gokan, Marco Taisch, Andrea Bettoni, Omid Maghazei, and Bojan Stahl. 2015. A New Human-centric Factory Model. *Procedia CIRP* 26: 103–108. https://www.sciencedirect.com/science/article/pii/S2212827114009251#!

Monbiot, George. 2019. Could This Local Experiment be the Start of a National Transformation? *The Guardian*, January 24. https://www.theguardian.com/commentisfree/2019/jan/24/neighbourhood-project-barking-dagenham

Mulvaney, Dustin. 2019. *Solar Power: Innovation, Sustainability, and Environmental Justice*. Orlando: University of California Press.

Perry, Elizabeth. 2019. A Proposal to Convert GM Oshawa to Electric Vehicle Production under Public and Worker Ownership. *Work and Climate Change Report*, October 30. https://workandclimatechangereport.org/2019/10/30/a-proposal-to-convert-gm-oshawa-to-electric-vehicle-production-under-public-and-worker-ownership/

Pizzolato, Nicola. 2013 The Making and Unmaking of Fordism. In *Challenging Global Capitalism*, N. Pizzolato. New York: Palgrave Macmillan.

Pollin, Robert. 2018. De-Growth vs A Green New Deal. *New Left Review*, 112, August. https://newleftreview.org/issues/II112/articles/robert-pollin-de-growth-vs-a-green-new-deal

Rifkin, Jeremy. 1995. *The End of Work: The Decline of the Global Labor Force and the Dawn of the Post-Market Era*. London: Putnam.

Roberts, Cary, Henry Parkes, Rachel Statham, and Lesley Rankin. 2019. The Future is Ours: Women, Automation and Equality in the Digital Age. Institute for Public Policy Research. London: IPPR. https://www.ippr.org/research/publications/women-automation-and-equality

Rosenbrock, Howard, ed. 1989. *Designing Human-Centred Technology*. London: Springer-Verlag.

Sanders, Bernie. 2019. The Green New Deal. Campaign blog. https://berniesanders.com/issues/the-green-new-deal

Semuels, A. 2019. Every Game You Like is Built on the Backs of Workers: Video Game Creators are Burned Out and Desperate for Change. *Time*, June 11. https://time.com/5603329/e3-video-game-creators-union/

Silver, Beverley. 2003. *Workers' Movements and Globalization since 1870: Forces of Labour*. Cambridge: Cambridge University Press.

Smith, Adrian. 2014. Socially Useful Production. STEPS, University of Sussex, Working Paper 58. https://steps-centre.org/wp-content/uploads/Socially-Useful-Production.pdf

———. 2015. Technology Networks for Socially Useful Production. *Journal of Peer Production* (5). http://peerproduction.net/issues/issue-5-shared-machine-shops/peer-reviewed-articles/technology-networks-for-socially-useful-production/

Srnicek, Nick, and Alex Williams. 2015. *Inventing the Future: Postcapitalism and a World Without Work*. London: Verso.

Stollery, Brad. 2018. Universal Basic Income is an Inevitable Part of Our Automated Future. *Medium.com*, February 21. https://medium.com/age-of-awareness/universal-basic-income-is-an-inevitable-part-of-our-automated-future-3cc181d4778d

Trainer, Ted. 2019. Why De Growth is Essential: A Rejection of Left Ecomodernists Phillips, Sharzer, Bastini, and Parenti. *Resilience*, October 17. https://www.resilience.org/stories/2019-10-17/why-de-growth-is-essential-a-rejection-of-left-ecomodernists-phillips-sharzer-bastini-and-parenti/

TUC (Trades Union Congress). 2008. A Green and Fair Future. London: TUC. https://www.tuc.org.uk/sites/default/files/documents/greenfuture.pdf

United Nations Secretariat, Department of Social and Economic Affairs. 2014. https://www.un.org/esa/socdev/documents/2014/coopsegm/grace.pdf. Accessed November 2020.

Urban, Janina, and Andrea Pürckhauer. 2016. Feminist Economics. *Exploring Economics*, December 18. https://www.exploring-economics.org/en/orientation/feminist-economics/

Wainwright, Hilary. 2019. The Harland and Wolff Workers Want to Make Renewable Energy. A Labour Government Would Help Them. *Red Pepper*, August 24. https://www.redpepper.org.uk/the-harland-and-wolff-workers-want-to-make-renewable-energy-a-labour-government-would-help-them/

Wainwright, Hilary, and David Elliott. 1982. *The Lucas Plan: A New Trade Unionism in the Making*. London: Allison and Busby. New ed. Nottingham: Spokesman Press, 2017.

WEF. 2018. The Future of Jobs. *World Economic Forum*. Centre for the New Economy and Society. http://www3.weforum.org/docs/WEF_Future_of_Jobs_2018.pdf

# Index[1]

**NUMBERS AND SYMBOLS**
3D' jobs, 576
15&Fairness, 22, 298, 301–305, 307, 309–311

**A**
Abstract labour/abstract labor, 686, 733, 774, 807
Accountability, 42, 43, 91, 407
Accumulation, 25, 62, 68, 87, 155, 178, 183, 192, 208, 215, 259, 261, 320–322, 367, 393, 395, 398, 402, 407, 408, 662, 678, 681, 682, 687, 691, 693, 695, 700, 722, 728, 731, 735–739, 754, 756, 757, 760–762, 774, 803, 806, 806n8, 807, 808n9, 808n10, 809, 810
Activism, 11, 132, 159, 164, 190, 213, 215–217, 261, 285, 320, 340, 344, 365, 374, 381, 396, 397, 403, 452, 453, 457, 525, 565, 573, 574, 577, 613, 744
Advocacy coalition, 415–434
Affective labour, 393
Affluence hypothesis, 443
Affordances, 670, 671
African National Congress (ANC), 60, 85, 91–93, 92n13, 96, 98
Agrarian, 52, 210, 211, 272, 274–282, 289, 290, 322, 324, 325, 327, 333, 341, 345, 347, 419, 431, 433
 protest, 340–341, 347–350
 reform, 241, 274–278, 323, 324, 327, 328, 333, 340, 343–349
Agriculture, 3, 23, 36, 37, 45, 50, 62–65, 77, 182, 210, 212, 213, 273–275, 277, 281, 289, 340, 344, 345, 357, 392–394, 401, 405, 409, 415, 416, 418–423, 426, 432, 434, 470, 505, 542, 564n1, 568, 653n3, 707, 709, 736, 752, 810
 agribusiness, 274–277, 281, 282, 289, 340, 343–347, 349, 354, 359, 360, 398, 425, 426, 433
 agricultural intensification, 568
 agricultural production units (UPA), 274
 agri-food sector, 416

---
[1] Note: Page numbers followed by 'n' refer to notes.

© The Author(s), under exclusive license to Springer Nature Switzerland AG 2021
N. Räthzel et al. (eds.), *The Palgrave Handbook of Environmental Labour Studies*,
https://doi.org/10.1007/978-3-030-71909-8

863

## Index

Agroecology, 9, 23, 279, 281, 323, 354, 408, 415–421, 423, 425–434
 transition, 417, 421, 430, 431
Alienation, 52, 84, 90, 322, 395, 396, 725–727, 747, 751, 800–804, 820, 826
Alliances, 3, 4, 6, 7, 16, 17, 19, 21–24, 27, 28, 76, 85, 86, 86n7, 88, 92, 93, 93n14, 95, 97, 98, 114, 119, 128, 129, 137, 143, 166, 178, 181, 182, 184, 185, 188, 193, 217, 226, 230, 237–239, 252, 258, 273, 279, 283, 285, 289, 290, 296, 298, 306, 309, 311, 326, 327, 330, 334, 335, 345, 356, 361, 365–382, 389, 390, 393, 397–400, 409, 429, 434, 454, 498, 510, 526, 566, 577, 598, 601, 607–611, 615, 616, 653, 660, 661, 668–670, 712, 713, 715, 735, 739, 747, 758, 759, 762, 782, 785, 797, 798, 819
 Alliance for the Climate (Alianza por el Clima), 668, 669
 Alliance Technique (*Allianztechnik*), 653
 BlueGreen/Blue-green, 299, 379, 390, 393, 397–400, 409, 591, 602, 607–611, 613, 712
 BlueGreen Working Group, 597–601, 597n5, 601n11, 604, 608, 611, 612
 Labour-Green Alliance, 160, 162
 red-green, 76, 88, 98, 184
Althusser, Luis, 360, 662
Amazon/Amazonas, 275, 276, 278, 282–284, 287, 288, 319, 324, 325, 328–330, 334, 340n1, 346, 351, 355, 358, 359, 398, 795, 817

American Federation of Labor (AFL), USA, 142, 231, 587, 594
American Federation of Labor and Congress of Industrial Organizations (AFL-CIO), USA, 106, 109, 114, 115, 127, 128, 136, 232, 253, 522, 587, 591, 592, 594, 597, 598, 598n7, 602–608, 610, 612, 615
Antagonism, 9, 217, 730
Anthropocene, 6, 677
Anti-Nuclear, 142, 599, 635, 670
Apartheid, 64, 89, 91, 180
Appropriation, 27, 28, 163, 321, 360, 367, 391, 417, 433, 680, 728, 733, 756, 760, 770n1, 772, 773, 775, 775n6, 776, 780, 786, 794, 807–811
Arcelor-Mittal, 66, 527, 527n1, 528, 531
Architecture, 404, 815, 821, 823, 831, 833
Ashley, Brian, 182, 254, 259, 261, 714
Australian Workers Union, 157
Autogestion, 395, 404, 406–409
Automation, 26, 47, 129, 596, 733, 806, 820, 828, 830, 839–843, 847, 849, 850, 854–856
Automobiles, 75, 540, 549, 555–558, 609, 784, 785
 industry, 547–557, 783, 784
Automobility, 704, 777, 781, 784
Autonomy, 90, 288, 302, 320, 322, 324, 326–329, 331, 333–335, 390, 393, 395, 403, 404, 417, 421, 427, 555, 594, 604n14, 687, 723, 726, 727, 734, 751, 754, 756, 761

## B

B-Team, 234, 236, 239, 297, 526
Bargaining for the Common Good (BCG), 121
Base-load, 60
BASF strike, 612
Baugh, Bob, 598n7, 602–606, 609, 615, 616
Baugh, Chris, 251
Being-in-common, 397, 409
Bhopal, 39–41, 44, 49
Bielefeld School, 745, 758
Bio-
    capitalism, 735–739
    diversity, 2, 4, 17, 45, 189, 279, 288, 321, 323, 331, 332, 621, 654, 661, 668, 810, 816
    economy, 736
    sphere, 677–695, 728, 736, 756, 800, 802
Black Panthers Party, 832
Blake, William, 4, 4n1
Bloch, Ernst, 6, 653, 660, 726n7, 796n1
Boff, Leonardo, 342, 356, 621, 622, 633
Bolsa Verde programme (Brazil), 478n14
Bookchin, Murray, 723n1, 822, 823, 840
Bourdieu, Pierre, 239n2, 359, 529
Brazil-nut (*castanha*), 327, 328
Brecher, Jeremy, 115, 116, 118, 120, 159, 259, 375, 612, 647
Bridge to the Future Alliance (Norway), 255, 258
Burrow, Sharan, 252
Business unionism, 374, 375, 547, 554, 558

## C

Camargo, João, 253, 257, 263
Campaigning, 10, 157, 160, 163, 165, 445, 446, 448, 450, 452–454, 629, 642, 853
Canadian, 253, 259, 263, 265, 296–298, 300, 301, 303, 305–307, 310, 367, 368, 423, 453, 584, 593, 593n1, 596, 596n3, 603
    Labour Congress, CLC, 232, 252, 298–300, 307
    Union of Postal Workers, CUPW, 304, 308
Capital, 14, 26, 41, 46, 48, 68, 72, 98n18, 120, 155, 212, 216, 232, 242, 259, 261, 320–322, 324, 328, 335, 339, 345, 350, 354, 380, 381, 389, 393, 401, 421, 423, 483, 541, 545–547, 571, 574, 576, 606, 607, 630, 653, 679, 681–683, 687, 687n16, 691, 695, 700, 703n3, 707, 710, 721, 726, 728–730, 732, 733, 735, 737–739, 749n3, 752–754, 756, 760–762, 774, 777, 782, 793, 794, 799–802, 806–811, 806n8, 808n9, 816, 822, 841, 847, 854
    capitalism, 7, 53, 83, 84n2, 85, 90, 93, 95, 97, 98n18, 99, 120, 121, 129, 183–184, 200, 204–206, 205n5, 208, 214, 217, 234, 243, 319, 322–325, 333, 342, 344, 346n3, 354, 360, 366–369, 371, 374, 378, 379, 381, 382, 391, 392, 394, 397, 398, 400, 401, 407, 415, 419, 589, 594, 679, 681, 686–689, 691, 693, 694, 700, 701, 704n4, 705, 706, 708, 715, 724, 727–738, 733n9, 736n10, 744, 744n1, 751, 754, 755, 757, 760, 762, 763, 774, 775, 799–802, 829, 848, 854
    capitalist growth imperative, 700
    cognitive capitalism, 722, 731–734, 736
    mobility, 128, 143
Capitalocene, 793

# 866  Index

Carbon, 64, 70–72, 74, 76, 93, 108, 118, 167, 178, 253, 260, 263, 319, 381, 400, 483, 485, 519, 521, 522, 525, 526, 532, 539, 544, 553, 577, 609, 816, 828, 831, 833, 852
  $CO_2$ emissions, 4, 24, 519, 543, 552, 555, 657, 668, 743
  democracy, 710
  efficiency, 260
  emissions, 13, 21, 63, 120, 154, 164, 179, 182, 194, 195, 231, 254, 519, 520, 532, 557, 624, 628, 643, 748, 816
  footprint, 520
  lock-in, 64, 70–73, 75–77
  low carbon futures, 237, 621, 828
Care, 8, 46, 85, 115, 143, 231, 254, 302, 304, 311, 353, 393, 395, 396, 402, 405, 409, 443, 451, 503, 505, 542, 569, 571n3, 572, 573, 597, 604, 607, 628, 635, 701, 714, 735, 752, 754, 756, 756n7, 759–763, 776n7, 786, 798, 805, 806, 809, 811, 817, 828, 834, 849, 852, 855, 856
  caring responsibilities, 569
  work, 8, 26, 760, 776, 780, 805, 805n7, 809, 810
  workers, 304, 852
Caste, 36, 38, 272, 818
Catholicism, 277, 632, 633, 724n2
Certification schemes, 420, 426, 427, 432
Chief Seattle, 637
Christian, 341, 342, 356, 522, 626, 632
Church of Norway (Protestant), 252
Civil society, 19, 237–239, 239n3, 280, 285, 286, 307, 343, 346n5, 418, 421, 431, 495, 496, 502, 508, 574, 667–669, 671, 732

Clarke, Tony, 253, 257–259, 265
Class
  alliances, 85, 86, 345
  compromise, 200, 705, 710
  consciousness, 156, 207n6, 242, 243, 759
  fragmentation of class interests, 206
  politics, 200, 217, 369, 738
  and social movements, 204
  struggle, 200, 204, 207n6, 208, 231, 276, 356, 357, 398, 400, 735
Clean coal technology, 186, 629
Clean Development Mechanism (CDM), 74
Clearcutting, 133, 134
Climate
  change, 14, 59, 70, 84, 105, 127, 152, 179, 215n10, 225, 249, 287, 304, 342, 366, 400, 415, 467, 493, 517, 539, 584, 621, 655, 707, 749, 773, 851
  change adaptation, 45, 416, 417
  change mitigation, 68, 106, 110, 115, 117, 167, 288, 484, 517–519, 521, 524, 528–531, 584
  commons, 400
  contrast with green jobs, 250
  crisis, 6, 13, 14, 17, 20, 99, 116, 120, 122, 166, 167, 177–180, 183, 184, 192, 194, 225, 241, 249, 255–257, 259, 261, 264, 265, 304, 305, 358, 365, 371, 382, 400, 401, 408, 503, 584, 621, 623, 624, 647, 649, 655, 666, 667, 670, 711, 713, 785, 794, 795, 810, 851 (*see also* Climate emergency)
  job quality of, 255
  mitigation, 105, 113, 400, 494–496, 503, 509, 510, 520, 523, 524

Index    867

origin of, 305
plans - coalitions for, 264
plans - inclusiveness of strike, 264
policy, 115, 308, 468, 472, 473, 482, 484, 503, 519, 521, 522, 525, 532, 542–545, 548, 550, 555, 558, 600, 605, 606, 612, 615, 643, 711, 713, 749, 782, 783
protection, 106, 109–117, 120, 770, 783
Climate Action Summit 2019 (UN, New York), 264
Climate emergency, 539, 668
Climate jobs
*One Million Climate Jobs (UK)*, 234, 250
*One Million Climate Jobs Campaign* (South Africa), 93, 93n14
Climate Justice Alliance, 166, 408, 598, 614, 615
Climate Resilient Green Economy (Ethiopia) (CRGE), 485
*Climáximo* (Portugal), 253, 257, 263, 264
Coal
fired power stations, 45, 65, 70, 164, 177, 186–188, 190, 191, 194, 194n2, 296, 715n13
mining, 21, 60, 177, 178, 186, 189–194, 298, 471, 477n13, 480–482, 484, 487, 550, 552, 567, 577, 634n4, 711, 851
phase-out, 298, 299, 548–556
Coalition, 3, 22, 37, 86, 86n8, 88, 89n12, 95, 121, 138, 150–152, 159–161, 182, 185, 186, 189, 190, 217, 236, 238, 239n3, 250–252, 256, 258, 260, 262, 264–266, 306–311, 320, 330, 365, 366, 369, 371, 374, 379, 381, 399, 405, 408, 417, 430, 431, 433, 451, 453, 533, 552, 577, 613
Coalition for Clean and Safe Ports, 160–162, 168
Co-becoming, 395–397, 409
Collective action
agreements, 639
bargaining, 156, 380
Colonialism
colonial job blackmail, 402–408
decolonisation, 240, 406
Comisiones Obreras (Spain) (CCOO), 586, 588, 589, 647–671
Commission on Sustainable Development (United Nations) (CSD), 165
Commodification, 180, 259, 322, 328, 331, 333, 334, 401, 425, 434, 677–695, 726, 733–735, 749, 751, 807
Commons
commoning, 8, 23, 322, 323, 326–331, 333, 334, 389–409, 761, 763
Communist, 136, 200, 207n6, 208, 326, 587, 593, 625, 626
communist party, 45, 88, 88n11, 207, 208, 214, 588, 651, 724
Community
-development, 441, 442, 445–452, 565, 822
unionism, 159, 442, 453–455, 457, 759
Company-level bargaining, 526
Compensation, 50, 51, 119, 138, 157, 158, 191, 192, 283, 328, 348, 426, 508, 523, 552, 571, 571n2, 572, 574–576, 779
Comunidades Eclesiais de Base (Basic Ecclesial Communities, Brazil), CEB, 341, 342

## 868  Index

Confederação Nacional dos Trabalhadores na Agricultura (National Confederation of Agricultural Workers, Brazil), CONTAG, 324
Confédération Française Démocratique du Travail, (FRANCE) CFDT, 524, 552, 727
Conference of the Parties (COP) of UNFCCC, 59, 473
Congress of Industrial Organizations (CIO), 134, 135, 142, 587, 594, 604n14, 611
Congress of South African Trade Unions (COSATU), 59, 60, 68, 70, 84–86, 86n7, 92, 93, 95–97, 99, 178n1, 181–183, 185, 187, 188, 193, 194, 194n2, 454, 483n21, 545
Consensus, 22, 89n12, 106, 109, 166, 225–244, 289, 480, 483, 484, 529, 530, 604, 655, 684
Conservation, 10, 45, 46, 86n8, 133–136, 180, 208, 320–323, 329–331, 334, 400, 420, 444, 486, 709, 834
Construction, 23, 25, 42, 50, 63, 64, 72, 74, 112, 113, 113n1, 113n2, 115, 135, 188, 207, 253, 259, 265, 309, 329, 333, 345, 348, 353, 365, 369, 372–378, 380, 381, 405, 429, 454, 468, 470, 483n21, 486, 520, 525, 530, 542, 565, 566, 568, 575, 576, 596n3, 600, 601, 604, 606, 610, 652, 663, 683–691, 694, 772, 807, 831, 834
Constructionist grounded theory, 206
Constructivism, 689, 694
Consumption, 15, 64, 71, 84, 87, 108, 183, 184, 194, 232, 242, 482, 493, 496, 497, 500, 504, 506, 539, 542, 557, 568, 578, 584, 668, 683, 683n6, 699–708, 729, 730, 743, 745, 747, 748, 750, 752, 754, 777, 778, 781, 785, 825, 830, 848, 853, 854, 856
Cooley, Mike, 26, 625n2, 636–639, 815, 827, 842–844, 847, 850, 851, 856
Coordinating Body of Social Movements (CMS), 285
Cosmopolitics, 282–288
Council for Citizen Participation and Social Control, 280
Countries
 Argentina, 232, 242, 399, 826
 Australia, 21, 61, 157, 158, 256, 444, 454, 455
 Austria, 700n1, 724n2, 771, 771n3, 775n6, 779, 781–786
 Bangladesh, 452, 479n17
 Bolivia, 87, 88, 397, 493, 494
 Brazil, 10, 14n2, 22, 23, 85, 184, 320, 322–326, 331, 335n2, 339–361, 390n1, 398, 418, 427, 470, 478n14, 493, 525, 584, 622n1, 649n1, 654, 659, 667
 Burkina Faso, 468
 Cambodia, 468
 Cameroon, 494
 Canada, 22, 251, 253, 257, 264, 265, 295–311, 366, 367, 369, 370, 379, 381, 451, 453, 455, 588, 845
 Chad, 468
 Chile, 825
 China, 15, 24, 45, 263, 470, 477n13, 493, 563–578, 706, 707n7, 822, 846
 Colombia, 479n17
 Cuba, 418
 Denmark, 74, 478n15, 525
 Dominican Republic, 233

## Index

Ecuador, 22, 271–274, 277, 279, 279n4, 282–284, 283n7, 286–288, 290, 291, 478n14
Egypt, 493, 693
Estonia, 478n15
Ethiopia, 24, 478, 480, 485–487
France, 24, 73, 74, 84, 478n15, 479n16, 525, 526, 548, 552, 554, 555, 557, 559, 681, 725, 745, 852
Germany, 74, 475n11, 478n15, 526, 543, 548, 549, 551, 554, 555, 557–559, 654, 712n11, 745, 771, 771n3, 775n6, 779, 781–786
Ghana, 24, 494, 498–509
Guyana, 477n12, 479n17
India, 2, 14, 15, 20, 35–53, 88, 199–218, 351, 390n1, 393, 470, 478n15, 493, 494, 584, 622n1, 649n1, 667, 706
Indonesia, 475n11, 494, 825
Iran, 494
Ireland, 73, 74, 640
Italy, 10, 73, 74, 397–399, 553, 681n2
Jamaica, 468
Jordan, 494
Kenya, 390n1, 398, 475n11, 478
Mexico, 392, 418, 425, 427
Nepal, 88, 88n11, 493
Netherlands, 158
Nigeria, 24, 398, 493, 494, 498–509
Norway, 22, 73, 74, 215n10, 251, 252, 254, 256, 258, 260, 264, 826, 827
Peru, 15
Philippines, 471n3, 477n12, 478n15, 479n17
Poland, 236, 473, 524, 525, 548–551, 554, 557, 559
Portugal, 251, 253, 256, 257, 263, 264
Puerto Rico, 23, 390, 402–409
Republic of Korea, 478n15
Senegal, 23, 417, 421, 423, 425–433, 433n1
Sierra Leone, 468
South Africa, 14n2, 21, 24, 51, 59–66, 68, 69, 70, 72, 76, 85n3, 89, 95, 96, 99, 177–195, 251, 252, 254, 263, 264, 266, 351, 390n1, 455, 475n11, 478, 478n15, 480, 482–484, 486, 487, 584, 622n1, 649n1, 659
Spain, 6, 12, 14n2, 24, 25, 73, 74, 351, 402, 480–482, 486, 487, 583–589, 622n1, 642, 649, 649n1, 655, 663, 669
Sudan, 24, 468, 494, 495, 498–505, 507–509
Sweden, 351, 524, 584, 622n1, 649n1, 667, 826, 827
Togo, 468
Trinidad and Tobago, 398, 493
United Kingdom, 6, 10, 25, 73, 93n14, 249–252, 255–258, 263, 264, 348, 351, 441–457, 496, 497, 526, 583–589, 622–625, 622n1, 629, 631n3, 649n1, 654, 659, 667, 842, 845, 848, 851
USA, 7, 9, 10, 21, 25, 40, 42, 61, 75, 86, 143, 144, 159, 160, 164, 165, 230, 232, 259, 262, 279, 299, 302, 306, 308, 311, 347, 402, 403, 405n11, 407, 444, 449, 451, 453–455, 468, 470, 473, 475n11, 497, 522, 541, 583–589, 591–616, 653n3, 654, 661, 712, 729, 829–831, 833, 846, 847, 851, 852
Venezuela, 456, 494
Yemen, 493
Zimbabwe, 24, 494, 498–505, 507–509

## 870  Index

COVID-19, 128, 456, 494n1, 639, 714, 715, 762, 783
Craft work, 842
Creativity, 636, 751, 752, 843
Crime rates, 470
Critical juncture, 71, 72, 76
Cry of the earth, 633
Cry of the poor, 633
The Cry of the Xcluded (South Africa), 266
*Cuencas Sagradas* (Amazon), 287, 288
Cultural diversity, 279, 288
Cuomo, Governor Andrew, 254, 255

### D

Dakota Access Pipeline (DAPL), 120, 365, 366, 371–379
Dams, 38, 42, 45, 49–51, 348, 395
Debt, 110, 115, 179, 283–285, 288, 321, 326, 328, 402, 407, 503, 505, 563–578, 736, 736n10, 737, 748
Decarbonisation, 61, 65, 112, 297, 402, 484, 517, 519–521, 523–526, 529–532, 545, 547, 549, 552, 554, 557–559, 714, 779, 824, 825, 828
Decent
  jobs, 93, 234, 243, 380, 475, 523, 780n10
  life, 20, 84, 99
  work, 20, 84, 84n2, 99, 165, 235, 304, 305, 309, 468–470, 474, 475, 477, 478, 522, 710, 713, 780, 855
Decision-making, 11, 43, 44, 71, 89, 118, 119, 235, 236, 240, 242, 334, 381, 401, 448, 508, 518, 529, 530, 532, 533, 541, 584, 660, 785

Deforestation, 15, 17, 37, 39, 84, 130, 133, 135, 344, 346, 347, 358, 523, 707
Degrowth, 8, 26, 87, 90, 419, 743–763, 819, 835
Deindustrialization, 112, 128, 129, 143, 376, 451
Democracy, 38, 42, 51, 52, 86, 86n5, 88, 89, 89n12, 92, 96, 98, 109, 115, 166, 183, 226, 239, 240, 341, 368, 394, 397, 401, 405, 498, 499, 502, 503, 510, 518, 532, 593, 651, 710, 727, 745, 816, 825–828
  democratisation, 232, 344, 456, 730, 761, 825, 827
Department of Environment, Food and Rural Affairs (UK), Defra, 623
Department of Mineral Resources and Energy (South Africa) (DME), 65
Depatriarchalisation, 761
Deprivation, 446, 450
Design, 26, 40, 43, 122, 254, 330, 483, 487, 503, 508, 597, 636, 638, 682, 691, 753, 771, 815–835, 842, 847
Developing countries, 52, 59, 67, 73, 76, 87, 486, 506, 522, 523, 525, 752
Development, 11–13, 42, 45, 48, 49, 51, 52, 62, 64–66, 72, 74, 76, 77, 83–85, 87, 88, 88n10, 90, 93–95, 108, 118, 128, 152, 165, 177, 189–191, 200, 201, 203, 204, 212, 213, 215, 217, 230, 232, 238, 240, 242, 258, 260, 261, 265, 275, 280, 283, 288, 290, 319, 323, 331, 334, 339–341, 344–346, 350, 355, 369–371, 379, 403, 404, 416–418, 422, 426, 427,

430–432, 444–452, 468–484,
472n4, 478n15, 486, 487,
494n1, 495–497, 525, 530, 544,
551, 553, 555, 556, 563–578,
584, 595, 596, 603–605, 610,
615, 622, 629, 637, 638, 653n3,
654, 662, 668, 669, 679, 683n6,
686, 688, 702, 703, 714, 723,
731–736, 745, 745n2, 746, 757,
762, 776n8, 779–781, 780n10,
783, 808, 822, 828, 839,
843–845, 847, 848, 855
Development Platform of the Americas
(PLADA), 231, 232
Displacement, 36, 45, 49–52, 122,
278, 321, 367, 404, 475, 476,
685n13, 688, 690, 821
Dispossession, 27, 28, 86, 177, 180,
191, 321, 323, 328, 367, 376,
380, 396, 398, 402, 407,
808n10, 812
Distribution of the sensible, 227, 228,
239, 240
Documentary films
*An Inconvenient Truth* (2006), 120
*Behemoth* (2015), 577
*Dying to Breath* (2015), 563, 576
*Miners, Groom, and Pneumoconiosis*
(2020), 576
*A Preservação da Floresta
Amazônica* (1988)
*Tears of Shuangxi Village* (2009), 576
*This Changes Everything* (2015), 120
*Toxic: Amazoni (2011)*, 332n1
Double internality, 800, 809

E
EarthFirst!, 142
Eco-
feminism, 776
socialism, 183

Ecological sustainability, 88, 93, 116,
282, 340, 509, 780, 781
Ecology, 26–28, 203, 273, 320, 321,
323, 335, 390, 393, 396–397,
403, 404, 408–409, 415, 418,
419, 630, 652, 654, 678, 679,
708, 712, 712n11, 721–739,
746, 770–773, 775, 775n6, 800,
816–819, 821–824, 827, 835
Economic Development Department
(South Africa), EDD, 60
Economic growth, 9, 20, 25, 85, 87,
95, 178, 180, 285, 375, 379,
478, 480, 485, 505, 564, 566,
578, 699, 708, 735, 743, 744,
746, 747, 749, 754, 755, 757,
763, 781–783, 853, 855
Economic surplus, 204
Eco-social state, 24, 539–559
Ecosystem, 65, 137, 144, 239, 275,
367, 372, 404, 415, 418, 429,
478, 478n14, 621, 679, 759, 853
Ecuador, 22, 271–274, 277, 279,
279n4, 279n5, 282–284, 283n7,
286–288, 290, 291, 478n14
Egan, Carolyn, 259
Emissions
Big Shift, 253
intensive industries, 62, 71, 77, 520,
532, 545, 852
reduction effect, 253
Employers, 25, 83, 85, 98, 110–114,
127, 128, 130, 131, 137–141,
143, 156, 157, 229, 231, 234,
241, 261, 262, 297, 301, 342,
353, 375, 471, 474–476,
475n11, 485, 486, 509, 518,
519, 523, 525–529, 531–533,
549–551, 554, 555, 557, 558,
571–576, 571n2, 576n4,
585–587, 594, 598, 600, 606,
652, 665, 700, 783, 827

Employment, 26, 63, 74, 76, 111–113, 117, 119, 133, 134, 136, 137, 139–142, 155, 187, 189, 191, 193, 212, 252, 255, 264, 266, 274, 277, 300, 375, 376, 378, 379, 406, 441, 444, 451, 455, 468, 470, 471, 474, 475, 477–479, 479n16, 481, 482, 484–486, 502, 504, 505, 508, 510, 519, 520, 525–529, 531–533, 542, 544–550, 552–556, 558, 563–565, 569, 572–575, 578, 587, 621, 653–656, 668, 711, 729, 730, 733, 745, 749, 750, 752, 778, 780, 780n10, 787, 844, 845, 848–850, 852, 854–857

Empty-belly environmentalism, 207

Enclosures, 320, 321, 325, 390–392, 394–396, 398–400, 402, 409, 734

Endangered Species Act, 140

Energy
 colonialism, 405
 efficiency, 66, 118, 478, 479, 542, 550, 609, 629
 humanities, 688
 justice, 231, 405, 540, 542, 543
 sector, 61, 62, 64, 68, 70, 72, 75, 93, 111, 112, 120, 121, 188, 232, 233, 255, 377, 470, 471, 480, 526, 550, 552, 782
 system, 61, 65, 66, 69, 71, 107, 108, 166, 167, 187, 235, 511, 543, 680–682, 845, 847, 848
 transition, 59–77, 117, 178, 400, 406–408, 480, 482, 484, 510, 553, 557, 667, 712, 782, 852

Energy Intensive Users Group (EIUG), 63, 75, 179

Energy Task Force, AFL-CIO, 602–607, 610, 615

Engineering knowledge of workers, 253

Environment, 1, 4–7, 9–20, 35–53, 60, 71, 90, 93, 109–119, 121, 129, 130, 132, 137–139, 143, 151, 152, 154, 159, 162, 167, 181, 183, 186, 189–191, 194, 202, 213, 214, 225, 228–230, 249, 255, 256, 271–274, 277, 278, 286, 288, 289, 329, 333, 334, 341, 344, 346, 349, 358, 375, 380, 390, 391, 394, 395, 402, 418, 419, 442–445, 457, 468, 470, 473–475, 477, 478, 487, 494, 496, 506, 510, 541, 545, 547, 550, 551, 563–578, 584, 585, 593, 597n5, 602, 603, 605n16, 621–631, 633–636, 638–641, 647, 650–652, 654–667, 669, 700, 701, 709, 711–713, 711n10, 715, 728, 730, 735, 736, 745, 756, 769, 773, 774, 779, 780, 782–784, 795, 827

Environmental
 conflicts, 283, 340, 341, 348, 349, 578, 608, 610, 804
 crisis, 4, 14, 99, 208, 242, 342, 379, 583, 642, 667, 708, 744, 746, 748, 749, 753, 757, 758, 762, 772
 Environmental Justice Atlas, 17, 348–350
 Environmental Protection Agency, 140
 Environmental Remediation Program, 287
 exploitation, 200, 205, 206, 208
 hazards, 41, 163, 541, 543, 816
 inequality, 200, 204–206, 217, 591
 justice, 3, 15–17, 21, 68, 99, 119, 131, 149–168, 177, 178, 180–182, 184–186, 190, 192–195, 203n3, 218, 232, 238, 240, 271–291, 323, 349, 350,

365, 369, 371, 375, 381, 389,
390, 397, 398, 400, 403, 404,
407, 408, 444, 445, 448, 449,
452, 540–543, 565, 598, 611,
615, 633, 641, 708, 758,
804, 817
justice movements/organisations,
16, 131, 163, 178, 181, 182,
184–186, 190, 192, 193, 195,
203n3, 238, 240, 272, 273, 340,
381, 397, 398, 444, 541, 598,
615, 758
labour, 2–4, 7, 15, 18–21, 23,
25–28, 227, 229, 230, 240, 241,
249, 250, 261, 265, 273, 320,
389–409, 540, 565, 584, 671,
770, 771, 803, 805, 816, 817,
823, 825, 831, 835
labour studies, field of, 35, 107,
122, 252, 389, 721
modernisation, 290
movements of the Global South,
199, 203, 207
policies, 1, 2, 6, 9, 10, 12, 18, 20,
21, 26, 117, 154, 296, 321, 330,
346, 441, 444, 477–479,
478n15, 583–585, 589, 603n13,
622, 628, 633, 642, 643,
647–671, 760, 782, 804, 851
protests, 128, 350, 374
Environmentalism
of the poor, 3, 13, 15–17, 20, 22,
27, 48, 201, 203n1, 271, 398,
443, 445, 709, 810
of the working class, 273
Environmentalist, 1, 3, 5, 6, 9–12, 15,
16, 21–23, 42, 45, 46, 110, 117,
127, 128, 137, 139–143,
159–161, 164, 166, 177, 200,
201, 211–214, 216, 218, 229,
239, 251, 252, 258, 282, 285,
290, 331, 333, 340, 347, 348,
351, 365, 366, 371–373, 379,
380, 404, 415, 416, 431, 433,
434, 456, 494, 546, 578,
583–589, 593, 594, 596–611,
603n13, 613–615, 626, 632,
640, 650–654, 656–658,
662–664, 668, 669, 671, 722,
745, 815, 817, 833, 851
Eskom (South Africa), 60, 62, 63, 66,
67, 69, 70, 94, 95, 98, 178–180,
186–188, 191, 193, 194, 194n2,
232, 261
Ethnicity, 164, 272, 276, 286,
288, 442
EU Emissions Trading System (EU
ETS), 520, 527–529, 630
European
European Patent Organisation
(EPO), 73
European Trade Union
Confederation (ETUC), 11, 230,
236, 524, 531, 545, 547, 558,
658, 659, 851
European Trade Union Institute
(ETUI), 19, 548
European Union (EU), 236, 347,
480, 520, 525, 526, 527n1, 531,
548–551, 553, 555, 557–559,
655, 657, 658, 661, 662,
709n9, 824
European Union Emissions Trading
System, EU ETS, 520,
527–529, 630
Everyday
life, 192, 201, 353, 402, 443, 449,
622, 701, 703, 726, 777, 784
practices, 391, 396n5, 707, 758,
772, 775, 777, 778, 781,
784, 785
Exchange value, 393, 409, 683n6, 685,
687n16, 692, 756, 774, 805,
807, 843
Expertise, 38, 40, 43, 44, 283, 289,
426, 522, 523, 532, 843

Exploitation, 23, 25, 27, 28, 119, 129, 133, 200, 205, 207, 208, 242, 271, 275, 282–286, 288, 290, 322, 342, 344, 351–360, 370, 378, 380, 394–396, 399, 402–408, 547, 653, 678, 681, 687, 702, 703n3, 704, 725–728, 732, 735, 736, 748, 752, 756, 760, 775–777, 776n7, 781, 794, 806–811, 816, 821, 854
  exploitative, 133, 274, 377, 395, 687, 702, 797
Extended case method, 206
Externalisation, 215, 290, 706, 708, 757, 807
Extraction
  extractivism, 177, 178, 191, 273, 287, 329, 331, 380, 408, 688
  extrativistas, 330–332, 335
Extra-union, 225, 229–231, 233–238, 239n3, 241–244, 450

F

False solutions, 243
Farmer, 8, 9, 15, 22–23, 37, 41, 42, 49, 50, 135, 201, 209, 210, 253, 274, 278, 331, 349, 415–434, 486, 804, 810
  suicides, 42
  unions, 161, 209, 425, 428, 429, 431–433
Feminist theory, 7, 807
Financial crisis, 265, 285, 427, 660, 663, 670, 743, 746
First Nation, 367, 368, 370, 381, 453
*Florestania*, 330
Food, 5, 8, 21, 27, 40, 41, 90, 95, 106, 130–132, 185, 191, 278, 281, 303, 304, 311, 341, 345, 346, 352, 354, 369, 390, 392, 393, 394n4, 401, 402, 404–409, 415–417, 422, 423, 426, 431, 434, 443, 445–448, 485, 486, 505, 633, 681, 691, 703, 776n7, 801, 806, 808, 810, 816, 821, 829, 832, 849
  sovereignty, 16, 241, 276–282, 289, 323, 408, 415–418, 427, 428
  systems, 239, 278, 282, 289, 419, 434
*Force Ouvrière* (FO),France, 526, 552
Ford, Gerald, 139
Forestry, 36–39, 49, 64, 133–136, 140, 299, 321, 398, 409
Fossil energy, 510, 678–680, 689n21, 693, 704
  fossil fuels, 17, 22, 24, 60–67, 71, 72, 75–77, 93, 94, 98, 106, 107, 110–115, 117–120, 164, 177, 180, 181, 188, 194, 236, 238, 249, 250, 252, 257, 260–262, 264, 266, 298, 299, 301, 303–305, 307–311, 366, 369, 371–381, 400, 401, 467, 471, 480, 481, 483n21, 486, 493–511, 520, 526, 532, 543, 545, 557, 592, 600, 604, 606, 609, 653, 656, 668, 688, 703, 705, 708, 774, 782, 852
  fossil fuel workers, 111, 256, 264, 300, 306, 308, 310, 828
Foster, David, 306, 607–610, 615
Foster, Ellery, 134–136
Foster, John Bellamy, 6, 84, 87, 199, 202–204, 205n5, 206, 686n15, 800–802
Fracking, 255, 365, 372, 630, 633, 635, 636
Frame analysis, 200
France, 24, 73, 74, 84, 478n15, 479n16, 525, 526, 548, 554, 555, 557, 559, 681, 725, 745, 852
Franco dictatorship, 588, 650

Freire, Paulo, 342, 351, 826
Fridays For Future (FFF), Portugal, 264
Friends of the Earth (FOE), 236, 239, 258, 599, 608, 626
Frontline Green New Deal, 409
Fuel subsidy, 24, 486, 493–511

G

G20, 70, 495–497
Gaia, 797, 798
Gender, 37, 38, 47, 86n8, 94, 95, 205, 231, 238, 242, 272, 279, 344, 356, 368, 380, 381, 392, 395, 409, 442, 477, 506, 569, 587, 702, 708, 710, 758, 759, 761, 776, 780, 786, 818, 823, 831, 855
  equality, 119, 415, 479, 593, 780n10
  relations, 8, 9, 26–28, 703, 705, 757, 772, 775, 780, 786, 793, 806n8, 808, 809, 811, 812
  relations as relations of production, 26, 794, 807, 808, 810, 811
General Confederation of the Portuguese Workers (CGTP), 233, 252, 256
Genetically modified organisms (GMO), 41
Geoengineering, 253
Germany, 74, 475n11, 478n15, 526, 543, 548, 549, 551, 554, 555, 557–559, 654, 706n5, 712n11, 745, 771, 771n3, 775n6, 779, 781–786
Ghana, 24, 494, 498–509
Global, 2, 3, 9, 13, 14, 16–18, 59, 71, 72, 76, 83, 84n2, 85, 95, 112, 116, 152, 162–167, 179, 184, 227, 238, 244, 249–252, 256, 258, 260, 264–266, 278, 290, 322, 323, 329, 334, 346, 360, 371, 382, 389, 397, 400, 402, 408, 419, 427, 443, 449, 452, 468, 474, 477, 478, 495–497, 503, 509, 519, 522, 527n1, 532, 539, 542, 543, 545, 554, 583, 584, 603n13, 605, 605n16, 607, 615, 622, 623, 647, 650, 654, 655, 668, 677–680, 682, 683n6, 687, 687n17, 689–693, 689n21, 695, 700–703, 705–707, 709, 710, 730, 736, 743, 746–749, 757, 758, 760, 762, 772, 773, 775–777, 775n6, 776n7, 798, 806, 810–812, 847, 853, 854
  asymmetries, 680, 690, 691, 693, 695, 702
  division of labour, 747–749, 775n6, 776, 780, 786, 828
  Global North, 3, 7, 11, 14, 15, 18, 19, 22, 25, 27, 61, 163, 166, 177, 178, 203n3, 203n4, 230–232, 234, 272, 298, 360, 361, 390, 400, 402, 434, 442, 454, 495, 510, 542, 622, 687n17, 702, 703, 705–708, 710, 732, 746, 783, 786, 804, 810, 812, 854
  Global South, 3, 7, 8, 11, 14, 15, 18, 19, 22, 25, 27, 59–77, 163, 166, 177, 178, 184, 194, 199, 203, 203n4, 207, 230, 232, 234, 237, 243, 273, 298, 341, 360, 361, 389, 398, 400, 402–409, 420, 434, 442, 454, 493, 494, 497, 498, 509–511, 525, 542, 622, 659, 687n17, 688, 702, 703, 705–708, 713, 730, 732, 746, 775n6, 783, 804, 810, 854, 855
Global Union Federation (GUF), 163, 165, 166, 230
Global Wind Energy Council (GWEC), 72
South, 822
temperatures, 252, 258, 468

Good Jobs for All (GJFA), Toronto, 258, 259
Good sense, concept of, 251
*See also* Gramsci, Antonio
Gorz, André, 10, 26, 87, 91, 687, 721–739, 745, 747, 751
Gramsci, Antonio, 91, 92, 251, 342, 352, 355, 360, 396n5, 583, 624, 637, 640, 648, 649, 666, 670
Great Depression, 118
Green, 4n1, 5, 7, 11, 16, 45, 70, 86n8, 93, 95, 98n18, 99, 111, 128, 136, 153, 182, 188, 199–213, 216, 217, 234, 243, 259, 260, 265, 319, 331, 381, 443, 445, 446, 448, 478, 478n15, 481, 483, 484, 486, 506, 510, 520, 524, 526, 529, 540, 542, 544, 545, 548, 550, 591, 592, 606, 609, 610, 628–630, 632, 652, 708, 712, 728–731, 753, 821, 823–825, 828, 839, 846, 847, 849, 855
  economy, 7, 180, 185, 250, 259–263, 265, 331, 334, 467, 479n16, 484, 485, 532, 541, 545, 605, 605n16, 607, 668, 708n8, 736, 749n3, 779, 780, 780n10
  growth, 162, 181, 185, 250, 485, 668, 743, 745, 749n3, 779, 783
  industrial policy, 592, 600–602, 605, 607, 615, 616
  jobs, 6, 13, 92, 98, 111, 162, 181, 185, 228, 230, 236, 243, 250, 260, 261, 263, 265, 377, 409, 471n3, 474, 475n11, 478n15, 484–487, 517, 605, 609, 708, 712, 779, 783, 828, 851, 853
  nationalism, 14
  reps, 629
  revolution, 40, 417, 420
  transition, 468, 471, 476, 478n15, 479, 486, 510, 544–546, 558, 592, 606, 610, 616

Green Economy Accord (South Africa), 59, 60, 68, 70
Green Economy Network (Canada), 257, 260, 265
Greenhouse gas (GHG) emissions, 63, 64, 70, 105, 108, 182–184, 253, 346, 472n6, 496, 503, 519, 551, 635, 707n7, 713
Green Jobs Initiative (ILO), 235, 377, 474
Green Jobs Programme (ILO), 474
Green New Deal (GND), 21, 22, 87, 105–122, 128, 239, 251, 259, 261–263, 265, 266, 298, 301, 305–309, 311, 390, 400–402, 408, 434, 591, 597, 601, 609, 611, 613, 614, 616, 739, 817, 828, 833–835, 852
Gross Domestic Product (GDP), 20, 63, 83, 83n1, 84, 98, 212, 321, 322, 468, 506, 507, 730, 744, 753, 780n10
Gross national happiness (GNH), 84

H

Happiness, 87, 90, 706n5, 747, 750–752
Haraway, Donna, 7, 689n20, 773n5, 796–799, 798n2, 804
Harvey, David, 98n18, 108, 110, 115, 206, 808n10
Haug, Frigga, 8, 668, 714, 760, 794, 806–808, 810–812
Health and safety, 2, 4, 10, 12, 16, 17, 137, 139, 150, 156, 159, 163, 281, 373, 403, 442, 443, 453–454, 457, 541, 594, 595, 623, 627, 647, 649, 654, 655, 657
Heat stress, 468, 468n1, 469
Hegemony, 216, 217, 241, 272, 352, 587, 639, 648, 649, 671, 685, 699–715, 731, 744, 746

High reliability, 43
Hook, Frank, 135, 136
*Huasipungueros*, 275
Human-centred technology, 842–843
Human Development Index (HDI), 84
Humanist, 90, 92, 96, 632, 723
Human rights, 36, 153, 183, 232, 323, 330, 349, 356, 365, 407, 408, 637
Hydrocarbon
　law, 286
　sectors, 273

Identity
　livelihood, 272, 403
　politics, 272, 284, 288, 332, 731
IG Bergbau, Chemie, Energie, Germany (IG BCE), 523, 551, 552, 554, 700, 712n11, 782
IG Metall, 526, 530, 554, 555, 712, 712n11, 714, 725, 745, 771, 779, 782–785
Illich, Ivan, 87n9, 678, 683, 687, 688, 692–694, 726, 819, 822, 823
Imperial mode of living, 25, 699–715
Independent Power Producers (IPP), 60, 62, 65, 94, 187, 188
Indigenous and tribal peoples
　land, 327, 340, 366–368, 371, 376, 380
　movement, 88, 272, 278, 284, 326, 349, 368, 402
Individual capabilities, 585
Industrial
　democracy, 825–828
　revolution, 154, 155, 586, 677, 678, 681, 682, 687n16, 695, 842
　Workers of the World, 129, 130
IndustriALL, 554
Industrial Union Council of the AFL-CIO (IUC), 604, 604n14, 605

Industrial Union Department of the AFL-CIO (IUD), 594, 595, 604, 604n14, 605, 612
Industry and Energy Union (Norway), 260
Inequality, 18, 27, 86, 90, 105, 119, 203–206, 226, 227, 233, 235, 237, 241–244, 285, 342, 344, 350–359, 378, 401, 432, 433, 449, 450, 467, 468, 470, 476, 477, 482, 483n21, 487, 496, 499, 505, 509, 540–542, 544, 557, 587, 591, 640, 643, 652, 681, 683, 683n6, 690, 693, 700, 702, 706, 708, 713, 730, 731, 746, 748, 756, 759, 772, 775, 776, 779, 794, 795, 855
Infrastructure, 36–38, 41, 42, 48, 71, 72, 73, 77, 113, 113n1, 118, 184, 207, 212, 276, 286, 287, 301, 309, 323, 365, 366, 371, 376, 377, 380, 394n4, 405, 406, 408, 427, 476, 481, 483, 519, 551, 592, 601, 604, 680, 682, 682n4, 687, 694, 702–705, 712–714, 748, 751, 777, 778, 785, 816, 830, 831, 833
Injury, 47, 138, 139, 572, 575
Injustice, 18, 38, 108, 119, 120, 131, 150, 152, 155, 228, 238, 271, 278, 283, 349, 367, 378, 379, 390, 400, 432, 445, 448, 452, 509–511, 541, 631, 640, 643, 652, 690, 693, 694
Inorganic energy, 678, 680
Irrigation, 425, 427, 429
Instituto Sindical de Trabajo, Ambiente y Salud (ISTAS), 651, 655, 656, 658, 661, 663, 665, 669
Insurance
　health, 110, 573
　social, 574
　unemployment, 110, 149

Integrated Resource Plan (IRP), 60–62, 64, 65
Interculturality, 278
Interdisciplinary, 18–19, 679, 771, 781
Interest groups, 108, 366, 368, 518
Intergovernmental Panel on Climate Change (IPCC), 468, 495, 504, 519, 612, 655
International Association of Machinists and Aerospace Workers, USA, (IAM), 142
International Brotherhood of Electrical Workers, USA (IBEW), 106, 299, 376, 605
International Brotherhood of Teamsters, USA (IBT), 376
International Confederation of Free Trade Unions (ICFTU), 11, 165, 230, 522, 605n16, 659
International Federation of Chemical, Energy, Mine and General Workers' Unions (now part of IndustriALL), ICEM, 230
International Labour Organisation (ILO), 12, 13, 19, 24, 41, 84n2, 118, 234–237, 284n8, 297, 452, 468–480, 483, 484, 486, 519, 522, 544–546, 659, 661, 712, 780n10, 855
International Longshore and Warehouse Union (USA, Canada), ILWU, 114
International Metalworkers' Federation (now part of IndustriALL) (IMF), 495–498, 501, 504, 506–508
International Trade Union Confederation (ITUC), 99, 115, 163, 165, 166, 229–233, 235–239, 243, 260, 295–297, 452, 456, 474, 499, 522, 523, 525, 526, 530, 531, 545, 547, 558, 605n16, 623, 659, 660, 664, 853

International Transport Workers' Federation, ITF, 230, 713
International Woodworkers of America (IWA), 132–143
International Workers' Memorial Day, 163, 454
Intersectional, 340, 360, 368
Ireland, 73, 74, 640
Irrigation, 36–39, 391
Irvin, Jim, 70, 188
Isle Royale National Park, 135


J

Jamaica, 468
Job blackmail, 110, 157, 229, 394, 402–408, 577, 774
Job demand, 470, 471
Jobs vs. environment, 109–115, 117, 121, 186, 202, 249, 255, 256, 375, 527, 577, 621, 655–657, 783
Johnson, Keith, 137, 138, 140, 393
Just energy transition, 527, 667
Justice, 3, 7, 11, 15–17, 21, 76, 86, 93, 97, 99, 106–108, 118, 119, 121, 131, 149–168, 177, 178, 180–186, 190, 192–195, 203n3, 218, 227, 228, 231, 235, 236, 238, 240, 243, 256, 257, 259, 261, 263, 271–291, 298, 305, 307, 309, 310, 323, 327, 333, 340, 342, 348, 349, 357, 365, 366, 368, 369, 371–375, 377, 380, 381, 389, 390, 397–405, 407, 408, 416, 417, 419, 421, 428, 431, 432, 434, 444, 445, 448, 449, 452, 470, 479, 509, 510, 540–544, 556, 565, 570–575, 578, 592, 598, 602, 611, 614–616, 621–644, 685, 708, 713, 758, 798n2, 804, 817, 832, 833

Just recovery, 407, 408
Just transition
    Guidelines for a just transition (ILO), 473–476, 475n11
    Just Transition Alliance (JTA), 309, 598, 615
    Task Force (Canada), 299–301, 310

## K

Katowice (Poland), 236, 473
*Kawsak Sacha* (The Living Forest), 287
Kelly, Petra, 652
Kerala Model, 200, 218
Keynesianism, 250, 262
Keystone XL Pipeline, 120, 127, 374, 379, 610, 613
Kinder Scout trespass, 5, 444
King, Martin Luther Jr., 149, 150, 168
Klein, Naomi, 120, 121, 258, 259, 261, 265, 298, 306, 399, 739
Knowledge, 28, 43, 71, 122, 137, 139, 190, 191, 193, 194, 204, 229, 253, 278, 279, 288, 326, 330, 355, 356, 381, 392, 418–420, 425, 432, 474, 479, 519, 602, 624, 638, 639, 664, 689, 693, 729, 732–736, 736n10, 794, 795, 810, 817, 825, 826, 834, 842, 843, 850, 855
Krenak, Ailton, 326, 328
Kyoto Protocol, 115, 472, 522, 598, 612

## L

Laborers' International Union of North America (LIUNA), 127, 375, 376, 378, 379
Labor Network for Sustainability (LNS), 109, 121, 542, 545, 611, 613–615
Labor Party Advocates (LPA), 596, 597
Labour
    climate movement, 107, 116–121, 447
    commons, 390, 392–395, 409
    environmental conflicts, 199–218, 608, 610
    environmentalism, 3, 9, 10, 23, 128, 165, 177, 178, 186, 203n3, 203n4, 217, 230, 320, 333, 380, 390, 392, 409, 416, 418–420, 546, 584, 592–595, 611, 613, 615, 699–715, 721, 804, 810
    labour movement, 1, 2, 11, 14, 21, 35, 41, 44, 59, 60, 61, 66, 69, 70, 75, 76, 84–86, 92, 93, 97, 105–122, 127, 128, 136, 142, 143, 149, 163, 177, 178n1, 181–184, 186, 187, 193, 194, 199, 202, 217, 242–244, 276–282, 297, 299, 302, 307–311, 322, 333, 334, 365, 366, 369, 371, 375, 377, 379, 381, 450, 473, 587, 592, 599, 604, 605n16, 607, 608, 610, 615, 627, 631, 640, 641, 685, 687, 724, 729, 758, 785, 827, 843
    nature relationship, 8, 26–28, 634, 650–654, 660, 661, 666, 793, 810
    politics, 136, 225, 589, 592, 615, 629, 759
    productivity, 468, 469, 702, 754, 841
    rights, 119, 236, 241, 302, 303, 333, 547, 853
Labour Convergence on Climate, 613
Labour Party, 451, 452, 586, 597, 612, 623, 626, 629, 635, 848, 850, 852
Labour's 'Other,' 1

Land
  Landless Workers Movement, 341
  Land Struggle Database (Brazil), 344, 347, 348
  occupations, 344, 347, 348
Landless Workers' Movement, *(Movimento dos Trabalhadores Rurais Sem Terra, MST)*, Brazil, 341, 343, 344, 348
*La Via Campesina*, 415, 418, 422
Leap Manifesto, 306, 307, 309
Legal, 38, 71, 92n13, 156, 157, 161, 189, 280, 282–284, 286, 287, 289, 290, 327, 328, 330, 343, 367, 367n1, 368, 372, 374, 380, 404, 472n4, 482, 487, 570–574, 643, 811, 852
Lévi-Strauss, Claude, 5, 6
Liberation Theology, 341–343, 351, 352, 355, 356, 359, 403, 633, 650
Life histories, 6–8, 19, 25, 340, 584–585, 622, 624, 625, 639, 648–654
Livelihood, 16, 17, 22, 35, 36, 42, 49, 119, 157, 177, 180, 186, 191, 195, 201, 209–212, 214, 271, 272, 275, 276, 283, 288, 310, 326, 327, 345, 357–359, 367, 368, 376, 381, 390, 391, 398, 399, 402, 403, 405, 406, 443, 470, 485, 498, 510, 521–523, 544, 568, 570, 575, 621, 626, 756, 760, 820
Local Agenda 21, 13, 165, 623
LO Norway, 260
Los Angeles Alliance for a New Economy (LAANE), 160, 161
Low-income countries, 426, 468, 472, 478, 480
Loyal Legion of Loggers and Lumbermen, 132

Lucas
  Lucas Aerospace, 456, 636, 642, 643, 827, 844
  Lucas Combine Committee, 851
  Lucas Plan, 13, 456, 637–639, 642, 643, 827, 844–845
  Lucas Plan Group (New), 639
Lula da Silva, Inácio, 331

M

Mabola Protected Environment, 189–191
Markey, Edward, 21, 105, 106, 128, 154, 306, 591
Martínez-Alier, Joan, 3, 15, 16, 162, 163, 180, 201, 203, 207, 271, 272, 323, 340, 341, 359, 443, 684, 684n8, 684n10, 707n7, 709, 744, 810
Marx, Karl, 84, 90, 205n5, 208, 391, 653, 653n3, 654, 684–686, 687n16, 695, 700, 704, 705, 733, 756, 773, 774, 800–806, 803n4, 803n5, 806n8, 808, 809, 854
Marxism
  Marx's ecology, 205n5
  Marxian economy, 205n5, 756
  Marxist-Leninism, 85n3, 95–97, 96n15, 99
Mazzocchi, Tony, 107, 116, 117, 541, 594–598, 595n2, 611
Mendes, Chico, 3, 319–335
Metabolic
  rift, 6, 84, 334, 758, 800, 801
  shift, 750, 801
Meta-industrial labour, 759, 762
Migrant workers, 304, 305, 307, 310, 326, 563–568, 564n1, 570–574, 576, 576n4
Migration, 38, 50, 51, 416, 457, 568, 576, 706, 706n6, 707, 737

Militancy, 184, 356, 565, 722, 846, 851
Military, 91, 132, 324, 341, 351, 352, 403, 456, 484, 500, 501, 682n5, 689n21
Minerals
  extractivism, 273
  minerals energy complex (MEC), 62, 63, 71, 177–182
Mining
  Mining Affected Communities United in action (MACUA), 192, 193
  Mining Council, 178
Modern Monetary Theory (MMT), 262
Money, 131, 183, 284, 299, 358, 402n9, 503, 507, 563, 567, 568, 578, 605, 613, 663, 664, 677–695, 738, 776n7, 810, 844, 849
Moore, Jason, 7, 350, 775, 776, 776n7, 793, 794, 799–803, 806–812, 808n9, 808n10, 818
Morris, William, 138, 637, 819–823, 830, 833, 835, 842, 843
Movements, 1, 35, 59, 84, 105–122, 127, 149, 177, 180–185, 199, 207–209, 225, 238–241, 250, 272, 276–282, 296, 365, 372–374, 390, 415–434, 442, 453–454, 473, 494, 520–523, 541, 585, 592, 624, 648, 685, 724, 743, 785, 803, 819, 842
  environmental justice movement, 16, 163, 181, 182, 184–186, 190, 193, 195, 203n3, 238, 240, 272, 340, 381, 398, 444, 541, 598, 615, 758
  Movement of People Affected by Dams *(Movimento dos Antingidos por Barragens, MAB)*, Brazil, 348
  Occupy, 498
  Rural Women Movement *(Movimento de Mulheres Camponesas,* MMC, 351, 356

  Rural Women's Workers' Movement, *(Movimiento de Trabajadores de las Mujeres Rurales, MMTR)*, Brazil, 351
Mumford, Lewis, 693, 822, 823

N

Naidoo, Kumi, 252
Nairobi conference, 659
National Association for the Advancement of Colored People (USA), NAACP, 160
National Climate Change Response White Paper, 482
National Confederation of Farmworkers (Brazil) (CONTAG), 324
National Development Plan (South Africa)/New Democratic Party (NDP), 430, 482
National Economic Development and Labour Council (South Africa) (NEDLAC), 60, 69
National Education Health and Allied Workers Union (South Africa) (NEHAWU), 69, 70
Nationalisation, 94, 278, 288, 290
National Labor Relations Act (USA), 132
National Rubber-Tappers' Council (Brazil) (CNS), 320, 322, 327, 329, 334, 335, 335n2
National Union of Metalworkers of South Africa (NUMSA), 19, 20, 59, 60, 66, 68, 84, 91–97, 96n15, 99, 99n19, 179, 181, 183, 185, 187–189, 193, 194, 232, 483n21, 713
National Union of Mineworkers (South Africa; UK) (NUM), 60, 70, 97, 179, 185, 186, 188, 193, 194n2

# Index

Nature, 1, 37, 60, 84, 108, 129, 163, 177, 205n5, 242, 271, 288, 320, 339, 367, 389, 455, 518, 541, 594, 621–644, 648, 678, 700, 721, 756, 793–812, 817, 851
Nature – labour alliance, 653, 660
Neale, Jonathan, 251
Neoliberalism, 115, 129, 240, 285, 449, 452, 586, 588, 630, 736, 737, 746, 754
Neo-Malthusian, 46
Nigeria, 24, 398, 493, 494, 498–509
Nisqually National Wildlife Refuge, 139
Non-governmental organizations (NGOs), 16, 22, 42, 95, 225, 230–232, 237, 239n3, 273, 279, 280, 280n6, 282, 283, 285, 287, 320, 401, 417, 420–423, 425, 428–432, 525, 533, 551, 564, 574, 575, 665, 712, 782
North American Free Trade Agreement (NAFTA), 109, 599
North Dakota Access Pipeline, 23
Northern Ireland/Ulster, 631, 632, 640, 845
Nuclear energy, 42, 43, 69, 188, 403, 603, 657
Nuclear power, 12, 45, 69, 70, 95, 253, 524, 592, 599, 626, 627, 629, 635, 680

## O

Obama, Barack, 265, 373, 374, 496, 592, 605, 609–611
Objectivism, 351, 689
Ocasio-Cortez, Alexandria, 21, 105, 106, 128, 306, 591
Occupational health and safety
 occupational disease, 563–565, 571n3, 572, 575, 576
 occupational health standards, 572

Occupational Safety and Health Administration, 137
O'Grady, Frances, 625, 851
Oikeios, 7, 799, 801, 810
Oil
 and gas workers, 254
 Oil Change International (Washington), 256
 Oil, Chemical and Atomic Workers' International Union (USA), (OCAW), 142, 159, 298, 661
Olympic National Park, 134, 136
One Million Climate Jobs (UK), 234, 250, 256
One Million Climate Jobs Campaign (South Africa), 93, 93n14
Oppression, 27, 152, 238, 381, 631, 747
Oregon
 environmental council, 138
 wilderness coalition, 140
Organic agriculture, 182, 210, 423, 479n17
Organic intellectuals, 25, 351–359, 583–589, 624, 637, 639, 640, 648, 649, 653, 664, 670, 671
Organisation for Economic Cooperation and Development (OECD), 495, 520, 586, 587, 703
Organization of Employers (IOE), 474
OSHA Environmental Network, 594, 595, 612
O'Sullivan, Terry, 127, 128, 376, 379
Over determination, 360, 662

## P

Paid labour/living labour, 730
Paid labour/wage labour, 187, 272, 275–277, 393, 394, 394n4, 399, 518, 630, 691n23, 701, 705, 712, 714, 725, 727, 729, 738, 744n1, 745, 747–754, 756, 758–759, 762, 774, 777, 780

Pain, 138, 569, 570, 774
Paper, Allied-Industrial, Chemical and Energy Workers (PACE),USA, 601
Paris Agreement, 234, 346, 472, 472n6, 473, 476, 482, 495, 504, 519, 525, 577
Participation, 38, 84, 85, 86n5, 89, 90, 94, 95, 122, 141, 233, 234, 242, 280, 306–309, 352, 355, 401, 407, 427, 428, 430, 449, 455, 467, 476, 479, 487, 523, 530, 541, 639, 649, 660, 701, 706, 707, 760, 826, 852, 856
  participatory, 12, 66, 89, 89n12, 96, 427, 428, 448, 480, 530, 641–644, 826, 835
  participatory-democratic, 89, 92, 96, 97
*Partido dos Trabalhadores* (PT, Workers' Party), Brazil, 92, 343, 345, 346, 355–357
Patriarchal society, 569
Payments for ecosystem services (PES), 478n14
Peace movement, 670
Peasant, 2, 3, 6, 8, 9, 15, 22, 23, 88n11, 240, 241, 272, 274–279, 276n2, 281, 282, 288–290, 320–322, 326, 330, 332, 343, 345–347, 349, 389–392, 394, 397–402, 415, 418, 422, 565–567, 569, 576, 681, 709, 758, 759, 810
  leagues, 343
  movements, 240, 272, 343, 415, 746
  peasant-proletarian, 403, 404, 408
Pelosi, Nancy, 105
Pentachlorophenol, 138
Perkins, Jane, 599, 600n10, 604, 615
Pesticides, 40, 49, 62, 139, 344, 347, 348, 354, 360, 423, 425, 454, 570

Petroleum, 10, 62, 403, 471, 499, 502, 504, 505, 658, 705
Physics, 683–685, 684n9, 687n16, 688, 691
Pinchot, Gifford, 133
Pinheiro, Wilson, 325
Pipeline, 23, 71, 112, 114, 127, 365, 372–376, 378, 379, 406, 610, 613, 705
*Plan der Carbon* (Spain), 481
Plurinational state, 276–282, 276n1, 289
Pneumoconiosis, 25, 563–578
Political, 7, 11, 19, 20, 36, 39, 41, 45, 46, 48, 51–53, 84–86, 86n7, 88, 89n12, 91, 93, 95, 98, 107, 108, 113–115, 120, 122, 128, 129, 143, 152, 156, 162, 164, 166, 167, 179, 182, 184, 202, 203, 205n5, 207, 208, 213, 215, 218, 225–235, 237–244, 239n2, 250, 257, 271–278, 276n1, 280, 282, 284–291, 296, 303, 309, 319, 321, 324–327, 330–335, 340, 346, 353, 355–361, 366, 368, 370, 374, 379–381, 391, 392, 407, 408, 416, 418–422, 428, 431–434, 433n1, 449, 452, 454, 455, 457, 470, 476, 487, 493, 494, 500, 501, 510, 518, 521, 523–525, 531–533, 539, 541, 542, 545–547, 550, 556, 558, 559, 570, 572, 576n4, 585–588, 596, 597, 599, 601, 602, 608, 609, 611, 614–616, 625, 626, 629, 631, 640, 647, 648, 650–654, 660, 661, 665, 678, 679, 686–690, 687n16, 692–695, 705, 706n5, 708, 710, 713, 714, 744, 745, 749, 750, 755, 775, 777, 780, 786, 804, 816, 819, 821, 823, 824, 827–829, 832, 844, 851, 856

Political (*cont.*)
  ecology, 26–28, 273, 321, 323, 335, 679, 708, 721–739, 770–772, 775, 775n6, 817, 824
  unionism, 85, 86
Politics of dissensus, 227, 241–244
Pollution, 2, 10, 35, 38, 41, 84, 86n8, 139, 153, 167, 184, 192, 195, 199, 201, 202, 204, 205, 207, 209–216, 283, 346, 347, 367, 371, 375, 403, 446, 454, 481, 504, 541, 543, 577, 578, 594, 602, 635, 706, 754, 816, 820, 847
  air, 4, 41, 160, 161, 191, 192, 378, 493, 543
  water, 139, 191, 621
Popular ecologism, 340
Popular environmentalism, 128, 271
Portelli, Alessandro, 351, 359, 625
Postcolonial social settings, 204
Posthumanist, 7, 686n14, 688–690, 689n20, 693, 798
Post-materialist values theory, 442
Poverty, 39, 40, 86, 113n1, 133, 166, 167, 181, 259, 278, 285, 286, 332, 342, 351–359, 372, 416, 420, 423, 432, 446, 452, 475, 482, 483n21, 499, 506, 507, 541, 558, 564, 565, 567, 568, 576, 577, 621, 730, 746
Power, 5, 38, 83, 108, 111–112, 128, 154, 177–182, 205, 226, 238–241, 253, 262–264, 277, 280, 296, 305, 340, 366, 393, 416, 446, 448, 482, 494, 524, 549, 572, 592, 626, 648, 680, 701, 724, 754, 772, 793, 816, 844
Power resources approach, 181
Precarious workers, 302, 303, 307, 310, 342
Primary exports, 274
Production of the means of life, 8, 16, 806, 808, 809, 811

Productivism, 95, 183, 653, 685, 724, 726, 747, 755, 776
Productivity, 236, 280, 321–323, 331, 333, 432, 468, 469, 486, 521, 653, 681, 685, 686, 691, 702, 734, 738, 748–750, 753, 754, 774, 841–843, 846, 847, 854, 855
Project labor agreements (community workforce agreement, PLA (USA)), 113, 114, 375
Protests, 5, 24, 36, 41, 43, 127–129, 141, 158, 178, 190–192, 199, 203, 207n6, 211, 216, 251, 256, 258, 328, 340–341, 347–350, 365, 366, 373, 374, 376–378, 389, 428–429, 443, 447, 481, 493, 494, 498–504, 507, 508, 510, 525, 543, 573–578, 608, 631n3, 708, 735, 804
Public and Commercial Services Union (UK, PCS), 251, 455, 623

Q
*Quilombolas*, 348, 348n6, 349

R
Race, 86n8, 205, 231, 238, 242, 272, 286, 288, 290, 380, 381, 541, 587, 593, 602, 608, 702, 817, 818, 823, 829
Racism, 16, 23, 27, 120, 240, 370, 372, 378, 379, 450, 703n3, 710, 807
Radiation, 12, 42, 635
Rancière, Jacques, 226–230, 233, 241
Räthzel, Nora, 1, 3, 5–7, 9, 11, 12, 18, 23, 25, 26, 59, 106, 111, 153, 165, 182, 183, 187, 202, 203, 229, 243, 249, 252, 255, 261, 265, 322, 333, 356, 380, 416, 480, 510, 517, 530, 532, 546,

584, 621–623, 637, 640n7, 649, 649n1, 655, 668, 700, 709, 713, 721, 776n8, 779, 783, 810, 827
Raw and the cooked, 5
Reagan, Ronald, 128, 136, 143
Real
  utopia, 243
  wages, 505, 507, 510
Redwood National Park, 141
Reformist, 86, 93, 131, 181, 586, 587, 747, 753, 754
Regional policy, 549, 557
Regulation, 13, 14, 18, 38, 108, 122, 128, 133–135, 137–139, 143, 152, 202, 213, 229, 237, 284, 375, 405, 444, 453, 478n15, 496, 520, 532, 545, 549, 554, 572–574, 586, 606–608, 624, 630, 643, 711, 729, 774–778
Rehabilitation, 40, 50, 483
Relations of production, 26, 204, 340, 360, 680, 757, 793, 794, 806–811, 825
Religion, 272, 632, 637
Renewable energy
  Renewable Energy Association (IRENA)
  Renewable Energy Independent Power Producer Procurement Programme (South Africa), REIPPPP, 65, 68, 71, 73, 74, 75, 483, 483n21, 484
  renewables, 45, 60, 64–66, 67–77, 94, 111, 118, 166, 187, 258, 301, 329, 479, 482, 483n21, 526, 542, 635, 642, 652, 699, 703, 823, 851
Repertoire
  of collective action, 347, 417, 423–425, 431
  of contention, 417, 423
Reproduction/social reproduction/(re) production, 8, 52, 152, 273, 334, 357, 381, 390, 393–395, 398, 399, 407, 409, 565, 700–702, 704, 705, 709, 713–715, 722, 725, 728–731, 733–738, 747, 751, 754–760, 755n6, 756n7, 762, 763, 775, 776, 778, 805, 807–809, 828, 829
Reproductive
  commons, 392, 403, 409
  labour, 390, 393–395, 399, 409, 705, 756, 757
Reservation, 69, 193, 365, 367, 368, 372, 373, 380, 669
Reskilling, 68, 194, 418, 470, 553
Resource
  efficiency, 260
  extraction, 15, 365–382, 624, 702
Restructuring, 68, 69, 155, 238, 407, 456, 467, 470, 474, 476, 477, 533, 544, 546, 547, 552–555, 745, 746, 760, 828
Reuters, 65, 501, 725n5
Right to know legislation, 594
Rio +20 Summit (2012), 265
Rio Tinto, 15
Roles
  gender roles, 569, 758
  generational roles, 569
  leadership roles, 142
  union roles, 151, 156, 162, 710
Roosevelt, Franklin D., 118, 133, 134
Rosemberg, Anabella, 166, 230, 243, 295, 456, 522, 542, 605n16, 660
Rousseff, Dilma, 345, 348, 359
Rubber, 320, 322–329, 332, 357, 398
Rural protest, 340
Rural Women's Workers' Movement, *(Movimiento de Trabajadores de las Mujeres Rurales, MMTR)*, Brazil, 355
Rural Workers Trade-Union Movement *(Movimento Sindical de Trabalhadores Rurais, MSTR)*, Brazil, 324–327

## S

Sagebrush Rebellion, 141
Samarco, 348
Sanders, Bernie, 261, 591, 597, 602n12, 852
Sassen, Sakia, 737
Scott, Denny, 137–140, 346
Scottish Green Party, 256
Self-management, 395, 726, 727, 751
Service Employees International Union (USA), SEIU, 105, 114, 143, 377, 378, 599, 602
Service work, 3, 8
Settler colonialism, 366–371, 374, 378, 379, 381, 382
Sexism, 353, 356
Sierra Club, 10, 140, 160, 379, 407, 608, 609
Silcox, Ferdinand, 133
Silesia Declaration on Solidarity and Just Transition, 473
Silicosis, 563, 564, 566–568
Sioux Tribe, 365, 372, 373, 377
Skaggs, Tim, 139
Skills
  development, 72, 418, 449, 470, 475–479, 478n15, 484, 553
  policies, 475–479, 478n15, 482, 487
Slavery, 680n1, 681, 682n5, 688n18, 690n22, 693, 695
Slum clearance, 46
Smallholder/small farmer, 281, 352, 355, 356, 358–360, 415, 416, 422, 425, 426, 428, 430, 431, 434
Social
  dialogue, 60, 118, 179, 231, 234, 236, 237, 240, 256, 296–298, 301, 308, 475, 477, 480, 483, 483n21, 485–487, 508, 525, 527, 541, 549, 551–553, 557, 558, 669

  ecology, 773, 817, 822, 823
  instability, 470
  insurance, 574
  justice, 11, 86, 93, 106–108, 160, 180, 193, 238, 243, 272, 273, 340, 365, 374, 377, 381, 400n8, 416, 417, 419, 421, 431, 432, 434, 470, 479, 541–544, 556, 565, 598, 602, 614, 615, 621, 685
  metabolism, 690, 735, 752, 759
  movements, 22, 24, 36, 37, 52, 92, 93n14, 95, 96, 99, 106, 107, 119, 121, 127, 143, 150, 156, 159, 184, 190, 203n2, 204, 206, 217, 225, 226, 228, 230–232, 238–242, 250, 272, 273, 281, 296, 304, 330, 340–350, 359, 366, 368, 373, 374, 381, 397, 417, 418, 421, 431, 445, 449, 450, 456, 510, 648, 652, 660–662, 665, 668, 669, 671, 713, 731, 744, 759, 832
  movement unionism, 85, 85n4, 86, 86n5, 86n7, 92, 159, 184, 377, 454, 546, 554, 558, 713, 759
  nature, 7, 794
  ownership, 20, 94, 167, 183
  power, 187, 226, 228, 238–242, 244, 256, 276n1, 296–298, 301, 303, 304, 309, 400, 712, 795
  practices, 351, 391n3, 584
  protection, 110, 181, 664, 730, 754
  reproduction, 8, 52, 381, 390, 393–395, 398, 407, 700, 701, 709, 713–715, 737, 747, 757, 759, 805, 809, 828, 829
Socialism, 85, 86, 88, 93, 98, 183, 335n2, 593, 616, 626–629, 631, 640, 731, 819
Socialist, 5, 53, 70, 89, 90, 98, 183, 262, 277, 403, 566, 587, 593, 600–602, 602n12, 615,

626–630, 634, 639, 643, 654, 662, 665, 713, 724, 818–820, 823, 828, 829
Socially
  owned, 68, 76, 85, 94, 95, 98, 99, 188, 189, 193
  production, 442, 455–457, 639, 843, 844, 846
  products, 637, 638, 844
  useful work, 637
Societal
  exchange relations, 680
  relations with nature, 6, 8, 775
  shift, 22, 296, 298, 301, 401
  society–nature relationship, 4–8, 18, 227, 714, 793–803
Socio-
  ecological transformation, 389, 399, 700, 710, 712, 713, 745, 770, 781, 782, 785, 786
  technical transitions, 826, 830
Socio Bosque programme (Ecuador), 478n14
Solar energy, 371–372, 678, 679
Solidarity, 14, 19, 23, 27, 114, 152–154, 156–158, 163, 166, 182, 186, 187, 207, 212, 265, 335, 342, 365, 366, 371, 377, 378, 380, 381, 391, 393, 395–397, 401, 406, 425, 449, 452, 454, 456, 457, 473, 521, 550, 591, 631, 750, 755, 761, 812, 825, 828
Solidarność (Poland), NSZZ, 524, 525
South, 187
South African Federation of Trade Unions (SAFTU), 92, 97, 178n1, 180–182, 187, 188, 194, 195, 232, 263
South African Food Sovereignty Campaign (SAFSC), 94, 258

South African Renewable Energy Initiative (SARI), 72
South African Synthetic Oil Limited (SASOL), 62, 63
Sovereignty, 16, 232, 240, 241, 276–282, 285, 289, 323, 366, 368–370, 372, 374, 376, 378, 397, 406, 408, 415–418, 427, 428, 434, 525, 751, 756
Spain, 481, 583–589
Spanish Socialist Workers' Party (PSOE), 588
Special Economic Zone, 566
Standing Rock Reservation, 127, 373
State
  state-capitalist nexus, 208
  state owned enterprises, 572
Steam
  engine, 680, 680n1, 682, 683, 688, 689, 690n22, 820
  technology, 680, 682, 686, 690, 693
Steel sector, 527
Steelworker Toronto Area Council, 259
Stevis, Dimitris, 2, 6, 10, 11, 14, 25, 68, 71, 75, 106, 107, 109, 121, 165, 166, 203n3, 203n4, 206, 217, 227, 236, 242, 249, 250, 256, 260, 263, 265, 298, 309, 322, 366, 375, 380, 389, 390n1, 396, 397, 416, 517, 522, 543, 545, 546, 584, 592, 597n4, 700, 708, 711, 711n10, 712, 721, 739, 771
Subsidy, 24, 486, 494–510, 700n1
Subsistence workers, 2, 19, 326, 328, 805
Sufficiency, 87, 87n9, 187, 729
Suicides, 42, 134, 141, 243, 265, 564, 570
Superfund for Workers (USA), 116, 596, 597, 597n4

Suquamish/Duwamish worldview, 637
Sustainability, 59, 65, 67, 88, 93, 107–109, 116, 119, 153, 180, 183, 194, 218, 232, 240, 243, 278, 282, 323, 330, 340, 391, 392, 397, 431, 434, 441–457, 467–487, 494, 509, 518, 528, 578, 611, 624, 636, 667, 745, 745n2, 746, 763, 769–772, 776, 778–781, 780n10, 786, 816, 822–825
    transformations, 397, 763, 770, 778–780, 786
Sustainable development
  mobility, 657, 658
  Sustainable Development Goals (SDG), 475, 544, 779, 780, 780n10
  work, 12, 474, 779, 780n10
Sustained yield forestry, 134, 135
Sustain labour, 623, 659, 661, 662, 664
Sweeney, John, 116, 119, 120, 167, 179, 187, 231, 237, 256, 261, 262, 296, 297, 303, 308, 366, 371, 379, 494, 496, 531, 545, 598–600, 708, 712, 715n13
Sympoiesis, 7, 796–798

T

Taft-Hartley Act (USA), 136
Távora, Euclides, 326
Teamsters, 114, 160–162
Technology, 26, 40, 41, 45, 48, 61, 62, 64, 67, 68, 71–74, 75, 83, 93, 108, 111, 130, 166, 180, 186, 234, 280, 283, 284, 345, 347, 392, 418, 425, 457, 528, 556, 604, 606, 609, 628, 637, 638, 642, 677–695, 702, 726, 732, 734–736, 738, 796, 819, 821–823, 826–828, 830, 832, 839–857

Territory, 16, 275, 278–280, 283, 284, 284n8, 287–289, 321–323, 326, 327, 332, 366–368, 367n1, 372, 392, 393, 397–399, 401, 402, 405n11, 408, 409, 433, 589, 653, 658, 667–669, 671, 681
Texaco-Gulf, 283
Thermodynamics, 680, 684, 684n9, 685, 689, 690, 693
Three Sisters Wilderness, 136
Thunberg, Greta, 264–266
Total work environment, 137, 139
Trade Adjustment Assistance (TAA), 109
Trade Union
  bureaucratization, 91
  confederations, 256, 263, 531, 545, 649, 659
  democracy, 86, 166, 826, 827
  identities, 518, 521, 524, 532, 661, 665, 670
  structure, 22, 152, 518, 530, 531
  as "swords of justice," 263 (*see also* Class-based strategy)
Trade Union Advisory Committee, TUAC, 165, 230, 660
Trade Union Confederation of the Americas, TUCA, 230–233, 240
Trades Union Congress (UK), TUC, 450, 455, 524, 585, 586, 586n1, 623–625, 636, 642, 851
Trade Unions for Energy Democracy, TUED, 95, 121, 166, 167, 231, 261, 297, 298, 545
Trade Union Sustainable Development Advisory Committee (UK), TUSDAC, 623, 624
Tragic choices, 40
Transformation, 2, 8, 11, 20, 21, 24, 38, 52, 86n6, 88n11, 89, 129, 141, 149, 184, 194, 228, 235, 242, 253, 261, 273, 275, 282, 296, 299, 301, 308, 333, 356, 360, 381, 389, 391n3, 396, 397, 399, 401, 407, 430, 449, 477, 485,

530, 539, 543–546, 548–559, 637, 639–644, 648, 650, 655, 656, 664, 668, 671, 680, 700, 709, 710, 712–715, 712n11, 723, 732, 735–738, 745–748, 753–755, 758, 762, 763, 770, 772, 773, 778–783, 785, 786, 798, 809, 815, 817, 829, 856
  transformative, 5, 12, 17, 19, 22, 68, 91, 93, 94, 116, 119–121, 156, 167, 177, 179, 181, 184, 191, 228, 250, 297, 306, 356, 381, 389, 392, 396, 400n8, 402, 417, 467, 487, 545, 586, 641, 642, 753, 754, 805
Transparency, 43, 392, 407, 508, 552
Tribe, 348, 365, 367, 367n1, 368, 370–377, 380, 637
Tripartism, 475
Turner, Rick, 89–92, 96

U

Uehlein, Joe, 594, 605, 611–613, 615
Underemployment, 45, 51
Unemployed, 113, 262, 304, 342, 441, 451, 455, 477n13
Unemployment, 14, 14n2, 83, 84, 84n2, 87, 110, 111, 113, 113n1, 149, 181, 190, 194, 195, 237, 254, 256, 264, 373, 406, 416, 446, 468, 477, 477n13, 478, 481, 482, 483n21, 484, 498, 499, 529, 642, 664, 733, 749–751, 779, 840, 854
Unequal exchange, 681, 687, 689, 693n25, 707, 776
UNIFOR (Canada), 259, 298, 299, 308, 310
Union
  -NGO cooperation, 236, 320
  plus, 225, 226, 228, 229, 231, 237–242, 244
  union hiring halls, 113

Unión General de Trabajadores (General Union of Workers, Spain), UGT, 232, 586, 588, 589
Union-plus, 225, 226, 228, 229, 231, 237–242, 239n3, 244
United
  United Auto Workers (USA), UAW, 142, 594
  United Brotherhood of Carpenters and Joiners of America (USA and Canada), UBC, 114, 143
  United Mineworkers of America, UMWA, 109
  United Nations, UN, 11, 84, 165, 166, 225, 228, 230, 235, 238, 240, 265, 468, 503, 522, 530, 532, 623, 659, 662, 771, 779, 786, 824
  United Nations Conference of the Parties, UN COP, 59, 228, 232, 233, 235, 236, 238, 288, 295, 523, 604, 605, 605n16, 664
  United Nations Environment Programme, UNEP, 73, 230, 235, 238, 265, 474, 519, 545, 609, 659, 661, 699, 708n8
  United Nations Framework Convention on Climate Change, UNFCCC, 59, 236, 472, 472n4, 472n6, 473, 476, 503, 504, 522, 539, 623
  United Paper International Union (Paperworkers), 596
  United States Forest Service, USFS, 133
  United Steelworkers Union (USA, Canada, Caribbean), USW, 159, 299, 310, 370, 594, 601, 607–609
University and College Union (UK), UCU, 623
Unpaid work, 9, 26, 360, 760, 780, 805, 807, 808, 811
Unsustainability, 690, 708, 769–787

# Index

Urbanization, 45–49, 51, 52, 403, 564, 565
Utopia, 243, 342, 344, 724, 726
Uzzell, David, 1, 5–7, 11, 18, 25, 106, 111, 153, 165, 182, 187, 202, 204, 229, 243, 249, 252, 255, 261, 265, 322, 333, 380, 416, 510, 517, 530, 532, 546, 584, 621, 640n7, 649, 649n1, 655, 668, 700, 709, 713, 721, 779, 783, 810

## V

Value, 11, 15, 41, 44, 47, 48, 72, 88, 91, 95, 135, 180, 203, 212, 334, 340, 381, 393, 396, 401, 405, 409, 430, 442, 449, 551, 553, 626, 631, 632, 637, 652, 683–685, 683n6, 684n10, 685n12, 687n16, 694, 702, 707, 726, 732–738, 748, 750, 751, 756, 759n8, 770, 773n5, 774, 776, 777, 786, 801, 805–808, 806n8, 810, 821, 843, 848, 850, 855–857
Vegetarian, 632, 640
Vestas, 74, 263, 631, 631n3
Via Campesina (VC), 278, 281, 289, 339, 340, 351, 416
Vibration White Finger, 138
Violence, 23, 51, 130, 278, 320, 321, 323–325, 332, 334, 335, 340, 340n1, 343, 345, 349, 354, 355, 367, 373, 376, 379, 575, 631, 703, 723, 739, 759, 797, 811

## W

Wages, Robert, 595–598, 601, 615
Washington Environmental Council, 139
Water, 15, 16, 36–38, 45, 74, 88, 91, 95, 120, 132, 134, 136, 139, 151, 185, 188–191, 208, 210, 211, 232, 277, 278, 280, 332, 339, 345, 348, 349, 356, 359, 365, 372–374, 378, 392, 394n4, 397, 399–401, 404, 407, 409, 418, 419, 422, 425, 429–431, 434, 453, 481, 486, 541, 594, 621, 635, 654, 661, 668, 756, 758, 794, 804, 820, 830
water-power, 681
Web of life, 350, 637, 799–801, 818
Welfare state, 24, 254, 342, 402n9, 556, 557, 729, 734, 737, 753, 762, 833
Wilderness, 136, 140, 180, 321
Winstanley, Gerrard, 634
Work/labour productivity, 2, 35, 60, 84, 111, 129, 152, 192, 207, 227, 250, 271, 298, 319, 341, 376, 390, 419, 443, 468, 469, 494, 520, 539, 565, 584, 593, 622, 649, 678, 702, 721, 743–763, 771, 794, 817, 839–857
Worker(s)
    control, 89, 90, 322, 453, 455–456, 724, 726, 727, 846
    focused approach, 296–301, 310
    subsistence, 2, 19, 26, 326, 328, 700, 805
    Workers' Action Centre, 302
    Workers' plans, 845, 846, 852
Working class
    conditions, 273
    hours, 302, 753, 848
    politics, 83–99, 371, 451
Working-class environmentalism, 10, 127–144, 271–291, 406, 441–457, 565, 709, 804
Workplace transitions, 455, 457, 544, 546, 547, 558
World Bank, 49, 284, 329, 480n18, 482, 485, 495, 497, 507

World markets, 678, 679, 682, 687, 690, 691, 693, 693n25
Worldview, 7, 20, 368, 430, 603, 621–644, 651, 660, 661, 665, 682, 687, 689, 824
Wright Mills, Charles, 640

X
Xapuri, 327

Y
Youth Pastoral (Brazil), 352, 353, 355, 357

Printed in the United States
by Baker & Taylor Publisher Services